Perimeter and Area of Plane Figures

1. Triangle

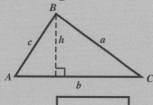

Perimeter: $P = a + b + c$

Area: $A = \dfrac{1}{2}bh$

2. Rectangle

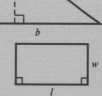

Perimeter: $P = 2l + 2w$

Area: $A = lw$

3. Parallelogram

Perimeter: $P = 2a + 2b$

Area: $A = bh$

4. Trapezoid

Perimeter: $P = a + b + c + d$

Area: $A = \dfrac{1}{2}h(a + b)$

5. Circle

Circumference: $C = 2\pi r$

Area: $A = \pi r^2$

Volume and Surface Area of Solid Figures

1. Rectangular prism

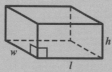

Volume: $V = lwh$

Surface area: $S = 2lw + 2lh + 2wh$

2. Right circular cylinder

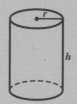

Volume: $V = \pi r^2 h$

Surface area: $S = 2\pi r^2 + 2\pi rh$

3. Sphere

Volume: $V = \dfrac{4}{3}\pi r^3$

Surface area: $S = 4\pi r^2$

4. Right circular cone

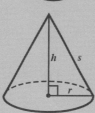

Volume: $V = \dfrac{1}{3}\pi r^2 h$

Surface area: $S = \pi r^2 + \pi rs$

Modeling, Functions, and Graphs

Prealgebra (softcover), by Katherine Yoshiwara
ISBN: 0-534-95448-0

Elementary Algebra: Equations and Graphs (hardcover), by Katherine
Yoshiwara, Bruce Yoshiwara, and Irving Drooyan
ISBN: 0-534-35823-3

Modeling, Functions, and Graphs: Algebra for College Students, Third Edition
(hardcover), by Katherine Yoshiwara, Bruce Yoshiwara
ISBN: 0-534-36832-8

3RD EDITION

Modeling, Functions, and Graphs

Algebra for College Students

Katherine Yoshiwara
Los Angeles Pierce College

Bruce Yoshiwara
Los Angeles Pierce College

BROOKS/COLE

™

THOMSON LEARNING

Australia • Canada • Mexico • Singapore • Spain • United Kingdom • United States

BROOKS/COLE

THOMSON LEARNING

Sponsoring Editor: *Bob Pirtle*
Marketing Team: *Leah Thomson,*
  Sheila Randazzo, Samantha Cabaluna
Editorial Assistant: *Erin Mettee-McCutchon*
Production Editor: *Laurel Jackson*
Production Service: *Matrix Productions*
Manuscript Editor: *Cheryl Smith*
Permissions Editor: *Sue Ewing*
Interior Design: *Carolyn Deacy*

Cover Design: *Roy R. Neuhaus*
Cover Illustration: *Charles Fuhrman,* Untitled,
  1982, Courtesy of Candice Fuhrman
Interior Illustration: *Asterisk*
Print Buyer: *Vena Dyer*
Typesetting: *Better Graphics*
Cover Printing: *Phoenix Color Corporation*
Printing and Binding: *Quebecor World Book Services*

For more information about this or any other Brooks/Cole product, contact:
BROOKS/COLE
511 Forest Lodge Road
Pacific Grove, CA 93950 USA
www.brookscole.com
1-800-423-0563 (Thomson Learning Academic Resource Center)

Printed in the United States of America

10   9   8   7   6   5   4   3   2   1

Library of Congress Cataloging-in-Publication Data

Yoshiwara, Katherine, 1953–
    Modeling, functions, and graphs: algebra for college students/
  Katherine Yoshiwara, Bruce Yoshiwara–3rd ed.
        p.   cm.
    Includes index.
    ISBN 0-534-36832-8
    1. Algebra. I. Yoshiwara, Bruce. II. Title.
QA154.2.Y67 2000   00-58575
512.9-dc21

Contents

Preface

This third edition of *Modeling, Functions, and Graphs* builds on earlier editions to incorporate the benefits of technology and the philosophy of the reform movement into algebra for college students. We have added several features to this edition, including Investigations for every chapter, Reading Questions for every section, and a variety of new application problems based on material in texts and research journals from many fields.

Modeling

The ability to model problems or phenomena by algebraic expressions and equations is the ultimate goal of any algebra course. With this end in mind, we motivate students to acquire the skills and techniques of algebra by placing them in the context of simple applications. Each chapter begins with an interactive Investigation that gives students an opportunity to explore open-ended modeling problems. These Investigations, which can be used in class either as guided explorations or as projects for small groups, are designed to show students how the mathematical techniques they are learning can be applied to study and to understand new situations.

Functions

The fundamental concept underlying calculus and related disciplines is the notion of function, and students should acquire a thorough understanding of functions before they embark on their study of calculus. Although the formal study of functions is usually the content of precalculus, it is not too early to begin building an intuitive understanding of functional relationships in the preceding algebra courses. Functions are useful not only in calculus but in nearly every field students may pursue. We begin working with functions in Chapter 1, "Linear Models." In Chapter 2 we explore applications of linear models, and Chapters 3 and 4 cover quadratic models. Formal function notation and terminology appear in Chapter 5, after students have had an opportunity to absorb the meaning behind the notation.

In our work with functions and modeling, we employ the now-celebrated "Rule of Four": that all problems should be considered using algebraic, numerical, graphical, and verbal methods. It is the connections among these approaches that we have endeavored to establish in this course. At this level, it is crucial that students learn to write an algebraic expression from a verbal description, to recognize trends in a table of data, and to extract and interpret information from the graph of a function.

Graphs

No tool for conveying information about a system is more powerful than a graph. However, many students have trouble progressing from a pointwise understanding of graphs to a more global view. By taking advantage of the useful features of graphing calculators, we examine a large number of examples and study them in more detail than is possible when every graph is plotted by hand. With graphing calculators, we can consider realistic models in which calculations by more traditional methods are difficult or impossible to perform.

We have incorporated graphing calculators into the text wherever they can enhance understanding. Calculator use is not simply an add-on; in many ways, it shapes the organization of the material. Although the instructions in this text are written for the TI-83 graphing calculator, they can easily be adapted to any other graphing utility. We have not attempted to use all the features of the calculator or to teach calculator use for its own sake but in all cases have let the mathematics suggest how technology should be used.

Content

Modeling, Functions, and Graphs includes the material found in a typical algebra course, along with introductions to curve-fitting and display of data. The order and presentation of topics are organized around families of functions and their applications.

Appendix A includes review material on polynomial products and factoring, laws of exponents, and facts from geometry. Appendix B provides a note on the structure of the real number system and an introduction to complex numbers.

Changes in the Third Edition

The following is a list of the major changes that appear in this edition:

- The text now begins with linear models. Review material from Chapter 1 of the second edition now appears in Appendix A.

- Each chapter begins with an Investigation of an applied problem. Investigations are guided explorations of a particular problem (such as population growth or the global positioning system) that illustrate the topic of the chapter. For example, the Investigation for Chapter 6 analyzes data from an award-winning experiment performed by two high school students.

- We have added dozens of new applications from biology, chemistry, geology, astronomy, and other fields. These problems contain real data and are referenced to source material.

- Expanded sections on curve fitting appear in Chapter 1, "Linear Models"; Chapter 4, "Applications of Quadratic Models"; and Chapter 7, "Exponential and Logarithmic Functions." We have also added an introduction to linearizing data by transformation in Chapter 6.

- In this edition, we have included more use of tables in Examples and Homework Problems and more problems on the properties of graphs.

- We have expanded Chapter 10 from the second edition to two chapters, "More About Functions" and "More About Graphs." These chapters include new sections on joint variation and displaying data.

- After most of the Examples in the text, we have added guided Exercises for students to complete. The Exercises can be used as either individual or group activities in class, or students can use them to work through a section independently. In addition, each section ends with a set of Reading Questions on the content of the section, and each chapter contains both Midchapter Review and Chapter Review exercises.

Exercise Sets

The Homework Problems in this book reflect our focus on modeling, functions, and graphs. We have provided a wide range of problems that emphasize mathematical modeling using tables of values, algebraic expressions, and graphs. We also include practice exercises for each new skill so that instructors can decide how much drill on manipulation is appropriate for their classes. In each chapter, we offer a set of Midchapter Review exercises and a set of Chapter Review exercises that students can use as a practice test. In addition, we have provided answers to the odd-numbered exercises in each section and in the chapter reviews.

Ancillaries

- *Student Solutions Manual.* The student solutions manual contains worked-out solutions to the odd-numbered problems.

- *BCA Tutorial.* This text-specific, interactive tutorial software is delivered via the Web at *http://bca.brookscole.com* and is offered in student and instructor versions. Like *BCA Testing*, the tutorial is browser-based, making it an intuitive mathematical guide even for students with little technological proficiency. So sophisticated that it's simple, *BCA Tutorial* allows students to work with real math notation in real time, providing instant analysis and feedback. The tracking program built into the instructor's version of the software enables instructors to monitor student progress closely.

- *Greatest Hits.* This interactive tutorial CD-ROM covers concepts in each chapter section that the text authors have found to be the most difficult for students. It features short videos of a professor working out specific problems.

Acknowledgments

We would like to thank the following reviewers for their valuable comments and suggestions: Debbie Garrison, Valencia Community College; Judith M. Jones, Valencia Community College; and Jolene Rhodes, Valencia Community College.

Katherine Yoshiwara
Bruce Yoshiwara

Modeling, Functions, and Graphs

Linear Models

You may have heard the saying that "mathematics is the language of science." Today, however, professionals not only in the sciences but in nearly every discipline take advantage of mathematical methods. Mathematical techniques are used to analyze data, identify trends, and predict the effects of change. At the heart of these quantitative methods are the concepts and skills of algebra. In this course, you will use skills you learned in elementary algebra to solve problems and to study a variety of phenomena.

In this chapter, we describe relationships between variables by using equations, graphs, and tables of values. This process is called **mathematical modeling.**

 INVESTIGATION 1

Sales on Commission

Delbert is offered a job as a salesman for a company that manufactures restaurant equipment. He will be paid $1000 per month plus a 6% commission on his sales. The sales manager informs Delbert that he can expect to sell about $8000 worth of equipment per month. To help him decide whether or not to accept the job, Delbert does a few calculations.

1. Based on the sales manager's estimate, what monthly income can Delbert expect from this job? What annual salary would that provide?

2. What would Delbert's monthly salary be if he sold, on average, only $5000 worth of equipment per month? What would his salary be if he sold $10,000 worth per month? Compute monthly salaries for the sales totals shown in Table 1.1.

TABLE 1.1

Sales	Income
5000	
8000	
10,000	
12,000	
15,000	
18,000	
20,000	
25,000	
30,000	
35,000	

3. Plot your data points on a graph, using sales figures (S) on the horizontal axis, and income (I) on the vertical axis, as shown in Figure 1.1. Connect the data points to show Delbert's monthly income for all possible monthly sales totals.

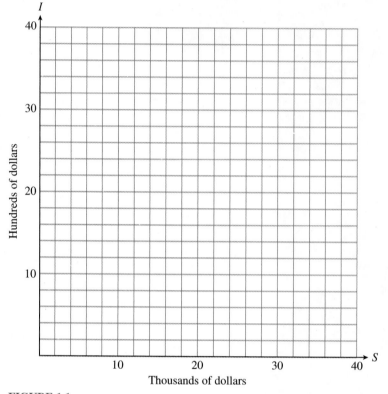

FIGURE 1.1

4. Add two new data points to Table 1.1 by reading values from your graph.

5. Write an algebraic expression for Delbert's monthly income, I, in terms of his monthly sales, S. Use the verbal description in the problem to help you:

   "He will be paid: $1000 . . . plus a 6% commission on his sales."

   Income = _____

6. Test your formula from part (5) to see whether it yields the same results as those you recorded in Table 1.1.

7. Use your formula to find out what monthly sales total Delbert would need in order to have a monthly income of $2500.

8. Each increase of $1000 in monthly sales increases Delbert's monthly income by _____.

9. Summarize the results of your work: In your own words, describe the relationship between Delbert's monthly sales and his monthly income. Include in your discussion a description of your graph.

1.1　Some Examples of Linear Models

A relationship between variables can be represented by

1. a table of values
2. a graph
3. an algebraic equation

EXAMPLE 1　Annelise is on vacation at a seaside resort. She can rent a bicycle from her hotel for $3 an hour, plus a $5 insurance fee. (A fraction of an hour is charged as the corresponding fraction of $3.)

a. Make a table of values showing the cost, C, of renting a bike for various lengths of time, t.

b. Plot the points on a graph. Draw a curve through the data points.

c. Write an equation for C in terms of t.

Solutions

a. To find the cost, multiply the time by $3, and add the $5 insurance fee. For example, the cost of a one-hour bike ride is

$$\text{Cost} = (\$5 \text{ insurance fee}) + \$3 \times (\text{one hour})$$
$$C = 5 + 3(1) = 8$$

A one-hour bike ride costs $8. Record the results in a table, as shown here:

Length of rental (hours)	Cost of rental (dollars)		(t, C)
0	5	$C = 5 + 3(0)$	$(0, 5)$
1	8	$C = 5 + 3(1)$	$(1, 8)$
2	11	$C = 5 + 3(2)$	$(2, 11)$
3	14	$C = 5 + 3(3)$	$(3, 14)$

b. Each pair of values represents a point on the graph. The first value gives the horizontal coordinate of the point, and the second value gives the vertical coordinate. (To review ideas about graphing, see Appendix A.4.) The points lie on a straight line, as shown in Figure 1.2. The line extends infinitely in only one direction, because negative values of t do not make sense here.

c. To find an equation, let C represent the cost of the rental, and use t for the number of hours:

$$\text{Cost} = (\$5 \text{ insurance fee}) + \$3 \times (\text{number of hours})$$
$$C = 5 + 3 \cdot t$$

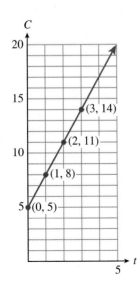

FIGURE 1.2

EXAMPLE 2

Use the equation $C = 5 + 3t$ you found in Example 1 to answer the following questions. Then show how the answers can be found by using the graph.

a. How much will it cost Annelise to rent a bicycle for 6 hours?

b. How long can Annelise bicycle for $18.50?

Solutions

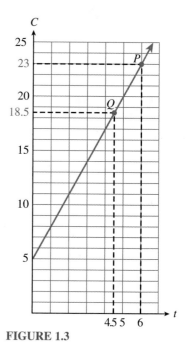

a. Substitute $t = 6$ into the equation to find

$$C = 5 + 3(6) = 23$$

A 6-hour bike ride will cost $23. The point P on the graph in Figure 1.3 represents the cost of a 6-hour bike ride. The value on the C-axis at the same height as point P is 23, so a 6-hour bike ride costs $23.

b. Substitute $C = 18.50$ into the equation and solve for t.

$$18.50 = 5 + 3t$$
$$13.50 = 3t$$
$$t = 4.5$$

For $18.50 Annelise can bicycle for $4\frac{1}{2}$ hours. The point Q on the graph represents an $18.50 bike ride. The value on the t-axis below point Q is 4.5, so $18.50 will buy a 4.5 hour bike ride.

FIGURE 1.3

EXERCISE 1 Frank plants a dozen corn seedlings, each 6 inches tall. With plenty of water and sunlight they will grow approximately 2 inches per day. Complete the table of values for the height, h, of the seedlings after t days.

t	0	5	10	15	20
h					

a. Write an equation for the height h of the seedlings in terms of the number of days t since they were planted.

b. Graph the equation.

c. How tall is the corn after 3 weeks?

d. How long will it be before the corn is 6 feet tall?

Choosing Scales for the Axes

To draw a useful graph, we must choose appropriate scales for the axes.

EXAMPLE 3

In 1980, a three-bedroom house in Midville cost $50,000. The price of the house increased by $2500 per year since 1980.

a. Make a table of values showing the price of the house in 1980, 1984, 1988, and 1992.

b. Choose suitable scales for the axes and plot the values you found in part (a) on a graph. Use t, the number of years since 1980, on the horizontal axis and the price of the house, P, on the vertical axis. Draw a curve through the points.

c. Use your graph to calculate the increase in the price of the house from 1990 to 1996. Illustrate the increase on your graph.

d. Write an equation that expresses P in terms of t.

Solutions

a. In 1980, the price of the house was $50,000. Four years later, in 1984, the price had increased by $4(2500) = 10,000$ dollars, so

$$P = 50,000 + 4(2500) = 60,000$$

In 1988 the price had increased by $8(2500) = 20,000$ dollars, so

$$P = 50,000 + 8(2500) = 70,000$$

You can verify the price of the house in 1992 by a similar calculation.

Year	Price of house		(t, P)
1980	50,000		(0, 50,000)
1984	60,000		(4, 60,000)
1988	70,000		(8, 70,000)
1992	80,000		(12, 80,000)

b. Let t stand for the number of years since 1980, so that $t = 0$ in 1980, $t = 4$ in 1984, and so on. To choose scales for the axes, look at the values in the table. For this graph, we scale the horizontal axis, or t-axis, in 2-year intervals and the vertical axis, or P-axis, in intervals of $10,000. The points in Figure 1.4 lie on a straight line.

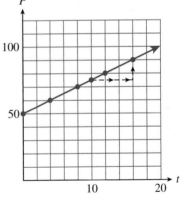

FIGURE 1.4

c. Find the points on the graph corresponding to 1990 and 1996. These points lie above $t = 10$ and $t = 16$ on the t-axis. Now find the values on the P-axis corresponding to the two points. The values are $P = 75,000$ in 1990 and $P = 90,000$ in 1996. The increase in price is the difference of the two P-values.

$$\text{increase in price} = 90,000 - 75,000$$
$$= 15,000$$

The price of the home increased $15,000 between 1990 and 1996. This increase is indicated by the arrows in Figure 1.4.

d. Look back at the calculations in part (a). The price of the house started at $50,000 in 1980 and increased by $t \times 2500$ dollars after t years. Thus,

$$P = 50,000 + 2500t \qquad (t \geq 0) \qquad \blacksquare$$

Linear Equations

The graphs in Examples 1–3 were all portions of straight lines. In fact, the graph of any equation

$$ax + by = c$$

where a and b are not both equal to zero, is a straight line. For this reason such equations are called **linear equations.** The equation in Example 1,

$$C = 5 + 3t$$

can be written equivalently as

$$-3t + C = 5$$

and the equation in Example 3,

$$P = 50,000 + 2500t$$

can be written as

$$-2500t + P = 50,000$$

The graphs in the preceding examples are **increasing graphs.** As we move along the graph from left to right (that is, in the direction of increasing t), the second coordinate of the points on the graph increases as well. Exercise 2 illustrates a **decreasing graph.**

EXERCISE 2 Silver Lake has been polluted by industrial waste products. The concentration of toxic chemicals in the water is currently 285 parts per million (ppm). Local environmental officials would like to reduce the concentration by 15 ppm each year.

a. Complete the table of values showing the desired concentration, C, of toxic chemicals t years from now. For each t-value, calculate the corresponding value for C. Write your answers as ordered pairs.

t	C		(t, C)
0		$C = 285 - 15(0)$	$(0, \ )$
5		$C = 285 - 15(5)$	$(5, \ )$
10		$C = 285 - 15(10)$	$(10, \ )$
15		$C = 285 - 15(15)$	$(15, \ )$

b. Graph the ordered pairs, and connect them with a straight line. Extend the graph until it reaches the horizontal axis, but no farther. Points with negative C-coordinates have no meaning for the problem.

c. Write an equation for the desired concentration, C, of toxic chemicals t years from now. The concentration is initially 285 ppm, and we *subtract* 15 ppm for each year that passes, or $15 \times t$.

Intercepts

Consider the graph of the equation

$$3y - 4x = 12$$

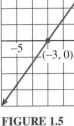

FIGURE 1.5

shown in Figure 1.5. The points at which the graph crosses the axes are called the **intercepts** of the graph. The coordinates of these points are easy to find.

The y-coordinate of the x-intercept is zero, so we set $y = 0$ in the equation to get

$$3(0) - 4x = 12$$
$$x = -3$$

The x-intercept is the point $(-3, 0)$. Also, the x-coordinate of the y-intercept is zero, so we set $x = 0$ in the equation to get

$$3y - 4(0) = 12$$
$$y = 4$$

The y-intercept is $(0, 4)$.

EXERCISE 3 Find the intercepts of the graph of

$$2y = -18 - 3x$$

To find the y-intercept, set $x = 0$ and solve for y.
To find the x-intercept, set $y = 0$ and solve for x.

The intercepts of a graph tell us something about the situation it models.

EXAMPLE 4

  a. Find the intercepts of the graph in Exercise 2.
  b. What is the significance of the intercepts to the problem?

Solutions

  a. An equation for the concentration of toxic chemicals is $C = 285 - 15t$. To find the C-intercept, set t equal to zero.

$$C = 285 - 15(0) = 285$$

  The C-intercept is the point $(0, 285)$, or simply 285. To find the t-intercept, set C equal to zero and solve for t.

$$0 = 285 - 15t \quad \text{Add } 15t \text{ to both sides.}$$
$$15t = 285 \qquad \text{Divide both sides by 15.}$$
$$t = 19$$

  The t-intercept is the point $(19, 0)$, or simply 19.

  b. The C-intercept represents the concentration of toxic chemicals in Silver Lake now: When $t = 0$, $C = 285$, so the concentration is currently

285 ppm. The t-intercept represents the number of years it will take for the concentration of toxic chemicals to drop to zero: When $C = 0$, $t = 19$, so it will take 19 years for the pollution to be eliminated entirely. ■

Intercept Method for Graphing Lines

Since we really only need two points to graph a linear equation, we might as well find the intercepts first and use them to draw the graph. It is always a good idea to find a third point as a check.

EXAMPLE 5

a. Find the x- and y-intercepts of the graph of $150x - 180y = 9000$.

b. Use the intercepts to graph the equation. Find a third point as a check.

Solutions

a. To find the x-intercept, set $y = 0$.

$$150x - 180(0) = 9000 \quad \text{Simplify.}$$
$$150x = 9000 \quad \text{Divide both sides by 150.}$$
$$x = 60$$

The x-intercept is the point $(60, 0)$. To find the y-intercept, set $x = 0$.

$$150(0) - 180y = 9000 \quad \text{Simplify.}$$
$$-180y = 9000 \quad \text{Divide both sides by } -180.$$
$$y = -50$$

The y-intercept is the point $(0, -50)$.

b. Scale both axes in intervals of 10, then plot the two intercepts, $(60, 0)$ and $(0, -50)$. Draw the line through them, as shown in Figure 1.6. Now find another point and check that it lies on this line. We choose $x = 20$ and solve for y.

$$150(20) - 180y = 9000$$
$$3000 - 180y = 9000$$
$$-180y = 6000$$
$$y = -33.\overline{3}$$

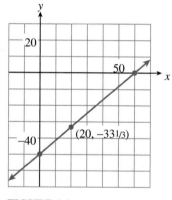

FIGURE 1.6

Plot the point $\left(20, -33\frac{1}{3}\right)$. Since this point lies on the line, we can be reasonably confident that our graph is correct. ■

EXERCISE 4

a. Find the intercepts of the graph of

$$60x - 13y = 390$$

b. Use the intercepts to help you choose appropriate scales for the axes, and graph the equation.

The examples in this section model simple linear relationships between two variables. Such relationships, in which the value of one variable is determined by the value of the other, are called *functions*. We will study various kinds of functions throughout the course.

READING QUESTIONS

1. Name three ways to represent a relationship between two variables.
2. What do the coordinates of the points on the graph in Exercise 2 represent?
3. What is the general form for a linear equation?
4. What is a decreasing graph?
5. What are the intercepts of a line? How can you find them?

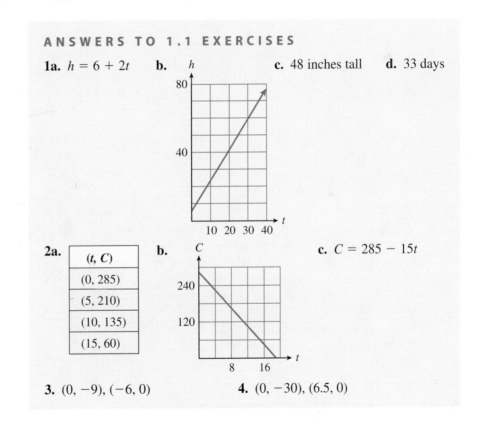

ANSWERS TO 1.1 EXERCISES

1a. $h = 6 + 2t$ **b.** h **c.** 48 inches tall   **d.** 33 days

2a.

(t, C)
(0, 285)
(5, 210)
(10, 135)
(15, 60)

b. C

c. $C = 285 - 15t$

3. $(0, -9), (-6, 0)$ **4.** $(0, -30), (6.5, 0)$

HOMEWORK 1.1

Write and graph a linear equation for each situation. Then answer the questions.

1. On October 31, Betty and Paul fill their 250-gallon heating fuel oil tank. Beginning in November they use an average of 15 gallons of heating fuel oil per week. Complete the table of values for the amount of oil, A, left in the tank after w weeks.

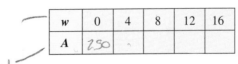

w	0	4	8	12	16
A	250				

   a. Write an equation that expresses the amount of oil, A, in the tank in terms of the number of weeks, w, since October 31.

b. Graph the equation.

c. How much did the amount of fuel oil in the tank decrease between the third week and the eighth week? Illustrate this amount on the graph.

d. When will the tank contain more than 175 gallons of fuel oil? Illustrate on the graph.

2. Leon's camper has a 20-gallon fuel tank, and he gets 12 miles to the gallon. (Note that getting 12 miles to the gallon is the same as using $\frac{1}{12}$ gallon per mile.) Complete the table of values for the amount of gasoline, g, left in Leon's tank after driving m miles.

m	0	48	96	144	192
g					

a. Write an equation that expresses the amount of gasoline, g, in Leon's fuel tank in terms of the number of miles, m, he has driven.

b. Graph the equation.

c. How much gasoline will Leon use between 8 a.m., when his odometer reads 96 miles, and 9 a.m., when the odometer reads 144 miles? Illustrate this amount on the graph.

d. If Leon has less than 5 gallons of gas left, how many miles has he driven? Illustrate on the graph.

a. Find the intercepts of the graph.

b. Graph the equation by the intercept method.

5. $x + 2y = 8$ **6.** $2x - y = 6$

9. $\dfrac{x}{9} - \dfrac{y}{4} = 1$ **10.** $\dfrac{x}{5} + \dfrac{y}{8} = 1$

a. Find the intercepts of the graph.

b. Use the intercepts to choose scales for the axes, and graph the equation by the intercept method.

11. $\dfrac{2x}{3} + \dfrac{3y}{11} = 1$ **12.** $\dfrac{8x}{7} - \dfrac{2y}{7} = 1$

13. $20x = 30y - 45,000$ **14.** $30x = 45y + 60,000$

15. $0.4x + 1.2y = 4.8$ **16.** $3.2x - 0.8y = 12.8$

3. Phil and Ernie buy a used photocopier for $800 and set up a copy service on their campus. For each hour that the copier runs continuously, Phil and Ernie make $40. Complete the table of values for their profit, P, after running the copier for t hours.

t	0	5	10	15	20
P					

a. Write an equation that expresses Phil and Ernie's profit (or loss), P, in terms of the number of hours, t, they run the copier.

b. Find the intercepts and sketch the graph.

c. What is the significance of the intercepts to Phil and Ernie's profit?

4. A deep-sea diver is taking some readings at a depth of 400 feet. He begins ascending at 20 feet per minute. (Consider a depth of 400 feet as an altitude of -400 feet.) Complete the table of values for the diver's altitude, h, after m minutes.

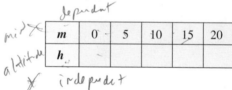

m	0	5	10	15	20
h					

a. Write an equation that expresses the diver's altitude, h, in terms of the number of minutes m elapsed.

b. Find the intercepts and sketch the graph.

c. What is the significance of the intercepts to the diver's depth?

7. $3x - 4y = 12$ **8.** $2x + 6y = 6$

17. The owner of a gas station has $4800 to spend on unleaded gas this month. Regular unleaded costs him $0.60 per gallon, and premium unleaded costs him $0.80 per gallon.

a. How much do x gallons of regular cost? How much do y gallons of premium cost?

b. Write an equation that relates the amount of regular unleaded gasoline, x, the owner can buy and the amount of premium unleaded, y. (Your equation will have the form $ax + by = c$.)

c. Find the intercepts and sketch the graph.

d. What is the significance of the intercepts to the amount of gasoline?

18. Five pounds of body fat is equivalent to 16,000 calories. Carol can burn 600 calories per hour bicycling and 400 calories per hour swimming.

a. How many calories will Carol burn in x hours of cycling? How many calories will she burn in y hours of swimming?

b. Write an equation that relates the number of hours, x, of cycling and y, of swimming Carol needs to perform in order to lose 5 pounds. (Your equation will have the form $ax + by = c$.)

c. Find the intercepts and sketch the graph.

d. What is the significance of the intercepts to Carol's exercise program?

19. A real estate agent receives a salary of $10,000 plus 3% of her total sales for the year. Complete the table of values. Total sales, s, and income, I, are expressed in thousands of dollars.

s	200	500	800	1200	1500
I					

a. Write an equation that expresses the agent's salary, I, in terms of her total annual sales, s.

b. Use your table to graph the equation. (*Hint:* Use increments of $100,000 for the horizontal axis.)

c. Use your equation to determine what annual sales result in salaries between $16,000 and $22,000 for the year. Illustrate your result on the graph.

d. Calculate the increase in the agent's salary if her sales increase from $500,000 to $700,000. Use arrows to illustrate the increase on your graph.

20. Under a proposed graduated income tax system, a single taxpayer whose taxable income is between $13,000 and $23,000 would pay $1500 plus 20% of the amount of income over $13,000. (For example, if your income is $18,000, you would pay $1500 plus 20% of $5000.) Complete the table of values for the tax, T. Income over $13,000, I, is expressed in thousands of dollars.

I	2	5	6	7.5	10
T					

a. Write an equation that expresses the tax, T, in terms of the amount of income over $13,000, I.

b. Use your table to graph the equation.

c. Use your equation to determine what incomes result in taxes between $2500 and $3100. Illustrate your result on the graph.

d. Calculate the increase in taxes corresponding to an increase in income from $20,000 to $22,000. Use arrows to illustrate the increase on your graph.

1.2 Using a Graphing Calculator

A linear equation can be written in the form

$$ax + by = c$$

where a, b, and c are constants, and a and b are not both zero. Some examples of linear equations are

$$2x + 5y = 10 \qquad 4x = 3 - 8y$$
$$y = \frac{2}{3}x - 1 \qquad 0 = x - 2 - y$$

The order of the terms is not important, as long as each variable has an exponent of one, and no variables appear in the denominators of fractions.

Solving for y in Terms of x

We can use a graphing calculator to graph linear equations if they are written in the form "$y = $ (expression in x)." Suppose we would like to use a graphing calculator to graph the equation

$$2x - y = 5$$

Before we can enter the equation, we must solve for y in terms of x.

$$2x - y = 5 \qquad \text{Subtract } 2x \text{ from both sides.}$$
$$-y = -2x + 5 \qquad \text{Divide both sides by } -1.$$
$$y = 2x - 5$$

After solving for y in terms of x, we can enter the equation. Press $\boxed{Y =}$ and enter $2X - 5$ after $Y_1 =$ by keying in 2 $\boxed{X, T, \theta, n}$ $\boxed{-}$ 5. The display should look like Figure 1.7a.

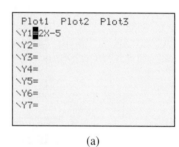

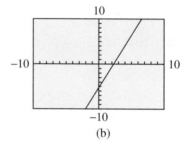

(a)   (b)

FIGURE 1.7

Next we select a "graphing window," which corresponds to drawing the x- and y-axes and choosing a scale for each when we graph by hand. Press the $\boxed{\text{ZOOM}}$ key, and then the number $\boxed{6}$ to select the "standard graphing window." This window displays values from -10 to 10 on both axes. The graph is shown in Figure 1.7b.

EXAMPLE 1

Use a graphing calculator to graph the equation $3x + 2y = 16$.

Solution

First solve the equation for y in terms of x.

$$3x + 2y = 16 \qquad \text{Subtract } 3x \text{ from both sides.}$$
$$2y = -3x + 16 \qquad \text{Divide both sides by 2.}$$
$$y = \frac{-3x + 16}{2} \qquad \text{Simplify.}$$
$$\text{or} \qquad y = -1.5x + 8$$

Now press $\boxed{\text{Y}=}$ and $\boxed{\text{CLEAR}}$ to clear any equations on the display. Enter $-1.5\text{X} + 8$ after "$Y_1 = .$" Use the negation key, $\boxed{(-)}$, when you enter $-1.5x$. (Do not use the $\boxed{-}$ key, which is for subtraction.) Choose the standard graphing window by pressing $\boxed{\text{ZOOM}}\ \boxed{6}$. The graph is shown in Figure 1.8.

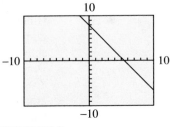

FIGURE 1.8

Here is a summary of the procedure for entering a graph in the standard window.

To Graph an Equation:

1. Press $\boxed{\text{Y}=}$ and enter the equation(s) you wish to graph.
2. Select the standard graphing window by pressing $\boxed{\text{ZOOM}}\ \boxed{6}$.

When you want to erase the graphs, press $\boxed{\text{Y}=}$ and then press $\boxed{\text{CLEAR}}$. This deletes the expression following Y_1. Use the down arrow key, $\boxed{\triangledown}$, and the $\boxed{\text{CLEAR}}$ key to delete the rest of the equations on the display.

EXERCISE 1

a. Solve the equation $7 - 2y = 4x$ for y in terms of x.

Subtract 7 from both sides.
Divide both sides by -2.
Simplify.

b. Graph the equation in the standard window.

Press $\boxed{\text{Y}=}$ and enter the equation.
Then press $\boxed{\text{ZOOM}}\ \boxed{6}$.

Choosing a Graphing Window

Now consider the equation

$$6x - 5y = 90 \qquad (1)$$

To solve for y, we first subtract $6x$ from both sides of the equation.

$$-5y = 90 - 6x \qquad \text{Divide both sides by } -5.$$

$$y = \frac{90}{-5} - \frac{6x}{-5} \qquad \text{Simplify.}$$

$$y = -18 + \frac{6}{5}x$$

If we graph this equation in the standard window, we see only a small piece of the graph in the lower corner of the screen, as shown in Figure 1.9. We should choose a window that shows more of the graph. How do we choose an appropriate window for a graph? The window should show the essential features of the graph and give the viewer a good idea of its overall shape. For the graph of a linear equation, the most informative points are the *x*- and *y*-intercepts. Both intercepts should be visible in the graphing window.

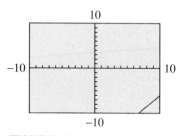

FIGURE 1.9

To graph a linear equation, we should choose scales for the axes so that both intercepts can be shown on the graph.

You can check that the intercepts of the graph of Equation (1) are (15, 0) and (0, −18). Thus, a good choice for the window might be

$$\text{Xmin} = -20 \qquad \text{Xmax} = 20$$
$$\text{Ymin} = -20 \qquad \text{Ymax} = 20$$

Press $\boxed{\text{WINDOW}}$ and enter the values above for the edges of the window, using the $\boxed{\triangledown}$ key to move through the menu items. Then press $\boxed{\text{GRAPH}}$. The resulting graph is shown in Figure 1.10. If we want to concentrate on the portion of the graph in the fourth quadrant, we can choose the window settings

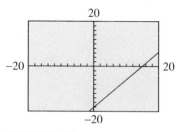

FIGURE 1.10

$$\text{Xmin} = 0 \qquad \text{Xmax} = 20$$
$$\text{Ymin} = -20 \qquad \text{Ymax} = 0$$

(Try this window for yourself.)

EXAMPLE 2

Choose an appropriate window and graph the equation $10y - 15x = 6$.

Solution

First solve for *y* in terms of *x*. Add 15*x* to both sides to find

$$10y = 15x + 6 \qquad \text{Divide both sides by 10.}$$
$$y = 1.5x + 0.6$$

Press $\boxed{\text{Y =}}$ and enter this equation.

To decide on a good window, we find the *x*- and *y*-intercepts of the graph. Setting *x* equal to zero, we find $y = 0.6$, and when $y = 0$, we find $x = -0.4$. The intercepts are thus (−0.4, 0) and (0, 0.6). For this graph we choose a very small window so that we can distinguish the intercepts from the origin. We will set *x*- and *y*-limits slightly larger than the intercept values. Figure 1.11 shows the graph in the window

$$\text{Xmin} = -1 \qquad \text{Xmax} = 1$$
$$\text{Ymin} = -1 \qquad \text{Ymax} = 1$$

We have also set Xscl and Yscl equal to 0.1, so that the tick marks occur at intervals of 0.1 unit.

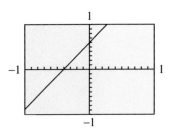

FIGURE 1.11

■

EXERCISE 2

a. Find the *x*- and *y*-intercepts of the graph of $2y - 1440 = 45x$.

x	y
0	
	0

b. Graph the equation on a graphing calculator. Choose a window that shows both of the intercepts.

$$\text{Xmin} = \qquad \text{Xmax} =$$
$$\text{Ymin} = \qquad \text{Ymax} =$$

Finding Coordinates with a Graphing Calculator

We can use the TRACE feature of the calculator to find the coordinates of points on a graph. For example, graph the equation $y = -2.6x - 5.4$ in the window

$$\text{Xmin} = -5 \qquad \text{Xmax} = 4.4$$
$$\text{Ymin} = -20 \qquad \text{Ymax} = 15$$

Press TRACE, and a "bug" begins flashing on the display. The coordinates of the bug appear at the bottom of the display, as shown in Figure 1.12. Use the left and right arrows to move the bug along the graph. (You can check that the coordinates of the point $(2, -10.6)$ do satisfy the equation $y = -2.6x - 5.4$.)

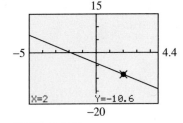

FIGURE 1.12

The points identified by the **Trace** bug depend on the window settings and on the type of calculator. If we want to find the *y*-coordinate for a particular *x*-value, we can type the *x*-value and press ENTER. After activating the **Trace**, try entering 2, and you should find that $y = -10.6$, as before.

Graphical Solution of Equations

The graph of an equation in two variables is just a picture of its solutions. When we read the coordinates of a point on the graph, we are reading a pair of x- and y-values that make the equation true.

EXAMPLE 3

a. Verify, algebraically and graphically, that the ordered pair $(2, 7)$ is a solution of the equation $y = 2x + 3$.

b. Verify that the ordered pair $(3, 6)$ is not a solution of $y = 2x + 3$.

Solutions

a. In Figure 1.13, the point $(2, 7)$ lies on the graph of $y = 2x + 3$. This confirms graphically that $(2, 7)$ is a solution. To verify algebraically that $(2, 7)$ is a solution, substitute $x = 2$ and $y = 7$ into the equation:

$$7 \overset{?}{=} 2\,(2) + 3 \quad \text{True}$$

b. The point $(3, 6)$ does not lie on the graph of $y = 2x + 3$, so $(3, 6)$ cannot be a solution of the equation. We can verify algebraically that $(3, 6)$ is not a solution:

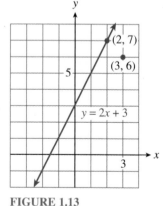

FIGURE 1.13

$$6 \overset{?}{=} 2\,(3) + 3 \quad \text{False.}$$

In Example 3, we can also say that $x = 2$ is a solution of the equation $2x + 3 = 7$. Thus, we can use graphs to find solutions to equations in one variable. For example, we'll use the graph of

$$y = 285 - 15x$$

shown in Figure 1.14, to find the solution of

$$150 = 285 - 15x \tag{2}$$

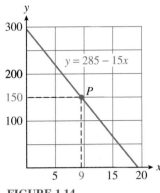

FIGURE 1.14

Begin by locating the point P on the graph for which $y = 150$. Now find the x-coordinate of point P by drawing an imaginary line from P straight down to the x-axis. The x-coordinate of P is $x = 9$. Thus, P is the point $(9, 150)$, and $x = 9$ when $y = 150$. The solution of Equation (2) is $x = 9$.

You can verify the solution algebraically by substituting $x = 9$ into Equation (2):

Does $150 = 285 - 15(9)$?

$$285 - 15(9) = 285 - 135 = 150 \quad \text{Yes.}$$

The relationship between an equation and its graph is an important one. For the previous example, make sure you understand that the following three statements are equivalent.

1. The point $(9, 150)$ lies on the graph of $y = 285 - 15x$.
2. The ordered pair $(9, 150)$ is a solution of the equation $y = 285 - 15x$.
3. $x = 9$ is a solution of the equation $150 = 285 - 15x$.

EXERCISE 4

a. Use the graph of $y = 30 - 8x$ to solve the equation $30 - 8x = 50$.

b. Verify your solution algebraically.

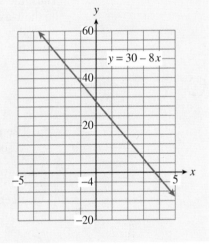

Graphical Solution of Inequalities

We can also use graphs to solve inequalities. Consider the inequality

$$285 - 15x \geq 150$$

To "solve" this inequality means to find all values of x that make the expression $285 - 15x$ greater than 150 or equal to 150. We could begin by trying some values of x. Here is a table obtained by evaluating $285 - 15x$.

TABLE 1.2

x	0	2	4	6	8	10	12
$285 - 15x$	285	255	225	195	165	135	105

From Table 1.2, it appears that values of x less than or equal to 8 are solutions of the inequality, but we have not checked all possible x-values. We can get a more complete picture from a graph. Figure 1.15 shows the graph of the equation $y = 285 - 15x$. To solve the inequality

$$285 - 15x \geq 150$$

we look for points on the graph with y-coordinates greater than or equal to 150. These points are shown in color.

Which x-values produced these points with y-values greater than or equal to 150? We can read the x-coordinates by dropping straight down to the x-axis, as shown by the arrows. For example, the x-value corresponding to $285 - 15x = 150$

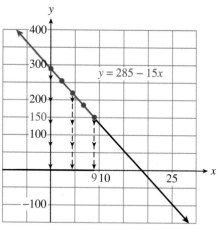

FIGURE 1.15

is $x = 9$. For larger values of $285 - 15x$ we must choose x-values less than 9. Thus, the solutions are all values of x less than or equal to 9, as shown in color on the x-axis in Figure 1.15. You may want to verify the solution by solving the inequality algebraically. (See Appendix A.3 to review algebraic solution of inequalities.)

EXERCISE 5

a. Use the graph of $y = 30 - 8x$ to solve the inequality $30 - 8x < 14$.

b. Verify your solution algebraically.

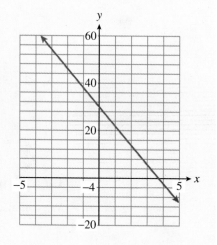

Using a Graphing Calculator to Solve Equations and Inequalities

With the aid of a graphing calculator, we can quickly find (at least approximate) solutions to equations and inequalities.

EXAMPLE 4

a. Use a graphing calculator to solve the equation $572 - 23x = 181$. Set the (WINDOW) to

$$\text{Xmin} = -40 \qquad \text{Xmax} = 54$$
$$\text{Ymin} = -100 \qquad \text{Ymax} = 900$$

b. Use the same graph to solve the inequality $572 - 23x \geq 181$.

Solutions

a. Because the graphing calculator does not display scales on the axes, it is often helpful to plot a horizontal line corresponding to the y-value of interest. After setting the (WINDOW), enter $Y_1 = 572 - 23X$ and $Y_2 = 181$. Press (GRAPH) to see the graphs shown in Figure 1.16. Activate the (TRACE), and use the arrow keys until the bug rests on the intersection of the two lines, that is, when the y-coordinate is 181. The corresponding value of x is $x = 17$. Thus, the solution to the equation $572 - 23x = 181$ is $x = 17$.

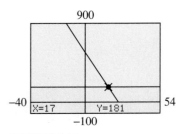

FIGURE 1.16

b. To solve the inequality

$$572 - 23x \geq 181$$

we want points with y-values greater than 181. Experiment by pressing the left and right arrow keys to move the bug to either side of the intersection point. Notice that points with y-coordinates *greater* than 181 have x-coordinates *less* than 17. Thus, the solution to the inequality is $x \leq 17$. ∎

EXERCISE 6

a. Graph the equation $y = 1.3x + 2.4$. Set the (WINDOW) to

$$\text{Xmin} = -4.6 \qquad \text{Xmax} = 4.8$$
$$\text{Ymin} = -10 \qquad \text{Ymax} = 10$$

b. Use your graph to solve the inequality $1.3x + 2.4 < 8.51$.

READING QUESTIONS

1. How can you decide on an appropriate graphing window for a linear equation?

2. How can you determine graphically whether a given ordered pair is a solution to an equation in two variables?

3. How can you determine algebraically whether a given ordered pair is a solution to an equation in two variables?

4. What equation could you graph to solve $1.2x - 8.4 = 6$ graphically?

5. Explain how to use a graph to solve the inequality $35x + 850 \leq 2400$.

ANSWERS TO 1.2 EXERCISES

1a. $y = -2x + \dfrac{7}{2}$ **2a.**

x	y
0	720
−32	0

3. −2.5   **4.** −2.5

5. $x > 2$ **6b.** $x < 4.7$

HOMEWORK 1.2

In Problems 1–8,

a. Solve each equation for y in terms of x,

b. Graph the equation on your calculator in the specified window,

c. Make a pencil and paper sketch of the graph. Label the scales on your axes, and show the coordinates of the intercepts.

1. $2x + y = 6$
   Standard window

2. $8 - y + 3x = 0$
   Standard window

3. $3x - 4y = 1200$
   Xmin = −1000
   Xmax = 1000
   Xscl = 100
   Ymin = −1000
   Ymax = 1000
   Yscl = 100

4. $x + 2y = 500$
   Xmin = −1000
   Xmax = 1000
   Xscl = 100
   Ymin = −1000
   Ymax = 1000
   Yscl = 100

5. $0.2x + 5y = 0.1$
   Xmin = −1
   Xmax = 1
   Xscl = 0.1
   Ymin = −0.1
   Ymax = 0.1
   Yscl = 0.01

6. $1.2x - 4.2y = 3.6$
   Xmin = −1
   Xmax = 4
   Xscl = 0.2
   Ymin = −1
   Ymax = 1
   Yscl = 0.1

7. $70x + 3y = y + 420$
   Xmin = 0
   Xmax = 10
   Xscl = 1
   Ymin = 0
   Ymax = 250
   Yscl = 25

8. $40y - 5x = 780 - 20y$
   Xmin = −200
   Xmax = 0
   Xscl = 20
   Ymin = 0
   Ymax = 20
   Yscl = 2

For each of the equations in Problems 9–16,

a. Find the x- and y-intercepts.

b. Solve the equation for y.

c. Use the intercepts to select a suitable graphing window. (Both intercepts should be visible and distinguishable).

d. Graph the resulting equation using the window you specified. It may be convenient to set both Xscl and Yscl to 0.

9. $x + y = 100$

10. $x - y = -72$

11. $25x - 36y = 1$

12. $43x + 71y = 1$

13. $\dfrac{y}{12} - \dfrac{x}{47} = 1$

14. $\dfrac{x}{8} + \dfrac{y}{21} = 1$

15. $-2x = 3y + 84$

16. $7x = 91 - 13y$

17. Figure 1.17 shows a graph of $y = -2x + 6$.

  a. Use the graph to find all values of x for which:

     i) $y = 12$ ii) $y > 12$ iii) $y < 12$

  b. Use the graph to solve:

     i) $-2x + 6 = 12$ ii) $-2x + 6 > 12$

     iii) $-2x + 6 < 12$

  c. Explain why your answers to parts (a) and (b) are the same.

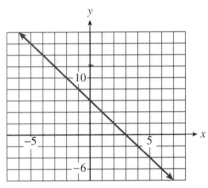

FIGURE 1.17

18. Figure 1.18 shows a graph of $y = \dfrac{-x}{3} - 6$.

  a. Use the graph to find all values of x for which:

     i) $y = -4$ ii) $y > -4$ iii) $y < -4$

  b. Use the graph to solve:

     i) $\dfrac{-x}{3} - 6 = -4$ ii) $\dfrac{-x}{3} - 6 > -4$

     iii) $\dfrac{-x}{3} - 6 < -4$

  c. Explain why your answers to parts (a) and (b) are the same.

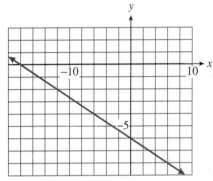

FIGURE 1.18

In Problems 19 and 20, use the graph to solve the equation or inequality, and then solve algebraically. (If you have forgotten how to solve linear inequalities algebraically, please review Appendix A.3.)

19. Figure 1.19 shows the graph of $y = 1.4x - 0.64$. Solve:

  a. $1.4x - 0.64 = 0.2$
  b. $-1.2 = 1.4x - 0.64$
  c. $1.4x - 0.64 > 0.2$
  d. $-1.2 > 1.4x - 0.64$

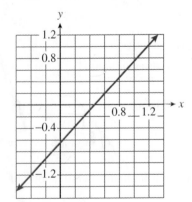

FIGURE 1.19

20. Figure 1.20 shows the graph of $y = -2.4x + 2.32$. Solve:

  a. $1.6 = -2.4x + 2.32$
  b. $-2.4x + 2.32 = 0.4$
  c. $-2.4x + 2.32 \geq 1.6$
  d. $0.4 \geq -2.4x + 2.32$

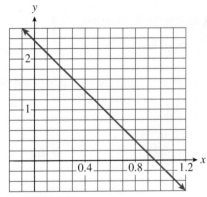

FIGURE 1.20

In Problems 21–24, graph each equation with the ZInteger setting. (Press [ZOOM] [6], and then [ZOOM] [8] [ENTER].) Use the graph to answer each question. Use the equation to verify your answers.

21. Graph $y = 2x - 3$.
  a. For what value of x is $y = 5$?
  b. For what value of x is $y = -13$?
  c. For what values of x is $y > -1$?
  d. For what values of x is $y < 25$?

23. Graph $y = 6.5 - 1.8x$.
  a. For what value of x is $y = -13.3$?
  b. For what value of x is $y = 24.5$?
  c. For what values of x is $y \leq 15.5$?
  d. For what values of x is $y \geq -7.9$?

22. Graph $y = 4 - 2x$.
  a. For what value of x is $y = 6$?
  b. For what value of x is $y = -4$?
  c. For what values of x is $y > -12$?
  d. For what values of x is $y < 18$?

24. Graph $y = 0.2x + 1.4$.
  a. For what value of x is $y = -5.2$?
  b. For what value of x is $y = 2.8$?
  c. For what values of x is $y \leq -3.2$?
  d. For what values of x is $y \geq 4.4$?

In Problems 25–28, graph each equation with the ZInteger setting. Use the graph to solve each equation or inequality. Check your solutions algebraically.

25. Graph $y = -0.4x + 3.7$.
  a. Solve $-0.4x + 3.7 = 2.1$.
  b. Solve $-0.4x + 3.7 > -5.1$.

26. Graph $y = 0.4(x - 1.5)$
  a. Solve $0.4(x - 1.5) = -8.6$.
  b. Solve $0.4(x - 1.5) < 8.6$.

27. Graph $y = \dfrac{2}{3}x - 24$.
  a. Solve $\dfrac{2}{3}x - 24 = -10\dfrac{2}{3}$.
  b. Solve $\dfrac{2}{3}x - 24 \leq -19\dfrac{1}{3}$.

28. Graph $y = \dfrac{80 - 3x}{5}$.
  a. Solve $\dfrac{80 - 3x}{5} = 22\dfrac{3}{5}$.
  b. Solve $\dfrac{80 - 3x}{5} \leq 9\dfrac{2}{5}$.

29. Kieran's resting blood pressure (in mm Hg) is 120 and it rises by 6 mm for each minute he jogs on a treadmill programmed to increase the level of intensity at a steady rate.
  a. Fill in the table.

Time on treadmill (minutes)	0	1	2	5	10
Blood pressure (mm Hg)					

  b. Find a formula for Kieran's blood pressure, p, in terms of time, t.
  c. Use your calculator to graph the equation for p in terms of t.
  d. What is Kieran's blood pressure after 3.5 minutes?
  e. Kieran's blood pressure should not exceed 165 mm Hg. When should the treadmill stop increasing his exercise intensity?

30. When Francine is at rest, her cardiac output is 5 liters per minute. The output increases by 3 liters per minute for each minute she spends on a cycling machine that gradually increases her exercise intensity.
  a. Fill in the table.

Time on cycle (minutes)	0	1	2	5	10
Cardiac output (liters/minute)					

  b. Find a formula for Francine's cardiac output, c, in terms of time, t.
  c. Use your calculator to graph the equation for c in terms of t.
  d. What will Francine's cardiac output be after 6 minutes?
  e. For how long will Francine's cardiac output be less than 14.5 liters/minute?

1.3 Slope

Using Ratios for Comparison

Which is more expensive: a 64-ounce bottle of Velvolux dish soap that costs $3.52, or a 60-ounce bottle of Rainfresh dish soap that costs $3.36?

You are probably familiar with the notion of "comparison shopping." To determine which dish soap is the better buy, we compute the unit price, or price per ounce, for each bottle. The unit price for Velvolux is

$$\frac{352 \text{ cents}}{64 \text{ ounces}} = 5.5 \text{ cents per ounce}$$

and the unit price for Rainfresh is

$$\frac{336 \text{ cents}}{60 \text{ ounces}} = 5.6 \text{ cents per ounce}$$

The Velvolux costs less per ounce, so it is the better buy. By computing the price of each brand for *the same amount of soap* it is easy to compare them.

In many situations a ratio, similar to a unit price, can provide a basis for comparison. Example 1 uses a ratio to measure a rate of growth.

EXAMPLE 1

Which grow faster: Hybrid A wheat seedlings, which grow 11.2 centimeters in 14 days, or Hybrid B seedlings, which grow 13.5 centimeters in 18 days?

Solution

We compute the growth rate for each strain of wheat. Growth rate is expressed as a ratio, $\dfrac{\text{centimeters}}{\text{days}}$, or centimeters per day. The growth rate for Hybrid A is

$$\frac{11.2 \text{ centimeters}}{14 \text{ days}} = 0.8 \text{ centimeter per day}$$

and the growth rate for Hybrid B is

$$\frac{13.5 \text{ centimeters}}{18 \text{ days}} = 0.75 \text{ centimeter per day}$$

Because their rate of growth is larger, the Hybrid A seedlings grow faster. ■

By computing the growth of each strain of wheat seedling over *the same unit of time,* a single day, we have a basis for comparison. We can use this same idea, finding a common basis for comparison, to measure and compare the steepness of an incline.

Measuring Steepness

Imagine you are an ant carrying a heavy burden along one of the two paths shown in Figure 1.21. Which path is more strenuous? Most ants would agree that the steeper path is more difficult. But what exactly is "steepness?" It is not merely the gain in altitude, because even a gentle incline will reach a great height eventually. Steepness measures how sharply the altitude increases. An ant finds the upper path more difficult, or steeper, because it rises five feet while the lower path rises only two feet *over the same horizontal distance*.

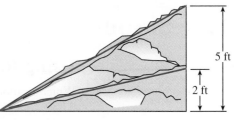

FIGURE 1.21

To compare the steepness of two inclined paths, we compute the ratio of change in altitude to change in horizontal distance for each path.

EXAMPLE 2

Which is steeper: Stony Point trail, which climbs 400 feet over a horizontal distance of 2500 feet, or Lone Pine trail, which climbs 360 feet over a horizontal distance of 1800 feet?

Solution

For each trail, we compute the ratio of vertical gain to horizontal distance. For Stony Point trail, the ratio is

$$\frac{400 \text{ feet}}{2500 \text{ feet}} = 0.16$$

and for Lone Pine trail, the ratio is

$$\frac{360 \text{ feet}}{1800 \text{ feet}} = 0.20$$

Lone Pine trail is steeper, since it has a vertical gain of 0.20 foot for every foot traveled horizontally. Or, in more practical units, Lone Pine trail rises 20 feet for every 100 feet of horizontal distance, whereas Stony Point trail rises only 16 feet over a horizontal distance of 100 feet. ∎

Definition of Slope

To compare the steepness of the two trails in Example 2 it is not enough to know which trail has the greater gain in elevation overall! Instead we compare their elevation gains over the *same horizontal distance*. Using the same horizontal distance provides a basis for comparison. The two trails are illustrated in Figure 1.22 as lines on a coordinate grid.

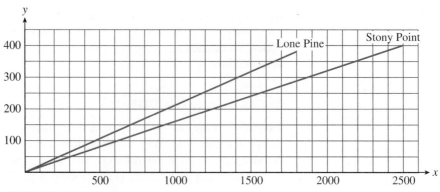

FIGURE 1.22

The ratio we computed in Example 2,

$$\frac{\text{change in elevation}}{\text{change in horizontal position}}$$

appears on the graphs in Figure 1.22 as

$$\frac{\text{change in } y\text{-coordinate}}{\text{change in } x\text{-coordinate}}$$

For example, as we travel along the line representing Stony Point trail, we move from the point $(0, 0)$ to the point $(2500, 400)$. The y-coordinate changes by 400 and the x-coordinate changes by 2500, giving the ratio 0.16 that we found in Example 2. We call this ratio the **slope** of the line.

> The **slope** of a line is the ratio
>
> $$\frac{\textbf{change in } y\textbf{-coordinate}}{\textbf{change in } x\textbf{-coordinate}}$$
>
> as we move from one point to another on the line.

EXAMPLE 3

Compute the slope of the line that passes through points A and B in Figure 1.23.

Solution

As we move along the line from $A(1, 3)$ to $B(5, 6)$, the y-coordinate changes by 3 units, and the x-coordinate changes by 4 units. The slope of the line is thus

$$\frac{\text{change in } y\text{-coordinate}}{\text{change in } x\text{-coordinate}} = \frac{3}{4}$$

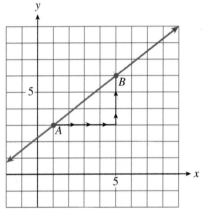

FIGURE 1.23   ■

EXERCISE 1   Compute the slope of the line through the indicated points. On both axes, one square represents one unit.

$$\frac{\text{change in } y\text{-coordinate}}{\text{change in } x\text{-coordinate}} =$$

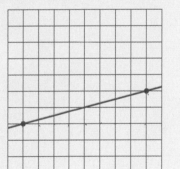

$\frac{2}{8}$

The slope of a line is a *number*. It tells us how much the *y*-coordinates of points on the line increases when we increase the *x*-coordinate by one unit. A larger slope indicates a greater increase in altitude (at least for increasing graphs), and hence a steeper line.

Notation for Slope

Because the ratio that defines slope $\dfrac{\text{change in } y\text{-coordinate}}{\text{change in } x\text{-coordinate}}$ is cumbersome to write out, we use a short-hand notation. The symbol Δ (the Greek letter "delta") is used in mathematics to denote "change in." In particular, Δy means "change in *y*-coordinate," and Δx means "change in *x*-coordinate." (See Figure 1.24.) We also use the letter **m** to stand for slope. With these symbols, we can write the definition of slope as follows.

The **slope** of a line is given by

$$m = \frac{\Delta y}{\Delta x}, \ (\Delta x \neq 0)$$

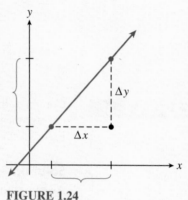

FIGURE 1.24

So far we have only considered examples in which Δx and Δy are positive numbers, but they can also be negative. We say that Δx is positive if we move to the right, and negative if we move to the left as we travel from one point to another on the line. We say that Δy is positive if we move up, and negative if we move down.

EXAMPLE 4

Compute the slope of the line that passes through the points $P(-4, 2)$ and $Q(5, -1)$ shown in Figure 1.25. Illustrate Δy and Δx on the graph.

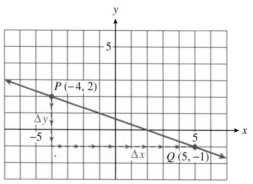

FIGURE 1.25

Solution

As we move from the point $P(-4, 2)$ to the point $Q(5, -1)$, we move 3 units *down* in the y-direction, so $\Delta y = -3$. We then move 9 units to the right in the x-direction, so $\Delta x = 9$. Thus the slope is

$$m = \frac{\Delta y}{\Delta x} = \frac{-3}{9} = \frac{-1}{3}$$

Δy and Δx are labeled on the graph. ∎

The line graphed in Example 4 *decreases* as we move from left to right and has a negative slope. The slope is the same if we move from point Q to point P, instead of from P to Q. In that case our computation looks like this:

$$m = \frac{\Delta y}{\Delta x} = \frac{3}{-9} = \frac{-1}{3}$$

(See Figure 1.26.) We can move from point to point in either direction to compute the slope.

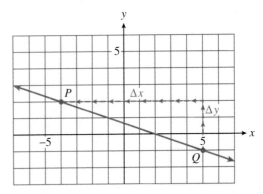

FIGURE 1.26

Lines Have Constant Slope

How do we know which two points to choose when we want to compute the slope of a line? It turns out that any two points on the line will do.

EXERCISE 2

a. Graph the line $4x - 2y = 8$ by finding the x- and y-intercepts.

x	y
0	~ 4
2	0

b. Compute the slope of the line using the x-intercept and y-intercept. Move from $(0, -4)$ to $(2, 0)$ along the line.

$$m = \frac{\Delta y}{\Delta y} = \qquad \frac{-4}{0}$$

c. Compute the slope of the line using the points $(4, 4)$ and $(1, -2)$.

$$m = \frac{\Delta y}{\Delta y} =$$

$$\frac{4-2}{4-1} \qquad \frac{2}{3}$$

Exercise 2 illustrates an important property of lines: They have constant slope. No matter what two points we use to calculate the slope, we will always get the same result. We will see later that lines are the only graphs that have this property.

Meaning of Slope

In Example 1 on page 3, we graphed the equation $C = 5 + 3t$ showing the cost of a bicycle rental in terms of the length of the rental. The graph is reproduced in Figure 1.27. We can choose any two points on the line to compute its slope. Using points P and Q as shown,

$$m = \frac{\Delta C}{\Delta t} = \frac{9}{3} = 3$$

The slope of the line is 3.

What does this value mean for the cost of renting a bicycle? The expression

$$\frac{\Delta C}{\Delta t} = \frac{9}{3}$$

stands for

$$\frac{\text{change in cost}}{\text{change in time}} = \frac{9 \text{ dollars}}{3 \text{ hours}}$$

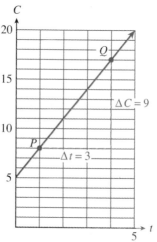

FIGURE 1.27

If we increase the length of the rental by 3 hours, the cost of the rental increases by \$9. The slope gives the *rate of increase* in the rental fee, 3 dollars per hour.

In general, the slope of a line measures the rate of change of y with respect to x. In different situations, this rate might be interpreted as a rate of growth or a speed. A negative slope might represent a rate of decrease or a rate of consumption. Interpreting the slope of a graph can give us valuable information about the variables involved.

EXAMPLE 5

The graph in Figure 1.28 shows the distance in miles traveled by a driver for a cross-country trucking firm in terms of the number of hours she has been on the road.

  a. Compute the slope of the graph.

  b. What is the significance of the slope to the problem?

Solutions

  a. Choose any two points on the line, say $G(2, 100)$ and $H(4, 200)$ in Figure 1.28. As we move from G to H, we find

$$m = \frac{\Delta D}{\Delta t} = \frac{100}{2} = 50$$

The slope of the line is 50.

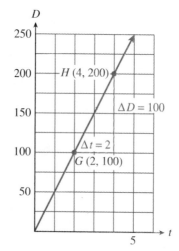

FIGURE 1.28

b. Consider the meanings of ΔD and Δt in the computation of the slope. The best way to understand these is to include the units in the calculation.

$$\frac{\Delta D}{\Delta t} \qquad \text{means} \qquad \frac{\text{change in distance}}{\text{change in time}}$$

or

$$\frac{\Delta D}{\Delta t} = \frac{100 \text{ miles}}{2 \text{ hours}} = 50 \text{ miles per hour}$$

The slope represents the trucker's average speed or velocity. ∎

EXERCISE 3 The graph shows the altitude a (in feet) of a skier t minutes after getting on a ski lift.

a. Choose two points and compute the slope (including units).

$$m = \frac{\Delta a}{\Delta t} =$$

b. Explain what the slope measures in the context of the problem.

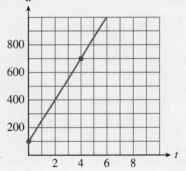

READING QUESTIONS

1. Explain how unit pricing works.
2. Explain how slope measures the steepness of an incline.
3. Explain the notation used for slope.
4. A classmate says that you must always use the intercepts to calculate the slope of a line. Do you agree? Explain.
5. In an application, what does the slope of the graph tell you about the situation?

ANSWERS TO 1.3 EXERCISES

1. $\frac{1}{4}$ **2a.**

x	y
0	-4
2	0

b. 2   **c.** 2

3a. 150   **b.** Altitude increases by 150 feet per minute.

Compute ratios to answer the questions in Problems 1–4.

1. Carl runs 100 meters in 10 seconds. Anthony runs 200 meters in 19.6 seconds. Who has the faster average speed?

2. On his 512-mile round trip to Las Vegas and back, Corey needed 16 gallons of gasoline. He used 13 gallons of gasoline on a 429-mile trip to Los Angeles. On which trip did he get better fuel economy?

3. Grimy Gulch Pass rises 0.6 mile over a horizontal distance of 26 miles. Bob's driveway rises 12 feet over a horizontal distance of 150 feet. Which is steeper?

4. Which is steeper: the truck ramp for Acme Movers, which rises 4 feet over a horizontal distance of 9 feet, or a toy truck ramp, which rises 3 centimeters over a horizontal distance of 7 centimeters?

In Problems 5–8, compute the slope of the line through the indicated points. (On both axes, one square represents one unit.)

5.

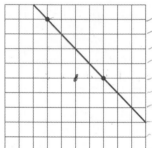

FIGURE 1.29

6.

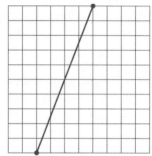

FIGURE 1.30

7.

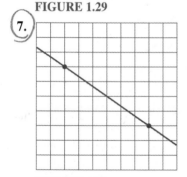

FIGURE 1.31

8.

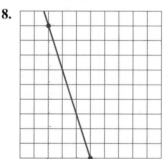

FIGURE 1.32

9. **a.** Compute the slope of the line in Figure 1.33.

   b. Start at point (0, 2) and move 4 units in the positive x-direction. How many units must you move in the y-direction to get back to the line? What is the ratio of Δy to Δx?

   c. Repeat part (b) starting at any other point on the line. Are your answers the same?

   d. Start at point (0, 2) and move -6 units in the x-direction. How many units must you move in the y-direction to get back to the line? What is the ratio of Δy to Δx?

   e. Suppose you start at any point on the line and move 18 units in the x-direction. How many units must you move in the y-direction to get back to the line? Use the equation $m = \dfrac{\Delta y}{\Delta x}$ to calculate your answer.

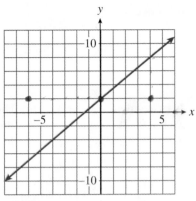

FIGURE 1.33

10. **a.** Compute the slope of the line in Figure 1.34.

    b. Start at point $(0, -6)$ and move -6 units in the y-direction (down). How many units must you move in the x-direction to get back to the line? What is the ratio of Δy to Δx?

    c. Repeat part (b) starting at any other point on the line. Are your answers the same?

    d. Start at point $(0, -6)$ and move 9 units in the y-direction. How many units must you move in the x-direction to get back to the line? What is the ratio of Δy to Δx?

    e. Suppose you start at any point on the line and move 20 units in the y-direction. How many units must you move in the x-direction to get back to the line? Use the equation $m = \dfrac{\Delta y}{\Delta x}$ to calculate your answer.

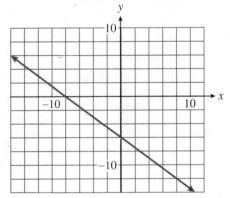

FIGURE 1.34

11. A line has slope $-\frac{4}{5}$. Find the horizontal change associated with each of the following vertical changes along the line. Use the equation $m = \dfrac{\Delta y}{\Delta x}$.

    a. $\Delta y = -4$ **b.** $\Delta y = 2$

    c. $\Delta y = -12$ **d.** $\Delta y = 5$

12. A line has slope $\frac{7}{3}$. Find the vertical change associated with each of the following horizontal changes along the line. Use the equation $m = \dfrac{\Delta y}{\Delta x}$.

    a. $\Delta x = 3$ **b.** $\Delta x = 6$

    c. $\Delta x = 1$ **d.** $\Delta x = -24$

13. Residential staircases are usually built with a slope of 70%, or $\frac{7}{10}$. If the vertical distance between stories is 10 feet, how much horizontal space does the staircase require?

14. A straight section of highway in the Midwest maintains a grade (slope) of 4%, or $\frac{1}{25}$, for 12 miles. How much does your elevation change as you travel the road?

15. Geologists can measure the depth of the ocean at different points using a technique called echo-sounding. Scientists on board a ship project a pulse of sound toward the ocean floor and measure the time interval until the echo returns to the ship. The speed of sound in seawater is about 1500 meters per second. If the echo returns in 4.5 seconds, what is the depth of the ocean at that point?

16. Niagara Falls was discovered by Father Louis Hennepin in 1682. In 1952 much of the water of the Niagara River was diverted for hydroelectric power, but until that time erosion caused the Falls to recede upstream by 3 feet per year.

    a. How far have the Falls receded since Father Hennepin first encountered them?

    b. The Falls were formed about 12,000 years ago during the end of the last ice age. How far downstream from their current position were they then? (Give your answer in miles.)

In Problems 17–22, (a) graph each line by the intercept method, and (b) use the intercepts to compute the slope.

17. $3x - 4y = 12$ **18.** $2y - 5x = 10$ **19.** $2y + 6x = -18$ **20.** $9x + 12y = 36$

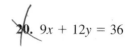

21. $\dfrac{x}{5} - \dfrac{y}{8} = 1$ **22.** $\dfrac{x}{7} - \dfrac{y}{4} = 1$

23. The table shows the amount of ammonium chloride salt, in grams, that can be dissolved in 100 grams of water at different temperatures.

Temperature, °C	10	12	15	21	25	40	52
Grams of salt	33	34	35.5	38.5	40.5	48	54

   a. If you plot the data, will it lie on a straight line? Why or why not?

   b. Calculate the rate of change of salt dissolved with respect to temperature.

24. A spring is suspended from the ceiling. The table shows the length of the spring in centimeters as it is stretched by hanging various weights from it.

Weight, kg	3	4	8	10	12	15	22
Length, cm	25.76	25.88	26.36	26.6	26.84	27.2	28.04

   a. If you plot the data, will it lie on a straight line? Why or why not?

   b. Calculate the rate of change of length with respect to weight.

Which of the following tables represent variables that are related by a linear equation? (Hint: Which relationships have constant slope?)

25. **a.**

x	y
2	12
3	17
4	22
5	27
6	32

b.

t	P
2	4
3	9
4	16
5	25
6	36

26. **a.**

h	w
−6	20
−3	18
0	16
3	14
6	12

b.

t	d
5	0
10	3
15	6
20	12
25	24

27. A temporary typist's paycheck (before deductions) in dollars is given by $S = 8t$, where t is the number of hours she worked.

   a. Make a table of values for the equation.

   b. Graph the equation.

   c. Using two points on the graph, compute the slope $\dfrac{\Delta S}{\Delta t}$, including units.

   d. What is the significance of the slope in terms of the typist's paycheck?

28. The distance in miles covered by a cross-country competitor is given by $d = 6t$, where t is the number of hours she runs.

   a. Make a table of values for the equation.

   b. Graph the equation.

   c. Using two points on the graph, compute the slope $\dfrac{\Delta d}{\Delta t}$, including units.

   d. What is the significance of the slope in terms of the cross-country runner?

For each graph in Problems 29–36,

a. Choose two points and compute the slope (including units).

b. Explain what the slope measures in the context of the problem.

29. Figure 1.35 shows the number of barrels of oil, B, that has been pumped at a drill site t days after a new drill is installed.

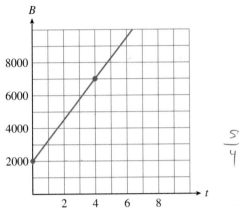

FIGURE 1.35

30. Figure 1.36 shows the amount of garbage, G, (in tons) that has been deposited at a dump t years after new regulations go into effect.

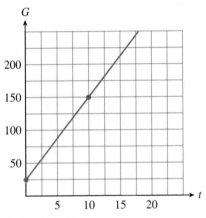

FIGURE 1.36

31. Figure 1.37 shows the amount of water, W, remaining (in liters) in a southern California household t days after an earthquake.

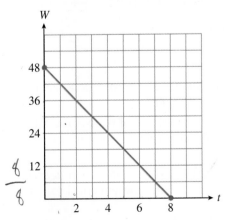

FIGURE 1.37

32. Figure 1.38 shows the amount of money, M, (in dollars) in Tammy's bank account w weeks after she loses all sources of income.

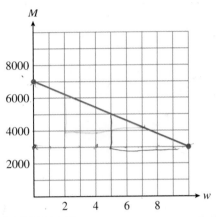

FIGURE 1.38

33. Figure 1.39 shows the length in inches, i, corresponding to various lengths in feet, f.

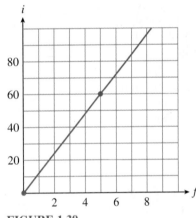

FIGURE 1.39

34. Figure 1.40 shows the number of ounces, z, that correspond to various weights measured in pounds, p.

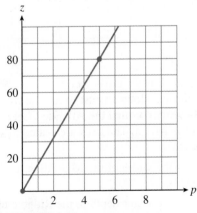

FIGURE 1.40

35. Figure 1.41 shows the cost, C, (in dollars) of coffee beans in terms of the amount of coffee, b (in kilograms).

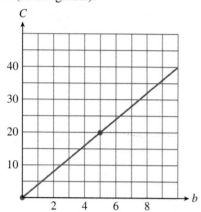

FIGURE 1.41

37. The table gives the radius and circumference of various circles, rounded to three decimal places.

  a. If we plot the data, will the points lie on a straight line?

  b. What familiar number does the slope turn out to be? (*Hint:* Recall a formula from geometry.)

r	C
4	25.133
6	37.699
10	62.832
15	94.248

36. Figure 1.42 shows Tracey's earnings, E, (in dollars) in terms of the number of hours, h, she babysits.

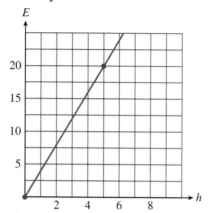

FIGURE 1.42

38. The table gives the side and the diagonal of various squares, rounded to three decimal places.

  a. If we plot the data, will the points lie on a straight line?

  b. What familiar number does the slope turn out to be? (*Hint:* Draw a picture of one of the squares and use the Pythagorean Theorem to compute its diagonal.)

s	d
3	4.243
6	8.485
8	11.314
10	14.142

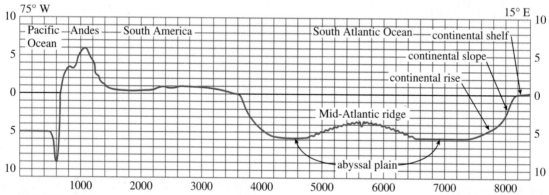

FIGURE 1.43

39. Figure 1.43 shows a cross section of the Earth's surface along an east–west line from South America through the Atlantic Ocean to the coast of Africa. The axes are scaled in kilometers.

  a. What is the width of the Mid-Atlantic Ridge?

  b. What is the water depth at the Ridge crest and at the adjacent abyssal plain? What is

the height of the Mid-Atlantic Ridge above the abyssal plain?

  c. Calculate the slopes of the continental shelf, the continental slope, and the continental rise.

  d. Why do these slopes look much steeper in Figure 1.43 than their numerical values suggest?

Source: Open University, 1998.

40. Geologists calculate the speed of seismic waves by plotting the travel times for waves to reach seismometers at stations a known distance from the epicenter. The speed of the wave can help them determine the nature of the material it passes through. Figure 1.44 shows a travel time graph for P-waves from a shallow earthquake.

a. Why do you think the graph is plotted with distance as the independent variable?

b. Use the graph to calculate the speed of the wave.

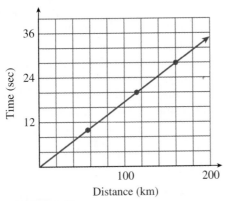

FIGURE 1.44

41. Naismith's Rule is used by runners and walkers to estimate journey times in hilly terrain. In 1892 Naismith wrote in the Scottish Mountaineering Club Journal that a person "in fair condition should allow for easy expeditions an hour for every three miles on the map, with an additional hour for every 2000 feet of ascent."

a. According to Naismith, one unit of ascent requires the same time as how many units of horizontal travel? (This is called Naismith's number.) Round your answer to one decimal place.

b. A walk in the Brecon Beacons in Wales covers 3.75 kilometers horizontally and climbs 582 meters. What is the equivalent flat distance?

c. If you can walk at a pace of 15 minutes per kilometer over flat ground, how long will the walk in the Brecon Beacons take?
Source: Scarf, 1998.

42. Empirical investigations have improved Naismith's number (see Problem 41) to 8.0 for men and 9.5 for women. Part of the Karrimor International Mountain Marathon in the Arrochar Alps in Scotland has a choice of two routes. Route A is 1.75 kilometers long with a 240-meter climb, and route B is 3.25 kilometers long with a 90-meter climb.

a. Which route is faster for women?

b. Which route is faster for men?

c. At a pace of 6 minutes per flat kilometer, how much faster is the preferred route for women?

d. At a pace of 6 minutes per flat kilometer, how much faster is the preferred route for men?
Source: Scarf, 1998.

43. A line passes through the point $(-5, 3)$ and has slope $\frac{2}{3}$. Find the coordinates of two more points on the line.

44. A line passes through the point $(-2, -6)$ and has slope $-\frac{8}{5}$. Find the coordinates of two more points on the line.

MIDCHAPTER REVIEW

1. In the desert the temperature at 6 a.m., just before sunrise, was 65°F. The temperature rose 5 degrees every hour until it reached its maximum value at about 5 p.m. Complete the table of values for the temperature, T, at h hours after 6 a.m.

h	0	3	6	9	10
T					

a. Write an equation for the temperature, T, in terms of h.

b. Graph the equation.

c. How hot is it at noon?

d. When will the temperature be 110°F?

2. The taxi out of Dulles Airport charges a traveler with one suitcase an initial fee of $2.00, plus $1.50 for each mile traveled. Complete the table of values showing the charge, C, for a trip of n miles.

n	0	5	10	15	20	25
C						

a. Write an equation for the charge, C, in terms of the number of miles traveled, n.

b. Graph the equation.

c. What is the charge for a trip to Mount Vernon, 40 miles from the airport?

d. If a ride to the National Institutes of Health costs $39.50, how far is it from the airport to the NIH?

3. Delbert's altitude is 300 meters when he begins to descend in his hot air balloon. He descends by 5 meters each minute.

t	0	10	20	30	40
A					

a. Complete the table of values showing Delbert's altitude, A, t minutes after starting his descent. Graph the data points and connect them with a straight line. Extend the graph until it reaches the horizontal axis.

b. Write an equation for Delbert's altitude, A, in terms of t.

c. Explain the significance of the intercepts to the problem situation.

4. Francine received 120 hours of free Internet connect time as an introductory membership offer from Yippee.com. She spends 1.5 hours per day connected to the Internet.

a. Complete the table of values showing the number of free hours, H, that Francine has left t days after starting her membership. Graph the data points and connect them with a straight line. Extend the graph until it reaches the horizontal axis.

t	0	20	40	60	80
H					

b. Write an equation for the remaining free hours, H, in terms of t.

c. Explain the significance of the intercepts for the problem situation.

a. Find the intercepts of the graph.

b. Use the intercepts to choose appropriate scales for the axes, and graph the equation on paper.

c. Solve the equation for y in terms of x.

d. Graph the equation using your calculator. Use the intercepts to help you choose an appropriate window.

5. $36x - 24y = 7200$

6. $1.7x + 0.09y = 0.0765$

7. $0.3y - 56x = 84$

8. $6y + 0.11x = 330$

9. **a.** Use the graph in Figure 1.45 to solve $0.24x - 3.44 = -2$.

b. Verify your answer algebraically.

c. Use the graph to solve $0.24x - 3.44 > -2$.

d. Solve the inequality in part (c) algebraically.

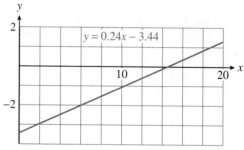

FIGURE 1.45

10. a. Use the graph in Figure 1.46 to solve
$5.2 - 1.4x = 1$.

   b. Verify your answer algebraically.

   c. Use the graph to solve $5.2 - 1.4x > 1$.

   d. Solve the inequality in part (c) algebraically.

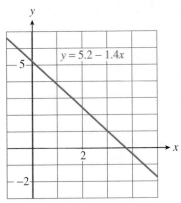

FIGURE 1.46

11. Graph the equation $y = 4.3x - 8.3$ in the
WINDOW Xmin $= -4.7$, Xmax $= 4.7$,
Ymin $= -30$, Ymax $= 15$. Use the graph to solve the following equations and inequalities.

   a. $8.9 = 4.3x - 8.3$

   b. $-25.5 > 4.3x - 8.3$

   c. $0.3 \le 4.3x - 8.3$

   d. $4.3x - 8.3 \le -12.6$

12. Graph the equation $y = 3.6 - 1.7x$ in the
WINDOW Xmin $= 0$, Xmax $= 9.4$,
Ymin $= -15$, Ymax $= 5$. Use the graph to solve the following equations and inequalities.

   a. $-5.24 = 3.6 - 1.7x$

   b. $-2.35 > 3.6 - 1.7x$

   c. $0.2 \le 3.6 - 1.7x$

   d. $3.6 - 1.7x < 1.9$

Compute the slope of the line through the indicated points. Notice the scales on the axes.

13.

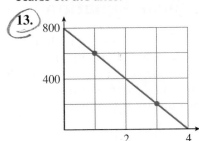

FIGURE 1.47

14.

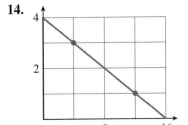

FIGURE 1.48

15.

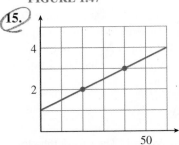

FIGURE 1.49

16.

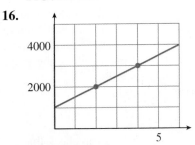

FIGURE 1.50

a. Find the intercepts of the graph.

b. Compute the slope of the graph by using the intercepts.

17. $3y - 5x = 30$

18. $7x + 4y = 28$

19. $y = 40x - 120$

20. $\dfrac{x}{3} + \dfrac{y}{2} = 1$

21. Francine is riding in a glass elevator in a shopping mall. Figure 1.51 shows the height,

h, of her eyes above ground level t seconds after the elevator begins climbing.

   a. Choose two points and compute the slope (including units).

   b. Explain what the slope measures in the context of the problem.

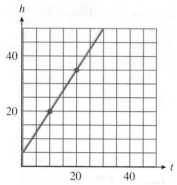

FIGURE 1.51

22. Figure 1.52 shows the cost, c, (in dollars) charged for a taxi ride of distance, d, traveled (in miles).

   a. Choose two points and compute the slope (including units).

b. Explain what the slope measures in the context of the problem.

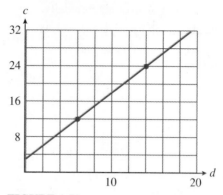

FIGURE 1.52

The Coefficients in a Linear Equation

What do the coefficients in a linear equation tell us about its graph? In this Investigation we'll use a graphing calculator to study some examples. For the following exercises, set your calculator's graphing window as follows: First press ZOOM 6, and then press ZOOM 8 and ENTER to obtain the **ZInteger** window centered at the origin. Or, enter the window settings

$$\text{Xmin} = -47 \qquad \text{Xmax} = 47 \qquad \text{Xscl} = 10$$
$$\text{Ymin} = -31 \qquad \text{Ymax} = 31 \qquad \text{Yscl} = 10$$

In this window you can find the slope of a line using the TRACE feature. Try this example: Graph the line $y = 3x - 10$ in the **ZInteger** window. Now press TRACE and use the arrow keys to move along the graph. Notice that the bug moves along the graph so that $\Delta x = 1$ between consecutive displayed points. Thus

$$m = \frac{\Delta y}{\Delta x} = \frac{\Delta y}{1} = \Delta y$$

You can check that $\Delta y = 3$ for each increment of $\Delta x = 1$. Therefore, the slope of the line is 3.

1. **a.** Graph the three equations below in the same graphing window.

$$y = x - 12 \qquad y = 2x - 12 \qquad y = 4x - 12$$

   b. What is the y-intercept for each graph?

   c. Use the TRACE feature to find the slope of each line.

   d. Which of the three lines is the steepest?

2. **a.** Graph the three equations below in the same graphing window.

$$y = -x + 5 \qquad y = -2.5x + 5 \qquad y = -5x + 5$$

b. What is the y-intercept for each graph?

c. Use the TRACE feature to find the slope of each line.

d. Which of the three lines is the steepest?

e. Compare the three lines in Part (2) with the three lines in Part (1). What can you say about lines with negative slopes?

3. **a.** Graph the three equations below in the same graphing window.

$$y = 1.5x - 18 \qquad y = 1.5x \qquad y = 1.5x + 18$$

b. What do the three equations have in common? How is this feature reflected in the graphs?

c. What is the y-intercept for each graph?

4. Repeat Part (3) for the three equations below:

$$y = -0.8x + 4 \qquad y = -0.8x + 10 \qquad y = -0.8x + 20$$

5. **a.** Graph the four equations below in the same graphing window.

$$y = \frac{1}{3}x - 12 \qquad y = \frac{2}{3}x - 12 \qquad y = x - 12 \qquad y = \frac{4}{3}x - 12$$

b. Use the TRACE feature to find the slope of each line.

c. Which of the three lines is the steepest?

6. Repeat Part 5 for the equations below:

$$y = \frac{-3}{2}x + 3 \qquad y = -x + 3 \qquad y = \frac{-1}{2}x + 3 \qquad y = \frac{-1}{4}x + 3$$

7. Write a paragraph summarizing your observations about the examples in this Investigation. Include answers to the following questions: What do the coefficients in a linear equation tell you about its graph? If two lines have the same slope but different y-intercepts, what is true about their graphs? What if they have different slopes but the same y-intercept?

8. Write a paragraph explaining what the slope tells you about the graph of a line. Include answers to the following questions: How are lines with negative slopes different from lines with positive slopes? If two lines have positive slopes, how can you tell which one is steeper? What if the two lines have negative slopes?

9. Match each equation with its graph in Figure 1.53.

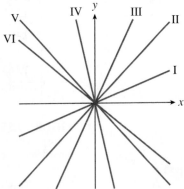

a. $y = \frac{2}{5}x$ **d.** $y = x$

b. $y = -4x$ **e.** $y = 2x$

c. $y = -x$ **f.** $y = \frac{-3}{4}x$ **FIGURE 1.53**

10. Match each equation with its graph in Figure 1.54.
   a. $y = 2x - 2$
   b. $y = 2x + 2$
   c. $y = -2x - 2$
   d. $y = -2x + 2$

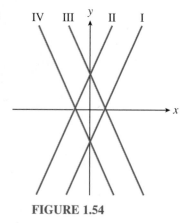

FIGURE 1.54

1.4 Equations of Lines

Slope-Intercept Form

In Investigation 2, we saw that the equation of a line written in the form

$$y = mx + b$$

gives us direct information about the graph of the line. The constant term in the equation is the y-intercept of the line, and the coefficient of x is the slope of the line. This should come as no surprise when we recall that slope measures how steep the line is. Think of the slope as a scale factor that tells us how many units y increases (or decreases) for each unit of increase in x.

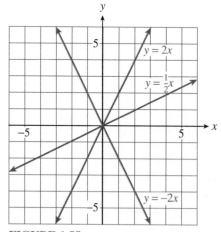

FIGURE 1.55
These lines have the same y-intercept but different slopes.

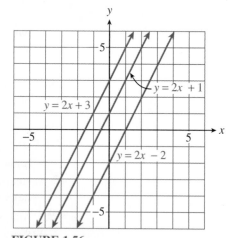

FIGURE 1.56
These lines have the same slope but different y-intercepts.

This form for the equation of a line is called the **slope-intercept** form.

Slope-Intercept Form

If we write the equation of a line in the form

$$y = mx + b$$

then m is the **slope** of the line, and b is the **y-intercept.**

Any linear equation (except equations of vertical lines) can be put into slope-intercept form simply by solving for y. Once that is done, we can find the slope and y-intercept without any further calculation by reading off their values from the equation.

EXAMPLE 1

a. Write the equation $3x + 4y = 6$ in slope-intercept form.

b. Specify the slope of the line and its y-intercept.

Solutions

a. Solve the equation for y.

$$3x + 4y = 6 \qquad \text{Subtract } 3x \text{ from both sides.}$$
$$4y = -3x + 6 \qquad \text{Divide both sides by 4.}$$
$$y = \frac{-3x}{4} + \frac{6}{4} \qquad \text{Simplify.}$$
$$y = \frac{-3}{4}x + \frac{3}{2}$$

b. The slope is the coefficient of x, and the y-intercept is the constant term. Thus,

$$m = \frac{-3}{4} \qquad \text{and} \qquad b = \frac{3}{2} \qquad \blacksquare$$

Slope-Intercept Method of Graphing

We can graph a line given in slope-intercept form without having to make a table of values.

1. First plot the y-intercept $(0, b)$.

2. Use the definition of slope to find a second point on the line: Starting at the y-intercept, move Δy units in the y-direction and Δx units in the x-direction. (See Figure 1.57.) The point at this location lies on the graph.

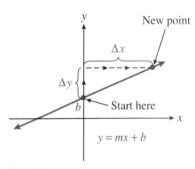

FIGURE 1.57

EXAMPLE 2

Graph the line $y = \dfrac{4}{3}x - 2$ by hand.

Solution

Since the equation is given in slope-intercept form, we see that the slope of the line is $m = \frac{4}{3}$ and its y-intercept is $b = -2$. We begin by plotting the y-intercept, $(0, -2)$. We then use the slope to find another point on the line. We have

$$m = \frac{\Delta y}{\Delta x} = \frac{4}{3}$$

so starting at $(0, -2)$ we move 4 units in the y-direction and 3 units in the x-direction, to arrive at the point $(3, 2)$. Finally, we draw the line through these two points. (See Figure 1.58.)

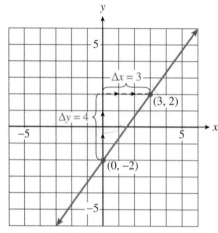

FIGURE 1.58

The slope of a line is a ratio and can be written in many equivalent ways. In Example 2, the slope is equal to $\frac{8}{6}$, $\frac{12}{9}$, and $\frac{-4}{-3}$. We can use any of these fractions to locate a third point on the line as a check. If we use $m = \dfrac{\Delta y}{\Delta x} = \dfrac{-4}{-3}$, we move *down* 4 units and *left* 3 units from the y-intercept to find the point $(-3, -6)$ on the line.

EXERCISE 1
a. Write the equation

$$2y + 3x + 4 = 0$$

in slope-intercept form.
b. Use the slope-intercept method to graph the line.

A Formula for Slope

In Section 1.3, we defined the slope of a line to be the ratio

$$m = \frac{\Delta y}{\Delta x}$$

as we move from one point to another on the line. So far we have computed Δy and Δx by counting squares on the graph, but this method is not always practical. All we really need are the coordinates of two points on the graph.

We will use *subscripts* to refer to the two points:

P_1 means "first point" and
P_2 means "second point"

We denote the coordinates of P_1 by (x_1, y_1) and the coordinates of P_2 by (x_2, y_2).

Now consider a specific example. The line through the two points P_1 (2, 9) and P_2 (7, −6) is shown in Figure 1.59. We can find Δx by subtracting the x-coordinate of the first point from the x-coordinate of the second point:

$$\Delta x = 7 - 2 = 5$$

In general, we have

$$\Delta x = x_2 - x_1$$

and similarly

$$\Delta y = y_2 - y_1$$

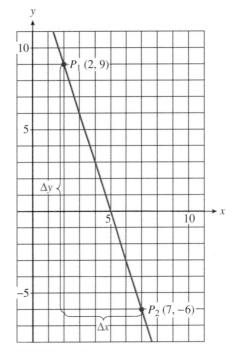

FIGURE 1.59

These formulas work even if some of the coordinates are negative; in our example

$$\Delta y = y_2 - y_1 = -6 - 9 = -15$$

By counting squares *down* from P_1 to P_2 we see that Δy is indeed -15. The slope of the line in Figure 1.59 is

$$m = \frac{\Delta y}{\Delta x} = \frac{y_2 - y_1}{x_2 - x_1}$$

$$= \frac{-15}{5} = -3$$

We now have a formula for the slope of a line that works even if we do not have a graph.

Slope Formula

The slope of the line passing through the points P_1 (x_1, y_1) and P_2 (x_2, y_2) is given by

$$m = \frac{\Delta y}{\Delta x} = \frac{y_2 - y_1}{x_2 - x_1}, \qquad (x_2 \neq x_1)$$

EXAMPLE 3

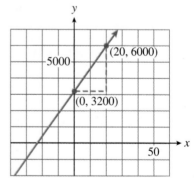

Compute the slope of the line in Figure 1.59 using the points Q_1 (6, −3) and Q_2 (4, 3).

Solution

Substitute the coordinates of Q_1 and Q_2 into the slope formula to find

$$m = \frac{y_2 - y_1}{x_2 - x_1}$$

$$= \frac{3 - (-3)}{4 - 6} = \frac{6}{-2} = -3$$

This value for the slope, −3, is the same value found above. ∎

EXERCISE 2

a. Find the slope of the line passing through the points (2, −3) and (−2, −1).

$m =$

b. Make a rough sketch of the line by hand.

Finding a Linear Equation from a Graph

We can use the slope-intercept form to find the equation of a line from its graph. First note the value of the y-intercept from the graph, then calculate the slope using two convenient points.

EXAMPLE 4

Find an equation for the line shown in Figure 1.60.

Solution

The line crosses the y-axis at the point (0, 3200), so the y-intercept is 3200. To calculate the slope of the line, locate another point, say (20, 6000), and compute:

$$m = \frac{\Delta y}{\Delta x} = \frac{6000 - 3200}{20 - 0}$$

$$= \frac{2800}{20} = 140$$

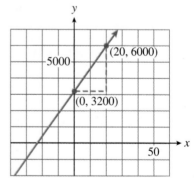

FIGURE 1.60

The slope-intercept form of the equation, with $m = 140$ and $b = 3200$, is $y = 140x + 3200$. ∎

EXERCISE 3   Find an equation for the line shown.

$b =$

$m =$

$y =$

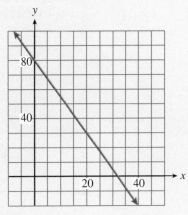

Point-Slope Form

There is only one line that passes through a given point and has a given slope. For example, we can graph the line of slope $\frac{-3}{4}$ that passes through the point $(1, -4)$. We first plot the given point, $(1, -4)$, as shown in Figure 1.61. Then we use the slope to find another point on the line. Since the slope is

$$m = \frac{-3}{4} = \frac{\Delta y}{\Delta x}$$

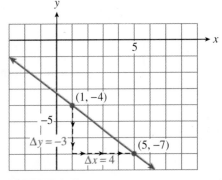

FIGURE 1.61

we move down 3 units and then 4 units to the right, starting from $(1, -4)$. This brings us to the point $(5, -7)$. We can then draw the line through these two points.

We can also find an equation for the line, as shown in Example 5.

EXAMPLE 5

Find an equation for the line that passes through $(1, -4)$ and has slope $\frac{-3}{4}$.

Solution

We will use the formula for slope,

$$m = \frac{y_2 - y_1}{x_2 - x_1}$$

We substitute $\frac{-3}{4}$ for the slope, m, and $(1, -4)$ for (x_1, y_1). For the second point, (x_2, y_2), we will use the variable point (x, y). Substituting these values into the slope formula gives us

$$\frac{-3}{4} = \frac{y - (-4)}{x - 1} = \frac{y + 4}{x - 1}$$

To solve for y, we first multiply both sides by $x - 1$.

$$(x - 1)\frac{-3}{4} = \frac{y - (-4)}{x - 1}(x - 1)$$

$$\frac{-3}{4}(x - 1) = y + 4 \qquad \text{Apply the distributive law.}$$

$$\frac{-3}{4}x + \frac{3}{4} = y + 4 \qquad \text{Subtract 4 from both sides.}$$

$$\frac{-3}{4}x - \frac{13}{4} = y \qquad\qquad \frac{3}{4} - 4 = \frac{3}{4} - \frac{16}{4} = \frac{-13}{4} \qquad \blacksquare$$

When we use the slope formula in this way to find the equation of a line, we substitute a variable point (x, y) for the second point. This version of the formula,

$$m = \frac{y - y_1}{x - x_1}$$

is called the **point-slope form** for a linear equation. It is sometimes stated in another form obtained by clearing the fraction to get

$$(x - x_1)\, m = \frac{y - y_1}{x - x_1}(x - x_1)$$

or

$$(x - x_1)\, m = y - y_1$$

Point-Slope Form

The equation of the line that passes through the point (x_1, y_1) and has slope m is

$$y - y_1 = m(x - x_1)$$

EXERCISE 4 Use the point-slope form to find the equation of the line that passes through the point $(-3, 5)$ and has slope -1.4.

$$y - y_1 = m(x - x_1) \qquad \text{Substitute } -1.4 \text{ for } m \text{ and } (-3, 5) \text{ for } (x_1, y_1).$$
Simplify: apply the distributive law.
Solve for y.

READING QUESTIONS

1. How can you put a linear equation into slope-intercept form?
2. What do the coefficients in the slope-intercept form tell you about the line?
3. Explain how to graph a line using the slope-intercept method.

4. State a formula for finding the slope of a line from two points on the line.

5. Explain how to find an equation for a line from its graph.

6. Explain how to use the point-slope form for a linear equation.

ANSWERS TO 1.4 EXERCISES

1a. $y = \dfrac{-3}{2}x - 2$ **b.**

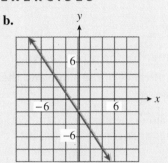

2a. $\dfrac{-1}{2}$ **b.**

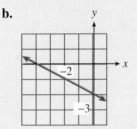

3. $b = 80$, $m = \dfrac{-5}{2}$, $y = \dfrac{-5}{2}x + 80$ **4.** $y = -1.4x + 0.8$

HOMEWORK 1.4

In Problems 1–10,

a. Write each equation in slope-intercept form.

b. State the slope and y-intercept of the line.

1. $3x + 2y = 1$ **2.** $5x - 4y = 0$ **3.** $\dfrac{1}{4}x + \dfrac{3}{2}y = \dfrac{1}{6}$ **4.** $\dfrac{7}{6}x - \dfrac{2}{9}y = 3$

5. $4.2x - 0.3y = 6.6$ **6.** $0.8x + 0.004y = 0.24$ **7.** $y + 29 = 0$ **8.** $y - 37 = 0$

9. $250x + 150y = 2450$ **10.** $80x - 360y = 6120$

In Problems 11–14,

a. Sketch by hand the graph of the line with the given slope and *y*-intercept.

b. Write an equation for the line.

c. Find the *x*-intercept of the line.

11. $m = 3$ and $b = -2$

12. $m = -4$ and $b = 1$

13. $m = -\dfrac{5}{3}$ and $b = -6$

14. $m = \dfrac{3}{4}$ and $b = -2$

a. Find an equation for the graph shown.

b. State the significance of the slope and the vertical intercept in terms of the problem.

15. Figure 1.62 shows the altitude, *a,* (in feet) of a skier *t* minutes after getting on a ski lift.

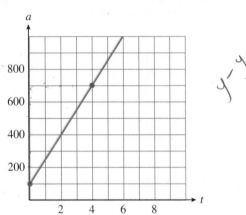

FIGURE 1.62

16. Figure 1.63 shows the distance, *d,* traveled (in meters) by a train *t* seconds after the train passes an observer.

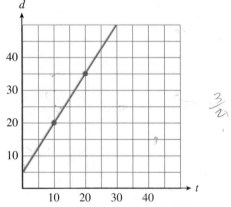

FIGURE 1.63

17. Figure 1.64 shows the amount of garbage, *G,* (in tons) that has been deposited at a dump site *t* years after new regulations go into effect.

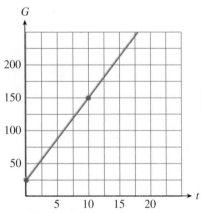

FIGURE 1.64

18. Figure 1.65 shows the number of barrels of oil, *B,* that has been pumped at a drill site *t* days afer a new drill is installed.

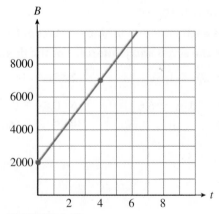

FIGURE 1.65

19. Figure 1.66 shows the amount of money, M, (in dollars) in Tammy's bank account w weeks after she loses all sources of income.

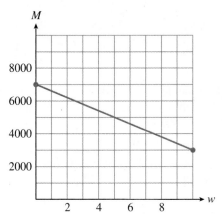

FIGURE 1.66

21. The formula $F = \dfrac{9}{5} C + 32$ converts the temperature in degrees Celsius to degrees Fahrenheit.

   a. What is the Fahrenheit temperature when it is 10°Celsius?

   b. What is the Celsius temperature when it is −4°Fahrenheit?

   c. Choose appropriate $\boxed{\textbf{WINDOW}}$ settings and graph the equation $y = \dfrac{9}{5} x + 32$.

   d. Find the slope and explain its meaning for this problem.

   e. Find the intercepts and explain their meanings for this problem.

23. In England, oven cooking temperatures are often given as Gas Marks rather than degrees Fahrenheit. The table shows the equivalent oven temperatures for various Gas Marks.

Gas mark	3	5	7	9
Degrees (F)	325	375	425	475

   a. Plot the data and draw a line through the data points.

   b. Calculate the slope of your line. Estimate the y-intercept from the graph.

   c. Find an equation that gives the temperature in degrees Fahrenheit in terms of the Gas Mark.

20. Figure 1.67 shows the amount of emergency water, W, remaining (in liters) in a southern Californian household t days after an earthquake.

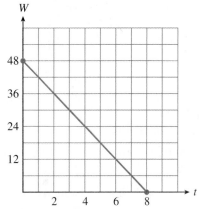

FIGURE 1.67

22. If the temperature on the ground is 70° Fahrenheit, the formula $T = 70 - \dfrac{3}{820} h$ gives the temperature at an altitude of h feet.

   a. What is the temperature at an altitude of 4100 feet?

   b. At what altitude is the temperature 34°?

   c. Choose appropriate $\boxed{\textbf{WINDOW}}$ settings and graph the equation $y = 70 - \dfrac{3}{820} x$.

   d. Find the slope and explain its meaning for this problem.

   e. Find the intercepts and explain their meanings for this problem.

24. European shoe sizes are scaled differently than American shoe sizes. The table shows the European equivalents for various American shoe sizes.

American shoe size	5.5	6.5	7.5	8.5
European shoe size	37	38	39	40

   a. Plot the data and draw a line through the data points.

   b. Calculate the slope of your line. Estimate the y-intercept from the graph.

   c. Find an equation that gives the European shoe size in terms of American shoe size.

25. A spring is suspended from the ceiling. The table shows the length of the spring in centimeters as it is stretched by hanging various weights from it.

Weight, kg	3	4	8	10	12	15	22
Length, cm	25.76	25.88	26.36	26.6	26.84	27.2	28.04

a. Plot the data and draw a straight line through the points. Estimate the y-intercept of your graph.

b. Find an equation for the line.

c. If the spring is stretched to 27.56 cm, how heavy is the attached weight?

26. The table shows the amount of ammonium chloride salt, in grams, that can be dissolved in 100 grams of water at different temperatures.

Temperature, °C	10	12	15	21	25	40	52
Grams of salt	33	34	35.5	38.5	40.5	48	54

a. Plot the data and draw a straight line through the points. Estimate the y-intercept of your graph.

b. Find an equation for the line.

c. At what temperature will 46 grams of salt dissolve?

Find the slope of each line in Problems 27–32.

27.

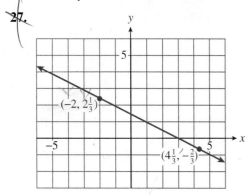

FIGURE 1.68

28.

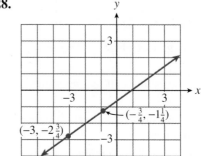

FIGURE 1.69

29.

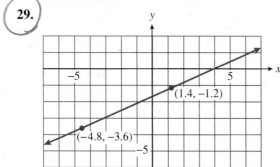

FIGURE 1.70

30.

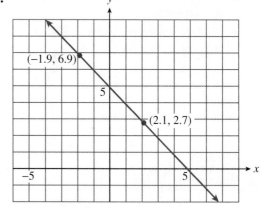

FIGURE 1.71

31.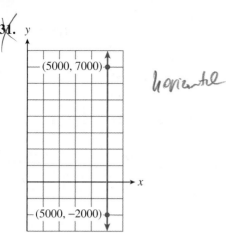

(5000, 7000)

(5000, −2000)

horizontal

FIGURE 1.72

32.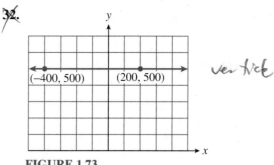

(−400, 500)   (200, 500)

vertical

FIGURE 1.73

In Problems 33–36,

a. Sketch by hand the graph of the line that passes through the given point and has the given slope.

b. Write an equation for the line in point-slope form.

c. Put your equation from part (b) into slope-intercept form.

33. $(2, -5)$, $m = -3$ **34.** $(-6, -1)$, $m = 4$ **35.** $(2, -1)$, $m = \dfrac{5}{3}$ **36.** $(-1, 2)$, $m = -\dfrac{3}{2}$

In Problems 37 and 38, match each graph with its equation.

37. a. $y = \dfrac{3}{4}x + 2$ **b.** $y = -\dfrac{3}{4}x + 2$

c. $y = \dfrac{3}{4}x - 2$ **d.** $y = -\dfrac{3}{4}x - 2$

I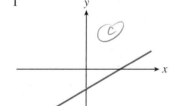

II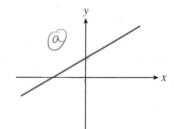

III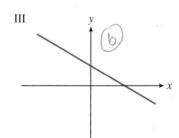

IV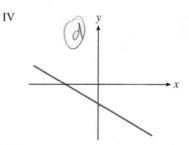

FIGURE 1.74

38. a. $y - 1 = 2(x + 3)$ **b.** $y + 1 = 2(x - 3)$
  c. $y + 1 = 2(x + 3)$ **d.** $y - 1 = 2(x - 3)$

I 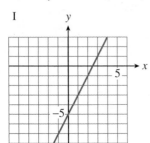   II 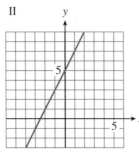   III 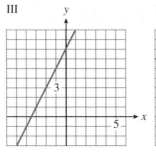   IV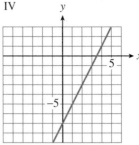

FIGURE 1.75

In Problems 39–42, find the slope of each line and the coordinates of one point on the line. (No calculation is necessary!)

39. $y + 1 = 2(x - 6)$

40. $2(y - 8) = 5(x + 2)$

41. $\dfrac{y - 3}{x + 5} = \dfrac{-4}{3}$

42. $7x = -3y$

In Problems 43–46,

  a. Write an equation in point-slope form for the line that passes through the given point and has the given slope.

  b. Put your equation from part (a) into slope-intercept form.

  c. Use your graphing calculator to graph the line.

43. $(-6.4, -3.5), m = -0.25$

44. $(7.2, -5.6), m = 1.6$

45. $(80, -250), m = 2.4$

46. $(-150, 1800), m = -24$

In Problems 47–52,

  a. Estimate the slope and vertical intercept of each line.

  b. Using your estimates from (a), write an equation for the line.

  (Hint: To calculate the slope, find two points on the graph that lie on the intersection of grid lines.)

47.

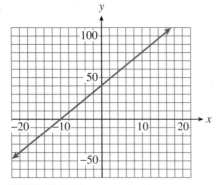

FIGURE 1.76

48.

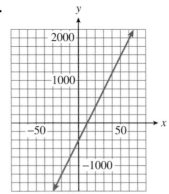

FIGURE 1.77

49.

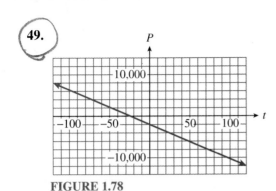

FIGURE 1.78

50.

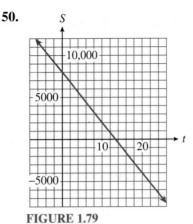

FIGURE 1.79

51.

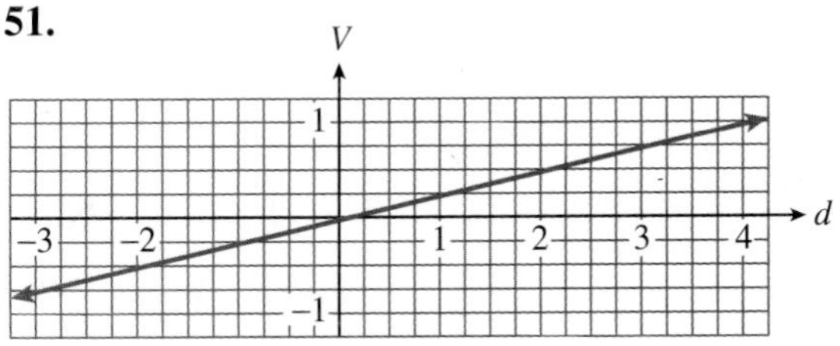

FIGURE 1.80

52.

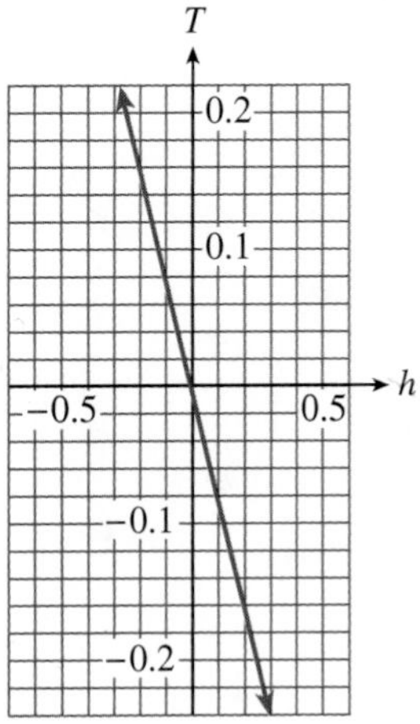

FIGURE 1.81

For Problems 53–56, set the graphing window as indicated and graph the equation. It may be convenient to set both Xscl and Yscl to 0.

a. Use ⬚TRACE or *value* to read off coordinates of two convenient points and compute the slope of the line.

b. Use ⬚TRACE or *value* again to find the coordinates of the *y*-intercept.

53. Xmin $= -4.7$, Xmax $= 4.7$, Ymin $= -10$,   Ymax $= 20$; $y = 2.5x + 6.25$

54. Xmin $= -4$,   Xmax $= 5.4$, Ymin $= -20$,   Ymax $= 15$; $y = -4.2x - 3.7$

55. Xmin $= -0.4$, Xmax $= 9$,   Ymin $= -10$,   Ymax $= 100$; $y = -8.4x + 63$

56. Xmin $= -7.2$, Xmax $= 2.2$, Ymin $= -200$, Ymax $= 10$; $y = -28x - 182$

1.5 Lines of Best Fit

An equation that relates two variables can be used to find values of one variable from values of the other. In this section we consider methods for fitting a linear equation to a collection of data points.

Scatterplots

Figure 1.82 is called a **scatterplot.** Each point on a scatterplot exhibits a pair of measurements about a single event. This plot shows the boiling points of various substances on the horizontal axis, and their heats of vaporization on the vertical axis. (The heat of vaporization is the energy needed to change a standard amount of a substance from liquid to gas at its boiling point.)

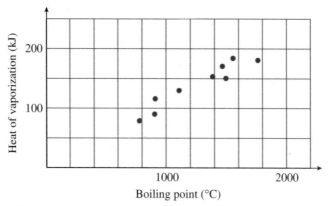

FIGURE 1.82

The points on a scatterplot may or may not show some sort of pattern. Consider the three plots in Figure 1.83. In Figure 1.83a the data points resemble a cloud of gnats; there is no apparent pattern to their locations. In Figure 1.83b the data follow a generally decreasing trend: as the values of x increase, most of the corresponding y-values decrease. The points in Figure 1.83c are even more organized. Although the points do not all lie on a straight line, they seem to be clustered around some imaginary line.

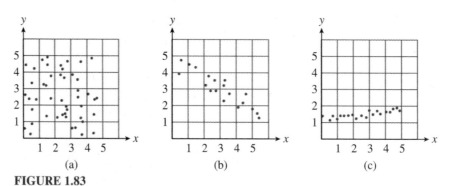

FIGURE 1.83

Linear Regression

If the data in a scatterplot are roughly linear, we can estimate the location of an imaginary "line of best fit" that passes as close as possible to the data points. We can then use this line to make predictions about the data.

EXAMPLE 1

a. Estimate a line of best fit for the scatterplot in Figure 1.84.

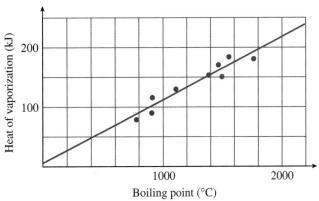

FIGURE 1.84

b. Use your line to predict the heat of vaporization of silver, whose boiling temperature is 2160°C.

Solutions

a. We draw a line that "fits" the data points as best we can, as shown in Figure 1.84. We try to end up with roughly equal numbers of data points above and below our line.

b. We see that when $x = 2160$ on this line, the y-value is approximately 235. We therefore predict that the heat of vaporization of silver is about 235 kilojoules. (It is actually 255.1 kilojoules.) ∎

The process of predicting a value of y based on a straight line that fits the data is called **linear regression,** and the line itself is called the **regression line.** The equation of the regression line is usually used (instead of a graph) to predict values.

EXAMPLE 2

a. Find the equation of the regression line in Example 1.
b. Use the regression line to predict the heat of vaporization of potassium bromide, whose boiling temperature is 1435°C.

Solutions

a. We first calculate the slope by choosing two points on the regression line we drew in Figure 1.84. The line appears to pass through the points

(900, 100) and (1560, 170). (Note that these points are not necessarily any of the original data points.) The slope of the line is then

$$m = \frac{170 - 100}{1560 - 900} = \frac{70}{660} \approx 0.106$$

Now we use the point-slope formula to find the equation of the line. We substitute $m = 0.106$ and use either of the two points for (x_1, y_1); we'll choose (900, 100). The equation of the regression line is

$$y - y_1 = m(x - x_1)$$
$$y - 100 = 0.106(x - 900) \quad \text{Simplify and solve for } y.$$
$$y = 0.106x + 4.6$$

b. We'll use the regression equation to make our prediction. For potassium bromide, $x = 1435$ and

$$y = 0.106(1435) + 4.6 = 156.71$$

We predict that the heat of vaporization for potassium bromide is about 156.7°C. ∎

EXERCISE 1 High-frequency radiation is harmful to living things because it can cause changes in their genetic material. The following data, collected by C. P. Oliver in 1930, show the frequency of genetic transmutations induced in fruit flies by doses of x-rays, measured in roentgens.

Dosage (roentgens)	285	570	1640	3280	6560
Percent of mutated genes	1.18	2.99	4.56	9.63	15.85

Source: Oliver, 1930.

a. Plot the data on the grid.

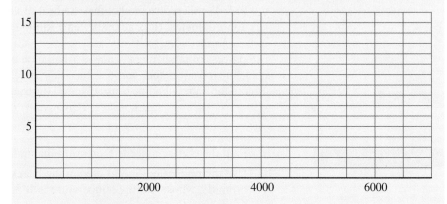

b. Draw a line of best fit through the data points.

c. Find the equation of your regression line.

d. Use the regression equation to predict the percent of mutations that might result from exposure to 5000 roentgens of radiation.

Linear Interpolation and Extrapolation

If we know that the relationship between two variables is linear, we need only two points to find an equation. Sometimes a variable relationship is not linear, but a portion of its graph can be approximated by a line. The graph in Figure 1.85 shows a child's height in centimeters each month. The graph is not linear because her rate of growth is not constant; her growth slows down as she approaches her adult height. However, over a short time interval the graph is close to a line, and that line can be used to approximate the coordinates of points on the curve.

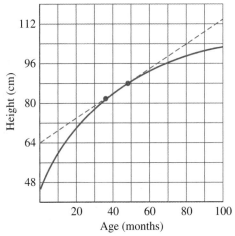

FIGURE 1.85

The process of estimating between known data points is called **interpolation**. Making predictions beyond the range of known data is called **extrapolation**.

EXAMPLE 3

Emily was 82 centimeters tall at age 36 months and 88 centimeters tall at age 48 months.

   a. Find a linear equation that approximates Emily's height in terms of her age over the given time interval.

   b. Use linear interpolation to estimate Emily's height when she was 38 months old, and extrapolate to predict her height at age 50 months.

Solutions

   a. Let y represent Emily's height at age x months, and use the point-slope formula to find a linear equation relating x and y. First calculate the slope using the given points, $(36, 82)$ and $(48, 88)$.

$$m = \frac{88 - 82}{48 - 36} = \frac{6}{12} = \frac{1}{2}$$

Now substitute $m = \frac{1}{2}$ and either of the two points into the formula; we'll choose $(x_1, y_1) = (36, 82)$.

$$y - y_1 = m(x - x_1)$$

$$y - 82 = \frac{1}{2}(x - 36) \quad \text{Simplify and solve for } y.$$

$$y = \frac{x}{2} + 64$$

   b. We use this equation to approximate Emily's height at other times. When Emily was 38 months old, $x = 38$ and

$$y = \frac{38}{2} + 64 = 83$$

When $x = 50$,

$$y = \frac{50}{2} + 64 = 89$$

We estimate that Emily was 83 centimeters tall at age 38 months, and we predict that she will be approximately 89 centimeters tall at age 50 months. ∎

If we try to extrapolate too far, we may get unreasonable results. For example, if we use our model to predict Emily's height at 300 months (25 years old), we get

$$y = \frac{300}{2} + 64 = 214$$

It is unlikely that Emily will be 214 centimeters, or over 7 feet tall, when she is 25 years old. Our linear model provides a fair approximation for Emily's height over a short time interval, but diverges from the actual graph as Emily gets older.

EXERCISE 2

a. The temperature in Delbert's apartment was 35°C at 1 P.M. when he turned on the air conditioning, and by 5 P.M. it had dropped to 29°C. Find a linear equation that approximates the temperature h hours after noon.

h	T

$m = \dfrac{T_2 - T_1}{h_2 - h_1} =$

$T - T_1 = m(h - h_1)$

$T =$

b. Use linear interpolation to estimate the temperature at 2 P.M., and extrapolate to predict the temperature at 8 P.M.

Using a Calculator for Linear Regression

Estimating a line of best fit is a subjective process. Rather than base their estimates on such a line, statisticians use the **least squares regression line.** This regression line minimizes the sum of the squares of all the vertical distances between the data points and the corresponding points on the line (see Figure 1.86). Many calculators are programmed to find the least squares regression line, using an algorithm that depends only on the data, and not on the appearance of the graph.

On the TI-83 calculator we use the statistics mode, which you can access by pressing ▢STAT▢. You will see a display that looks like Figure 1.87a. Choose ▢1▢ to **Edit** (enter or alter) data.

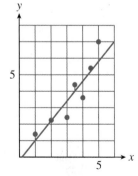

FIGURE 1.86

If there are data in column L_1 or L_2, clear them out: Use the △ key to select L_1, press **CLEAR** **ENTER**, then do the same for L_2. Now follow the instructions in Example 4 for using your calculator's statistics features.

(a)   (b)

FIGURE 1.87

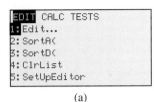

EXAMPLE 4

a. Find the equation of the least squares regression line for the following data.

$$(10, 12), \quad (11, 14), \quad (12, 14), \quad (12, 16), \quad (14, 20)$$

b. Plot the data points and the least squares regression line on the same axes.

c. Use the least squares regression line to predict the value of y when $x = 9$.

Solutions

a. We must first enter the data. Press **STAT** **ENTER** to select **Edit.** Enter the x-coordinates of the data points in the L_1 column and the y-coordinates in the L_2 column, as shown in Figure 1.88a.

Now we are ready to find the regression equation for our data. Press **STAT** ▷ **4** to select linear regression, or **LinReg($ax + b$),** then press **ENTER**. The calculator will display the equation $y = ax + b$ and values for a and b as shown in Figure 1.88b. You should find that your regression line is approximately

$$y = 1.95x - 7.86$$

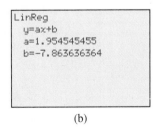

(a)   (b)

FIGURE 1.88

b. To plot the data and the regression line, we first clear out any old definitions in the **Y =** list. Then copy the regression equation into the equation list by pressing **VARS** 5 ▷ ▷ **ENTER**. To draw the scatterplot, press **2nd** **Y =** **1** and set the **Plot1** menu as shown in Figure 1.89a.

Finally, press [ZOOM] [9] to see the scatterplot of data and the least squares regression line. The graph is shown in Figure 1.89b.

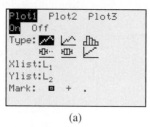

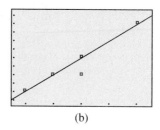

(a)   (b)

FIGURE 1.89

c. Press [2nd] [GRAPH] to see a table of values for the regression equation. Use the arrow keys to scroll the table and find that $y \approx 9.7273$ when $x = 9$. ■

When you are through with the scatterplot, press [Y =] [△] [ENTER] to turn off the **Stat Plot.** If you neglect to do this, the calculator will continue to show the scatterplot even after you ask it to plot a new equation. ■

EXERCISE 3

a. Use your calculator's statistics features to find the least squares regression equation for the data in Exercise 1.

b. Plot the data and the graph of the regression equation.

c. Use the regression equation to predict the number of mutations expected from a 5000 roentgen dose of radiation.

READING QUESTIONS

1. What is a regression line?
2. State two formulas you will need to calculate the equation of a line through two points.
3. Explain the difference between interpolation and extrapolation.
4. In general, should you have more confidence in figures obtained by interpolation or by extrapolation? Why?

ANSWERS TO 1.5 EXERCISES

1a, b.

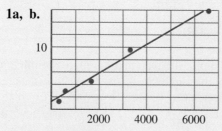

c. $y = 0.0023x + 1.18$ **d.** 13%

2a. $T = -1.5h + 36.5$ **b.** 33.5°C, 24.5°C

3a. $y = 0.0022937101x + 1.183417216$ **c.** 12.65%

Use information from the graphs to answer the questions.

1. The scatterplot in Figure 1.90 shows the ages of ten army drill sergeants and the time it took each to run 100 meters, in seconds.

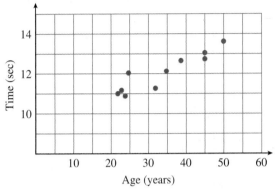

FIGURE 1.90

    a. What was the hundred-meter time for the 25-year-old drill sergeant?

    b. How old was the drill sergeant whose hundred-meter time was 12.6 seconds?

    c. Use a straightedge to draw a line of best fit through the data points.

    d. Use your line of best fit to predict the hundred-meter time of a 28-year-old drill sergeant.

    e. Choose two points on your regression line and find its equation.

    f. Use the equation to predict the hundred-meter time of a 40-year-old drill sergeant and a 12-year-old drill sergeant. Are these predictions reasonable?

2. The scatterplot in Figure 1.91 shows the outside temperature and the number of cups of hot chocolate sold at an outdoor skating rink snack bar on 13 consecutive nights.

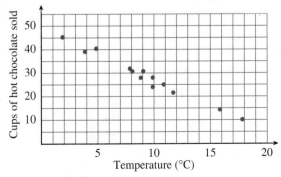

FIGURE 1.91

a. How many cups of hot chocolate were sold when the temperature was 2° C?

b. What was the temperature on the night when 25 cups of hot chocolate were sold?

c. Use a straightedge to draw a line of best fit through the data points.

d. Use your line of best fit to predict the number of cups of hot chocolate that will be sold at the snack bar if the temperature is 7° C.

e. Choose two points on your regression line and find its equation.

f. Use the equation to predict the number of cups of hot chocolate that will be sold when the temperature is 10°C and when the temperature is 24°C. Are these predictions reasonable?

3. The scatterplot in Figure 1.92 shows weights (in pounds) and heights (in inches) of a team of distance runners.

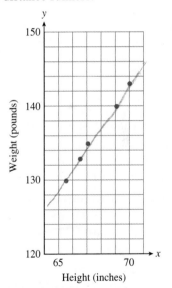

FIGURE 1.92

a. Use a straightedge to draw a line that fits the data.

b. Use your line to predict the weight of a 65-inch-tall runner and the weight of a 71-inch-tall runner.

c. Use your answers from part (b) to approximate the equation of a regression line.

d. Use your answer to part (c) to predict the weight of a runner who is 68 inches tall.

4. The scatterplot in Figure 1.93 shows best times for various women running 400 meters and 100 meters.

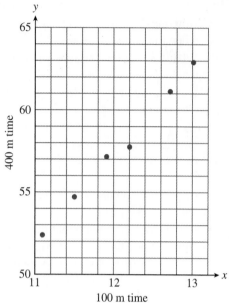

FIGURE 1.93

a. Use a straightedge to draw a line that fits the data.

b. Use your line to predict the 400-meter time of a woman who runs the 100-meter dash in 11.2 seconds, and the 400-meter time of a woman who runs the 100-meter dash in 13.2 seconds.

c. Use your answers from part (b) to approximate the equation of a regression line.

d. Use your answer to part (c) to predict the 400-meter time of a woman who runs the 100-meter dash in 12.1 seconds.

5. The points on the scatterplot in Figure 1.92 are (65.5, 130), (66.5, 133), (67, 135), (69, 140), and (70, 143).

a. Use your calculator to find the least squares regression line.

b. Use the regression line to predict the weight of a runner who is 68 inches tall. Compare the weight this equation gives with the prediction you made in part (d) of Problem 3.

6. The points on the scatterplot in Figure 1.93 are (11.1, 52.4), (11.5, 54.7), (11.9, 57.4), (12.2, 57.9), (12.7, 61.3), and (13.0, 63.0).

a. Use your calculator to find the least squares regression line.

b. Use the regression line to predict the 400-meter time of a woman who runs the 100-meter dash in 12.1 seconds. Compare the time this equation gives with the prediction you made in part (d) of Problem 4.

7. Bracken, a type of fern, is one of the most successful plants in the world, growing on every continent except Antarctica. New plants, which are genetic clones of the original, spring from a network of underground stems, or rhizomes, to form a large circular colony. Figure 1.94 shows the age of various colonies plotted against their diameters.

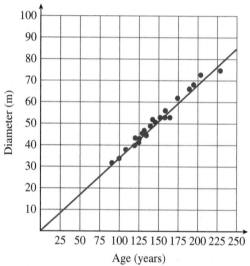

FIGURE 1.94

Source: From Chapman, J. L. and Reiss, M. J., *Ecology: Principles and Applications,* 1992. Copyright © 1992 Cambridge University Press. Reprinted with the permission of Cambridge University Press. Data from E. Oionen, "Sporal regeneration of bracken in Finland," *Acta Forestalia Fennica,* 83, (1) and (2), 1967.

a. Calculate the rate of growth of a bracken colony, in meters per year.

b. Find an equation for the line of best fit. (What should the vertical intercept of the line be?)

c. In Finland, bracken colonies over 450 meters in diameter have been found. How old are these colonies?

8. The European sedge warbler has an intricate song consisting of a long stream of trills, whistles and buzzes. Males with the most elaborate songs are the first to acquire mates in the spring. The following data show the number of different songs sung by several male warblers and the day on which they acquired mates, where Day 1 corresponds to April 20. (Pairing can be dated because the male stops singing when it pairs.)

Number of songs	41	38	34	32	30	25	24	24	23	14
Pairing day	20	24	25	21	24	27	31	35	40	42

a. Plot the data points. A regression line for the data is $y = -0.85x + 53$. Graph this line on the same axes with the data.

b. What does the slope of the regression line represent?

c. When can a sedge warbler that knows 10 songs expect to find a mate?

d. What do the intercepts of the regression line represent? Do these values make sense in context?

Source: Krebs and Davies, 1993.

9. Astronomers use a numerical scale called magnitude to measure the brightness of a star, with brighter stars assigned smaller magnitudes. When we view a star from earth, molecules and dust in the air scatter and absorb some of the light, making the star appear fainter than it really is. Thus, the observed magnitude of a star depends on the distance its light rays must travel through the earth's atmosphere. The observed magnitude, m, is given by

$$m = m_0 + kx$$

where m_0 is the apparent magnitude of the star outside the atmosphere, x is the air mass (a measure of the distance through the atmosphere), and k is called the extinction coefficient. To calculate m_0, astronomers observe the same object several times during the night at different positions in the sky, and hence different values of x. Here are data from such observations.

Altitude	Air mass, x	Magnitude, m
50°	1.31	0.90
35°	1.74	0.98
25°	2.37	1.07
20°	2.92	1.17

a. Plot magnitude against air mass, and draw a line of best fit through the data.

b. Find the equation of your line of best fit, or use a calculator to find the regression line for the data.

c. What is the value of the extinction coefficient? What is the apparent magnitude of the star outside the earth's atmosphere?

Source: Karttunen, 1987.

10. One of the factors that determines the strength of a muscle is its cross-sectional area. The data below show the cross-sectional area of the arm flexor muscle for several men and women, and their strength, measured by the maximum force they exerted against a resistance.

Women		Men	
Area (sq cm)	Strength (kg)	Area (sq cm)	Strength (kg)
11.5	11.3	13.5	15.0
10.8	13.2	13.8	17.3
11.7	13.2	15.4	19.0
12.0	14.5	15.4	19.8
12.5	15.6	17.7	20.6
12.7	14.8	18.6	20.8
14.4	15.6	20.8	26.3
14.4	16.1	—	—
15.7	18.4	—	—

a. Plot the data for both men and women on the same graph, using different symbols for the data points for men and the data points for women.

b. Are the data for both men and women described reasonably well by the same regression line? Draw a line of best fit through the data.

c. Find the equation of your line of best fit, or use a calculator to find the regression line for the data.

d. What is the significance of the slope in this context?

Source: Davis, Kimmet, and Autry, 1986.

11. The Mid-Atlantic Ridge is a mountain range on the sea floor beneath the Atlantic Ocean. It was discovered in the late nineteenth century during the laying of trans-Atlantic telephone cables. The ridge is volcanic, and the ocean floor is moving away from the ridge on either side. We can estimate the speed of this sea-floor spreading by recording the age of the rocks on the sea floor and their distance from the ridge. (The age of the rocks is calculated by measuring their magnetic polarity. At known intervals over the last four million years, the earth reversed its polarity, and this information is encoded in the rocks.) Plot the data in the table, with age on the horizontal axis and separation distance on the vertical axis.

Age (millions of years)	Distance (km)
0.78	17
0.99	18
1.07	21
1.79	32
1.95	39
2.60	48
3.04	58
3.11	59
3.22	62
3.33	65
3.58	66

Source: Open University, 1998.

a. What was the speed of spreading since the most recent polarity reversal? (This is the rate of spreading closest to the ridge.)

b. Draw a line of best fit through the data and calculate its slope in kilometers per million years.

c. Convert the spreading rate to millimeters per year.

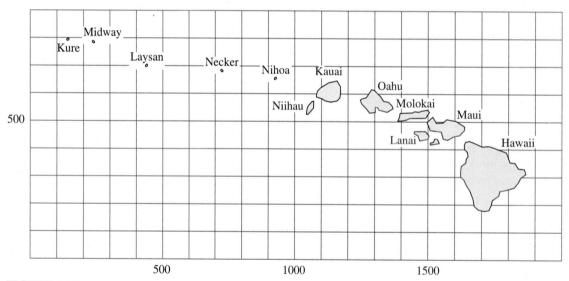

FIGURE 1.95

Island	Midway	Laysan	Necker	Nihoa	Niihau	Kauai	Oahu	Molokai	Lanai	Maui	Hawaii
Age	27	20	10	7.5	4.9	5.1	3.8	1.8	1.3	0.8	0.5

12. The Big Island of Hawaii is the last island in a chain of islands and submarine mountain peaks that stretch almost 6000 km across the Pacific Ocean. All are extinct volcanoes except for the Big Island itself, which is still active. The age of the extinct peaks is roughly proportional to their distance from the Big Island. Geologists believe that the volcanic islands were formed as the tectonic plate drifted across a "hot spot" in the earth's mantle. Figure 1.95 shows a map of the islands.

a. The table gives the ages of the islands in the Hawaiian chain, in millions of years. Plot the ages of the islands against their distance from Hawaii along the chain. (Use Figure 1.95 to determine the distances. The grid is scaled in kilometers.)

b. Draw a line of best fit through the data.

c. Use your graph to estimate the speed of the Pacific plate.

Source: Open University, 1998.

13. Six students are trying to identify an unknown chemical compound by heating the compound and measuring the density of the gas that evaporates. (To find the density of a gas, we divide its mass by its volume.) The students record the mass lost by the solid substance and the volume of the gas that evaporated from it. They know that the mass lost by the solid must be the same as the mass of the gas that evaporated.

Student	A	B	C	D	E	F
Loss in mass (mg)	64	81	32	107	88	72
Volume of gas (cm³)	48	60	24	81	76	54

a. Plot the data with mass on the horizontal axis. Which student made an error in the experiment?

b. Ignoring the incorrect data point, draw a line of best fit through the other points.

c. Find an equation of the form $y = mx$ for the data. Why should you expect the regression line to pass through the origin?

d. Use your equation to calculate the mass of 1000 cm^3 of the gas.

e. Here are the densities of some gases at room temperature.

Hydrogen	8 mg/liter
Nitrogen	1160 mg/liter
Oxygen	1330 mg/liter
Carbon dioxide	1830 mg/liter

Which of these might have been the gas that evaporated from the unknown compound? (*Hint:* Use your answer to part (d) to calculate the density of the gas. $1 \text{ cm}^3 = 1$ milliliter.)

Source: © Andrew Hunt and Alan Sykes, *Chemistry*, 1984. Reprinted by permission of Pearson Education Limited.

14. The formulas for many chemical compounds involve ratios of small integers. For example, the formula for water, H_2O, means that two atoms of hydrogen combine with one atom of oxygen to make one water molecule. Magnesium and oxygen combine to produce magnesium oxide. Twenty-four grams of magnesium contain the same number of atoms as sixteen grams of oxygen.

a. Complete the table showing the amount of oxygen needed if the formula for magnesium oxide is MgO, Mg_2O, or MgO_2.

Grams of Mg	Grams of O (if MgO)	Grams of O (if Mg₂O)	Grams of O (if MgO₂)
24	16		
48			
12			
6			

b. Graph three lines on the same axes to represent the three possibilities, with grams of magnesium on the horizontal axis and grams of oxygen on the vertical axis.

c. Here are the results of some experiments synthesizing magnesium oxide.

Experiment	Grams of magnesium	Grams of oxygen
1	15	10
2	22	14
3	30	20
4	28	18
5	10	6

Plot the data on your graph from part (b). Which is the correct formula for magnesium oxide?

Source: © Andrew Hunt and Alan Sykes, *Chemistry*, 1984. Reprinted by permission of Pearson Education Limited.

15. In this problem we'll calculate the efficiency of swimming as a means of locomotion. A swimmer generates power to maintain a constant speed in the water. If she must swim against an opposing force, the power needed increases. The table shows the power generated by a swimmer maintaining a speed of 0.4 meter per second against various forces. (A negative force is helping the swimmer.)

Force (newtons)	−3.5	0	0	6	8	10	17	17
Metabolic power (watts)	100	190	230	320	380	450	560	600

a. Plot the data and find the least squares regression line.

b. Estimate the power needed to overcome an opposing force of 15 newtons.

c. Estimate the force necessary to tow a resting swimmer at 0.4 meter per second. (If she is resting she is not expending any power).

d. The swimmer's *mechanical* power is computed by multiplying her speed times the force needed to tow her at rest. Estimate the mechanical power for swimming at 0.4 meter per second.

e. The ratio of mechanical power to metabolic power is a measure of the swimmer's efficiency. Compute the efficiency of the swimmer when there is no external force hindering or helping her.

Source: diPrampero et al., 1974 and Alexander, 1992.

16. An athlete uses oxygen slowly when resting, but more quickly during physical exertion. The table shows the oxygen intake, in liters per minute, for several trained cyclists in terms of the power each generated on a bicycle ergometer.

Work rate (watts)	40	100	180	220	280	300	320	410
Oxygen use (l/min)	1	1.7	2	3.3	3.9	3.6	4.3	5

a. Plot the data and find the least squares regression line.

b. What does the vertical intercept of the regression line tell you about this situation?

c. Estimate the power produced by a cyclist consuming oxygen at 5.9 liters per minute.

d. One watt of power is an energy output of one joule per second. How many joules of energy can be extracted from each cubic centimeter of oxygen used?

Source: Pugh, 1974.

In Problems 17–22, we'll find a linear model from two data points.

a. Make a table showing the coordinates of two data points for the model. (Which variable should be plotted on the horizontal axis?)

b. Find a linear equation relating the variables.

c. Graph the equation.

d. State the slope of the line, including units, and explain its meaning in the context of the problem.

17. It cost a bicycle company $9000 to make 50 touring bikes in its first month of operation and $15,000 to make 125 bikes during its second month. Express the company's monthly production cost, C, in terms of the number x of bikes it makes.

18. Flying lessons cost $645 for an 8-hour course and $1425 for a 20-hour course. Both prices include a fixed insurance fee. Express the cost, C, of flying lessons in terms of the length, h, of the course in hours.

19. Under ideal conditions Andrea's Porsche can travel 312 miles on a full tank (12 gallons of gasoline) and 130 miles on 5 gallons. Express the distance, d, Andrea can drive in terms of the amount of gasoline, g, she buys.

20. On an international flight a passenger may check two bags each weighing 70 kilograms, or 154 pounds, and one carry-on bag weighing 50 kilograms, or 110 pounds. Express the weight, p, of a bag in pounds in terms of its weight, k, in kilograms.

21. A radio station in Detroit, Michigan, reports the high and low temperatures in the Detroit/Windsor area as 59°F and 23°F, respectively. A station in Windsor, Ontario, reports the same temperatures as 15°C and −5°C. Express the Fahrenheit temperature, F, in terms of the Celsius temperature, C.

22. Ms. Randolph bought a new car in 1990. In 1992 the car was worth $9000, and in 1995 it was valued at $4500. Express the value, V, of Ms. Randolph's car in terms of the number of years, t, she has owned it.

a. Use linear interpolation to give approximate answers.

b. What is the meaning of the slope in the context of the problem?

23. The temperature in Encino dropped from 81° at 1 A.M. to 73° at 5 A.M. Estimate the temperature at 4 A.M.

24. Newborn blue whales are about 24 feet long and weigh 3 tons. The young whale nurses for 7 months, at which time it is 53 feet long. Estimate the length of a one-year-old blue whale.

25. A car starts from a standstill and accelerates to a speed of 60 miles per hour in 6 seconds. Estimate the car's speed 2 seconds after it began to accelerate.

26. A truck is moving at 24 feet per second on a slippery road when the driver steps on the brakes. The truck needs 3 seconds to come to a stop. Estimate the truck's speed 2 seconds after the brakes were applied.

Use linear interpolation or extrapolation to answer the questions.

27. The temperature of an automobile engine is 9° Celsius when the engine is started and is 51° seven minutes later. Use a linear model to predict the engine temperature for both two minutes and two hours after it started. Are your predictions reasonable?

28. The temperature in Death Valley is 95° at 5 A.M. and rises to 110° by noon. Use a linear model to predict the temperature at 2 P.M. and at midnight. Are your predictions reasonable?

29. Ben weighed 8 pounds at birth and 20 pounds at age 1 year. How much will he weigh at age 10 if his weight increases at a constant rate?

30. The elephant at the City Zoo becomes ill and loses weight. She weighed 10,012 pounds when healthy and only 9,641 pounds a week later. Predict her weight after 10 days of illness.

31. Birds' nests are always in danger from predators. If there are other nests close by, the chances of predators finding the nest increase. The table shows the probability of a nest being found by predators and the distance to the nearest neighboring nest.

Distance to nearest neighbor (meters)	20	40	60	80	100
Probability of loss to predators (%)	47	34	32	17	1.5

a. Plot the data and the least squares regression line.

b. Use the regression line to estimate the probability of predators finding a nest if its nearest neighbor is 50 meters away.

c. If the probability of predators finding a nest is 10%, how far away is its nearest neighbor?

d. What is the probability of predators finding a nest if its nearest neighbor is 120 meters away? Is your answer reasonable?
Source: Perrins, 1979.

32. A trained cyclist pedals faster as he increases his cycling speed, even with a multiple-gear bicycle. The table gives the pedal frequency, p, (in revolutions per minute) and the cycling speed, c, (in kilometers per hour) of one cyclist.

Speed (km/hr)	8.8	12.5	16.2	24.4	31.9	35.0
Pedal frequency (rpm)	44.5	50.7	60.6	77.9	81.9	95.3

a. Plot the data and the least squares regression line.

b. Estimate the pedal frequency at a speed of 20 kilometers per hour.

c. Estimate the cyclist's speed when he is pedaling at 70 revolutions per minute.

d. Does your regression line give a reasonable prediction for the pedaling frequency when the cyclist is not moving? Explain.
Source: Pugh, 1974.

1.6 Additional Properties of Lines

Horizontal and Vertical Lines

Two special cases of linear equations are worth noting. First, an equation such as $y = 4$ can be thought of as an equation in two variables,

$$0x + y = 4$$

For each value of x, this equation assigns the value 4 to y. Thus, any ordered pair of the form $(x, 4)$ is a solution of the equation. For example,

$$(-1, 4), \quad (2, 4) \quad \text{and} \quad (4, 4)$$

are all solutions of the equation. If we draw a straight line through these points, we obtain the horizontal line shown in Figure 1.96.

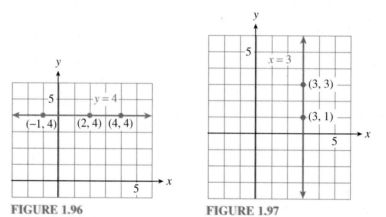

FIGURE 1.96   **FIGURE 1.97**

The other special case of a linear equation is of the type $x = 3$, or

$$x + 0y = 3$$

Here, only one value is permissible for x, namely 3, while any value may be assigned to y. Any ordered pair of the form $(3, y)$ is a solution of this equation. If we choose two solutions, say $(3, 1)$ and $(3, 3)$, and draw a straight line through these two points, we have the vertical line shown in Figure 1.97. In general, we have the following results.

> The graph of $x = k$ (k a constant) is a **vertical** line.
> The graph of $y = k$ (k a constant) is a **horizontal** line.

EXAMPLE 1

a. Graph $y = 2$

b. Graph $x = -3$

Solutions

a.

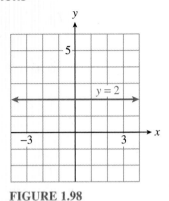

FIGURE 1.98

b.

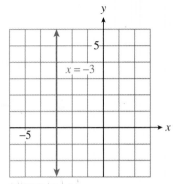

FIGURE 1.99 ■

Now let's compute the slopes of the two lines in Example 1. Choose two points on the graph of $y = 2$, say, $(-5, 2)$ and $(4, 2)$. Use these points to compute the slope.

$$m = \frac{y_2 - y_1}{x_2 - x_1} = \frac{2 - 2}{4 - (-5)}$$

$$= \frac{0}{9} = 0$$

The slope of the horizontal line $y = 2$ is zero. In fact, the slope of any horizontal line is zero, because the y-coordinates of all the points on the line are equal. Thus

$$m = \frac{y_2 - y_1}{x_2 - x_1} = \frac{0}{x_2 - x_1} = 0$$

On a vertical line, the x-coordinates of all the points are equal. For example, two points on the line $x = -3$ are $(-3, 1)$ and $(-3, 6)$. Using these points to compute the slope, we find

$$m = \frac{y_2 - y_1}{x_2 - x_1} = \frac{6 - 1}{-3 - (-3)}$$

$$= \frac{5}{0}$$

which is undefined. The slope of any vertical line is undefined because the expression $x_2 - x_1$ equals zero.

The slope of a **horizontal** line is **zero.**
The slope of a **vertical** line is **undefined.**

Parallel and Perpendicular Lines

Consider the graphs of the equations

$$y = \frac{2}{3}x - 4$$

$$y = \frac{2}{3}x + 2$$

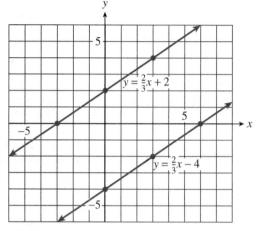

shown in Figure 1.100. The lines have the same slope, $\frac{2}{3}$, but different *y*-intercepts. Because slope measures the "steepness" or inclination of a line, lines with the same slope are parallel.

FIGURE 1.100

> Two lines with slopes m_1 and m_2 are **parallel** if and only if $\boldsymbol{m_1 = m_2}$.

EXAMPLE 2

Are the graphs of the equations $3x + 6y = 6$ and $y = -\frac{1}{2}x + 5$ parallel?

Solution

The lines are parallel if their slopes are equal. We can find the slope of the first line by putting its equation into slope-intercept form. Solve for *y*:

$$3x + 6y = 6 \qquad \text{Subtract } 3x \text{ from both sides.}$$
$$6y = -3x + 6 \qquad \text{Divide both sides by 6.}$$
$$y = \frac{-3x}{6} + \frac{6}{6} \qquad \text{Simplify.}$$
$$y = -\frac{1}{2}x + 1$$

The slope of the first line is $m_1 = -\frac{1}{2}$. The equation of the second line is already in slope-intercept form, and its slope is $m_2 = -\frac{1}{2}$. Thus, $m_1 = m_2$, so the lines are parallel. ∎

Now consider the graphs of the equations

$$y = \frac{2}{3}x - 2$$

$$y = -\frac{3}{2}x + 3$$

shown in Figure 1.101. The lines appear to be perpendicular. The relationship between the slopes of perpendicular lines is not as easy to see as the relationship for parallel lines. However, for this example, $m_1 = \frac{2}{3}$ and $m_2 = -\frac{3}{2}$. Note that

$$m_2 = -\frac{3}{2} = \frac{-1}{\frac{2}{3}} = \frac{-1}{m_1}$$

This relationship holds for any two perpendicular lines with slopes m_1 and m_2, as long as $m_1 \neq 0$ and $m_2 \neq 0$.

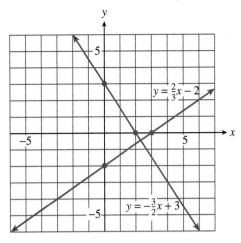

FIGURE 1.101

Two lines with slopes m_1 and m_2 are **perpendicular** if

$$m_2 = \frac{-1}{m_1}$$

We say that m_2 is the *negative reciprocal* of m_1.

EXERCISE 2 Are the graphs of $3x - 5y = 5$ and $2y = \frac{10}{3}x + 3$ perpendicular?

Put each equation into slope-intercept form by solving for y.

$3x - 5y = 5$ $\qquad\qquad\qquad 2y = \frac{10}{3}x + 3$

$y =$ $\qquad\qquad\qquad\qquad\quad y =$

Read the slope of each line from its equation.

$m_1 =$ $\qquad\qquad\qquad\qquad\quad m_2 =$

Compute the negative reciprocal of m_1: $\qquad \frac{-1}{m_1} =$

Does $m_2 = \frac{-1}{m_1}$?

Applications to Geometry

These relationships for the slopes of parallel and perpendicular lines can help us solve numerous geometric problems.

EXAMPLE 3   Show that the triangle with vertices $A(0, 8)$, $B(6, 2)$, and $C(-4, 4)$ is a right triangle.

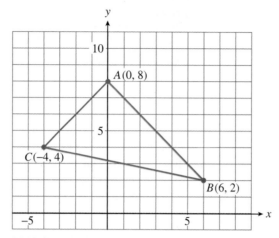

FIGURE 1.102

Solution

We will show that two of the sides of the triangle are perpendicular. The line segment $\overline{AB}$ has slope

$$m_1 = \frac{2 - 8}{6 - 0} = \frac{-6}{6} = -1$$

and the line segment $\overline{AC}$ has slope

$$m_2 = \frac{4 - 8}{-4 - 0} = \frac{-4}{-4} = 1$$

Since

$$\frac{-1}{m_1} = \frac{-1}{-1} = 1 = m_2$$

the sides $\overline{AB}$ and $\overline{AC}$ are perpendicular, and the triangle is a right triangle.   ■

Consider the graph of $4x - 2y = 6$ shown in Figure 1.103. Can we find the equation of the line that is parallel to this line but passes through the point $(1, 4)$?

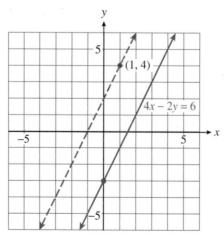

FIGURE 1.103

If we can find the slope of the desired line, we can use the slope-intercept formula to find its equation. Now, since the line we want is parallel to the given line, they must have the same slope. To find the slope of the given line, we write its equation in slope-intercept form:

$$4x - 2y = 6 \qquad \text{Subtract } 4x \text{ from both sides.}$$
$$-2y = -4x + 6 \qquad \text{Divide both sides by } -2.$$
$$y = \frac{-4x}{-2} + \frac{6}{-2} \qquad \text{Simplify.}$$
$$y = 2x - 3$$

The slope of the given line is $m_1 = 2$. Because the unknown line is parallel to this line, its slope is also 2. We know the slope of the desired line, $m = 2$, and one point on the line $(1, 4)$. Substituting these values into the point-slope formula will give us the equation.

$$y - y_1 = m(x - x_1)$$
$$y - 4 = 2(x - 1) \qquad \text{Apply the distributive law.}$$
$$y - 4 = 2x - 2 \qquad \text{Add 4 to both sides.}$$
$$y = 2x + 2$$

EXAMPLE 4

Find an equation for the line that passes through the point $(1, 4)$ and is perpendicular to the line $4x - 2y = 6$.

Solution

We follow the same strategy as in the discussion above: First find the slope of the desired line, then use the point-slope formula to write its equation. The line we want is perpendicular to the given line, so its slope is the negative reciprocal of $m_1 = 2$, the slope of the given line. Thus

$$m_2 = \frac{-1}{m_1} = \frac{-1}{2}$$

Now use the point-slope formula with $m = \frac{-1}{2}$ and $(x_1, y_1) = (1, 4)$.

$$y - y_1 = m(x - x_1)$$
$$y - 4 = \frac{-1}{2}(x - 1) \qquad \text{Apply the distributive law.}$$
$$y - 4 = \frac{-1}{2}x + \frac{1}{2} \qquad \text{Add 4 to both sides.}$$
$$y = \frac{-1}{2}x + \frac{9}{2} \qquad \frac{1}{2} + 4 = \frac{1}{2} + \frac{8}{2} = \frac{9}{2}$$

The given line and the perpendicular line are shown in Figure 1.104.

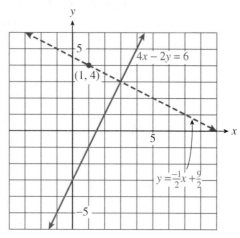

FIGURE 1.104

EXERCISE 3 Find an equation for the altitude of the triangle shown.

The altitude of a triangle is perpendicular to its base. Find the slope of the base.

$$m_1 =$$

Find the slope of the altitude.

$$m_2 =$$

Use the point-slope formula to find the equation of the altitude. Use m_2 for the slope, and the vertex of the triangle for (x_1, y_1).

$$y - y_1 = m(x - x_1)$$

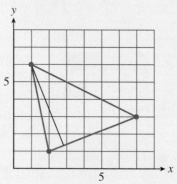

1. Give an example of an equation for a vertical line and for a horizontal line.
2. Why is the slope of a vertical line undefined?
3. What is the best way to determine whether two lines are parallel?
4. Suppose you know the equation of a certain line. Explain how to find the slope of a second line perpendicular to the first line.

ANSWERS TO 1.6 EXERCISES

1. a. 0   **b.** undefined   **2.** $m_1 = \dfrac{3}{5}$, $m_2 = \dfrac{5}{3}$; no

3. $m_1 = \dfrac{2}{5}$, $m_2 = \dfrac{-5}{2}$; $y = \dfrac{-5}{2}x + \dfrac{17}{2}$

a. Sketch a rough graph of each equation, and label its intercept.

b. State the slope of each line.

1. $y = -3$ **2.** $x = -2$ **3.** $2x = 8$ **4.** $3y = 15$

5. $x = 0$ **6.** $y = 0$

In Problems 7–12, find the equation of the line described.

7. A vertical line through the point $(-5, 8)$.

8. A horizontal line through the point $(2, -4)$.

9. The x-axis.

10. The y-axis.

11. Perpendicular to $x = 3$ and intersecting it at $(3, 9)$.

12. Parallel to the y-axis and including the point $(-1, -2)$.

In Problems 13 and 14,

a. Determine whether the slope of each line is positive, negative, zero, or undefined.

b. List the lines in order of increasing slope.

13.

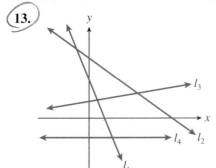

FIGURE 1.105

14.

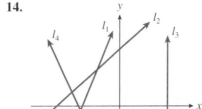

FIGURE 1.106

15. a. Use your calculator to graph the equations $y = 3x + 8$ and $y = 3.1x + 6$ together in the standard window. Do you think the lines are parallel?

   b. Find the slope of each line in part (a). Are the lines parallel?

   c. Find the y-value for each equation when $x = 20$. What do your answers tell you about the two lines?

16. a. Use your calculator to graph the equation $y = 0.001x + 4$ in the standard window. Do you think the line is horizontal?

   b. Find the slope and the x-intercept of the line in part (a). Is the line horizontal?

   c. Graph the equation in part (a) in the window

   $$Xmin = -5000 \qquad Xmax = 5000$$
   $$Ymin = -10 \qquad Ymax = 10$$

   Find the coordinates of two convenient points on the line and compute its slope using the slope formula.

17. The slopes of several lines are given below. Which of the lines are parallel to the graph of $y = 0.75x + 2$, and which are perpendicular to it?

   a. $m = \frac{3}{4}$ **c.** $m = \frac{-20}{15}$ **e.** $m = \frac{4}{3}$ **g.** $m = \frac{36}{48}$

   b. $m = \frac{8}{6}$ **d.** $m = \frac{-39}{52}$ **f.** $m = \frac{-16}{12}$ **h.** $m = \frac{9}{12}$

18. The slopes of several lines are given below. Which of the lines are parallel to the graph of $y = 2.5x - 3$, and which are perpendicular to it?

   a. $m = \frac{2}{5}$ **c.** $m = \frac{-8}{20}$ **e.** $m = \frac{40}{16}$ **g.** $m = \frac{-1}{25}$

   b. $m = \frac{25}{10}$ **d.** $m = \frac{-45}{18}$ **f.** $m = 25$ **h.** $m = \frac{-5}{10}$

19. In each part, determine whether the two lines are parallel, perpendicular, or neither.

  a. $y = \dfrac{3}{5}x - 7$; $3x - 5y = 2$

  b. $y = 4x + 3$; $y = \dfrac{1}{4}x - 3$

  c. $6x + 2y = 1$; $x = 1 - 3y$

  d. $2y = 5$; $5y = -2$

20. In each part, determine whether the two lines are parallel, perpendicular, or neither.

  a. $2x - 7y = 14$; $7x - 2y = 14$

  b. $x + y = 6$; $x - y = 6$

  c. $x = -3$; $3y = 5$

  d. $\dfrac{1}{4}x - \dfrac{3}{4}y = \dfrac{2}{3}$; $\dfrac{1}{6}x = \dfrac{1}{2}y + \dfrac{1}{3}$

21. a. Sketch the triangle with vertices $A(2, 5)$, $B(5, 2)$, and $C(10, 7)$.

  b. Show that the triangle is a right triangle. (*Hint:* What should be true about the slopes of the two sides that form the right angle?)

22. a. Sketch the triangle with vertices $P(-1, 3)$, $Q(-3, 8)$, and $R(4, 5)$.

  b. Show that the triangle is a right triangle. (See the hint for Problem 21.)

23. a. Sketch the quadrilateral with vertices $P(2, 4)$, $Q(3, 8)$, $R(5, 1)$, and $S(4, -3)$.

  b. Show that the quadrilateral is a parallelogram. (*Hint:* What should be true about the slopes of the opposite sides of the parallelogram?)

24. a. Sketch the quadrilateral with vertices $A(-5, 4)$, $B(7, -11)$, $C(12, 25)$, and $D(0, 40)$.

  b. Show that the quadrilateral is a parallelogram. (See the hint for Problem 23.)

25. Show that the line passing through the points $A(0, -3)$ and $B\left(3, \frac{1}{2}\right)$ also passes through the point $C(-6, -10)$.

26. Do the points $P\left(-5, -3\frac{1}{2}\right)$, $Q(4, -2)$, and $R\left(9\frac{1}{2}, -1\right)$ lie on the same line? Why or why not?

27. a. Put the equation $x - 2y = 5$ into slope-intercept form, and graph the equation on paper.

  b. What is the slope of any line that is parallel to $x - 2y = 5$?

  c. On your graph for part (a), sketch by hand a line that is parallel to $x - 2y = 5$ and passes through the point $(2, -1)$.

  d. Use the point-slope formula to write an equation for the line that is parallel to the graph of $x - 2y = 5$ and passes through the point $(2, -1)$.

28. a. Put the equation $2y - 3x = 5$ into slope-intercept form, and graph the equation on paper.

  b. What is the slope of any line that is parallel to $2y - 3x = 5$?

  c. On your graph for part (a), sketch by hand a line that is parallel to $2y - 3x = 5$ and passes through the point $(-3, 2)$.

  d. Use the point-slope formula to write an equation for the line that is parallel to the graph of $2y - 3x = 5$ and passes through the point $(-3, 2)$.

29. a. Put the equation $2y - 3x = 5$ into slope-intercept form, and graph the equation on paper.

  b. What is the slope of any line that is perpendicular to $2y - 3x = 5$?

  c. On your graph for part (a), sketch by hand a line that is perpendicular to $2y - 3x = 5$ and passes through the point $(1, 4)$.

  d. Use the point-slope formula to write an equation for the line that is perpendicular to the graph of $2y - 3x = 5$ and passes through the point $(1, 4)$.

30. a. Put the equation $x - 2y = 5$ into slope-intercept form, and graph the equation on paper.

  b. What is the slope of any line that is perpendicular to $x - 2y = 5$?

  c. On your graph for part (a), sketch by hand a line that is perpendicular to $x - 2y = 5$ and passes through the point $(4, -3)$.

  d. Use the point-slope formula to write an equation for the line that is perpendicular to the graph of $x - 2y = 5$ and passes through the point $(4, -3)$.

31. Two of the vertices of rectangle *ABCD* are $A(-5, 2)$ and $B(-2, -4)$.

   a. Find an equation for the line that includes side $\overline{AB}$.

   b. Find an equation for the line that includes side $\overline{BC}$.

32. Two of the vertices of rectangle *PQRS* are $P(-2, -6)$ and $Q(4, -4)$.

   a. Find an equation for the line that includes side $\overline{PQ}$.

   b. Find an equation for the line that includes side $\overline{QR}$.

In Problems 33 and 34, recall from geometry that the altitude from one vertex of a triangle to the opposite side is perpendicular to that side.

33. **a.** Sketch the triangle with vertices $A(-6, -3)$, $B(-6, 3)$, and $C(4, 5)$.

   b. Find the slope of the side $\overline{AC}$.

   c. Find the slope of the altitude from point B to side $\overline{AC}$.

   d. Find an equation for the line that includes the altitude from point B to side $\overline{AC}$.

34. **a.** Sketch the triangle with vertices $A(-5, 12)$, $B(4, -2)$, and $C(1, -6)$.

   b. Find the slope of the side $\overline{AC}$.

   c. Find the slope of the altitude from point B to side $\overline{AC}$.

   d. Find an equation for the line that includes the altitude from point B to side $\overline{AC}$.

In Problems 35 and 36, recall from geometry that the tangent line to a circle is perpendicular to the radius to the point of tangency.

35. The center of a circle is the point $C(2, 4)$, and $P(-1, 6)$ is a point on the circle. Find the equation of the line tangent to the circle at the point P. (See Figure 1.107.)

36. The center of a circle is the point $D(-2, 1)$, and $Q(1, -3)$ is a point on the circle. Find the equation of the line tangent to the circle at the point Q. (See Figure 1.108.)

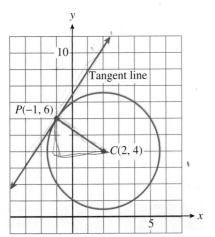

FIGURE 1.107

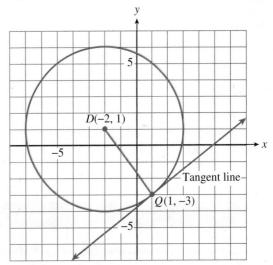

FIGURE 1.108

37. In this problem, we will show that parallel lines have the same slope. In Figure 1.109, l_1 and l_2 are two parallel lines that are neither horizontal nor vertical. Their y-intercepts are A and B. The segments $\overline{AC}$ and $\overline{CD}$ are constructed parallel to the x- and y-axes, respectively. Explain why each of the following statements is true.

a. Angle ACD equals angle CAB.

b. Angle DAC equals angle ACB.

c. Triangle ACD is similar to triangle CAB.

d. $m_1 = \dfrac{CD}{AC} \qquad m_2 = \dfrac{AB}{AC}.$

e. $m_1 = m_2.$

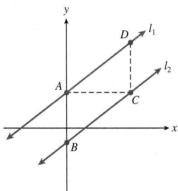

FIGURE 1.109

38. In this problem, we will show that if two lines with slopes m_1 and m_2 (where neither line is vertical) are perpendicular, then m_2 is the negative reciprocal of m_1. In Figure 1.110, lines l_1 and l_2 are perpendicular. Their y-inter-cepts are B and C. The segment $\overline{AP}$ is constructed through the point of intersection of l_1 and l_2 parallel to the x-axis. Explain why each of the following statements is true.

a. Angle ABC and angle ACB are complementary.

b. Angle ABC and angle BAP are complementary.

c. Angle BAP equals angle ACB.

d. Angle CAP and angle ACB are complementary.

e. Angle CAP equals angle ABC.

f. Triangle ABP is similar to triangle CAP.

g. $m_1 = \dfrac{BP}{AP} \qquad m_2 = -\dfrac{CP}{AP}.$

h. $m_2 = \dfrac{-1}{m_2}.$

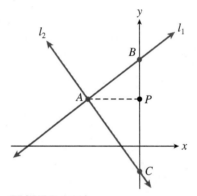

FIGURE 1.110

<hr>

CHAPTER 1 REVIEW

Write and graph a linear equation for each situation. Then answer the questions.

1. Last year Pinwheel Industries introduced a new model calculator. It cost \$2000 to develop the calculator and \$20 to manufacture each one.

a. Complete the table of values showing the total cost, C, of producing n calculators.

n	100	500	800	1200	1500
C					

b. Write an equation that expresses C in terms of n.

c. Graph the equation by hand.

d. What is the cost of producing 1000 calculators? Illustrate this as a point on your graph.

e. How many calculators can be produced for \$10,000? Illustrate this as a point on your graph.

2. Megan weighed 5 pounds at birth and gained 18 ounces per month during her first year.

  a. Complete the table of values that shows Megan's weight, w, in terms of her age, m, in months.

m	2	4	6	9	12
w					

  b. Write an equation that expresses w in terms of m.

  c. Graph the equation by hand.

  d. How much did Megan weigh at 9 months? Illustrate this as a point on your graph.

  e. When did Megan weigh 9 pounds? Illustrate this as a point on your graph.

3. The world's oil reserves were 1660 billion barrels in 1976; total annual consumption is 20 billion barrels.

  a. Complete the table of values that shows the remaining oil reserves, R, in terms of time, t, (in years since 1976).

t	5	10	15	20	25
R					

  b. Write an equation that expresses R in terms of t.

  c. Find the intercepts and graph the equation by hand.

  d. What is the significance of the intercepts to the world's oil supply?

4. The world's copper reserves were 500 million tons in 1976; total annual consumption is 8 million tons.

  a. Complete the table of values that shows the remaining copper reserves, R, in terms of time, t, (in years since 1976).

t	5	10	15	20	25
R					

  b. Write an equation that expresses R in terms of t.

  c. Find the intercepts and graph the equation by hand.

  d. What is the significance of the intercepts to the world's copper supply?

5. The owner of a movie theater needs to bring in $1000 at each screening in order to stay in business. He sells adult tickets at $5 apiece and children's tickets at $2 each.

  a. Write an equation that relates the number of adult tickets, A, he must sell and the number of children's tickets, C.

  b. Find the intercepts and graph the equation by hand.

  c. If the owner sells 120 adult tickets, how many children's tickets must he sell?

  d. What is the significance of the intercepts to the sale of tickets?

6. Alida plans to spend part of her vacation in Atlantic City and part in Saint-Tropez. She estimates that after airfare her vacation will cost $60 per day in Atlantic City and $100 per day in Saint-Tropez. She has $1200 to spend after airfare.

  a. Write an equation that relates the number of days, C, Alida can spend in Atlantic City and the number of days, T, in Saint-Tropez.

  b. Find the intercepts and graph the equation by hand.

  c. If Alida spends 10 days in Atlantic City, how long can she spend in Saint-Tropez?

  d. What is the significance of the intercepts to Alida's vacation?

Graph each equation on graph paper. Use the most convenient method for each problem.

7. $4x - 3y = 12$

8. $\dfrac{x}{6} - \dfrac{y}{12} = 1$

9. $50x = 40y - 20{,}000$

10. $1.4x + 2.1y = 8.4$

11. $3x - 4y = 0$

12. $x = -4y$

13. $4x = -12$

14. $2y - x = 0$

15. A spiked volleyball travels 6 feet in 0.04 second. A pitched baseball travels 66 feet in 0.48 second. Which ball travels faster?

16. Kendra needs $4\frac{1}{2}$ gallons of Luke's Brand primer to cover 1710 square feet of wall. She uses $5\frac{1}{3}$ gallons of Slattery's Brandprimer for 2040 square feet of wall. Which brand covers more wall per gallon?

17. Which is steeper: Stone Canyon Drive, which rises 840 feet over a horizontal distance of 1500 feet, or Highway 33, which rises 1150 feet over a horizontal distance of 2000 feet?

18. The top of Romeo's ladder is on Juliet's window sill that is 11 feet above the ground, and the bottom of the ladder is 5 feet from the base of the wall. Is the incline of this ladder as steep as a firefighter's ladder that rises a height of 35 feet over a horizontal distance of 16 feet?

19. The table shows the amount of oil, B, (in thousands of barrels) left in a tanker t minutes after it hits an iceberg and springs a leak.

t	0	10	20	30
B	800	750	700	650

  a. Write a linear equation expressing B in terms of t.

  b. Choose appropriate window settings on your calculator and graph your equation.

  c. Give the slope of the graph, including units, and explain the meaning of the slope in terms of the oil leak.

20. A traditional first experiment for chemistry students is to make 98 observations about a burning candle. Delbert records the height, h, of the candle in inches at various times t minutes after he lit it.

t	0	10	30	45
h	12	11.5	10.5	9.75

  a. Write a linear equation expressing h in terms of t.

  b. Choose appropriate window settings on your calculator and graph your equation.

  c. Give the slope of the graph, including units, and explain the meaning of the slope in terms of the candle.

21. An interior decorator bases her fee on the cost of a remodeling job. The table below shows her fee, F, for jobs of various costs, C, both given in dollars.

C	5000	10,000	20,000	50,000
F	1000	1500	2500	5500

  a. Write a linear equation for F in terms of C.

  b. Choose appropriate window settings on your calculator and graph your equation.

  c. Give the slope of the graph, and explain the meaning of the slope in terms of the decorator's fee.

22. Auto registration fees in Connie's home state depend on the value of the automobile. The table below shows the registration fee, R, for a car whose value is V, both given in dollars.

V	5000	10,000	15,000	20,000
R	135	235	335	435

  a. Write a linear equation for R in terms of V.

  b. Choose appropriate window settings on your calculator and graph your equation.

  c. Give the slope of the graph, and explain the meaning of the slope in terms of the registration fee.

Find the slope of the line segment joining each pair of points.

23. $(-1, 4), (3, -2)$

24. $(5, 0), (2, -6)$

25. $(6.2, 1.4), (-2.1, 4.8)$

26. $(0, -6.4), (-5.6, 3.2)$

Which of the following tables represent variables that are related by a linear equation?

27. a.

r	E
1	5
2	$\frac{5}{2}$
3	$\frac{5}{3}$
4	$\frac{5}{4}$
5	1

b.

s	T
10	6.2
20	9.7
30	12.6
40	15.8
50	19.0

28. a.

w	A
2	−13
4	−23
6	−33
8	−43
10	−53

b.

x	C
0	0
2	5
4	10
8	20
16	40

Each table gives values for a linear equation in two variables. Fill in the missing values.

29.

d	V
−5	−4.8
−2	−3
	−1.2
6	1.8
10	

30.

q	S
−8	−8
−4	56
3	
	200
9	264

31. The planners at AquaWorld want the small water slide to have a slope of 25%. If the slide is 20 feet tall, how far should the end of the slide be from the base of the ladder?

32. In areas with heavy snowfall, the pitch (or slope) of the roof of an A-frame house should be at least 1.2. If a small ski chalet is 40 feet wide at its base, how tall is the center of the roof?

Find the slope and y-intercept of each line.

33. $2x - 4y = 5$

34. $\frac{1}{2}x + \frac{2}{3}y = \frac{5}{6}$

35. $8.4x + 2.1y = 6.3$

36. $y - 3 = 0$

a. Graph by hand the line that passes through the given point with the given slope.

b. Find an equation for the line.

37. $(-4, 6)$ $m = -\frac{2}{3}$

38. $(2, -5)$ $m = \frac{3}{2}$

39. The rate at which air temperature decreases with altitude is called the lapse rate. In the troposphere, the layer of atmosphere that extends from the earth's surface to a height of about 7 miles, the lapse rate is about 3.6°F for every 1000 feet.

   a. If the temperature on the ground is 62°F, write an equation for the temperature, T, at an altitude of h feet.

   b. What is the temperature outside an aircraft flying at an altitude of 30,000 feet? How much colder is that than the ground temperature?

   c. What is the temperature at the top of the troposphere?

Source: Ahrens, C. Donald, 1998.

40. In his television program, *Notes from a Small Island,* aired in February of 1999, Bill Bryson discussed the future of the British aristocracy. Because not all families produce an heir, 4 or 5 noble lines die out each year. At this rate, Mr. Bryson says, and if no more peers are created, there will be no titled families left by the year 2175.

   a. Assuming that on average 4.5 titled families die out each year, write an equation for the number, N, of noble houses left in year t, where $t = 0$ in the year 1999.

   b. Graph your equation.

   c. According to your graph, how many noble families existed in 1999? Which point on the graph corresponds to this information?

Find an equation for the line passing through the two given points.

41. $(3, -5), (-2, 4)$

42. $(0, 8), (4, -2)$

  a. Make a table of values showing two data points.

  b. Find a linear equation relating the variables.

  c. State the slope of the line, including units, and explain its meaning in the context of the problem.

43. The population of Maple Rapids was 4800 in 1982 and had grown to 6780 by 1997. Assume that the population increases at a constant rate. Express the population, P, of Maple Rapids in terms of the number of years, t, since 1982.

44. Cicely's odometer read 112 miles when she filled up her 14-gallon gas tank and 308 when the gas gauge read half full. Express her odometer reading, m, in terms of the amount of gas, g, she used.

  a. Find the slope and y-intercept of each line.

  b. Write an equation for the line.

45.

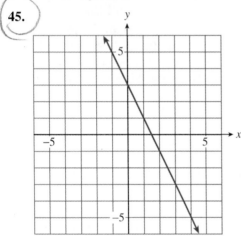

FIGURE 1.111

46.

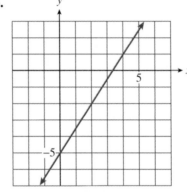

FIGURE 1.112

47. What is the slope of the line whose intercepts are $(-5, 0)$ and $(0, 3)$?

48. a. Find the x- and y-intercepts of the line
$$\frac{x}{4} - \frac{y}{6} = 1.$$

  b. What is the slope of the line in part (a)?

49. a. What is the slope of the line
$$y = 2 + \frac{3}{2}(x - 4)?$$

  b. Find the point on the line whose x-coordinate is 4. Can there be more than one such point?

  c. Use your answers from parts (a) and (b) to find another point on the line:
$(x_2, y_2) = (x_1 + \triangle x, y_1 + \triangle y)$.

50. Find an equation in point-slope form for the line of slope $\frac{6}{5}$ that passes through $(-3, -4)$.

Decide whether the following lines are parallel, perpendicular, or neither.

51. $y = \frac{1}{2}x + 3$; $x - 2y = 8$

52. $4x - y = 6$; $x + 4y = -2$

53. Write an equation for the line that is parallel to the graph of $2x + 3y = 6$ and passes through the point $(1, 4)$.

54. Write an equation for the line that is perpendicular to the graph of $2x + 3y = 6$ and passes through the point $(1, 4)$.

55. Two vertices of the rectangle $ABCD$ are $A(3, 2)$ and $B(7, -4)$. Find an equation for the line that includes side $\overline{BC}$.

56. One leg of the right triangle PQR has vertices $P(-8, -1)$ and $Q(-2, -5)$. Find an equation for the line that includes the leg $\overline{QR}$.

57. In 1986, the space shuttle *Challenger* exploded because of "O-ring" failure due to cold temperatures. On the morning of the disaster, the temperature was about 30°F. Before that date, there had been 1 incident of O-ring failure when the temperature was 70°F and 3 incidents when the temperature was 54°F. Use linear extrapolation to estimate the number of incidents of O-ring failure you would expect when the temperature is 30°F.

58. Thelma typed a 19-page technical report in 40 minutes. She required only 18 minutes for an 8-page technical report. Use linear interpolation to estimate how long Thelma would require to type a 12-page technical report.

59. *Archaeopteryx* is an extinct beast with characteristics of both birds and reptiles. Only six fossil specimens are known, and only five of those include both a femur (leg bone) and a humerus (forearm bone) The scatterplot in Figure 1.113 shows the lengths of femur and humerus for the five Archaeopteryx specimens.

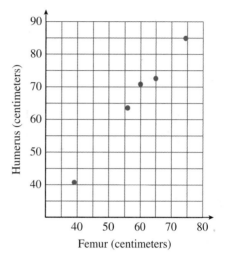

FIGURE 1.113

a. Predict the humerus length of an Archaeopteryx whose femur is 40 centimeters.

b. Predict the humerus length of an Archaeopteryx whose femur is 75 centimeters.

c. Use your answers from parts (a) and (b) to approximate the equation of a regression line.

d. Use your answer to part (c) to predict the humerus length of an Archaeopteryx whose femur is 60 centimeters.

e. Use your calculator and the given points on the scatterplot to find the least squares regression line. Compare the score this equation gives for part (d) with what you predicted earlier. The ordered pairs defining the data are $(38, 41)$, $(56, 63)$, $(59, 70)$, $(64, 72)$, $(74, 84)$.

60. The scatterplot in Figure 1.114 shows the metabolic rate of various members of the women's swim team compared with the athlete's body mass in kilograms.

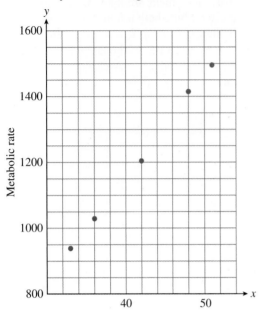

FIGURE 1.114

a. Predict the metabolic rate of a swimmer whose body mass is 30 kilograms.

b. Predict the metabolic rate of a swimmer whose body mass is 50 kilograms.

c. Use your answers from parts (a) and (b) to approximate the equation of a regression line.

d. Use your answer to part (c) to predict the metabolic rate of a swimmer with a body mass of 45 kilograms.

e. Use your calculator and the given points on the scatterplot to find the least squares regression line. Compare the score this equation gives for part (d) with what you predicted earlier. The ordered pairs defining the data are (33, 935), (36, 1027), (42, 1202), (48, 1409), (51, 1489).

Applications of Linear Models

Linear models describe variable quantities that increase or decrease at a constant rate. In this chapter we consider some linear techniques used in applications.

INVESTIGATION 3

Water Level

When sailing upstream in a canal or a river that has rapids, ships must sometimes negotiate locks to raise them to a higher water level. Suppose your ship is in one of the lower locks, at an elevation of 20 feet. The next lock is at an elevation of 50 feet. Water begins to flow from the higher lock to the lower one, raising your level by one foot per minute, and simultaneously lowering the water level in the next lock by 1.5 feet per minute.

1. Fill in the table.

t (minutes)	Lower lock water level	Upper lock water level
0		
2		
4		
6		
8		
10		

2. Let t stand for the number of minutes the water has been flowing.

   a. Write an equation for L, the water level in the lower lock after t minutes.

   b. Write an equation for U, the water level in the upper lock after t minutes.

3. Graph both your equations on the grid in Figure 2.1.

4. When will the water levels in the two locks be 10 feet apart?

5. When will the water level in the two locks be the same?

6. Write an equation you can use to verify your answer to part (5), and solve it.

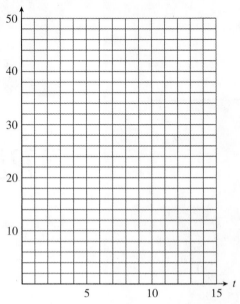

FIGURE 2.1

2.1 Systems of Linear Equations in Two Variables

Solving Systems by Graphing

A biologist wants to know the average weights of two species of birds in a wildlife preserve. She sets up a feeder whose platform is actually a scale, and mounts a camera to monitor the feeder. She waits until the feeder is occupied only by members of the two species she is studying, robins and thrushes. Then she takes a picture, which records the number of each species on the scale, and the total weight registered.

From her two best pictures, she obtains the following information. The total weight of three thrushes and six robins is 48 ounces, and the total weight of five thrushes and two robins is 32 ounces. Using these data, the biologist estimates the average weight of a thrush and of a robin. She begins by assigning variables to the two unknown quantities:

Average weight of a thrush: t

Average weight of a robin: r

Because there are two variables, the biologist must write two equations about the weights of the birds. In each of the two photos,

(weight of thrushes) + (weight of robins) = total weight

Thus,

$$3t + 6r = 48$$
$$5t + 2r = 32$$

This pair of equations is an example of a **linear system of two equations in two unknowns** (or a 2×2 linear system, for short). A **solution** to the system is an ordered pair of numbers (t, r) that satisfies both equations in the system.

Recall that every point on the graph of an equation represents a solution to that equation. A solution to *both* equations thus corresponds to a point on *both* graphs. Therefore, a solution to the system is a point where the two graphs intersect. From Figure 2.2 it appears that the intersection point is $(4, 6)$, so we expect that the values $t = 4$ and $r = 6$ are

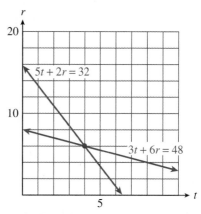

FIGURE 2.2

the solution to the system. We can check the solution by verifying that these values satisfy *both* equations in the system.

$$3(\mathbf{4}) + 6\,(\mathbf{6}) \overset{?}{=} 48 \quad \text{True}$$
$$5(\mathbf{4}) + 2\,(\mathbf{6}) \overset{?}{=} 32 \quad \text{True}$$

Both equations are true, so we conclude that the average weight of a thrush is 4 ounces, and the average weight of a robin is 6 ounces.

We can obtain graphs for the equations in a system quickly and easily using a calculator.

EXAMPLE 1

Use your calculator to solve the system

$$y = 1.7x + 0.4$$
$$y = 4.1x + 5.2$$

by graphing.

Solution

Set the graphing window to

$$\text{Xmin} = -9.4 \qquad \text{Xmax} = 9.4$$
$$\text{Ymin} = -10 \qquad \text{Ymax} = 10$$

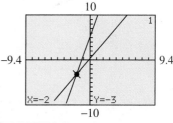

FIGURE 2.3

and enter the two equations. We can see in Figure 2.3 that the two lines intersect in the third quadrant. Use the [**TRACE**] key to find the coordinates of the intersection point. Check that the point $(-2, -3)$ lies on both graphs. (Use the up and down arrow keys to move the bug from one graph to the other.) The solution to the system is $x = -2, y = -3$. ∎

The values we obtain from the calculator may be only approximations, so it is a good idea to check the solution algebraically. In Example 1 we find that both equations are true when we substitute $x = -2$ and $y = -3$.

$$-3 = 1.7\,(-2) + 0.4 \quad \text{True}$$
$$-3 = 4.1\,(-2) + 5.2 \quad \text{True}$$

EXERCISE 1 **a.** Solve the system of equations

$$y = -0.7x + 6.9$$
$$y = 1.2x - 6.4$$

by graphing. Use the "friendly" window

$$\text{Xmin} = -9.4 \qquad \text{Xmax} = 9.4$$
$$\text{Ymin} = -10 \qquad \text{Ymax} = 10$$

b. Verify algebraically that your solution satisfies both equations.

Using the "Intersect" Feature

Because the **Trace** feature does not show every point on the graph, we may not find the exact solution to a system by tracing the graphs. Consider the system

$$3x - 2.8y = 21.06$$
$$2x + 1.2y = 5.3$$

We can graph this system in the standard window by solving each equation for y. Enter

$$Y_1 = (21.06 - 3X)/-2.8$$
$$Y_2 = (5.3 - 2x)/1.2$$

and press [ZOOM] [6].

Trace along the first line to find the intersection point. It appears to be at $x = 4.468051$, $y = -2.734195$, as shown in Figure 2.4(a). However, if we press the up or down arrow to read the coordinates off the second line, we see that for the same x-coordinate we obtain a different y-coordinate, as in Figure 2.4(b). The different y-coordinates indicate that we have *not* found an intersection point, although we are close. We can use another feature of the calculator to find the exact coordinates of the intersection point.

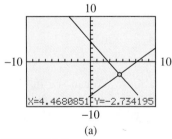
(a)

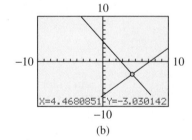
(b)

FIGURE 2.4

Use the arrow keys to position the **Trace** bug as close to the intersection point as you can, as in Figure 2.4(a) or (b). Then press [2ND] [CALC] to see the **Calculate** menu. Press [5] for **intersect**, and respond to the calculator's questions, **First curve?**, **Second curve?**, and **Guess?** by pressing [ENTER]. The calculator will then display the intersection point, $x = 4.36$, $y = -2.85$.

We can substitute these values into the original system to check that they satisfy both equations.

$$3(4.36) - 2.8(-2.85) = 21.06$$
$$2(4.36) + 1.2(-2.85) = 5.3$$

Thus $(4.36, -2.85)$ is the exact solution to the system. How does the calculator find the exact coordinates of the intersection point? In the next section we'll learn how to find the solution of a system using algebra.

EXERCISE 2 Solve the system of equations

$$y = 47x - 1930$$
$$y + 19x = 710$$

by graphing. Find the intercepts of each graph to help you choose a suitable window, and use the **intersect** feature to locate the solution.

Inconsistent and Dependent Systems

Because two straight lines do not always intersect at a single point, a 2×2 system of linear equations does not always have a unique solution. In fact, there are three possibilities, as illustrated in Figure 2.5.

1. The graphs may be the same line, as shown in Figure 2.5(a).

2. The graphs may be parallel but distinct lines, as shown in Figure 2.5(b).

3. The graphs may intersect in one and only one point, as shown in Figure 2.5(c).

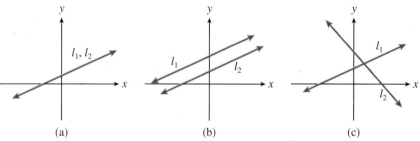

FIGURE 2.5

EXAMPLE 2 Solve the system $y = -x + 5$
$$2x + 2y = 3$$

Solution

We will use the calculator to graph both equations on the same axes, as shown in Figure 2.6. First, rewrite the second equation in slope-intercept form by solving for y.

$$2x + 2y = 3 \qquad \text{Subtract } 2x \text{ from both sides.}$$
$$2y = -2x + 3 \qquad \text{Divide both sides by 2.}$$
$$y = -x + 1.5$$

Now enter the equations as

$$Y_1 = -X + 5$$
$$Y_2 = -X + 1.5$$

The lines do not intersect within the viewing window; they appear to be parallel. If we look again at the equations of the lines we recognize that both have slope -1, so they *are* parallel. Since parallel lines never meet, there is no solution to the system.

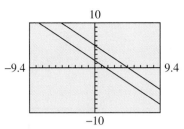

FIGURE 2.6

A system with no solutions, such as the system in Example 2, is called **inconsistent.** A 2×2 system of linear equations is inconsistent when the two equations correspond to parallel lines. This situation occurs when the lines have the same slope but different y-intercepts.

EXAMPLE 3

Solve the system $\qquad x = \dfrac{2}{3}y + 3$

$$3x - 2y = 9$$

Solution

We begin by putting each equation in slope-intercept form.

$$x = \frac{2}{3}y + 3 \quad \text{Subtract 3.}$$

$$x - 3 = \frac{2}{3}y \qquad \text{Multiply by } \frac{3}{2}.$$

$$\frac{3}{2}x - \frac{9}{2} = y$$

For the second equation:

$$3x - 2y = 9 \qquad \text{Subtract } 3x.$$
$$-2y = -3x + 9 \quad \text{Divide by } -2.$$
$$y = \frac{3}{2}x - \frac{9}{2}$$

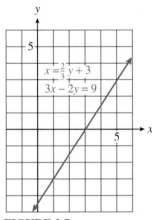

FIGURE 2.7

The two equations are actually different forms of the same equation. Because they are equivalent, they share the same line as a graph, as shown in Figure 2.7. Every point on the first line is also a point on the second line, so every solution to the first equation is also a solution of the second equation. Thus, the system has infinitely many solutions.

A linear system with infinitely many solutions, such as the system in Example 3, is called **dependent.** A 2×2 system is dependent when the two equations actually describe the same line. This situation occurs when the two lines have the same slope *and* the same y-intercept.

Here is a summary of the three cases for a 2×2 system of linear equations.

1. **Dependent system.** All the solutions of one equation are also solutions to the second equation, and hence are solutions of the system. The

graphs of the two equations are the same line. A dependent system has infinitely many solutions.

2. **Inconsistent system.** The graphs of the equations are parallel lines and hence do not intersect. An inconsistent system has no solutions.

3. **Consistent and independent system.** The graphs of the two lines intersect in exactly one point. The system has exactly one solution.

EXERCISE 3

a. Graph the system

$$y = -3x + 6$$
$$6x + 2y = 15$$

by hand, using either the intercept method or the slope-intercept method. (See Sections 1.1 and 1.4 to review these methods.)

b. Identify the system as dependent, inconsistent, or consistent and independent.

Applications

Many practical problems involve two or more unknown quantities.

EXAMPLE 4

A cup of rolled oats provides 11 grams of protein. A cup of rolled wheat flakes provides 8.5 grams of protein. Francine wants to combine oats and wheat to make a cereal with 10 grams of protein per cup. How much of each grain will she need in one cup of her mixture?

Solution

Step 1   Fraction of a cup of oats needed: x
                            Fraction of a cup of wheat needed: y

Step 2 Because we have two variables we must find two equations that describe the problem. It may be helpful to organize the information into a table.

	Cups	Grams of protein per cup	Grams of protein
Oats	x	11	$11x$
Wheat	y	8.5	$8.5y$
Mixture	1	—	10

The wheat and oats together will make one cup of mixture, so the first equation is

$$x + y = 1$$

The 10 grams of protein must come from the protein in the oats plus the protein in the wheat. This gives us a second equation:

$$11x + 8.5y = 10$$

We now have a system of equations.

Step 3 We will solve the system by graphing. First, solve each equation for y in terms of x to get

$$y = -x + 1$$
$$y = (10 - 11x)/8.5$$

Although we could simplify the second equation, the calculator can graph both equations as they are. We know that x and y represent fractions of one cup, so we set the WINDOW (as shown in Figure 2.8) with

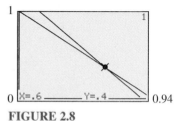

FIGURE 2.8

$$Xmin = 0 \qquad Xmax = 0.94$$
$$Ymin = 0 \qquad Ymax = 1$$

The lines intersect at (0.6, 0.4), which we can verify by substituting these values into the original two equations of our system.

Step 4 Francine needs 0.6 cups of oats and 0.4 cups of wheat.   ■

Being unable to read exact coordinates from a graph is not always a disadvantage. In many situations fractional values of the unknowns are not acceptable.

EXAMPLE 5 The mathematics department has $40,000 to set up a new computer lab. They will need one printer for every four terminals they purchase. If a printer costs $560 and a terminal costs $1520, how many of each should they buy?

Solution

Step 1   Number of printers: p
Number of terminals: t

Step 2 Since the math department needs four times as many terminals as printers,

$$t = 4p$$

The total cost of the printers will be $560p$ dollars and the total cost of the terminals will be $1520t$ dollars, so we have

$$560p + 1520t = 40,000$$

Step 3 Solve the second equation for t to get

$$t = \frac{(40,000 - 560p)}{1520}$$

Now we graph the equations

$$Y_1 = 4X$$
$$Y_2 = (40000 - 560X)/1520$$

on the same set of axes. The second graph is not visible in the standard graphing window, but with a little experimentation we can find an appropriate window setting. The WINDOW values used for Figure 2.9 are

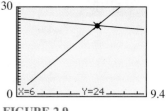

FIGURE 2.9

$$\text{Xmin} = 0 \qquad \text{Xmax} = 9.4$$
$$\text{Ymin} = 0 \qquad \text{Ymax} = 30$$

The lines intersect at approximately $(6, 24)$. These values satisfy the first equation, but not the second.

$$560(6) + 1520(24) \overset{?}{=} 40{,}000$$
$$39{,}840 \neq 40{,}000$$

Step 4 The exact solution to the system is $\left(\frac{500}{83}, \frac{2000}{83}\right)$. (We will explore two methods for finding exact solutions in Section 2.2.) But this solution is not of practical use, since the math department cannot purchase fractions of printers or terminals. The department *can* purchase 6 printers and 24 terminals (with some money left over). ∎

EXERCISE 4 The manager for Books for Cooks plans to spend $300 stocking a new diet cookbook. The paperback version costs her $5, and the hardback costs $10. She finds that she will sell three times as many paperbacks as hardbacks. How many of each should she buy?

a. Let x represent the number of hardbacks and y the number of paperbacks she should buy. Write an equation about the cost of the books.

b. Write a second equation about the number of each type of book.

c. Graph both equations and solve the system. (Find the intercepts of the graphs to help you choose a window.) Answer the question in the problem.

An Application from Economics

The owner of a retail business must try to balance the demand for his product from consumers with the supply he can obtain from manufacturers. Supply and demand both vary with the price of the product: Consumers usually buy fewer items if the price increases, but manufacturers will be willing to supply more units of the product if its price increases.

The **demand equation** gives the number of units of the product that consumers will buy in terms of the price per unit. The **supply equation** gives the number of units that the producer will supply in terms of the price per unit. The price at which the supply and demand are equal is called the **equilibrium price.** This is the price at which the consumer and the producer agree to do business.

EXAMPLE 6

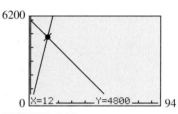

A woolen mill can produce $400x$ yards of fine suit fabric if they can charge x dollars per yard. Their clients in the garment industry will buy $6000 - 100x$ yards of wool fabric at a price of x dollars per yard. Find the equilibrium price and the amount of fabric that will change hands at that price.

Solution

Step 1   Price per yard: x
              Number of yards: y

Step 2 The supply equation tells us how many yards of fabric the mill will produce for a price of x dollars per yard.

$$y = 400x$$

The demand equation tells us how many yards of fabric the garment industry will buy at a price of x dollars per yard.

$$y = 6000 - 100x$$

Step 3 Graph the two equations on the same set of axes, as shown in Figure 2.10. Set the [WINDOW] values to

$$\text{Xmin} = 0 \qquad \text{Xmax} = 94$$
$$\text{Ymin} = 0 \qquad \text{Ymax} = 6200$$

and use the [TRACE] to locate the solution. (Or use the **intersect** command.) The graphs intersect at the point (12, 4800).

FIGURE 2.10

Step 4 The equilibrium price is $12 per yard, and the mill will sell 4800 yards of fabric at that price. ∎

READING QUESTIONS

1. What is a solution to a system of two linear equations?
2. Explain how to solve a 2×2 linear system by graphing.
3. What is an inconsistent system?
4. What is a dependent system?
5. Explain the terms demand equation, supply equation, and equilibrium price.

ANSWERS FOR 2.1 EXERCISES

1. $(7,2)$ **2.** $(40, -50)$ **3a.**

3b. inconsistent

4a. $10x + 5y = 300$ **b.** $y = 3x$ **c.** $(12, 36)$. She should buy 12 hardbacks and 36 paperbacks.

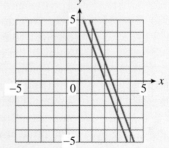

In Problems 1–4 solve each system of equations using the graphs given. Verify algebraically that your solution satisfies both equations.

1. $2.3x - 3.7y = 6.9$
$1.1x + 3.7y = 3.3$

2. $-2.3x + 5.9y = 38.7$
$9.3x + 7.4y = -0.2$

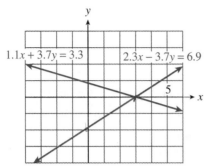

FIGURE 2.11

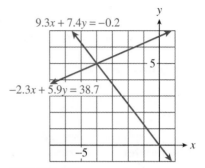

FIGURE 2.12

3. $35s - 17t = 560$
$24s + 15t = 2250$

4. $56a + 32b = -880$
$23a - 7b = 1250$

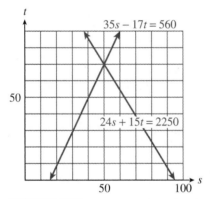

FIGURE 2.13

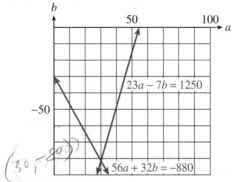

FIGURE 2.14

5. According to the Bureau of the Census, the average age of the U.S. population is steadily rising. The table gives data for two types of average, the mean and the median, for the ages of women.

Date	July 1990	July 1992	July 1994	July 1996	July 1998
Median age	34.0	34.6	35.2	35.8	36.4
Mean age	36.6	36.8	37.0	37.3	37.5

a. Which is growing more rapidly, the mean age or the median age?

b. Plot the data for median age versus the date, using July 1990 as $t = 0$. Draw a line through the data.

c. What is the meaning of the slope of the line in part (b)?

d. Plot the data for mean age versus date on the same axes. Draw a line that fits the data.

e. Use your graph to estimate when the mean age and the median age will be the same.

f. For the data given, the mean age of women in the United States is greater than the median age. What does this tell you about the U.S. population of women?

6. Repeat Exercise 5 for the mean and median ages of U.S. men, given in the adjacent table.

Date	July 1990	July 1992	July 1994	July 1996	July 1998
Median age	31.6	32.2	32.9	33.5	34.0
Mean age	33.8	34.0	34.3	34.5	34.9

In Problems 7–10 solve each system of equations by graphing. Use the "friendly" window:

$$Xmin = -9.4 \qquad Xmax = 9.4$$
$$Ymin = -10 \qquad Ymax = 10$$

Verify algebraically that your solution satisfies both equations.

7. $y = 2.6x + 8.2$
   $y = 1.8 - 0.6x$

8. $y = 5.8x - 9.8$
   $y = 0.7 - 4.7x$

9. $y = 7.2 - 2.1x$
   $-2.8x + 3.7y = 5.5$

10. $y = -2.3x - 5.5$
   $3.1x + 2.4y = -1.1$

In Problems 11–16, graph each system by hand, using either the intercept method or the slope-intercept method. (See Sections 1.1 and 1.4 to review these methods.) Identify each system as dependent, inconsistent, or consistent and independent.

11. $2x = y + 4$
   $8x - 4y = 8$

12. $2t + 12 = -6s$
   $12s + 4t = 24$

13. $w - 3z = 6$
   $2w + z = 8$

14. $2u + v = 5$
   $u - 2v = 3$

15. $2L - 5W = 6$
   $\dfrac{15W}{2} + 9 = 3L$

16. $-3A = 4B + 12$
   $\dfrac{1}{2}A + 2 = \dfrac{-2}{3}B$

Solve Problems 17–24 by graphing a system of equations. If you need to review writing equations, please see Appendices A.1 and A.2.

17. Dash Phone Company charges a monthly fee of $10 plus $0.09 per minute for long-distance calls. Friendly Phone Company charges $15 per month plus $0.05 per minute for long-distance calls.

a. Write an equation for Dash Phone Company's monthly bill if you talk long-distance for x minutes.

b. Write an equation for Friendly Phone Company's monthly bill if you talk long-distance for x minutes.

c. Fill in the table of values for the bills from each company.

x	0	30	60	90	120	150
Dash						
Friendly						

Estimate when (after how many long-distance minutes) Dash's bill is larger than Friendly's.

d. Graph your system. Use the table to help you choose a suitable window.

e. How many minutes of long-distance calls would result in equal bills from the two companies?

18. The Olympus Health Club charges an initial fee of $230 and $13 monthly dues. The Valhalla Health Spa charges $140 initially and $16 per month.

a. Write an equation for the cost of belonging to Olympus Health Club for x months.

b. Write an equation for the cost of belonging to Valhalla Health Spa for x months.

c. Fill in the table of values for the cost of membership in both clubs.

x	0	6	12	18	24	30
Olympus						
Valhalla						

x	5	10	15	20	25	30
C						
R						

For how many months is membership in Valhalla cheaper than in Olympus?

d. Graph your system. Use the table to help you choose a suitable window.

e. After how many months of membership would the costs of belonging to the two clubs be equal?

19. Yasuo can afford to produce $50x$ bushels of wheat if he can sell them at x cents per bushel, and the market will buy $2100 - 20x$ bushels at x cents per bushel.

a. What is the supply equation?

b. What is the demand equation?

c. Find the intercepts of the graphs to help you choose a window, then graph both equations.

d. Find the equilibrium price and the number of bushels of wheat Yasuo can sell at that price.

20. Mel's Pool Service can clean $1.5x$ pools per week if it charges x dollars per pool, and the public will book $120 - 2.5x$ pool cleanings at x dollars per pool.

a. What is the supply equation?

b. What is the demand equation?

c. Find the intercepts of the graphs to help you choose a window, then graph both equations.

d. Find the equilibrium price and the number of pools Mel will clean at that price.

21. The Aquarius jewelry company determines that each production run to manufacture a pendant involves an initial setup cost of $200 and $4 for each pendant produced. The pendants sell for $12 each.

a. Express the cost C of production in terms of the number x of pendants produced.

b. Express the revenue R in terms of the number x of pendants sold.

c. Complete the table of values for the cost and revenue.

d. Graph the revenue and cost on the same set of axes.

e. How many pendants must be sold for the Aquarius company to break even on a particular production run?

22. The Bread Alone Bakery has a daily overhead of $90. It costs $0.60 to bake each loaf of bread, and the bread sells for $1.50 per loaf.

a. Express the cost C in terms of the number x of loaves baked.

b. Express the revenue R in terms of the number x of loaves sold.

c. Complete the table of values for the cost and revenue.

x	20	40	60	80	100	120
C						
R						

d. Graph the revenue and cost on the same set of axes.

e. How many loaves must the bakery sell to break even on a given day?

23. The admissions at a Bengals' baseball game was $7.50 for adults and $4.25 for students. The ticket office took in $465.50 for 82 paid admissions. How many adults and how many students attended the game?

a. Write algebraic expressions to fill in the table.

	Number of tickets	Cost per ticket	Revenue
Adults	x		
Students	y		
Total			

b. Write an equation about the number of tickets sold.

c. Write a second equation about the revenue from the tickets.

d. Graph both equations and solve the system.

24. There were 42 passengers on an airplane flight for which first-class fare was $400 and tourist fare was $320. If the revenue for the flight totaled $14,400, how many first-class and how many tourist passengers paid for the flight?

a. Write algebraic expressions to fill in the table.

b. Write an equation about the number of tickets sold.

	Number of tickets	Cost per ticket	Revenue
First-class	x		
Tourist	y		
Total			

c. Write a second equation about the revenue from the tickets.

d. Graph both equations and solve the system.

In Problems 25–28 solve each system of equations by graphing. Find the intercepts of each graph to help you choose a suitable window, and use the *intersect* feature to locate the solution.

25. $38x + 2.3y = -55.2$
    $y = 15x + 121$

26. $25x - 1.7y = 10.5$
    $y + 5x = 49$

27. $64x + 58y = 707$
    $82x - 21y = 496$

28. $35x - 76y = 293$
    $15x + 44y = -353$

2.2 Solution of Systems by Algebraic Methods

Although calculators and computers make it fairly easy to solve systems by graphing, it may be faster to use algebraic methods. In this section we will review two algebraic methods: substitution and elimination.

Solving Systems by Substitution

You are probably familiar with the method of substitution. The basic strategy can be described as follows.

Steps for Solving a 2 × 2 System by Substitution

1. Solve one of the equations for one of the variables in terms of the other.

2. Substitute this expression into the second equation, which yields an equation in one variable.

3. Solve the new equation.

4. Use the result of Step 1 to find the other variable.

EXAMPLE 1

Staci stocks two kinds of sleeping bags in her sporting goods store, a standard model and a down-filled model for colder temperatures. From past experience she estimates that she will sell twice as many of the standard variety as of the down-filled. She has room to stock 60 sleeping bags at a time. How many of each variety should Staci order?

Solution

Step 1
Number of standard sleeping bags: x

Number of down-filled sleeping bags: y

Step 2 Write two equations about the variables. Staci needs twice as many standard model as down-filled, so

$$x = 2y \tag{1}$$

Also, the total number of sleeping bags is 60, so

$$x + y = 60 \tag{2}$$

These two equations give us a system.

Step 3 We will solve this system using substitution. Notice that Equation (1) is already solved for x in terms of y: $x = 2y$. Substitute $2y$ for x in Equation (2) to obtain

$$2y + y = 60$$
$$3y = 60$$

Solving for y we find $y = \mathbf{20}$. Finally, substitute this value into Equation (1) to find

$$x = 2(\mathbf{20}) = 40$$

The solution to the system is $x = 40$, $y = 20$.

Step 4 Staci should order 40 standard sleeping bags and 20 down-filled bags. ∎

EXERCISE 1 Solve the system by substitution.

$$2x - 3y = 6$$
$$x + 3y = 3$$

Solving Systems by Elimination

The method of substitution is convenient if one of the variables in the system has a coefficient of 1 or -1, because it is easy to solve for that variable. If none of the coefficients is 1 or -1, then a second method, called **elimination,** is usually more efficient.

The method of elimination is based on the following properties of linear equations.

Properties of Linear Systems

1. Multiplying a linear equation by a (nonzero) constant does not change its solutions. That is, any solution of the equation

$$ax + by = c$$

is also a solution of the equation

$$kax + kby = kc$$

2. Adding (or subtracting) two linear equations does not change their common solutions. That is, any solution of the system

$$a_1x + b_1y = c_1$$
$$a_2x + b_2y = c_2$$

is also a solution of the equation

$$(a_1 + a_2)x + (b_1 + b_2)y = c_1 + c_2$$

EXAMPLE 2 Rani kayaks downstream for 45 minutes and travels a distance of 6000 meters. On the return journey upstream she covers only 4800 meters in 45 minutes. How fast is the current in the river, and how fast would Rani kayak in still water? (Give your answers in meters per minute.)

Solution
Step 1 Rani's speed in still water: r
Speed of the current: s

Step 2 We must write two equations using the variables r and s. First, organize the information into a table. Note that when Rani travels downstream, the current in the river helps her, so her effective speed is $r + s$. When she travels upstream, she is fighting the current, so her speed is actually $r - s$.

	Rate	Time	Distance
Downstream	$r + s$	45	6000
Upstream	$r - s$	45	4800

Using the formula *Rate × Time = Distance,* we can write one equation describing Rani's journey downstream and a second equation for the journey upstream.

$$(r + s) \cdot 45 = 6000$$
$$(r - s) \cdot 45 = 4800$$

Apply the distributive law to write each equation in standard form.

$$45r + 45s = 6000 \tag{1}$$
$$45r - 45s = 4800 \tag{2}$$

Step 3 To solve the system, we will eliminate the variable s by adding the two equations together vertically. (Property 2 of linear systems allows us to take this step.)

$$
\begin{array}{r}
45r + 45s = 6000 \\
+45r - 45s = 4800 \\
\hline
90r = 10{,}800
\end{array}
$$

Now we have an equation in one variable only, which we can solve for r.

$$90r = 10{,}800 \quad \text{\small Divide both sides by 90.}$$
$$r = 120$$

To solve for s we substitute $r = 120$ into any previous equation involving both r and s. We'll use Equation (1).

$$45(120) + 45s = 6000 \quad \text{\small Simplify the left side.}$$
$$5400 + 45s = 6000 \quad \text{\small Subtract 5400 from both sides.}$$
$$45s = 600 \quad \text{\small Divide both sides by 45; reduce.}$$
$$s = \frac{40}{3}$$

Step 4 The speed of the current is $\frac{40}{3}$, or $13\frac{1}{3}$ meters per minute, and Rani's speed in still water is 120 meters per minute. ∎

In Example 2 we eliminated the variable s by adding the two equations together. This worked because the coefficients of s in the two equations were opposites, namely 45 and -45. If the coefficients of the chosen variable are not opposites, we must multiply each equation by a suitable factor so that the new coefficients are opposites. We choose the "suitable factors" the same way we choose building factors to obtain a least common denominator when adding fractions.

EXAMPLE 3

Solve the system

$$2x + 3y = 8 \tag{1}$$
$$3x - 4y = -5 \tag{2}$$

by the method of elimination.

Solution

We first decide which variable to eliminate, x or y. We can choose whichever looks easier. In this problem there is no advantage to either choice, so we'll choose to eliminate x. We next look for the smallest number that both coefficients, 2 and 3, divide into evenly. This number is 6. We want the coefficients of x to become 6 and -6, so we'll multiply Equation (1) by 3 and Equation (2) by -2 to obtain

$$6x + 9y = 24 \qquad \text{(1a)}$$
$$-6x + 8y = 10 \qquad \text{(2a)}$$

Now add the corresponding terms of (1a) and (2a). The x-terms are eliminated, yielding an equation in one variable.

$$\begin{array}{r} 6x + 9y = 24 \\ -6x + 8y = 10 \\ \hline 17y = 34 \end{array} \qquad \text{(3)}$$

Solve this equation for y to find $y = 2$. We can substitute this value of y into any of our equations involving both x and y. If we choose Equation (1), then

$$2x + 3(2) = 8$$

and solving this equation yields $x = 1$. The ordered pair $(1, 2)$ is a solution to the system. You should verify that these values satisfy both original equations. ∎

We summarize the strategy for solving a linear system by elimination.

Steps for Solving a 2 × 2 Linear System by Elimination

1. Choose one of the variables to eliminate. Multiply each equation by a suitable factor so that the coefficients of that variable are opposites.
2. Add the two new equations termwise.
3. Solve the resulting equation for the remaining variable.
4. Substitute the value found in Step 3 into either of the original equations and solve for the other variable.

In Example 3, we added 3 times the first equation to -2 times the second equation. The result from adding a constant multiple of one equation to a constant multiple of another equation is called a **linear combination** of the two equations. The method of elimination is also called the method of linear combinations.

EXERCISE 2 Solve the system by linear combinations.

$$3x - 4y = -11$$
$$2x + 6y = -3$$

If either equation in a system has fractional coefficients, it is helpful to clear the fractions before applying the method of linear combinations.

EXAMPLE 4

Solve the system by linear combinations.

$$\frac{2}{3}x - y = 2 \qquad (1)$$

$$x + \frac{1}{2}y = 7 \qquad (2)$$

Solution

Multiply each side of Equation (1) by 3 and each side of Equation (2) by 2 to clear the fractions:

$$2x - 3y = 6 \qquad (1a)$$
$$2x + \ y = 14 \qquad (2a)$$

To eliminate the variable x, multiply Equation (2a) by -1 and add the result to Equation (1a) to get

$$-4y = -8 \quad \text{Divide both sides by } -4.$$
$$y = 2$$

Substitute 2 for y in one of the original equations and solve for x. Using Equation (2) we find

$$x + \frac{1}{2}(2) = 7 \quad \text{Subtract 1 from both sides.}$$
$$x = 6$$

Verify that $x = 6$ and $y = 2$ satisfy both Equations (1) and (2). The solution to the system is the ordered pair $(6, 2)$. ∎

Systems of linear equations are some of the most useful and widely used mathematical tools for solving problems. Systems involving hundreds of variables and equations are not uncommon in applications such as scheduling airline flights or routing telephone calls. The method of substitution does not generalize easily to larger linear systems, but the method of elimination is the basis for a number of sophisticated algorithms that can be implemented on computers.

EXERCISE 3 It took Leon 7 hours to fly the same distance that Marlene drove in 21 hours. Leon flies 120 miles per hour faster than Marlene drives. At what speed did each travel?

a. Choose variables for the unknown quantities, and fill in the table.

	Rate	Time	Distance
Leon			
Marlene			

b. Write one equation about Leon's and Marlene's speeds.

c. Write a second equation about distances.

d. Solve the system and answer the question in the problem.

Inconsistent and Dependent Systems

Recall that a system of equations may not have a unique solution. It is not always easy to tell from the equations themselves whether there is one solution, no solution, or infinitely many solutions. However, the method of elimination will reveal which of the three cases applies.

EXAMPLE 5

Solve each system.

a. $2x = 2 - 3y$
$6y = 7 - 4x$

b. $3x - 4 = y$
$2y + 8 = 6x$

Solutions

a. First, rewrite the system in standard form as

$$2x + 3y = 2 \tag{1}$$
$$4x + 6y = 7 \tag{2}$$

Multiply Equation (1) by -2 and add the result to Equation (2) to obtain

$$
\begin{array}{r}
-4x - 6y = -4 \\
4x + 6y = 7 \\
\hline
0x + 0y = 3
\end{array}
$$

This equation has no solutions. The system is inconsistent. (Notice that both lines have slope $\frac{-2}{3}$, so their graphs are parallel.)

b. Rewrite the system in standard form as

$$3x - y = 4 \tag{3}$$
$$-6x + 2y = -8 \tag{4}$$

Multiply Equation (3) by 2 and add the result to Equation (4) to obtain

$$
\begin{array}{r}
6x - 2y = 8 \\
-6x + 2y = -8 \\
\hline
0x + 0y = 0
\end{array}
$$

This equation has infinitely many solutions. The system is dependent. (Notice that both original equations have the same slope-intercept form: $y = 3x - 4$, so they have the same graph. Thus, every point on the graph is a solution of the system.) ■

We can generalize the results from Example 5 as follows.

1. If an equation of the form

$$0x + 0y = k \qquad (k \neq 0)$$

is obtained as a linear combination of the equations in a system, the system is **inconsistent.**

2. If an equation of the form

$$0x + 0y = 0$$

is obtained as a linear combination of the equations in a system, the system is **dependent.**

EXERCISE 4 Identify the system as dependent, inconsistent, or consistent and independent.

$$x + 3y = 6$$
$$2x - 12 = -6y$$

READING QUESTIONS

1. Name two algebraic techniques for solving a linear system.
2. When is the substitution method easier to use?
3. When using elimination, how do you decide what to multiply each equation by?
4. If the system is inconsistent, what is the result of the elimination method?

ANSWERS FOR 2.2 EXERCISES

1. $(3, 0)$ **2.** $\left(-3, \frac{1}{2}\right)$

3a.

	Rate	Time	Distance
Leon	x	7	$7x$
Marlene	y	21	$21y$

b. $x = y + 120$ **c.** $7x = 21y$
d. $(180, 60)$ Leon flies at 180 mph, Marlene drives 60 mph.

4. dependent

In Problems 1–6 solve each system by substitution or by linear combinations.

1. $3m + n = 7$
$2m = 5n - 1$

2. $2r = s + 7$
$2s = 14 - 3r$

3. $2u - 3v = -4$
$5u + 2v = 9$

4. $3x + 5y = 1$
$2x - 3y = 7$

5. $3y = 2x - 8$
$4y + 11 = 3x$

6. $4L - 3 = 3W$
$25 + 5L = -2W$ $+25$

In Problems 7–12, clear the fractions in each equation first, then solve the system by substitution or by linear combinations.

7. $\dfrac{2}{3}A - B = 4$
$A - \dfrac{3}{4}B = 6$

8. $\dfrac{1}{8}w - \dfrac{3}{8}z = 1$
$\dfrac{1}{2}w - \dfrac{1}{4}z = -1$

9. $\dfrac{M}{4} = \dfrac{N}{3} - \dfrac{5}{12}$
$\dfrac{N}{5} = \dfrac{1}{2} - \dfrac{M}{10}$

10. $\dfrac{R}{3} = \dfrac{2S}{3} + 2$
$\dfrac{S}{3} = \dfrac{R}{6} - 1$

11. $\dfrac{s}{2} = \dfrac{7}{6} - \dfrac{t}{3}$
$\dfrac{s}{4} = \dfrac{3}{4} - \dfrac{t}{4}$

12. $\dfrac{2p}{3} + \dfrac{8q}{9} = \dfrac{4}{3}$
$\dfrac{p}{3} = 2 + \dfrac{q}{2}$

Use linear combinations to identify each system in Problems 13–18 as dependent, inconsistent, or consistent and independent.

13. $2m = n + 1$
$8m - 4n = 3$

14. $6p = 1 - 2q$
$12p + 4q = 2$

15. $r - 3s = 4$
$2r + s = 6$

16. $2u + v = 4$
$u - 3v = 2$

17. $2L - 5W = 6$
$\dfrac{15W}{2} + 9 = 3L$

18. $-3A = 4B + 8$
$\dfrac{1}{2}A + \dfrac{4}{3} = \dfrac{-2}{3}B$

19. Refer to Problem 5 of Section 2.1.

   a. Find an equation for the line you graphed in part (b), median age versus date.

   b. Find an equation for the line you graphed in part (d), mean age versus date.

   c. Solve the system of equations algebraically. How does your answer compare to the solution you found by graphing?

20. Refer to Problem 6 of Section 2.1.

   a. Find an equation for the line you graphed in part (b), median age versus date.

   b. Find an equation for the line you graphed in part (d), mean age versus date.

   c. Solve the system of equations algebraically. How does your answer compare to the solution you found by graphing?

21. Water slowly contracts as it cools, causing it to become denser, until just before freezing, when the density decreases slightly. The salinity of water affects its freezing point, and also the temperature at which it reaches its maximum density: the higher the salinity, the lower the temperature at which the maximum density occurs. Pure water is densest at 4°C and freezes at 0°C. Water that is 15% saline is densest at 0.8°C and freezes at −0.8°C.

   a. Write an equation for the water's temperature of maximum density in terms of its salinity in percent.

   b. Write an equation for the water's freezing point in terms of its salinity in percent.

   c. Graph both equations on the same set of axes.

   d. Solve the system to find the salinity of water that freezes when it is densest. What is the freezing point?

22. Charles' law says that the temperature of a gas is related to its volume by a linear equation. A gas that occupied one liter at 0°C expanded to 1.2 liters at a temperature of 54.6°C. A second sample of gas that occupied 2 liters at 0°C expanded to 2.8 liters at 109.2°C.

a. Find a linear equation for temperature, T, in terms of volume, V, for the first sample of gas, and a second linear equation for the second sample of gas.

b. Graph both equations on the same set of axes.

c. Solve the system. What is the significance of the intersection point?

d. Show that both equations from part (a) can be written in the form $T = c\left(\dfrac{V}{V_0} - 1\right)$, where V_0 is the volume of the gas at 0°C.

In Problems 23–34 write a system of equations for each problem, and solve algebraically.

23. Francine has $2000, part of it invested in bonds paying 10% and the rest in a certificate account at 8%. Her annual income from the two investments is $184. How much did Francine invest at each rate?

a. Choose variables for the unknown quantities, and fill in the table.

	Principal	Interest rate	Interest
Bonds			
Certificate			
Total		—	

b. Write one equation about the amount Francine invested.

c. Write a second equation about Francine's annual interest.

d. Solve the system and answer the question in the problem.

24. Carmella has $1200 invested in two stocks; one returns 8% per year and the other returns 12% per year. The income from the 8% stock is $3 more than the income from the 12% stock. How much did she invest in each stock?

a. Choose variables for the unknown quantities, and fill in the table.

	Principal	Interest rate	Interest
First stock			
Second stock			
Total		—	—

b. Write one equation about the amount Carmella invested.

c. Write a second equation about Carmella's annual interest.

d. Solve the system and answer the question in the problem.

25. Paul needs 40 pounds of 48% silver alloy to finish a collection of jewelry. How many pounds of 45% silver alloy should he melt with 60% silver alloy to obtain the alloy he needs?

a. Choose variables for the unknown quantities, and fill in the table.

	Pounds	% silver	Amount of silver
First alloy			
Second alloy			
Mixture			

b. Write one equation about the amount of alloy Paul needs.

c. Write a second equation about the amount of silver in the alloys.

d. Solve the system and answer the question in the problem.

26. Amal plans to make 10 liters of a 17% acid solution by mixing a 20% acid solution with a 15% acid solution. How much of each should she use?

a. Choose variables for the unknown quantities, and fill in the table on page 108.

	Liters	% acid	Amount of acid
First solution			
Second solution			
Mixture			

b. Write one equation about the amount of solution Amal needs.

c. Write a second equation about the amount of acid in the solutions.

d. Solve the system and answer the question in the problem.

27. Delbert answered 13 true-false and 9 fill-in questions correctly on his last test and got a score of 71. If he had answered 9 true-false and 13 fill-ins correctly, he would have made an 83. How many points was each type of problem worth?

28. In a recent election 7179 votes were cast for the two candidates. If 6 votes had been switched from the winner to the loser, the loser would have won by one vote. How many votes were cast for each candidate?

29. Because of prevailing winds a flight from Detroit to Denver, a distance of 1120 miles, takes 4 hours on Econoflite, while the return trip takes 3.5 hours. What were the speed of the airplane and the speed of the wind?

a. Choose variables for the unknown quantities, and fill in the table.

	Rate	Time	Distance
Detroit to Denver			
Denver to Detroit			

b. Write one equation about the trip from Detroit to Denver.

c. Write a second equation about the return trip.

d. Solve the system and answer the question in the problem.

30. On a breezy day Bonnie propelled her human-powered aircraft 100 meters in 15 seconds going into the wind and made the return trip in 10 seconds with the wind. What were the speed of the wind and Bonnie's speed in still air?

a. Choose variables for the unknown quantities, and fill in the table.

b. Write one equation about Bonnie's initial flight.

	Rate	Time	Distance
Against the wind			
With the wind			

c. Write a second equation Bonnie's return trip.

d. Solve the system and answer the question in the problem.

31. A cup of rolled oats provides 310 calories. A cup of rolled wheat flakes provides 290 calories. A new breakfast cereal combines wheat and oats to provide 302 calories per cup. How much of each grain does 1 cup of the cereal include?

a. Choose variables for the unknown quantities, and fill in the table.

b. Write one equation about the amounts of each grain.

	Cups	Calories per cup	Calories
Oat flakes			
Wheat flakes			
Mixture		—	

c. Write a second equation about the number of calories.

d. Solve the system and answer the question in the problem.

32. Acme Motor Company is opening a new plant to produce chassis for two of its models, a sports coupe and a wagon. Each sports coupe requires a riveter for 3 hours and a welder for 4 hours; each wagon requires a riveter for 4 hours and a welder for 5 hours. The plant has available 120 hours of riveting and 155 hours of welding per day. How many of each model of chassis can it produce in a day?

a. Choose variables for the unknown quantities and fill in the table.

b. Write one equation about the hours of riveting.

	Sports coupes	Wagons	Total
Hours of riveting			
Hours of welding			

c. Write a second equation about the hours of welding.

d. Solve the system and answer the question in the problem.

Use a calculator to solve each system.

35. $4.8x - 3.5y = 5.44$
$2.7x + 1.3y = 8.29$

37. $0.9x = 25.78 + 1.03y$
$0.25x + 0.3y = 85.7$

33. Sanaz can afford to produce $35x$ pairs of sunglasses if she can sell them at x dollars per pair, and the market will buy $1700 - 15x$ at x dollars a pair.

a. Write the supply and demand equations for the sunglasses.

b. Find the equilibrium price and the number of sunglasses Sanaz will produce and sell at that price.

34. Benham will service $2.5x$ copy machines per week if he can charge x dollars per machine, and the public will pay for $350 - 4.5x$ jobs at x dollars per machine.

a. Write the supply and demand equations for servicing copy machines.

b. Find the equilibrium price and the number of copy machines Benham will service at that price.

36. $6.4x + 2.3y = -14.09$
$-5.2x - 3.7y = -25.37$

38. $0.02x = 0.6y - 78.72$
$1.1y = -0.4x + 108.3$

2.3 Systems of Linear Equations in Three Variables

Some problems involve more than two unknown quantities. Efficient techniques for solving linear systems in many variables are available. In this section, we solve systems of three linear equations in three variables.

3 × 3 Linear Systems

A solution to an equation in three variables, such as

$$x + 2y - 3z = -4 \tag{1}$$

is an **ordered triple** of numbers that satisfies the equation. For example, $(0, -2, 0)$ and $(-1, 0, 1)$ are solutions to the equation above, but $(1, 1, 1)$ is not. You can verify this by substituting the coordinates into the equation to see if a true statement results.

For $(0, -2, 0)$: $\quad 0 + 2(-2) - 3(0) = -4 \quad$ True

For $(-1, 0, 1)$: $\quad -1 + 2(0) - 3(1) = -4 \quad$ True

For $(1, 1, 1)$: $\quad 1 + 2(1) - 3(1) \;\; = -4 \quad$ Not true

As with the two-variable case, a single linear equation in three variables has infinitely many solutions.

An ordered triple (x, y, z) can be represented geometrically as a point in space using a three-dimensional Cartesian coordinate system (see Figure 2.15). In this coordinate system the graph of a linear equation in three variables, such as Equation (1), is a plane, and the fact that there are infinitely many solutions to the equation tells us that that there are infinitely many points in the corresponding plane.

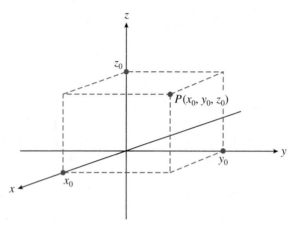

FIGURE 2.15

A solution to a *system* of three linear equations in three variables is an ordered triple that satisfies each equation in the system. That triple represents a point that must lie on all three graphs. Figure 2.16 shows different ways in which three planes may intersect in space.

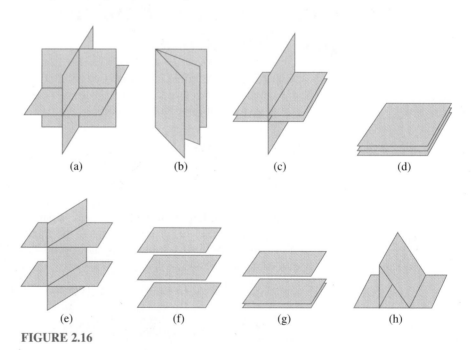

FIGURE 2.16

In Figure 2.16(a) the three planes intersect in a single point, so the corresponding system of three equations has a unique solution. In Figure 2.16(b), (c), and (d) the intersection is either a line or an entire plane, so the corresponding system has infinitely many solutions. Such a system is called **dependent.** In Figure 2.16(e), (f), (g), and (h) the three planes have no common intersection, so the corresponding system has no solution. In this case the system is said to be **inconsistent.**

It is impractical to solve 3×3 systems by graphing. Even when technology for producing three-dimensional graphs is available, we cannot read coordinates on such graphs with any confidence. Thus we will restrict our attention to algebraic methods of solving such systems.

Back-Substitution

When we review our algebraic methods for 2×2 systems—substitution and linear combinations—we find features common to both. In both methods we obtain an equation in one variable. Once we have solved for that variable, we substitute the value into an earlier equation to find the other variable.

One strategy for solving a 3×3 system extends this idea to include a third variable and a third equation. We would like to get an equation in a single variable. Once we have found the value for that variable, we can substitute its value into the other equations to find the remaining unknowns. The following special case illustrates the substitution part of the procedure.

EXAMPLE 1

Solve the system

$$x + 2y + 3z = 2$$
$$-2y - 4z = -2$$
$$3z = -3$$

Solution

The third equation involves only the variable z, so we solve that equation to find $z = -1$. Then we substitute -1 for z in the second equation and solve for y.

$$-2y - 4(-1) = -2$$
$$-2y + 4 = -2$$
$$-2y = -6$$
$$y = 3$$

Finally, we substitute -1 for z and 3 for y into the first equation to find x.

$$x + 2(3) + 3(-1) = 2$$
$$x + 6 - 3 = 2$$
$$x = -1$$

The solution is the ordered triple $(-1, 3, -1)$. You should verify that this triple satisfies all three equations of the system. ■

The technique used in Example 1 is called **back-substitution.** It works in the special case when one of the equations involves exactly one variable, and a second equation involves that same variable and just one other variable. A 3×3 linear system with these properties is said to be in **triangular** form. If we can transform a system into triangular form, we can use back-substitution to complete the solution.

EXERCISE 1 Use back-substitution to solve the system

$$2x + 2y + z = 10$$
$$y - 4z = 9$$
$$3z = -6$$

Gaussian Reduction

On page 100 we stated two properties of linear systems. These properties allow us to form linear combinations of the equations in a system without changing the system's solution. They apply to linear systems with any number of variables. We can use linear combinations to reduce a 3×3 system to triangular form, and then use back-substitution to find the solutions.

Our strategy will be to eliminate one of the variables from each of the three equations by considering them in pairs. This results in a 2×2 system that we can solve using elimination, as we did in Section 2.2. As an example, consider the system

$$x + 2y - 3z = -4 \tag{1}$$
$$2x - y + z = 3 \tag{2}$$
$$3x + 2y + z = 10 \tag{3}$$

We can choose any one of the three variables to eliminate first. For this example we will eliminate x. We then choose two of the equations, say (1) and (2), and use a linear combination: we multiply Equation (1) by -2 and add the result to Equation (2) to produce Equation (4).

$$-2x - 4y + 6z = 8 \tag{1a}$$
$$\underline{2x - y + z = 3} \tag{2}$$
$$-5y + 7z = 11 \tag{4}$$

Now we have an equation involving only two variables. But we need *two* equations in two unknowns to specify a unique solution. So we choose a different pair of equations, say (1) and (3), and eliminate x again. We multiply Equation (1) by -3 and add the result to Equation (3) to obtain Equation (5).

$$-3x - 6y + 9z = 12 \tag{1b}$$
$$\underline{3x + 2y + z = 10} \tag{3}$$
$$-4y + 10z = 22 \tag{5}$$

We now form a 2×2 system with our new Equations (4) and (5).

$$-5y + 7z = 11 \tag{4}$$
$$-4y + 10z = 22 \tag{5}$$

Finally, we eliminate either y or z to obtain an equation in a single variable. If we choose to eliminate y, we add 4 times Equation (4) to -5 times Equation (5) to obtain Equation (6).

$$-20y + 28z = 44 \qquad (4a)$$
$$\underline{20y - 50z = -110} \qquad (5a)$$
$$-22z = -66 \qquad (6)$$

Now we can start solving for the variables. To keep things organized, we choose one of our original equations (in three variables), one of the equations from our 2×2 system, and our final equation in one variable. We choose Equations (1), (4), and (6).

$$x + 2y - 3z = -4 \qquad (1)$$
$$-5y + 7z = 11 \qquad (4)$$
$$-22z = -66 \qquad (6)$$

This new system is in triangular form, and it has the same solutions as the original system. We can complete its solution by back-substitution. We first solve Equation (6) to find $z = 3$. Substituting 3 for z in Equation (4), we find

$$-5y + 7(3) = 11$$
$$-5y + 21 = 11$$
$$-5y = -10$$
$$y = 2$$

Next, we substitute 3 for z and 2 for y into Equation (1) to find

$$x + 2(2) - 3(3) = -4$$
$$x + 4 - 9 = -4$$
$$x = 1$$

The solution to the system is the ordered triple $(1, 2, 3)$. You should verify that this triple satisfies all three of the original equations.

The method described above for putting a linear system into triangular form is called **Gaussian reduction,** after the German mathematician Carl Gauss. We summarize our method for solving a 3×3 linear system as follows.

Steps for Solving a 3 × 3 Linear System

1. Clear each equation of fractions and put it in standard form.
2. Choose two of the equations and eliminate one of the variables by forming a linear combination.
3. Choose a different pair of equations and eliminate the *same* variable.
4. Form a 2×2 system with the equations found in Steps (2) and (3). Eliminate one of the variables from this 2×2 system by using a linear combination.
5. Form a triangular system by choosing among the previous equations. Use back-substitution to solve the triangular system.

EXAMPLE 2   Solve the system

$$x + 2y + z = -3 \tag{1}$$

$$\frac{1}{3}x - y + \frac{1}{3}z = 2 \tag{2}$$

$$x + \frac{1}{2}y + z = \frac{5}{2} \tag{3}$$

Solution

Follow the steps outlined above.

Step 1 Multiply each side of Equation (2) by 3 and each side of Equation (3) by 2 to obtain the equivalent system

$$x + 2y - z = -3 \tag{1}$$
$$x - 3y + z = 6 \tag{2a}$$
$$2x + y + 2z = 5 \tag{3a}$$

Step 2 Eliminate z from Equations (1) and (2a) by adding.

$$
\begin{array}{rl}
x + 2y - z = -3 & \quad (1) \\
\underline{x - 3y + z = 6} & \quad (2a) \\
2x - y = 3 & \quad (4)
\end{array}
$$

Step 3 Eliminate z from Equations (1) and (3a): multiply Equation (1) by 2 and add the result to Equation (3a).

$$
\begin{array}{rl}
2x + 4y - 2z = -6 & \quad (1a) \\
\underline{2x + y + 2z = 5} & \quad (3a) \\
4x + 5y = -1 & \quad (5)
\end{array}
$$

Step 4 Form the 2×2 system consisting of Equations (4) and (5).

$$2x - y = 3 \tag{4}$$
$$4x + 5y = -1 \tag{5}$$

Eliminate x by adding -2 times Equation (4) to Equation (5).

$$
\begin{array}{rl}
-4x + 2y = -6 & \quad (4a) \\
\underline{4x + 5y = -1} & \quad (5) \\
7y = -7 & \quad (6)
\end{array}
$$

Step 5 Form a triangular system using Equations (1), (4), and (6).

$$
\begin{array}{rl}
x + 2y - z = -3 & \quad (1) \\
2x - y = 3 & \quad (4) \\
7y = -7 & \quad (6)
\end{array}
$$

Use back-substitution to find the solution. Divide both sides of Equation (6) by 7 to find $y = -1$. Substitute -1 for y in Equation (4) and solve for x.

$$2x - (-1) = 3$$
$$2x = 2$$
$$x = 1$$

Finally, substitute 1 for x and -1 for y in Equation (1) and solve for z.

$$1 + 2(-1) - z = -3$$
$$-1 - z = -3$$
$$-z = -2$$
$$z = 2$$

The solution is the ordered triple $(1, -1, 2)$. You should verify that this triple satisfies all three of the original equations of the system. ∎

EXERCISE 2 Use Gaussian reduction to solve the system

$$x - 2y + z = -1 \qquad (1)$$
$$\frac{2}{3}x + \frac{1}{3}y - z = 1 \qquad (2)$$
$$3x + 3y - 2z = 10 \qquad (3)$$

Follow the steps suggested below:

1. Clear the fractions from Equation (2).
2. Eliminate z from Equations (1) and (2).
3. Eliminate z from Equations (1) and (3).
4. Eliminate x from your new 2 × 2 system.
5. Form a triangular system and solve by back-substitution.

Inconsistent and Dependent Systems

The results on page 105 for identifying dependent and inconsistent systems can be extended to 3 × 3 linear systems. If at any step in forming linear combinations we obtain an equation of the form

$$0x + 0y + 0z = k \qquad (k \neq 0)$$

then the system is inconsistent and has no solution. If we obtain an equation of the form

$$0x + 0y + 0z = 0$$

then the system is dependent and has infinitely many solutions.

EXAMPLE 3

Solve the system

$$3x + y - 2z = 1 \qquad (1)$$
$$6x + 2y - 4z = 5 \qquad (2)$$
$$-2x - y + 3z = -1 \qquad (3)$$

Solution

To eliminate y from Equations (1) and (2), multiply Equation (1) by -2 and add the result to Equation (2).

$$-6x - 2y + 4z = -2$$
$$6x + 2y - 4z = 5$$
$$\overline{0x + 0y + 0y = 3}$$

Since the resulting equation has no solution, the system is *inconsistent*. ■

EXAMPLE 4

Solve the system

$$-x + 3y - z = -2 \tag{1}$$
$$2x + y - 4z = 6 \tag{2}$$
$$2x - 6y + 2z = 4 \tag{3}$$

Solution

To eliminate x from Equations (1) and (3), multiply Equation (1) by 2 and add Equation (3).

$$-2x + 6y - 2z = -4$$
$$2x - 6y + 2z = 4$$
$$\overline{0x + 0y + 0z = 0}$$

Since the resulting equation vanishes, the system is *dependent* and has infinitely many solutions. ■

> **EXERCISE 3**   Decide whether the system is inconsistent, dependent, or consistent and independent.
>
> $$x + 3y - z = 4$$
> $$-2x - 6y + 2z = 1$$
> $$x + 2y - z = 3$$

Applications

Here are some problems that can be modeled by a system of three linear equations. When writing such systems we must be careful to find three *independent* equations describing the conditions of the problem.

EXAMPLE 5

One angle of a triangle measures 4° less than twice the second angle, and the third angle is 20° greater than the sum of the first two. Find the measure of each angle.

Solution

Step 1 Represent the measure of each angle by a separate variable.

First angle: x
Second angle: y
Third angle: z

Step 2 Write the conditions stated in the problem as three equations.

$$x = 2y - 4$$
$$z = x + y + 20$$
$$x + y + z = 180$$

(The third equation states that the sum of the angles of a triangle is 180°.)

Step 3 We follow the steps outlined for solving a 3 × 3 linear system.

1. Write the three equations in standard form.

$$x - 2y = -4 \tag{1}$$
$$x + y - z = -20 \tag{2}$$
$$x + y + z = 180 \tag{3}$$

2–3. Since Equation (1) has no z term, it will be most efficient to eliminate the variable z from Equations (2) and (3). Add these two equations.

$$
\begin{array}{rcll}
x + y - z &=& -20 & (2) \\
x + y + z &=& 180 & (3) \\
\hline
2x + 2y &=& 160 & (4)
\end{array}
$$

4. Form a 2 × 2 system from Equations (1) and (4). Add the two equations to eliminate the variable y, yielding

$$
\begin{array}{rcll}
x - 2y &=& -4 & (1) \\
2x + 2y &=& 160 & (4) \\
\hline
3x &=& 156 & (5)
\end{array}
$$

5. Form a triangular system using Equations (3), (1), and (5). Use back-substitution to complete the solution.

$$x + y + z = 180 \tag{3}$$
$$x - 2y = -4 \tag{1}$$
$$3x = 156 \tag{5}$$

Divide both sides of Equation (5) by 3 to find $x = 52$. Substitute 52 for x in Equation (1) and solve for y to find

$$52 - 2y = -4$$
$$2y = 56$$
$$y = 28$$

Substitute 52 for x and 28 for y in Equation (3) to find

$$52 + 28 + z = 180$$
$$z = 100$$

Step 4 The angles measure 52°, 28°, and 100°. ■

EXERCISE 4 A manufacturer of office supplies makes three types of file cabinet: two-drawer, four-drawer, and horizontal. The manufacturing process is divided into three phases: assembly, painting, and finishing. A two-drawer cabinet requires 3 hours to assemble, 1 hour to paint, and 1 hour to finish. The four-drawer model takes 5 hours to assemble, 90 minutes to paint, and 2 hours to finish. The horizontal cabinet takes 4 hours to assemble, 1 hour to paint, and 3 hours to finish. The manufacturer employs enough workers for 500 hours of assembly time, 150 hours of painting, and 230 hours of finishing per week. How many of each type of file cabinet should he make in order to use all the hours available?

Step 1 Represent the number of each model of file cabinet by a different variable.

Number of two-drawer cabinets: x

Number of four-drawer cabinets: y

Number of horizontal cabinets: z

Step 2 Organize the information into a table. (Convert all times to hours.)

	2-Drawer	4-Drawer	Horizontal	Total available
Assembly				
Painting				
Finishing				

Write three equations describing the time constraints in each of the three manufacturing phases. For example, the assembly phase requires $3x$ hours for the two-drawer cabinets, $5y$ hours for the four-drawer cabinets, and $4z$ hours for the horizontal cabinets, and the sum of these times should be the time available, 500 hours.

(Assembly time)　　　　　　　　　　(1)

(Painting time)　　　　　　　　　　(2)

(Finishing time)　　　　　　　　　　(3)

Step 3 Solve the system. Follow the steps suggested below.

1. Clear the fractions from the second equation.
2. Subtract Equation (1) from 3 times Equation (3) to obtain a new Equation (4).
3. Subtract Equation (2) from twice Equation (3) to obtain a new Equation (5).
4. Equation (4) and (5) form a 2×2 system in y and z. Subtract Equation (5) from Equation (4) to obtain a new Equation (6).
5. Form a triangular system with equations (3), (4), and (6). Use back-substitution to complete the solution.

Step 4 You should have found the following solution: The manufacturer should make 60 two-drawer cabinets, 40 four-drawer cabinets, and 30 horizontal cabinets.

1. Explain how to solve a triangular system by back-substitution.
2. Explain how to obtain a triangular system using Gaussian reduction.

ANSWERS FOR 2.3 EXERCISES

1. $(5, 1, -2)$ **2.** $(2, 2, 1)$ **3.** inconsistent

4.

	2-Drawer	4-Drawer	Horizontal	Total available
Assembly	3	5	4	500
Painting	1	1.5	1	150
Finishing	1	2	3	230

$$3x + 5y + 4z = 500$$
$$x + 1.5y + z = 150$$
$$x + 2y + 3z = 230$$

HOMEWORK 2.3

Use back-substitution to solve the systems in Problems 1–6.

1. $x + y + z = 2$
$3y + z = 5$
$-4y = -8$

2. $2x + 3y - z = -7$
$y - 2z = -6$
$5z = 15$

3. $2x - y - z = 6$
$5y + 3z = -8$
$13y = -13$

4. $x + y + z = 1$
$x + 4y = 1$
$3x = 3$

5. $2x + z = 5$
$3y + 2z = 6$
$5x = 20$

6. $3x - y = 6$
$x - 2z = -7$
$13x = 13$

Use Gaussian reduction to solve the systems in Problems 7–20.

7. $x + y + z = 0$
$2x - 2y + z = 8$
$3x + 2y + z = 2$

8. $x - 2y + 4z = -3$
$3x + y - 2z = 12$
$2x + y - 3z = 11$

9. $4x + z = 3$
$2x - y = 2$
$3y + 2z = 0$

10. $3y + z = 3$
$-2x + 3y = 7$
$3x + 2z = -6$

11. $2x + 3y - 2z = 5$
$3x - 2y - 5z = 5$
$5x + 2y + 3z = -9$

12. $3x - 4y + 2z = 20$
$4x + 3y - 3z = -4$
$2x - 5y + 5z = 24$

13. $4x + 6y + 3z = -3$
$2x - 3y - 2z = 5$
$-6x + 6y + 2z = -5$

14. $3x + 4y + 6z = 2$
$-2x + 2y - 3z = 1$
$4x - 10y + 9z = 0$

15. $x - \dfrac{1}{2}y - \dfrac{1}{2}z = 4$
$x - \dfrac{3}{2}y - 2z = 3$
$\dfrac{1}{4}x + \dfrac{1}{4}y - \dfrac{1}{4}z = 0$

16. $x + 2y + \dfrac{1}{2}z = 0$
$x + \dfrac{3}{5}y - \dfrac{2}{5}z = \dfrac{1}{5}$
$4x - 7y - 7z = 6$

17. $x + y - z = 2$
$\dfrac{1}{2}x - y + \dfrac{1}{2}z = -\dfrac{1}{2}$
$x + \dfrac{1}{3}y - \dfrac{2}{3}z = \dfrac{4}{3}$

18. $x + y - 2z = 3$
$x - \dfrac{1}{3}y + \dfrac{1}{3}z = \dfrac{5}{3}$
$\dfrac{1}{2}x - \dfrac{1}{2}y - z = \dfrac{3}{2}$

19.
$$x = -y$$
$$x + z = \frac{5}{6}$$
$$y - 2z = -\frac{7}{6}$$

20.
$$x = y + \frac{1}{2}$$
$$y = z + \frac{5}{4}$$
$$2z = x - \frac{7}{4}$$

Solve the systems in Problems 21–30. If the system is inconsistent or dependent, say so.

21.
$$3x - 2y + z = 6$$
$$2x + y - z = 2$$
$$4x + 2y - 2z = 3$$

22.
$$x - 2y + z = 5$$
$$-x + y = -2$$
$$y - z = -3$$

23.
$$2x + 3y - z = -2$$
$$x - y + \frac{1}{2}z = 2$$
$$4x - \frac{1}{3}y + 2z = 8$$

24.
$$3x + 6y + 2z = -2$$
$$\frac{1}{2}x - 3y - z = 1$$
$$4x + y + \frac{1}{3}z = -\frac{1}{3}$$

25.
$$x = 2y - 7$$
$$y = 4z + 3$$
$$z = 3x + y$$

26.
$$x = y + z$$
$$y = 2x - z$$
$$z = 3x - y$$

27.
$$\frac{1}{2}x + y = \frac{1}{2}z$$
$$x - y = -z - 2$$
$$-x - 2y = -z + \frac{4}{3}$$

28.
$$x = \frac{1}{2}y - \frac{1}{2}z + 1$$
$$x = 2y + z - 1$$
$$x = \frac{1}{2}y - \frac{1}{2}z + \frac{1}{4}$$

29.
$$x - y = 0$$
$$2x + 2y + z = 5$$
$$2x + y - \frac{1}{2}z = 0$$

30.
$$x + y = 1$$
$$2x - y + z = -1$$
$$x - 3y - z = -\frac{2}{3}$$

Solve Problems 31–40 by using a system of equations.

  a. Identify three unknown quantities and choose variables to represent them.

  b. If appropriate, make a table organizing the information in the problem.

  c. Write three equations about the variables in the problem.

31. A box contains $6.25 in nickels, dimes, and quarters. There are 85 coins in all, with 3 times as many nickels as dimes. How many coins of each kind are there?

32. Vanita has $446 in 10-dollar, 5-dollar, and 1-dollar bills. There are 94 bills in all and 10 more 5-dollar bills than 10-dollar bills. How many bills of each kind does she have?

33. The perimeter of a triangle is 155 inches. Side x is 20 inches shorter than side y, and side y is 5 inches longer than side z. Find the lengths of the sides of the triangle.

34. One angle of a triangle measures 10° more than a second angle, and the third angle is 10° more than six times the measure of the smallest angle. Find the measure of each angle.

35. Vegetable Medley is made of carrots, green beans, and cauliflower. The package says that 1 cup of Vegetable Medley provides 29.4 milligrams of vitamin C and 47.4 milligrams of calcium. One cup of carrots contains 9 milligrams of vitamin C and 48 milligrams of calcium. One cup of green beans contains 15 milligrams of vitamin C and 63 milligrams of calcium. One cup of cauliflower contains 69 milligrams of vitamin C and 26 milligrams of calcium. How much of each vegetable is in 1 cup of Vegetable Medley?

36. The Java Shoppe sells a house brand of coffee that is only 2.25% caffeine for $6.60 per pound. The house brand is a mixture of Colombian coffee that sells for $6 per pound and is 2% caffeine, French roast that sells for $7.60 per pound and is 4% caffeine, and Sumatran at $6.80 per pound and 1% caffeine. How much of each variety is in a pound of house brand?

37. The ABC Psychological Testing Service offers three types of reports on test results: score only, evaluation, and narrative report. Each score-only test takes 3 minutes to score using an optical scanner and 1 minute to print the interpretation. Each evaluation takes 3 minutes to score, 4 minutes to analyze, and 2 minutes to print. Each narrative report takes 3 minutes to score, 5 minutes to analyze, and 8 minutes to print. If ABC Services uses its optical scanner 7 hours per day, has 8 hours in which to analyze results, and has 12 hours of printer time available per day, how many of each type of report can it complete each day when it is using all its resources?

38. Reliable Auto Company wants to ship 1700 Status Sedans to three major dealers in Los Angeles, Chicago, and Miami. From past experience Reliable figures that it will sell twice as many sedans in Los Angeles as in Chicago. It costs $230 to ship a sedan to Los Angeles, $70 to Chicago, and $160 to Miami. If Reliable Auto has $292,000 to pay for shipping costs, how many sedans should it ship to each city?

39. Ace, Inc. produces three kinds of wooden rackets: tennis rackets, Ping-Pong paddles, and squash rackets. After the pieces are cut each racket goes through three phases of production: gluing, sanding, and finishing. A tennis racket takes 3 hours to glue, 2 hours to sand, and 3 hours to finish. A Ping-Pong paddle takes 1 hour to glue, 1 hour to sand, and 1 hour to finish. A squash racket takes 2 hours to glue, 2 hours to sand, and $2\frac{1}{2}$ hours to finish. Ace has available 95 man-hours in its gluing department, 75 man-hours in sanding, and 100 man-hours in finishing per day. How many of each racket should it make in order to use all the available manpower?

40. A farmer has 1300 acres on which to plant wheat, corn, and soybeans. The seed costs $6 for an acre of wheat, $4 for an acre of corn, and $5 for an acre of soybeans. An acre of wheat requires 5 acre-feet of water during the growing season, while an acre of corn requires 2 acre-feet and an acre of soybeans requires 3 acre-feet. If the farmer has $6150 to spend on seed and can count on 3800 acre-feet of water, how many acres of each crop should he plant in order to use all his resources?

MIDCHAPTER REVIEW

a. Solve the system of equations by graphing. Use the "friendly" window suggested.

b. Verify algebraically that your solution satisfies both equations.

1. $y = 3.7x - 6.1$ Xmin $= -9.4$, Xmax $= 9.4$ **2.** $y = 7.5x$ Xmin $= 0$, Xmax $= 94$
 $y = -1.6x + 9.8$ Ymin $= -10$, Ymax $= 10$ $y = 1.5x + 408$ Ymin $= 0$, Ymax $= 800$

Solve the system of equations by graphing. Find the intercepts of each graph to help you choose a suitable window, and use the *intersect* feature to locate the solution.

3. $y - 32x = 2630$
$y = -21x - 1610$

4. $y = 1200 - 49x$
$y - 14x = 129$

a. Graph the system by hand, using either the intercept method or the slope-intercept method.

b. Identify the system as dependent, inconsistent, or consistent and independent.

5. $3a - b = 6$
$6a - 12 = 2b$

6. $3r - 12 = 4s$
$12s - 9r = 72$

7. Etienne plans to open a coffee house, and he has $7520 to spend on furniture. A table costs $460 and a chair costs $120. Etienne will buy four chairs for each table. How many tables can he buy?

    a. Let x represent the number of tables Etienne should buy and y the number of chairs. Write an equation about the cost of the furniture.

    b. Write a second equation about the number of tables and chairs.

    c. Graph both equations, solve the system, and answer the question in the problem. (Find the intercepts of the graphs to help you choose a window.)

8. Cynelle is arranging a field trip for 300 students from her elementary school. The tour company offers two types of buses, a 40-seat model and a newer 50-seat model. It costs $100 to rent a 40-seat bus for the day, and $140 to rent a 50-seat bus. Cynelle has $780 for transportation. How many of each size bus should she reserve?

    a. Let x represent the number of 40-seat buses and y the number of 50-seat buses. Write an equation about the number of buses needed.

    b. Write a second equation about the cost of renting the buses.

    c. Graph both equations, solve the system, and answer the question in the problem. (Find the intercepts of the graphs to help you choose a window.)

Solve the system by substitution.

9. $a + 2b = -6$
$2a - 3b = 16$

10. $7r - 4s = 1$
$3r + s = 14$

Solve the system by elimination.

11. $5x - 2y = -4$
$-6x + 3y = 5$

12. $2p + 3q = 38$
$6p - 5q = 2$

Identify the system as dependent, inconsistent, or consistent and independent.

13. $3a = 6 + 2b$
$4b = 6a - 8$

14. $5u = 4v + 20$
$8v = 10u - 1$

15. Earthquakes simultaneously send out two types of waves called "P" waves and "S" waves, but the two types travel at different speeds. A seismograph records arrival of P waves from an earthquake, and ninety seconds later the seismograph receives S waves from the same earthquake. The P waves travel at 5.4 miles per second, and S waves at 3 miles per second. How far is the seismograph from the earthquake?

a. Let x represent the time in seconds for the P wave to arrive at the seismograph and y the distance in miles between the earthquake and seismograph. Fill in the table.

	Rate	Time	Distance
P waves			
S waves			

b. Write an equation about how far the S waves travel.

c. Write a second equation about how far the P waves travel.

d. Solve the system and answer the question in the problem.

16. Thelma and Louise start together and drive in the same direction. Thelma is driving twice as fast as Louise. At the end of 3 hours they are 96 miles apart. How fast is each traveling?

a. Choose variables for the unknown quantities, and fill in the table.

b. Write one equation about Thelma's and Louise's speeds.

	Rate	Time	Distance
Thelma			
Louise			

c. Write a second equation about distances.

d. Solve the system and answer the question in the problem.

Use back-substitution to solve the system.

17. $-3x + 2y + 3z = -3$
$4x + z = 5$
$-2z = 6$

18. $2a + b + 3c = 3$
$5b - 8c = 16$
$4c = 12$

Use Gaussian reduction to solve the system.

19. $x - 5y - z = 2$
$3x - 9y + 3z = 6$
$x - 3y - z = -6$

20. $2u - v + 3w = 16$
$u + 3v - 2w = 1$
$u + 5v + w = 62$

In problems 21 and 22, decide whether the system is inconsistent, dependent, or consistent and independent.

21. $2x + y = 6$
$x - z = 4$
$3x + y - z = 10$

22. $a - c = 2$
$2a + b = 5$
$a + b + c = 3$

23. Juan is preparing a quart of fruit punch made of cranberry juice, apricot nectar, and club soda. Each quart of cranberry juice has 1200 calories and costs $1.25. The apricot nectar has 1600 calories and costs $1.00 per quart.

Club soda has no calories and costs $0.25 per quart. How much of each ingredient should Juan use to make a quart of fruit punch that contains 800 calories and costs $0.80?

24. The sum of the two smaller angles in a triangle is 12° less than the largest angle. The middle size angle is 1° less than half the largest angle. What are the three angles?

2.4 Solution of Linear Systems Using Matrices

In this section we consider a mathematical tool called a matrix (plural: matrices) that has wide application in mathematics, business, science, and engineering. A **matrix** is a rectangular array of numbers or **entries.** These entries are ordinarily displayed in rows and columns, and the entire matrix is enclosed in brackets or parentheses. Thus

$$\begin{bmatrix} 1 & 2 & 3 \\ 4 & 5 & 6 \\ 7 & 8 & 9 \end{bmatrix}, \qquad \begin{bmatrix} 2 & -1 & 3 \\ 4 & 0 & 2 \end{bmatrix}, \qquad \text{and} \qquad \begin{bmatrix} 4 \\ 5 \\ 6 \end{bmatrix}$$

are matrices. A matrix of **order,** or **dimension,** $n \times m$ (read "n by m") has n (horizontal) rows and m (vertical) columns. The matrices above are 3×3, 2×3, and 3×1, respectively. The first matrix—in which the number of rows is equal to the number of columns—is an example of a **square matrix.**

Coefficient Matrix and Augmented Matrix of a System

For a system of linear equations of the form

$$a_1 x + b_1 y + c_1 z = d_1$$
$$a_2 x + b_2 y + c_2 z = d_2$$
$$a_3 x + b_3 y + c_3 z = d_3$$

the matrices

$$\begin{bmatrix} a_1 & b_1 & c_1 \\ a_2 & b_2 & c_2 \\ a_3 & b_3 & c_3 \end{bmatrix} \qquad \text{and} \qquad \begin{bmatrix} a_1 & b_1 & c_1 & d_1 \\ a_2 & b_2 & c_2 & d_2 \\ a_3 & b_3 & c_3 & d_3 \end{bmatrix}$$

are called the **coefficient matrix** and the **augmented matrix,** respectively. Each *row* of the augmented matrix represents one of the equations of the system. For example, the augmented matrix of the system

$$\begin{aligned} 3x - 4y + z &= 2 \\ -x + 2y &= -1 \qquad \text{is} \\ 2x - y - 3z &= 4 \end{aligned} \qquad \begin{bmatrix} 3 & -4 & 1 & 2 \\ -1 & 2 & 0 & -1 \\ 2 & -1 & -3 & 4 \end{bmatrix}$$

and the augmented matrix of the system

$$\begin{aligned} x - 3y + 2z &= 5 \\ 2y - z &= 4 \qquad \text{is} \\ 4z &= 8 \end{aligned} \qquad \begin{bmatrix} 1 & -3 & 2 & 5 \\ 0 & 2 & -1 & 4 \\ 0 & 0 & 4 & 8 \end{bmatrix}$$

The augmented matrix of this last system, which has all zero entries in the lower left corner (below the diagonal), is said to be in **upper triangular form.** As we saw in Section 2.3, it is easy to find the solution of such a system by back-substitution.

Elementary Row Operations

We used the two properties on page 100 to change a given linear system into an equivalent one, that is, one that has the same solutions as the original system. The properties allowed us to perform the following operations on the equations of a system.

1. Multiply both sides of an equation by nonzero real number.
2. Add a constant multiple of one equation to another equation.

We can also perform the following operation without changing the solution of the system.

3. Interchange two equations.

Because each equation of the system corresponds to a row in the augmented matrix, the three operations above correspond to three **elementary row operations** for the augmented matrix. We can perform one or more of these operations on an augmented matrix without changing the solution of the system it represents.

> ## Elementary Row Operations
> 1. Multiply the entries of any row by a nonzero real number.
> 2. Add a constant multiple of one row to another row.
> 3. Interchange two rows.

EXAMPLE 1

a. $A = \begin{bmatrix} 1 & 3 & -1 & | & -1 \\ 2 & 1 & 4 & | & 5 \\ 6 & 2 & -1 & | & -12 \end{bmatrix}$ and $B = \begin{bmatrix} 1 & 3 & -1 & | & -1 \\ 6 & 3 & 12 & | & 15 \\ 6 & 2 & -1 & | & -12 \end{bmatrix}$

represent equivalent systems because we can multiply each entry in the second row of A by 3 to obtain B.

b. $A = \begin{bmatrix} 3 & -1 & 2 & | & 7 \\ 2 & 1 & 4 & | & -5 \\ 3 & 1 & 9 & | & -16 \end{bmatrix}$ and $B = \begin{bmatrix} 3 & 1 & 9 & | & -16 \\ 2 & 1 & 4 & | & -5 \\ 3 & -1 & 2 & | & 7 \end{bmatrix}$

represent equivalent systems because we can interchange the first and third rows of A to obtain B.

c. $A = \begin{bmatrix} 1 & 2 & 1 & | & -3 \\ 2 & 0 & -1 & | & 7 \\ 3 & 1 & 2 & | & 10 \end{bmatrix}$ and $B = \begin{bmatrix} 1 & 2 & 1 & | & -3 \\ 0 & -4 & -3 & | & 13 \\ 3 & 1 & 2 & | & 10 \end{bmatrix}$

represent equivalent systems because we can add -2 times each entry of the first row of A to the corresponding entry of the second row of A to obtain B. ■

EXERCISE 1   Subtract 4 times row 2 from row 3.

$$\begin{bmatrix} 3 & 2 & -5 & | & -3 \\ 2 & -3 & 2 & | & 6 \\ 8 & -4 & 2 & | & 12 \end{bmatrix}$$

EXAMPLE 2

Use row operations to obtain an equivalent matrix with the given elements:

$$\begin{bmatrix} 1 & -4 & | & -5 \\ 3 & 6 & | & 3 \end{bmatrix} \rightarrow \begin{bmatrix} 1 & -4 & | & -5 \\ 0 & ? & | & ? \end{bmatrix}$$

Solution

To obtain 0 as the first entry in the second row, we can multiply the first row by -3 and add the result to the second row. That is, we add -3(row 1) to row 2:

$$-3(\text{row } 1) + \text{row } 2 \quad \begin{bmatrix} 1 & -4 & | & -5 \\ 3 & 6 & | & 3 \end{bmatrix} \rightarrow \begin{bmatrix} 1 & -4 & | & -5 \\ 0 & 18 & | & 18 \end{bmatrix}$$ ∎

EXAMPLE 3

Use row operations on the first matrix to obtain an equivalent matrix in upper triangular form.

$$\begin{bmatrix} 1 & -3 & 1 & | & -4 \\ 3 & -1 & -1 & | & 8 \\ 2 & -2 & 3 & | & -1 \end{bmatrix} \rightarrow \begin{bmatrix} 1 & -3 & 1 & | & -4 \\ 0 & ? & ? & | & ? \\ 0 & 0 & ? & | & ? \end{bmatrix}$$

Solution

We perform the transformation in two steps. First, we obtain zeros in the lower two entries of the first column. We get these zeros by adding suitable multiples of the first row to the second and third rows:

$$\begin{matrix} -3(\text{row } 1) + \text{row } 2 \\ -2(\text{row } 1) + \text{row } 3 \end{matrix} \quad \begin{bmatrix} 1 & -3 & 1 & | & -4 \\ 3 & -1 & -1 & | & 8 \\ 2 & -2 & 3 & | & -1 \end{bmatrix} \rightarrow \begin{bmatrix} 1 & -3 & 1 & | & -4 \\ 0 & 8 & -4 & | & 20 \\ 0 & 4 & 1 & | & 7 \end{bmatrix}$$

Next we obtain a zero as the second entry of the third row by adding a suitable multiple of the *second* row. For this matrix, we add $-\frac{1}{2}$(row 2) to row 3:

$$-\frac{1}{2}(\text{row } 2) + \text{row } 3 \quad \begin{bmatrix} 1 & -3 & 1 & | & -4 \\ 0 & 8 & -4 & | & 20 \\ 0 & 4 & 1 & | & 7 \end{bmatrix} \rightarrow \begin{bmatrix} 1 & -3 & 1 & | & -4 \\ 0 & 8 & -4 & | & 20 \\ 0 & 0 & 3 & | & -3 \end{bmatrix}$$

The last matrix is in upper triangular form. ∎

$$\begin{bmatrix} 1 & -2 & 4 & | & 3 \\ 5 & -7 & 8 & | & 6 \\ -2 & 6 & -7 & | & 6 \end{bmatrix} \rightarrow \begin{bmatrix} 1 & -2 & 4 & | & 3 \\ 0 & ? & ? & | & ? \\ 0 & 0 & ? & | & ? \end{bmatrix}$$

Matrix Reduction

We use elementary row operations to adapt the method of Gaussian reduction to the augmented matrix of a linear system. This matrix method can be used to solve linear systems of any size and is well suited for implementation by a computer program. The method has three steps.

Solving a Linear System with Matrix Reduction

1. Write the augmented matrix for the system.
2. Using elementary row operations, transform the matrix into an equivalent one in upper triangular form.
3. Use back-substitution to find the solution to the system.

EXAMPLE 4 Use matrix reduction to solve the system.

$$x - 2y = -5$$
$$2x + 3y = 11$$

Solution

The augmented matrix is

$$\begin{bmatrix} 1 & -2 & | & -5 \\ 2 & 3 & | & 11 \end{bmatrix}$$

We use row operations to obtain 0 in the first entry of the second row.

$$-2(\text{row 1}) + \text{row 2} \quad \begin{bmatrix} 1 & -2 & | & -5 \\ 2 & 3 & | & 11 \end{bmatrix} \rightarrow \begin{bmatrix} 1 & -2 & | & -5 \\ 0 & 7 & | & 21 \end{bmatrix}$$

The last matrix is upper triangular, and corresponds to the system

$$x - 2y = -5 \qquad (1)$$
$$7y = 21 \qquad (2)$$

Now we use back-substitution to solve the system. From Equation (2), $y = 3$. Substitute 3 for y in Equation (1) to find

$$x - 2(3) = -5$$
$$x = 1$$

The solution is the ordered pair $(1, 3)$. ∎

EXERCISE 3   Use matrix reduction to solve the system

$$x + 4y = 3$$
$$3x + 8y = 1$$

Follow the suggested steps:

1. Write the augmented matrix for the system.
2. Add -3(row 1) to (row 2).
3. Solve the resulting system by back-substitution.

Reducing a 3 × 3 Matrix

Although there are many ways to reduce an augmented 3×3 matrix to upper triangular form, the following two-step procedure may help you organize the row operations.

1. Obtain zeros in the *first* entries of the second and third rows by adding suitable multiples of the *first* row to the second and third rows.

2. Obtain a zero in the *second* entry of the third row by adding a suitable multiple of the *second* row to the third row.

EXAMPLE 5

Use matrix reduction to solve the system.

$$2x - 4y = 6$$
$$3x - 4y + z = 8$$
$$2x - 3z = -11$$

Solution

The augmented matrix is

$$\begin{bmatrix} 2 & -4 & 0 & | & 6 \\ 3 & -4 & 1 & | & 8 \\ 2 & 0 & -3 & | & -11 \end{bmatrix}$$

Obtain 1 as the first entry of the first row by multiplying each entry of the first row by $\frac{1}{2}$. (This will make it easier to obtain zeros in the first entries of the second and third rows):

$$\tfrac{1}{2}(\text{row 1}) \quad \begin{bmatrix} 2 & -4 & 0 & | & 6 \\ 3 & -4 & 1 & | & 8 \\ 2 & 0 & -3 & | & -11 \end{bmatrix} \qquad \begin{array}{l} 2x - 4y + 0z = 6 \\ 3x - 4y + z = 8 \\ 2x + 0y - 3z = -11 \end{array}$$

$$\downarrow$$

$$\begin{bmatrix} 1 & -2 & 0 & | & 3 \\ 3 & -4 & 1 & | & 8 \\ 2 & 0 & -3 & | & -11 \end{bmatrix} \qquad \begin{array}{l} x - 2y + 0z = 3 \\ 3x - 4y + z = 8 \\ 2x + 0y - 3z = -11 \end{array}$$

Next, obtain zeros in the first entries of the second and third rows by adding suitable multiples of the first row:

$$\begin{array}{l} -3(\text{row 1}) + \text{row 2} \\ -2(\text{row 1}) + \text{row 3} \end{array} \begin{bmatrix} 1 & -2 & 0 & | & 3 \\ 3 & -4 & 1 & | & 8 \\ 2 & 0 & -3 & | & -11 \end{bmatrix} \qquad \begin{array}{l} x - 2y + 0z = 3 \\ 3x - 4y + z = 8 \\ 2x + 0y - 3z = -11 \end{array}$$

$$\downarrow$$

$$\begin{bmatrix} 1 & -2 & 0 & | & 3 \\ 0 & 2 & 1 & | & -1 \\ 0 & 4 & -3 & | & -17 \end{bmatrix} \qquad \begin{array}{l} x - 2y + 0z = 3 \\ 0x + 2y + z = -1 \\ 0x + 4y - 3z = -17 \end{array}$$

Finally, obtain a zero in the second entry of the third row by adding $-2(\text{row 2})$ to row 3:

$$-2(\text{row 2}) + \text{row 3} \begin{bmatrix} 1 & -2 & 0 & | & 3 \\ 0 & 2 & 1 & | & -1 \\ 0 & 4 & -3 & | & -17 \end{bmatrix} \qquad \begin{array}{l} x - 2y + 0z = 3 \\ 0x + 2y + z = -1 \\ 0x + 4y - 3z = -17 \end{array}$$

$$\downarrow$$

$$\begin{bmatrix} 1 & -2 & 0 & | & 3 \\ 0 & 2 & 1 & | & -1 \\ 0 & 0 & -5 & | & -15 \end{bmatrix} \qquad \begin{array}{l} x - 2y + 0z = 3 \\ 0x + 2y + z = -1 \\ 0x + 0y - 5z = -15 \end{array}$$

The system is now in upper triangular form, and we use back-substitution to find the solution. Solve the last equation to get $z = 3$ and substitute 3 for z in the second equation to find $y = -2$. Finally, substitute 3 for z and -2 for y in the first equation to find $x = -1$. The solution is the ordered triple $(-1, -2, 3)$. ∎

EXERCISE 4 Use matrix reduction to solve the system

$$\begin{array}{l} x + 3z = -11 \\ 2x + y + z = 1 \\ -3x - 2y = 3 \end{array}$$

Follow the suggested steps.

1. Write the augmented matrix for the system.
2. Add $-2(\text{row 1})$ to (row 2)
3. Add $3(\text{row 1})$ to (row 3)
4. Add $-2(\text{row 2})$ to (row 3)
5. Solve the resulting system by back-substitution.

It is a good idea to make the first entry in the first row equal to 1, as we did in Example 5, to simplify the calculations that follow. However, if that entry is zero we can interchange two rows. For example, to reduce the matrix

$$\begin{bmatrix} 0 & 1 & 2 & 3 \\ 1 & 2 & 3 & 4 \\ 2 & 5 & 1 & 4 \end{bmatrix}$$

we begin by interchanging the first and second rows to obtain

$$\begin{bmatrix} 1 & 2 & 3 & 4 \\ 0 & 1 & 2 & 3 \\ 2 & 5 & 1 & 4 \end{bmatrix}$$

Then we follow the two-step procedure described above to find

$$\begin{bmatrix} 1 & 2 & 3 & 4 \\ 0 & 1 & 2 & 3 \\ 2 & 5 & 1 & 4 \end{bmatrix} \underset{-2(\text{row 1}) + \text{row 3}}{\longrightarrow} \begin{bmatrix} 1 & 2 & 3 & 4 \\ 0 & 1 & 2 & 3 \\ 0 & 1 & -5 & -4 \end{bmatrix} \underset{-(\text{row 2}) + \text{row 3}}{\longrightarrow} \begin{bmatrix} 1 & 2 & 3 & 4 \\ 0 & 1 & 2 & 3 \\ 0 & 0 & -7 & -7 \end{bmatrix}$$

READING QUESTIONS

1. What are the coefficient matrix and the augmented matrix for a system of equations?
2. What does it mean for a matrix to be in upper triangular form?
3. State the three elementary row operations.
4. Describe a two-step procedure for reducing a 3×3 matrix to upper triangular form.
5. Once the matrix is in upper triangular form, how do we complete the solution of the system?

ANSWERS FOR 2.4 EXERCISES

1. $\begin{bmatrix} 3 & 2 & -5 & -3 \\ 2 & -3 & 2 & 6 \\ 0 & 8 & -6 & -12 \end{bmatrix}$

2. $\begin{bmatrix} 1 & -2 & 4 & 3 \\ 0 & 1 & -4 & -3 \\ 0 & 0 & 9 & 18 \end{bmatrix}$

3. $(-5, 2)$

4. $(37, -57, -16)$

HOMEWORK 2.4

Perform the given elementary row operation on the matrices in Problems 1–8.

1. Multiply row 2 by -3:

$$\begin{bmatrix} -2 & 1 & 0 \\ 3 & -1 & 2 \end{bmatrix}$$

2. Multiply row 1 by $\frac{1}{4}$:

$$\begin{bmatrix} 2 & 0 & 3 \\ -1 & 5 & 4 \end{bmatrix}$$

3. Add 2 (row 1) to row 2:

$$\begin{bmatrix} 1 & -3 & | & 6 \\ -2 & 4 & | & -1 \end{bmatrix}$$

4. Add -3(row 1) to row 2:

$$\begin{bmatrix} 1 & -4 & | & 8 \\ 3 & -2 & | & 10 \end{bmatrix}$$

5. Interchange row 1 and row 3:

$$\begin{bmatrix} 0 & -3 & 2 & | & -3 \\ 2 & 6 & -1 & | & 3 \\ 1 & 0 & -2 & | & 5 \end{bmatrix}$$

6. Interchange row 2 and row 3:

$$\begin{bmatrix} 1 & 6 & 0 & | & -2 \\ 0 & 0 & 5 & | & -10 \\ 0 & 3 & -2 & | & 8 \end{bmatrix}$$

7. Add -4(row 1) to row 3:

$$\begin{bmatrix} 1 & 2 & 1 & | & -5 \\ 0 & 4 & -2 & | & 3 \\ 4 & -1 & 6 & | & -8 \end{bmatrix}$$

8. Add 2(row 2) to row 3:

$$\begin{bmatrix} 1 & -7 & 5 & | & 2 \\ 0 & 1 & -3 & | & -1 \\ 0 & -2 & -3 & | & 4 \end{bmatrix}$$

In Problems 9–18 use row operations on the first matrix to form an equivalent matrix with the given entries.

9. $\begin{bmatrix} 1 & -3 & | & 2 \\ 2 & 1 & | & 4 \end{bmatrix} \rightarrow \begin{bmatrix} 1 & -3 & | & 2 \\ 0 & ? & | & ? \end{bmatrix}$

10. $\begin{bmatrix} -2 & 3 & | & 0 \\ 4 & 1 & | & 6 \end{bmatrix} \rightarrow \begin{bmatrix} -2 & 3 & | & 0 \\ 0 & ? & | & ? \end{bmatrix}$

11. $\begin{bmatrix} 2 & 6 & | & -4 \\ 5 & 3 & | & 1 \end{bmatrix} \rightarrow \begin{bmatrix} 2 & 6 & | & -4 \\ ? & 0 & | & ? \end{bmatrix}$

12. $\begin{bmatrix} 6 & 4 & | & -2 \\ -1 & -2 & | & -3 \end{bmatrix} \rightarrow \begin{bmatrix} 6 & 4 & | & -2 \\ ? & 0 & | & ? \end{bmatrix}$

13. $\begin{bmatrix} 1 & -2 & 2 & | & 1 \\ 2 & 3 & -1 & | & 6 \\ 4 & 1 & -3 & | & 3 \end{bmatrix} \rightarrow \begin{bmatrix} 1 & -2 & 2 & | & 1 \\ 0 & ? & ? & | & ? \\ 0 & ? & ? & | & ? \end{bmatrix}$

14. $\begin{bmatrix} 2 & -1 & 3 & | & -1 \\ -4 & 0 & 4 & | & 5 \\ 6 & 2 & -1 & | & -2 \end{bmatrix} \rightarrow \begin{bmatrix} 2 & -1 & 3 & | & -1 \\ 0 & ? & ? & | & ? \\ 0 & ? & ? & | & ? \end{bmatrix}$

15. $\begin{bmatrix} -1 & 4 & 3 & | & 2 \\ 2 & -2 & -4 & | & 6 \\ 1 & 2 & 3 & | & -3 \end{bmatrix} \rightarrow \begin{bmatrix} -1 & 4 & 3 & | & 2 \\ ? & 0 & ? & | & ? \\ ? & 0 & ? & | & ? \end{bmatrix}$

16. $\begin{bmatrix} 3 & -2 & 4 & | & -4 \\ 2 & 2 & 1 & | & 2 \\ -1 & 1 & 5 & | & -1 \end{bmatrix} \rightarrow \begin{bmatrix} 3 & -2 & 4 & | & -4 \\ ? & ? & 0 & | & ? \\ ? & ? & 0 & | & ? \end{bmatrix}$

17.

$$\begin{bmatrix} -2 & 1 & -3 & | & -2 \\ 4 & 2 & 0 & | & 2 \\ 6 & -1 & 2 & | & 0 \end{bmatrix} \rightarrow \begin{bmatrix} -2 & 1 & -3 & | & -2 \\ 0 & ? & ? & | & ? \\ 0 & 0 & ? & | & ? \end{bmatrix}$$

18. $\begin{bmatrix} -1 & 2 & 3 & | & 3 \\ 4 & 0 & 1 & | & -6 \\ 2 & 2 & -3 & | & -2 \end{bmatrix} \rightarrow \begin{bmatrix} -1 & 2 & 3 & | & 3 \\ 0 & ? & ? & | & ? \\ 0 & 0 & ? & | & ? \end{bmatrix}$

Use matrix reduction on the augmented matrix to solve each system in Problems 19–26.

19. $x + 3y = 11$
$2x - y = 1$

20. $x - 5y = 11$
$2x + 3y = -4$

21. $x - 4y = -6$
$3x + y = -5$

22. $x + 6y = -14$
$5x + 3y = -4$

23. $2x + y = 5$
$3x - 5y = 14$

24. $3x - 2y = 16$
$4x + 2y = 12$

25. $x - y = -8$
$x + 2y = 9$

26. $4x - 3y = 16$
$2x + y = 8$

Use matrix reduction on the augmented matrix to solve each system in Problems 27–34.

27. $x + 3y - z = 5$
$3x - y + 2z = 5$
$x + y + 2z = 7$

28. $x - 2y + 3z = 6$
$2x + 3y - z = 6$
$3x - y - z = 3$

29. $2x - y + z = 5$
$x - 2y - 2z = 2$
$3x + 3y - z = 4$

30. $x - 2y - 2z = 4$
$2x + y - 3z = 7$
$x - y - z = 3$

31. $2x - y - z = -4$
$x + y + z = -5$
$x + 3y - 4z = 12$

32. $x - 2y - 5z = 2$
$2x + 3y + z = 11$
$3x - y - z = 11$

33. $2x - y = 0$
$3y + z = 7$
$2x + 3z = 1$

34. $3x - z = 7$
$2x + y = 6$
$3y - z = 7$

2.5 Linear Inequalities in Two Variables

In this section we study linear inequalities in two variables and how they arise in applications.

Graphs of Inequalities in Two Variables

Ivana is investing in the hotel business. She has bought two hotels and will expand her investments when her total profit from the two hotels is at least $10,000. If we let x represent the profit from one hotel and let y represent the profit from the other, then Ivana will expand her investments when

$$x + y \geq 10,000 \tag{1}$$

Notice that the equation $x + y = 10,000$ is not appropriate to model our situation, since Ivana will be delighted if her profits are not exactly equal to $10,000 but actually exceed that amount.

A **solution** to an inequality in two variables is an ordered pair of numbers that satisfies the inequality. The graph of the inequality must show all the points whose coordinates are solutions. As an example, let us graph inequality (1).

We first rewrite the inequality by subtracting x from both sides to get

$$y \geq -x + 10,000 \tag{2}$$

Inequality (2) says that for each x-value, we must choose points with y-values greater than or equal to $-x + 10,000$. For example, when $x = 2000$, we must choose points with y-values greater than or equal to 8000. Solutions for several choices of x are shown in Figure 2.17a.

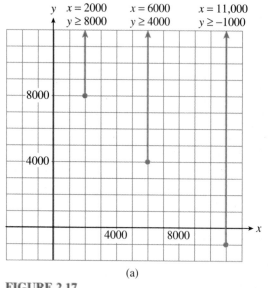

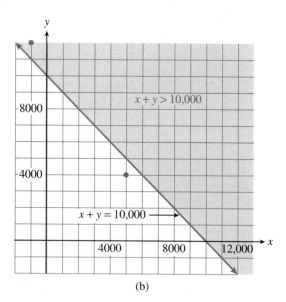

(a)

(b)

FIGURE 2.17

A more efficient way to find all the solutions of inequality (2) is to start with the graph of the corresponding equation $y = -x + 10,000$. The graph is a straight line, as illustrated in Figure 2.17b. Observe that any point *above* this line has a *y*-coordinate greater than $-x + 10,000$ and hence satisfies (2). Thus, the graph of inequality (2) includes all the points on or above the line $y = -x + 10,000$, as shown by the shaded region in Figure 2.17b.

You can check that the shaded points are also solutions to inequality (1). Consider the point $(-1,000, 12,000)$, which lies in the shaded region above the line. This pair *does* satisfy (1) since

$$-1,000 + 12,000 \geq 10,000$$

(Ivana will expand her investments if her first hotel loses \$1,000 and her second has a profit of \$12,000.) On the other hand, the point $(5000, 4000)$ does *not* lie in the graph of (1) because the coordinates do not satisfy (1).

EXERCISE 1

a. Find one *y*-value that satisfies the inequality $y - 3x < 6$ for each of the *x*-values in the table.

x	1	0	-2
y			

b. Graph the line $y - 3x = 6$. Then plot your solutions from part (a) on the same grid.

Linear Inequalities

A **linear inequality** can be written in the form

$$ax + by + c \leq 0 \qquad \text{or} \qquad ax + by + c \geq 0$$

The solutions consist of the line $ax + by + c = 0$ and a **half-plane** on one side of that line. We shade the half-plane to show that all its points are included in the solution set. If the inequality is strict, then the graph includes only the half-plane and not the line. In that case we use a dashed line for the graph of the equation $ax + by + c = 0$ to show that it is not part of the solution.

To decide which side of the line to shade, we can solve the inequality for *y* in terms of *x*. If we obtain

$$y \geq mx + b \qquad \text{(or} \qquad y > mx + b)$$

then we shade the half-plane *above* the line. If the inequality is equivalent to

$$y \leq mx + b \qquad \text{(or} \qquad y < mx + b)$$

then we shade the half-plane *below* the line. Be careful when isolating *y*: We must remember to reverse the direction of the inequality whenever we multiply or divide by a negative number. (See Appendix A.3 if you would like to review solving inequalities.)

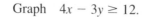

EXAMPLE 1

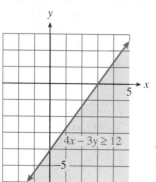

FIGURE 2.18

Graph $4x - 3y \geq 12$.

Solution

Solve the inequality for y.

$$4x - 3y \geq 12 \qquad \text{Subtract } 4x \text{ from both sides.}$$
$$-3y \geq -4x + 12 \qquad \text{Divide both sides by } -3.$$
$$y \leq \frac{4}{3}x - 4$$

Graph the corresponding line $y = \frac{4}{3}x - 4$. Note that the y-intercept is -4 and the slope is $\frac{4}{3}$. (See Section 2.1 to review the slope-intercept method of graphing.) Finally, shade the half-plane below the line. The completed graph is shown in Figure 2.18. ∎

Using a Test Point

A second method for graphing inequalities does not require us to solve for y. Once we have graphed the boundary line, we can decide which half-plane to shade by using a "test point." The test point can be any point that is not on the boundary line itself.

EXAMPLE 2

Graph $3x - 2y < 6$.

Solution

First, graph the line $3x - 2y = 6$, as shown in Figure 2.19. We will use the intercept method. The intercepts are $(2, 0)$ and $(0, -3)$, so we sketch the boundary line through those points. Next, choose a test point. Since $(0, 0)$ does not lie on the line, we choose it as our test point. Substitute the coordinates of the test point into the inequality to obtain

$$3(0) - 2(0) < 6$$

Since this is a true statement, $(0, 0)$ *is* a solution of the inequality. Since *all* the solutions lie on the same side of the boundary line, we shade the half-plane that contains the test point. In this example the boundary line is a dashed line because the original inequality was strict.

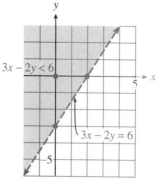

FIGURE 2.19

∎

We can choose *any* point for the test point, as long as it does not lie on the boundary line. We chose $(0, 0)$ in Example 2 because the coordinates are easy to substitute into the inequality. If the test point *is* a solution to the inequality, then the half-plane including that point should be shaded. If the test point is *not* a solution to the inequality, then the *other* half-plane should be shaded. For example, suppose we had chosen (5, 0) as the test point in Example 2. When we substitute its coordinates into the inequality, we find,

$$3(5) - 2(0) < 6$$

which is a *false* statement. This tells us that (5, 0) is not a solution to the inequality, so the solutions must lie on the other side of the boundary line. Using (5, 0) as the test point gives us the same solutions we found in Example 2.

Here is a summary of our test point method for graphing inequalities.

Graphing an Inequality Using a Test Point

1. Graph the corresponding equation to obtain the boundary line.
2. Choose a test point that does not lie on the boundary line.
3. Substitute the coordinates of the test point into the inequality.
   a. If the resulting statement is true, shade the half-plane that includes the test point.
   b. If the resulting statement is false, shade the half-plane that does not include the test point.
4. If the inequality is strict, make the boundary line a dashed line.

EXERCISE 2 Graph the solutions of the inequality

$$y > \frac{-3}{2}x$$

1. Graph the line $y = \frac{-3}{2}x$. (Use the slope-intercept method.)
2. Choose a test point. (Do not choose (0, 0)!)
3. Decide which side of the line to shade.
4. Should the boundary line be dashed or solid?

Recall that the equation of a vertical line has the form

$$x = k$$

where k is a constant, and a horizontal line has an equation of the form

$$y = k$$

Similarly, the inequality $x \geq k$ may represent the inequality in two variables.

$$x + 0y \geq k$$

Its graph is then a region in the plane.

EXAMPLE 3

Graph $x \geq 2$ in the plane.

Solution

First graph the equation $x = 2$; its graph is a vertical line. Since the origin does not lie on this line, we can use it as a test point. Substitute 0 for x (there is no y) into the inequality to obtain

$$0 \geq 2.$$

Since this is a false statement, shade the half-plane that does not contain the origin. We see in Figure 2.20 that the graph of the inequality contains all points whose x-coordinates are greater than or equal to 2. ∎

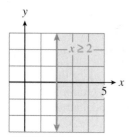

FIGURE 2.20

Systems of Inequalities

Some applications are best described by a system of two or more inequalities. The solutions to a system of inequalities include all points that are solutions to each inequality in the system. The graph of the system is the intersection of the shaded regions for each inequality in the system. For example, Figure 2.21 shows the solutions of the system

$$y > x \quad \text{and} \quad y > 2$$

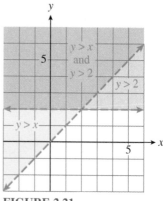

FIGURE 2.21

EXAMPLE 4

Laura is a finicky eater, and dislikes most foods that are high in calcium. Her morning cereal satisfies some of her calcium requirements, but she needs an additional 500 milligrams of calcium, which she will get from a combination of broccoli, at 160 milligrams per serving, and zucchini, at 30 milligrams per serving. Draw a graph representing the possible combinations of broccoli and zucchini that fulfill Laura's calcium requirements.

Solution

Step 1 Number of servings of broccoli: $\quad x$

Number of servings of zucchini: $\quad y$

Step 2 To consume at least 500 milligrams of calcium, Laura must choose x and y so that

$$160x + 30y \geq 500$$

It makes no sense to consider negative values of x or of y, since Laura cannot eat a negative number of servings. Thus we have two more inequalities to satisfy:

$$x \geq 0 \quad \text{and} \quad y \geq 0$$

Step 3 Graph all three inequalities on the same axes. The inequalities $x \geq 0$ and $y \geq 0$ restrict the solutions to lie in the first quadrant. The solutions common to all three inequalities are shown in Figure 2.22.

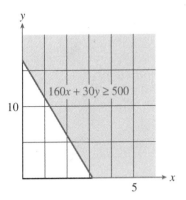

FIGURE 2.22

Step 4 Laura can choose any combination of broccoli and zucchini represented by points in the shaded region. For example, the point (3, 1) is a solution to the system of inequalities, so Laura could choose to eat 3 servings of broccoli and 1 serving of zucchini.

To describe the solutions of a system of inequalities, it is useful to locate the **vertices,** or corner points, of the boundary.

EXAMPLE 5 Graph the solution set of the system below and find the coordinates of its vertices.

$$x - y - 2 \leq 0$$
$$x + 2y - 6 \leq 0$$
$$x \geq 0, \quad y \geq 0$$

Solution

The last two inequalities, $x \geq 0$ and $y \geq 0$, restrict the solutions to the first quadrant. Graph the line $x - y - 2 = 0$, and use the test point (0, 0) to decide to shade the half-plane including the origin. Finally, graph the line $x + 2y - 6 = 0$, and again use the test point (0, 0) to shade the half-plane below the line. The intersection of the shaded regions is shown in Figure 2.23.

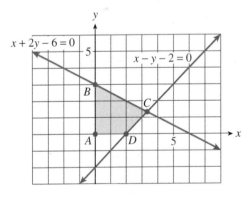

FIGURE 2.23

To find the coordinates of the vertices *A, B, C,* and *D,* solve simultaneously the equations of the two lines that intersect at the vertex. (See Section 2.2 to review solving 2 × 2 linear systems.) Thus,

for *A*, solve the system $x = 0$ to find (0, 0);
$$y = 0$$

for *B*, solve the system $x = 0$ to find (0, 3);
$$x + 2y = 6$$

for *C*, solve the system $x + 2y = 6$ to find $\left(\frac{10}{3}, \frac{4}{3}\right)$;
$$x - y = 2$$

for *D*, solve the system $y = 0$ to find (2, 0).
$$x - y = 2$$

The vertices are the points (0, 2), (0, 3), $\left(\frac{10}{3}, \frac{4}{3}\right)$ and (2, 0).   ■

EXERCISE 3

a. Graph the system of inequalities

$$5x + 4y < 40$$
$$-3x + 4y < 12$$
$$x < 6, \quad y > 2$$

b. Find the coordinates of the vertices of the solution set.

---

1. The solutions of a linear inequality in two variables form what sort of set?
2. How can you find the boundary of the solution set?
3. Explain how to use a test point to solve a linear inequality.
4. How can you find the vertices of the solution set of a system of inequalities?

ANSWERS FOR 2.5 EXERCISES

1a.

x	1	0	−2
y	5	0	−3

b.

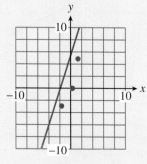

2.

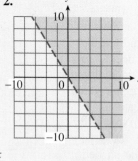

3a.

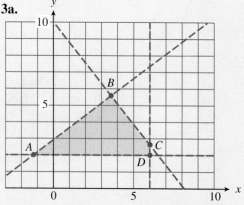

b. A (−1.3, 2)
B (3.5, 5.625)
C (6, 2.5)
D (6, 2)

Graph each inequality.

1. $y > 2x + 4$
2. $y < 9 - 3x$
3. $3x - 2y \leq 12$
4. $2x + 5y \geq 10$

5. $x + 4y \geq -6$
6. $3x - y \leq -2$
7. $x > -3y + 1$
8. $x > 2y - 5$

9. $x \geq -3$
10. $y < 4$
11. $y < \frac{1}{2}x$
12. $y > \frac{4}{3}x$

13. $0 \geq x - y$
14. $0 \geq x + 3y$
15. $-1 < y \leq 4$
16. $-2 \leq y < 0$

Graph each system of inequalities.

17. $y > 2$
    $x \geq -2$

18. $y \leq -1$
    $x > 2$

19. $y < x$
    $y \geq -3$

20. $y \geq -x$
    $y < 2$

21. $x + y \leq 6$
    $x + y \geq 4$

22. $x - y < 3$
    $x - y > -2$

23. $2x - y \leq 4$
    $x + 2y > 6$

24. $2y - x < 2$
    $x + y \leq 4$

25. $3y - 2x < 2$
    $y > x - 1$

26. $2x + y < 4$
    $y > 1 - x$

Graph each system of inequalities and find the coordinates of the vertices.

27. $2x + 3y - 6 < 0$
    $x \geq 0, y \geq 0$

28. $3x + 2y < 6$
    $x \geq 0, y \geq 0$

29. $5y - 3x \leq 15$
    $x + y \leq 11$
    $x \geq 0, y \geq 0$

30. $y - 2x \geq -4$
    $x + y \leq 5$
    $x \geq 0, y \geq 0$

31. $2y \leq x$
    $2x \leq y + 12$
    $x \geq 0, y \geq 0$

32. $y \geq 3x$
    $2y + x \leq 14$
    $x \geq 0, y \geq 0$

33. $x + y \geq 3$
    $2y \leq x + 8$
    $2y + 3x \leq 24$
    $x \geq 0, y \geq 0$

34. $2y + 3x \geq 6$
    $2y + x \leq 10$
    $y \geq 3x - 9$
    $x \geq 0, y \geq 0$

35. $3y - x \geq 3$
    $y - 4x \geq -10$
    $y - 2 \leq x$
    $x \geq 0, y \geq 0$

36. $2y + x \leq 12$
    $4y \leq 2x + 8$
    $x \leq 4y + 4$
    $x \geq 0, y \geq 0$

Graph the set of solutions to each problem. Two of the inequalities in each system are $x \geq 0$ and $y \geq 0$.

37. The math club is selling tickets for a show by a "mathemagician." Student tickets will cost $1 and faculty tickets will cost $2. The ticket receipts must be at least $250 to cover the fee for the performer. Write a system of three inequalities for the number of student tickets and the number of faculty tickets that must be sold, and graph the solutions.

38. The math department is selling old textbooks to raise at least $300 for scholarships. Paperback textbooks will cost $2 and the hardcover textbooks will cost $5. Write a system of three inequalities for the number of paperback and hardback textbooks that must be sold, and graph the solutions.

39. Vassilis plans to invest at most $10,000 in two banks. One bank pays 6% annual interest and the other pays 5%. Vassilis wants at least $540 total annual interest from his two investments. Write a system of four inequalities for the amount Vassilis can invest in the two accounts, and graph the system.

40. Jeannette has 180 acres of farmland for growing wheat or soy. She can get a profit of $36 per acre for wheat and $24 per acre for soy. She wants to have a profit of at least $5400 from her crops. Write a system of four inequalities for the number of acres she can use for each crop, and graph the solutions.

41. Gary's pancake recipe includes corn meal and whole wheat flour. Corn meal has 2.4 grams of linoleic acid and 2.5 milligrams of niacin per cup. Whole wheat flour has 0.8 grams of linoleic acid and 5 milligrams of niacin per cup. These two ingredients should not exceed 3 cups total. The mixture should provide at least 3.2 grams of linoleic acid and at least 10 milligrams of niacin. Write a system of five inequalities for the amount of corn meal and the amount of whole wheat flour Gary can use, and graph the solutions.

42. Cho and his brother go into business making comic book costumes. They need 1 hour of cutting and 2 hours of sewing to make a Batman costume. They need 2 hours of cutting and 1 hour of sewing to make a Wonder Woman costume. They have available at most 10 hours per day for cutting and at most 8 hours per day for sewing. They must make at least one costume each day to stay in business. Write a system of five inequalities for the number of each type of costume Cho can make, and graph the solutions.

Problems 43 and 44 investigate how the dietary requirements of animals affect their survival strategy. Grazing animals spend most of their time foraging for food. A small animal such as a ground squirrel must also be alert for predators. Which foraging strategy favors survival: Should the animal look for foods that satisfy its dietary requirements in minimum time, thus minimizing its exposure to predators and the elements, or should it try to maximize its intake of nutrients? (See Section 11.4 for more on this topic.)

43. The Columbian ground squirrel eats forb, a type of flowering weed, and grass. One gram of forb provides 2.44 kilocalories, and one gram of grass provides 2.26 kilocalories. For survival, the squirrel needs at least 100 kilocalories per day. However, the squirrel can digest no more than 314 wet grams of food daily. Each dry gram of forb becomes 2.67 wet grams in the squirrel's stomach, and each dry gram of grass becomes 1.64 wet grams. In addition, the squirrel has at most 342 minutes available for grazing each day. It takes the squirrel 2.05 minutes to eat one gram of forb and 5.21 minutes to eat one gram of grass.
Source: Belovsky, 1986.

   a. Write a system of inequalities for the number of grams of grass and forb the squirrel can eat and survive.

   b. Graph the system.

   c. Give two possible combinations of grass and forb that satisfy the squirrel's feeding requirements.

   d. If the squirrel can find only grass on a given day, how much grass will it need to eat? If the squirrel can find only forb, how much forb will it need?

44. The Isle Royale moose eats deciduous leaves that provide 3.01 kilocalories of energy per gram and aquatic plants that provide 3.82 kilocalories per gram. The moose needs at least 10,946 kilocalories of food per day, and its daily sodium requirement must be met by consuming at least 453 grams of aquatic plants. It takes the moose 0.05 minute to eat one gram of aquatic plants and 0.06 minute for each gram of leaves. To maintain its thermal balance, the moose cannot spend more than 150 minutes each day standing in water and eating aquatics plants, or more than 256 minutes eating leaves. Each gram of aquatic plants becomes 20 wet grams in the moose's stomach, and each gram of leaves becomes 4 wet grams, and the stomach can hold no more than 32,900 wet grams of food daily.
Source: Belovsky, 1978.

   a. Write a system of inequalities for the number of grams of aquatic plants and of leaves the moose can eat.

   b. Graph the system.

   c. Give two possible combinations of aquatic plants and leaves that satisfy the moose's feeding requirements.

   d. Can the moose survive by eating only aquatic plants?

Solve each system by graphing. Use the *ZDecimal* window.

1. $y = -2.9x - 0.9$
   $y = 1.4 - 0.6x$

2. $y = 0.6x - 1.94$
   $y = -1.1x + 1.29$

Solve each system using substitution or elimination.

3. $x + 5y = 18$
   $x - y = -3$

4. $x + 5y = 11$
   $2x + 3y = 8$

5. $\dfrac{2}{3}x - 3y = 8$
   $x + \dfrac{3}{4}y = 12$

6. $3x = 5y - 6$
   $3y = 10 - 11x$

Decide whether each system is inconsistent, dependent, or consistent and independent.

7. $2x - 3y = 4$
   $x + 2y = 7$

8. $2x - 3y = 4$
   $6x - 9y = 4$

9. $2x - 3y = 4$
   $6x - 9y = 12$

10. $x - y = 6$
    $x + y = 6$

Solve each system using Gaussian reduction.

11. $x + 3y - z = 3$
    $2x - y + 3z = 1$
    $3x + 2y + z = 5$

12. $x + y + z = 2$
    $3x - y + z = 4$
    $2x + y + 2z = 3$

13. $x + z = 5$
    $y - z = -8$
    $2x + z = 7$

14. $x + 4y + 4z = 0$
    $3x + 2y + z = -4$
    $2x - 4y + z = -11$

15. $\dfrac{1}{2}x + y + z = 3$

    $x - 2y - \dfrac{1}{3}z = -5$

    $\dfrac{1}{2}x - 3y - \dfrac{2}{3}z = -6$

16. $\dfrac{3}{4}x - \dfrac{1}{2}y + 6z = 2$

    $\dfrac{1}{2}x + y - \dfrac{3}{4}z = 0$

    $\dfrac{1}{4}x + \dfrac{1}{2}y - \dfrac{1}{2}z = 0$

Use matrix reduction to solve each system.

17. $x - 2y = 5$
    $2x + y = 5$

18. $4x - 3y = 16$
    $2x + y = 8$

21. $x + 2y - z = -3$
    $2x - 3y + 2z = 2$
    $x - y + 4z = 7$

22. $x + y + z = 1$
    $2x - y - z = 2$
    $2x - y + 3z = 2$

19. $2x - y = 7$
    $3x + 2y = 14$

20. $2x - y + 3z = -6$
    $x + 2y - z = 7$
    $3x + y + z = 2$

Solve each problem by writing and solving a system of linear equations in two or three variables.

23. A math contest exam has 40 questions. A contestant scores 5 points for each correct answer, but loses 2 points for each wrong answer. Lupe answered all the questions and her score was 102. How many questions did she answer correctly?

24. A game show contestant wins $25 for each correct answer he gives but loses $10 for each incorrect response. Roger answered 24 questions and won $355. How many answers did he get right?

25. Barbara wants to earn $500 a year by investing $5000 in two accounts, a savings plan that pays 8% annual interest and a high-risk option that pays 13.5% interest. How much should she invest in each account?

26. An investment broker promises his client a 12% return on her funds. If the broker invests $3000 in bonds paying 8% interest, how much must he invest in stocks paying 15% interest to keep his promise?

27. The perimeter of a triangle is 30 centimeters. The length of one side is 7 centimeters shorter than the second side, and the third side is 1 centimeter longer than the second side. Find the length of each side.

28. A company ships its product to three cities: Boston, Chicago, and Los Angeles. The cost of shipping is $10 per crate to Boston, $5 per crate to Chicago, and $12 per crate to Los Angeles. The company's shipping budget for April is $445. It has 55 crates to ship, and demand for their product is twice as high in Boston as in Los Angeles. How many crates should the company ship to each destination?

Graph each inequality.

29. $3x - 4y < 12$

30. $x > 3y - 6$

31. $y < -\dfrac{1}{2}$

32. $-4 \leq x < 2$

Graph the solutions to each system of inequalities.

33. $y > 3, x \leq 2$

34. $y \geq x, x > 2$

35. $3x - y < 6, x + 2y > 6$

36. $x - 3y > 3, y < x + 2$

a. Graph the solutions to the system of inequalities.

b. Find the coordinates of the vertices.

37. $3x - 4y \leq 12$
$x \geq 0, y \leq 0$

38. $x - 2y \leq 6$
$y \leq x$
$x \geq 0, y \geq 0$

39. $x + y \leq 5$
$y \geq x$
$y \geq 2, x \geq 0$

40. $x - y \leq -3$
$x + y \leq 6$
$x \leq 4$
$x \geq 0, y \geq 0$

41. Ruth wants to provide cookies for the customers at her video rental store. It takes 20 minutes to mix the ingredients for each batch of peanut butter cookies and 10 minutes to bake them. Each batch of granola cookies takes 8 minutes to mix and 10 minutes to bake. Ruth does not want to use the oven more than 2 hours a day, or to spend more than 2 hours a day mixing ingredients. Write a system of inequalities for the number of batches of peanut butter cookies and of granola cookies that Ruth can make in one day, and graph the solutions.

42. A vegetarian recipe calls for no more than 32 ounces of a combination of tofu and tempeh. Tofu provides 2 grams of protein per ounce and tempeh provides 1.6 grams of protein per ounce. Graham would like the dish to provide at least 56 grams of protein. Write a system of inequalities for the amount of tofu and the amount of tempeh for the recipe, and graph the solutions.

Quadratic Models

A **quadratic** equation involves the square of the variable. It has the form

$$y = ax^2 + bx + c$$

where *a, b,* and *c* are constants, and *a* is not equal to zero. (If *a* is zero, there is no *x*-squared term, so the equation is not quadratic.) In this chapter we investigate problems that can be described by quadratic equations.

INVESTIGATION 4

Falling

Suppose you drop a small object from a height and let it fall under the influence of gravity. Does it fall at the same speed throughout its descent? The diagram in Figure 3.1 shows a sequence of photographs of a steel ball falling onto a table. The photographs were taken using a stroboscopic flash at intervals of 0.05 seconds, and a scale on the left side shows the distance the ball has fallen, in feet, at each interval.

1. Complete Table 3.1 showing the distance the ball has fallen at each 0.05-second interval. Measure the distance at the bottom of the ball in each image. The first photo was taken at time $t = 0$.

2. On the grid provided, plot the points in the table. (Notice that the scale on the vertical axis increases from top to bottom instead of the usual way, from bottom to top. This is done to reflect the motion of the ball.) Connect the points with a smooth curve to sketch a graph of distance fallen versus time elapsed. Is the graph linear?

3. Use your graph to estimate the time elapsed when the ball has fallen 0.5 feet, and when it has fallen 3 feet.

4. How far did the ball fall during the first quarter second, from $t = 0$ to $t = 0.25$? How far did the ball fall from $t = 0.25$ to $t = 0.5$?

5. Add a line segment to your graph connecting the points at $t = 0$ and $t = 0.25$, and a second line segment connecting the points at $t = 0.25$ and $t = 0.5$. Compute the slope of each line segment.

6. What do the slopes in part (5) represent in terms of the problem?

7. Use your answers to part (4) to verify algebraically that the graph is not linear.

TABLE 3.1

t	0	0.05	0.1	0.15	0.2	0.25	0.3	0.35	0.4	0.45	0.5
d											

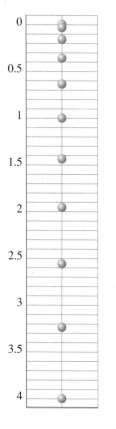
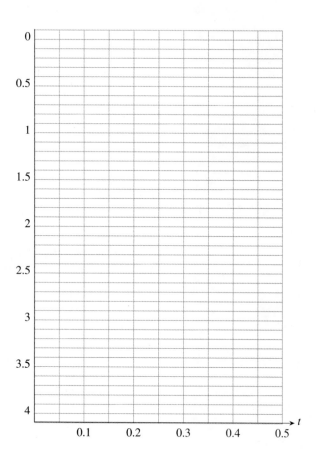

FIGURE 3.1

3.1 Extraction of Roots

Solving Quadratic Equations

Figure 3.2 shows a table and graph for the quadratic equation $y = 2x^2 - 5$.

x	-3	-2	-1	0	1	2	3
y	13	3	-3	-5	-3	3	13

You can see that there are two points on the graph for each y-value greater than -5. For example, the two points with y-coordinate 7 are shown. To solve the quadratic equation

$$2x^2 - 5 = 7$$

we need only find the x-coordinates of these points. (See Section 1.2 if you would like to review solving equations by graphing.) From the graph, the solutions appear to be about 2.5 and -2.5.

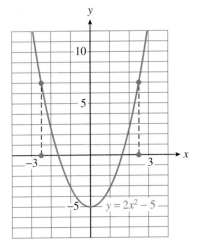

FIGURE 3.2

How can we solve quadratic equations algebraically? The opposite operation for squaring a number is taking a square root. Consequently, we can "undo" the operation of squaring by extracting square roots. To solve the equation above,

$$2x^2 - 5 = 7$$

we first solve for x^2 to get

$$2x^2 = 12$$
$$x^2 = 6$$

and then take square roots to find

$$x = \pm\sqrt{6}$$

Don't forget that every positive number has *two* square roots. The symbol $\pm$ (read "plus or minus") is a shorthand notation used here to indicate both square roots of 6. The *exact* solutions are thus $\sqrt{6}$ and $-\sqrt{6}$. We can also find decimal *approximations* for the solutions using a calculator. To two decimal places, the solutions are 2.45 and -2.45.

In general, we can solve quadratic equations of the form

$$ax^2 + c = 0$$

where the linear term bx is missing, by isolating x^2 on one side of the equation, and then taking the square root of each side. This method for solving quadratic equations is called **extraction of roots.**

EXAMPLE 1

If a cat falls off a tree branch 20 feet above the ground, its height t seconds later is given by $h = 20 - 16t^2$.

   a. What is the height of the cat 0.5 second later?
   b. How long does the cat have to get in position to land on its feet before it reaches the ground?

Solutions

a. In this question, we are given the value of t and asked to find the corresponding value of h. To do this, we evaluate the formula for $t = 0.5$. We
substitute **0.5** for t into the formula, and simplify.

$$h = 20 - 16(\mathbf{0.5})^2 \quad \text{Compute the power.}$$
$$= 20 - 16(0.25) \quad \text{Multiply, then subtract.}$$
$$= 20 - 4 = 16$$

The cat is 16 feet above the ground after 0.5 second.

b. We would like to find the value of t
when the height, h, is known. We
substitute $h = 0$ into the equation to
obtain

$$\mathbf{0} = 20 - 16t^2$$

To solve this equation, we use extraction of roots. First "isolate" t^2
on one side of the equation.

$$16t^2 = 20 \qquad \text{Divide by 16.}$$

$$t^2 = \frac{20}{16} = 1.25$$

Now take the square root of both
sides of the equation to find

$$t = \pm\sqrt{1.25} \approx \pm 1.118$$

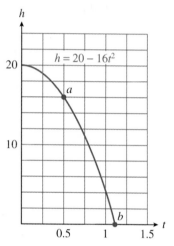

FIGURE 3.3

Only the positive solution makes sense here, so the cat has approximately 1.12 seconds to get into position for landing. A graph of the cat's
height after t seconds is shown in Figure 3.3. The points corresponding
to parts (a) and (b) are labeled. ∎

EXERCISE 1 Solve by extracting roots $\dfrac{3x^2 - 8}{4} = 10$

First, "isolate" x^2.
Take the square root of both sides.

Solving Formulas

Extraction of roots can be used to solve many formulas with quadratic terms.

EXAMPLE 2

The formula $V = \frac{1}{3}\pi r^2 h$ gives the volume of a cone in terms of its height and radius. Solve the formula for r in terms of V and h.

Solution

Because the variable we want is squared, we use extraction of roots. First, multiply both sides by 3 to clear the fraction.

$$3V = \pi r^2 h \quad \text{Divide both sides by } \pi h.$$

$$\frac{3V}{\pi h} = r^2 \quad \text{Take square roots.}$$

$$\pm\sqrt{\frac{3V}{\pi h}} = r$$

Because the radius of a cone must be a positive number, we use only the positive square root: $r = \sqrt{\dfrac{3V}{\pi h}}$ ∎

EXERCISE 2 Find a formula for the radius of a circle in terms of its area.

Start with the formula for the area of a circle. $\qquad A =$

Solve for r in terms of A.

More Extraction of Roots

Equations of the form

$$a(x - p)^2 = q$$

can also be solved by extraction of roots after isolating the squared expression, $(x - p)^2$.

EXAMPLE 3 Solve the equation $3(x - 2)^2 = 48$.

Solution

First, isolate the perfect square, $(x - 2)^2$.

$$3(x - 2)^2 = 48 \qquad \text{Divide both sides by 3.}$$

$$(x - 2)^2 = 16 \qquad \text{Take the square root of each side.}$$

$$x - 2 = \pm\sqrt{16} = \pm 4$$

This gives us two equations for x,

$$x - 2 = 4 \quad \text{or} \quad x - 2 = -4 \quad \text{Solve each equation.}$$

$$x = 6 \quad \text{or} \quad x = -2$$

The solutions are 6 and -2. ∎

EXERCISE 3 Solve by extracting roots:

$$2(5x + 3)^2 = 38$$

(handwritten: $5x + 3 = 19$)

a. Give your answers as exact values.

b. Find approximations for the solutions to two decimal places.

Compound Interest

Many savings institutions offer accounts on which the interest is *compounded annually*. At the end of each year the interest earned that year is added to the principal, and the interest for the next year is computed on this larger sum of money. After n years, the amount of money in the account is given by the formula

$$A = P(1 + r)^n$$

where P is the original principal and r is the interest rate, expressed as a decimal fraction.

EXAMPLE 4

Carmella invests $3000 in an account that pays an interest rate, r, compounded annually.

a. Write an expression for the amount of money in Carmella's account after 2 years.

b. What interest rate would be necessary for Carmella's account to grow to $3500 in 2 years?

Solutions

a. Use the formula above with $P = 3000$ and $n = 2$. Carmella's account balance will be

$$A = 3000(1 + r)^2$$

b. Substitute 3500 for A in the equation.

$$3500 = 3000(1 + r)^2$$

This is a quadratic equation in the variable r, which we can solve by extraction of roots. First, isolate the perfect square.

$$3500 = 3000(1 + r)^2 \qquad \text{Divide both sides by 3000}$$
$$1.1\overline{6} = (1 + r)^2 \qquad \text{Take the square root of both sides.}$$
$$\pm 1.0801 \approx 1 + r \qquad \text{Subtract 1 from both sides.}$$
$$r \approx 0.0801 \qquad \text{or} \qquad r \approx -2.0801$$

Because the interest rate must be a positive number, we discard the negative solution. Carmella needs an account with interest rate $r \approx 0.0801$, or over 8%, to achieve an account balance of $3500 in 2 years. ■

The same formula that we used for compound interest also applies to calculating the effects of inflation. For instance, if there is a steady inflation rate of 4% per year, an item that now costs $100 will cost

$$A = P(1 + r)^2$$
$$= 100(1 + 0.04)^2 = \$108.16$$

2 years from now.

EXERCISE 4 The average cost of dinner and a movie two years ago was $24. This year the average cost is $25.44. What was the rate of inflation over the past two years?

READING QUESTIONS

1. Under what conditions would you use extraction of roots to solve a quadratic equation?
2. How many square roots does a positive number have?
3. What is the first step in solving the equation $a(x - p)^2 = q$ by extraction of roots?
4. State the formula for the amount of an account on which interest is compounded annually.

ANSWERS TO 3.1 EXERCISES

1. $x = \pm 4$ **2.** $r = \sqrt{\dfrac{A}{\pi}}$ **3a.** $x = \dfrac{-3 \pm \sqrt{19}}{5}$

3b. $x \approx -1.47$ or $x \approx 0.27$ **4.** $r \approx 2.96\%$

HOMEWORK 3.1

Solve by extracting roots. Give exact values for your answers.

1. $9x^2 = 25$ **2.** $4x^2 = 9$ **3.** $4x^2 - 24 = 0$ **4.** $3x^2 - 9 = 0$

5. $\dfrac{2x^2}{3} = 4$ **6.** $\dfrac{3x^2}{5} = 6$

Solve by extracting roots. Round your answers to two decimal places.

7. $2x^2 = 14$ **8.** $3x^2 = 15$ **9.** $1.5x^2 = 0.7x^2 + 26.2$ **10.** $0.4x^2 = 2x^2 - 8.6$

11. $5x^2 - 97 = 3.2x^2 - 38$ **12.** $17 - \dfrac{x^2}{4} = 43 - x^2$

Solve the formulas for the specified variable.

13. $F = \dfrac{mv^2}{r}$, for v **14.** $A = \dfrac{\sqrt{3}}{4}s^2$, for s **15.** $s = \dfrac{1}{2}gt^2$, for t **16.** $S = 4\pi r^2$, for r

For Problems 17 and 18, refer to the geometric formulas at the front of the book.

17. A conical coffee filter is 8.4 centimeters tall.

   a. Write a formula for the filter's volume in terms of its widest radius (at the top of the filter).

   b. Complete the table of values for the volume equation. If you double the radius of the filter, by what factor does the volume increase?

r	1	2	3	4	5	6	7	8
V								

c. If the volume of the filter is 302.4 cubic centimeters, what is its radius?

d. Use your calculator to graph the volume equation. Locate the point on the graph that corresponds to the filter in part (c).

18. A large bottle of shampoo is 20 centimeters tall and cylindrical in shape.

a. Write a formula for the volume of the bottle in terms of its radius.

b. Complete the table of values for the volume equation. If you halve the radius of the bottle, by what factor does the volume decrease?

r	1	2	3	4	5	6	7	8
V								

c. What radius should the bottle have if it must hold 240 milliliters of shampoo? (A milliliter is equal to 1 cubic centimeter.)

d. Use your calculator to graph the volume equation. Locate the point on the graph that corresponds to the bottle in part (c).

a. **Make a sketch of the situation described, and label a right triangle.**

b. **Use the Pythagorean theorem to solve each problem. (See Appendix A.3 to review the Pythagorean theorem.)**

19. The size of a TV screen is the length of its diagonal. If the width of a 35-inch TV screen is 28 inches, what is its height? (See Figure 3.4.)

FIGURE 3.4

20. How high on a building will a 25-foot ladder reach if its foot is 15 feet away from the base of the wall? (See Figure 3.5.)

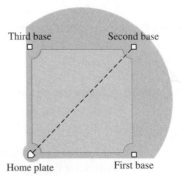

FIGURE 3.5

21. If a 30-meter pine tree casts a shadow of 30 meters, how far is the tip of the shadow from the top of the tree? (See Figure 3.6.)

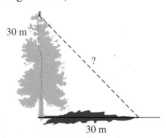

FIGURE 3.6

22. A baseball diamond is a square whose sides are 90 feet in length. Find the straight-line distance from home plate to second base. (See Figure 3.7.)

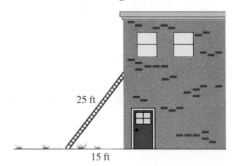

FIGURE 3.7

23. What size square can be inscribed in a circle of radius 8 inches? (See Figure 3.8.)

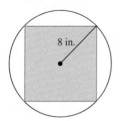

8 in.

FIGURE 3.8

24. What size rectangle can be inscribed in a circle of radius 30 feet if the length of the rectangle must be three times its width? (See Figure 3.9.)

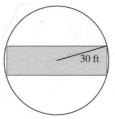

30 ft

$A = \pi r^2$
$A = l \times w$
$3L + W = A$

FIGURE 3.9

In Problems 25–30, (a) use a calculator to graph the quadratic equation in the suggested window; (b) use your graph to find two solutions for the given equation. (If you would like to review graphical solution of equations, see Section 1.2.); and (c) check your solutions algebraically, using mental arithmetic.

25. a. $y = \dfrac{1}{4}x^2$

Xmin = −15   Ymin = −10
Xmax = 15   Ymax = 50

    b. $\dfrac{1}{4}x^2 = 36$

26. a. $y = 8x^2$

Xmin = −15   Ymin = −50
Xmax = 15   Ymax = 450

    b. $8x^2 = 392$

27. a. $y = (x − 5)^2$
Xmin = −5   Ymin = −5
Xmax = 15   Ymax = 30

    b. $(x − 5)^2 = 16$

28. a. $y = (x + 2)^2$
Xmin = −10   Ymin = −2
Xmax = 10   Ymax = 12

    b. $(x + 2)^2 = 9$

29. a. $y = 3(x − 4)^2$

Xmin = −5   Ymin = −20
Xmax = 15   Ymax = 150

    b. $3(x − 4)^2 = 108$

30. a. $y = \dfrac{1}{2}(x + 3)^2$

Xmin = −15   Ymin = −5
Xmax = 5   Ymax = 15

    b. $\dfrac{1}{2}(x + 3)^2 = 8$

Solve by extraction of roots.

31. $(x − 2)^2 = 9$ **32.** $(x + 3)^2 = 4$ **33.** $(2x − 1)^2 = 16$ **34.** $(3x + 1)^2 = 25$

35. $4(x + 2)^2 = 12$ **36.** $6(x − 5)^2 = 42$ **37.** $\left(x − \dfrac{1}{2}\right)^2 = \dfrac{3}{4}$ **38.** $\left(x − \dfrac{2}{3}\right)^2 = \dfrac{5}{9}$

39. $81\left(x + \dfrac{1}{3}\right)^2 = 1$ **40.** $16\left(x + \dfrac{1}{2}\right)^2 = 1$ **41.** $3(8x − 7)^2 = 24$ **42.** $2(5x − 12)^2 = 48$

43. Cyril plans to invest $5000 in a money market account paying interest compounded annually.

    a. Write a formula in terms of the interest rate, r, for the balance, B, in Cyril's account after two years.

    b. Complete the table showing Cyril's account balance after two years for the given interest rates.

r	0.02	0.04	0.06	0.08
B				

    c. If Cyril would like to have $6250 in two years, what interest rate must the account pay?

    d. Use your calculator to graph the formula for Cyril's account balance. Locate the point on the graph that corresponds to the amount in part (c).

44. You plan to deposit your savings of $1600 in an account that compounds interest annually.

   a. Write a formula in terms of the interest rate, r, for the amount in your savings account after two years.

   b. Complete the table showing your account balance after 2 years for the given interest rates.

r	0.02	0.04	0.06	0.08
A				

   c. To the nearest tenth of a percent, what interest rate will you require if you want your $1600 to grow to $2000 in two years?

   d. Use your calculator to graph the formula for the account balance. Locate the point on the graph that corresponds to the amount in part (c).

45. Two years ago Carol's living expenses were $1200 per month. This year the same items cost Carol $1400 per month. What was the annual inflation rate for the past two years?

46. Two years ago the average price of a house in the suburbs was $88,600. This year the average price is $93,996. What was the annual percent increase in the cost of a house?

47. A machinist wants to make a metal section of pipe that is 80 millimeters long and has an interior volume of 9000 cubic millimeters. If the pipe is 2 millimeters thick, its interior volume is given by the formula

$$V = \pi(r - 2)^2 h$$

where h is the length of the pipe and r is its radius. What should the radius of the pipe be?

48. A storage box for sweaters is constructed from a square sheet of corrugated cardboard measuring x inches on a side. The volume of the box, in cubic inches, is

$$V = 10(x - 20)^2$$

If the box should have a volume of 1960 cubic inches, what size cardboard square is needed?

49. The area of an equilateral triangle is given by the formula $A = \dfrac{\sqrt{3}}{4}s^2$, where s is the length of the side.

   a. Find the areas of equilateral triangles with sides of length 2 centimeters, 4 centimeters, and 10 centimeters.

   b. Graph the area equation in the window

       Xmin = 0   Xmax = 14.1
       Ymin = 0   Ymax = 60

   (Make sure the equation is entered correctly in your calculator.) Use the ⟨TRACE⟩ or ⟨VALUE⟩ to verify your answers to part (a).

   c. **Trace** along the curve to the point (5.1, 11.26266). What do the coordinates of this point represent?

   d. Use your graph to estimate the side of an equilateral triangle whose area is 20 square centimeters.

   e. Write and solve an equation to answer part (d).

   f. If the area of an equilateral triangle is $100\sqrt{3}$ square centimeters, what is the length of its side?

50. The area of the ring in Figure 3.10 is given by the formula $A = \pi R^2 - \pi r^2$, where R is the radius of the outer circle and r is the radius of the inner circle.

   a. Suppose the inner radius of the ring is kept fixed at $r = 4$ centimeters, but the radius of the outer ring, R, is allowed to vary. Find the area of the ring when the outer radius is 6 centimeters, 8 centimeters, and 12 centimeters.

   b. Graph the area equation, with $r = 4$, in the window

       Xmin = 0   Xmax = 14.1
       Ymin = 0,   Ymax = 400

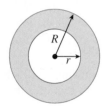

FIGURE 3.10

Use the [TRACE] to verify your answers to part (a).

c. **Trace** along the curve to the point (9.75, 248.38217). What do the coordinates of this point represent?

d. Use your graph to estimate the outer radius of the ring when its area is 100 square centimeters.

e. Write and solve an equation to answer part (d).

f. If the area of the ring is 11π square centimeters, what is the radius of the outer circle?

Solve for x in terms of a, b, and c.

51. $\dfrac{ax^2}{b} = c$

52. $\dfrac{bx^2}{c} - a = 0$

53. $(x - a)^2 = 16$

54. $(x + a)^2 = 36$

55. $(ax + b)^2 = 9$

56. $(ax - b)^2 = 25$

57. The jump height, J, in meters, achieved by a pole vaulter is given approximately by $J = \dfrac{v^2}{2g}$, where v is the vaulter's speed in meters per second at the end of his run, and $g = 9.8$ is the gravitational acceleration.

a. Fill in the table of values for jump heights achieved with values of v from 0 to 11 meters per second.

v	0	1	2	3	4	5	6	7	8	9	10	11
J												
H												

b. Graph the jump height versus final speed. (Use the table values to help you choose a window for the graph.)

c. The jump height should be added to the height of the vaulter's center of gravity (at about hip level) to give the maximum height, H, he can clear. For a typical pole vaulter, his center of gravity at the end of the run is 0.9 meters from the ground (slightly lower than when standing erect). Complete the table in part (a) with values of H, and graph H on your graph of J.

d. A good pole vaulter can reach a final speed of 9.5 meters per second. What height will he clear?

e. The world record in pole vaulting is 6.15 meters. What was the vaulter's speed at the end of his run?

Source: Alexander, 1992.

58. To be launched into space, a satellite must travel fast enough to escape from earth's gravity. This escape velocity, v, satisfies the equation

$$\frac{1}{2}mv^2 = \frac{GMm}{R}$$

where m is the mass of the satellite, M is the mass of the earth, R is the radius of the earth, and G is the universal gravitational constant.

a. Solve the equation for v in terms of the other constants.

b. The equation

$$mg = \frac{GMm}{R^2}$$

gives the force of gravity at the earth's surface. We can use this equation to simplify the expression for v as follows: First, multiply both sides of the equation by $\dfrac{R}{m}$. You now have an expression for $\dfrac{GM}{R}$. Substitute this new expression into your formula for v.

c. The radius of the earth is about 6400 km, and $g = 0.0098$. Calculate the escape velocity from earth in kilometers per second. Convert your answer to miles per hour. (One kilometer is 0.621 miles.)

d. The radius of the moon is 1740 km, and the value of g at the moon's surface is 0.0016. Calculate the escape velocity from the moon in kilometers per second, and convert to miles per hour.

3.2 Some Examples of Quadratic Models

Not all quadratic equations can be solved by extraction of roots. Here are some problems that require new algebraic techniques.

INVESTIGATION 5

Perimeter and Area

Do all rectangles with the same perimeter, say 36 inches, have the same area? Two different rectangles with perimeter 36 inches are shown in Figure 3.11. The first rectangle has base 10 inches and height 8 inches, and its area is 80 square inches. The second rectangle has base 12 inches and height 6 inches. Its area is 72 square inches.

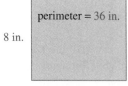

FIGURE 3.11

1. Table 3.2 shows the bases of various rectangles, in inches. Each rectangle has a perimeter of 36 inches. Fill in the height and the area of each rectangle. (To find the height of the rectangle, reason as follows: The base plus the height makes up half of the rectangle's perimeter.)

2. What happens to the area of the rectangle when we change its base? On the grid in Figure 3.12, plot the points with coordinates *(Base, Area)*. (For this graph we will not use the heights of the rectangles.) The first two points, (10, 80) and (12, 72), are shown. Connect your data points with a smooth curve.

3. What are the coordinates of the highest point on your graph?

4. Each point on your graph represents a particular rectangle with perimeter 36 inches. The first coordinate of the point gives the base of the rectangle, and the second coordinate gives the area of the rectangle. What is the largest area you found among rectangles with perimeter 36 inches? What is the base for that rectangle? What is its height?

5. Describe the rectangle that corresponds to the point (13, 65).

6. Find two points on your graph with vertical coordinate 80.

7. If the rectangle has area 80 square inches, what is its base? Why are there 2 different answers here? Describe the rectangle corresponding to each answer.

TABLE 3.2

Base	Height	Area
10	8	80
12	6	72
3		
14		
5		
17		
9		
2		
11		
4		
16		
15		
1		
6		
8		
13		
7		

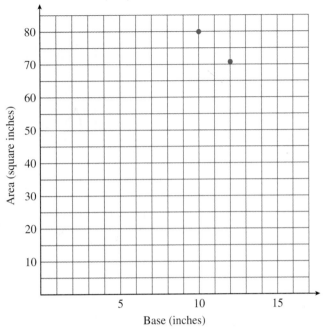

FIGURE 3.12

8. Now we'll write an algebraic expression for the area of the rectangle in terms of its base. Let x represent the base of the rectangle. First, express the height of the rectangle in terms of x. (*Hint:* If the perimeter of the rectangle is 36 inches, what is the sum of the base and the height?) Now write an expression for the area of the rectangle in terms of x.

9. Use your formula from part (8) to compute the area of the rectangle when the base is 5 inches. Does your answer agree with the values in your table and the point on your graph?

10. Use your formula to compute the area of the rectangle when $x = 0$ and when $x = 18$. Describe the "rectangles" that correspond to these data points.

11. Continue your graph to include the points corresponding to $x = 0$ and $x = 18$.

 **INVESTIGATION 6**

Height of a Baseball

Suppose a baseball player "pops up," that is, hits the baseball straight up into the air. The height, h, of the baseball t seconds after it leaves the bat can be calculated using a formula from physics. This formula takes into account the initial speed of the ball (64 feet per second), and its height when it was hit (4 feet). The formula for the height of the ball (in feet) is

$$h = -16t^2 + 64t + 4$$

1. Evaluate the formula to complete the table of values for the height of the baseball.

t	0	1	2	3	4
h					

2. On the grid in Figure 3.13, graph the height of the baseball versus time. Plot data points from your table, and connect the points with a smooth curve.

3. What are the coordinates of the highest point on the graph? When does the baseball reach its maximum height, and what is that height?

4. Use the formula to find the height of the baseball after 1/2 second.

5. Check that your answer to part (4) corresponds to a point on your graph. Approximate from your graph another time at which the baseball is at the same height as your answer to part (4).

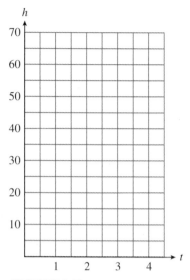

FIGURE 3.13

6. Use your graph to find two times when the baseball is at a height of 64 feet.

7. Use your graph to approximate two times when the baseball is at a height of 20 feet. Then use the formula to find the actual heights at those times.

8. Suppose the catcher catches the baseball at a height of 4 feet, before it strikes the ground. At what time was the ball caught?

Using the Table Feature on a Graphing Calculator

We can use a graphing calculator to make a table of values for an equation in two variables. Consider again the quadratic model from Investigation 6:

$$h = -16t^2 + 64t + 4$$

Begin by entering the equation: Press the [Y =] key, then define $Y_1 = -16X^2 + 64X + 4$, and clear out any other definitions.

Next we choose the x-values for the table. Press [2nd] [WINDOW] to access the **"TblSet"** (Table Set-up) menu and set it to look like Figure 3.14a. This setting will give us an initial x-value of 0 (**TblStart = 0**), and an increment of 1 unit in the x-values, (**ΔTbl = 1**). It also fills in values of both variables automatically. Now press [2nd] [GRAPH] to see the table of values, as shown in Figure 3.14b. From this table we can check the heights we found in part (1) of Investigation 6.

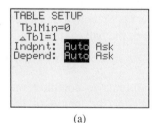

(a)   (b)

FIGURE 3.14

9. Use your calculator to make a table of values for the equation

$$h = -16t^2 + 64t + 4$$

with **TblStart = 0** and **ΔTbl = 0.5**. Use the table to check your answers to parts (4) and (6) of Investigation 6. Use the △ and ▽ keys to "scroll" up and down the table.

10. Use your calculator to graph the equation for the height of the ball, with WINDOW settings to match Figure 3.13:

$$\text{Xmin} = 0 \qquad \text{Xmax} = 4.5 \qquad \text{Xscl} = 0.5$$
$$\text{Ymin} = 0 \qquad \text{Ymax} = 70 \qquad \text{Yscl} = 5$$

11. We can use the **intersect** command to answer part (7): Estimate two times when the baseball is at a height of 20 feet. Press Y = and enter $Y_2 = 20$, then press GRAPH. You should see the horizontal line $y = 20$ superimposed on your graph of the height equation. Move the **Trace** bug close to one of the intersection points of the two graphs and activate the **intersect** feature. (If you would like to review the **intersect** command, see Section 2.1.) Does the result agree with your answer to part (7)? Now move the **Trace** bug to the second intersection point, and find its coordinates using **intersect**.

12. Use the **intersect** command to verify your answer to part (8): At what time was the ball caught, if it was caught at a height of 4 feet?

To help you analyze quadratic models, you will need to find products of monomials and binomials, and to factor certain quadratic expressions. Homework 3.2 includes some review problems on multiplying and factoring polynomials. If you would like to review these topics in more detail, see Appendix A.6.

READING QUESTIONS

1. Give an example of two rectangles with the same perimeter but different areas.

2. The perimeter of a rectangle is 50 meters. Write an expression for the length of the rectangle in terms of its width.

3. The height of a ball at time t is given by

$$h = -16t^2 - 6t + 32$$

What was the height of the ball at time $t = 0$?

4. Factor:
   a. $x^2 - 4$ b. $x^2 - 4x$
   c. $x^2 - 4x + 4$ d. $x^2 + 4$

1. Suppose you have 36 feet of rope and you want to enclose a rectangular display area against 1 wall of an exhibit hall, as shown in Figure 3.15. The area enclosed depends on the dimensions of the rectangle you make. Because the wall makes 1 side of the rectangle, the length of the rope accounts for only 3 sides. Thus

$$Base + 2 \, (Height) = 36$$

FIGURE 3.15

a. Complete the table showing the base and the area of the rectangle for the given heights.

Height	Base	Area	Height	Base	Area
1	34	34	10		
2	32	64	11		
3			12		
4			13		
5			14		
6			15		
7			16		
8			17		
9			18		

b. Make a graph with "Height" on the horizontal axis and "Area" on the vertical axis. Draw a smooth curve through your data points. (Use your table to help you decide on appropriate scales for the axes.)

c. What is the area of the largest rectangle you can enclose in this way? What are its dimensions? On your graph, label the point that corresponds to this rectangle with the letter M.

d. Let x stand for the height of a rectangle, and write algebraic expressions for the base and the area of the rectangle.

e. Enter your algebraic expression for the area in your calculator, and use the **Table** feature to verify the entries in your table in part (a).

f. Graph your formula for area on your graphing calculator. Use your table of values and your hand-drawn graph to help you choose appropriate WINDOW settings.

g. Use the **intersect** command to find the height of the rectangle whose area is 149.5 square feet.

2. We are going to make an open box from a square piece of cardboard by cutting 3-inch squares from each corner, and then turning up the edges, as shown in Figure 3.16.

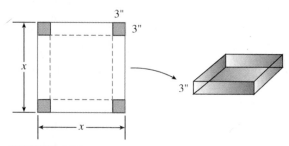

FIGURE 3.16

a. Complete the table showing the side of the original sheet of cardboard, the dimensions of the box created from it, and the volume of the box.

Side	Length of box	Width of box	Height of box	Volume of box
7	1	1	3	3
8	2	2	3	12
9				
10				
11				
12				
13				
14				
15				

Explain why the side of the cardboard square cannot be smaller than 6 inches. What happens if the cardboard is exactly 6 inches on a side?

b. Make a graph with "Side" on the horizontal axis and "Volume" on the vertical axis. Draw a smooth curve through your data points. (Use your table to help you decide on appropriate scales for the axes.)

c. Let x represent the side of the original sheet of cardboard. Write algebraic expressions for the dimensions of the box and for its volume.

d. Enter your expression for the volume of the box in your calculator, and use the **Table** feature to verify the values in your table in part (a).

e. Graph your formula for volume on your graphing calculator. Use your table of values and your hand-drawn graph to help you choose appropriate WINDOW settings.

f. Use the **intersect** command to find out how large a square of cardboard you need to make a box with volume 126.75 cubic inches.

g. Does your graph have a highest point? What happens to the volume of the box as you increase x?

3. Delbert stands at the top of a 300-foot cliff and throws his algebra book directly upward with a velocity of 20 feet per second. The height of his book above the ground t seconds later is given by the formula

$$h = -16t^2 + 20t + 300$$

where h is in feet.

a. Use your calculator to make a table of values for the height formula, with increments of 0.5 second.

b. Graph the height formula on your calculator. Use your table of values to help you choose appropriate WINDOW settings.

c. What is the highest altitude Delbert's book reaches? When does it reach that height? Use the TRACE feature to find approximate answers first. Then use the **Table** feature to improve your estimate.

d. When does Delbert's book pass him on its way down? (Delbert is standing at a height of 300 feet.) Use the **intersect** command.

e. How long will it take Delbert's book to hit the ground at the bottom of the cliff?

4. James Bond stands on top of a 240-foot building and throws a film canister upward to a fellow agent in a helicopter 16 feet above the building. The height of the film above the ground t seconds later is given by the formula

$$h = -16t^2 + 32t + 240$$

where h is in feet.

a. Use your calculator to make a table of values for the height formula, with increments of 0.5 second.

b. Graph the height formula on your calculator. Use your table of values to help you choose appropriate WINDOW settings.

c. How long will it take the film canister to reach the agent in the helicopter? (What is the agent's altitude?) Use the TRACE feature to find approximate answers first. Then use the **Table** feature to improve your estimate.

d. If the agent misses the canister, when will it pass James Bond on the way down? Use the **intersect** command.

e. How long will it take to hit the ground?

Write each product as a polynomial in simplest form. (If you would like to review the skills in Problems 5–34, consult Appendix A.6, Review of Products and Factoring.)

5. $3x(x - 5)$

6. $-5a(2a + 3)$

7. $(b + 6)(2b - 3)$

8. $(3z - 8)(4z - 1)$

9. $(4w - 3)^2$

10. $(2d + 8)^2$

11. $3p(2p - 5)(p - 3)$

12. $2v(v + 4)(3v - 4)$

13. $-50(1 + r)^2$

14. $12(1 - t)^2$

15. $3q^2(2q - 3)^2$

16. $-5m^2(3m + 4)^2$

Factor completely.

17. $x^2 - 7x + 10$ **18.** $x^2 - 7x + 12$ **19.** $x^2 - 225$ **20.** $x^2 - 121$

21. $w^2 - 4w - 32$ **22.** $w^2 + 5w - 150$ **23.** $2z^2 + 11z - 40$ **24.** $5z^2 - 28z - 12$

25. $9n^2 + 24n + 16$ **26.** $4n^2 - 28n + 49$ **27.** $3a^4 + 6a^3 + 3a^2$ **28.** $2a^3 - 12a^2 + 18a$

29. $4h^4 - 36h^2$ **30.** $80h - 5h^3$ **31.** $-10u^2 - 100u + 390$

32. $270 - 15u + 5u^2$ **33.** $24t^4 + 6t^2$ **34.** $27t^3 + 75t$

3.3 Solving Quadratic Equations by Factoring

In Investigation 6 the height of a baseball t seconds after being hit was given by

$$h = -16t^2 + 64t + 4 \tag{1}$$

We used a graph to find two times when the baseball was 64 feet high. Can we solve the same problem algebraically?

We are looking for values of t that produce $h = 64$ in the height equation. So, if we substitute $h = 64$ into Equation (1), we would like to solve the quadratic equation

$$64 = -16t^2 + 64t + 4$$

This equation cannot be solved by extraction of roots. There are two terms containing the variable t, and they cannot be combined. To solve this equation, we will appeal to a property of our number system.

Zero-Factor Principle

Can you multiply two numbers together and obtain a product of zero? Only if one of the two numbers happens to be zero. This property of numbers is called the **zero-factor principle.**

> ### Zero-Factor Principle
>
> The product of two factors equals zero if and only if one or both of the factors equals zero. In symbols,
>
> $$ab = 0 \qquad \text{if and only if} \qquad a = 0 \quad \text{or} \quad b = 0$$

The principle is true even if the numbers a and b are represented by algebraic expressions, such as $x - 5$ or $2x + 1$. For example, if

$$(x - 5)(2x + 1) = 0$$

then it must be true that either $x - 5 = 0$ or $2x + 1 = 0$. Thus, we can use the zero-factor principle to solve equations.

EXAMPLE 1

a. Solve the equation $(x - 6)(x + 2) = 0$.

b. Find the x-intercepts of the graph of $y = x^2 - 4x - 12$.

Solutions

a. Apply the zero-factor principle to the product $(x - 6)(x + 2)$.

$$(x - 6)(x + 2) = 0 \qquad \text{Set each factor equal to zero.}$$
$$x - 6 = 0 \quad \text{or} \quad x + 2 = 0 \qquad \text{Solve each equation.}$$
$$x = 6 \quad \text{or} \quad x = -2$$

There are two solutions, 6 and -2. You should check that both of these values satisfy the original equation.

b. To find the x-intercepts of the graph, we set $y = 0$, and solve the equation

$$0 = x^2 - 4x - 12$$

But this is the equation we solved in part (a), because

$$(x - 6)(x + 2) = x^2 - 4x - 12$$

The solutions of that equation were 6 and -2, so the x-intercepts of the graph are 6 and -2. You can see this by graphing the equation on your calculator, as shown in Figure 3.17.

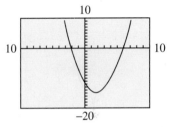

FIGURE 3.17 ∎

EXERCISE 1 Graph the equation

$$y = (x - 3)(2x + 3)$$

on a calculator, and use your graph to solve the equation $y = 0$. (Use Xmin $= -9.4$, Xmax $= 9.4$.) Check your answer with the zero-factor principle.

Solving Quadratic Equations by Factoring

Before we apply the zero-factor principle, we must first write the quadratic equation so that one side of the equation is zero. We then factor the other side and set each factor equal to zero separately. (If you need to refresh your factoring skills, see Appendix A.6.)

EXAMPLE 2

Solve $\quad 3x(x + 1) = 2x + 2$

Solution

First, write the equation in standard form.

$$3x(x + 1) = 2x + 2 \qquad \text{Apply the distributive law to the left side.}$$
$$3x^2 + 3x = 2x + 2 \qquad \text{Subtract } 2x + 2 \text{ from both sides.}$$
$$3x^2 + x - 2 = 0$$

Next, factor the left side to obtain

$$(3x - 2)(x + 1) = 0$$

Apply the zero-factor principle: Set each factor equal to zero.

$$3x - 2 = 0 \qquad \text{or} \qquad x + 1 = 0$$

Finally, solve each equation to find

$$x = \frac{2}{3} \qquad \text{or} \qquad x = -1$$

The solutions are $\frac{2}{3}$ and -1. ∎

COMMON ERROR

When we apply the zero-factor principle, one side of the equation *must* be zero. For example, to solve the equation

$$(x - 2)(x - 4) = 15$$

it is *incorrect* to set each factor equal to 15! (There are many ways that the product of two numbers can equal 15; it is not necessary that one of the numbers be 15.) We must first simplify the left side and write the equation in standard form. (The correct solutions are 7 and -1; make sure you can find these solutions.) ∎

> **EXERCISE 2**   Solve by factoring $(t - 3)^2 = 3(9 - t)$
>
> Multiply out each side of the equation.
> Obtain zero on one side of the equation.
> Factor the other side.
> Apply the zero-factor principle.

We can use factoring to solve the equation from Investigation 6.

EXAMPLE 3

The height, h, of a baseball t seconds after being hit is given by

$$h = -16t^2 + 64t + 4$$

When will the baseball reach a height of 64 feet?

Solution
Substitute 64 for h in the formula, and solve for t.

$64 = -16t^2 + 64t + 4$	Write the equation in standard form.
$16t^2 - 64t + 60 = 0$	Factor 4 from the left side.
$4(4t^2 - 16t + 15) = 0$	Factor the quadratic expression.
$4(2t - 3)(2t - 5) = 0$	Set each variable factor equal to zero.
$2t - 3 = 0 \qquad \text{or} \qquad 2t - 5 = 0$	Solve each equation.
$t = \dfrac{3}{2} \qquad \text{or} \qquad t = \dfrac{5}{2}$	

There are two solutions to the quadratic equation. At $t = \frac{3}{2}$ seconds, the ball reaches a height of 64 feet on the way up, and at $t = \frac{5}{2}$ seconds, the ball is 64 feet high on its way down. (See Figure 3.18.)

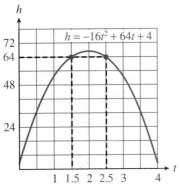

FIGURE 3.18

In the solution to Example 3, the factor 4 does not affect the solutions of the equation at all. You can understand why this is true by looking at some graphs. First, check that the two equations

$$x^2 - 4x + 3 = 0$$
$$4(x^2 - 4x + 3) = 0$$

have the same solutions, $x = 1$ and $x = 3$. Then use your graphing calculator to graph the equation

$$Y_1 = X^2 - 4X + 3$$

in the window

$$\text{Xmin} = -2 \quad \text{Xmax} = 8$$
$$\text{Ymin} = -5 \quad \text{Ymax} = 10$$

Notice that when $y = 0$, $x = 3$ or $x = 1$. These two points are the x-intercepts of the graph. Now in the same window graph

$$Y_2 = 4(X^2 - 4X + 3)$$

(See Figure 3.19.) This graph has the same x-values when $y = 0$. The factor of 4 makes the graph "skinnier," but does not change the location of the x-intercepts.

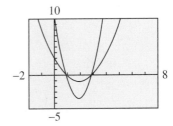

FIGURE 3.19

Applications

Here is another example of how quadratic equations arise in applications.

EXAMPLE 4

The size of a rectangular computer monitor screen is given by the length of its diagonal. (See Figure 3.20.) If the length of the screen should be 3 inches greater than its width, what are the dimensions of a 15-inch monitor?

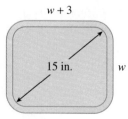

$w + 3$

15 in.

w

FIGURE 3.20

Solution

Express the two dimensions of the screen in terms of a single variable:

Width of screen: w

Length of screen: $w + 3$

Use the Pythagorean theorem to write an equation:

$$w^2 + (w + 3)^2 = 15^2$$

(See Appendix A.3 if you would like to review the Pythagorean theorem.) Solve the equation. Begin by simplifying the left side.

$$w^2 + w^2 + 6w + 9 = 225 \qquad \text{Write the equation in standard form.}$$
$$2w^2 + 6w - 216 = 0 \qquad \text{Factor 2 from the left side.}$$
$$2(w^2 + 3w - 108) = 0 \qquad \text{Factor the quadratic expression.}$$
$$2(w - 9)(w + 12) = 0 \qquad \text{Set each factor equal to zero.}$$
$$w - 9 = 0 \quad \text{or} \quad w + 12 = 0 \qquad \text{Solve each equation.}$$
$$w = 9 \quad \text{or} \quad w = -12$$

Because the width of the screen cannot be a negative number, the width is 9 inches and the length is 12 inches. ∎

EXERCISE 3 Francine is designing the layout for a botanical garden. The plan includes a square herb garden, with a path five feet wide through the center of the garden. To include all the species of herbs, the planted area must be 300 square feet. Find the dimensions of the herb garden.

Solutions of Quadratic Equations

A quadratic equation in one variable always has two solutions. However, in some cases, the solutions may be equal. For example, the equation

$$x^2 - 2x + 1 = 0$$

can be solved by factoring as follows:

$$(x - 1)(x - 1) = 0 \qquad \text{Apply the zero-factor principle.}$$
$$x - 1 = 0 \quad \text{or} \quad x - 1 = 0$$

Both of these equations have solution 1. We say that 1 is a solution *of multiplicity two,* meaning that it occurs twice as a solution of the quadratic equation. Notice that the solutions of the quadratic equation

$$(x - r_1)(x - r_2) = 0 \tag{2}$$

are r_1 and r_2. If we know the two solutions of a quadratic equation we can work backward and reconstruct the equation, starting from its factored form, as in

Equation (2). We can then write the equation in standard form by multiplying together the two factors.

EXAMPLE 5

Find a quadratic equation whose solutions are $\frac{1}{2}$ and -3.

Solution
The quadratic equation is

$$\left(x - \frac{1}{2}\right)[x - (-3)] = 0$$

or

$$\left(x - \frac{1}{2}\right)(x + 3) = 0$$

To write the equation in standard form, multiply the factors together.

$$x^2 + \frac{5}{2}x - \frac{3}{2} = 0$$

We can also find an equation with integer coefficients if we clear the equation of fractions. Multiply both sides by 2.

$$2\left(x^2 + \frac{5}{2}x - \frac{3}{2}\right) = 2(0)$$
$$2x^2 + 5x - 3 = 0$$

Check that the solutions of this last equation are in fact $\frac{1}{2}$ and -3. Multiplying by a constant factor does not change an equation's solutions. ∎

EXERCISE 4 Find a quadratic equation with integer coefficients whose solutions are $\frac{2}{3}$ and -5.

READING QUESTIONS

1. What is a quadratic equation?
2. What is the zero-factor principle?
3. Explain how to solve a quadratic equation by factoring.
4. How many solutions does a quadratic equation have?
5. If you know the solutions of $ax^2 + bx + c = 0$, how can you find the solutions of $k(ax^2 + bx + c) = 0$?

ANSWERS TO 3.3 EXERCISES

1. $x = -\frac{3}{2}, x = 3$ 2. $x = -3, x = 6$

3. 20 feet by 20 feet   4. $3x^2 + 13x - 10 = 0$

In Problems 1–8 graph each equation on a graphing calculator, and use your graph to solve the equation $y = 0$. (Use Xmin $= -9.4$, Xmax $= 9.4$.) Check your answers with the zero-factor principle.

1. $y = (2x + 5)(x - 2)$ **2.** $y = (x + 1)(4x - 1)$ **3.** $y = x(3x + 10)$ **4.** $y = x(3x - 7)$

5. $y = (4x + 3)(x + 8)$ **6.** $y = (x - 2)(x - 9)$ **7.** $y = (x - 4)^2$ **8.** $y = (x + 6)^2$

Solve the equations in Problems 9–22 by factoring.

9. $2a^2 + 5a - 3 = 0$ **10.** $3b^2 - 4b - 4 = 0$ **11.** $2x^2 = 6x$

12. $5z^2 = 5z$ **13.** $3y^2 - 6y = -3$ **14.** $4y^2 + 4y = 8$

15. $x(2x - 3) = -1$ **16.** $2x(x - 2) = x + 3$ **17.** $t(t - 3) = 2(t - 3)$

18. $5(t + 2) = t(t + 2)$ **19.** $z(3z + 2) = (z + 2)^2$ **20.** $(z - 1)^2 = 2z^2 + 3z - 5$

21. $(v + 2)(v - 5) = 8$ **22.** $(w + 1)(2w - 3) = 3$

Use a graphing calculator to graph each set of equations in Problems 23–26. Use the standard graphing window. What do you notice about the x-intercepts? Try to generalize your observation, and test your idea with a few examples.

23. **a.** $y = x^2 - x - 20$

   b. $y = 2(x^2 - x - 20)$

   c. $y = 0.5(x^2 - x - 20)$

24. **a.** $y = x^2 + 2x - 15$

   b. $y = 3(x^2 + 2x - 15)$

   c. $y = 0.2(x^2 + 2x - 15)$

25. **a.** $y = x^2 + 6x - 16$

   b. $y = -2(x^2 + 6x - 16)$

   c. $y = -0.1(x^2 + 6x - 16)$

26. **a.** $y = x^2 - 16$

   b. $y = -1.5(x^2 - 16)$

   c. $y = -0.4(x^2 - 16)$

In Problems 27–34 write a quadratic equation whose solutions are given. The equation should be in standard form with integer coefficients.

27. -2 and 1 **28.** -4 and 3 **29.** 0 and -5 **30.** 0 and 5

31. -3 and $\dfrac{1}{2}$ **32.** $\dfrac{-2}{3}$ and 4 **33.** $\dfrac{-1}{4}$ and $\dfrac{3}{2}$ **34.** $\dfrac{-1}{3}$ and $\dfrac{-1}{2}$

Graph each equation in Problems 35–38 on a graphing calculator. (Use the ZInteger setting.) Locate the x-intercepts of the graph. Use the x-intercepts to write the quadratic expression in factored form.

35. $y = 0.1(x^2 - 3x - 270)$ **36.** $y = 0.1(x^2 + 9x - 360)$

37. $y = -0.08(x^2 + 14x - 576)$ **38.** $y = -0.06(x^2 - 22x - 504)$

Use the Pythagorean theorem to solve Problems 39 and 40. (See Appendix A.3 to review the Pythagorean theorem.)

39. One end of a ladder is 10 feet from the base of a wall, and the other end reaches a window in the wall. The ladder is 2 feet longer than the height of the window. (See Figure 3.21.)

a. Write a quadratic equation about the height of the window.

b. Solve your equation to find the height of the window.

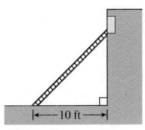

FIGURE 3.21

40. The diagonal of a rectangle is 20 inches. One side of the rectangle is 4 inches shorter than the other side.

   a. Write a quadratic equation about the length of the rectangle.

   b. Solve your equation to find the dimensions of the rectangle.

If an object is thrown into the air from a height s_0 above the ground with an initial velocity v_0, then its height t seconds later is given by the formula

$$h = -\frac{1}{2}gt^2 + v_0t + s_0$$

where g is a constant that measures the force of gravity. Use this formula to answer Problems 41 and 42.

41. A tennis ball is thrown into the air with an initial velocity of 16 feet per second from a height of 8 feet. The value of g is 32.

   a. Write a quadratic equation that gives the height of the tennis ball at time t.

   b. Find the height of the tennis ball at $t = \frac{1}{2}$ second and at $t = 1$ second.

   c. Write and solve an equation to answer the question: At what time is the tennis ball 11 feet high?

   d. Use the **Table** feature on your calculator to verify your answers to parts (b) and (c). (What value of **ΔTbl** is useful for this problem?)

   e. Graph your equation from part (a) on your calculator. Use your table to help you choose an appropriate window.

   f. If nobody hits the tennis ball, approximately how long will it be in the air?

42. A mountain climber stands on a ledge 80 feet above the ground and tosses a rope down to a companion clinging to the rock face below the ledge. The initial velocity of the rope is -8 feet per second, and the value of g is 32.

   a. Write a quadratic equation that gives the height of the rope at time t.

   b. What is the height of the rope after $\frac{1}{2}$ second? After 1 second?

   c. Write and solve an equation to answer the question: How long does it take the rope to reach the second climber, who is 17 feet above the ground?

   d. Use the **Table** feature on your calculator to verify your answers to parts (b) and (c). (What value of **ΔTbl** is useful for this problem?)

   e. Graph your equation from part (a) on your calculator. Use your table to help you choose an appropriate window.

   f. If the second climber misses the rope, approximately how long will the rope take to reach the ground?

43. A rancher has 360 yards of fence to enclose a rectangular pasture. If the pasture should be 8000 square yards in area, what should its dimensions be? We'll use 3 methods to solve this problem: a table of values, a graph, and an algebraic equation.

   a. Make a table by hand that shows the areas of pastures of various widths, as shown here.

Width	Length	Area
10	170	1700
20	160	3200
⋮	⋮	⋮

(To find the length of each pasture, ask yourself: What is the sum of the length plus the width if there are 360 yards of fence?) Continue the table until you find the pasture whose area is 8000 square yards.

b. Write an expression for the length of the pasture if its width is x. Next write an expression for the area, A, of the pasture if its width is x. Graph the equation for A on your calculator, and use the graph to find the pasture of area 8000 square yards.

c. Write an equation for the area, A, of the pasture in terms of its width x. Solve your equation algebraically for $A = 8000$. Explain why there are two solutions.

44. If the rancher in Problem 43 uses a riverbank to border 1 side of the pasture as shown in Figure 3.22, he can enclose 16,000 square yards with 360 yards of fence. What will the dimensions of the pasture be then? We'll use 3 methods to solve this problem: a table of values, a graph, and an algebraic equation.

FIGURE 3.22

a. Make a table by hand that shows the areas of pastures of various widths, as shown here.

Width	Length	Area
10	340	3400
20	320	6400
⋮	⋮	⋮

(Be careful computing the length of the pasture: Remember that one side of the pasture does not need any fence!) Continue the table until you find the pasture whose area is 16,000 square yards.

b. Write an expression for the length of the pasture if its width is x. Next write an expression for the area, A, of the pasture if its width is x. Graph the equation for A, and use the graph to find the pasture of area 16,000 square yards.

c. Write an equation for the area, A, of the pasture in terms of its width x. Solve your equation algebraically for $A = 16,000$.

45. A box is made from a square piece of cardboard by cutting 2-inch squares from each corner and turning up the edges.

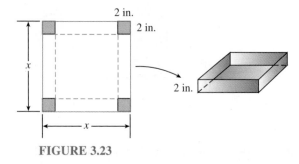

FIGURE 3.23

a. If the piece of cardboard is x inches square, write expressions for the length, width, and height of the box. Then write an expression for the volume, V, of the box in terms of x. (See Figure 3.23.)

b. Use your calculator to make a table of values showing the volumes of boxes made from cardboard squares of side 4 inches, 5 inches, and so on.

c. Graph your expression for the volume on your calculator. What happens to V as x increases?

d. Use your table or your graph to find what size cardboard you need to make a box with volume 50 cubic inches.

e. Write and solve a quadratic equation to answer part (d).

46. A length of rain gutter is made from a piece of aluminum 6 feet long and 1 foot wide.

a. If a strip of width x is turned up along each long edge, write expressions for the length, width and height of the gutter. Then write an expression for the volume, V, of the gutter in terms of x. (See Figure 3.24.)

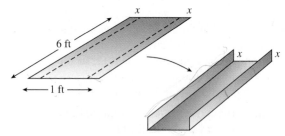

FIGURE 3.24

b. Use your calculator to make a table of values showing the volumes of various rain gutters formed by turning up edges of 0.1 foot, 0.2 foot, and so on.

c. Graph your expression for the volume. What happens to V as x increases?

d. Use your table or your graph to discover how much metal should be turned up along each long edge so that the gutter has a capacity of $\frac{3}{4}$ cubic foot of rainwater.

e. Write and solve a quadratic equation to answer part (d).

47. How far can you throw a baseball? The distance depends on the initial speed of the ball, v, and on the angle at which you throw it. For maximum range, you should throw the ball at $45°$.

a. If there were no air resistance, the height, h, of the ball t seconds after its release would be given in meters by the formula

$$h = \frac{vt}{\sqrt{2}} - \frac{gt^2}{2}$$

where g is the acceleration due to gravity. Find an expression for the total time the ball is in the air. (*Hint:* Set $h = 0$ and solve for t in terms of the other variables.)

b. At time t, the ball has traveled a horizontal distance, d, given by

$$d = \frac{vt}{\sqrt{2}}$$

Find an expression for the range of the ball in terms of its velocity, v. (*Hint:* In part (a) you found an expression for t when $h = 0$. Use that value of t to calculate d when $h = 0$.)

c. The fastest baseball pitch on record was 45 meters per second, or about 100 miles per hour. Use your formula from part (b) to calculate the theoretical range of such a pitch. The value of g is 9.8.

d. The maximum distance a baseball has actually been thrown is 136 meters. Can you explain the discrepancy between this figure and your answer to part (c)?

48. The annual increase, I, in the deer population in a national park depends on the size, x, of the population that year, according to the formula

$$I = kCx - kx^2$$

where k and C are constants related to the fertility of the population and the availability of food. Suppose that $k = 0.0002$ and $C = 6000$.

a. Make a table of values for I for $0 \leq x \leq 7000$. Use increments of 500 in x.

b. How much will a population of 2000 deer increase? A population of 5000 deer? A population of 7000 deer?

c. Use your calculator to graph the annual increase, I, versus the size of the population, x, for $0 \leq x \leq 7000$.

d. What is the physical significance of the x-intercepts of the graph?

e. Estimate the population size that results in the largest annual increase. What is that increase?

MIDCHAPTER REVIEW

Solve by extracting roots.

1. $x^2 + 7 = 13 - 2x^2$ **2.** $\dfrac{2x^2}{5} - 7 = 9$ **3.** $3(x + 4)^2 = 60$ **4.** $(3.5 - 0.2x)^2 = 1.44$

Solve each formula for the indicated variable.

5. $A = \dfrac{3\sqrt{3}}{2}s^2$, for s **6.** $A = P(1 + r)^2$, for r

7. What happens to the volume of a cylinder if you double its radius?

8. The perimeter of an equilateral triangle is 36 inches. Find its altitude. (*Hint:* The altitude is the perpendicular bisector of the base.)

9. a. State the formulas for the area and perimeter of a rectangle.

b. Give the dimensions of two different rectangles with perimeter 60 inches. Compute the areas of the two rectangles.

10. a. A rectangle has a perimeter of 48 inches. If the width of the rectangle is w inches, write an expression for its length.

b. Write an expression for the area of the rectangle.

Write each expression as a polynomial by computing the products.

11. $-3b(6b - 2)$ **12.** $(2q + 5)(3q - 2)$ **13.** $2(t - 8)(2t - 1)$ **14.** $4n(n + 3)^2$

Factor completely.

15. $2x^2 + 8x - 42$ **16.** $3w^2 + 13w - 10$ **17.** $49a^2 + 28ab + 4b^2$ **18.** $75 - 48t^2$

19. State the zero-factor principle.

20. Explain how to use the zero-factor principle to solve a quadratic equation.

21. Graph the equation

$$y = (2x - 7)(x + 1)$$

on a graphing calculator, and use your graph to solve the equation $y = 0$. (Use Xmin = -9.4, Xmax = 9.4.) Check your answers with the zero-factor principle.

22. a. Find a quadratic equation whose solutions are -4 and $\frac{3}{2}$.

b. Find another quadratic equation with the same solutions.

Solve.

23. $x^2 + x = 4 - (x + 2)^2$

24. $(n - 3)(n + 2) = 6$

25. A car traveling at 50 feet per second (about 34 miles per hour) can stop in 2.5 seconds after applying the brakes hard. The distance the car travels, in feet, t seconds after applying the brakes is

$$d = 50t - 10t^2$$

How long does it take the car to travel 40 feet?

26. You have 300 feet of wire fence to mark off a rectangular Christmas tree lot with a center-divider, using a brick wall as one side of the lot. The divider will be perpendicular to the wall. If you would like to enclose a total area of 7500 square feet, what should be the dimensions of the lot?

3.4 Graphing Parabolas: Special Cases

The graph of a quadratic equation is called a **parabola.** The simplest quadratic equation in two variables is

$$y = x^2$$

We can sketch its graph by plotting a few points, as shown in Figure 3.25.

x	-2	-1	0	1	2
y	4	1	0	1	4

We will call the graph of $y = x^2$ the "basic" parabola. The graphs of some other quadratic equations are shown in Figure 3.26.

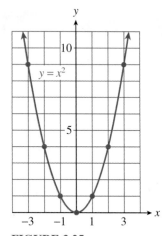

FIGURE 3.25

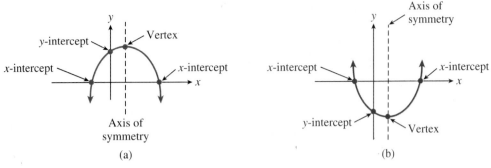

FIGURE 3.26

All parabolas share certain features. The graph has either a highest point (if the parabola opens downward, as in Figure 3.26a) or a lowest point (if the parabola opens upward, as in Figure 3.26b). This high or low point is called the **vertex** of the graph. The parabola is symmetric about a vertical line, called the **axis of symmetry,** that runs through the vertex. The **y-intercept** is the point where the parabola intersects the y-axis. The parabola may intersect the x-axis at zero, one, or two points, called the **x-intercepts.** If there are two x-intercepts, they are equidistant from the axis of symmetry.

The values of the constants a, b, and c in the equation $y = ax^2 + bx + c$ determine the location and orientation of the parabola. We'll begin by considering each of these constants separately.

The Graph of $y = ax^2$

First we'll investigate the effect of the constant a on the shape of the graph. Use your calculator to graph the following three equations on the same screen:

$$y = x^2$$
$$y = 3x^2$$
$$y = 0.1x^2$$

The graphs are shown in Figure 3.27. You can see that, compared to the basic parabola, the graph of $y = 3x^2$ is narrower, and the graph of $y = 0.1x^2$ is wider. We can also describe the graphs by saying that as x increases, the graph of $y = 3x^2$ increases more quickly than the basic parabola, and the graph of $y = 0.1x^2$ increases more slowly. This is clearer if we compare the corresponding y-values for the three graphs, as shown in Table 3.3.

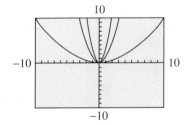

FIGURE 3.27

TABLE 3.3

x	$y = x^2$	$y = 3x^2$	$y = 0.1x^2$
-2	4	12	0.4
1	1	3	0.1
3	9	27	0.9

For each x-value the points on the graph of $y = 3x^2$ are higher than the points on the basic parabola, while the points on the graph of $y = 0.1x^2$ are lower. Multiplying x^2 by a positive constant greater than 1 stretches the graph vertically, while multiplying by a positive constant less than 1 squashes the graph vertically.

What about negative values for a? Consider the graphs of

$$y = x^2$$
$$y = -x^2$$
$$y = -0.5x^2$$

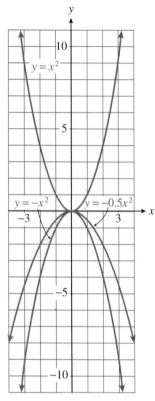

shown in Figure 3.28. We see that multiplying x^2 by a negative constant reflects the graph about the x-axis. These parabolas open downward. In general, the graph of $y = ax^2$ opens upward if $a > 0$ and opens downward if $a < 0$. The magnitude of a determines how "wide" or "narrow" the parabola is. For equations of the form $y = ax^2$, the vertex, the x-intercepts, and the y-intercept all coincide at the origin.

FIGURE 3.28

EXAMPLE 1

Sketch graphs of each quadratic equation by hand.

a. $y = 2x^2$ **b.** $y = -\dfrac{1}{2}x^2$

Solutions

Both equations are of the form $y = ax^2$. The graph of $y = 2x^2$ opens upward because $a = 2 > 0$, and the graph of $y = -\frac{1}{2}x^2$ opens downward because $a = -\frac{1}{2} < 0$. To make a reasonable sketch by hand, it is enough to plot a few "guide points;" the points with x-coordinates 1 and -1 are easy to compute.

x	$y = 2x^2$	$y = -\frac{1}{2}x^2$
-1	2	$-\frac{1}{2}$
0	0	0
1	2	$-\frac{1}{2}$

Sketch parabolas through each set of guidepoints, as shown in Figure 3.29.

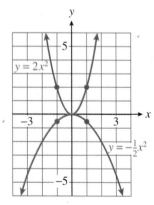

FIGURE 3.29

EXERCISE 1 Match each parabola with its equation. The basic parabola is shown in red.

a. $y = -\dfrac{3}{4}x^2$

b. $y = \dfrac{1}{4}x^2$

c. $y = \dfrac{5}{2}x^2$

d. $y = -\dfrac{5}{4}x^2$

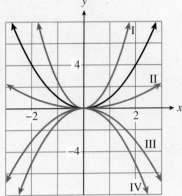

The Graph of $y = x^2 + c$

Next we consider the effect of the constant term, c, on the graph. Consider the graphs of

$$y = x^2$$
$$y = x^2 + 4$$
$$y = x^2 - 4$$

shown in Figure 3.30. The graph of $y = x^2 + 4$ is shifted *upward* four units compared to the basic parabola, and the graph of $y = x^2 - 4$ is shifted *downward* four units. Look at Table 3.4, which shows the y-values for the three graphs.

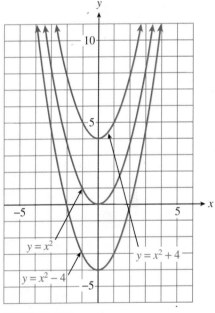

FIGURE 3.30

TABLE 3.4

x	$y = x^2$	$y = x^2 + 4$	$y = x^2 - 4$
-1	1	5	-3
0	0	4	-4
2	4	8	0

Each point on the graph of $y = x^2 + 4$ is four units higher than the corresponding point on the basic parabola, and each point on the graph of $y = x^2 - 4$ is four units lower. In particular, the vertex of the graph of $y = x^2 + 4$ is the point $(0, 4)$, and the vertex of the graph of $y = x^2 - 4$ is the point $(0, -4)$. In general, adding a positive constant c shifts the graph upward by c units, and adding a negative constant c shifts the graph downward.

The x-intercepts of the graph of $y = x^2 - 4$ can be found by setting y equal to zero and solving for x:

$$x^2 - 4 = 0$$
$$x^2 = 4 \qquad \text{Extract roots.}$$
$$x = 2 \qquad x = -2$$

so the x-intercepts are the points $(2, 0)$ and $(-2, 0)$. The graph of $y = x^2 + 4$ has no x-intercepts, because the equation

$$0 = x^2 + 4$$

(obtained by setting $y = 0$) has no real-number solutions.

EXAMPLE 2

Sketch graphs for the following quadratic equations.

a. $y = x^2 - 2$ **b.** $y = -x^2 + 4$

Solutions

a. The graph of $y = x^2 - 2$ is shifted downward by two units, compared to the basic parabola. The vertex is the point $(0, -2)$, and the x-intercepts are the solutions of the equation

$$0 = x^2 - 2$$

or $\sqrt{2}$ and $-\sqrt{2}$. The graph is shown in Figure 3.31.

b. The graph of $y = -x^2 + 4$ opens downward and is shifted 4 units up, compared to the basic parabola. Its vertex is the point $(0, 4)$. Its x-intercepts are the solutions of the equation

$$0 = -x^2 + 4$$

or 2 and -2. (See Figure 3.31.) You can verify both graphs with your graphing calculator.

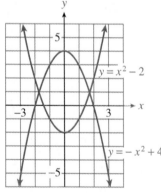

FIGURE 3.31

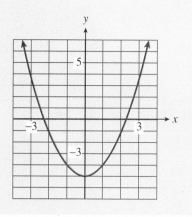

EXERCISE 2

a. Find the equation of the parabola shown.

b. Give the x- and y-intercepts of the graph.

The Graph of $y = ax^2 + bx$

How does the presence of a linear term, bx, affect the graph? Let's begin by considering an example. Graph the equation

$$y = 2x^2 + 8x$$

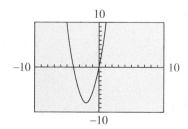

FIGURE 3.32

on your calculator. The graph is shown in Figure 3.32. Note that $a = 2$ and $2 > 0$, so the parabola opens upward. We can find the x-intercepts of the graph by setting y equal to zero:

$$0 = 2x^2 + 8x \quad \text{Factor.}$$
$$= 2x(x + 4)$$

The solutions of this equation are 0 and -4, so the x-intercepts are the points $(0, 0)$ and $(-4, 0)$.

Recall that the parabola is symmetric about a vertical line through its vertex. The two x-intercepts are equidistant from this line of symmetry, so the x-coordinate of the vertex lies exactly halfway between the x-intercepts. We can average their values to find

$$x_v = \frac{1}{2}[0 + (-4)] = -2$$

(We use (x_v, y_v) to denote the coordinates of the vertex.) To find the y-coordinate of the vertex, substitute $x = -2$ into the equation for the parabola.

$$y_v = 2(-2)^2 + 8(-2)$$
$$= 8 - 16 = -8$$

Thus, the vertex is the point $(-2, -8)$.

a. Find the x-intercepts and the vertex of the parabola

$$y = 6x - x^2$$

x-intercepts: Solve $6x - x^2 = 0$

Vertex: Compute the average of the x-intercepts:

$$x_v =$$

Evaluate y at $x = x_v$:

$$y_v =$$

b. Verify your answers by graphing the equation in the window

$$\text{Xmin} = -9.4 \qquad \text{Xmax} = 9.4$$
$$\text{Ymin} = -10 \qquad \text{Ymax} = 10$$

Finding the Vertex

We can use the same method to find a formula for the vertex of any parabola of the form

$$y = ax^2 + bx$$

We proceed as we did in the example above. First, find the x-intercepts of the graph by setting y equal to zero and solving for x.

$$0 = ax^2 + bx \qquad \text{Factor.}$$
$$= x(ax + b)$$

Thus,

$$x = 0 \qquad \text{or} \qquad ax + b = 0$$

so

$$x = 0 \qquad \text{or} \qquad x = \frac{-b}{a}$$

The x-intercepts are the points $(0, 0)$ and $\left(\dfrac{-b}{a}, 0\right)$.

Next, find the x-coordinate of the vertex by taking the average of the two x-intercepts found above.

$$x_v = \frac{1}{2}\left[0 + \left(\frac{-b}{a}\right)\right] = \frac{-b}{2a}$$

This gives us a formula for the x-coordinate of the vertex:

$$x_v = \frac{-b}{2a}$$

Also, the axis of symmetry is the vertical line $x = \dfrac{-b}{2a}$. (See Figure 3.33.) Finally, we find the y-coordinate of the vertex by substituting its x-coordinate into the equation for the parabola.

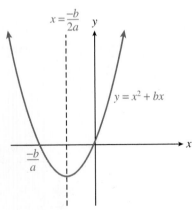

FIGURE 3.33

EXAMPLE 3

a. Find the vertex of the graph of $y = -1.8x^2 - 16.2x$.

b. Find the x-intercepts of the graph.

Solutions

a. The x-coordinate of the vertex is

$$x_v = \frac{-b}{2a} = \frac{-(-16.2)}{2(-1.8)} = -4.5$$

To find the y-coordinate of the vertex, evaluate y at $x = -4.5$.

$$y_v = -1.8(-4.5)^2 - 16.2(-4.5) = 36.45$$

The vertex is $(-4.5 \; 36.45)$

b. To find the x-intercepts of the graph, set $y = 0$ and solve

$$-1.8x^2 - 16.2x = 0 \quad \text{Factor.}$$

$$-x(1.8x + 16.2) = 0 \quad \text{Set each factor equal to zero.}$$

$$-x = 0 \qquad 1.8x + 16.2 = 0 \quad \text{Solve each equation.}$$

$$x = 0 \qquad\qquad x = -9$$

The x-intercepts of the graph are $(0, 0)$ and $(-9, 0)$. The graph is shown in Figure 3.34.

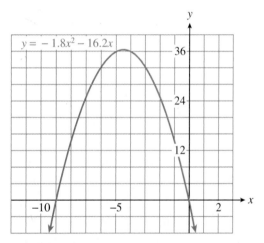

FIGURE 3.34

■

READING QUESTIONS

1. Sketch a parabola that opens downward. Show the location of the x-intercepts, the y-intercept, the vertex, and the axis of symmetry.

2. Describe how the value of a in $y = ax^2$ alters the graph of the basic parabola.

3. Describe how the value of c in $y = x^2 + c$ alters the graph of the basic parabola.

4. Explain how to find the coordinates of the vertex of $y = ax^2 + bx$.

HOMEWORK 3.4

Describe what each graph will look like compared to the basic parabola. Then sketch the graph by hand, and label the coordinates of three points on the graph.

1. $y = 2x^2$

2. $y = 4x^2$

3. $y = \dfrac{1}{2}x^2$

4. $y = 0.6x^2$

5. $y = -x^2$

6. $y = -3x^2$

7. $y = -0.2x^2$

8. $y = \dfrac{-3}{4}x^2$

Describe what each graph will look like compared to the basic parabola. Then sketch each graph by hand. Label the vertex and the x-intercepts (if there are any) with their coordinates.

9. $y = x^2 + 2$

10. $y = x^2 + 5$

11. $y = x^2 - 1$

12. $y = x^2 - 9$

13. $y = x^2 - 5$

14. $y = x^2 - 3$

15. $y = 100 - x^2$

16. $y = 225 - x^2$

Find the x-intercepts and the vertex of each graph. Then sketch the graph by hand.

17. $y = x^2 - 4x$

18. $y = x^2 - 2x$

19. $y = x^2 + 2x$

20. $y = x^2 + 6x$

21. $y = 3x^2 + 6x$

22. $y = 2x^2 - 6x$

23. $y = -2x^2 + 5x$

24. $y = -3x^2 - 8x$

Graph each set of equations in the same window on your calculator. Describe how the constants in each equation affect the graph.

25. a. $y = x^2$
  b. $y = 3x^2$
  c. $y = \dfrac{1}{4}x^2$
  d. $y = -2x^2$

26. a. $y = x^2$
  b. $y = x^2 + 1$
  c. $y = x^2 + 3$
  d. $y = x^2 - 6$

27. a. $y = x^2 - 4x$
  b. $y = x^2 + 4x$
  c. $y = 4x - x^2$
  d. $y = -x^2 - 4x$

28. a. $y = x^2 - 4x$
  b. $y = 2x^2 - 8x$
  c. $y = \dfrac{1}{2}x^2 - 2x$
  d. $y = -x^2 + 4x$

29. Match each equation with its graph in Figure 3.35. In each equation, $a > 0$.

a. $y = x^2 + a$ **b.** $y = x^2 + ax$ **c.** $y = ax^2$
d. $y = ax$ **e.** $y = x + a$ **f.** $y = x^2 - a$

I

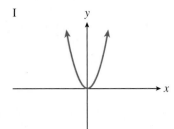

II

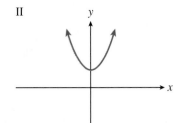

III

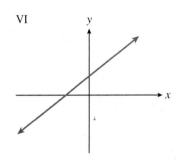

IV

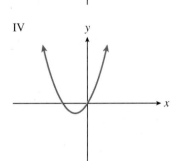

V

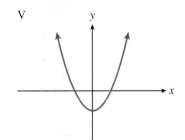

VI

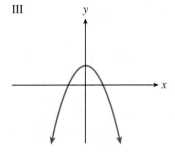

FIGURE 3.35

30. Match each equation with its graph in Figure 3.36. In each equation, $b > 0$.

a. $y = -bx$ **b.** $y = -bx^2$ **c.** $y = b - x^2$
d. $y = x - b$ **e.** $y = b - x$ **f.** $y = x^2 - bx$

I

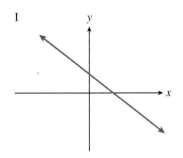

II

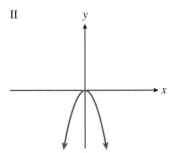

III

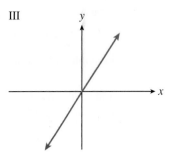

IV

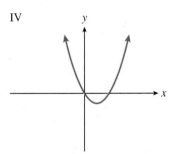

V

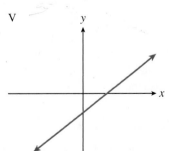

VI

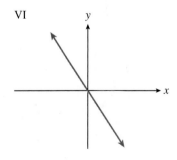

FIGURE 3.36

31. Commercial fishermen rely on a steady supply of fish in their area. To avoid overfishing, they adjust their harvest to the size of the population. The equation

$$y = 0.4x - 0.0001x^2$$

gives the annual rate of growth, in tons per year, of a fish population of biomass x tons.

   a. Find the x-intercepts of the graph of y. What is their significance in terms of the fish population?

   b. Find the vertex of the graph. What is its significance in the context of the problem?

   c. Sketch the graph for $0 \le x \le 5000$.

   d. For what values of x does the fish population decrease rather than increase? Suggest a reason why the population might decrease.

32. Many animals live in groups. A species of marmot found in Colorado lives in "harems" composed of a single adult male and several females with their young. The average number of offspring each female can raise depends on the number of females in the harem. On average, if there are x females in the harem, each female can raise $y = 2 - 0.4x$ young marmots each year.

   a. Complete the table of values for the average number of offspring per female, and the total number of young marmots, A, produced by the entire harem in one year.

x	1	2	3	4	5
y					
A					

   b. Write a formula for A in terms of x.

   c. Graph the equation for A in terms of x.

   d. What is the maximum number of young marmots a harem can produce (on average)? What is the optimal number of female marmots per harem?

   Source: Chapman and Reiss, 1992.

3.5 Completing the Square

Not every quadratic equation can be solved by factoring or by extraction of roots. For example, $x^2 + x - 1$ cannot be factored, so the equation $x^2 + x - 1 = 0$ cannot be solved by factoring. For some equations, factoring may be difficult. In this section and the next we will learn two methods that can be used to solve any quadratic equation.

Squares of Binomials

In Section 3.3 we used extraction of roots to solve equations of the form

$$(x - p)^2 = q \tag{1}$$

where the left side of the equation is the square of a binomial, or a "perfect square." We can write any quadratic equation in the form of Equation (1) by completing the square.

Consider the following squares of binomials.

1. $(x + 5)^2 = x^2 + 10x + 25$ $2(5) = 10$ and $5^2 = 25$
2. $(x - 3)^2 = x^2 - 6x + 9$ $2(-3) = -6$ and $(-3)^2 = 9$
3. $(x - 12)^2 = x^2 - 24x + 144$ $2(-12) = -24$ and $(-12)^2 = 144$

In each case the square of the binomial is a quadratic trinomial,

$$(x + p)^2 = x^2 + 2px + p^2$$

The coefficient of the linear term, $2p$, is twice the constant in the binomial, and the constant term of the trinomial, p^2, is its square.

We'd like to reverse the process and write a quadratic trinomial as the square of a binomial. For example, what constant term can we add to

$$x^2 - 16x$$

to produce a perfect square? Compare the expression to the formula above:

$$x^2 + 2px + p^2 = (x + p)^2$$
$$x^2 - 16x + \underline{\ ?\ } = (x + \underline{\ ?\ })^2$$

We see that $2p = -16$, so $p = \frac{1}{2}(-16) = -8$, and $p^2 = (-8)^2 = 64$. Substitute these values for p^2 and p into the equation to find

$$x^2 - 16x + 64 = (x - 8)^2$$

Notice that in the resulting trinomial, the constant term is equal to *the square of one-half the coefficient of x*. In other words, we can find the constant term by taking one-half the coefficient of x, and then squaring the result. Obtaining the constant term in this way is called **completing the square.**

EXAMPLE 1

Complete the square by adding an appropriate constant, and write the result as the square of a binomial.

    a. $x^2 - 12x +$ _____   **b.** $x^2 + 5x +$ _____

Solutions

    a. One-half of -12 is -6, so the constant term is $(-6)^2$ or 36. Add 36 to obtain

$$x^2 - 12x + 36 = (x - 6)^2 \quad \begin{aligned} p &= \frac{1}{2}(-12) = -6 \\ p^2 &= (-6)^2 = 36 \end{aligned}$$

    b. One-half of 5 is $\frac{5}{2}$, so the constant term is $\left(\frac{5}{2}\right)^2$, or $\frac{25}{4}$. Add $\frac{25}{4}$ to obtain

$$x^2 + 5x + \frac{25}{4} = \left(x + \frac{5}{2}\right)^2 \quad \begin{aligned} p &= \frac{1}{2}(5) = \frac{5}{2} \\ p^2 &= \left(\frac{5}{2}\right)^2 = \frac{25}{4} \end{aligned}$$

    ∎

Solving Quadratic Equations by Completing the Square

Now we'll use completing the square to solve quadratic equations. First we'll solve equations in which the coefficient of the squared term is 1. Consider the equation

$$x^2 - 6x - 7 = 0 \qquad (2)$$

We move the constant term to the other side of the equation to get

$$x^2 - 6x \quad = 7$$

and complete the square on the left. Because

$$p = \frac{1}{2}(-6) = -3 \qquad \text{and} \qquad p^2 = (-3)^2 = 9$$

we add 9 to *both* sides of our equation to get

$$x^2 - 6x + 9 = 7 + 9$$

The left side of the equation is now the square of a binomial, namely $(x - 3)^2$, so we have

$$(x - 3)^2 = 16$$

(You can check that this equation is equivalent to the original one; if you expand the left side and collect like terms, you will return to the original form of the equation.) We can now use extraction of roots to find the solutions. Taking square roots of both sides we get

$$x - 3 = 4 \qquad \text{or} \qquad x - 3 = -4$$

Solve each equation.

$$x = 7 \qquad \text{or} \qquad x = -1$$

The solutions are 7 and -1.

The graph of $y = x^2 - 6x - 7$ is shown in Figure 3.37. Note that the x-intercepts of the graph are $x = 7$ and $x = -1$, and the axis of symmetry is half-way between the intercepts, at $x = 3$.

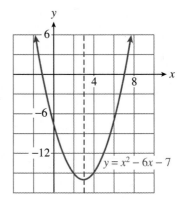

FIGURE 3.37

We could solve Equation (2) by factoring instead of completing the square. Of course, we get the same solutions by either method. In Example 2 we'll solve an equation that cannot be solved by factoring.

EXAMPLE 2

Solve $x^2 - 4x - 3 = 0$ by completing the square.

Solution

First, write the equation with the constant term on the right side.

$$x^2 - 4x \underline{\hspace{1cm}} = 3$$

Now complete the square on the left side. The coefficient of x is -4, so

$$p = \frac{1}{2}(-4) = -2, \quad \text{and} \quad p^2 = (-2)^2 = 4$$

We add 4 to both sides of the equation.

$$x^2 - 4x + 4 = 3 + 4$$

Write the left side as the square of a binomial, and combine terms on the right side.

$$(x - 2)^2 = 7$$

Finally, use extraction of roots to obtain

$$x - 2 = \sqrt{7} \qquad \text{or} \qquad x - 2 = -\sqrt{7} \qquad \text{Solve each equation.}$$
$$x = 2 + \sqrt{7} \qquad \text{or} \qquad x = 2 - \sqrt{7}$$

These are the exact values of the solutions. We can use a calculator to find decimal approximations for each solution:

$$2 + \sqrt{7} \approx 4.646 \qquad \text{and} \qquad 2 - \sqrt{7} \approx -0.646$$

The graph of $y = x^2 - 4x - 3$ is shown in Figure 3.38.

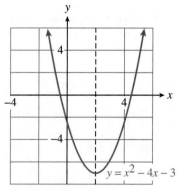

FIGURE 3.38

The General Case: $ax^2 + bx + c = 0, a \neq 1$

Our method for completing the square works only if the coefficient of x^2 is 1. If we want to solve a quadratic equation whose lead coefficient is not 1, we first divide each term of the equation by the lead coefficient.

EXAMPLE 3

Solve $2x^2 - 6x - 5 = 0$.

Solution

Because the coefficient of x^2 is 2, we must divide each term of the equation by 2.

$$x^2 - 3x - \frac{5}{2} = 0$$

Now, proceed as before. Rewrite the equation with the constant on the right side.

$$x^2 - 3x \underline{\qquad} = \frac{5}{2}$$

Complete the square:

$$p = \frac{1}{2}(-3) = \frac{-3}{2} \qquad \text{and} \qquad p^2 = \left(\frac{-3}{2}\right)^2 = \frac{9}{4}$$

Add $\frac{9}{4}$ to both sides of the equation.

$$x^2 - 3x + \frac{9}{4} = \frac{5}{2} + \frac{9}{4}$$

Rewrite the left side as the square of a binomial and simplify the right side to get

$$\left(x - \frac{3}{2}\right)^2 = \frac{19}{4}$$

Finally, extract roots and solve each equation for x.

$$x - \frac{3}{2} = \sqrt{\frac{19}{4}} \qquad \text{or} \qquad x - \frac{3}{2} = -\sqrt{\frac{19}{4}}$$

The solutions are $\frac{3}{2} + \sqrt{\frac{19}{4}}$ and $\frac{3}{2} - \sqrt{\frac{19}{4}}$. Using a calculator, we can find decimal approximations for the solutions: 3.679 and -0.679. ∎

COMMON ERROR

In Example 3, it is essential that we first divide each term of the equation by 2, the coefficient of x^2. The following attempt at a solution is *incorrect*.

$$2x^2 - 6x = 5$$
$$2x^2 - 6x + 9 = 5 + 9$$
$$(2x - 3)^2 = 14 \qquad \leftarrow \text{Incorrect!}$$

You can check that $(2x - 3)^2$ is *not* equal to $2x^2 - 6x + 9$. We have not written the left side of the equation as a perfect square, so the solutions we obtain by extracting roots will not be correct. ∎

EXERCISE 3

a. Solve by completing the square $-4x^2 - 36x - 65 = 0$.

Divide each term by -4.
Write the equation with the constant on the right.
Complete the square on the left:

$$p = \frac{1}{2}(9) = \underline{}, p^2 = \underline{}$$

Add p^2 to both sides.
Write the left side as a perfect square; simplify the right side.
Solve by extracting roots.

b. Graph $y = -4x^2 - 36x - 65$ in the window

$$\text{Xmin} = -9.4 \qquad \text{Xmax} = 0$$
$$\text{Ymin} = -10 \qquad \text{Ymax} = 20$$

Here is a summary of the steps for solving quadratic equations by completing the square.

To Solve a Quadratic Equation by Completing the Square:

1. Write the equation in standard form.
2. Divide both sides of the equation by the coefficient of the quadratic term, and subtract the constant term from both sides.
3. Complete the square on the left side:

   a. Multiply the coefficient of the first-degree term by one-half, then square the result.

   b. Add the value obtained in (a) to both sides of the equation.
4. Write the left side of the equation as the square of a binomial. Simplify the right side.
5. Use extraction of roots to finish the solution.

READING QUESTIONS

1. Name three algebraic methods for solving a quadratic equation.
2. Give an example of a quadratic trinomial that is the square of a binomial.
3. What number must be added to $x^2 - 26x$ to make it the square of a binomial?
4. After completing the square, how do we finish solving the quadratic equation?
5. What is the first step in solving the equation $2x^2 - 6x = 5$ by completing the square?

ANSWERS TO 3.5 EXERCISES

1a. $x - 18x + 81 = (x - 9)^2$ **b.** $x^2 + 9x + \dfrac{81}{4} = \left(x + \dfrac{9}{2}\right)^2$

2a. $x = \dfrac{3}{2} \pm \sqrt{\dfrac{13}{4}}$ **b.** $x \approx -0.30$ or $x \approx 3.30$

3a. $x = \dfrac{-13}{2}, x = \dfrac{-5}{2}$

HOMEWORK 3.5

Complete the square and write the result as the square of a binomial.

1. $x^2 + 8x$
2. $x^2 - 14x$
3. $x^2 - 7x$
4. $x^2 + 3x$

5. $x^2 + \dfrac{3}{2}x$
6. $x^2 - \dfrac{5}{2}x$
7. $x^2 - \dfrac{4}{5}x$
8. $x^2 + \dfrac{2}{3}x$

Solve by completing the square.

9. $x^2 - 2x + 1 = 0$
10. $x^2 + 4x + 4 = 0$
11. $x^2 + 9x + 20 = 0$
12. $x^2 - x - 20 = 0$

13. $x^2 = 3 - 3x$
14. $x^2 = 5 - 5x$
15. $2x^2 + 4x - 3 = 0$
16. $3x^2 + 12x + 2 = 0$

17. $3x^2 + x = 4$
18. $4x^2 + 6x = 3$
19. $4x^2 - 3 = 2x$
20. $2x^2 - 5 = 3x$

21. $3x^2 - x - 4 = 0$
22. $2x^2 - x - 3 = 0$
23. $5x^2 + 8x = 4$
24. $9x^2 - 12x + 5 = 0$

Solve by completing the square. Your answers will involve *a*, *b*, or *c*.

25. $x^2 + 2x + c = 0$
26. $x^2 - 4x + c = 0$
27. $x^2 + bx + 1 = 0$
28. $x^2 + bx - 4 = 0$

29. $ax^2 + 2x - 4 = 0$
30. $ax^2 - 4x + 9 = 0$

Solve each formula for the indicated variable.

31. $V = \pi(r - 3)^2 h$, for r
32. $A = P(1 + r)^2$, for r

33. $E = \dfrac{1}{2}mv^2 + mgh$, for v
34. $h = \dfrac{1}{2}gt^2 + dl$, for t

35. $V = 2(s^2 + t^2)w$, for t
36. $V = \pi(r^2 + R^2)h$, for R

Solve for *y* in terms of *x*. Use whichever method of solution seems easiest.

37. $x^2y - y^2 = 0$
38. $x^2y^2 - y = 0$
39. $(2y + 3x)^2 = 9$
40. $(3y - 2x)^2 = 4$

41. $4x^2 - 9y^2 = 36$
42. $9x^2 + 4y^2 = 36$
43. $4x^2 - 25y^2 = 0$
44. $(2x - 5y)^2 = 0$

45. Complete the square to find the solutions of the equation $x^2 + bx + c = 0$. (Your answers will be expressions in b and c.)

46. Complete the square to find the solutions of the equation $ax^2 + bx + c = 0$. (Your answers will be expressions in a, b, and c.)

47. a. Write an expression for the area of the square in Figure 3.39.

    b. Express the area as a polynomial.

    c. Divide the square into four pieces whose areas are given by the terms of your answer to (b).

48. a. Write an expression for the area of the shaded region in Figure 3.40.

    b. Express the area in factored form.

    c. By making one cut in the shaded region, rearrange the pieces into a rectangle whose area is given by your answer to (b).

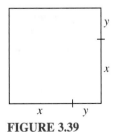

FIGURE 3.39

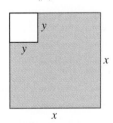

FIGURE 3.40

49. 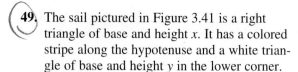 The sail pictured in Figure 3.41 is a right triangle of base and height x. It has a colored stripe along the hypotenuse and a white triangle of base and height y in the lower corner.

a. Write an expression for the area of the colored stripe.

b. Express the area of the stripe in factored form.

c. If the sail is $7\frac{1}{2}$ feet high and the white strip is $4\frac{1}{2}$ feet high, use your answer to (b) to calculate mentally the area of the stripe.

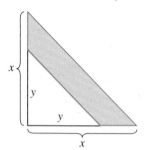

FIGURE 3.41

50. An hors d'oeuvres tray has radius x, and the dip container has radius y, as shown in Figure 3.42.

a. Write an expression for the area for the chips (shaded region).

b. Express the area in factored form.

c. If the tray has radius $8\frac{1}{2}$ inches and the space for the dip has radius $2\frac{1}{2}$ inches, use your answer to part (b) to calculate mentally the area for chips. (Express your answer as a multiple of π.)

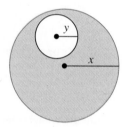

FIGURE 3.42

3.6 Quadratic Formula

Instead of completing the square every time we solve a new quadratic equation, we can complete the square on the general quadratic equation,

$$ax^2 + bx + c = 0, \qquad a \neq 0$$

and obtain a formula for the solutions of any quadratic equation.

The Quadratic Formula

The solutions of the equation $ax^2 + bx + c = 0$, $a \neq 0$, are

$$x = \frac{-b \pm \sqrt{b^2 - 4ac}}{2a}$$

This formula expresses the solutions of a quadratic equation in terms of its coefficients. (The proof of the formula is considered in Homework problems 45 and 46 in Section 3.5.) The symbol $\pm$, read "plus or minus," is used to combine the two equations

$$x = \frac{-b + \sqrt{b^2 - 4ac}}{2a} \qquad \text{and} \qquad x = \frac{-b - \sqrt{b^2 - 4ac}}{2a}$$

into a single equation.

To solve a quadratic equation using the quadratic formula, all we have to do is substitute the coefficients a, b, and c into the formula.

EXAMPLE 1

Solve $2x^2 + 1 = 4x$.

Solution

Write the equation in standard form as

$$2x^2 - 4x + 1 = 0$$

Substitute 2 for a, -4 for b, and 1 for c into the quadratic formula, and simplify.

$$x = \frac{-(-4) \pm \sqrt{(-4)^2\, 4(2)(1)}}{2(2)}$$

$$= \frac{4 \pm \sqrt{8}}{4}$$

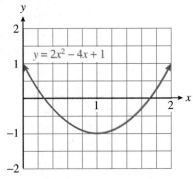

FIGURE 3.43

Using a calculator, we find that the solutions are approximately 1.707 and 0.293. (If you would like to review the order of operations with a calculator, please see Appendix A.1.) We can also verify that the x-intercepts of the graph of $y = 2x^2 - 4x + 1$ are approximately 1.707 and 0.293, as shown in Figure 3.43. ∎

EXERCISE 1 Use the quadratic formula to solve $x^2 - 3x = 1$.

Write the equation in standard form.

Substitute $a = 1$, $b = -3$, $c = -1$ into the quadratic formula.

Simplify.

Note that the solutions to this equation are the same as the solutions to Exercise 2 in Section 3.5.

Applications

We have now seen four different algebraic methods for solving quadratic equations:

1. Factoring
2. Extraction of Roots
3. Completing the Square
4. Quadratic Formula

Factoring and extraction of roots are relatively fast and simple, but they don't work on all quadratic equations. The quadratic formula will work on any quadratic equation.

EXAMPLE 2

The owners of a day-care center plan to enclose a divided play area against the back wall of their building as shown in Figure 3.44. They have 300 feet of picket fence and would like the total area of the playground to be 6000 square feet. Can they enclose the playground with the fence they have, and if so what should the dimensions of the playground be?

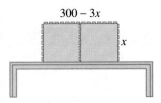

FIGURE 3.44

Solution

Suppose the width of the play area is x feet. Because there are three sections of fence along the width of the play area, that leaves $300 - 3x$ feet of fence for its length. (See Figure 3.44.) The area of the play area should be 6000 square feet, so we have the equation

$$x(300 - 3x) = 6000$$

This is a quadratic equation. In standard form,

$$3x^2 - 300x + 6000 = 0 \quad \text{Divide each term by 3.}$$

$$x^2 - 100x + 2000 = 0$$

The left side cannot be factored, so we use the quadratic formula with $a = 1$, $b = -100$, and $c = 2000$.

$$\begin{aligned} x &= \frac{-(-100) \pm \sqrt{(-100)^2 - 4(1)(2000)}}{2(1)} \\ &= \frac{100 \pm \sqrt{2000}}{2} \\ &\approx \frac{100 \pm 44.7}{2} \end{aligned}$$

Thus, $x \approx 72.35$ or $x \approx 27.65$. Both values give solutions to the problem. If the width of the play area is 72.35 feet, then the length is $300 - 3(72.35)$, or 82.95 feet. If the width is 27.65 feet, the length is $300 - 3(27.65)$, or 217.05 feet. ∎

EXERCISE 2 In Investigation 6, we considered the height of a baseball, given by the equation

$$h = -16t^2 + 64t + 4.$$

Find two times when the ball is at a height of 20 feet. Give your answers to two decimal places.

Set $h = 20$, then write the equation in standard form.

Divide each term by -16.

Use the quadratic formula to solve.

Sometimes it is useful to solve a quadratic equation for one variable in terms of the others.

EXAMPLE 3   Solve $x^2 - xy + y = 2$ for x in terms of y.

Solution

We first write the equation in standard form as a quadratic equation in the variable x.

$$x^2 - yx + (y - 2) = 0$$

Expressions in y are treated as constants with respect to x, so that $a = 1$, $b = -y$, $c = y - 2$. Substitute these expressions into the quadratic formula.

$$x = \frac{-(-y) \pm \sqrt{(-y)^2 - 4(1)(y - 2)}}{2(1)}$$

$$= \frac{y \pm \sqrt{y^2 - 4y + 8}}{2}$$

∎

EXERCISE 3 Solve $2x^2 + kx + k^2 = 1$ for x in terms of k.

Write in standard form, treating k as a constant.

Use the quadratic formula to solve.

Complex Numbers

Not all quadratic equations have solutions that are real numbers.

EXAMPLE 4

Solve $x^2 - \frac{x}{2} + 1 = 0$.

Solution

Notice that for this equation, b is a fraction, $-\frac{1}{2}$. It is easier to apply the quadratic formula if the coefficients are integers, so we will first multiply both sides of the equation by 2 in order to clear the fraction. (Recall that multiplying both sides of an equation by the same number does not change the solutions of the equation.) This gives us

$$2x^2 - x + 2 = 0$$

For this new equation, $a = 2$, $b = -1$, and $c = 2$. Substitute these values into the quadratic formula to obtain

$$x = \frac{-(-1) \pm \sqrt{(-1)^2 - 4(2)(2)}}{2(2)} = \frac{1 \pm \sqrt{-15}}{4}$$

Because $\sqrt{-15}$ is not a real number, this equation does not have real-number solutions. Consequently, the graph of $y = x^2 + \frac{x}{2} + 1$ has no x-intercepts, as shown in Figure 3.45. ∎

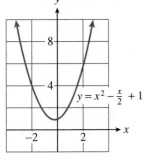

FIGURE 3.45

Although square roots of negative numbers such as $\sqrt{-15}$ are not real numbers, they occur often in mathematics and its applications. (See Appendix B.1 if you would like to review the definition of real number.) Mathematicians began working with square roots of negative numbers in the sixteenth century, in their attempts to solve quadratic and cubic equations. (In a cubic equation the highest power is x^3.) René Descartes gave them the name "imaginary numbers," which reflected the mistrust with which mathematicians regarded them at the time. Today, however, such numbers are well understood and used routinely by scientists and engineers.

We'll begin by defining a new number, i, whose square is -1.

$$i^2 = -1 \qquad \text{or} \qquad i = -1$$

With this new number we can define the principal square root of any negative real number as follows.

For $a > 0$,

$$\sqrt{-a} = \sqrt{-1}\,\sqrt{a} = i\sqrt{a}$$

EXAMPLE 5

a. $\sqrt{-4} = \sqrt{-1}\,\sqrt{4}$ **b.** $\sqrt{-3} = \sqrt{-1}\,\sqrt{3}$
$\quad = i\sqrt{4} = 2i$ $\quad = i\sqrt{3}$ ∎

Thus, the square root of any negative real number can be written as the product of a real number and i. Such numbers are still called **imaginary numbers.** Each negative real number has two imaginary square roots, $i\sqrt{a}$ and $-i\sqrt{a}$, because

$$\left(i\sqrt{a}\right)^2 = i^2\left(\sqrt{a}\right)^2 = -1 \cdot a = -a$$

and

$$\left(-i\sqrt{a}\right)^2 = (-i)^2\left(\sqrt{a}\right)^2 = i^2 \cdot a = -a$$

For example, the two square roots of -9 are $3i$ and $-3i$.

Using this new notation, we can write the solutions to Example 3, $\dfrac{1 \pm \sqrt{-15}}{4}$, as $\dfrac{1 \pm i\sqrt{15}}{4}$ or $\dfrac{1}{4} \pm \dfrac{\sqrt{15}}{4}i$. The sum of a real number and an imaginary number is called a **complex number.** You can read more about complex numbers in Appendix B.2.

> **EXERCISE 4** Use extraction of roots to solve $(2x + 1)^2 + 9 = 0$. Write your answers as complex numbers.

The Discriminant

A closer look at the quadratic formula reveals useful information about the solutions of quadratic equations. The solutions of a quadratic equation always occur in **conjugate pairs,**

$$\frac{-b}{2a} + \frac{\sqrt{b^2 - 4ac}}{2a} \qquad \text{and} \qquad \frac{-b}{2a} - \frac{\sqrt{b^2 - 4ac}}{2a}$$

For example, if we know that one solution of a particular quadratic equation with rational coefficients is $3 + \sqrt{2}$, then the other solution must be $3 - \sqrt{2}$.

The expression $b^2 - 4ac$, which appears under the radical in the quadratic formula, is called the **discriminant** of the equation. The value of the discriminant determines the nature of the solutions of the equation:

The **discriminant** of the quadratic equation $ax^2 + bx + c = 0$ is

$$D = b^2 - 4ac.$$

1. If $D > 0$, the equation has two unequal real solutions;
2. If $D = 0$, the equation has one real solution of multiplicity two;
3. If $D < 0$, the equation has two complex (conjugate) solutions.

EXAMPLE 6

Use the discriminant to determine the nature of the solutions of each equation.

a. $x^2 - x - 3 = 0$ b. $2x^2 + x + 1 = 0$ c. $x^2 - 2\sqrt{3}x + 3 = 0$

Solutions

Compute the discriminant for each equation.

a. $D = b^2 - 4ac = (-1)^2 - 4(1)(-3) = 13 > 0$. The equation has two real, unequal solutions.
b. $D = b^2 - 4ac = 1^2 - 4(2)(1) = -7 < 0$. The equation has two complex solutions.
c. $D = b^2 - 4ac = \left(-2\sqrt{3}\right)^2 - 4(1)(3) = 0$. The equation has one real solution of multiplicity two.

You can verify the conclusions above by solving each equation. ■

We can also use the discriminant to decide whether a quadratic equation with rational coefficients can be solved by factoring. First clear the equation of fractions. If the discriminant is a perfect square, that is, the square of an integer, the solutions are rational numbers. This means that the equation can be solved by factoring.

EXERCISE 5 Use the discriminant to determine the nature of the solutions of the equation

$$6x^2 + 13x = 240.$$

Can the equation be solved by factoring?

Write the equation in standard form, and compute the discriminant.

READING QUESTIONS

1. The quadratic formula can solve any quadratic equation. When might you prefer to use factoring or extraction of roots?
2. What is an imaginary number? What is a complex number?
3. What are conjugate pairs? Give an example.
4. Explain how to use the discriminant.

HOMEWORK 3.6

Solve using the quadratic formula. Round your answers to three decimal places.

1. $x^2 - x - 1 = 0$

2. $x^2 + x - 1 = 0$

3. $y^2 + 2y = 5$

4. $y^2 - 4y = 4$

5. $3z^2 = 4.2z + 1.5$

6. $2z^2 = 7.5z - 6.3$

7. $0 = x^2 - \dfrac{5}{3}x + \dfrac{1}{3}$

8. $0 = -x^2 + \dfrac{5}{2}x - \dfrac{1}{2}$

9. $-5.2z^2 + 176z + 1218 = 0$

10. $15z^2 - 18z - 2750 = 0$

11. A car traveling at s miles per hour on a dry road surface will require approximately d feet to stop, where d is given by

$$d = \dfrac{s^2}{24} + \dfrac{s}{2}$$

   a. Make a table showing the stopping distance, d, for speeds of 10, 20, . . . , 100 miles per hour. (Use the **Table** feature of your calculator.)

   b. Graph the equation for d in terms of s. Use your table values to help you choose appropriate window settings.

   c. Write and solve an equation to answer the question: If a car must be able to stop in 50 feet, what is the maximum safe speed it can travel? Verify your answer on your graph.

12. A car traveling at s miles per hour on a wet road surface will require approximately d feet to stop, where d is given by

$$d = \dfrac{s^2}{12} + \dfrac{s}{2}$$

   a. Make a table showing the stopping distance, d, for speeds of 10, 20, . . . , 100 miles per hour. (Use the **Table** feature of your calculator.)

   b. Graph the equation for d in terms of s. Use your table values to help you choose appropriate window settings.

   c. Insurance investigators at the scene of an accident find skid marks 100 feet long leading up to the point of impact. Write and solve an equation to discover how fast the car was traveling when it put on the brakes. Verify your answer on your graph.

13. A skydiver jumps out of an airplane at 11,000 feet. While she is in free-fall, her altitude in feet t seconds after jumping is given by

$$h = -16t^2 - 16t + 11{,}000$$

   a. Make a table of values showing the skydiver's altitude at 5-second intervals after she jumps from the airplane. (Use the **Table** feature of your calculator.)

   b. Graph the equation. Use your table of values to choose appropriate window settings.

   c. If the skydiver must open her parachute at an altitude of 1000 feet, how long can she free-fall? Write and solve an equation to find the answer.

   d. If the skydiver drops a marker just before she opens her parachute, how long will it take the marker to hit the ground? (*Hint:* The marker continues to fall according to the equation given above.)

   e. Find points on your graph that correspond to your answers to parts (c) and (d).

14. A high diver jumps from the 10-meter springboard. His height in meters above the water t seconds after leaving the board is given by

$$h = -4.9t^2 + 8t + 10$$

a. Make a table of values showing the diver's altitude at 0.25-second intervals after he jumps from the springboard. (Use the **Table** feature of your calculator.)

b. Graph the equation. Use your table of values to choose appropriate window settings.

c. How long is it before the diver passes the board on the way down?

d. How long is it before the diver hits the water?

e. Find points on your graph that correspond to your answers to parts (c) and (d).

15. A dog trainer has 100 meters of chain link fence. She wants to enclose 250 square meters in three pens of equal size as shown in Figure 3.46.

a. Let l and w represent the length and width of the entire area. Write an equation about the amount of chain link fence.

b. Solve your equation for l in terms of w.

c. Write and solve an equation in w for the total area enclosed.

d. Find the dimensions of each pen.

FIGURE 3.46

16. An architect is planning to include a rectangular window topped by a semicircle in his plans for a new house, as shown in Figure 3.47. In order to admit enough light the window should have an area of 120 square feet. The architect wants the rectangular portion of the window to be 2 feet wider than it is tall.

a. Let x stand for the horizontal width of the window. Write expressions for the height of the rectangular portion and for the radius of the semi-circular portion.

FIGURE 3.47

b. Write an expression for the total area of the window.

c. Write and solve an equation to find the width and over-all height of the window.

17. When you look down from a height, say a tall building or a mountain peak, your line of sight is tangent to the earth at the horizon. (See Figure 3.48.)

a. Suppose you are standing on top of the World Trade Center in New York, 1350 feet high. How far can you see on a clear day? (You will need to use the Pythagorean theorem, and the fact that the radius of the earth is 3960 miles. Don't forget to convert the height of the World Trade Center to miles.)

b. How tall a building should you stand on in order to see 100 miles?

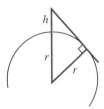

FIGURE 3.48

18. a. If the radius of the earth is 6370 kilometers, how far can you see from an airplane at an altitude of 10,000 meters? (*Hint:* See Problem 17.)

b. How high would the airplane have to be in order for you to see a distance of 10 kilometers?

In Problems 19–22, (a) given one solution of a quadratic equation with rational coefficients, find the other solution, (b) write a quadratic equation that has those solutions.

19. $2 + \sqrt{5}$ **20.** $3 - \sqrt{2}$ **21.** $4 - 3i$ **22.** $5 + i$

23. a. Graph the three equations $y = x^2 - 6x + 5$, $y = x^2 - 6x + 9$, $y = x^2 - 6x + 12$ in the window

$$\text{Xmin} = -2 \qquad \text{Xmax} = 7.4$$
$$\text{Ymin} = -5 \qquad \text{Ymax} = 15$$

Use the **Trace** to locate the x-intercepts of each graph.

b. Calculate the discriminant for each of the equations.

$$x^2 - 6x + 5 = 0$$
$$x^2 - 6x + 9 = 0$$
$$x^2 - 6x + 12 = 0$$

What does the discriminant tell you about the solutions of the equation? How does your answer relate to the graphs in part (a)?

24. a. Graph the three equations $y = 3 - 2x - x^2$, $y = -1 - 2x - x^2$, $y = -4 - 2x - x^2$ in the window

$$\text{Xmin} = -6.4 \qquad \text{Xmax} = 3$$
$$\text{Ymin} = -10 \qquad \text{Ymax} = 5$$

Use the **Trace** to locate the x-intercepts of each graph.

b. Calculate the discriminant for each of the equations.

$$3 - 2x - x^2 = 0$$
$$-1 - 2x - x^2 = 0$$
$$-4 - 2x - x^2 = 0$$

What does the discriminant tell you about the solutions of the equation? How does your answer relate to the graphs in part (a)?

Use the discriminant to determine the nature of the solutions of each equation.

25. $3x^2 + 26 = 17x$

26. $4x^2 + 23x = 19$

27. $16x^2 - 712x + 7921 = 0$

28. $121x^2 + 1254x + 3249 = 0$

29. $65.2x = 13.2x^2 + 41.7$

30. $0.03x^2 = 0.05x - 0.12$

Use the discriminant to decide if we can use factoring to solve the equation.

31. $3x^2 - 7x + 6 = 0$

32. $6x^2 - 11x - 7 = 0$

33. $15x^2 - 52x - 32 = 0$

34. $17x^2 + 65x - 12 = 0$

Use the quadratic formula to solve each equation for the indicated variable.

35. $A = 2w^2 + 4lw$,　for w

36. $A = \pi r^2 + \pi rs$,　for r

37. $h = 4t - 16t^2$,　for t

38. $P = IE - RI^2$,　for I

39. $s = vt - \dfrac{1}{2}at^2$,　for t

40. $S = \dfrac{n^2 + n}{2}$,　for n

41. $3x^2 + xy + y^2 = 2$,　for y

42. $y^2 - 3xy + x^2 = 3$,　for x

43. What is the sum of the two solutions of the quadratic equation $ax^2 + bx + c = 0$? (*Hint:* The two solutions are given by the quadratic formula.)

44. What is the product of the two solutions of the quadratic equation $ax^2 + bx + c = 0$? (*Hint:* Do *not* try to multiply the two solutions given by the quadratic formula! Think about the factored form of the equation.)

45. Explain why the value of the discriminant determines the nature of the solutions of a quadratic equation. Consider the three cases discussed in the text.

46. Explain why a quadratic equation with integer coefficients has rational solutions if and only if its discriminant is the square of an integer.

Solve by extraction of roots.

1. $(2x - 5)^2 = 9$

2. $(7x - 1)^2 = 15$

Solve by factoring.

3. $x(3x + 2) = (x + 2)^2$

4. $6y = (y + 1)^2 + 3$

5. $4x - (x + 1)(x + 2) = -8$

6. $3(x + 2)^2 = 15 + 12x$

Write a quadratic equation with integer coefficients and with the given solutions.

7. $\dfrac{-3}{4}$ and 8

8. $\dfrac{5}{3}$ and $\dfrac{5}{3}$

Graph each equation using the ZDecimal setting. Locate the x-intercepts, and use them to write the quadratic expression in factored form.

9. $y = x^2 - 0.6x - 7.2$

10. $y = -x^2 + 0.7x + 2.6$

a. Find the coordinates of the vertex and the intercepts.

b. Sketch the graph.

11. $y = \dfrac{1}{2}x^2$

12. $y = x^2 - 4$

13. $y = x^2 - 9x$

14. $y = -2x^2 - 4x$

Solve by completing the square.

15. $x^2 - 4x - 6 = 0$

16. $x^2 + 3x = 3$

17. $2x^2 + 3 = 6x$

18. $3x^2 = 2x + 3$

Solve by using the quadratic formula.

19. $\dfrac{1}{2}x^2 + 1 = \dfrac{3}{2}x$

20. $x^2 - 3x + 1 = 0$

21. $x^2 - 4x + 2 = 0$

22. $2x^2 + 2x = 3$

Solve each formula for the indicated variable.

23. $K = \dfrac{1}{2}mv^2,$ for v

24. $a^2 + b^2 = c^2,$ for b

25. $h = 6t - 3t^2,$ for t

26. $D = \dfrac{n^2 - 3n}{2},$ for n

Use the discriminant to determine the nature of the solutions of each equation.

27. $4x^2 - 12x + 9 = 0$

28. $2t^2 + 6t + 5 = 0$

29. $2y^2 = 3y - 4$

30. $\dfrac{x^2}{4} = x + \dfrac{5}{4}$

31. In a tennis tournament among n competitors, $\dfrac{n(n - 1)}{2}$ matches must be played. If the organizers can schedule 36 matches, how many players should they invite?

32. The formula $S = \dfrac{n(n + 1)}{2}$ gives the sum of the first n positive integers. How many consecutive integers must be added to make a sum of 91?

33. Lewis invested $2000 in an account that compounds interest annually. He made no deposits or withdrawals after that. Two years later he closed the account, withdrawing $2464.20. What interest rate did Lewis earn?

34. Earl borrowed $5500 from his uncle for 2 years with interest compounded annually. At the end of 2 years he owed his uncle $6474.74. What was the interest rate on the loan?

35. Irene wants to enclose 2 adjacent chicken coops of equal size against the henhouse wall. She has 66 feet of chicken wire fencing and would like the total area of the two coops to be 360 square feet. What should the dimensions of the chicken coops be?

37. The height, h, of an object t seconds after being thrown from ground level is given by

$$h = v_0 t - \frac{1}{2}gt^2$$

where v_0 is its starting velocity and g is a constant that depends on gravity. On the moon, the value of g is approximately 5.6. Suppose you hit a golf ball on the moon with an upward velocity of 100 feet per second.

a. Write an equation for the height of the golf ball t seconds after you hit it.

b. Graph your equation in the window

$$Xmin = 0 \qquad Xmax = 47$$
$$Ymin = 0 \qquad Ymax = 1000$$

c. Use the ⟨ TRACE ⟩ key to estimate the maximum height the golf ball reaches.

d. Use your equation to calculate when the golf ball will reach a height of 880 feet.

36. The base of an isosceles triangle is one inch shorter than the equal sides, and the altitude of the triangle is two inches shorter than the equal sides. What is the length of the equal sides?

38. An acrobat is catapulted into the air from a springboard at ground level. Her height h in meters is given by the formula

$$h = -4.9t^2 + 14.7t$$

where t is the time in seconds from launch. Use your calculator to graph the acrobat's height versus time. Set the ⟨ WINDOW ⟩ values on your calculator to

$$Xmin = 0 \qquad Xmax = 4.7$$
$$Ymin = 0 \qquad Ymax = 12$$

a. Use the ⟨ TRACE ⟩ key to find the coordinates of the highest point on the graph. When does the acrobat reach her maximum height, and what is that height?

b. Use the formula to find the height of the acrobat after 2.4 seconds.

c. Use the ⟨ TRACE ⟩ key to verify your answer to part (b). Find another time when the acrobat is at the same height.

d. Use the formula to find two times when the acrobat is at a height of 6.125 meters. Verify your answers on the graph.

e. What are the coordinates of the horizontal intercepts of your graph? What do these points have to do with the acrobat?

Show that the shaded areas are equal.

39.

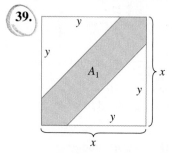

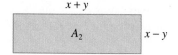

FIGURE 3.49

40.

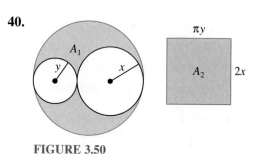

FIGURE 3.50

Applications of Quadratic Models

In this chapter we learn more about the graphs of quadratic equations and consider some applications of quadratic models.

INVESTIGATION 7

Revenue from Theater Tickets

The local theater group sold tickets to its opening night performance for $5 and drew an audience of 100 people. The next night they reduced the ticket price by $0.25 and 10 more people attended; that is, 110 people bought tickets at $4.75 apiece. In fact, for each $0.25 reduction in ticket price, 10 additional tickets can be sold.

1. Complete Table 4.1.

   TABLE 4.1

Number of Price reductions	Price of ticket	Number of Tickets sold	Total revenue
0	5.00	100	500
1	4.75	110	522.50
2			
3			
4			
5			
6			
7			
8			
9			
10			
11			

2. Use your table to make a graph. Plot *Total Revenue* on the vertical axis versus *Number of price reductions* on the horizontal axis, as shown in Figure 4.1.

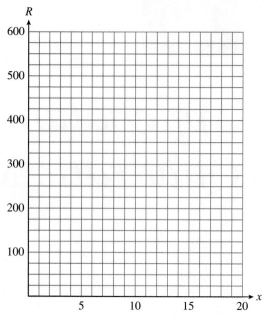

FIGURE 4.1

3. Let *x* represent the *number of price reductions,* as in the first column of the table. Write algebraic expressions in terms of *x* for

The *price of a ticket* after *x* price reductions:

$$Price =$$

The *number of tickets* sold at that price:

$$Number =$$

The *total revenue* from ticket sales:

$$Revenue =$$

4. Enter your expressions for the price of a ticket, the number of tickets sold, and the total revenue into the calculator as Y_1, Y_2, and Y_3. Use the **Table** feature to verify that your algebraic expressions agree with your table from part (1).

5. Use your calculator to graph your expression for total revenue in terms of *x*. Use your table to choose appropriate window settings that show the high point of the graph and both *x*-intercepts.

6. What is the maximum revenue possible from ticket sales? What price should the theater group charge for a ticket to generate that revenue? How many tickets will they sell at that price?

4.1 Graphing Parabolas: The General Case

We have already graphed parabolas of several special forms. We now consider graphs of any quadratic equation

$$y = ax^2 + bx + c$$

The constants a, b, and c determine the relative size and position of the graph. If we can locate the vertex of the parabola and a few other points, perhaps the x- and y-intercepts, we can sketch a fairly accurate graph.

Finding the Vertex

In Section 3.4, we found that the x-coordinate of the vertex of the graph of $y = ax^2 + bx$ is given by

$$x_v = \frac{-b}{2a}$$

Now we'll see that this same formula holds for any parabola.

As an example of the general case, $y = ax^2 + bx + c$, consider the equation

$$y = 2x^2 + 8x + 6 \qquad (1)$$

In Section 3.4 we graphed the equation $y = 2x^2 + 8x$, and that graph is shown again in Figure 4.2. Adding 6 to $2x^2 + 8x$ will have the effect of shifting each point on the graph 6 units upward, as shown in Figure 4.2. Notice that the x-coordinate of the vertex will not be affected by an upward shift. Thus, the formula

$$x_v = \frac{-b}{2a}$$

for the x-coordinate of the vertex still holds. We have

$$x_v = \frac{-8}{2(2)} = -2$$

and from Equation (1),

$$y_v = 2(-2)^2 + 8(-2) + 6 \quad \text{Substitute} \atop -2 \text{ for } x.$$
$$= 8 - 16 + 6 = -2$$

so the vertex is the point $(-2, -2)$. (Notice that this point is shifted six units upward from the vertex of $y = 2x^2 + 8x$.)

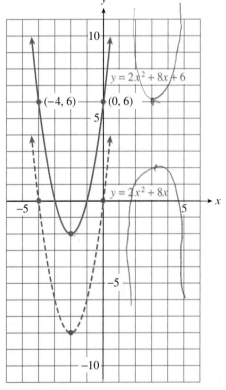

FIGURE 4.2

We find the x-intercepts of the graph by setting y equal to zero.

$$0 = 2x^2 + 8x + 6 \qquad \text{Factor the right side.}$$
$$= 2(x + 1)(x + 3) \qquad \text{Set each factor equal to zero.}$$
$$x + 1 = 0 \qquad \text{or} \qquad x + 3 = 0$$
$$x = -1 \qquad\qquad x = -3$$

The x-intercepts are the points $(-1, 0)$ and $(-3, 0)$.

The y-intercept of the graph is found by setting x equal to zero.

$$y = 2\,(0)^2 + 8(0) + 6 = 6$$

You can see that the y-intercept, 6, is just the constant term of the quadratic equation. The completed graph is shown in Figure 4.2.

EXAMPLE 1

Find the vertex of the graph of $y = -2x^2 + x + 1$.

Solution

For this equation, $a = -2$, $b = 1$, and $c = 1$. The x-coordinate of the vertex is given by

$$x_v = \frac{-b}{2a} = \frac{-1}{2(-2)} = \frac{1}{4}$$

To find the y-coordinate of the vertex, we substitute $x = \frac{1}{4}$ into the equation. We can do this by hand to find

$$y_v = -2\left(\frac{1}{4}\right)^2 + \frac{1}{4} + 1$$

$$= -2\left(\frac{1}{16}\right) + \frac{4}{16} + \frac{16}{16} = \frac{18}{16} = \frac{9}{8}$$

So the coordinates of the vertex are $\left(\frac{1}{4}, \frac{9}{8}\right)$. Or we can use the calculator to evaluate $-2x^2 + x + 1$ for $x = 0.25$. Enter

$$\boxed{(-)}\; 2 \;\boxed{\times}\; 0.25 \;\boxed{x^2}\; \boxed{+}\; 0.25 \;\boxed{+}\; 1$$

and press $\boxed{\text{ENTER}}$. The calculator returns the y-value, 1.125. Thus, the vertex is the point $(0.25, 1.125)$, which is the decimal equivalent of $\left(\frac{1}{4}, \frac{9}{8}\right)$. ∎

EXERCISE 1 Find the vertex of the graph of $y = 3x^2 - 6x + 4$. Decide whether the vertex is a maximum point or a minimum point of the graph.

$x_v =$

$y_v =$

Number of x-Intercepts

The graph of the quadratic equation $y = ax^2 + bx + c$ may have two, one, or no x-intercepts, according to the number of real-valued solutions of the equation $ax^2 + bx + c = 0$. Consider the three equations graphed in Figure 4.3. The graph of

$$y = x^2 - 4x + 3$$

has two x-intercepts, because the equation

$$x^2 - 4x + 3 = 0$$

has two real-valued solutions, $x = 1$ and $x = 3$. The graph of

$$y = x^2 - 4x + 4$$

has only one x-intercept, because the equation

$$x^2 - 4x + 4 = 0$$

has only one real-valued solution, $x = 2$.

Because the solutions of the equation

$$x^2 - 4x + 6 = 0$$

are complex numbers, they do not appear on the graph. The graph of

$$y = x^2 - 4x + 6$$

has no x-intercepts.

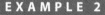

FIGURE 4.3

EXAMPLE 2

Use the discriminant to determine how many x-intercepts the graph of each equation has. (See Section 3.6 to review the discriminant.)

a. $y = x^2 - 4x + 6$ **b.** $y = -\frac{1}{2}x^2 + 4x - 8$

Solutions

a. For this equation, $a = 1$, $b = -4$, and $c = 6$, so

$$D = b^2 - 4ac = (-4)^2 - 4(1)(6) = -8$$

The discriminant is negative, so the equation $x^2 - 4x + 6 = 0$ has no real-valued solutions, and the graph has no x-intercepts.

b. For this equation, $a = -\frac{1}{2}$, $b = 4$, and $c = -8$, so

$$D = b^2 - 4ac = 4^2 - 4\left(\frac{-1}{2}\right)(-8) = 0$$

The discriminant is zero, so the equation $\frac{1}{2}x^2 + 4x - 8 = 0$ has one real-valued solution of multiplicity two, and the graph has one x-intercept. ∎

In Example 2b, you should check that the single x-intercept is also the vertex of the parabola.

EXERCISE 2 Use the discriminant of $y = 3x^2 - 6x + 4$ to determine how many x-intercepts the graph has.

$$D = b^2 - 4ac =$$

Sketching a Parabola

Once we have located the vertex of the parabola, the x-intercepts, and the y-intercept, we can sketch a reasonably accurate graph. Recall that the graph should be symmetric about a vertical line through the vertex. We summarize the procedure as follows.

> ### To graph the quadratic equation $y = ax^2 + bx + c$:
>
> 1. Determine whether the parabola opens upward (if $a > 0$) or downward (if $a < 0$).
> 2. Locate the vertex of the parabola.
>   a. The x-coordinate of the vertex is $x_v = \dfrac{-b}{2a}$.
>   b. Find the y-coordinate of the vertex by substituting x_v into the equation of the parabola.
> 3. Locate the x-intercepts (if any) by setting $y = 0$ and solving for x.
> 4. Locate the y-intercept by evaluating y for $x = 0$.
> 5. Locate the point symmetric to the y-intercept across the axis of symmetry.

EXAMPLE 3

Sketch a graph of the equation $y = x^2 + 3x + 1$, showing the significant points.

Solution
We follow the steps outlined above.

Step 1 Because $a = 1 > 0$, we know that the parabola opens upward.

Step 2 Compute the coordinates of the vertex.

$$x_v = \frac{-b}{2a} = \frac{-3}{2(1)} = -1.5$$

$$y_v = (-1.5)^2 + 3(-1.5) + 1 = -1.25$$

The vertex is the point $(-1.5, -1.25)$.

Step 3 Set y equal to zero to find the x-intercepts.

$$0 = x^2 + 3x + 1$$

$$x = \frac{-3 \pm \sqrt{3^2 - 4(1)(1)}}{2(1)} \qquad \text{Use the quadratic formula.}$$

$$= \frac{-3 \pm \sqrt{5}}{2}$$

Rounding to the nearest tenth, we find that the x-intercepts are approximately $(-2.6, 0)$ and $(-0.4, 0)$.

Step 4 Substitute $x = 0$ to find the y-intercept, $(0, 1)$.

Step 5 The axis of symmetry is the vertical line $x = -1.5$, so the y-intercept lies 1.5 units to the right of the axis of symmetry. There must be another point on the parabola with the same y-coordinate as the y-intercept, but 1.5 units to the left of the axis of symmetry. The coordinates of this point are $(-3, 1)$.

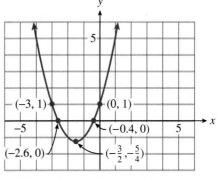

FIGURE 4.4

Finally, plot the x-intercepts, the vertex, the y-intercept and its symmetric point, and draw a parabola through them. The finished graph is shown in Figure 4.4. ∎

EXERCISE 3

a. Find the intercepts and the vertex of the graph of $y = x^2 - 5x + 4$.

b. Sketch the graph by hand.

> y-intercept: Set $x = 0$:
> x-intercepts: Set $y = 0$:
> vertex: $x_v =$
> $\qquad$ $y_v =$

c. Use your calculator to verify your graph.

The Vertex Form for a Parabola

Consider the quadratic equation

$$y = 2(x + 2)^2 - 2 \qquad (2)$$

By expanding the squared expression and collecting like terms, we can rewrite the equation in standard form as

$$y = 2(x^2 + 4x + 4) - 2$$
$$y = 2x^2 + 8x + 6$$

Now we see that Equation (2) is just another form of Equation (1), which we graphed at the start of this section. (See Figure 4.2 on page 201.) However, Equation (2) has one advantage over Equation (1): The coordinates of the vertex are apparent in Equation (2); we don't need to do any computation to find the vertex.

Recall that the vertex of the graph of Equation (1) is $(-2, -2)$. The coordinates of the vertex appear in Equation (2) as follows:

$$y = 2[x - (-2)]^2 - 2$$
$$\qquad\quad x_v \qquad\quad y_v$$

Equation (2) is an example of the **vertex form** for a quadratic equation.

Vertex Form for a Quadratic Equation

A quadratic equation $y = ax^2 + bx + c$, $a \neq 0$, can be written in the form

$$y = a(x - x_v)^2 + y_v$$

where the vertex of the graph is (x_v, y_v).

EXAMPLE 4

Find the vertex of the graph of $y = -3(x - 4)^2 + 6$. Is the vertex a maximum or a minimum point of the graph?

Solution

Compare the equation to the vertex form to see that the coordinates of the vertex are $(4, 6)$. For this equation $a = -3 < 0$ so the parabola opens downward. The vertex is the maximum point of the graph. ■

To understand why the vertex form works, substitute $x_v = 4$ into $y = -3(x - 4)^2 + 6$ from Example 4 to find

$$y = -3(4 - 4)^2 + 6 = 6$$

which confirms that when $x = 4$, $y = 6$. Next, notice that if x is any number except 4, the expression $-3(x - 4)^2$ is negative, so $y < 6$. Therefore, 6 is the maximum value for y on the graph, so $(4, 6)$ is the high point or vertex.

You can also rewrite $y = -3(x - 4)^2 + 6$ in standard form and use the formula $x_v = \dfrac{-b}{2a}$ to confirm that the vertex is the point $(4, 6)$. Any quadratic equation in vertex form can be written in standard form by expanding, and any quadratic equation in standard form can be put into vertex form by completing the square.

EXERCISE 4

a. Find the vertex of the graph of

$$y = 5 - \tfrac{3}{2}(x + 2)^2$$

b. Write the equation of the parabola in standard form.

c. Complete the table and sketch the graph.

x	-4	0	2
y			

READING QUESTIONS

1. State a formula for the x-coordinate of the vertex of a parabola. How can you find the y-coordinate of the vertex?

2. Explain how to use the discriminant to predict the number of x-intercepts of a parabola.

3. Suppose that a given parabola has only one x-intercept. What can you say about the vertex of the parabola?

4. Describe five points to look for when sketching a parabola.

5. What is the vertex form for a quadratic equation?

ANSWERS TO 4.1 EXERCISES

1. $(1, 1)$, minimum **2.** none

3a. $(0, 4)$; $(1, 0)$, $(4, 0)$; vertex $\left(\frac{5}{2}, \frac{-9}{4}\right)$

b.

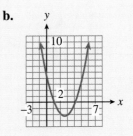

4a. $(-2, 5)$ **b.** $y = \frac{-3}{2}x^2 - 6x - 1$

c.

x	-4	0	2
y	-1	-1	-19

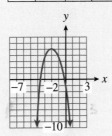

HOMEWORK 4.1

Find the coordinates of the vertex. Decide whether the vertex is a maximum point or a minimum point of the graph.

1. $y = 2 + 3x - x^2$

2. $y = 3 - 5x + x^2$

3. $y = \frac{1}{2}x^2 - \frac{2}{3}x + \frac{1}{3}$

4. $y = \frac{-3}{4}x^2 + \frac{1}{2}x - \frac{1}{4}$

5. $y = 2.3 - 7.2x - 0.8x^2$

6. $y = 5.1 - 0.2x + 4.6x^2$

In Problems 7–16, (a) find the coordinates of the vertex and the intercepts; (b) sketch the graph by hand; (c) use your calculator to verify your graph.

7. $y = -2x^2 + 7x + 4$

8. $y = -3x^2 + 2x + 8$

9. $y = 0.6x^2 + 0.6x - 1.2$

10. $y = 0.5x^2 - 0.25x - 0.75$

11. $y = x^2 + 4x + 7$

12. $y = x^2 - 6x + 10$

13. $y = x^2 + 2x - 1$

14. $y = x^2 - 6x + 2$

15. $y = -2x^2 + 6x - 3$

16. $y = -2x^2 - 8x - 5$

For Problems 17 and 18, match each equation with one of the eight graphs shown in Figure 4.5.

Use vertex $-\frac{b}{2a}$ formula (handwritten)

17. a. $y = 1 - x^2$ **b.** $y = (x + 2)^2$ **18. a.** $y = -2 - (x - 2)^2$ **b.** $y = x - x^2$

c. $y = 2x^2$ **d.** $y = (x - 4)(x + 2)$ **c.** $y = x^2 - 4$ **d.** $y = -0.5x^2$

I
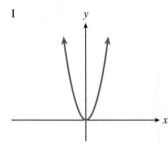

II
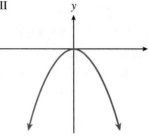

III
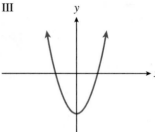

IV
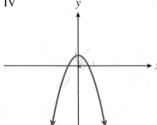

V
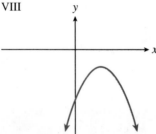

VI
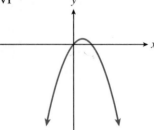

VII
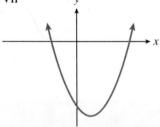

VIII

FIGURE 4.5

19. a. Write an equation for a parabola that has x-intercepts at $(2, 0)$ and $(-3, 0)$.

b. Write an equation for another parabola that has the same x-intercepts.

20. a. Write an equation for a parabola that opens upward with x-intercepts $(-1, 0)$ and $(4, 0)$.

b. Write an equation for a parabola that opens downward with x-intercepts $(-1, 0)$ and $(4, 0)$.

In Problems 21–24, (a) find the vertex of the parabola, (b) write the equation in standard form.

21. $y = 2(x - 3)^2 + 4$

22. $y = -3(x + 1)^2 - 2$

23. $y = -\frac{1}{2}(x + 4)^2 - 3$

24. $y = 4(x - 2)^2 - 6$

(handwritten) $x = 2, -3$

(handwritten) $y = -72(x - 2)(x \times 3)$

(handwritten) 2, -6 Gil ant

25. a. Write an equation for a parabola whose vertex is the point $(-2, 6)$. (Many answers are possible.)

b. Find the value of a if the y-intercept of the parabola in part (a) is 18.

27. a. Write an equation for a parabola with vertex at $(0, -3)$ and one of its x-intercepts at $(2, 0)$.

b. Write an equation for a parabola with vertex at $(0, -3)$ and no x-intercepts. *Solve form*

26. a. Write an equation for a parabola whose vertex is the point $(5, -10)$. (Many answers are possible.)

b. Find the value of a if the y-intercept of the parabola in part (a) is -5.

28. Write an equation for a parabola with vertex at $(4, 0)$ and y-intercept at $(0, 4)$. How many x-intercepts does the parabola have?

x_v y_v

vertex $(0, 3)$

y-intercept $(0, 3)$

$b = 0$

Find an equation for each parabola. Use the vertex form or the factored form of the equation, whichever is more appropriate.

29.

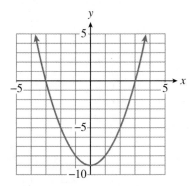

FIGURE 4.6

30.

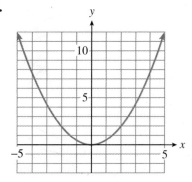

FIGURE 4.7

31.

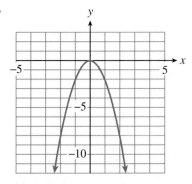

FIGURE 4.8

32.

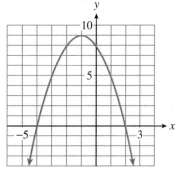

FIGURE 4.9

33.

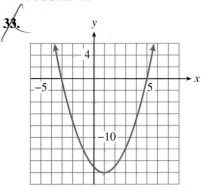

FIGURE 4.10

34.

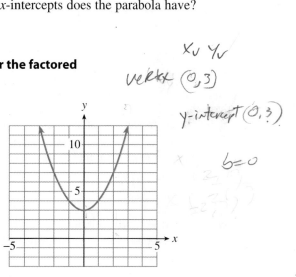

FIGURE 4.11

35.

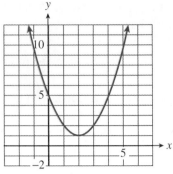

FIGURE 4.12

37. Because of air resistance, the path of a kicked soccer ball is not actually parabolic. However, both the horizontal and vertical coordinates of points on its trajectory can be approximated by quadratic equations. For a soccer ball kicked from the ground, these equations are

$$x = 17t - 1.1t^2$$
$$y = 17.28t - 4.8t^2$$

where x and y are given in meters and t is the number of seconds since the ball was kicked.

a. Fill in the table.

t	0	0.5	1.0	1.5	2.0	2.5	3.0	3.5
x								
y								

b. Plot the points (x, y) from your table and connect them with a smooth curve to represent the path of the ball.

c. Use your graph to estimate the maximum height of the ball.

d. Estimate the horizontal distance traveled by the ball before it strikes the ground.

e. Using the formula given for y, determine how long the ball is in the air.

f. Use your answer from part (e) and the formula for x to find the horizontal distance traveled by the ball before it strikes the ground.

g. Use the formula given for y to find the maximum height for the ball.

38. Figure 4.14 shows the typical weight of two species of birds each day after hatching.

36.

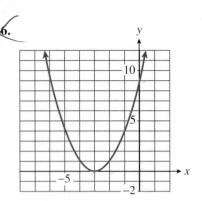

FIGURE 4.13

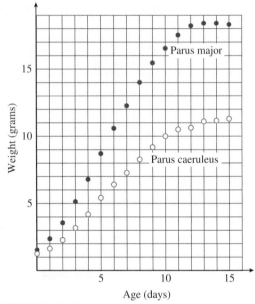

FIGURE 4.14

Source: Perrins, 1979.

a. Describe the rate of growth for each species over the first 15 days of life. How are the growth rates for the two species similar, and how are they different?

b. Complete the tables showing the weight and the daily rate of growth for each species.

Parus major

Day	0	1	2	3	4	5	6	7
Weight								
Growth rate								

Day	8	9	10	11	12	13	14	15
Weight								
Growth rate								

Parus caeruleus

Day	0	1	2	3	4	5	6	7
Weight								
Growth rate								

Day	8	9	10	11	12	13	14	15
Weight								
Growth rate								

c. For each species, plot the rate of growth against weight in grams. What type of curve does the growth rate graph appear to be?

d. For each species, at what weight did the maximum growth rate occur? Locate the corresponding point on each original curve in Figure 4.14.

Problems 39 and 40 use ideas from Investigation 7 on page 199.

39. The owner of a motel has 60 rooms to rent. She finds that if she charges $20 per room per night, all the rooms will be rented. For every $2 that she increases the price of a room, three rooms will stand vacant.

a. Complete the table. The first two rows are filled in for you.

No. of price increases	Price of room	No. of rooms rented	Total revenue
0	20	60	1200
1	22	57	1254
2			
3			
4			
5			
6			
7			
8			
10			
12			
16			
20			

b. Let x stand for the number of $2 price increases the owner makes. Write algebraic expressions for the price of a room, the number of rooms that will be rented, and the total revenue earned at that price.

c. Use your calculator to make a table of values for your algebraic expressions. Let Y_1 stand for the price of a room, Y_2 for the number of rooms rented, and Y_3 for the total revenue. Verify the values you calculated in part (a).

d. Use your table to find a value of x that causes the total revenue to be zero.

e. Use your graphing calculator to graph your formula for total revenue.

f. What is the lowest price that the owner can charge for a room to make her revenue be at least $1296 per night? What is the highest price she can charge to obtain this revenue?

g. What is the maximum revenue the owner can earn in one night? How much should she charge for a room to maximize her revenue? How many rooms will she rent at that price?

40. The owner of a video store sells 96 blank tapes per week if he charges $6 per tape. For every $0.50 he increases the price, he sells four fewer tapes per week.

a. Complete the table. The first two rows are filled in for you.

No. of price increases	Price of tape	No. of tapes sold	Total revenue
0	6	96	576
1	6.50	92	598
2			
3			
4			
5			
6			
7			
8			
12			
16			
20			
24			

b. Let x stand for the number of $0.50 price increases the owner makes. Write algebraic expressions for the price of a tape, the number of tapes sold, and the total revenue.

c. Use your calculator to make a table of values for your algebraic expressions. Let Y_1 stand for the price of a tape, Y_2 for the number of tapes sold, and Y_3 for the total revenue. Verify the values you calculated in part (a).

d. Use your table to find a value of x for which the total revenue is zero.

e. Use your graphing calculator to graph your formula for total revenue.

f. How much should the owner charge for a tape in order to bring in $630 per week from tapes? (You should have two answers.)

g. What is the maximum revenue the owner can earn from tapes in one week? How much should he charge for a tape to maximize his revenue? How many tapes will he sell at that price?

4.2 Curve Fitting

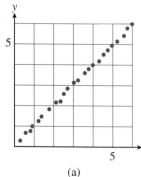

(a)

(b)

FIGURE 4.15

In Section 1.5 we used linear regression to "fit" a line through a collection of data points. If the data points do not cluster around a line, it does not make sense to describe them by a linear equation. Compare the data points in Figures 4.15a and 4.15b. The points in Figure 4.15a are roughly linear in appearance, but the points in Figure 4.15b are not. However, we can visualize a parabola that would approximate the data. In this section we'll see how to fit a quadratic equation to a collection of data points.

Finding a Quadratic Equation through Three Points

Every linear equation can be written in the form

$$y = mx + b$$

To find a specific line, we must find values for the two parameters (constants) m and b. We need two data points in order to find those two parameters. A quadratic equation, however, has three parameters, a, b, and c:

$$y = ax^2 + bx + c$$

To find these parameters, we need three data points. Here is an example of how to find the parabola that passes through three given points.

EXAMPLE 1   Find values for a, b, and c so that the points $(1, 3)$, $(3, 5)$, and $(4, 9)$ lie on the graph of $y = ax^2 + bx + c$.

Solution

Substitute the coordinates of each of the three points into the equation of the parabola to obtain three equations:

$$3 = a(1)^2 + b(1) + c$$
$$5 = a(3)^2 + b(3) + c$$
$$9 = a(4)^2 + b(4) + c$$

or, equivalently,

$$a + b + c = 3 \qquad (1)$$
$$9a + 3b + c = 5 \qquad (2)$$
$$16a + 4b + c = 9 \qquad (3)$$

This is a system of three equations in the three unknowns a, b, and c. (See Section 2.3 to review systems of three linear equations.) To solve the system, we first eliminate c. Subtract Equation (1) from Equation (2) to obtain

$$8a + 2b = 2 \qquad (4)$$

and subtract Equation (1) from Equation (3) to get

$$15a + 3b = 6 \qquad (5)$$

Now eliminate b from Equations (4) and (5): Add -3 times Equation (4) to 2 times Equation (5) to get

$$-24a - 6b = -6 \qquad \text{-3 times Equation (4)}$$
$$\underline{30a + 6b = 12} \qquad \text{2 times Equation (5)}$$
$$6a = 6$$

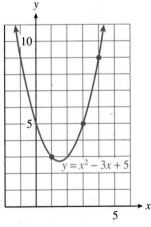

or $a = 1$. Substitute 1 for a in Equation (4) to find

$$8(1) + 2b = 2 \quad \text{Solve for b.}$$
$$b = -3$$

Finally, substitute -3 for b and 1 for a in Equation (1) to find

$$1 + (-3) + c = 3 \quad \text{Solve for c.}$$
$$c = 5$$

Thus, the equation of the parabola is

$$y = x^2 - 3x + 5$$

The parabola and the three points are shown in Figure 4.16.

FIGURE 4.16

■

The simplest way to fit a parabola to a set of data points is to pick three of the points and find the equation of the parabola that passes through those three points.

EXAMPLE 2

Major Motors Corporation is testing a new car designed for in-town driving. The data in Table 4.2 show the cost of driving the car at different speeds. The speeds, v, are given in miles per hour, and the cost, C, includes fuel and maintenance for driving the car 100 miles at that speed.

TABLE 4.2

v	30	40	50	60	70
C	6.50	6.00	6.20	7.80	10.60

Find a possible quadratic model,

$$C = av^2 + bv + c$$

that expresses C in terms of v.

Solution

When the data are plotted, it is clear that the relationship between v and C is not linear, but it may be quadratic. (See Figure 4.17.) We will use the last three data points, $(50, 6.20)$, $(60, 7.80)$, and $(70, 10.60)$, to fit a parabola to the data. We would like to find the coefficients a, b, and c of a parabola $C = av^2 + bv + c$ that includes the three data points. This gives us a system of equations:

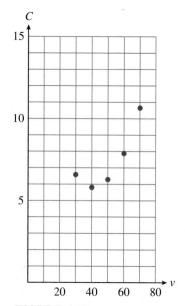

FIGURE 4.17

$$2500a + 50b + c = 6.20 \tag{1}$$
$$3600a + 60b + c = 7.80 \tag{2}$$
$$4900a + 70b + c = 10.60 \tag{3}$$

Eliminating c from Equations (1) and (2) yields Equation (4), and eliminating c from Equations (2) and (3) yields Equation (5).

$$1100a + 10b = 1.60 \tag{4}$$
$$1300a + 10b = 2.80 \tag{5}$$

Eliminating b from Equations (4) and (5) gives us

$$200a = 1.20$$
$$a = 0.006$$

Substitute this value into Equation (4) to find $b = -0.5$, then substitute both values into Equation (1) to find $c = 16.2$. Thus, our quadratic model is

$$C = 0.006v^2 - 0.5v + 16.2$$

The graph of this equation, along with the data points, is shown in Figure 4.18.

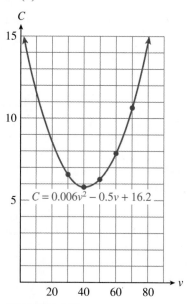

$C = 0.006v^2 - 0.5v + 16.2$

FIGURE 4.18

a. Find the equation of a parabola

$$y = ax^2 + bx + c$$

that passes through the points (0, 80), (15, 95), and (25, 55).

b. Plot the data points and sketch the graph on the same grid.

Using the Vertex Form

If one of the data points happens to be the vertex of the parabola, we can use the vertex form to find its equation.

EXAMPLE 3

When Andre practices free-throws at the park, the ball leaves his hands at a height of 7 feet, and reaches the vertex of its trajectory 10 feet away at a height of 11 feet.

    a. Find a quadratic equation for the ball's trajectory.

    b. Do you think Andre's free-throw would score on a basketball court where the hoop is 15 feet from the shooter and 10 feet high?

Solution

    a. If Andre's feet are at the origin, then the vertex of the ball's trajectory is the point (10, 11), and its y-intercept is (0, 7). Start with the vertex form for a parabola.

$$y = a(x - x_v)^2 + y_v$$
$$y = a(x - 10)^2 + 11$$

Then use the point (0, 7) to find the value of a.

$$7 = a(0 - 10)^2 + 11$$
$$7 = 100a + 11$$
$$a = -0.04$$

The equation of the trajectory is $y = -0.04(x - 10)^2 + 11$.

    b. We'd like to know if the point (15, 10) is on the trajectory of Andre's free-throw. Substitute $x = 15$ into the equation.

$$y = -0.04(15 - 10)^2 + 11$$
$$= -0.04(25) + 11 = 10$$

Andre's shot will score.   ■

EXERCISE 2 Francine is designing a synchronized fountain display for a hotel in Las Vegas. For each fountain, water emerges in a parabolic arc from a nozzle three feet above the ground. Francine would like the vertex of the arc to be eight feet high, and two feet horizontally from the nozzle.

a. Choose a coordinate system and write an equation for the path of the water.

b. How far from the base of the nozzle will the stream of water hit the ground?

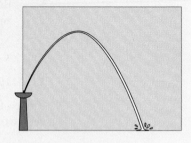

Using a Calculator for Quadratic Regression

We can use a graphing calculator to find an approximate "quadratic fit" for a set of data. The procedure for quadratic regression closely follows the steps outlined in Section 2.5 for linear regression.

EXAMPLE 4

a. Use your calculator to find a quadratic fit for the data in Example 2.

b. How many of the given data points actually lie on the graph of the quadratic approximation?

Solution

a. Press [STAT] [ENTER] and enter the data under columns L_1 and L_2, as shown in Figure 4.19(a). Calculate the quadratic regression equation and store it in Y_1 by pressing [STAT] [▷] [5] [VARS] [▷] [1] [1] [ENTER]. The regression equation has the form $y = ax^2 + bx + c$ where $a \approx 0.0057$, $b \approx -0.47$, and $c \approx 15.56$ as shown in Figure 4.19(b). Notice that a, b, and c are all close to the values we computed in Example 2.

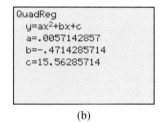

(a)   (b)

FIGURE 4.19

b. Next we'll graph the data and the regression equation. Press [Y =] and select **Plot1**, then press [ZOOM] [9] to see the graph shown in Figure 4.20(a). The parabola seems to pass through or very close to all the data points. However, we can use either the **value** feature or a table with

$x = 30, 40, 50, 60, 70$ to find the y-coordinates of the corresponding points on the graph of the regression equation. By comparing these y-coordinates with our original data points, we find that none of the given data points lies precisely on the parabola. (See Figure 4.20(b).)

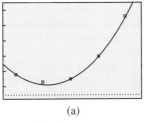

| (a) | (b) |

FIGURE 4.20   ∎

EXERCISE 3   To test the effects of radiation, a researcher irradiated male mice with various dosages and bred them to unexposed female mice. The table below shows the fraction of fertilized eggs that survived, in terms of the radiation dosage.

Radiation (1000 rems)	0.1	0.3	0.5	0.7	0.9	1.1	1.5
Relative survival of eggs (%)	94	70	54.4	42.4	36.6	27.7	19.5

a. Enter the data into your calculator and create a scatterplot. Does the graph appear to be linear? Does it appear to be quadratic?

b. Fit a quadratic regression equation to the data, and graph the equation on the scatterplot.

Source: Strickberger, Monroe W., *Genetics*, Macmillan, 1976

We must be careful that our data set gives a complete picture of the situation we want to model. A regression equation may fit a particular collection of data and still be a poor model if the rest of the data diverge from the regression graph. In Example 2, suppose Major Motors had collected only the first three data points and fit a line through them, as shown in Figure 4.21. This regression line gives poor predictions for the cost of driving at 60 or 70 miles per hour.

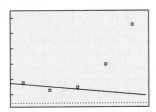

FIGURE 4.21

EXAMPLE 5   Francine records the height of the tip of the minute hand on the classroom's clock at different times. The data are shown in the table, where time is measured in minutes since noon. (A negative time indicates a number of minutes before noon.) Find a quadratic regression equation for the data and use it to predict the height of the minute hand's tip at 40 minutes past noon. Do you believe this prediction is valid?

Time (minutes)	−25	−20	−15	−10	−5	0	5	10	15	20	25
Height (feet)	7.13	7.50	8.00	8.50	8.87	9.00	8.87	8.50	8.00	7.50	7.13

Solution

Enter the time data under L_1 and the height data under L_2. Calculate and store the quadratic regression equation in Y_1 as you did in Example 4. The regression equation is

$$y \approx -0.00297x^2 + 0x + 8.834$$

From the graph of the regression equation or from the table (see Figure 4.22) you can see that the fit is not perfect, although the curve certainly fits the data better than any straight line could.

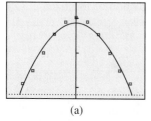
(a)

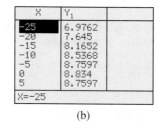
(b)

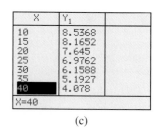
(c)

FIGURE 4.22

Scroll down the table to find that this equation predicts a height of approximately 4.08 feet at time 40 minutes. (See Figure 4.22(c).) This is a preposterous estimate! The position of the minute hand at 40 minutes after noon should be the same as it was exactly one hour earlier (at 20 minutes before noon), when it was 7.50 feet. ∎

Using the wrong type of equation to fit the data is a common error in making predictions. We know that the position of the minute hand of a clock will repeat itself every 60 minutes. The graph of the height of its tip oscillates up and down, repeating the same pattern over and over. We cannot describe such a graph using a straight line or a quadratic equation. The graph of the height is shown in Figure 4.23, along with the graph of our quadratic regression equation. You can see that the regression equation fits the actual curve only on a small interval.

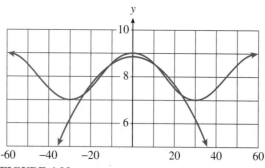

FIGURE 4.23

Your calculator can always compute a regression equation, but that equation is not necessarily appropriate for your data. Choosing a reasonable type of regression equation for a particular data set requires knowledge of different kinds of models and the physical or natural laws that govern the situation at hand.

1. How can you find the equation of a parabola that passes through three given points?

2. If you know the vertex of a parabola, how many more points do you need in order to find its equation?

3. How can you decide whether linear regression, quadratic regression, or neither one is appropriate for a collection of data?

ANSWERS TO 4.2 EXERCISES

1. $y = \frac{-1}{5}x^2 + 4x + 80$

2a. With the origin on the ground directly below the nozzle,
$y = \frac{-5}{4}x^2 + 5x + 3$.

b. Approximately 4.53 feet.

3a. The graph appears to be quadratic.

b. $y = 36.5x^2 - 108.8x + 101.9$

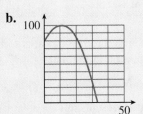

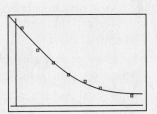

HOMEWORK 4.2

Find a quadratic equation that fits the data points.

1. Find values for a, b, and c so that the graph of the parabola $y = ax^2 + bx + c$ includes the points $(-1, 0)$ $(2, 12)$, and $(-2, 8)$.

2. Find values for a, b, and c so that the graph of the parabola $y = ax^2 + bx + c$ includes the points $(-1, 2)$ $(1, 6)$, and $(2, 11)$.

3. A survey to determine what percent of different age groups regularly use marijuana collected the following data.

Age	15	20	25	30
Percent	4	13	11	7

a. Use the percentages for ages 15, 20, and 30 to fit a quadratic equation to the data,

$$P = ax^2 + bx + c$$

where x represents age.

b. What does your equation predict for the percentage of 25-year-olds who use marijuana?

c. Sketch the graph of your quadratic equation and the given data on the same axes.

4. The following data show the number of people of certain ages who were victims of homicide in a large city last year.

Age	10	20	30	40
Number of victims	12	62	72	40

a. Use the first three data points to fit a quadratic equation to the data,

$$N = ax^2 + bx + c$$

where x represents age.

b. What does your equation predict for the number of 40-year-olds who were the victims of homicide?

c. Sketch the graph of your quadratic equation and the given data on the same axes.

5. The data below show Americans' annual per capita consumption of chicken for several years since 1970.

Year	Pounds of chicken
1970	27.7
1975	26.4
1980	32.5
1985	36.1
1989	40.5
1990	42.1
1991	43.9
1992	45.9

Source: *Statistical Abstract of the United States, 1994–1995*

a. Use the values for 1970, 1985, and 1990 to fit a quadratic equation to the data,

$$C = at^2 + bt + c$$

where t is measured in years since 1970.

b. What does your equation predict for per capita chicken consumption in 1992?

c. Sketch the graph of your equation and the given data on the same axes.

6. The data show sales of in-line skates at a sporting goods store at the beach.

Year	1990	1991	1992	1993	1994
Skates sold	54	82	194	446	726

a. Use the values for 1991 through 1993 to fit a quadratic equation to the data,

$$S = at^2 + bt + c$$

where t is measured in years since 1990.

b. What does your equation predict for the number of pairs of skates sold in 1994?

c. Sketch the graph of your equation and the given data on the same axes.

7. Find a quadratic formula for the number of diagonals that can be drawn in a polygon of n sides. Some data are provided.

Sides	4	5	6	7
Diagonals	2	5	9	14

8. You are driving at 60 miles per hour when you step on the brakes. Find a quadratic formula for the distance in feet that your car travels in t seconds after braking. Some data are provided.

Seconds	1	2	3	4
Feet	81	148	210	240

9. Find the equation for a parabola that has a vertex at (30, 280) and passes through the point (20, 80).

10. Find the equation for a parabola that has a vertex at $(-12, -40)$ and passes through the point (6, 68).

11. In skeet shooting, the clay pigeon is launched from a height of 4 feet and reaches a maximum height of 164 feet at a distance of 80 feet from the launch site.

a. Write an equation for the height of the clay pigeon in terms of the horizontal distance it has traveled.

b. If the shooter misses the clay pigeon, how far from the launch site will it hit the ground?

12. The batter in a softball game hits the ball when it is 4 feet above the ground. The ball reaches the greatest height on its trajectory, 35 feet, directly above the head of the left-fielder, who is 200 feet from home plate.

    a. Write an equation for the height of the softball in terms of its horizontal distance from home plate.

    b. Will the ball clear the left field wall, which is 10 feet tall and 375 feet from home plate?

13. The cables on a suspension bridge hang in the shape of parabolas. Imagine a coordinate system superimposed on a diagram of a suspension bridge, as shown in Figure 4.24. Each of the towers is 500 feet high, and the span between the towers is 4000 feet long. At its lowest point, the cable hangs 20 feet above the roadway. Find the coordinates of the vertex and one other point on the cable, and use them to find an equation for the shape of the cable in vertex form.

14. Some comets move about the sun in parabolic orbits. In 1973 the comet Kohoutek passed within 0.14 AU (astronomical units), or 21 million kilometers, of the sun. Imagine a coordinate system superimposed on a diagram of the comet's orbit, as shown in Figure 4.25. In that system the comet's coordinates at perihelion (its closest approach to the sun) were (0, 0.14). When the comet was first discovered, its coordinates were (1.68, −4.9). Find an equation for comet Kohoutek's orbit in vertex form.

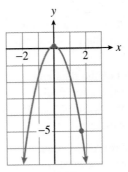

FIGURE 4.25

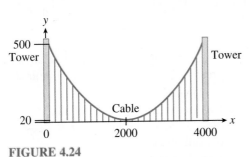

FIGURE 4.24

Use your calculator's statistics features for Problems 15–20.

15. The table shows the height of a projectile in meters at different times after it was fired.

Time (seconds)	2	4	6	8	10	12	14
Height (meters)	39.2	71.8	98.0	117.8	131.0	137.8	138.0

    a. Find the equation of the least-squares regression line for height in terms of time.

    b. Use the linear regression equation to predict the height of the projectile 15 seconds after it was fired.

    c. Make a scatterplot of the data and draw the regression line on the same axes.

    d. Find the quadratic regression equation for height in terms of time.

    e. Use the quadratic regression equation to predict the height of the projectile 15 seconds after it was fired.

    f. Draw the quadratic regression curve on the graph from part (c).

    g. Which model is more appropriate for the height of the projectile, linear or quadratic? Why?

16. The table shows the height in kilometers of a star-flare at different times after it exploded from the surface of a star.

Time (seconds)	0.2	0.4	0.6	0.8	1.0	1.2
Height (kilometers)	6.8	12.5	17.1	20.5	22.8	23.9

a. Find the equation of the least-squares regression line for height of the flare in terms of time.

b. Use the linear regression equation to predict the height of the flare 1.4 seconds after it exploded.

c. Make a scatterplot of the data and draw the regression line on the same axes.

d. Find the quadratic regression equation for height in terms of time.

e. Use the quadratic regression equation to predict the height of the flare 1.4 seconds after it exploded.

f. Draw the quadratic regression curve on the graph from part (c).

g. Which model is more appropriate for the height of the star-flare, linear or quadratic? Why?

17. The number of daylight hours increases each day from the beginning of winter until the beginning of summer, and then begins to decrease. The table below gives the number of daylight hours in Delbert's home town last year in terms of the number of days since January 1.

Days since January 1	0	50	100	150	200	250	300
Hours of daylight	9.8	10.9	12.7	14.1	13.9	12.5	10.7

a. Find the equation of the least-squares regression line for the number of daylight hours in terms of the number of days since January 1.

b. Use the linear regression equation to predict the number of daylight hours 365 days after January 1.

c. Make a scatterplot of the data and draw the regression line on the same axes.

d. Find the quadratic regression equation for the number of daylight hours in terms of the number of days since January 1.

e. Use the quadratic regression equation to predict the number of daylight hours 365 days after January 1.

f. Draw the quadratic regression curve on the graph from part (c).

g. Predict the number of daylight hours 365 days since January 1 without using any regression equation. What does this tell you about the linear and quadratic models you found?

18. To observers on earth, the moon looks like a disk that is completely illuminated at full moon and completely dark at new moon. The table below shows what fraction of the moon is illuminated at 5-day intervals after the last full moon.

Days since full moon	0	5	10	15	20	25
Fraction illuminated	1.000	0.734	0.236	0.001	0.279	0.785

a. Find the equation of the least-squares regression line for the fraction illuminated in terms of days.

b. Use the linear regression equation to predict the fraction illuminated 30 days after the full moon.

c. Make a scatterplot of the data and draw the regression line on the same axes.

d. Find the quadratic regression equation for the fraction illuminated in terms of days.

e. Use the quadratic regression equation to predict the fraction illuminated 30 days after the full moon.

f. Draw the quadratic regression curve on the graph from part (c).

g. Predict the fraction of the disk that is illuminated 30 days after the full moon without using any regression equation. What does this tell you about the linear and quadratic models you found?

19. a. Find a quadratic regression equation for the growth rate of *Parus major* in terms of its weight, using the data from Problem 38 of Section 4.1.

b. Make a scatterplot of the data and draw the regression curve on the same axes.

c. Find the vertex of the graph of the regression equation. How does this estimate for the maximum growth rate compare with your estimate in Problem 38?

20. a. Find a quadratic regression equation for the growth rate of *Parus caeruleus* in terms of its weight, using the data from Problem 38 of Section 4.1.

b. Make a scatterplot of the data and draw the regression curve on the same axes.

c. Find the vertex of the graph of the regression equation. How does this estimate for the maximum growth rate compare with your estimate in Problem 38?

4.3 Problem Solving

Maximum or Minimum Values

Finding the maximum or minimum value for a variable expression is a common problem in applications. For example, if you own a company that manufactures blue jeans, you might like to know how much to charge for your jeans in order to maximize your revenue. Recall that

Revenue = (price of one item) (number of items sold)

As you increase the price of your jeans, your revenue may increase for a while. But if you charge too much for your jeans, consumers will not buy as many pairs, and your revenue may actually start to decrease. Is there some optimum price you should charge for a pair of jeans in order to achieve the greatest revenue?

EXAMPLE 1

Late Nite Blues finds that it can sell $600 - 15x$ pairs of jeans per week if they charge x dollars per pair. (Notice that as the price increases, the number of pairs of jeans sold decreases.)

a. Write an equation for the revenue in terms of the price of a pair of jeans.

b. Graph the equation.

c. How much should Late Nite Blues charge for a pair of jeans in order to maximize their revenue?

Solutions

a. Using the formula for revenue stated above, we find

Revenue = (price of one item) (number of items sold)
$$R = x(600 - 15x)$$
$$R = 600x - 15x^2$$

b. We recognize the equation as quadratic, so the graph is a parabola. You can use your calculator to verify the graph in Figure 4.26.

c. The maximum value of R occurs at the vertex of the parabola. Thus,

$$x_v = \frac{-b}{2a} = \frac{-600}{2(-15)} = 20$$

$$R_v = 600(20) - 15(20)^2 = 6000$$

The revenue takes on its maximum value when $x = 20$, and the maximum value is $R = 6000$. This means that Late Nite Blues should charge $20 for a pair of jeans in order to maximize revenue at $6000 a week.

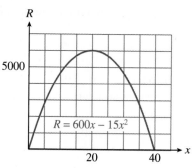

FIGURE 4.26

If the equation relating two variables is quadratic, then the maximum or minimum value is easy to find: It is the value at the vertex. If the parabola opens downward, as in Example 1, there is a maximum value at the vertex. If the parabola opens upward, there is a minimum value at the vertex. In Exercise 1 you'll minimize a quadratic expression to find a line of best fit.

EXERCISE 1   The figure shows a set of three data points and a line of best fit. For this example, the regression line passes through the origin, so its equation is $y = mx$ for some positive value of m. How shall we choose m to give the best fit for the data? We want the data points to lie as close to the line as possible. One way to achieve this is to minimize the sum of the squares of the vertical distances shown in the figure.

a. The data points are $(1, 2)$, $(2, 6)$, and $(3, 7)$. Verify that the sum S we want to minimize is

$$S = (2 - m)^2 + (6 - 2m)^2 + (7 - 3m)^2$$
$$= 14m^2 - 70m + 89$$

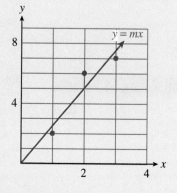

b. Graph the formula for S in the window

Xmin = 0	Xmax = 9.4
Ymin = 0	Ymax = 100

c. Find the vertex of the graph of S.

$$m_v =$$
$$S_v =$$

d. Use the value of m to write the equation of the regression line $y = mx$.

e. Graph the three data points and your regression line on the same axes.

Systems Involving Quadratic Equations

Recall that the solution to a 2×2 system of linear equations is the intersection point of the graphs of the equations. This is also true of systems in which one or both of the equations is quadratic. Such a system may have either one solution, two solutions, or no solution, as you can see in Figure 4.27.

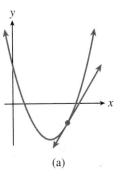
(a)

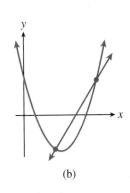
(b)

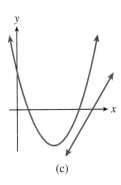
(c)

FIGURE 4.27

In Example 2 we will use both graphical and algebraic techniques to solve the system.

EXAMPLE 2

The Pizza Connection calculates that the cost in dollars of producing x pizzas per day is given by

$$C = 0.15x^2 + 0.75x + 180$$

The Pizza Connection charges \$15 per pizza, so the revenue from selling x pizzas is

$$R = 15x$$

How many pizzas per day must the Pizza Connection sell in order to break even?

Solution

Recall that to "break even" means to make zero profit. Since

$$Profit = Revenue - Cost$$

the break-even point occurs when the revenue equals the cost. In mathematical terms, we would like to find any values of x for which $R = C$. If we graph the revenue and cost equations on the same axes, these values correspond to points where the two graphs intersect. Use the WINDOW settings

$$Xmin = 0, \quad Xmax = 94$$
$$Ymin = 0, \quad Ymax = 1400$$

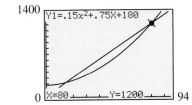

FIGURE 4.28

on your calculator to obtain the graph shown in Figure 4.28. You can verify that the two intersection points are $(15, 225)$ and $(80, 1200)$.

Thus, the Pizza Connection must sell either 15 or 80 pizzas in order to break even. On the graph we see that the revenue is greater than cost for x-values between 15 and 80, so the Pizza Connection will make a profit if they sell between 15 and 80 pizzas.

We can also solve algebraically for the break-even points. The intersection points of the two graphs correspond to the solutions of the system of equations

$$y = 0.15x^2 + 0.75x + 180$$
$$y = 15x$$

We equate the two expressions for y, and solve for x.

$$0.15x^2 + 0.75x + 180 = 15x \qquad \text{Subtract } 15x \text{ from both sides.}$$
$$0.15x^2 - 14.25x + 180 = 0 \qquad \text{Use the quadratic formula.}$$
$$x = \frac{14.25 \pm \sqrt{14.25^2 - 4(0.15)(180)}}{2(0.15)} \qquad \text{Simplify.}$$
$$= \frac{14.25 \pm 9.75}{0.3}$$

The solutions are 15 and 80, as we found from the graph in Figure 4.28. ∎

EXERCISE 2

a. Solve the system algebraically.

$$y = x^2 - 6x - 7$$
$$y = 13 - x^2$$

b. Graph both equations, and show the solutions on the graph.

Solving Systems with the Graphing Calculator

In Chapter 2 we used the **intersect** feature of the graphing calculator to solve systems of linear equations, and we can solve systems of quadratic equations in the same way. Consider the system of two quadratic equations

$$y = (x + 1.1)^2$$
$$y = 7.825 - 2x - 2.5x^2$$

We will graph these two equations in the standard window. The two intersection points are visible in the window, but we do not find their exact coordinates when we trace the graphs. Use the **intersect** command to locate one of the solutions, as shown in Figure 4.29. (See Section 2.1 to review the steps for using the **intersect** feature.) You can check that the point $(0.9, 4)$ is an exact solution to the system by substituting $x = 0.9$ and $y = 4$ into *each* equation of the system. (The calculator is not always able to find the exact coordinates, but it usually gives a very good approximation.)

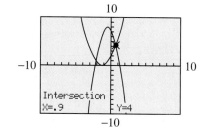

FIGURE 4.29

You can find the other solution of the system by following the same steps and moving the bug close to the other intersection point. You should verify that the other solution is the point $(-2.1, 1)$.

1. How can you tell whether a variable given by a quadratic equation has a maximum value or a minimum value?

2. What is wrong with this statement: The maximum or minimum value given by a quadratic equation is the average of the x-intercepts?

3. How many solutions does a system of one linear equation and one quadratic equation have? Illustrate with sketches.

4. How many solutions does a system of two quadratic equations have? Illustrate with sketches.

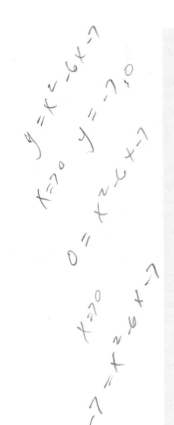

ANSWERS TO 4.3 EXERCISES

1b.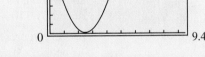

c. $\left(\frac{5}{2}, \frac{3}{2}\right)$ **d.** $y = \frac{5}{2}x$

e.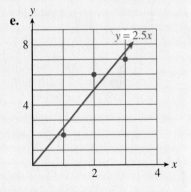

2a. $(-2, 9), (5, -12)$

b.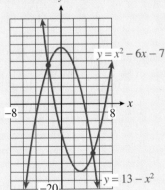

For each problem, (a) find the maximum or minimum value algebraically, (b) obtain a good graph on your calculator, and verify your answer. (Use the coordinates of the vertex and the vertical intercept to help you choose an appropriate window for the graph.)

1. Delbert launches a toy water rocket from ground level. Its distance above the ground t seconds after launch is given in feet by $d = 96t - 16t^2$. When will the rocket reach its greatest height, and what will that height be?

2. Francine throws a wrench into the air from the bottom of a trench 12 feet deep. Its height t seconds later is given in feet by $h = -12 + 32t - 16t^2$. When will the wrench reach its greatest height, and what will that height be?

   $lt = 2400 \times -100$

3. As part of a collage for her art class, Sheila wants to enclose a rectangle with 100 inches of yarn.

   a. Let w represent the width of the rectangle, and write an expression for its length. Then write an expression in terms of w for the area A of the rectangle.

   b. What is the area of the largest rectangle that Sheila can enclose with 100 inches of yarn?

4. Gavin has rented space for a booth at the county fair. As part of his display, he wants to rope off a rectangular area with 80 yards of rope.

   a. Let w represent the width of the roped-off rectangle, and write an expression for its length. Then write an expression in terms of w for the area A of the roped-off space.

   b. What is the largest area that Gavin can rope off? What will the dimensions of the rectangle be?

5. A farmer plans to fence a rectangular grazing area along a river with 300 yards of fence as shown in Figure 4.30.

   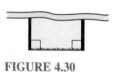

   FIGURE 4.30

   a. Write an expression for the area A of the grazing land in terms of the width w of the rectangle.

   b. What is the largest area he can enclose?

6. A breeder of horses wants to fence two rectangular grazing areas along a river with 600 meters of fence as shown in Figure 4.31.

   a. Write an expression for the total area A of the grazing land in terms of the width w of the rectangles.

   b. What is the largest area she can enclose?

   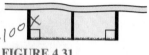

   FIGURE 4.31

7. A travel agent offers a group rate of $2400 per person for a week in London if 16 people sign up for the tour. For each additional person who signs up, the price per person is reduced by $100.

   a. Let x represent the number of additional people who sign up. Write expressions for the total number of people signed up, the price per person, and the total revenue.

   b. How many people must sign up for the tour in order for the travel agent to maximize her revenue?

8. An entrepreneur buys an apartment building with 40 units. The previous owner charged $240 per month for a single apartment and on average rented 32 apartments at that price. The entrepreneur discovers that for every $20 he raises the price another apartment stands vacant.

   a. Let x represent the number of $20 price increases. Write expressions for the new price, the number of rented apartments, and the total revenue.

   b. What price should the entrepreneur charge for an apartment in order to maximize his revenue?

9. The owners of a small fruit orchard decide to produce gift baskets as a sideline. The cost per basket for producing x baskets is

$$C = 0.01x^2 - 2x + 120$$

How many baskets should they produce in order to minimize the cost per basket? What will their total cost be at that production level?

10. A new electronics firm is considering marketing a line of telephones. The cost per phone for producing x telephones is

$$C = 0.001x^2 - 3x + 2270$$

How many telephones should they produce in order to minimize the cost per phone? What will their total cost be at that production level?

11. Starlings often feed in flocks, and their rate of feeding depends on the size of the flock. If the flock is too small, the birds are nervous and spend a lot of time watching for predators. If the flock is too large, the birds become over-crowded and fight each other, which interferes with feeding. Here are some data gathered at a feeding station. The data show the number of starlings in the flock and the total number of pecks per minute recorded at the station while the flock was feeding.

Number of starlings	Pecks per minute	Pecks per starling per minute
1	9	
2	26	
3	48	
4	80	
5	120	
6	156	
7	175	
8	152	
9	117	
10	180	
12	132	

Source: Chapman and Reiss, 1992.

a. For each flock size, calculate the number of pecks per starling per minute. For purposes of efficient feeding, what flock size appears to be optimum? How many pecks per minute would each starling make in a flock of optimal size?

b. Plot the number of pecks per starling per minute against flock size. Do the data points appear to lie on (or near) a parabola?

c. The quadratic regression equation for the data is $y = -0.45x^2 + 5.8x + 3.9$. Graph this parabola on the same axes with the data points.

d. What are the optimum flock size and the maximum number of pecks per starling per minute predicted by the regression equation?

12. Biologists conducted a four-year study of the nesting habits of the species *Parus major* in an area of England called Wytham Woods. The bar graph in Figure 4.32 shows the clutch size (the number of eggs) in 433 nests.

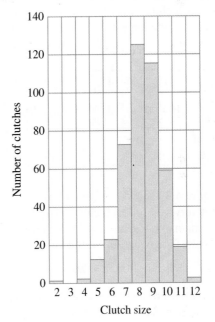

FIGURE 4.32

Source: Perrins and Moss, 1975.

a. Which clutch size was observed most frequently? Fill in the table showing the total number of eggs produced in each clutch size.

Clutch size	2	3	4	5	6	7
Number of clutches	1	0	2	12	23	73
Number of eggs						

Clutch size	8	9	10	11	12
Number of clutches	126	116	59	19	3
Number of eggs					

b. The average weight of the nestlings declines as the size of the brood increases, and the survival of individual nestlings is linked to their weight. A hypothetical (and simplified) model of this phenomenon is described by the table below. Calculate the number of surviving nestlings for each clutch size. Which clutch size produces the largest average number of survivors?

Clutch size	1	2	3	4	5	6	7	8	9	10
Percent survival	100	90	80	70	60	50	40	30	20	10
Number of survivors										

c. Figure 4.33 shows the percent of nestlings that survived for each clutch size in Wytham Woods, along with the curve of best fit. The equation for the curve is

$$y = -1.05x^2 + 20x - 3.5$$

Find the optimal clutch size for maximizing the number of surviving nestlings. How does this optimum clutch size compare with the most frequently observed clutch size in part (a)?

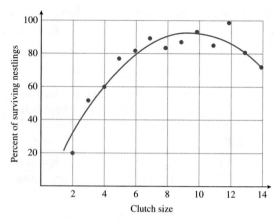

FIGURE 4.33

13. During a statistical survey, a public interest group obtains two estimates for the average monthly income of young adults aged 18 to 25. The first estimate is \$860 and the second estimate is \$918. To refine their estimate, they will take a weighted average of these two figures:

$$I = 860a + 918(1 - a) \qquad \text{where } 0 \le a \le 1$$

To get the best estimate, they must choose a to minimize the expression

$$V = 576a^2 + 5184(1 - a)^2$$

(The numbers that appear in this expression reflect the **variance** of the data, which measures how closely the data cluster around the mean, or average.) Find the value of a that minimizes V, and use this value to get a refined estimate for the average income.

14. The rate at which an antigen precipitates during an antigen–antibody reaction depends on the amount of antigen present. For a fixed quantity of antibody, the time required for a particular antigen to precipitate is given in minutes by

$$t = 2w^2 - 20w + 54$$

where w is the quantity of antigen present, in grams. For what quantity of antigen will the reaction proceed most rapidly, and how long will the precipitation take?

Solve each system algebraically, and verify your solutions with a graph.

15. $y = x^2 - 4x + 7$
$y = 11 - x$

16. $y = x^2 + 6x + 4$
$y = 3x + 8$

17. $y = -x^2 - 2x + 7$
$y = 2x + 11$

18. $y = x^2 - 8x + 17$
$y + 4x = 13$

19. $y = x^2 + 8x + 8$
$3y + 2x = -36$

20. $y = -x^2 + 4x + 2$
$4y - 3x = 24$

21. $y = x^2 - 9$
$y = -2x^2 + 9x + 21$

22. $y = 4 - x^2$
$y = 3x^2 - 12x - 12$

23. $y = x^2 - 0.5x + 3.5$
$y = -x^2 + 3.5x + 1.5$

24. $y = x^2 + 10x + 22$
$y = -0.5x^2 - 8x - 32$

25. $y = x^2 - 4x + 4$
$y = x^2 - 8x + 16$

26. $y = 0.5x^2 + 3x + 5.5$
$y = 2x^2 + 12x + 4$

Problems 27–30 deal with wildlife management and sustainable yield.

27. In Problem 31 of Section 3.4 you graphed the annual growth rate of a population of fish

$$y = 0.4x - 0.0001x^2$$

where x is the current biomass of the population, in tons.

a. Suppose that fishermen harvest 300 tons of fish each year. Sketch the graph of $H = 300$ on the same axes with your graph of y.

b. If the biomass is currently 2500 tons and 300 tons are harvested, will the population be larger or smaller next year? By how much? What if the biomass is currently 3500 tons?

c. What sizes of biomass will remain stable from year to year if 300 tons are harvested annually?

d. If the biomass ever falls below 1000 tons, what will happen after several years of harvesting 300 tons annually?

28. In Problem 48 of Section 3.3 you graphed the annual increase, I, in the deer population in a national park

$$I = kCx - kx^2$$

where $k = 0.0002$, $C = 6000$, and x is the current population.

a. Suppose hunters are allowed to kill 1000 deer per year. Sketch the graph of $H = 1000$ on the same axes with a graph of y.

b. What sizes of deer population will remain stable from year to year if 1000 deer are hunted annually?

c. Suppose 1600 deer are killed annually. What sizes of deer population will remain stable?

d. What is the largest annual harvest that still allows for a stable population? (This harvest is called the maximum sustainable yield.) What is the stable population?

e. What eventually happens if the population falls below the stable value but hunting continues at the maximum sustainable yield?

29. The annual increase, N, in a bear population of size x is given by

$$N = 0.0002x(2000 - x)$$

if the bears are not hunted. The number of bears killed each year by hunters is related to the bear population by the equation $K = 0.2x$. (Notice that in this model, hunting is adjusted to the size of the bear population.)

a. Graph N and K on the same axes for $0 \le x \le 2000$.

b. When the bear population is 1200, which is greater, N or K? Will the population increase or decrease in the next year? By how many bears?

c. When the bear population is 900, will the population increase or decrease in the next year? By how many bears?

d. What sizes of bear population will remain stable after hunting?

e. What sizes of bear population will increase despite hunting? What sizes will decrease?

f. Toward what size will the population tend over time?

g. Suppose hunting limits are raised so that $K = 0.3x$. Toward what size will the population tend over time?

30. The annual increase in the biomass of a whale population is given in tons by

$$w = 0.001x(1000 - x)$$

where x is the current population in tons.

a. Sketch a graph of w for $0 \le x \le 1100$. What size biomass remains stable?

b. Each year hunters are allowed to harvest a biomass given by $H = 0.6x$. Sketch H on the same graph with w. What is the stable biomass with hunting?

c. What sizes of population will increase despite hunting? What sizes will decrease?

d. What size will the population approach over time? What biomass are hunters allowed to harvest for that size population?

e. Find a value of k so that the graph of $H = kx$ will pass through the vertex of $w = 0.001x(1000 - x)$.

f. For the value of k found in part (e), what size will the population approach over time? What biomass are hunters allowed to harvest for that size population?

g. Explain why the whaling industry should prefer hunting quotas of kx rather than $0.6x$ for a long-term strategy, even though $0.6x > kx$ for any positive value of x.

For Problems 31–34,
a. Find the break-even points by solving a system of equations.
b. Graph the equations for Revenue and Cost in the same window, and verify your solutions on the graph.
c. Use the fact that

$$\textit{Profit} = \textit{Revenue} - \textit{Cost}$$

to find the value of x for which profit is maximum.

31. Writewell, Inc. makes fountain pens. It costs Writewell $C = 0.0075x^2 + x + 2100$ dollars to manufacture x pens, and the company receives $R = 13x$ dollars in revenue from the sale of the pens.

32. It costs The Sweetshop $C = 0.01x^2 + 1836$ dollars to produce x pounds of chocolate creams. The company brings in $R = 12x$ dollars revenue from the sale of the chocolates.

33. It costs an appliance manufacturer $C = 1.625x^2 + 33{,}150$ dollars to produce x front-loading washing machines, which will then bring in revenues of $R = 650x$ dollars.

34. A company can produce x lawn mowers for a cost of $C = 0.125x^2 + 100{,}000$ dollars. The sale of the lawn mowers will generate $R = 300x$ dollars in revenue.

MIDCHAPTER REVIEW

1. Find the vertex of the graph of the equation $y = -2x^2 + 5x - 1$. Decide whether the vertex is a maximum point or a minimum point of the graph.

2. Find the vertex of the graph of the equation $y = -12 - 3x + 4x^2$. Decide whether the vertex is a maximum point or a minimum point of the graph.

3. Use the discriminant to determine how many x-intercepts the graph of $y = -2x^2 + 5x - 1$ has.

4. Use the discriminant to determine how many x-intercepts the graph of $y = -12 - 3x + 4x^2$ has.

5. a. Find the intercepts and the vertex of the graph of $y = x^2 + x - 6$.

  b. Sketch the graph by hand.

  c. Use your calculator to verify your graph.

6. a. Find the intercepts and the vertex of the graph of $y = 8 - x - 2x^2$.

  b. Sketch the graph by hand.

  c. Use your calculator to verify your graph.

7. a. Find the vertex of the graph of $y = \frac{1}{3}(x + 3)^2 - 2$.

  b. Write the equation of the parabola in standard form.

  c. Complete the table and sketch the graph.

x	-6	0	1
y			

8. a. Find the vertex of the graph of $y = -2(x - 1)^2 + 5$.

  b. Write the equation of the parabola in standard form.

  c. Complete the table and sketch the graph.

x	-1	0	2
y			

9. a. Find the equation of a parabola

$$y = ax^2 + bx + c$$

that passes through the points $(0, -2)$, $(-6, 1)$ and $(4, 6)$.

  b. Plot the data points and sketch the graph.

10. a. Find the equation of a parabola

$$y = ax^2 + bx + c$$

that passes through the points $(0, 8)$, $(-3, -46)$ and $(4, -60)$.

  b. Plot the data points and sketch the graph.

11. The Tallahatchee bridge is 42 feet above the river below, and the railing on the bridge is 4 feet tall. Billy Joe is standing 4 feet from the railing, and tosses something over in a parabolic path. The object is 3 feet above the bridge when it leaves his hand, and reaches the vertex of its trajectory 3 feet directly above the railing.

  a. Make a sketch illustrating the path of the object. Use a coordinate system with the origin at water level directly below the railing, where h is the height of the object and x is its horizontal displacement. Find the coordinates (x, h) of the object when it leaves Billy Joe's hand, and at the vertex of the parabolic path.

  b. Find an equation for the height, h, of the object.

  c. How far from the origin will the object be when it hits the water?

12. Magda Nold bought a restaurant with a yellow parabolic archway. The arch is 20 feet high at the vertex, and 16 feet wide at its base. Magda wants to enlarge the restaurant so that cars can drive under the arch to a take-out window. This will mean cutting one side of the arch so that the cut end rests on the 8.75-foot high roof of the take-out window.

  a. Choose a coordinate system and write an equation for the arch.

  b. What is the horizontal clearance between the take-out window and the base of the other side of the arch?

  c. What is the horizontal clearance between the take-out window and the arch at a height of five feet from the ground?

13. Max took a sequence of photographs of an explosion spaced at equal time intervals. From the photographs he was able to estimate the height and vertical velocity of some debris from the explosion, as shown in the table. (Negative velocities indicate that the debris is falling back to earth.)

Velocity (meters/second)	67	47	27	8	-12	-31
Height (meters)		8	122	196	232	228	185

  a. Enter the data into your calculator and create a scatterplot. Fit a quadratic regression equation to the data, and graph the equation on the scatterplot.

b. Use your regression equation to find the vertex of the parabola. What do the coordinates represent, in terms of the problem? What should the velocity of the debris be at its maximum height?

14. A speeding motorist slams on the brakes when she sees an accident directly ahead of her. The distance she has traveled t seconds after braking is shown in the table.

Time (seconds)	0	0.5	1.0	1.5	2.0	2.5
Distance (feet)	0	51	95	131	160	181

a. Enter the data into your calculator and create a scatterplot. Fit a quadratic regression equation to the data, and graph the equation on the scatterplot.

b. Use your regression equation to find the vertex of the parabola. What do the coordinates represent, in terms of the problem?

15. The total profit Kiyoshi makes from producing and selling x floral arrangements is

$$P = -0.4x^2 + 36x$$

a. How many floral arrangements should Kiyoshi produce and sell to maximize his profit?

b. What is his maximum profit?

c. Verify your answers on a graph.

16. The Metro Rail service sells $1200 - 80x$ fares each day when it charges x dollars per fare.

a. Write an equation for the revenue in terms of the price of a fare.

b. What fare will return the maximum revenue?

c. What is the maximum revenue?

d. Verify your answers on a graph.

17. a. Solve the system algebraically.

$$y = \frac{1}{2}x^2 - \frac{3}{2}x$$

$$y = -\frac{1}{2}x^2 + \frac{1}{2}x + 3$$

b. Graph both equations, and show the solutions on the graph.

18. a. Solve the system algebraically.

$$y = 2x^2 + 5x - 3$$
$$y = x^2 + 4x - 1$$

b. Graph both equations, and show the solutions on the graph.

4.4 Quadratic Inequalities

In Chapter 1 we used graphs to help us solve equations and inequalities in one variable. For instance, to solve the inequality

$$1.3x - 4.2 \geq 2.3$$

we look for x-values that make $1.3x - 4.2$ greater than or equal to 2.3. We can graph the equation

$$y = 1.3x - 4.2$$

as shown in Figure 4.34, and look for points whose y-coordinates are greater than or equal to 2.3. These points are shown in color on the graph. The x-coordinates of these points are the solutions of the inequal-

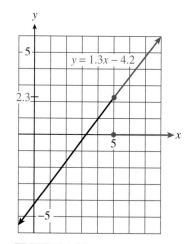

FIGURE 4.34

ity. From the graph, we see that all these points have x-coordinates greater than or equal to 5, so the solutions to the inequality are $x \geq 5$.

The graphing technique is also helpful for solving quadratic inequalities.

EXAMPLE 1

The Chamber of Commerce in River City wants to put on a Fourth of July fireworks display. City ordinance requires that fireworks at public gatherings explode at least 800 feet above the ground. The mayor particularly wants to include the Freedom Starburst model, which is launched from the ground so that its height after t seconds is given by

$$h = 256t - 16t^2$$

When should the Starburst explode in order to satisfy the safety ordinance?

Solution

We can get an approximate answer to this question by looking at the graph of the rocket's height shown in Figure 4.35. We would like to know when the rocket's height is greater than 800 feet, or, in mathematical terms, for what values of t is $h > 800$? The answer to this question is the solution of the inequality

$$256t - 16t^2 > 800$$

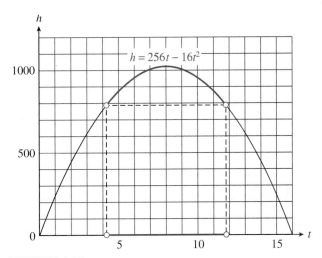

FIGURE 4.35

Points on the graph with $h > 800$ are shown in color, and the t-coordinates of those points are marked on the horizontal axis. If the Freedom Starburst explodes at any of these times, it will satisfy the safety regulation. From the graph, the safe time interval runs from approximately 4.25 seconds to 11.75 seconds after launch. The solution of the inequality is the set of all t-values greater than 4.25 but less than 11.75. ∎

The solution set in Example 1 is called a **compound inequality,** because it involves more than one inequality symbol. We write this set as

$$4.25 < t < 11.75$$

and read "t greater than 4.25 but less than 11.75."

You can also use your graphing calculator to solve the problem in Example 1. Graph the two equations

$$Y_1 = 256X - 16X^2$$
$$Y_2 = 800$$

on the same screen. (See Figure 4.36.) Use [WINDOW] settings to match the graph in Figure 4.35. Then use the [TRACE] or **intersect** feature to find the coordinates of the points (there are two of them) where the two graphs intersect. These points will have y-coordinates of 800. To two decimal places, you can see that $4.26 < t < 11.74$.

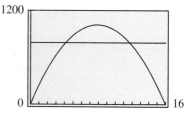

FIGURE 4.36

In Example 1, we solved the inequality $256t - 16t^2 > 800$ by comparing points on the graph of $h = 256 - 16t^2$ with points on the line $h = 800$. If one side of an inequality is zero, we can compare points on the graph with the x-axis.

EXAMPLE 2

Consider the graph of $y = x^2 - 4$. Find the solutions of
  a. $x^2 - 4 = 0$ **b.** $x^2 - 4 < 0$ **c.** $x^2 - 4 > 0$

Solutions

Look at the graph of the equation $y = x^2 - 4$ shown in Figure 4.37. When we substitute a value of x into the expression $x^2 - 4$, the result is either positive, negative, or zero. (You can see this more clearly if you compute a few values yourself to complete the table below. Your table should agree with the coordinates of points on the graph.)

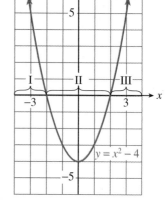

x	-3	-2	-1	0	1	2	3
y							

We'll use the graph to solve the given equation and inequalities.

FIGURE 4.37

  a. First locate the two points on the graph where $y = 0$. These points are $(-2, 0)$ and $(2, 0)$. Their x-coordinates, -2 and 2, are the solutions of the equation $x^2 - 4 = 0$. The two points divide the x-axis into three sections, which are labeled on the graph. On each of these sections, the value of $x^2 - 4$ is either always positive or always negative.

  b. To solve $x^2 - 4 < 0$, find the points on the graph where $y < 0$, that is, below the x-axis. These points have x-coordinates between -2 and 2 (labeled section II on the figure). Thus, the solution to the inequality $x^2 - 4 < 0$ is $-2 < x < 2$.

  c. To solve $x^2 - 4 < 0$, locate the points on the graph where $y > 0$, or above the x-axis. Points with positive y-values correspond to two sections of the x-axis, labeled I and III on the figure. In section I, $x < -2$, and in section III, $x > 2$. Thus, the solution to the inequality $x^2 - 4 > 0$ includes all values of x for which $x < -2$ or $x > 2$. ■

In Example 2c, the solution of the inequality $x^2 - 4 > 0$ is the set

$$x < -2 \quad \text{or} \quad x > 2$$

This set is another type of compound inequality, and its graph consists of two pieces, as shown in Figure 4.38.

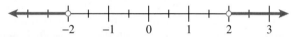

FIGURE 4.38

The left piece of the set is $x < -2$, and the right piece is $x > 2$. It would be incorrect to describe the solution set as $-2 > x > 2$, because this notation implies that $-2 > 2$. We must write the solution as two parts: $x < -2$ or $x > 2$. ■

Because it is relatively easy to decide whether the y-coordinate of a point on a graph is positive or negative, we often rewrite a given inequality so that one side is zero.

E X E R C I S E 1   Use a graph to solve the inequality $x^2 - 2x - 9 \geq 6$.

Rewrite the inequality so that the right side is zero.
Sketch the graph of $y = x^2 - 2x - 15$
    y-intercept:
    x-intercepts:
    vertex:
Darken the points on the graph with $y \geq 0$.
Darken the portions of the x-axis corresponding to those points.
Solution:

Interval Notation

The solution to the inequality $x^2 - 4 < 0$ in Example 2b, $-2 < x < 2$, is called an interval. An **interval** is a set that consists of all the real numbers between two numbers a and b. If the set includes both of the endpoints a and b, so that $a \leq x \leq b$, then the set is called a **closed interval** and is denoted with square brackets: $[a, b]$. If the set does not include its endpoints, so that $a < x < b$, then it is called an **open interval,** and is denoted with round brackets: (a, b).

The solution to Example 2b is an open interval, and we denote it by $(-2, 2)$. (See Figure 4.39a.) Do not confuse the open interval $(-2, 2)$ with the point

$(-2, 2)$! The notation is the same, so you must decide from the context whether an interval is being discussed, or a point.

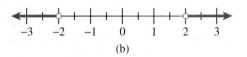

FIGURE 4.39

The solution to Example 2c consists of two **infinite intervals**, $x < -2$ or $x > 2$, shown in Figure 4.39b. We denote the interval $x < -2$ by $(-\infty, -2)$, and the interval $x > 2$ by $(2, \infty)$. The symbol ∞, "infinity," does not represent a specific real number; it indicates that the interval continues forever along the real line. The set consisting of both these intervals is called the **union** of the two intervals, and is denoted by $(-\infty, -2) \cup (2, \infty)$.

Many solutions of inequalities are intervals or unions of intervals.

EXAMPLE 4

Write each of the solution sets with interval notation, and graph the solution set on a number line.

a. $3 \leq x < 6$ **c.** $x \leq 1$ or $x > 4$

b. $x \geq -9$ **d.** $-8 < x \leq -5$ or $-1 \leq x < 3$

Solutions

a. $[3, 6)$. This is called a **half-open** or **half-closed** interval. (See Figure 4.40.)

FIGURE 4.40

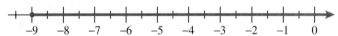

FIGURE 4.41

b. $[-9, \infty)$. We always use round brackets next to the symbol ∞ because ∞ is not a specific number and is not included in the set. (See Figure 4.41.)

c. $(-\infty, 1] \cup (4, \infty)$. The word "or" describes the union of two sets. (See Figure 4.42.)

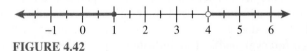
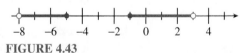

FIGURE 4.42   **FIGURE 4.43**

d. $(-8, -5] \cup [-1, 3)$. (See Figure 4.43.) ∎

EXERCISE 2   Solve the inequality

$$36 + 6x - x^2 \le 20$$

Write your answer with interval notation.

a. Rewrite the inequality so that the right side is zero.

b. Graph the equation $y = 16 + 6x - x^2$.

  y-intercept:

  x-intercepts: Solve $16 + 6x - x^2 = 0$

  Vertex:

  $$x_v = \frac{-b}{2a} =$$

  $y_v =$

c. Locate the points on the graph with *y*-coordinate less than zero, and mark the *x*-coordinates of the points on the *x*-axis. Write the solution in interval notation.

READING QUESTIONS

1. If $ax^2 + bx > c$ for a particular value of *x*, what can you say about the graph of $y = ax^2 + bx - c$ at that *x*-value?

2. What are the only *x*-values at which the graph of $y = ax^2 + bx + c$ can change sign?

3. Explain the difference between an open interval and a closed interval.

4. How do we indicate the union of two intervals?

ANSWERS TO 4.4 EXERCISES

1. $x \le -3$ or $x \ge 5$ **2a.** $16 + 6x - x^2 \le 0$

b, c.

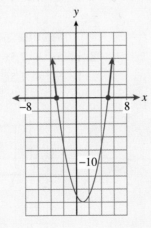

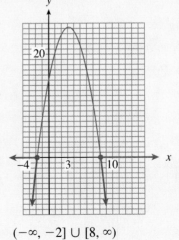

$$(-\infty, -2] \cup [8, \infty)$$

1. a. Graph the equation $y = x^2$ by hand on graph paper.

   b. Darken the portion of the x-axis for which $y > 9$.

   c. Solve the inequality $x^2 > 9$. Explain why $x > 3$ is incorrect as the answer.

3. a. Graph the equation $y = x^2 - 2x - 3$ by hand on graph paper.

   b. Darken the portion of the x-axis for which $y > 0$.

   c. Solve the inequality $x^2 - 2x - 3 > 0$.

2. a. Graph the equation $y = x^2$ by hand on graph paper.

   b. Darken the portion of the x-axis for which $y < 1$.

   c. Solve the inequality $x^2 < 1$. Explain why $x < 1$ is incorrect as the answer.

4. a. Graph the equation $y = x^2 + 2x - 8$ by hand on graph paper.

   b. Darken the portion of the x-axis for which $y < 0$.

   c. Solve the inequality $x^2 + 2x - 8 < 0$.

Use the graphs provided to estimate the solutions to each equation and inequality.

5. a. $x^2 - 3x - 180 = 0$

   b. $x^2 - 3x - 180 > 0$

6. a. $175 - 18x - x^2 = 0$

   b. $175 - 18x - x^2 < 0$

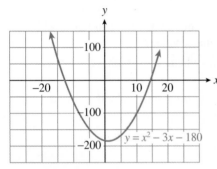

FIGURE 4.44

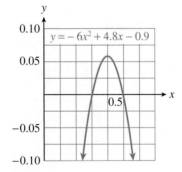

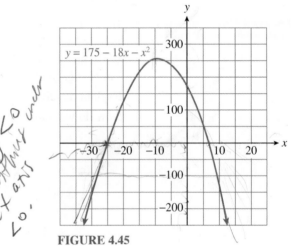

FIGURE 4.45

7. a. $-6x^2 + 4.8x - 0.9 = 0$

   b. $-6x^2 + 4.8x - 0.9 \geq 0$

8. a. $5x^2 + 7.5x + 1.8 = 0$

   b. $5x^2 + 7.5x + 1.8 \leq 0$

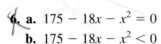

FIGURE 4.46

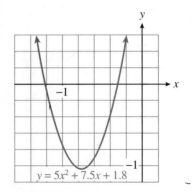

FIGURE 4.47

Write each set with interval notation, and graph the set on a number line.

9. $-5 < x \le 3$

10. $0 \le x < 4$

11. $0 \ge x \ge -4$

12. $8 > x > 5$

13. $x > -6$

14. $x \le 1$

15. $x < -3$ or $x \ge -1$

16. $x \ge 3$ or $x \le -3$

17. $-6 \le x < -4$ or $-2 < x \le 0$

18. $x < 2$ or $2 < x < 3$

Graph the parabola in the window

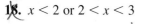

$$\text{Xmin} = -9.4 \qquad \text{Xmax} = 9.4$$
$$\text{Ymin} = -2.5 \qquad \text{Ymax} = 25$$

Then use the graph to solve the inequalities. Write your answers with interval notation.

19. $y = x^2 - 3x - 18$

    a. $x^2 - 3x - 18 > 0$

    b. $x^2 - 3x - 18 < 0$

    c. $x^2 - 3x - 18 \le -8$

    d. $x^2 - 3x - 18 \ge -8$

For parts (c) and (d), it may be helpful to graph $Y_2 = -8$ as well.

20. $y = 16 - 6x - x^2$

    a. $16 - 6x - x^2 \ge 0$

    b. $16 - 6x - x^2 \le 0$

    c. $16 - 6x - x^2 < 21$

    d. $16 - 6x - x^2 > 21$

For parts (c) and (d), it may be helpful to graph $Y_2 = 21$ as well.

21. $y = 16 - x^2$

    a. $16 - x^2 > 0$

    b. $16 - x^2 < 0$

    c. $16 - x^2 \le 7$

    d. $16 - x^2 \ge 7$

For parts (c) and (d), it may be helpful to graph $Y_2 = 7$ as well.

22. $y = x^2 - 9$

    a. $x^2 - 9 \ge 0$

    b. $x^2 - 9 \le 0$

    c. $x^2 - 9 > 16$

    d. $x^2 - 9 < 16$

For parts (c) and (d), it may be helpful to graph $Y_2 = 16$ as well.

Solve each inequality by graphing. Use the following [WINDOW] **settings on your calculator:**

$$\text{Xmin} = -9.4 \qquad \text{Xmax} = 9.4$$
$$\text{Ymin} = -15 \qquad \text{Ymax} = 15$$

23. $(x - 3)(x + 2) > 0$

24. $(x + 1)(x - 5) > 0$

25. $k(4 - k) \ge 0$

26. $-m(7 + m) \ge 0$

27. $6 + 5p - p^2 < 0$

28. $q^2 + 9q + 18 < 0$

Solve each inequality by graphing. Use the following [WINDOW] **settings on your calculator:**

$$\text{Xmin} = -9.4 \qquad \text{Xmax} = 9.4$$
$$\text{Ymin} = -62 \qquad \text{Ymax} = 62$$

29. $x^2 - 1.4x - 20 < 9.76$

30. $-x^2 + 3.2x + 20 > 6.56$

31. $5x^2 + 39x + 27 \ge 5.4$

32. $-6x^2 - 36x - 20 \le 25.36$

33. $-8x^2 + 112x - 360 < 6.08$

34. $10x^2 + 96x + 180 > 17.2$

Solve each inequality by graphing. Choose a suitable window for each problem. Use the *intersect* feature to estimate your solutions accurate to one decimal place.

35. $x^2 > 12.2$

36. $x^2 \le 45$

37. $-3x^2 + 7x - 25 \le 0$

38. $2.4x^2 - 5.6x + 18 \le 0$

39. $0.4x^2 - 54x < 620$

40. $-0.05x^2 - 3x > 76$

4.5 Solving Quadratic Inequalities Algebraically

Although a graph is very helpful in solving inequalities, it is not completely necessary. Every quadratic inequality can be put into one of the forms

$$ax^2 + bx + c < 0 \qquad ax^2 + bx + c > 0$$
$$ax^2 + bx + c \le 0 \qquad ax^2 + bx + c \ge 0$$

All we really need to know is whether the corresponding parabola

$$y = ax^2 + bx + c$$

opens upward or downward. Consider the parabolas shown in Figure 4.48.

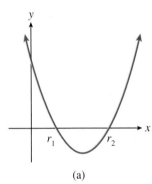
(a)

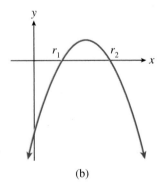
(b)

FIGURE 4.48

The parabola in Figure 4.48a opens upward. It crosses the x-axis at two points, $x = r_1$ and $x = r_2$. At these points, $y = 0$. The solutions to the inequality $y < 0$ lie between r_1 and r_2, because y is negative on that portion of the graph. The solutions to the inequality $y > 0$ are the x-values less than r_1 or greater than r_2, because y is positive in those regions.

If the parabola opens downward, as in Figure 4.48b, the situation is reversed. The solutions to the inequality $y > 0$ lie between the x-intercepts, and the solutions to $y < 0$ lie outside the x-intercepts.

From the graphs in Figure 4.48, we see that the x-intercepts are the "boundary points" between the portions of the graph with positive y-coordinates, and the portions with negative y-coordinates. To solve a quadratic inequality, we need only locate the x-intercepts of the corresponding graph, and then decide which intervals of the x-axis result in the correct sign for y.

To solve a quadratic inequality algebraically:

1. Write the inequality in standard form: One side is 0, and the other has the form $ax^2 + bx + c$.

2. Find the x-intercepts of the graph of $y = ax^2 + bx + c$ by setting $y = 0$ and solving for x.

3. Make a rough sketch of the graph, using the sign of a to determine whether the parabola opens upward or downward.

4. Decide which intervals on the x-axis give the correct sign for y.

EXAMPLE 1

Solve the inequality $36 + 6x - x^2 \le 20$ algebraically.

Solution

Step 1 Subtract 20 from both sides of the inequality so that we have zero on the right side.

$$16 + 6x - x^2 \le 0$$

Step 2 Consider the equation $y = 16 + 6x - x^2$. To locate the x-intercepts, set $y = 0$ and solve for x.

$$16 + 6x - x^2 = 0 \quad \text{Multiply each term by } -1.$$
$$x^2 - 6x - 16 = 0 \quad \text{Factor the left side.}$$
$$(x - 8)(x + 2) = 0 \quad \text{Apply the zero-factor principle.}$$

$$x - 8 = 0 \quad \text{or} \quad x + 2 = 0$$
$$x = 8 \quad \text{or} \quad x = -2$$

The x-intercepts are $x = -2$ and $x = 8$.

Step 3 Make a rough sketch of the graph of $y = 16 + 6x - x^2$, as shown in Figure 4.49. Because $a = -1 < 0$, the graph is a parabola that opens downward.

Step 4 We are interested in points on the graph for which $y \le 0$. The points with negative y-coordinates lie outside the x-intercepts of the graph, so the solution of the inequality is $x \le -2$ or $x \ge 8$. Or, using interval notation, the solution is $(-\infty, -2] \cup [8, \infty)$.

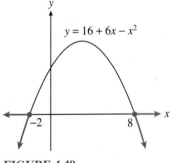

FIGURE 4.49

COMMON ERROR

Many people think that the inequality signs in the solution should point in the same direction as the sign in the original problem, and hence would incorrectly write the solution to Example 1 as $x \le -2$ or $x \le 8$. However, you can see from the graph that this is incorrect. Remember that the graph of a quadratic equation is a parabola, not a straight line!

If we cannot find the x-intercepts of the graph by factoring or extraction of roots, we can use the quadratic formula.

EXAMPLE 3

TrailGear, Inc. manufactures camping equipment. They find that the profit from producing and selling x alpine parkas per month is given in dollars by

$$P = -0.8x^2 + 320x - 25{,}200$$

How many parkas should they produce and sell each month if they must keep the profits above $2000?

Solution

Step 1 We would like to solve the inequality

$$-0.8x^2 + 320x - 25{,}200 > 2000$$

or, subtracting 2000 from both sides,

$$-0.8x^2 + 320x - 27{,}200 > 0$$

Step 2 Consider the equation

$$y = -0.8x^2 + 320x - 27{,}200$$

We locate the x-intercepts of the graph by setting $y = 0$ and solving for x. We will use the quadratic formula to solve the equation

$$-0.8x^2 + 320x - 27{,}200 = 0$$

so $a = -0.8$, $b = 320$, and $c = -27{,}200$.

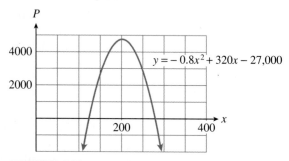

FIGURE 4.50

$$x =$$

$$\frac{-(320) \pm \sqrt{(320)^2 - 4(-0.8)(-27{,}200)}}{2(-0.8)}$$

$$= \frac{-320 \pm \sqrt{102{,}400 - 87{,}040}}{-1.6}$$

$$= \frac{-320 \pm \sqrt{15{,}360}}{-1.6}$$

To two decimal places, the solutions to the equation are 122.54 and 277.46.

Step 3 The graph of the equation is a parabola that opens downward, because the coefficient of x^2 is negative. (See Figure 4.50.)

Step 4 Positive values of y correspond to x-values between the two x-intercepts, that is, for $122.54 < x < 277.46$. Because we cannot produce a fraction of a parka, we restrict the interval to the closest whole number x-values included, namely 123 and 277. Thus, TrailGear can produce as few as 123 parkas or as many as 277 parkas per month to keep their profit above \$2000. ∎

EXERCISE 2 Solve the inequality $10 - 8x + x^2 > 4$

Step 1 Write the inequality in standard form.

Step 2 Find the x-intercepts of the corresponding graph. Use the quadratic formula.

Step 3 Make a rough sketch of the graph.

Step 4 Decide which intervals on the x-axis give the correct sign for y.

Write your answer with interval notation.

READING QUESTIONS

1. The parabola $y = x^2 + bx + c$ has x-intercepts at r_1 and r_2, with $r_1 < r_2$. What are the solutions of the inequality $x^2 + bx + c > 0$?

2. The parabola $y = x^2 + bx + c$ has x-intercepts at r_1 and r_2, with $r_1 < r_2$. What are the solutions of the inequality $x^2 + bx + c \le 0$?

3. The parabola $y = -x^2 + bx + c$ has x-intercepts at r_1 and r_2, with $r_1 < r_2$. What are the solutions of the inequality $x^2 + bx + c \le 0$?

4. The parabola $y = -x^2 + bx + c$ has x-intercepts at r_1 and r_2, with $r_1 < r_2$. What are the solutions of the inequality $x^2 + bx + c > 0$?

ANSWERS TO 4.5 EXERCISES

1. $-\sqrt{20} < x < \sqrt{20}$ **2.** $(-\infty, 4 - \sqrt{10}] \cup [4 + \sqrt{10}, \infty)$

HOMEWORK 4.5

Solve each inequality algebraically. Write your answers in interval notation, and round to two decimal places if necessary.

1. $(x + 3)(x - 4) < 0$

2. $(x + 2)(x - 5) > 0$

3. $28 - 3x - x^2 \ge 0$

4. $32 + 4x - x^2 \le 0$

5. $2z^2 - 7z > 4$

6. $6h^2 + 13h < 15$

7. $64 - t^2 > 0$

8. $121 - y^2 < 0$

9. $v^2 < 5$ **10.** $t^2 \geq 7$ **11.** $5a^2 - 32a + 12 \geq 0$

12. $6b^2 + 16b - 9 < 0$ **13.** $4x^2 + x \geq -2x^2 + 2$ **14.** $2x^2 + 8x \leq -x^2 + 3$

15. $x^2 - 4x + 1 \geq 0$ **16.** $x^2 + 4x + 2 \leq 0$ **17.** $-3 - m^2 < 0$

18. $11 + n^2 < 0$ **19.** $w^2 - w + 4 \leq 0$ **20.** $-z^2 + z - 1 \leq 0$

In Problems 21–28, (a) solve each problem by writing and solving an inequality, (b) graph the equation and verify your solution on the graph.

21. A fireworks rocket is fired from ground level. Its height in feet t seconds after launch is given by $h = 320t - 16t^2$. During what time interval is the rocket higher than 1024 feet?

22. A baseball thrown vertically reaches a height h in feet given by $h = 56t - 16t^2$, where t is measured in seconds. During what time intervals is the ball between 40 and 48 feet high?

23. The cost in dollars of manufacturing x pairs of garden shears is given by

$$C = -0.02x^2 + 14x + 1600$$

for $0 \leq x \leq 700$. How many pairs of shears can be produced if the total cost must be kept under $2800?

24. The cost in dollars of producing x cashmere[1] sweaters is given by

$$C = x^2 + 4x + 90$$

How many sweaters can be produced if the total cost must be kept under $1850?

25. The Locker Room finds that it sells $1200 - 30p$ sweatshirts each month when it charges p dollars per sweatshirt. It would like its revenue from sweatshirts to be over $9000 per month. In what range should it keep the price of a sweatshirt? *Hint:* Recall that *Revenue = (number of items sold) · (price per item).*

26. Green Valley Nursery sells $120 - 10p$ boxes of rose food per month at a price of p dollars per box. It would like to keep its monthly revenue from rose food over $350. In what range should it price a box of rose food?

27. A group of cylindrical storage tanks must be 20 feet tall. If the volume of each tank must be between 500π and 2880π cubic feet, what are the possible values for the radius of a tank?

28. The volume of a cylindrical can should be between 21.2 and 21.6 cubic inches. If the height of the can is 5 inches, what values for the radius (to the nearest hundredth of an inch) will produce an acceptable can?

The models in Problems 29 and 30 are similar to Investigation 7.

29. A travel agency offers a group rate of $600 per person for a weekend in Lake Tahoe if 20 people sign up. For each additional person who signs up, the price for all participants is reduced by $10 per person.

    a. Make a table as shown, and continue in increments of five additional participants until the total income is reduced to zero.

Additional people	Size of group	Price per person	Total income
0			
5			
10			

    b. Write algebraic expressions for the size of the group and the price per person if x additional people sign up.

    c. Write a quadratic equation for the travel agency's total income if x additional people sign up for the trip.

    d. Use your calculator's **Table** feature and your answer to part (c) to verify your values for the total income from part (a).

    e. What is the maximum income the travel agency can earn on the Lake Tahoe weekend? How many people should they enroll to achieve this income?

    f. How many people must sign up in order for the agency to bring in at least $15,750?

g. Graph your equation for total income on your calculator in the window

$$\text{Xmin} = 0 \quad \text{Xmax} = 94$$
$$\text{Ymin} = 0 \quad \text{Ymax} = 18{,}000$$

and use the graph to verify your answers to parts (e) and (f).

30. A farmer inherits an apple orchard on which 60 trees are planted per acre. Each tree yields 12 bushels of apples. Experimentation has shown that for each tree removed per acre, the yield per tree increases by $\frac{1}{2}$ bushel.

a. Make a table as shown, and continue in increments of two trees removed until the total yield (in bushels per acre) is reduced to zero.

Trees removed	Trees per acre	Bushels per tree	Total yield
0			
2			
4			

b. Write algebraic expressions for the number of trees per acre and for the yield per tree if x trees per acre are removed.

c. Write a quadratic equation for the total yield per acre if x trees are removed per acre.

d. Use your calculator's **Table** feature and your answer to part (c) to verify your values for the total yield from part (a).

e. What is the maximum yield per acre that can be achieved by removing trees? How many trees per acre should be removed to achieve this yield?

f. How many trees should be removed per acre in order to harvest at least 850 bushels per acre?

g. Graph your equation for total yield on your calculator in the window

$$\text{Xmin} = 0 \quad \text{Xmax} = 94$$
$$\text{Ymin} = 0 \quad \text{Ymax} = 1000$$

and use your graph to verify your answers to parts (e) and (f).

CHAPTER 4 REVIEW

In Problems 1–4, (a) find the coordinates of the vertex and the intercepts, (b) sketch the graph.

1. $y = x^2 - x - 12$

2. $y = -2x^2 + x - 4$

3. $y = -x^2 + 2x + 4$

4. $y = x^2 - 3x + 4$

Solve each inequality algebraically, and give your answers in interval notation. Verify your solutions by graphing.

5. $(x - 3)(x + 2) > 0$

6. $y^2 - y - 12 \le 0$

7. $2y^2 - y \le 3$

8. $3z^2 - 5z > 2$

9. $s^2 \le 4$

10. $4t^2 > 12$

11. The Sub Station sells $220 - \frac{1}{4}p$ submarine sandwiches at lunchtime if it sells them at p cents each.

a. Write a formula for the Sub Station's daily revenue in terms of p.

b. What range of prices can the Sub Station charge if it wants to keep its daily revenue from subs over $480? (Remember to convert $480 to cents.)

12. When it charges p dollars for an electric screwdriver, Handy Hardware will sell $30 - \frac{1}{2}p$ screwdrivers per month.

a. Write a formula in terms of p for Handy Hardware's monthly revenue from screwdrivers.

b. How much should Handy charge per screwdriver if it wants the monthly revenue from the screwdrivers to be over $400?

13. A beekeeper has beehives distributed over 60 square miles of pastureland. When she places four hives per square mile, each hive produces about 32 pints of honey per year. For each additional hive per square mile, honey production drops by 4 pints per hive.

    a. Write an equation for the total production of honey, in pints, in terms of the number of additional hives per square mile.

    b. How many additional hives per square mile should the beekeeper install in order to maximize honey production?

Solve each system algebraically, and verify your solution with a graph.

15. $y + x^2 = 4$
    $y = 3$

16. $y = 3 - x^2$
    $5x + y = 7$

17. $y = x^2 - 5$
    $y = 4x$

18. $y = x^2 - 2x + 1$
    $y = 3 - x$

19. $y = x^2 - 6x + 20$
    $y = 2x^2 - 2x - 25$

20. $y = x^2 - 5x - 28$
    $y = -x^2 + 4x + 28$

21. Find values of a, b, and c so that the graph of the parabola $y = ax^2 + bx + c$ contains the points $(-1, -4)$, $(0, -6)$, and $(4, 6)$.

22. Find a parabola that fits the following data points.

x	-8	-4	2	4
y	10	18	0	-14

23. Find the equation for a parabola that has a vertex at $(15, -6)$ and passes through the point $(3, 22.8)$.

14. A small company manufactures radios. When it charges $20 for a radio, it sells 500 radios per month. For each dollar the price is increased, 10 fewer radios are sold per month.

    a. Write an equation for the monthly revenue in terms of the price increase over $20.

    b. What should the company charge for a radio in order to maximize its monthly revenue?

24. The height of a cannonball was observed at 0.2-second intervals after the cannon was fired, and the data recorded in the table.

Time (seconds)	Height (meters)
0.2	10.2
0.4	19.2
0.6	27.8
0.8	35.9
1.0	43.7
1.2	51.1
1.4	58.1
1.6	64.7
1.8	71.0
2.0	76.8

    a. Find the equation of the least-squares regression line for height in terms of time.

    b. Use the linear regression equation to predict the height of the cannonball at 3 seconds and at 4 seconds after it was fired.

    c. Make a scatterplot of the data and draw the regression line on the same axes.

    d. Find the quadratic regression equation for height in terms of time.

    e. Use the quadratic regression equation to predict the height of the cannonball at 3 seconds and at 4 seconds after it was fired.

    f. Draw the quadratic regression curve on the graph from part (c).

    g. Which model is more appropriate for the height of the cannonball, linear or quadratic? Why?

CHAPTER

5

Functions and Their Graphs

We often want to predict values of one variable from the values of a related variable. For example, when a physician prescribes a drug in a certain dosage, she needs to know how long the dose will remain in the bloodstream. A pilot needs to know how the flight time for a trip is affected by the speed of the prevailing winds. A relationship between variables that allows us to make predictions is called a *function*.

INVESTIGATION 8

Epidemics

A contagious disease whose spread is unchecked can devastate a confined population. For example, in the early sixteenth century Spanish troops introduced smallpox into the Aztec population in Central America, and the resulting epidemic contributed significantly to the fall of Montezuma's empire.

Suppose that an outbreak of cholera follows severe flooding in an isolated town of 5000 people. Initially (Day 0), 40 people are infected. Every day after that, 25% of those still healthy fall ill.

1. At the beginning of the first day (Day 1), how many people are still healthy? How many will fall ill during the first day? What is the total number of people infected after the first day?

2. Check your results against Table 5.1. Subtract the total number of infected residents from 5000 to find the number of healthy residents at the beginning of the second day. Then fill in the rest of the table for 10 days. (Round off decimal results to the nearest whole number.)

TABLE 5.1

Day	Number healthy	New patients	Total infected
0	5000	40	40
1	4960	1240	1280
2			
3			
4			
5			
6			
7			
8			
9			
10			

3. Use the last column of Table 5.1 to plot the total number of infected residents (I) against time (t) in Figure 5.1. Connect your data points with a smooth curve.

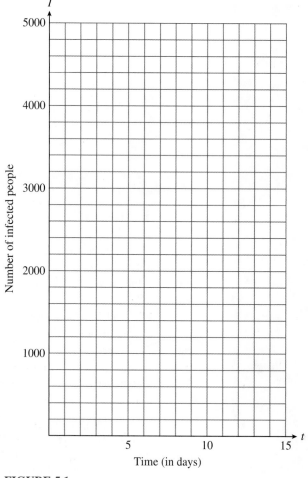

FIGURE 5.1

4. Do the values of I approach some largest value? Draw a dotted horizontal line at that value of I. Will the values of I ever exceed that value?

5. What is the first day on which at least 95% of the population is infected?

6. Look back at Table 5.1. What is happening to the number of new patients each day as time goes on? How is this phenomenon reflected in the graph? How would your graph look if the number of new patients every day were a constant?

7. Summarize your work: In your own words, describe how the number of residents infected with cholera changes with time. Include a description of your graph.

5.1 Definitions and Notation

We have already seen some examples of functions. For instance, suppose it costs $800 for flying lessons plus $30 per hour to rent a plane. If we let C represent the total cost for t hours of flying lessons, then

$$C = 30t + 800 \qquad (t \geq 0) \tag{1}$$

Thus, for example,

$$\text{when } t = 0, \quad C = 30(0) + 800 = 800$$
$$\text{when } t = 4, \quad C = 30(4) + 800 = 920$$
$$\text{when } t = 10, \quad C = 30(10) + 800 = 1100$$

The variable t in Equation (1) is called the **independent** variable, and C is the **dependent** variable, because its values are determined by the value of t. We can display an association between two variables by a table (such as Table 5.2) or by ordered pairs. The independent variable is the first component of the ordered pair, and the dependent variable is the second component. For the example above we have

TABLE 5.2

t	C		(t, C)
0	800		(0, 800)
4	920		(4, 920)
10	1100		(10, 1100)

For this relationship we can determine the value of C associated with any given value of t. All we have to do is substitute the value of t into Equation (1) and solve for C. The result has no ambiguity: Only *one* value for C corresponds to each value of t. This type of relationship between variables is called a **function.** In general, we make the following definition.

> A **function** is a relationship between two variables for which a *unique* value of the **dependent** variable can be determined from a value of the **independent** variable.

What distinguishes functions from other variable relationships? The definition of a function calls for a *unique value,* that is, *exactly one value* of the dependent variable corresponding to each value of the independent variable. This property makes functions useful in applications because they can often be used to make predictions. The linear and quadratic models we have already studied are examples of functions.

EXAMPLE 1

a. The distance, d, traveled by a car in 2 hours is a function of its speed, r. If we know the speed of the car, we can determine the distance it travels by the formula $d = r \cdot 2$.

b. The price of a fill-up with unleaded gasoline is a function of the number of gallons purchased. The gas pump represents the function by displaying the corresponding values of the independent variable (number of gallons) and the dependent variable (total cost).

c. Score on the Scholastic Aptitude Test is *not* a function of score on an IQ test, because two people with the same score on an IQ test may score differently on the SAT; that is, a person's score on the SAT is not uniquely determined by his or her score on an IQ test. ∎

EXERCISE 1

a. As part of a project to improve the success rate of freshmen, the counseling department studied the grades earned by a group of students in English and algebra. Do you think that a student's grade in algebra is a function of his or her grade in English? Explain why or why not.

b. Phatburger features a soda bar where you can serve your own soft drinks in any size. Do you think that the number of calories in a serving of Zap Kola is a function of the number of fluid ounces? Explain why or why not.

Functions Defined by Tables

A function can be described in several different ways. In the following examples we consider functions defined by tables, by graphs, and by equations.

EXAMPLE 2

Table 5.3 shows data on sales compiled over several years by the accounting office for Eau Claire Auto Parts, a division of Major Motors.

TABLE 5.3

Year (t)	1990	1991	1992	1993	1994
Total sales (S)	$612,000	$663,000	$692,000	$749,000	$904,000

In this example, the year is considered the independent variable, and we say that total sales, *S, is a function of t.* ▪

EXAMPLE 3

Table 5.4 gives the cost of sending printed material by first-class mail.

TABLE 5.4

Weight in ounces (w)	Postage (p)
$0 < w \le 1$	$0.33
$1 < w \le 2$	$0.55
$2 < w \le 3$	$0.77
$3 < w \le 4$	$0.99
$4 < w \le 5$	$1.21
$5 < w \le 6$	$1.43
$6 < w \le 7$	$1.65

If we know the weight of the article being shipped, we can determine the required postage from Table 5.4. For instance, a catalog weighing 4.5 ounces would require $1.21 in postage. In this example w is the independent variable and p is the dependent variable. We say that *p is a function of w.* ▪

EXAMPLE 4

Table 5.5 records the age and cholesterol count for 20 patients tested in a hospital survey.

TABLE 5.5

Patient number	Age	Cholesterol count	Patient number	Age	Cholesterol count
301	53	217	332	51	209
308	48	232	336	53	241
312	55	198	339	49	186
313	56	238	340	51	216
316	51	227	343	57	208
320	52	264	347	52	248
322	53	195	356	50	214
324	47	203	359	56	271
325	48	212	362	53	193
328	50	234	370	48	172

According to these data, cholesterol count is *not* a function of age, because several patients who are the same age have different cholesterol levels. For example, patients 316, 332, and 340 are all 51 years old but have cholesterol counts of 227, 209, and 216, respectively. Thus, we cannot determine a *unique* value of the dependent variable (cholesterol count) from the value of the independent variable (age). Other factors besides age must influence a person's cholesterol count. ■

EXERCISE 2 Decide whether each table describes y as a function of x. Explain your choice.

a.

x	3.5	2.0	2.5	3.5	2.5	4.0	2.5	3.0
y	2.5	3.0	2.5	4.0	3.5	4.0	2.0	2.5

b.

x	-3	-2	-1	0	1	2	3
y	17	3	0	-1	0	3	17

Functions Defined by Graphs

A graph may also be used to define one variable as a function of another.

EXAMPLE 5

Figure 5.2 shows the number of hours, H, that the sun is above the horizon in Peoria, Illinois, on day t, where January 1 corresponds to $t = 0$.

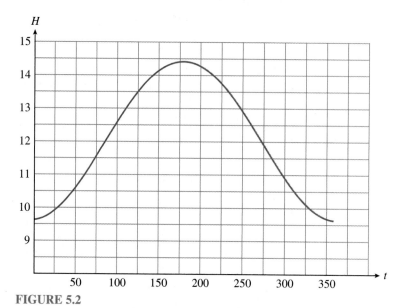

FIGURE 5.2

   a.   Which variable is independent, and which is dependent?

   b.   Approximately how many hours of sunlight are there in Peoria on day 150?

   c.   On which days are there 12 hours of sunlight?

d. What are the maximum and minimum values of H, and when do these values occur?

Solutions

   a. The independent variable, t, appears on the horizontal axis. The number of daylight hours, H, depends on the date. The dependent variable appears on the vertical axis.

   b. The point on the curve where $t = 150$ has $H \approx 14.1$, so Peoria has about 14.1 hours of daylight when $t = 150$, at the end of May.

   c. $H = 12$ at the two points where $t \approx 85$ (in late March) and $t \approx 270$ (late September).

   d. The maximum value of 14.4 hours is the longest day of the year, and it occurs when $t \approx 170$, about three weeks into June. The minimum of 9.6 hours is the shortest day, and it occurs when $t \approx 355$, about three weeks into December. ∎

Functions Defined by Equations

Example 6 illustrates a function defined by an equation.

EXAMPLE 6

The Sears Tower in Chicago is the nation's tallest building at 1454 feet. If an algebra book is dropped from the top of the Sears Tower, its height above the ground after t seconds is given by the equation

$$h = 1454 - 16t^2 \tag{2}$$

Thus, after 1 second the book's height is

$$h = 1454 - 16(1)^2 = 1438 \text{ feet}$$

After 2 seconds its height is

$$h = 1454 - 16(2)^2 = 1390 \text{ feet}$$

For this function, t is the independent variable and h is the dependent variable. For any value of t the corresponding unique value of h can be determined from Equation (2). We say that *h is a function of t*.

Function Notation

There is a convenient notation we use when discussing functions. First we choose a letter, such as f, g, or h, or F, G, or H, to name a particular function. (We can use any letter, but these are the most common choices.) For instance, in Example 6 we expressed the height h of an algebra book falling from the top of the Sears Tower as a function of the time t that it has been falling. We will call this function f. That is, f is the name of the relationship between the variables h and t. We can then write

$$h = f(t)$$

which means "h is a function of t, and f is the name of the function."

The new symbol $f(t)$, read "f of t," is another name for the height h. The parentheses in the symbol $f(t)$ do *not* indicate multiplication. (It would not make sense to multiply the name of a function by a variable.) Think of the symbol $f(t)$ as a single entity that represents the dependent variable of the function.

With this new notation we may write

$$h = f(t) = 1454 - 16t^2$$

or just

$$f(t) = 1454 - 16t^2$$

instead of

$$h = 1454 - 16t^2$$

to describe the function.

Perhaps it may seem a needless complication to introduce a new symbol for h, but the notation $f(t)$ is very useful for showing the correspondence between specific values of the variables h and t. In Example 6 we saw that

$$\text{when } t = 1 \qquad h = 1438$$
$$\text{when } t = 2 \qquad h = 1390$$

Using function notation, these relationships can be expressed more concisely as

$$f(1) = 1438 \qquad \text{and} \qquad f(2) = 1390$$

which we read as "f of 1 equals 1438" and "f of 2 equals 1390." The values for the independent variable, t, appear *inside* the parentheses, and the values for the dependent variable, h, appear on the other side of the equation.

$$f(t) = h$$

independent variable ⎯⎯⎯⎯⎯⎯↑   ↑⎯⎯⎯⎯ dependent variable

EXERCISE 3 Let F be the name of the function defined by the graph in Example 5.

a. Use function notation to state that H is a function of t.

b. What does the statement $F(15) = 9.7$ mean in the context of the problem?

Evaluating a Function

Finding the value of the dependent variable that corresponds to a particular value of the independent variable is called **evaluating the function.**

EXAMPLE 7

Let g be the name of the postage function defined by Table 5.5 in Example 3. Find $g(1)$, $g(3)$, and $g(6.75)$.

Solution

According to the table,

when $w = 1$	$p = 0.33$	so	$g(1) = 0.33$
when $w = 3$	$p = 0.77$	so	$g(3) = 0.77$
when $w = 6.75$	$p = 1.65$	so	$g(6.75) = 1.65$

Thus, a letter weighing 1 ounce costs \$0.33 to mail; a letter weighing 3 ounces costs \$0.77, and a letter weighing 6.75 ounces costs \$1.65. ■

EXERCISE 4 When you exercise, your heart rate should increase until it reaches your target heart rate. The table shows target heart rate, $r = f(a)$, as a function of age.

a	20	25	30	35	40	45	50	55	60	65	70
r	150	146	142	139	135	131	127	124	120	116	112

a. Find $f(25)$ and $f(50)$. Does $f(50) = 2f(25)$?
b. Find a value of a for which $f(a) = 135$.
c. Find a value of a for which $f(a) = 2a$. Is $f(a) = 2a$ for all values of a?

If a function is described by an equation, we simply substitute the given value into the equation to find the corresponding function value.

EXAMPLE 8

The function H is defined by $H(s) = \dfrac{\sqrt{s + 3}}{s}$. Evaluate the function at the given values.

a. $s = 6$ **b.** $s = -1$

Solutions

Substitute the given values for s into the equation defining H.

a. $H(6) = \dfrac{\sqrt{6 + 3}}{6} = \dfrac{\sqrt{9}}{6} = \dfrac{3}{6} = \dfrac{1}{2}$

b. $H(-1) = \dfrac{\sqrt{-1 + 3}}{-1} = \dfrac{\sqrt{2}}{-1} = -\sqrt{2}$ ■

EXERCISE 5 Complete the table displaying ordered pairs for the function $f(x) = 5 - x^3$.

Evaluate the function to find the corresponding $f(x)$-value for each value of x.

x	$f(x)$
-2	
0	
1	
3	

$f(-2) = 5 - (-2)^3 =$

$f(0) = 5 - 0^3 =$

$f(1) = 5 - 1^3 =$

$f(3) = 5 - 3^3 =$

To simplify the notation, we sometimes use the same letter for the dependent variable and for the name of the function. In the next example C is used in this way.

EXAMPLE 9

TrailGear decides to market a line of backpacks. The cost, C, of manufacturing backpacks is a function of the number, x, of backpacks produced, given by the equation

$$C(x) = 3000 + 20x$$

where $C(x)$ is measured in dollars. Find the cost of producing 500 backpacks.

Solution

To find the value of C that corresponds to $x = 500$, evaluate $C(500)$.

$$C(500) = 3000 + 20(500) = 13,000$$

The cost of producing 500 backpacks is $13,000. ∎

Using the Table Feature to Evaluate Functions

The table feature on a graphing calculator gives us a convenient tool for evaluating functions. We will demonstrate using the function of Exercise 5, $f(x) = 5 - x^3$.

Press $\boxed{Y =}$, clear any old functions, and enter

$$Y_1 = 5 - X \boxed{\wedge} 3$$

Then press $\boxed{\text{TblSet}}$ ($\boxed{\text{2nd}}$ $\boxed{\text{WINDOW}}$) and choose **Ask** after **Indpnt,** as shown in Figure 5.3a, and press $\boxed{\text{ENTER}}$. This setting allows you to enter any x-values you like. Next press $\boxed{\text{TABLE}}$ (using $\boxed{\text{2nd}}$ $\boxed{\text{GRAPH}}$).

To follow Exercise 5, key in (−) 2 ENTER for the *x*-value, and the calculator will fill in the *y*-value. Continue by entering 0, 1, 3, or any other *x*-values you choose. One such table is shown in Figure 5.3b.

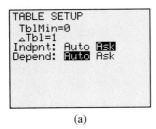

(a)   (b)

FIGURE 5.3

If you would like to evaluate a new function, you do not have to return to the Y = screen. Use the ▷ and △ keys to highlight Y_1 at the top of the second column. The definition of Y_1 will appear at the bottom of the display, as shown in Figure 5.3b. You can key in a new definition here, and the second column will be updated automatically to show the *y*-values of the new function.

READING QUESTIONS

1. What property makes a relation between two variables a function?
2. Name three ways to define a function.
3. Give an example of a function in which two distinct values of the independent variable correspond to the same value of the dependent variable.
4. Use function notation to write the statement "*G* defines *w* as a function of *p*."

ANSWERS TO 5.1 EXERCISES

1a. No, students with the same grade in English can have different grades in algebra.

  b. Yes, the number of calories is proportional to the number of fluid ounces.

2a. No, for example, $x = 3.5$ corresponds both to $y = 2.5$ and also $y = 4.0$.

  b. Yes, each value of *x* has exactly one value of *y* associated with it.

3a. $D = F(t)$

  b. The sun is above the horizon in Peoria for 9.7 hours on January 15.

4a. $f(25) = 146, f(50) = 127$; no

  b. $a = 40$

  c. $a = 60$; no

5.

x	$f(x)$
−2	13
0	5
1	4
3	−22

For which of the following pairs is the second quantity a function of the first? Explain your answers.

1. Price of an item; sales tax on the item at 4%

2. Time traveled at constant speed; distance traveled

3. Number of years of education; annual income

4. Distance flown in an airplane; price of the ticket

5. Volume of a container of water; its weight

6. Amount of a paycheck; amount of Social Security tax withheld

Each of the following objects establishes a correspondence between two variables. Suggest appropriate independent and dependent variables and decide whether the relationship is a function.

7. An itemized grocery receipt

8. An inventory list

9. An index

10. A will

11. An instructor's grade book

12. An address book

13. A bathroom scale

14. A radio dial

Which of the following tables define the second variable as a function of the first variable? Explain why or why not.

15.

x	t
−1	2
0	9
1	−2
0	−3
−1	5

16.

y	w
0	8
1	12
3	7
5	−3
7	4

17.

x	y
−3	8
−2	3
−1	0
0	−1
1	0
2	3
3	8

18.

s	t
2	5
4	10
6	15
8	20
6	25
4	30
2	35

19.

r	−4	−2	0	2	4
v	6	6	3	6	8

20.

p	−5	−4	−3	−2	−1
d	−5	−4	−3	−2	−1

21.

Pressure (p)	Volume (v)
15	100.0
20	75.0
25	60.0
30	50.0
35	42.8
40	37.5
45	33.3
50	30.0

22.

Frequency (f)	Wavelength (w)
5	60.0
10	30.0
20	15.0
30	10.0
40	7.5
50	6.0
60	5.0
70	4.3

23.

Temperature (*T*)	Humidity (*h*)
Jan. 1 34°F	42%
Jan. 2 36°F	44%
Jan. 3 35°F	47%
Jan. 4 29°F	50%
Jan. 5 31°F	52%
Jan. 6 35°F	51%
Jan. 7 34°F	49%

24.

Inflation rate (*I*)	Unemployment rate (*U*)
1972 5.6%	5.1%
1973 6.2%	4.5%
1974 10.1%	4.9%
1975 9.2%	7.4%
1976 5.8%	6.7%
1977 5.6%	6.8%
1978 6.7%	7.4%

25.

Adjusted gross income (*I*)	Tax bracket (*T*)
0–2479	0%
2480–3669	11%
3670–4749	12%
4750–7009	14%
7010–9169	15%
9170–11,649	16%
11,650–13,919	18%

26.

Cost of merchandise (*M*)	Shipping charge (*C*)
$0.01–10.00	$2.50
10.01–20.00	3.75
20.01–30.00	4.85
30.01–50.00	5.95
50.01–75.00	6.95
75.01–100.00	7.95
Over 100.00	8.95

27. The function described in Problem 21 is called *g*, so that $v = g(p)$. Find the following.

   a. $g(25)$ **b.** $g(40)$

   c. *x* so that $g(x) = 50$

28. The function described in Problem 22 is called *h*, so that $w = h(f)$. Find the following.

   a. $h(20)$ **b.** $h(60)$

   c. *x* so that $h(x) = 10$

29. The function described in Problem 25 is called *T*, so that $T = T(I)$. Find the following.

   a. $T(8750)$ **b.** $T(6249)$

   c. *x* so that $T(x) = 15\%$

30. The function described in Problem 26 is called *C*, so that $C = C(M)$. Find the following.

   a. $C(11.50)$ **b.** $C(47.24)$

   c. *x* so that $C(x) = 7.95$

In Problems 31–34 use the graph of the function to answer the questions.

31. Figure 5.4 on page 262 shows the graph of *C* as a function of *t*. *C* stands for the number of students (in thousands) at State University who consider themselves computer literate, and *t* represents time, measured in years since 1990.

   a. When did 2000 students consider themselves computer literate?

   b. How long did it take that number to double?

   c. How long did it take for the number to double again?

   d. How many students became computer literate between January 1992 and June 1993?

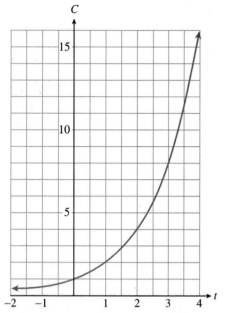

FIGURE 5.4

32. The graph in Figure 5.5 shows the graph of P as a function of t. P is the number of people in Cedar Grove who owned compact disc (CD) players t years after 1980.

   a. When did 3500 people own CD players?

   b. How many people owned CD players in 1986?

   c. The number of owners of CD players in Cedar Grove seems to be leveling off at what number?

   d. How many people acquired CD players between 1981 and 1984?

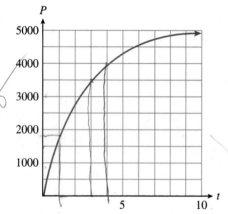

FIGURE 5.5

33. The graph in Figure 5.6 shows the revenue, R, a movie theater collects as a function of the price, d, it charges for a ticket.

   a. What is the revenue if the theater charges $6.50 for a ticket?

   b. What should the theater charge for a ticket in order to collect $250 in revenue?

   c. For what values of d is $R > 350$?

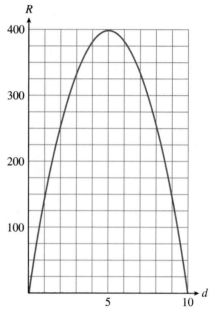

FIGURE 5.6

34. Figure 5.7 shows the graph of S as a function of w. S represents the weekly sales of a best-selling book, in thousands of dollars, w weeks after it is released.

   a. In which weeks were sales over $7000?

   b. In which week did sales fall below $5000 on their way down?

   c. For what values of w is $S > 3.4$?

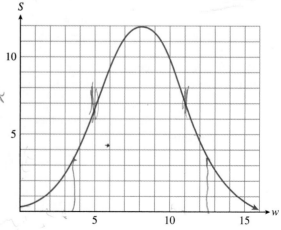

FIGURE 5.7

35. The number of eggs (clutch size) that a bird lays varies greatly. Is there an optimal clutch size for birds of a given species, or does it depend on the individual? In 1980 biologists reduced or enlarged the natural clutch size of magpies in Sweden by adding eggs to or removing eggs from the nests. They then recorded the number of fledglings successfully raised by the parent birds, and computed the average in each case. The graph in Figure 5.8 shows the results for magpies that initially laid 5, 6, 7, or 8 eggs.

Average Number of Fledglings Raised

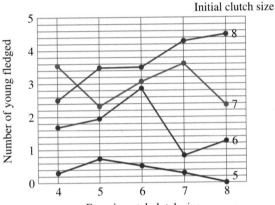

FIGURE 5.8

Source: Data from Hogstedt, G., "Evolution of clutch size in birds: adaptive variation in relation to territory quality," *Science*, 210, 1148–50, 1980. Used by permission of the author. Found in *An Introduction to Behavioural Biology* by J. R. Krebs and N. B. Davies, Blackwell Science, Ltd., 1993.

a. Use the graph to fill in the table of values for the average number of fledglings raised in each situation.

Initial clutch size laid	Experimental clutch size				
	4	5	6	7	8
5					
6					
7					
8					

b. For each initial clutch size, what was the optimal experimental clutch size, in terms of the number of fledglings raised? Record your answers in the table.

Initial clutch size	5	6	7	8
Optimum clutch size				

c. What conclusions can you draw in response to the question in the problem?

36. Energy is required to raise the temperature of a substance, and it is also needed to melt a solid substance to a liquid. The table shows data from heating a solid sample of stearic acid. Heat was applied at a constant rate throughout the experiment.

Time (minutes)	0	0.5	1.5	2	2.5	3	4	5
Temperature (°C)	19	29	40	48	53	55	55	55

Time (minutes)	6	7	8	8.5	9	9.5	10
Temperature (°C)	55	55	55	64	70	73	74

a. Did the temperature rise at a constant rate? Describe the temperature as a function of time.

b. Graph temperature as a function of time.

c. What is the melting point of stearic acid? How long did it take the sample to melt?

Source: © Andrew Hunt and Alan Sykes, *Chemistry*, 1984. Reprinted by permission of Pearson Education Limited.

37. The bar graph in Figure 5.9 shows the percent of earth's surface that lies at various altitudes or depths below the surface of the oceans. (Depths are given as negative altitudes.)

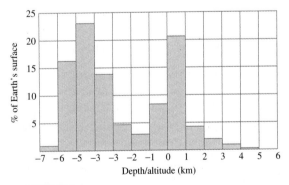

FIGURE 5.9

Source: Open University, 1998

a. Read the graph to complete the table.

Altitude (km)	Percent of earth's surface	Altitude (km)	Percent of earth's surface
−7 to −6		−1 to 0	
−6 to −5		0 to 1	
−5 to −4		1 to 2	
−4 to −3		2 to 3	
−3 to −2		3 to 4	
−2 to −1		4 to 5	

b. Is the level of the earth's surface evenly distributed between the highest and lowest altitudes?

c. What are the two most common altitudes?

d. In a few sentences, describe the distribution of earth's elevation.

e. There is a "relative minimum" value between the two high points on the graph. What altitude is represented by this value? (Refer to Figure 5.9.)

f. Where should Mount Everest appear on the graph? (*Hint:* What is the altitude of Mount Everest?) Why doesn't it show up?

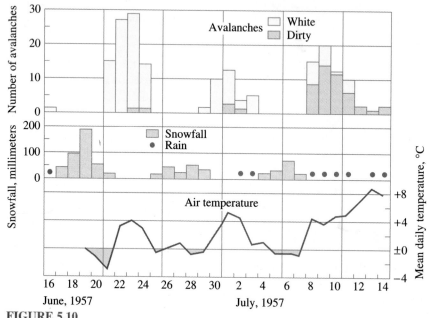

FIGURE 5.10
Source: Leopold, Wolman and Miller, 1992

38. Figure 5.10 gives data about snowfall, air temperature, and number of avalanches on the Mikka glacier in Sarek, Lapland, in 1957.

a. During June and July avalanches occurred over three separate time intervals. What were they?

b. Over what three time intervals did snow fall?

c. When was the temperature above freezing (0°C)?

d. According to the data, under what conditions do avalanches occur?

Evaluate each function for the given values.

39. $f(x) = 6 - 2x$

   a. $f(3)$ **b.** $f(-2)$ **c.** $f(12.7)$ **d.** $f\left(\dfrac{2}{3}\right)$

40. $g(t) = 5t - 3$

   a. $g(1)$ **b.** $g(-4)$ **c.** $g(14.1)$ **d.** $g\left(\dfrac{3}{4}\right)$

41. $h(v) = 2v^2 - 3v + 1$

   a. $h(0)$ **b.** $h(-1)$ **c.** $h\left(\dfrac{1}{4}\right)$ **d.** $h(-6.2)$

42. $r(s) = 2s - s^2$

   a. $r(2)$ **b.** $r(-4)$ **c.** $r\left(\dfrac{1}{3}\right)$ **d.** $r(-1.3)$

43. $H(z) = \dfrac{2z - 3}{z + 2}$

    a. $H(4)$ **b.** $H(-3)$ **c.** $H\left(\dfrac{4}{3}\right)$ **d.** $H(4.5)$

44. $F(x) = \dfrac{1 - x}{2x - 3}$

    a. $F(0)$ **b.** $F(-3)$ **c.** $F\left(\dfrac{5}{2}\right)$ **d.** $F(9.8)$

45. $E(t) = \sqrt{t - 4}$

    a. $E(16)$ **b.** $E(4)$ **c.** $E(7)$ **d.** $E(4.2)$

46. $D(r) = \sqrt{5 - r}$

    a. $D(4)$ **b.** $D(-3)$ **c.** $D(-9)$ **d.** $D(4.6)$

47. A sport utility vehicle costs \$28,000 and depreciates according to the formula

$$V(t) = 28,000(1 - 0.08t)$$

where V is the value of the vehicle after t years.

  a. Complete the table of values.

t	2	5	8	10
V				

  b. Evaluate $V(12)$ and explain what it means.

48. In a profit-sharing plan an employee receives a salary of

$$S(x) = 20,000 + 0.01x$$

where x represents the company's profit for the year.

  a. Complete the table of values.

x	50,000	100,000	500,000	1,000,000
S				

  b. Evaluate $S(850,000)$ and explain what it means.

49. The number of compact cars that a large dealership can sell at price, p, is given by

$$N(p) = \dfrac{12,000,000}{p}$$

  a. Complete the table of values.

p	5000	8,000	10,000	12,000
N				

  b. Evaluate $N(6,000)$ and explain what it means.

50. A department store finds that the market value of its Christmas-related merchandise is given by

$$M(t) = \dfrac{600,000}{t} \quad (t \le 30)$$

where t is the number of weeks after Christmas.

  a. Complete the table of values.

t	2	5	10	20
M				

  b. Evaluate $M(12)$ and explain what it means.

51. The velocity of a car that brakes suddenly can be determined from the length of its skid marks, d, by

$$v(d) = \sqrt{12d}$$

where d is in feet and v is in miles per hour.

  a. Complete the table of values.

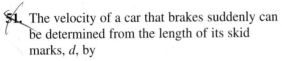

d	20	50	80	100
v				

  b. Evaluate $v(250)$ and explain what it means.

52. The distance, d, in miles that a person can see on a clear day from a height, h, in feet is given by

$$d(h) = 1.22 \sqrt{h}$$

  a. Complete the table of values.

h	200	1000	5000	10,000
d				

  b. Evaluate $d(20,320)$ and explain what it means.

5.2 Graphs of Functions

Reading Function Values from a Graph

The graph in Figure 5.11 shows the Dow-Jones Industrial Average (the average value of the stock prices of 500 major companies) during the stock market "correction" of October 1987.

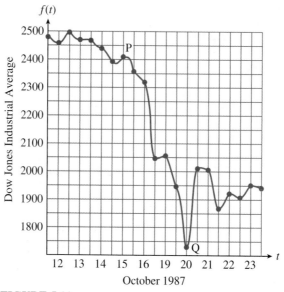

FIGURE 5.11

The Dow-Jones Industrial Average (DJIA) is given as a function of time during the 8 days from October 15 to October 22; that is, $f(t)$ is the DJIA recorded at time t. The values of the independent variable, time, are displayed on the horizontal axis, and values of the dependent variable, DJIA, are displayed on the vertical axis. There is no formula that gives the DJIA for a particular day; instead, $f(t)$ is defined for each value of t by the vertical coordinate of the point with the given t-coordinate.

EXAMPLE 1

  a. The coordinates of point P in Figure 5.11 are $(15, 2412)$. What do the coordinates tell you about the function f?

  b. If the DJIA was 1726 at noon on October 20, what can you say about the graph of f?

Solutions

  a. The coordinates of point P tell us that $f(15) = 2412$, so the DJIA was 2412 at noon on October 15.

  b. We can say that $f(20) = 1726$, so the point $(20, 1726)$ lies on the graph of f. This point is labeled Q in Figure 5.11. ∎

Thus, the coordinates of each point on the graph of the function represent a pair of corresponding values of the two variables. In general, we can make the following statement.

The point (a, b) lies on the graph of the function f if and only if $f(a) = b$.

Another way of saying this is:

Each point on the graph of the function f has coordinates $(x, f(x))$ for some value of x.

EXAMPLE 2

Consider the graph of the function g shown in Figure 5.12.

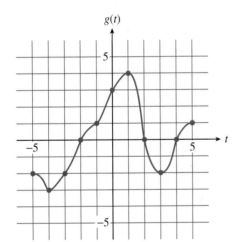

FIGURE 5.12

a. Find $g(-2)$, $g(0)$, and $g(5)$.

b. For what value(s) of t is $g(t) = -2$? For what value(s) of t is $g(t) = 0$?

c. What is the largest, or maximum, value of $g(t)$? For what value of t does the function take on its maximum value?

d. On what intervals is g increasing? Decreasing?

Solutions

a. The points $(-2, 0)$, $(0, 3)$, and $(5, 1)$ lie on the graph of g. Therefore, $g(-2) = 0$, $g(0) = 3$, and $g(5) = 1$.

b. Because the points $(3, -2)$ and $(-3, -2)$ lie on the graph, $g(-3) = -2$ and $g(3) = -2$. (The t-values are -3 and 3.) Because the points $(-2, 0)$, $(2, 0)$, and $(4, 0)$ lie on the graph, $g(-2) = 0$, $g(2) = 0$, and $g(4) = 0$. (The t-values are -2, 2, and 4).

c. The maximum value of $g(t)$ is the second coordinate of the highest point on the graph, $(1, 4)$. Thus, the maximum value of $g(t)$ is 4, and it occurs when $t = 1$.

d. A graph is increasing if the y-values get larger as we read from left to right. The graph of g is increasing for t-values between -4 and 1, and between 3 and 5. Thus, g is increasing on the intervals $(-4, 1)$ and $(3, 5)$, and decreasing on $(-5, -4)$ and $(1, 3)$. ∎

Constructing the Graph of a Function

Although some functions are *defined* by their graphs, we can also construct graphs for functions described by tables or equations. We make these graphs in the same way that we graph equations in two variables: by plotting points whose coordinates satisfy the equation.

EXAMPLE 3

Graph the function $f(x) = \sqrt{x + 4}$.

Solution

Choose several convenient values for x and evaluate the function to find the corresponding $f(x)$ values.(See Table 5.6.) For this function we cannot choose x-values less than -4, because the square root of a negative number is not a real number.

TABLE 5.6

x	$f(x)$
-4	0
-3	1
0	2
2	$\sqrt{6}$
5	3

$f(-4) = \sqrt{-4 + 4} = \sqrt{0} = 0$
$f(-3) = \sqrt{-3 + 4} = \sqrt{1} = 1$
$f(0) = \sqrt{0 + 4} = \sqrt{4} = 2$
$f(2) = \sqrt{2 + 4} = \sqrt{6} \approx 2.45$
$f(5) = \sqrt{5 + 4} = \sqrt{9} = 3$

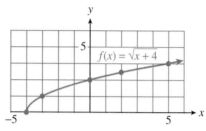

FIGURE 5.13

Points on the graph have coordinates $(x, f(x))$, so the vertical coordinate of each point is given by the value of $f(x)$. Plot the points and connect them with a smooth curve to obtain the graph shown. Notice that no points on the graph have x-coordinates less than -4. ∎

We can also use a graphing calculator to obtain the table and the graph in Example 3. Choose the standard window by pressing ZOOM 6 , then enter the equation that describes the function. Press Y = and type in

$$\sqrt{} \quad \boxed{\text{X, T, }\theta, n} \quad \boxed{+} \quad 4 \quad \boxed{)}$$

(Your calculator does not use the $f(x)$ notation for graphs, so we will continue to use Y_1, Y_2, etc. for the dependent variable.) Don't forget to enclose $x + 4$ in parentheses, because it appears under a radical. The calculator's picture of the graph is shown in Figure 5.14.

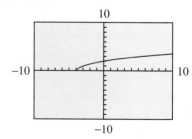

FIGURE 5.14

EXERCISE 1 $f(x) = x^3 - 2$

a. Complete the table of values and sketch a graph of the function.

x	-2	-1	$-\frac{1}{2}$	0	$\frac{1}{2}$	1	2
$f(x)$							

b. Use your calculator to make a table of values and graph the function.

The Vertical Line Test

If a variable relationship is a function, two different y-values cannot be related to the same x-value. This restriction means that two different ordered pairs cannot have the same first coordinate. What does it mean in terms of the graph of the function?

Consider the graph shown in Figure 5.15(a). Every vertical line intersects the graph in at most one point, so there is only one point on the graph for each x-value. This graph represents a function. In Figure 5.15b, however, the line $x = 2$ intersects the graph at two points, $(2, 1)$ and $(2, 4)$. Two different y-values, 1 and 4, are related to the same x-value, 2. This graph cannot be the graph of a function.

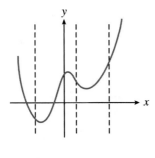

(a)

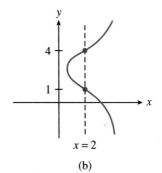

(b)

FIGURE 5.15

We summarize these observations as follows.

The Vertical Line Test

A graph represents a function if and only if every vertical line intersects the graph in at most one point.

EXAMPLE 4   Use the vertical line test to determine which of the graphs in Figure 5.16 represent functions.

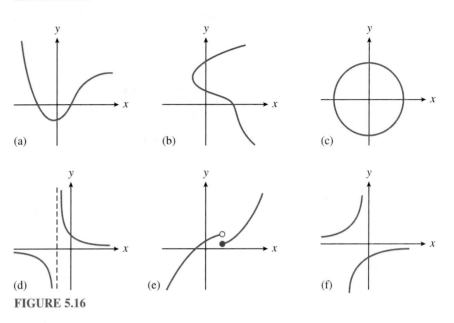

FIGURE 5.16

Solution

Graphs (a), (d), and (e) represent functions, because no vertical line intersects the graph in more than one point. Graphs (b), (c), and (f) do not represent functions. ∎

Linear and Quadratic Functions

You already know quite a bit about the graphs of two kinds of functions: linear and quadratic functions. All lines except vertical lines pass the vertical line test and hence are the graphs of functions. (See Figure 5.17.) Functions of the form

$$f(x) = mx + b$$

are therefore called **linear functions.** **Quadratic functions** can be written as

$$f(x) = ax^2 + bx + c \qquad a \neq 0$$

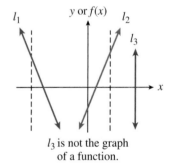

l_3 is not the graph of a function.

FIGURE 5.17

and their graphs are parabolas. These parabolas all pass the vertical line test.

By letting the vertical axis, or y-axis, represent the values of $f(x)$, we can graph linear functions the same way we graphed linear equations $y = mx + b$, and we can graph quadratic functions the same way we graphed quadratic equations $y = ax^2 + bx + c$.

EXERCISE 2 Sketch the graph of each function.

a. $f(x) = -2x + 5$ **b.** $g(x) = 6 - x^2$

a. The graph is a line.

    Slope: $m =$

    y-intercept: $b =$

Plot the y-intercept and use the definition of slope

    $m = \dfrac{\Delta y}{\Delta x} =$

to find a second point on the line.

b. The graph is a parabola.

    x-intercepts:

    Solve $0 = 6 - x^2$

Vertex:

    $x_v =$

    $y_v =$

READING QUESTIONS

1. How can you find the value of $f(3)$ from a graph of f?
2. If $f(8) = 2$, what point lies on the graph of f?
3. Explain how to construct the graph of a function from its equation.
4. Explain how to use the vertical line test.

ANSWERS TO 5.2 EXERCISES

1.

x	-2	-1	$\frac{-1}{2}$	0	$\frac{1}{2}$	1	2
$f(x)$	-10	-3	$\frac{-17}{8}$	-2	$\frac{-15}{8}$	-1	6

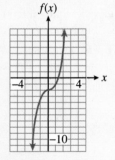

2a.

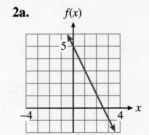

b.

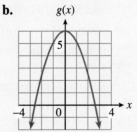

Use the graphs to answer the questions about the functions.

1. a. Find $h(-3)$, $h(1)$, and $h(3)$ in Figure 5.18.

   b. For what value(s) of z is $h(z) = 3$?

   c. Find the intercepts of the graph. List the function values given by the intercepts.

   d. What is the maximum value of $h(z)$?

   e. For what value(s) of z does h take on its maximum value?

   f. On what intervals is the function increasing? Decreasing?

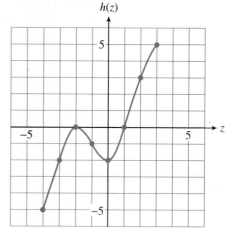

FIGURE 5.18

2. a. Find $G(-4)$, $G(-1)$, and $G(2)$ in Figure 5.19.

   b. For what value(s) of s is $G(s) = 3$?

   c. Find the intercepts of the graph. List the function values given by the intercepts.

   d. What is the minimum value of $G(s)$?

   e. For what value(s) of s does G take on its minimum value?

   f. On what intervals is the function increasing? Decreasing?

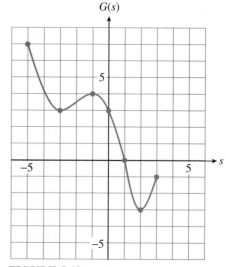

FIGURE 5.19

3. a. Find $R(1)$ and $R(3)$ in Figure 5.20.

   b. For what value(s) of p is $R(p) = 2$?

   c. Find the intercepts of the graph. List the function values given by the intercepts.

   d. Find the maximum and minimum values of $R(p)$.

   e. For what value(s) of p does R take on its maximum and minimum values?

   f. On what intervals is the function increasing? Decreasing?

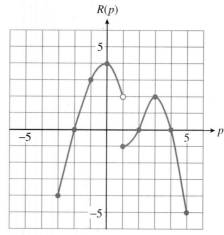

FIGURE 5.20

4. a. Find $f(-1)$ and $f(3)$ in Figure 5.21.

b. For what value(s) of t is $f(t) = 5$?

c. Find the intercepts of the graph. List the function values given by the intercepts.

d. Find the maximum and minimum values of $f(t)$.

e. For what value(s) of t does f take on its maximum and minimum values?

f. On what intervals is the function increasing? Decreasing?

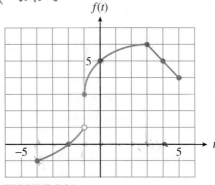

FIGURE 5.21

5. a. Find $S(0)$, $S\left(\frac{1}{6}\right)$, and $S(-1)$ in Figure 5.22.

b. Estimate the value of $S\left(\frac{1}{3}\right)$ from the graph.

c. For what value(s) of x is $S(x) = -\frac{1}{2}$?

d. Find the maximum and minimum values of $S(x)$.

e. For what value(s) of x does S take on its maximum and minimum values?

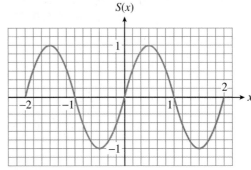

FIGURE 5.22

6. a. Find $C(0)$, $C\left(-\frac{1}{3}\right)$, and $C(1)$ in Figure 5.23.

b. Estimate the value of $C\left(\frac{1}{6}\right)$ from the graph.

c. For what value(s) of x is $C(x) = \frac{1}{2}$?

d. Find the maximum and minimum values of $C(x)$.

e. For what value(s) of x does C take on its maximum and minimum values?

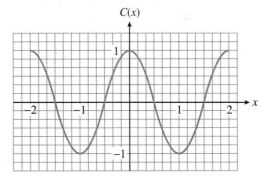

FIGURE 5.23

7. a. Find $F(-3)$, $F(-2)$, and $F(2)$ in Figure 5.24.

b. For what value(s) of s is $F(s) = -1$?

c. Find the maximum and minimum values of $F(s)$.

d. For what value(s) of s does F take on its maximum and minimum values?

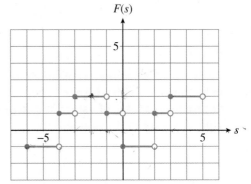

FIGURE 5.24

8. a. Find $P(-3)$, $P(-2)$, and $P(1)$ in Figure 5.25.

b. For what value(s) of n is $P(n) = 0$?

c. Find the maximum and minimum values of $P(n)$.

d. For what value(s) of n does P take on its maximum and minimum values?

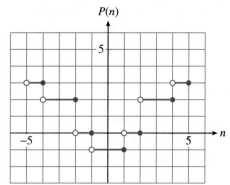

FIGURE 5.25

It's a function if the x value only touches in one place.

Which of the following graphs represent functions?

9.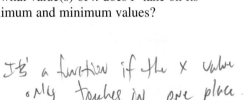

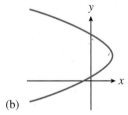

(a)

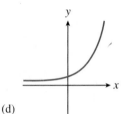

(d)

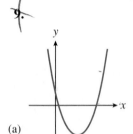

(b)

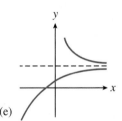

(e)

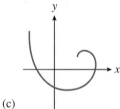

(c)

FIGURE 5.26

10.

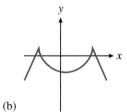

(a)

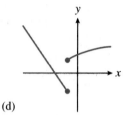

(d)

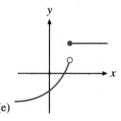

(b)

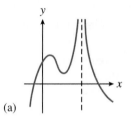

(e)

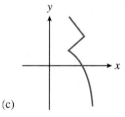

(c)

FIGURE 5.27

In Problems 11–16, (a) make a table of values and sketch a graph of the function by plotting points. (Use the suggested x-values.), and (b) use your calculator to make a table of values and to graph the function. Compare the calculator's graph with your sketch.

11. $g(x) = x^3 + 4$; $x = -2, -1, \ldots, 2$

12. $h(x) = 2 + \sqrt{x}$; $x = 0, 1, \ldots, 9$

13. $G(x) = \sqrt{4 - x}$; $x = -5, -4, \ldots, 4$

14. $F(x) = \sqrt{x - 1}$; $x = 1, 2, \ldots, 10$

15. $v(x) = 1 + 6x - x^3$; $x = -3, -2, \ldots, 3$

16. $w(x) = x^3 - 8x$; $x = -4, -3, \ldots, 4$

Use the graphs to solve the equations and inequalities.

17. Figure 5.28 shows the graph of
$$g(x) = \frac{12}{2 + x^2}.$$

a. Solve $\dfrac{12}{2 + x^2} = 4$

b. For what values of x is $\dfrac{12}{2 + x^2}$ between 1 and 2?

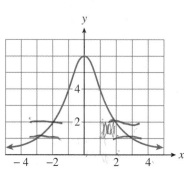

FIGURE 5.28

18. Figure 5.29 shows the graph of $f(x) = \dfrac{30\sqrt{x}}{1 + x}$.

a. Solve $\dfrac{30\sqrt{x}}{1 + x} = 15$.

b. For what values of x is $\dfrac{30\sqrt{x}}{1 + x}$ less than 12?

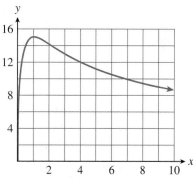

FIGURE 5.29

19. Figure 5.30 shows a graph of $M = g(q)$. Find all values of q for which

a. $g(q) = 0$

b. $g(q) = 16$

c. $g(q) < 6$

d. For what values of q is $g(q)$ increasing?

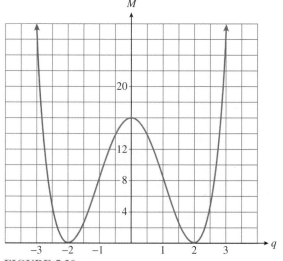

FIGURE 5.30

20. Figure 5.31 shows a graph of $P = f(t)$. Find all values of t for which

a. $f(t) = 3$

b. $f(t) > 4.5$

c. $2 \leq f(t) \leq 4$

d. For what values of t is $f(t)$ decreasing?

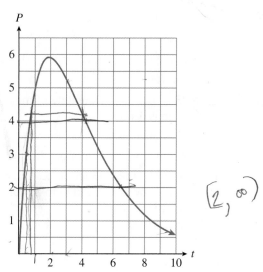

FIGURE 5.31

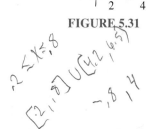

In Problems 21–24, graph each function in the "friendly" window

$$Xmin = -9.4 \qquad Xmax = 9.4$$
$$Ymin = -10 \qquad Ymax = 10$$

then answer the questions about the graph.

21. $g(x) = \sqrt{36 - x^2}$. Find all points on the graph that have y-coordinate 3.6.

22. $f(x) = \sqrt{x^2 - 6}$. Find all points on the graph that have y-coordinate -2.

23. $F(x) = 0.5x^3 - 4x$. Estimate the coordinates of the "turning points" of the graph, where the graph changes from increasing to decreasing or vice versa.

24. $G(x) = 2 + 4x - x^3$. Estimate the coordinates of the "turning points" of the graph, where the graph changes from increasing to decreasing or vice versa.

25. Figure 5.32 shows the relationship between annual precipitation, p, in a region and the amount of erosion, measured in tons of sediment per square mile, s, removed by runoff.

   a. Is the sediment yield an increasing function of annual precipitation?

   b. At what annual precipitation is sediment yield a maximum, and what is that maximum?

   c. For what interval of annual precipitation does sediment yield decrease?

   d. An increase in vegetation inhibits erosion, and precipitation encourages vegetation. Describe the effect of increased precipitation in each of these three environments:

desert shrub	$0 < p < 12$
grassland	$12 < p < 30$
forest	$30 < p < 60$

26. Figure 5.33 shows the temperature of the ocean at various depths.

   a. Is depth a function of temperature?

   b. Is temperature a function of depth?

   c. The axes in Figure 5.33 are scaled in an unusual way. Why is it useful to present the graph in this way?

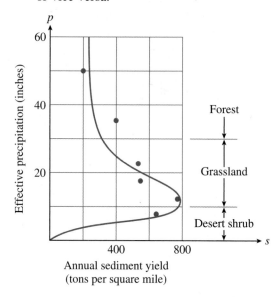

FIGURE 5.32
Source: Leopold, Wolman, and Miller, 1992

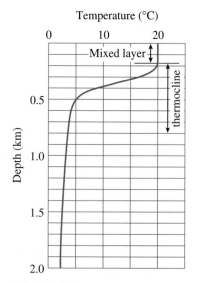

FIGURE 5.33
Source: Open University, 1998.

27. Refer to Figure 5.33 to answer the questions.

  a. What is the difference in temperature at the surface of the ocean and at the deepest level shown?

  b. Over what depths does the temperature change most rapidly?

  c. What is the average rate of change of temperature with respect to depth in the region called the thermocline?

  Source: Open University, 1998.

28. The graph in Figure 5.34 shows the average profile of air temperature above the earth's surface. (Figure b is an enlargement of the indicated region of Figure a.)

a. Is temperature a decreasing function of altitude?

b. The lapse rate is the rate at which the temperature changes with altitude. In which regions of the atmosphere is the lapse rate positive?

c. The region where the lapse rate is zero is called the isothermal zone. Give an interval of altitudes that describes the isothermal zone.

d. What is the lapse rate in the mesosphere?

e. Describe the temperature for altitudes greater than 90 kilometers.

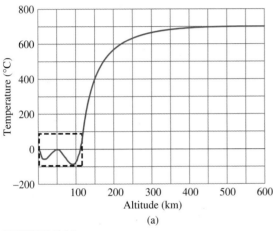

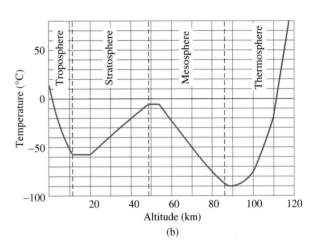

FIGURE 5.34
Source: Ahrens, 1998.

Graph each function (a) first using the standard window, (b) then using the suggested window. Explain how the window alters the appearance of the graph in each case.

29. a. $h(x) = \dfrac{1}{x^2 + 10}$

  b. Xmin $= -5$ Xmax $= 5$
      Ymin $= 0$ Ymax $= 0.5$

30. a. $H(x) = \sqrt{1 - x^2}$

  b. Xmin $= -2$ Xmax $= 2$
      Ymin $= -2$ Ymax $= 2$

31. a. $P(x) = (x - 8)(x + 6)(x - 15)$

  b. Xmin $= -10$ Xmax $= 20$
      Ymin $= -250$ Ymax $= 750$

32. a. $p(x) = 200x^3$

  b. Xmin $= -5$ Xmax $= 5$
      Ymin $= -10{,}000$ Ymax $= 10{,}000$

Sketch graphs of the following linear and quadratic functions by hand, and label the significant points.

33. $f(x) = 3x - 4$

34. $g(x) = -2x + 5$

35. $G(s) = -\dfrac{5}{3}s + 50$

36. $F(s) = -\dfrac{3}{4}s + 60$

37. $h(t) = t^2 - 8$

38. $p(t) = 3 - t^2$

39. $g(w) = (w + 2)^2$

40. $f(w) = (w - 1)^2$

41. $F(z) = -2z^2$

42. $G(z) = 0.5z^2$

For each pair of functions in Problems 43–46, do the following:

a. Compute $f(0)$ and $g(0)$.

b. Find all values of x for which $f(x) = 0$.

c. Find all values of x for which $g(x) = 0$.

d. Find all values of x for which $f(x) = g(x)$.

e. Graph each pair of functions in the same window, then sketch the graph on paper. Illustrate your answers to (a)–(d) as points on the graph.

43. $f(x) = 2x^2 + 3x \qquad g(x) = 5 - 6x$

44. $f(x) = 3x^2 - 6x \qquad g(x) = 8 + 4x$

45. $f(x) = 2x^2 - 2x \qquad g(x) = x^2 + 3$

46. $f(x) = x^2 + 4x + 6 \qquad g(x) = 4 - x^2$

5.3　Some Basic Graphs

In this section we will study the graphs of some important basic functions. Many functions fall into families or classes of similar functions—recognizing the appropriate family for a given situation is often the first step in modeling.

We'll need two new algebraic operations.

Cube Roots

You are familiar with square roots. Every non-negative number has two square roots, defined as follows.

$$s \text{ is a square root of } n \text{ if } s^2 = n$$

In this book we will use several other kinds of roots, one of which is called the **cube root.** We define the cube root as follows.

$$b \text{ is the \textbf{cube root} of } a \text{ if } b \text{ cubed equals } a$$

We use the symbol $\sqrt[3]{a}$ for the cube root of a, so in symbols we can write

$$b = \sqrt[3]{a} \qquad \text{if} \qquad b^3 = a$$

Although we cannot take the square root of a negative number, we can take the cube root of any real number. For example,

$$\sqrt[3]{64} = 4 \qquad \text{because} \qquad 4^3 = 64$$

absolute value sign

whatever is inside is positive

| |

and

$$\sqrt[3]{-27} = -3 \quad \text{because} \quad (-3)^3 = -27$$

Simplifying radicals occupies the same position in the order of operations as computing powers: after parentheses, and before products and quotients.

EXAMPLE 1

Simplify each expression.

a. $3\sqrt[3]{-8}$ **b.** $2 - \sqrt[3]{-125}$

Solutions

a. $3\sqrt[3]{-8} = 3(-2)$
$= -6$

b. $2 - \sqrt[3]{-125} = 2 - (-5)$
$= 7$ ∎

EXERCISE 1   Simplify each expression.

a. $5 - 3\sqrt[3]{64}$ **b.** $\dfrac{6 - \sqrt[3]{-27}}{2}$

We can use the calculator to find cube roots as follows. Press the MATH key to get a menu of options. Option **4,** labeled $\sqrt[3]{}$, is the cube root key. To find the cube root of, say, 15.625, we key in

MATH 4 15.625 ENTER

and the calculator returns the result, 2.5. Thus, $\sqrt[3]{15.625} = 2.5$. You can check this by verifying that $2.5^3 = 15.625$.

Absolute Value

The absolute value is used to discuss problems involving distance. For example, consider the number line in Figure 5.35. Starting at the origin, we travel in opposite *directions* to reach the two numbers 6 and −6, but the *distance* we travel in each case is the same.

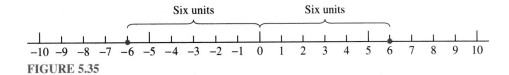

FIGURE 5.35

The distance from a number c to the origin is called the **absolute value** of c, denoted by $|c|$. Because distance is always positive, the absolute value of a number is always positive. Thus, $|6| = 6$ and $|-6| = 6$. In general, we define the absolute value of a number x as follows.

$$|x| = \begin{cases} x & \text{if } x \geq 0 \\ -x & \text{if } x < 0 \end{cases}$$

This definition says that the absolute value of a positive number (or zero) is the same as the number. To find the absolute value of a negative number, we take the opposite of the number, with a positive number as the result. For instance,

$$|-6| = -(-6) = 6$$

Absolute value bars act like grouping devices in the order of operations: You should complete any operations that appear inside absolute value bars before you compute the absolute value.

EXAMPLE 2

Simplify each expression.
    a. $|3 - 8|$ **b.** $|3| - |8|$

Solutions
    a. Simplify the expression inside the absolute value bars first.

$$|3 - 8| = |-5| = 5$$

    b. Simplify each absolute value, then subtract.

$$|3| - |8| = 3 - 8 = -5$$ ■

EXERCISE 2   Simplify each expression.

  a. $12 - 3|-6|$ **b.** $-7 - 3|2 - 9|$

Most of the graphs in this section will be new to you. Because they are fundamental to further study of mathematics and its applications, you should become familiar with the properties of each graph, so that you can sketch them easily from memory.

INVESTIGATION 9

Eight Basic Functions

Part I Some Powers and Roots

1. Complete Table 5.7, the table of values for the squaring function, $f(x) = x^2$, and the cubing function, $g(x) = x^3$. Then sketch each function on graph paper, using the table values to help you scale the axes.

2. Verify both graphs with your graphing calculator.

TABLE 5.7

x	$f(x) = x^2$	$g(x) = x^3$
-3		
-2		
-1		
$-\frac{1}{2}$		
0		
$\frac{1}{2}$		
1		
2		
3		

3. State the intervals on which each graph is increasing.

4. Write a few sentences comparing the two graphs.

5. Complete Table 5.8 and Table 5.9 for the functions $f(x) = \sqrt{x}$ and $g(x) = \sqrt[3]{x}$. (Round your answers to two decimal places.) Then sketch each function on graph paper, using the table values to help you scale the axes.

TABLE 5.8

x	$f(x) = \sqrt{x}$
0	
$\frac{1}{2}$	
1	
2	
3	
4	
5	
7	
9	

TABLE 5.9

x	$g(x) = \sqrt[3]{x}$
-8	
-4	
-1	
$-\frac{1}{2}$	
0	
$\frac{1}{2}$	
1	
4	
8	

6. Verify both graphs with your graphing calculator.

7. State the intervals on which each graph is increasing.

8. Write a few sentences comparing the two graphs.

Part II Asymptotes

9. Complete Table 5.10 for the functions $f(x) = \dfrac{1}{x}$ and $g(x) = \dfrac{1}{x^2}$. What is true about $f(0)$ and $g(0)$?

TABLE 5.10

x	$f(x) = \dfrac{1}{x}$	$g(x) = \dfrac{1}{x^2}$
-4		
-3		
-2		
-1		
$-\frac{1}{2}$		
0		
$\frac{1}{2}$		
1		
2		
3		
4		

10. Prepare a standard grid on graph paper for each function (scale both axes from -10 to 10). Plot the points from Table 5.10 and connect them with smooth curves.

11. As x increases through larger and larger values, what happens to the values of $f(x)$? Extend your graph to reflect your answer. What happens to $f(x)$ as x decreases through negative values (that is, for $x = -5, -6, -7, \ldots$)? Extend your graph for these x-values.

    As the values of x get larger in absolute value, the graph approaches the x-axis. However, because $\dfrac{1}{x}$ never *equals* zero for any x-value, the graph never actually touches the x-axis. We say that the x-axis is a **horizontal asymptote** for the graph.

12. Repeat (11) for the graph of $g(x)$.

13. Next we'll examine the graphs of f and g near $x = 0$. Use your calculator to evaluate f for several x-values close to zero and complete Table 5.11.

    TABLE 5.11

x	$f(x) = \dfrac{1}{x}$
-2	
-1	
-0.1	
-0.01	
-0.001	

x	$f(x) = \dfrac{1}{x}$
2	
1	
0.1	
0.01	
0.001	

    What happens to the values of $f(x)$ as x approaches zero? Extend your graph of f to reflect your answer.

    As x approaches zero from the left (through negative values), the function values decrease toward $-\infty$. As x approaches zero from the right (through positive values), the function values increase toward ∞. The graph approaches but never touches the vertical line $x = 0$ (the y-axis.) We say that the graph of f has a **vertical asymptote** at $x = 0$.

14. Repeat (13) for the graph of $g(x)$.

15. The functions $f(x) = \dfrac{1}{x}$ and $g(x) = \dfrac{1}{x^2}$ are examples of **rational functions.** Verify both graphs with your graphing calculator. Use the window

    $$\text{Xmin} = -4 \qquad \text{Xmax} = 4$$
    $$\text{Ymin} = -4 \qquad \text{Ymax} = 4$$

16. State the intervals on which each graph is increasing.

17. Write a few sentences comparing the two graphs.

Part III Absolute Value

18. Complete Table 5.12 for the two functions $f(x) = x$ and $g(x) = |x|$. Then sketch each function on graph paper, using the table values to help you scale the axes.

TABLE 5.12

x	$f(x) = x$	$g(x) = |x|$
-4		
-2		
-1		
$-\frac{1}{2}$		
0		
$\frac{1}{2}$		
1		
2		
4		

19. Verify both graphs with your graphing calculator. Your calculator uses the notation abs (x) instead of $|x|$ for the absolute value of x. Access the absolute value function through the Catalog: After opening the [**Y =**] screen, press [**2nd**] [**0**] to see the Catalog, then press [**ENTER**] for abs (x).

20. State the intervals on which each graph is increasing.

21. Write a few sentences comparing the two graphs.

Graphs of Eight Basic Functions

The graphs of the eight basic functions considered in the investigation are shown in Figures 5.36 to 5.40.

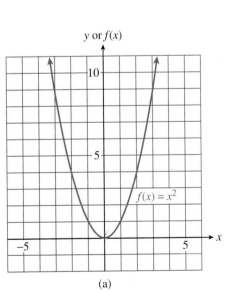

(a)

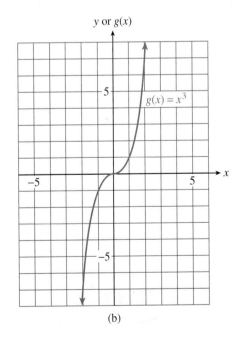

(b)

FIGURE 5.36

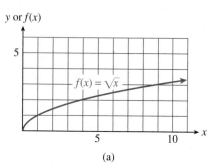

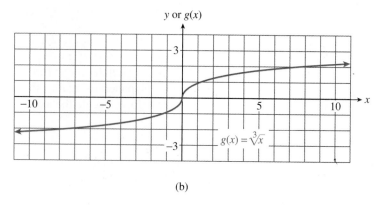

(a) (b)

FIGURE 5.37

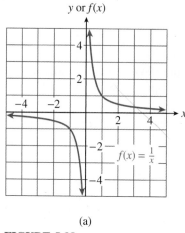

(a)

FIGURE 5.38

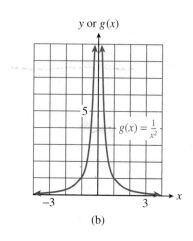

(b)

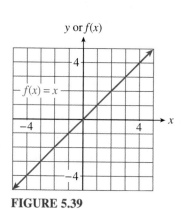

FIGURE 5.39

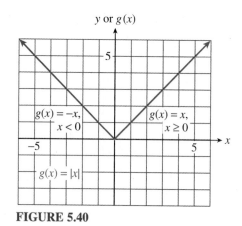

FIGURE 5.40

READING QUESTIONS

1. If $x > 1$, which is larger, $\sqrt{x}$ or $\sqrt[3]{x}$?

2. Give a geometric interpretation for the absolute value of a number.

3. What is a horizontal asymptote?

4. What is a vertical asymptote?

HOMEWORK 5.3

Compute each cube root. Round your answers to three decimal places if necessary. Verify your answers by cubing them.

1. a. $\sqrt[3]{512}$ b. $\sqrt[3]{-125}$ 2. a. $\sqrt[3]{9}$ b. $\sqrt[3]{258}$

   c. $\sqrt[3]{-0.064}$ d. $\sqrt[3]{1.728}$ c. $\sqrt[3]{-0.002}$ d. $\sqrt[3]{-3.1}$

Simplify each expression according to the order of operations.

3. a. $\dfrac{4 - 3\sqrt[3]{64}}{2}$ b. $\dfrac{4 + \sqrt[3]{-216}}{8 - \sqrt[3]{8}}$ 4. a. $\sqrt[3]{3^3 + 4^3 + 5^3}$ b. $\sqrt[3]{9^3 + 10^3 - 1^3}$

Simplify each expression according to the order of operations.

5. a. $-|-9|$ 6. a. $2 - (-6)$ 7. a. $|-8| - |12|$ 8. a. $|-3| + |-5|$
   b. $-(-9)$ b. $2 - |-6|$ b. $|-8 - 12|$ b. $|-3 + (-5)|$

9. $4 - 9|2 - 8|$ 10. $2 - 5|-6 - 3|$ 11. $|-4 - 5||1 - 3(-5)|$

12. $|-3 + 7||-2(6 - 10)|$ 13. $||-5| - |-6||$ 14. $||4| - |-6||$

Sketch each function by hand, paying attention to the shape of the graph. Carefully plot at least three "guide points" for each graph to ensure accuracy. If possible, plot the points with x-coordinates -1, 0, and 1.

15. $f(x) = x^3$ 16. $f(x) = |x|$ 17. $f(x) = \sqrt{x}$ 18. $f(x) = \sqrt[3]{x}$

19. $f(x) = \dfrac{1}{x}$ 20. $f(x) = \dfrac{1}{x^2}$

Use the graphs to solve the equations and inequalities. Estimate your answers to one decimal point.

21. Refer to the graph of $f(x) = x^3$ in Figure 5.41 on page 286.

   a. Estimate the value of $(1.4)^3$.

   b. Find all numbers whose cube is -20.

   c. Estimate the values of $\sqrt[3]{24}$.

   d. Find all solutions of the equation $x^3 = 6$.

   e. Find all solutions of the inequality $-12 \le x^3 \le 15$.

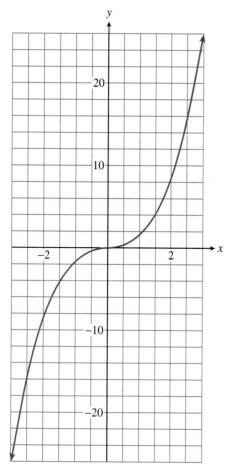

FIGURE 5.41

22. Refer to the graph of $f(x) = x^2$ in Figure 5.42.

   a. Estimate the value of $(-2.5)^2$.

   b. Find all numbers whose square is 12.

   c. Estimate the values of $\sqrt{10.5}$.

   d. Find all solutions of the equation $x^2 = 15$.

   e. Find all solutions of the inequality $4 \le x^2 \le 10$.

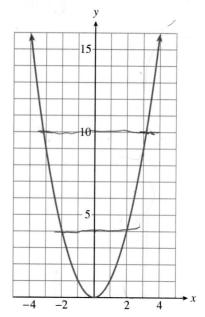

FIGURE 5.42

23. Refer to the graph of $y = \dfrac{1}{x}$ in Figure 5.43.

   a. Estimate the values of $\dfrac{1}{3.4}$.

   b. Find all numbers whose reciprocal is -2.5.

   c. Find all solutions of the equation $\dfrac{1}{x} = 4.8$.

   d. Find all solutions of the inequality $0.3 \le \dfrac{1}{x} \le 4.5$.

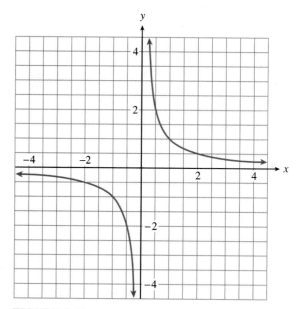

FIGURE 5.43

24. Refer to the graph of $y = |x - 2|$ in Figure 5.44.

  a. Estimate the values of $|1.6 - 2|$.

  b. Find all values of x for which $|x - 2| = 3$.

  c. Find all solutions of the equation $|x - 2| = 0.4$.

  d. Find all solutions of the inequality $|x - 2| > 1$.

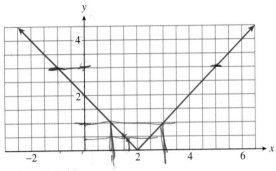

FIGURE 5.44

Graph each set of functions in the same window on your calculator. Use the ZDecimal window. Describe how graphs b. and c. are different from the basic graph.

25. a. $f(x) = x^3$
  b. $g(x) = x^3 - 2$
  c. $h(x) = x^3 + 1$

26. a. $f(x) = |x|$
  b. $g(x) = |x - 2|$
  c. $h(x) = |x + 1|$

27. a. $f(x) = \dfrac{1}{x}$
  b. $g(x) = \dfrac{1}{x + 1.5}$
  c. $h(x) = \dfrac{1}{x - 1}$

28. a. $f(x) = \dfrac{1}{x^2}$
  b. $g(x) = \dfrac{1}{x^2} + 2$
  c. $h(x) = \dfrac{1}{x^2} - 1$

29. a. $f(x) = \sqrt{x}$
  b. $g(x) = -\sqrt{x}$
  c. $h(x) = \sqrt{-x}$

30. a. $f(x) = \sqrt[3]{x}$
  b. $g(x) = -\sqrt[3]{x}$
  c. $h(x) = \sqrt[3]{-x}$

Each of the graphs in Problems 31 and 32 (Figures 5.45 and 5.46) is a variation of one of the basic graphs listed below. Identify the basic graph for each problem.

$$f(x) = x^3 \qquad f(x) = |x| \qquad f(x) = \sqrt{x} \qquad f(x) = \sqrt[3]{x}$$

$$f(x) = x \qquad f(x) = x^2 \qquad f(x) = \frac{1}{x} \qquad f(x) = \frac{1}{x^2}$$

31.

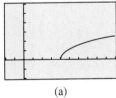

(a)

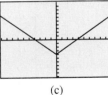

(c)

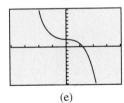

(e)

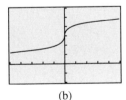

(b)

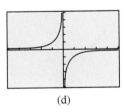

(d)

(f)

FIGURE 5.45

32.

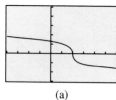

(a)

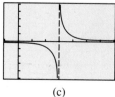

(c)

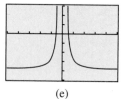

(e)

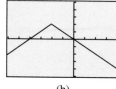

(b)

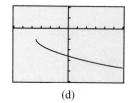

(d)

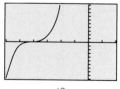

(f)

FIGURE 5.46

In Problems 33–38, use the graph to estimate the solution to the equation or inequality. Then check your answers algebraically.

33. Figure 5.47 shows a graph of $f(x) = \sqrt{x} - 2$, for $x > 0$. Solve:

a. $\sqrt{x} - 2 = 1.5$ **c.** $\sqrt{x} - 2 < 1$

b. $\sqrt{x} - 2 = 2.25$ **d.** $\sqrt{x} - 2 > -0.25$

34. Figure 5.48 shows a graph of $g(x) = \dfrac{4}{x + 2}$, for $x > -2$. Solve:

a. $\dfrac{4}{x + 2} = 4$ **b.** $\dfrac{4}{x + 2} = 0.8$

c. $\dfrac{4}{x + 2} > 1$ **d.** $\dfrac{4}{x + 2} < 3$

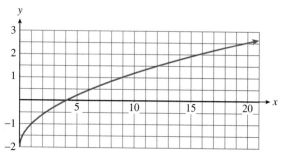

FIGURE 5.47

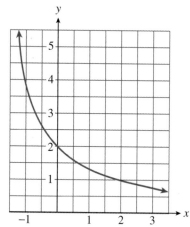

FIGURE 5.48

35. Figure 5.49 shows a graph of $w(t) = -10(t + 1)^3 + 10$. Solve:

a. $-10(t + 1)^3 + 10 = 100$

b. $-10(t + 1)^3 + 10 = -140$

c. $-10(t + 1)^3 + 10 > -50$

d. $-20 < -10(t + 1)^3 + 10 < 40$

36. Figure 5.50 shows a graph of $H = 4\sqrt[3]{z - 4} + 6$. Solve:

a. $4\sqrt[3]{z - 4} + 6 = 2$

b. $4\sqrt[3]{z - 4} + 6 = 12$

c. $4\sqrt[3]{z - 4} + 6 > 14$

d. $4\sqrt[3]{z - 4} + 6 < 6$

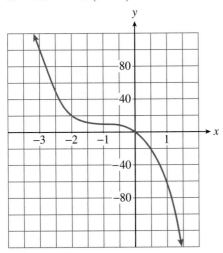

FIGURE 5.49

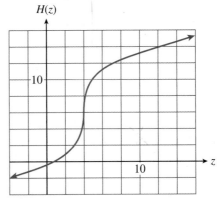

FIGURE 5.50

Match each graph in Figures 5.51 and 5.52 with its equation.

37.

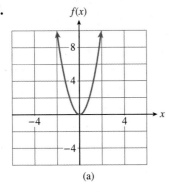

(a)

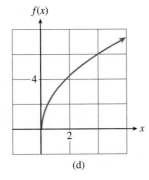

(c)

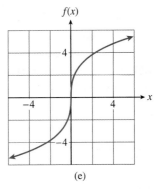

(e)

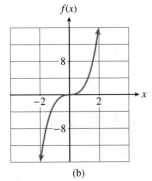

(b)

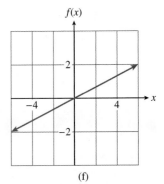

(d)

(f)

FIGURE 5.51

i. $f(x) = 3\sqrt{x}$ ii. $f(x) = 2x^3$ iii. $f(x) = \dfrac{x}{3}$ iv. $f(x) = \dfrac{3}{x}$ v. $f(x) = 3\sqrt[3]{x}$ vi. $f(x) = 3x^2$

38.

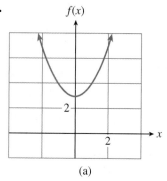

(a)

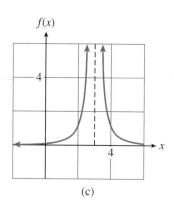

(c)

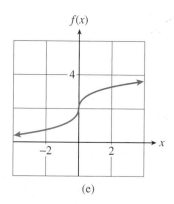

(e)

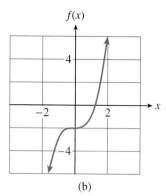

(b)

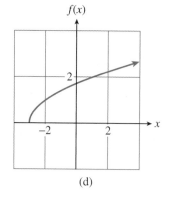

(d)

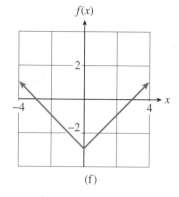

(f)

FIGURE 5.52

i. $f(x) = x^3 - 2$

ii. $f(x) = \sqrt[3]{x} + 2$

iii. $f(x) = \dfrac{1}{(x-3)^2}$

iv. $f(x) = |x| - 3$

v. $f(x) = x^2 + 3$

vi. $f(x) = \sqrt{x+3}$

In Problems 39–42, graph each function with the *ZInteger* setting. Use the graph to solve each equation or inequality. Check your solutions algebraically.

39. Graph $F(x) = 4\sqrt{x-25}$.
  a. Solve $4\sqrt{x-25} = 16$.
  b. Solve $8 < 4\sqrt{x-25} \le 24$.

41. Graph $H(x) = 24 - 0.25(x-6)^2$.
  a. Solve $25 - 0.25(x-6)^2 = -6.25$.
  b. Solve $24 - 0.25(x-6)^2 > 11.75$.

40. Graph $G(x) = 15 - 0.01(x-2)^3$.
  a. Solve $15 - 0.01(x-2)^3 = -18.75$.
  b. Solve $15 - 0.01(x-2)^3 \le 25$.

42. Graph $R(x) = 0.1(x+12)^2 - 18$.
  a. Solve $0.1(x+12)^2 - 18 = 14.4$.
  b. Solve $0.1(x+12)^2 - 18 < 4.5$.

MIDCHAPTER REVIEW

1. Do you believe that a person's age is a function of his or her social security number? Explain why or why not.

2. Professional baseball players often demand a salary increase after a successful season. Do you believe that a professional baseball player's salary is a function of the number of home runs hit the previous season? Explain why or why not.

3. Decide whether the table describes y as a function of x. Explain why or why not.

x	1	2	3	1	2
y	9	8	7	6	5

4. Decide whether the table describes y as a function of x. Explain why or why not.

x	-1	0	1	1	3
y	-5	-5	-5	-5	-5

5. The graph in Figure 5.53 defines a function, h, which shows the height, s, in meters of a duck t seconds after it was flushed out of the bushes.

   a. Use function notation to state that s is a function of t.

   b. What does the statement $h(3) = 7$ mean in this context?

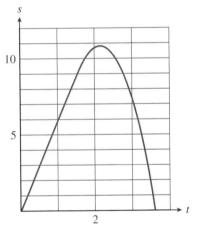

FIGURE 5.53

6. The graph in Figure 5.54 defines a function, f, which shows the atmospheric pressure, P, in inches of mercury at an altitude of a feet.

   a. Use function notation to state that s is a function of t.

   b. What does the statement

$$f(1500) = 28.3$$

mean in this context?

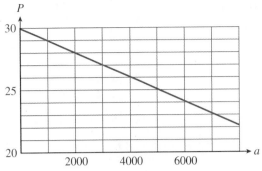

FIGURE 5.54

7. The table shows the revenue, R, from sales of the Miracle Mop as a function of the number of dollars spent on advertising, A. Let f be the name of the function defined by the table, so $R = f(A)$.

A (thousands of dollars)	100	150	200	250	300
R (thousands of dollars)	250	280	300	310	315

   a. Evaluate $f(300)$.

   b. What does the equation $f(0) = 20$ tell you in this context?

8. The table shows the systolic blood pressure, S, of a patient as a function of the dosage, d, of medication he receives. Let g be the name of the function defined by the table, so $S = g(d)$.

d (mg)	10	20	30	40	50
S (mm Hg)	220	200	190	185	183

   a. Evaluate $g(30)$.

   b. What does the equation $g(0) = 230$ tell you in this context?

9. $f(x) = 2 - \sqrt{x}$

   a. Complete the table of values and sketch a graph of the function.

x	0	1	2	3	4	5	6	7	8	9
$f(x)$										

   b. Use your calculator to make a table of values and graph the function.

10. $f(x) = 3 - |x|$

   a. Complete the table of values and sketch a graph of the function.

x	-4	-3	-2	-1	0	1	2	3	4
$f(x)$									

   b. Use your calculator to make a table of values and graph the function.

11. Sketch a graph of the function $p(x) = 3 + 2x$.

12. Sketch a graph of the function $q(x) = 1 - 2x - x^2$.

Simplify.

13. $-8 + 5\sqrt[3]{27}$

14. $\dfrac{4 - 7\sqrt[3]{-8}}{3}$

15. $10 - 5|5 - 10|$

16. $-1 + 2|9 - 6|$

5.4 Domain and Range

In Example 3 of Section 5.2 we graphed the function

$$f(x) = \sqrt{x + 4}$$

and observed that $f(x)$ is undefined for x-values less than -4. For this function we must choose x-values in the interval $[-4, \infty)$, and all the points on the graph have x-coordinates greater than or equal to -4, as shown in Figure 5.55. The set of all permissible values of the independent variable is called the **domain** of the function f.

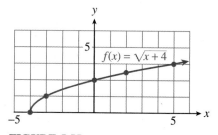

FIGURE 5.55

We also see that there are no points with negative $f(x)$-values on the graph of f: All the points have $f(x)$-values greater than or equal to zero. The set of all function values corresponding to the domain is called the **range** of the function. Thus, the domain of the function $f(x) = \sqrt{x + 4}$ is the set $[-4, \infty)$, and its range is the set $[0, \infty)$. In general we make the following definitions.

> The **domain** of a function is the set of permissible values for the independent variable. The **range** is the set of function values (that is, values of the dependent variable) that correspond to the domain values.

Using the notions of domain and range, we restate the definition of a function as follows.

> A relationship between two variables is a **function** if each element of the domain is paired with only one element of the range.

Finding Domain and Range from a Graph

We can identify the domain and range of a function from its graph. The domain is the set of *x*-values of all points on the graph, and the range is the set of *y*-values.

EXAMPLE 1

a. Determine the domain and range of the function *h* graphed in Figure 5.56.

b. For the indicated points, show the domain values and their corresponding range values in the form of ordered pairs.

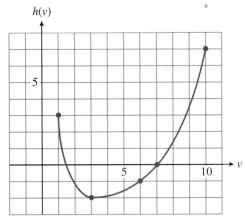

FIGURE 5.56

Solutions

a. All the points on the graph have *v*-coordinates between 1 and 10, inclusive, so the domain of the function *h* is the interval [1, 10]. The *h*(*v*)-coordinates have values between −2 and 7, inclusive, so the range of the function is the interval [−2, 7].

b. Recall that the points on the graph of a function have coordinates (*v*, *h*(*v*)). In other words, the coordinates of each point are made up of a domain value and its corresponding range value. Read the coordinates of the indicated points to obtain the ordered pairs (1, 3), (3, −2), (6, −1), (7, 0) and (10, 7). ■

Figure 5.57 shows the graph of the function *h* in Example 1 with the domain values marked on the horizontal axis and the range values marked on the vertical axis. Imagine a rectangle whose length and width are determined by those segments as shown in Figure 5.57. All the points (*v*, *h*(*v*)) on the graph of the function lie within this rectangle. This rectangle is a convenient "window" in the

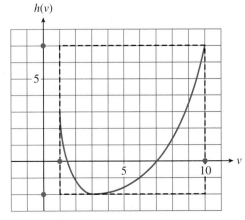

FIGURE 5.57

plane for viewing the function. Of course, if the domain or range of the function is an infinite interval, we can never include the whole graph within a viewing rectangle, and must be satisfied with studying only the "important" parts of the graph.

EXERCISE 1

a. Draw the smallest possible viewing window around the graph shown.

b. Find the domain and range of the function.

Domain:

Range:

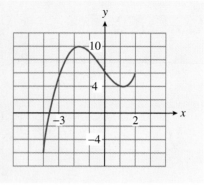

Sometimes the domain is given as part of the definition of a function.

EXAMPLE 2

Graph the function $f(x) = x^2 - 6$ on the domain $0 \leq x \leq 4$ and give its range.

Solution

The graph is part of a parabola that opens upward. Obtain several points on the graph by evaluating the function at convenient x-values in the domain.

$$f(0) = 0^2 - 6 = -6$$
$$f(1) = 1^2 - 6 = -5$$
$$f(2) = 2^2 - 6 = -2$$
$$f(3) = 3^2 - 6 = 3$$
$$f(4) = 4^2 - 6 = 10$$

TABLE 5.13

x	$f(x)$
0	-6
1	-5
2	-2
3	3
4	10

The range of the function is the set of all $f(x)$-values that appear on the graph. We can see in Figure 5.58 that the lowest point on the graph is $(0, -6)$, so the smallest $f(x)$-value is -6. The highest point on the graph is $(4, 10)$, so the largest $f(x)$-value is 10. Thus, the range of the function f is the interval $[-6, 10]$.

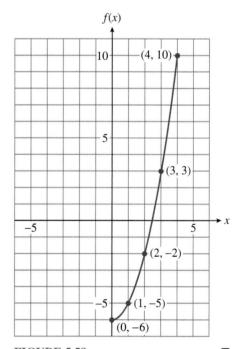

FIGURE 5.58 ∎

EXERCISE 2 Graph the function $g(x) = x^3 - 4$ on the domain $[-2, 3]$ and give its range.

Not all functions have domains and ranges that are intervals.

EXAMPLE 3

a. Graph the postage function given in Example 3 of Section 5.1.
b. Determine the domain and range of the function.

Solutions

a. From Table 5.4 (see page 253), we see that articles of any weight up to 1 ounce require \$0.33 postage. This means that for all w-values greater than 0 but less than or equal to 1, the p-value is 0.33. Thus, the graph of $p = g(w)$ between $w = 0$ and $w = 1$ looks like a small piece of the horizontal line $p = 0.33$. Similarly, for all w-values greater than 1 but less than or equal to 2 the p-value is 0.55, so the graph on this interval looks like a small piece of the line $p = 0.55$. Continue in this way to obtain the graph shown in Figure 5.59.

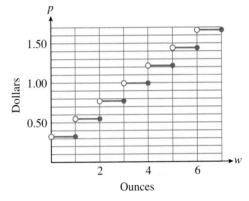

FIGURE 5.59

The open circles at the left endpoint of each horizontal segment indicate that that point is not included in the graph; the closed circles are points on the graph. For instance, if $w = 3$, the postage p is \$0.77, not \$0.99. Consequently, the point $(3, 0.77)$ is part of the graph of g but the point $(3, 0.99)$ is not.

b. Postage rates are given for all weights greater than 0 ounces up to and including 7 ounces, so the domain of the function is the half-open interval $(0, 7]$. (The domain is an interval because there is a point on the graph for *every* w-value from 0 to 7.) The range of the function is *not* an interval, however, because the possible values for p do not include *all* the real numbers between 0.33 and 1.65. The range is the set of discrete values 0.33, 0.55, 0.77, 0.99, 1.21, 1.43, and 1.65. ∎

Finding the Domain from a Formula

If the domain of a function is not given as part of its definition, we assume that the domain is as large as possible. We include in the domain all x-values that "make sense" when substituted into the function's formula. For example, the domain of $f(x) = \sqrt{x - 4}$ is the interval $[4, \infty)$, because x-values less than 4 result in square roots of negative numbers.

EXAMPLE 4

Find the domain of the function $g(x) = \dfrac{1}{x - 3}$.

Solution

We must omit any x-values that do not make sense in the function's formula. Because division by zero is undefined, we cannot allow the denominator of $\dfrac{1}{x - 3}$ to be zero. Now, $x - 3 = 0$ when $x = 3$, so we exclude $x = 3$ from the domain of g. Thus, the domain of g is the set of all real numbers except 3. ■

EXERCISE 3

a. Find the domain of the function $h(x) = \dfrac{1}{(x - 4)^2}$

b. Graph the function in the window

$$\text{Xmin} = -2 \qquad \text{Xmax} = 8$$
$$\text{Ymin} = -2 \qquad \text{Ymax} = 8$$

Use your graph and the function's formula to find its range.

For the functions we have studied so far, we need only avoid division by zero and square roots of negative numbers when finding the domain. Many common functions have as their domain the entire set of real numbers. For example, a linear function $f(x) = mx + b$ can be evaluated at any real number value of x, so its domain is the set of all real numbers. This set is represented in interval notation as $(-\infty, \infty)$.

The range of the linear function $f(x) = mx + b$ (if $m \neq 0$) is also the set of all real numbers, because the graph continues infinitely at both ends. (See Figure 5.60a.) If $m = 0$, then $f(x) = b$, and the graph of f is a horizontal line. In this case the range consists of a single number, b. (See Figure 5.60b.)

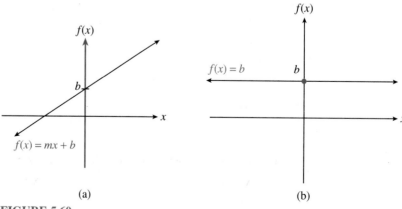

(a)   (b)

FIGURE 5.60

In many applications, we may restrict the domain and range of a function to suit the situation at hand. Consider the quadratic function

$$h = f(t) = 1454 - 16t^2$$

This function appeared in Example 6 of Section 5.1. It gives the height of an algebra book dropped from the top of the Sears Tower as a function of time. The graph of f is a parabola that opens downward. Its vertex occurs at the point (0, 1454). You can use the window

Xmin = −10   Xmax = 10
Ymin = −100   Ymax = 1500

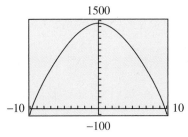

FIGURE 5.61

to obtain the graph shown in Figure 5.61.

Because t represents the time in seconds after the book was dropped, only positive t-values make sense for the problem. The book stops falling when it hits the ground, at $h = 0$. You can verify that this happens at approximately $t = 9.5$ seconds. Thus, only t-values between 0 and 9.5 are realistic for this application. During that time period, the height h of the book decreases from 1454 feet to 0 feet. The conditions of the problem prompt us to restrict the domain of the function f to the interval $[0, 9.5]$ and its range to $[0, 1454]$. The graph of this function is shown in Figure 5.62.

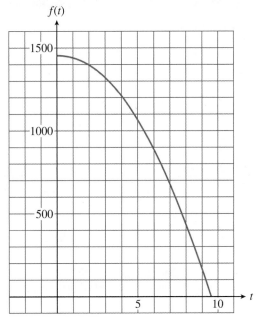

FIGURE 5.62

1. What is the domain of a function? What is its range?
2. Explain how to find the domain and range of a function from its graph.

3. What is the domain of the function $f(x) = 4$? What is its range?

4. Explain how to find the range of a quadratic function.

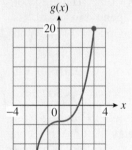
HOMEWORK 5.4

Find the domain and range of each function from its graph.

1.

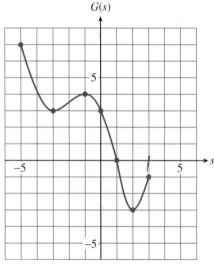

FIGURE 5.63

$[-5, 3) \cup$

$[-3, 7]$

2.

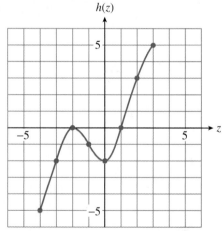

FIGURE 5.64

Domain is every x value

range is every y value

3.

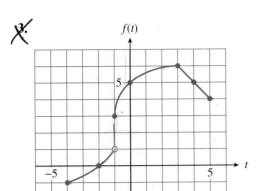

FIGURE 5.65

$[-4, 5)$ $[-1, 6)$

4.

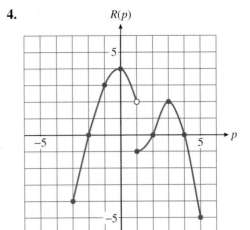

FIGURE 5.66

5.

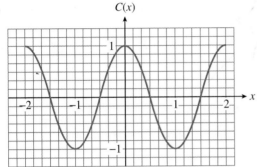

FIGURE 5.67

6.

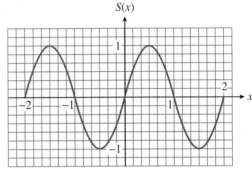

FIGURE 5.68

7.

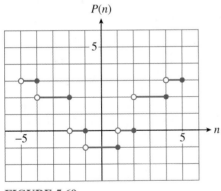

FIGURE 5.69

8.

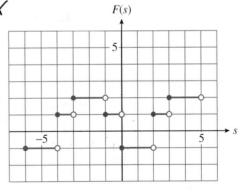

FIGURE 5.70

$[-6, 5)$

Use a graphing calculator to graph each function on the given domain. Adjust Ymin and Ymax until you can estimate the range of the function, using the [TRACE] key. Verify your answer algebraically by evaluating the function. State the domain and range in interval notation.

9. $f(x) = x^2 - 4x$ $-2 \le x \le 5$

10. $g(x) = 6x - x^2$ $-1 \le x \le 5$

11. $g(t) = -t^2 - 2t$ $-5 \le t \le 3$

12. $f(t) = -t^2 - 4t$ $-6 \le t \le 2$

13. $h(x) = x^3 - 1$ $-2 \le t \le 2$

14. $q(x) = x^3 + 4$ $-3 \le x \le 2$

15. $F(t) = \sqrt{8 - t}$ $-1 \le t \le 8$

16. $G(t) = \sqrt{t + 6}$ $-6 \le t \le 3$

17. $G(x) = \dfrac{1}{3 - x}$ $-1.25 \le x \le 2.75$

18. $H(x) = \dfrac{1}{x - 1}$ $3.25 \le x \le -1.25$

19. $G(x) = \dfrac{1}{3 - x}$ $3 < x \le 6$

20. $H(x) = \dfrac{1}{x - 1}$ $1 < x \le 4$

State the domain and range of each function.

21. **a.** $f(x) = x^3$ **b.** $g(x) = x^2$

22. **a.** $F(x) = |x|$ **b.** $G(x) = x$

23. **a.** $H(x) = \dfrac{1}{x^2}$ **b.** $M(x) = \dfrac{1}{x}$

24. **a.** $p(x) = \sqrt[3]{x}$ **b.** $q(x) = \sqrt{x}$

In Problems 25–30, (a) find the domain of each function algebraically, (b) graph the function, and use the graph to find the range.

25. **a.** $f(x) = \dfrac{1}{(x - 4)^2}$ **b.** $h(x) = \dfrac{1}{x^2} - 4$

26. **a.** $g(t) = \dfrac{1}{t} + 2$ **b.** $F(t) = \dfrac{1}{t + 2}$

27. **a.** $G(v) = v^3 + 5$ **b.** $H(v) = (v + 5)^3$

28. **a.** $h(n) = 3 + (n - 1)^2$

    b. $g(n) = 3 - (n + 1)^2$

29. **a.** $T(z) = \sqrt{z - 2}$ **b.** $S(z) = \sqrt{z} - 2$

30. **a.** $Q(x) = 4 - |x|$ **b.** $P(x) = |4 - x|$

Use a graphing calculator to explore some properties of the basic functions.

31. **a.** Graph $f(x) = x^2$ and $g(x) = x^3$ on the domain $[0, 1]$ and the range $[0, 1]$. On the interval $(0, 1)$, which is greater, $f(x)$ or $g(x)$?

    b. Graph $f(x) = x^2$ and $g(x) = x^3$ on the domain $[1, 10]$ and the range $[1, 100]$. On the interval $(1, 10)$, which is greater, $f(x)$ or $g(x)$?

32. **a.** Graph $f(x) = \sqrt{x}$ and $g(x) = \sqrt[3]{x}$ on the domain $[0, 1]$ and the range $[0, 1]$. On the interval $(0, 1)$, which is greater, $f(x)$ or $g(x)$?

    b. Graph $f(x) = \sqrt{x}$ and $g(x) = \sqrt[3]{x}$ on the domain $[1, 100]$ and the range $[1, 10]$. On the interval $(1, 100)$, which is greater, $f(x)$ or $g(x)$?

33. **a.** Graph $f(x) = \dfrac{1}{x}$ and $g(x) = \dfrac{1}{x^2}$ on the domain $[0.01, 1]$ and the range $(0, 10]$. On the interval $(0, 1)$, which is greater, $f(x)$ or $g(x)$?

    b. Graph $f(x) = \dfrac{1}{x}$ and $g(x) = \dfrac{1}{x^2}$ on the domain $[1, 10]$ and the range $(0, 1]$. On the interval $(1, \infty)$, which is greater, $f(x)$ or $g(x)$?

34. **a.** Graph $F(x) = |x^3|$ in the **ZDecimal** window. How does the graph compare to the graph of $y = x^3$?

    b. Graph $G(x) = \left| \dfrac{1}{x} \right|$ in the **ZDecimal** window. How does the graph compare to the graph of $y = \dfrac{1}{x}$?

For Problems 35 through 38, refer to the graphs of eight basic functions in Section 5.3.

35. Which of the eight basic functions are increasing on their entire domain? Which are decreasing on their entire domain?

36. Which of the eight basic functions can be evaluated at any real number? Which can take on any real number as a function value?

37. Which of the eight basic functions can be graphed in one piece, without lifting the pencil from the paper?

38. Which of the eight basic functions have no negative numbers in their range?

5.5 Variation

Two types of functions are widely used in the sciences and are known by special names: **direct variation** and **inverse variation.**

Direct Variation

Two variables are **directly proportional** (or just **proportional**) if the ratios of their corresponding values are always equal. Consider the functions described in Tables 5.14 and 5.15.

The first table shows the price of gasoline as a function of the number of gallons purchased. The ratio $\dfrac{\text{total price}}{\text{number of gallons}}$, or price per gallon, is the same for each pair of values in Table 5.14. This agrees with everyday experience: The price per gallon of gasoline is the same no matter how many gallons you buy. Thus, the total cost is directly proportional to the number of gallons purchased.

TABLE 5.14

Gallons of gasoline	Total price	$\dfrac{\text{Price}}{\text{gallons}}$
4	$4.60	$\dfrac{4.60}{4} = 1.15$
6	$6.90	$\dfrac{6.90}{6} = 1.15$
8	$9.20	$\dfrac{9.20}{8} = 1.15$
12	$13.80	$\dfrac{13.80}{12} = 1.15$
15	$17.25	$\dfrac{17.25}{15} = 1.15$

TABLE 5.15

Years	Population	$\dfrac{\text{People}}{\text{years}}$
10	432	$\dfrac{432}{10} \approx 43$
20	932	$\dfrac{932}{20} \approx 47$
30	2013	$\dfrac{2013}{30} \approx 67$
40	4345	$\dfrac{4345}{40} \approx 109$
50	9380	$\dfrac{9380}{50} \approx 188$
60	20,251	$\dfrac{20,251}{60} \approx 338$

Table 5.15 shows the population of a small town as a function of the town's age. The ratios $\dfrac{\text{number of people}}{\text{number of years}}$ give the average rate of growth of the town's population in people per year. You can see that this ratio is *not* constant; it increases as time goes on. The population of the town is *not* proportional to its age. The graphs of these two functions are shown in Figure 5.71.

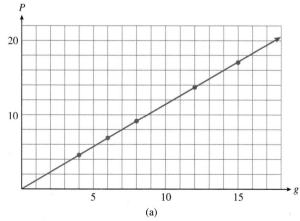

(a)

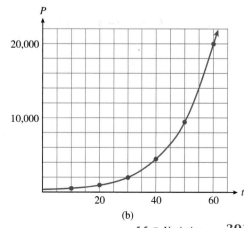

(b)

FIGURE 5.71

The price, P, of a fill-up is a linear function of the number of gallons, g, purchased. This should not be surprising if we consider an equation relating the variables g and P. Because the ratio of their values is a constant, we can write

$$\frac{P}{g} = k$$

where k is a constant. In this example the constant k is 1.15, the price of gasoline per gallon. Solving for P in terms of g we have,

$$P = kg = 1.15g$$

which we recognize as the equation of a line through the origin.
In general, we make the following definition.

> y **varies directly** with x if
>
> $$y = kx$$
>
> where k is a positive constant called the **constant of variation.**

From the discussion above, we see that "vary directly" means exactly the same thing as "are directly proportional." The two phrases are interchangeable.

EXAMPLE 1

a. The circumference C of a circle varies directly with its radius r, because

$$C = 2\pi r$$

The constant of variation is 2π, or about 6.28.

b. The amount of interest I earned in one year on an account paying 7% simple interest varies directly with the principal P invested, because

$$I = 0.07\,P$$ ■

Direct variation defines a linear function of the form

$$y = f(x) = kx$$

The fact that the constant term is zero is significant: If we double the value of x, then the value of y will double also. In fact, increasing x by any factor will result in y increasing by the same factor. For example, in Table 5.14 doubling the number of gallons of gas purchased, say, from 4 gallons to 8 gallons or from 6 gallons to 12 gallons, causes the total price to double also. As another example, consider investing $800 for one year at 7% simple interest, as in Example 1b. The interest earned is

$$I = 0.07(800) = \$56$$

If we increase the investment by a factor of 1.6 to 1.6(800) or $1280, the interest will be

$$I = 0.07(1280) = \$89.60$$

You can check that multiplying the original interest of $56 by a factor of 1.6 does give the same figure for the new interest, $89.60.

EXERCISE 1 Which of the following graphs could represent direct variation? Explain why.

a.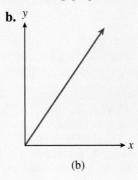

(a)

b.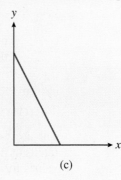

(b)

c.

(c)

EXAMPLE 2

a. Tuition at Woodrow University is $400 plus $30 per unit. Is the tuition proportional to the number of units you take?

b. Imogen makes a 15% commission on her sales of environmentally friendly products marketed by her co-op. Do her earnings vary directly with her sales?

Solutions

a. Let u represent the number of units you take, and let $T(u)$ represent your tuition. Then

$$T(u) = 400 + 30u$$

Thus $T(u)$ is a linear function of u, but the T-intercept is 400, not zero. Your tuition is *not* proportional to the number of units you take, so this is not an example of direct variation. You can check that doubling the number of units does not double the tuition. For example,

$$T(6) = 400 + 30(6) = 580$$

and

$$T(12) = 400 + 30(12) = 760$$

Tuition for 12 units is not double the tuition for 6 units. The graph of $T(u)$ in Figure 5.72a does not pass through the origin.

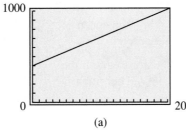

(a)

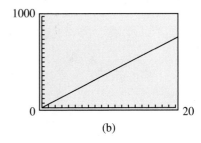

(b)

FIGURE 5.72

b. Let S represent Imogen's sales, and let $C(S)$ represent her commission. Then

$$C(S) = 0.15S$$

Thus, $C(S)$ is a linear function of S with a C-intercept of zero, so Imogen's earnings do vary directly with her sales. This is an example of direct variation. (See Figure 5.72b.) ∎

Finding a Variation Equation

The positive constant k in the equation $y = kx$ is just the slope of the graph, so it tells us how rapidly the graph increases. If we know any one pair of associated values for the variables, we can find the constant of variation. We can then use the constant to express one of the variables as a function of the other.

EXAMPLE 3

If an object is dropped from a great height, say, off the rim of the Grand Canyon, its speed, v, varies directly with the time, t, the object has been falling. A rock kicked off the edge of the Canyon is falling at a speed of 39.2 meters per second when it passes a lizard on a ledge 4 seconds later.

a. Express v as a function of t.

b. What is the speed of the rock after it has fallen for 6 seconds?

c. Sketch a graph of $v(t)$.

Solutions

a. Because v varies directly with t, there is a positive constant k for which

$$v = kt$$

Substitute $v = 39.2$ when $t = 4$ and solve for k to find

$$39.2 = k(4)$$

or

$$k = 9.8$$

Thus, $v(t) = 9.8t$.

b. Substitute $t = 6$ into the equation you found in part (a).

$$v = 9.8(6) = 58.8$$

The rock is falling at a speed of 58.8 meters per second.

c. Use your calculator to graph the function $v = 9.8t$. The graph is shown in Figure 5.73.

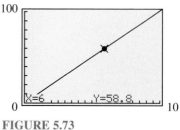

FIGURE 5.73 ∎

Other Types of Direct Variation

We can generalize the notion of direct variation to include situations in which y is proportional to a power of x, instead of x itself. This type of variable relation is modeled by a function of the form

$$y = f(x) = kx^n$$

where k and n are positive numbers.

EXAMPLE 4

a. The area of a circle varies directly with the *square* of the radius, according to the formula

$$A = \pi r^2$$

b. The volume of a sphere varies directly with the *cube* of the radius, according to the formula

$$V = \frac{4}{3}\pi r^3$$ ■

In any example of direct variation, as the independent variable increases through positive values, the dependent variable increases also. Thus, a direct variation is an example of an increasing function, as we can see when we consider the graphs of some typical direct variations in Figure 5.74.

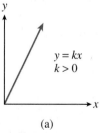
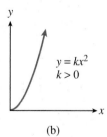
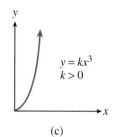

(a)   (b)   (c)

FIGURE 5.74

The graph of a direct variation always passes through the origin, so when the independent variable is zero, the dependent variable is zero also. Thus the functions $y = 3x + 2$ and $y = 0.4x^2 - 2.3$, for example, are *not* direct variations, even though they are increasing functions for positive x.

Inverse Variation

How long does it take to travel a distance of 600 miles? The answer depends on the average speed at which you travel. If you are on a bicycle trip, your average speed might be 15 miles per hour. In that case, your traveling time will be

$$T = \frac{D}{R} = \frac{600}{15} = 40 \text{ hours}$$

(Of course, you will have to add time for rest stops; the 40 hours are *just* your travel time.) If you are driving your car, you might average 50 miles per hour. Your travel time is then

$$T = \frac{D}{R} = \frac{600}{50} = 12 \text{ hours}$$

If you take a commercial air flight, the plane's speed might be 400 miles per hour, and the flight time would be

$$T = \frac{D}{R} = \frac{600}{400} = 1.5 \text{ hours}$$

TABLE 5.16

R	T
10	60
15	40
20	30
50	12
200	3
400	1.5

You can see that for higher average speeds, the travel time is shorter. In other words, the time needed for a 600-mile journey is a decreasing function of average speed. In fact, a formula for the function is

$$T(R) = \frac{600}{R}$$

This function is an example of **inverse variation**. A table of values and a graph of the function are shown in Table 5.16 and Figure 5.75.

In general, we say that

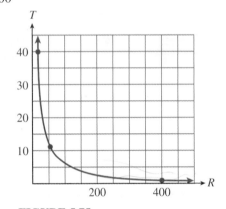

FIGURE 5.75

---

y **varies inversely** with x^n if

$$y = \frac{k}{x^n}, \qquad x \neq 0$$

where *k* is a positive constant and $n > 0$.

We may also say that *y* is **inversely proportional** to x^n.

EXAMPLE 5

a. The weight w of an object varies inversely with the square of its distance d from the center of the earth. Thus,

$$w = \frac{k}{d^2}$$

b. The amount of force F (in pounds) needed to lift a heavy object with the help of a lever is inversely proportional to the length l of the lever. Thus,

$$F = \frac{k}{l}$$

∎

In each case in Example 5, as the independent variable increases through positive values, the dependent variable decreases. An inverse variation is an example of a decreasing function. The graphs of some typical inverse variations are shown in Figure 5.76.

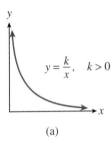
(a)

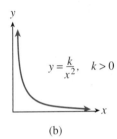
(b)

FIGURE 5.76

If we know that two variables vary inversely and we can find one pair of corresponding values for the variables, we can determine k, the constant of variation.

EXAMPLE 6

The intensity of electromagnetic radiation, such as light or radio waves, varies inversely with the square of the distance from its source. Radio station KPCC broadcasts a signal that is measured at 0.016 watt per square meter by a receiver one kilometer away.

a. Write a formula that gives signal strength as a function of distance.

b. If you live five kilometers from the station, what is the strength of the signal you will receive?

Solutions

a. Let I stand for the intensity of the signal in watts per square meter, and d for the distance from the station in kilometers. Then

$$I = \frac{k}{d^2}$$

To find the constant k, we substitute 0.016 for I and 1 for d. Solving for k gives us

$$0.016 = \frac{k}{1^2}$$

$$k = 0.016(1^2) = 0.016$$

Thus, $I = \dfrac{0.016}{d^2}$.

b. Now we can substitute 5 for d and solve for I.

$$I = \frac{0.016}{5^2} = 0.00064$$

At a distance of five kilometers from the station, the signal strength is 0.00064 watts per square meter. ∎

EXERCISE 3 Delbert's officemates want to buy a $120 gold watch for a colleague who is retiring. The cost per person is inversely proportional to the number of people who contribute.

a. Express the cost per person, C, as a function of the number of people who contribute, p.

b. Sketch the function on the domain $0 \leq p \leq 20$.

1. Describe the graph of $y = f(x)$ if y varies directly with x.
2. What is true about the ratio of two variables if they are directly proportional?
3. How can you tell from a table of values if two variables vary inversely?
4. If y is inversely proportional to x, then the graph of y versus x is a variation of which basic graph?
5. If y varies directly with a power of x, write a formula for y as a function of x.
6. If y varies inversely with a power of x, write a formula for y as a function of x.

ANSWERS TO 5.5 EXERCISES

1. (b): The graph is a straight line through the origin.

2a. $V = 1.75w$ **b.** 26.25

3a. $C = \dfrac{120}{p}$ **b.**

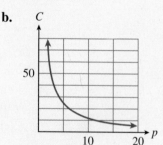

1. Delbert's credit card statement lists three purchases he made while on a business trip in the Midwest. His company's accountant would like to know the sales tax rate on the purchases.

Price of item	18	28	12
Tax	1.17	1.82	0.78

   a. Compute the ratio of the tax to the price of each item. Is the tax proportional to the price? What is the tax rate?
   b. Express the tax, T, as a function of the price, p, of the item.
   c. Sketch a graph of the function by hand, and label the scales on the axes.

3. The marketing department for a paper company is testing wrapping paper rolls in various dimensions to see which shape consumers prefer. All the rolls contain the same amount of wrapping paper.

Width (feet)	2	2.5	3
Length (feet)	12	9.6	8

   a. Compute the product of the length and width for each roll of wrapping paper. What is the constant of inverse proportionality?
   b. Express the length, L, of the paper as a function of the width, w, of the roll.
   c. Sketch a graph of the function by hand, and label the scales on the axes.

2. At constant acceleration from rest, the distance traveled by a race car is proportional to the square of the time elapsed. The highest recorded road-tested acceleration is 0 to 60 miles per hour in 3.07 seconds, which produces the following data.

Time (seconds)	2	2.5	3
Distance (feet)	57.32	89.563	128.97

   a. Compute the ratios of the distance traveled to the square of the time elapsed. What was the constant of proportionality.
   b. Express the distance traveled, d, as a function of time in seconds, t.
   c. Sketch a graph of the function by hand, and label the scales on the axes.

4. The force of gravity on a 1-kilogram mass is inversely proportional to the square of the object's distance from the center of the earth. The table shows the force on the object, in newtons, at distances that are multiples of the earth's radius.

Distance (earth radii)	1	2	4
Force (newtons)	9.8	2.45	0.6125

   a. Compute the products of the force and the square of the distance. What is the constant of inverse proportionality?
   b. Express the gravitational force, F, on a 1-kilogram mass as a function of its distance, r, from the earth's center, measured in earth radii.
   c. Sketch a graph of the function by hand, and label the scales on the axes.

5. The length of a rectangle is 10 inches, and its width is 8 inches. Suppose we increase the length of the rectangle while holding the width constant.

a. Fill in the table.

Length	Width	Perimeter	Area
10	8		
12	8		
15	8		
20	8		

b. Does the perimeter vary directly with the length?

c. Write a formula for the perimeter of the rectangle in terms of its length.

d. Does the area vary directly with the length?

e. Write a formula for the area of the rectangle in terms of its length.

6. The base of an isosceles triangle is 12 centimeters, and the equal sides have length 15 centimeters. Suppose we increase the base of the triangle while holding the sides constant.

a. Fill in the table.

Base	Sides	Height	Perimeter	Area
12	15			
15	15			
18	15			
20	15			

b. Does the perimeter vary directly with the base?

c. Write a formula for the perimeter of the triangle in terms of its base.

d. Write a formula for the area of the triangle in terms of its base.

e. Does the area vary directly with the base?

7. Which of the graphs in Figure 5.77 could describe direct variation? Explain.

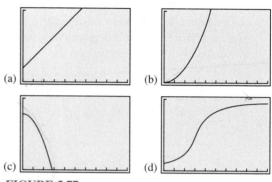

FIGURE 5.77

8. Which of the graphs in Figure 5.78 could describe direct variation? Explain.

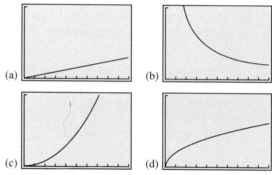

FIGURE 5.78

9. Which of the graphs in Figure 5.79 could describe inverse variation? Explain.

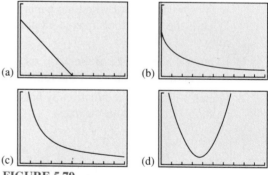

FIGURE 5.79

10. Which of the graphs in Figure 5.80 could describe inverse variation? Explain

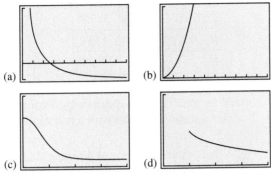

(a)   (b)
(c)   (d)

FIGURE 5.80

11. The weight of an object on the moon varies directly with its weight on earth. A person who weighs 150 pounds on earth would weigh only 24.75 pounds on the moon.

a. Find a function that gives the weight m of an object on the moon in terms of its weight w on earth. Complete the table and graph your function in a suitable window.

w	100	150	200	400
m				

b. How much would a person weigh on the moon if she weighs 120 pounds on earth?

c. A piece of rock weighs 50 pounds on the moon. How much will it weigh back on earth?

d. Locate the points on your graph that correspond to your answers for (b) and (c).

12. Hubble's law says that distant galaxies are receding from us at a rate that varies directly with their distance. (The speeds of the galaxies are measured using a phenomenon called redshifting.) A galaxy in the constellation Ursa Major is 980 million light-years away and is receding at a speed of 15,000 kilometers per second.

a. Find a function that gives the speed, v, of a galaxy in terms of its distance, d, from earth. Complete the table and graph your function in a suitable window. (Distances are given in millions of light-years.)

d	500	980	2500	5000
v				

b. How far away is a galaxy in the constellation Hydra that is receding at 61,000 kilometers per second?

c. A galaxy in Leo is 1240 million light-years away. How fast is it receding from us?

d. Locate the points on your graph that correspond to your answers for (b) and (c).

13. The length, L, of a pendulum varies directly with the square of its period, T, the time required for the pendulum to make one complete swing back and forth. The pendulum on a grandfather clock is 3.25 feet long and has a period of 2 seconds.

a. Express L as a function of T. Complete the table and graph your function in a suitable window.

T	1	5	10	20
L				

b. How long is the Foucault pendulum in the Pantheon in Paris, which has a period of 17 seconds?

c. A hypnotist uses a gold pendant as a pendulum to mesmerize his clients. If the chain on the pendant is 9 inches long, what is the period of its swing?

d. Locate the points on your graph that correspond to your answers for (b) and (c).

14. The load, L, that a beam can support varies directly with the square of its vertical thickness, h. A beam that is 4 inches thick can support a load of 2000 pounds.

   a. Express L as a function of h. Complete the table and graph your function in a suitable window.

h	2	4	5	8
L				

   b. What size load can be supported by a beam that is 6 inches thick?

   c. How thick a beam is needed to support a load of 100 pounds?

   d. Locate the points on your graph that correspond to your answers for (b) and (c).

15. Computer monitors produce a magnetic field. The effect of the field, B, on the user varies inversely with his or her distance, d, from the screen. The field from a certain 13-inch color monitor was measured at 22 milligauss 4 inches from the screen.

   a. Express the field strength as a function of distance from the screen. Complete the table and graph your function in a suitable window.

d	2	4	12	24
B				

   b. What is the field strength 10 inches from the screen?

   c. An elevated risk of cancer can result from exposure to field strengths of 4.3 milligauss. How far from the screen should the computer user sit to keep the field level below 4.3 milligauss?

   d. Locate the points on your graph that correspond to your answers for (b) and (c).

16. The amount of current, I, that flows through a circuit varies inversely with the resistance, R, on the circuit. An iron with a resistance of 12 ohms draws 10 amps of current.

   a. Express the current as a function of the resistance. Complete the table and graph your function in a suitable window.

R	1	5	10	20
I				

   b. How much current is drawn by a compact fluorescent light bulb with a resistance of 533.3 ohms?

   c. What is the resistance of a toaster that draws 12.5 amps of current?

   d. Locate the points on your graph that correspond to your answers for (b) and (c).

17. The amount of power, P, generated by a windmill varies directly with the cube of the wind speed, w. A windmill on Oahu, Hawaii, produces 7300 kilowatts of power when the wind speed is 32 miles per hour.

   a. Express the power as a function of wind speed. Complete the table and graph your function in a suitable window.

w	10	20	40	80
P				

   b. How much power would the windmill produce in a light breeze of 15 miles per hour?

   c. What wind speed is needed to produce 10,000 kilowatts of power?

   d. Locate the points on your graph that correspond to your answers for (b) and (c).

18. A crystal form of pyrite (a compound of iron and sulfur) has the shape of a regular solid with 12 faces. Each face is a regular pentagon. This compound is called pyritohedron, and its mass, M, varies directly with the cube of the length, L, of one edge. If each edge is 1.1 centimeters, then the mass is 51 grams.

   a. Express the mass of pyritohedron as a function of the length of one edge. Complete the table and graph your function in a suitable window.

L	0.5	1	2	4
M				

c. How long would each edge be for a 408-gram piece of pyritohedron?

d. Locate the points on your graph that correspond to your answers for (b) and (c).

b. What is the weight of a chunk of pyritohedron if each edge is 2.2 centimeters?

For each function described below, (a) use the values in the table to find the constant of variation, k, and write y as a function of x; (b) fill in the rest of the table with the correct values; (c) what happens to y when you double the value of x?

19. y varies directly with x

x	y
2	
5	1.5
	2.4
12	
	4.5

20. y varies directly with x.

x	y
0.8	
1.5	54
	108
	126
6	

21. y varies directly with the square of x.

x	y
3	
6	24
	54
12	
	150

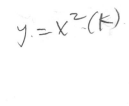

$y = x^2 (k)$

22. y varies directly with x.

x	y
2	120
3	
	1875
6	
	15,000

23. y varies inversely with the square of x.

x	y
4	
	15
20	6
30	
	3

24. y varies inversely with the cube of x.

x	y
0.2	
	80
2	
4	1.25
	0.8

$y = \dfrac{k}{x^3}$

For each table, decide whether

a. y varies directly with x,

b. y varies directly with x^2,

c. y does not vary directly with a power of x.

Explain why your choice is correct. If your choice is (a) or (b), find the constant of variation, k.

25.

x	y
2	2.0
3	4.5
5	12.5
8	32.0

26.

x	y
2	12
4	28
6	44
9	68

27.

x	y
1.5	3.0
2.4	7.2
5.5	33.0
8.2	73.8

28.

x	y
1.2	7.20
2.5	31.25
6.4	204.80
12.0	720.00

For each table, decide whether

a. *y* varies inversely with *x*,

b. *y* varies inversely with x^2,

c. *y* does not vary inversely with a power of *x*.

Explain why your choice is correct. If your choice is (a) or (b), find the constant of variation, *k*.

29.

x	*y*
0.5	288
2.0	18
3.0	8
6.0	2

30.

x	*y*
0.5	100.0
2.0	25.0
4.0	12.5
5.0	10.0

31.

x	*y*
1.0	4.0
1.3	3.7
3.0	2.0
4.0	1.0

32.

x	*y*
0.5	180.00
2.0	11.25
3.0	5.00
5.0	1.80

Each of the following functions described by a table of data or by a graph is an example of direct or inverse variation. For each problem, (a) find an algebraic formula for the function, including the constant of variation, *k*, (b) answer the question in the problem.

33. The faster a car moves, the more difficult it is to stop. The graph in Figure 5.81 shows the distance, *d*, required to stop a car as a function of its velocity, *v*, before the brakes were applied. What distance is needed to stop a car moving at 100 kilometers per hour?

d

Distance (meters)

20   40

v

Velocity (kilometers per hour)

FIGURE 5.81

34. A wide pipe can handle a greater water flow than a narrow pipe. The graph in Figure 5.82 shows the water flow through a pipe, *w*, as a function of its radius, *r*. How great is the water flow through a pipe of radius of 10 inches?

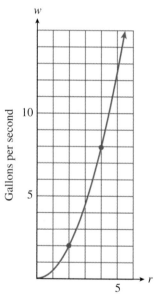

w

Gallons per second

10

5

5

r

Radius (inches)

FIGURE 5.82

35. If the price of mushrooms goes up, the amount consumers are willing to buy goes down. The graph in Figure 5.83 shows the number of tons of shiitake mushrooms, *m*, sold in California each week as a function of their price, *p*. If the price of shiitake mushrooms rises to $10 per pound, how many tons will be sold?

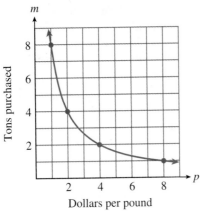

FIGURE 5.83

36. When an adult plays with a small child on a see-saw, the adult must sit closer to the pivot point to balance the see-saw. The graph in Figure 5.84 shows this distance, d, as a function of the adult's weight, w. How far from the pivot must Kareem sit if he weighs 280 pounds?

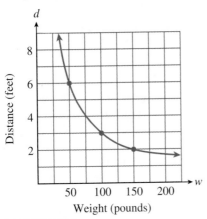

FIGURE 5.84

37. Ocean temperatures are generally colder at the greater depths. Table 5.17 shows the temperature of the water as a function of depth. What is the ocean temperature at a depth of 6 kilometers?

TABLE 5.17

Depth (km)	Temperature (°C)
0.5	12
1.0	6
2.0	3
3.0	2

38. The shorter the length of a vibrating guitar string, the higher the frequency of the vibra-

tions. Table 5.18 shows frequency as a function of effective length. The fifth string is 65 centimeters long and is tuned to A (with a frequency of 220 vibrations per second). The placement of the fret relative to the bridge changes the effective length of the guitar string. How far from the bridge should the fret be placed for the note C (256 vibrations per second)?

TABLE 5.18

Length (cm)	Frequency
55.0	260
57.2	250
65.0	220
71.5	200

39. The strength of a cylindrical rod depends on its diameter. The greater the diameter of the rod, the more weight it can support before collapsing. Table 5.19 shows the maximum weight supported by a rod as a function of its diameter. How much weight can a 1.2-centimeter rod support before collapsing?

TABLE 5.19

Diameter (cm)	Weight (newtons)
0.5	150
1.0	600
1.5	1350
2.0	2400

40. The maximum height attained by a cannonball depends on the speed at which it was shot. Table 5.20 shows maximum height as a function of initial speed. What height is attained by a cannonball whose initial upward speed was 100 feet per second?

TABLE 5.20

Speed (ft/sec)	Height (ft)
40	200.0
50	312.5
60	450.0
70	612.5

41. The intensity of illumination, I, from a lamp varies inversely with the square of your distance, d, from the lamp. If you double your distance from a reading lamp, what happens to the illumination?

42. The resistance, R, of an electrical wire varies inversely with the square of its diameter, d. If you replace an old wire with a new one whose diameter is two-thirds of the old one, what happens to the resistance?

43. The wind resistance, W, experienced by a vehicle on the freeway varies directly with the square of its speed, v. If you decrease your speed by 10%, what happens to the wind resistance?

44. The weight, w, of a bronze statue varies directly with the cube of its height, h. If you increase the height of a statue by 50%, what happens to its weight?

45. y varies directly with x. Show that if you multiply x by any constant c, then y will be multiplied by the same constant.

46. y varies inversely with x. Show that if you multiply x by any constant c, then y will be divided by the same constant.

47. Explain why the ratio $\dfrac{y}{x^2}$ is a constant when y varies directly with x^2.

48. Explain why the product yx^2 is a constant when y varies inversely with x^2.

49. If x varies directly with y and y varies directly with z, does x vary directly with z?

50. If x varies inversely with y and y varies inversely with z, does x vary inversely with z?

5.6 Functions as Mathematical Models

The Shape of the Graph

Creating a good model for a situation often begins with deciding what kind of function to use. An appropriate model can depend on very qualitative considerations, such as the general shape of the graph. What sort of function has the right shape to describe the process we want to model? Should it be increasing or decreasing, or some combination? Is the slope constant or is it changing? In Examples 1 and 2 we investigate how the shape of a graph illustrates the nature of the process it models.

EXAMPLE 1   Forrest leaves his house to go to school. Sketch a possible graph of Forrest's distance from the house versus the time since he left for each of the following situations.

    a. Forrest walks at a constant speed until he reaches the bus stop.

    b. Forrest walks at a constant speed until he reaches the bus stop, then waits there until the bus arrives.

    c. Forrest walks at a constant speed until he reaches the bus stop, waits there until the bus arrives, and then the bus drives him to school at a constant speed.

Solutions

    a. The graph could look like the straight line segment in Figure 5.85(a). It begins at the origin because at the instant that Forrest leaves the house, his distance from home is 0. (In other words, when $t = 0$, $y = 0$.) The graph is a straight line because Forrest has a constant speed. The slope of the line is equal to Forrest's walking speed.

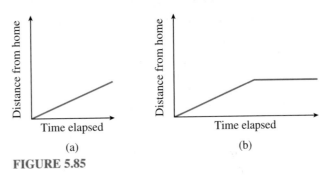

FIGURE 5.85

b. The graph begins exactly as the graph in Figure 5.85(a) does. But while Forrest waits for the bus, his distance from home remains constant, so the graph at that time must be a horizontal line as shown in Figure 5.85(b). The line has slope 0 because while Forrest is waiting for the bus, his speed is 0.

c. The graph begins just as the graph in Figure 5.85(b) does. The section of the graph that represents the bus ride should have a constant slope because the bus is moving at a constant speed. Because the bus (probably) moves much faster than Forrest walks, the slope of the line should be greater for the bus section of the graph than it was for the walking section. The graph is shown in Figure 5.85(c).

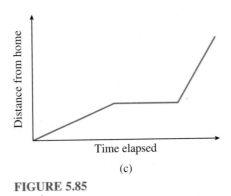

FIGURE 5.85

EXERCISE 1 Erin walks from her home to a convenience store, where she buys some cat food, and then walks back home. Sketch a possible graph of her distance from home as a function of time.

EXAMPLE 2 The two functions described here are both examples of increasing functions, but they increase in different ways. Match each function to its graph in Figure 5.86 and to the appropriate table of values.

a. The number of flu cases reported at an urban medical center during an epidemic is an increasing function of time, and it is growing at a faster and faster rate.

b. The temperature of a potato placed in a hot oven increases rapidly at first, then slows down as it approaches the temperature of the oven.

(1)

x	0	2	5	10	15
y	70	89	123	217	383

(2)

x	0	2	5	10	15
y	70	219	341	419	441

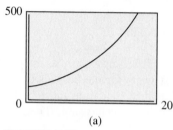

(a)

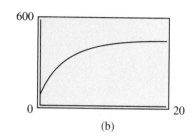

(b)

FIGURE 5.86

Solutions

a. The number of flu cases is described by graph (a) and table (1). The function values in table (1) increase at an increasing rate. We can see this by computing the rate of change over successive time intervals.

$$x = 0 \text{ to } x = 5: \qquad m = \frac{\Delta y}{\Delta x} = \frac{123 - 70}{5 - 0} = 10.6$$

$$x = 5 \text{ to } x = 10: \qquad m = \frac{\Delta y}{\Delta x} = \frac{217 - 123}{10 - 5} = 18.8$$

$$x = 10 \text{ to } x = 15: \qquad m = \frac{\Delta y}{\Delta x} = \frac{383 - 217}{15 - 10} = 33.2$$

The increasing rates are reflected in graph (a): The graph bends upward as the slopes increase.

b. The temperature of the potato is described by graph (b) and table (2). The function values in table (2) increase, but at a decreasing rate.

$$x = 0 \text{ to } x = 5: \qquad m = \frac{\Delta y}{\Delta x} = \frac{341 - 70}{5 - 0} = 54.2$$

$$x = 5 \text{ to } x = 10: \qquad m = \frac{\Delta y}{\Delta x} = \frac{419 - 341}{10 - 5} = 15.6$$

$$x = 10 \text{ to } x = 15: \qquad m = \frac{\Delta y}{\Delta x} = \frac{441 - 419}{15 - 10} = 4.4$$

The decreasing slopes are illustrated by graph (b): The graph is increasing but bends downward. ■

EXERCISE 2 Francine bought a cup of cocoa at the cafeteria. The cocoa cooled off rapidly at first, and then gradually approached room temperature. Which graph more accurately reflects the temperature of the cocoa as a function of time? Explain why.

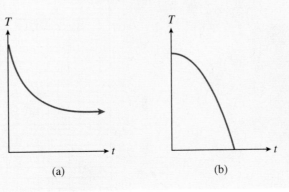

(a)   (b)

Using the Basic Functions as Models

In earlier chapters we considered situations that can be modeled by linear and quadratic functions. Direct and inverse variation, which we studied in Section 5.5, use functions of the form $y = kx^n$ and $y = \dfrac{k}{x^n}$. In this section we'll look at a few of the other basic functions. Example 3 illustrates an application of the function $f(x) = \sqrt{x}$.

EXAMPLE 3

The speed of sound is a function of the temperature of the air in degrees Kelvin. (The temperature, T, in degrees Kelvin is given by $T = C + 273$, where C is the temperature in degrees Celsius.) Table 5.21 shows the speed of sound, s, in meters per second, at various temperatures, T.

TABLE 5.21

T	0	20	50	100	200	400
s	0	89.7	141.8	200.6	283.7	401.2

    a. Plot the data to obtain a graph. Which of the basic functions does your graph most resemble?

    b. This function is an example of direct variation. Find a value of k that fits the data.

    c. On a summer night when the temperature is 20° Celsius, you see a flash of lightning, and 6 seconds later you hear the thunderclap. Use your function to estimate your distance from the thunderstorm.

Solutions

a. The graph of the data is shown in Figure 5.87. The shape of the graph reminds us of the square root function, $y = \sqrt{x}$.

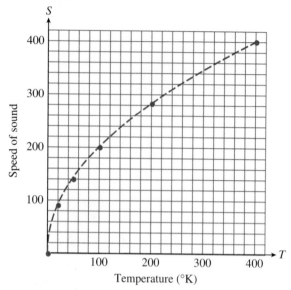

FIGURE 5.87

b. We are looking for a value of k so that the function $f(T) = k\sqrt{T}$ fits the data. We solved this type of problem in Section 5.4 on variation: We substitute one of the data points into the formula and solve for k. If we choose the point $(100, 200.6)$, we obtain

$$200.6 = k\sqrt{100}$$

and solving for k yields $k = 20.06$. We can check that the formula $s = 20.06\sqrt{T}$ is a good fit for the rest of the data points as well. Thus, we suggest the function

$$f(T) = 20.06\sqrt{T}$$

as a model for the speed of sound.

c. First, use the model to calculate the speed of sound at a temperature of $20°$ Celsius. The Kelvin temperature is

$$T = 20 + 273 = 293$$

so we evaluate $s = f(T)$ for $T = 293$.

$$f(293) = 20.06\sqrt{293} \approx 343.4$$

so s is approximately 343.4 meters per second.

Next we note that the speed of light is fast enough (approximately 30,000,000 meters per second) that for distances on earth we can consider the transmission of light to be instantaneous. We know that the lightning flash and the thunderclap actually occur simultaneously, so if the sound of the thunderclap takes 6 seconds to reach us after we see the lightning, we can use our calculated speed of sound to find our distance from the storm.

$$\text{distance} = \text{speed} \times \text{time}$$
$$= (343.4 \text{ m/sec})(6 \text{ sec}) = 2060.4 \text{ meters}$$

The thunderstorm is 2060 meters, or about 1.3 miles, away. ∎

Using Absolute Value to Model Distance

Next we'll consider the absolute value function. Recall that the absolute value of a number gives the distance from the origin to that number on the number line. More generally,

The **distance** between two points x and a is given by $|x - a|$.

For example, the equation $|x - 2| = 6$ means that "the distance between x and 2 is 6 units." The number x could be to the left or the right of 2 on the number line. Thus, the equation has two solutions, 8 and -4, as shown in Figure 5.88.

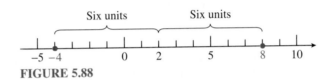

FIGURE 5.88

EXAMPLE 4

Write each statement using absolute-value notation.

    a. x is 3 units from the origin.

    b. p is 2 units from 5.

    c. a is within 4 units of -2.

Solutions

First, restate each sentence in terms of distance.

    a. The distance between x and the origin is 3 units, or $|x| = 3$.

    b. The distance between p and 5 is 2 units, or $|p - 5| = 2$.

    c. The distance between a and -2 is less than 4 units, or $|a - (-2)| < 4$, or $|a + 2| < 4$. ∎

EXERCISE 3 Write each statement using absolute value notation.

a. x is 5 units away from -3.

b. x is at least 6 units away from 4.

In Example 5 we use a graph to solve equations and inequalities involving absolute values.

EXAMPLE 5

Marlene is driving to a new outlet mall on Highway 17. There is a gas station at Marlene's on-ramp. She buys gas there and resets her odometer to 0 before getting on the highway. The mall is only 15 miles from Marlene's on-ramp, but she mistakenly drives past the mall and continues down the highway. Marlene's distance from the mall is a function of how far she has driven on Highway 17. (See Figure 5.89.)

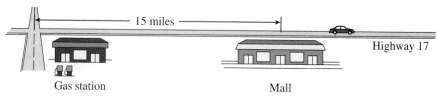

Gas station   Mall

FIGURE 5.89

   a. Make a table of values showing how far Marlene has driven on Highway 17 and how far she is from the mall.

   b. Make a graph of Marlene's distance from the mall versus the number of miles she has driven on the highway. Which of the basic graphs from Section 5.3 does your graph most resemble?

   c. Find a piecewise defined formula that describes Marlene's distance from the mall as a function of the distance she has driven on the highway.

   d. Use your graph to determine how far Marlene has driven when she is within 5 miles of the mall.

   e. Determine how far Marlene has driven when she is at least 10 miles from the mall.

Solutions

   a. Marlene gets closer to the mall for each mile that she has driven on the highway until she has driven 15 miles, and after that she gets farther from the mall.

   TABLE 5.22

Miles on highway	0	5	10	15	20	25	30
Miles from mall	15	10	5	0	5	10	15

   b. Plot the points in Table 5.22 to obtain the graph shown in Figure 5.90. This graph looks like the absolute value function defined in Section 5.3, except that the vertex is the point $(15, 0)$ instead of the origin.

   c. Let x represent the number of miles on the highway and $f(x)$ the number of miles from the mall. For x values less than 15, the graph is a straight line with slope -1 and y-intercept at $(0, 15)$, so its equation is $y = -x + 15$. Thus,

   $$f(x) = -x + 15 \qquad \text{when} \qquad 0 \le x < 15$$

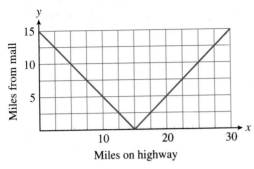

FIGURE 5.90

On the other hand, when $x \geq 15$, the graph of f is a straight line with slope 1 that passes through the point $(15, 0)$. The point-slope form of this line is

$$y - 0 = 1(x - 15)$$

so $y = x - 15$. Thus,

$$f(x) = x - 15 \qquad \text{when} \qquad 15 \leq x$$

Combining our two pieces together, we obtain

$$f(x) = \begin{cases} -x + 15, & \text{when } 0 \leq x < 15 \\ x - 15, & \text{when } x \geq 15 \end{cases}$$

The graph of $f(x)$ is a part of the graph of $y = |x - 15|$. (See Figure 5.90.) If we think of the highway as a portion of the real line, with Marlene's on-ramp located at the origin, then the outlet mall is located at 15. Marlene's coordinate as she drives along the highway is x, and the distance from Marlene to the mall is given by $|x - 15|$.

d. Marlene is within 5 miles of the mall when she has driven between 10 and 20 miles or when $10 < x < 20$. (See Figure 5.91.) We can restate this problem as "find Marlene's position when the distance between Marlene and the mall is less than 5," or, in mathematical terms, solve the inequality

$$|x - 15| < 5$$

The solution is $10 < x < 20$.

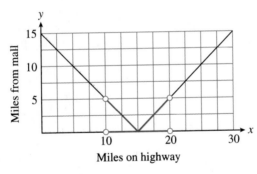

FIGURE 5.91

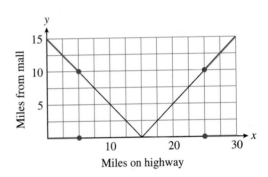

FIGURE 5.92

e. Marlene is at least 10 miles from the mall when she has driven less than 5 miles or more than 25 miles. (See Figure 5.92.) In mathematical terms, we are solving the inequality

$$|x - 15| \geq 10$$

The solution is $x \leq 5$ or $x \geq 25$. ∎

a. Sketch a graph of $y = |2x + 7|$.

b. Use your graph to solve the inequality $|2x + 7| < 13$.

1. Describe the graph of a function whose slope is positive and increasing.

2. Describe the graph of a function whose slope is positive and decreasing.

3. Which basic function is increasing but bending downward?

4. What function can be used to model the distance between x and a fixed point c?

ANSWERS TO 5.6 EXERCISES

1.

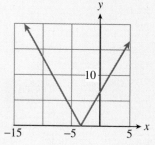

2. (a): The graph has a steep negative slope at first, corresponding to an initial rapid drop in the temperature of the cocoa. The graph becomes closer to a horizontal line, corresponding to the cocoa approaching room temperature.

3a. $|x + 3| = 5$ **b.** $|x - 4| \geq 6$

4a.   **b.** $(-\infty, -10) \cup (3, \infty)$

Which graph best illustrates each of the following situations?

1. Your pulse rate during an aerobics class.

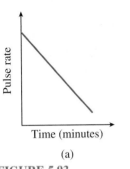 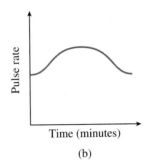

(a)   (b)

FIGURE 5.93

2. The stopping distances for cars traveling at various speeds.

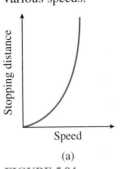 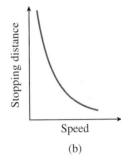

(a)   (b)

FIGURE 5.94

3. Your income in terms of the number of hours you worked.

(a)   (b)

FIGURE 5.95

4. Your temperature during an illness.

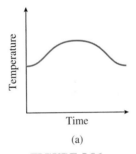 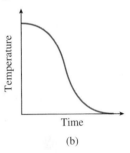

(a)   (b)

FIGURE 5.96

Sketch graphs to illustrate the following situations.

5. The height of your head above the ground during a ride on a Ferris wheel.

6. The height above the ground of a rubber ball dropped from the top of a 10-foot ladder.

7. Halfway from your English class to your math class, you realize that you left your math book in the classroom. You retrieve the book, then walk to your math class. Graph the distance between you and your English classroom as a function of time, from the moment you originally leave the English classroom until you reach the math classroom.

8. After you leave your math class, you start off toward your music class. Halfway there you meet an old friend, so you stop and chat for awhile. Then you continue to the music class. Graph the distance between you and your math classroom as a function of time, from the moment you originally leave the math classroom until you reach the music classroom.

9. Toni drives from home to meet her friend at the gym, which is halfway between their homes. They work out together at the gym, then they both go to the friend's home for a snack. Finally Toni drives home. Graph the distance between Toni and her home as a function of time, from the moment she leaves home until she returns.

10. While bicycling from home to school, Greg gets a flat tire. He repairs the tire in just a few minutes, but decides to backtrack a few miles to a service station, where he cleans up. Finally he bicycles the rest of the way to school. Graph the distance between Greg and his home as a function of time, from the moment he leaves home until he arrives at school.

Choose the graph that depicts the function described.

11. Inflation is still rising, but by less each month.

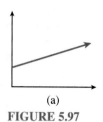 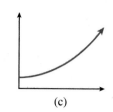 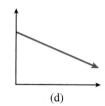 (d)

FIGURE 5.97

12. The price of wheat was rising more rapidly in 1996 than at any time during the previous decade.

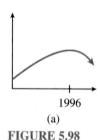 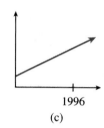 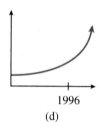

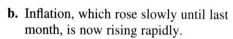

(a)　　　　(b)　　　　(c)　　　　(d)

FIGURE 5.98

Match each graph with the function it illustrates.

13. **a.** The volume of a cylindrical container of constant height as a function of its radius.

　　c. The simple interest earned at a given interest rate as a function of the investment.

　　b. The time it takes to travel a fixed distance as a function of average speed.

　　d. The number of Senators present versus the number absent in the U.S. Senate.

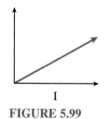 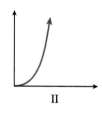 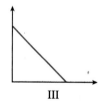 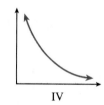

I　　　　II　　　　III　　　　IV

FIGURE 5.99

14. **a.** Unemployment was falling but is now steady.

　　c. The birthrate rose steadily until 1990, but is now beginning to fall.

　　b. Inflation, which rose slowly until last month, is now rising rapidly.

　　d. The price of gasoline has fallen steadily over the past few months.

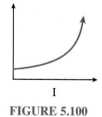 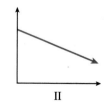 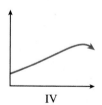 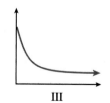

I　　　　II　　　　III　　　　IV

FIGURE 5.100

15. Four different functions are described below. Match each description with the appropriate table of values and with its graph in Figure 5.101.

a. As a chemical pollutant pours into a lake, its concentration is a function of time. The concentration of the pollutant initially increases quite rapidly, but due to the natural mixing and self-cleansing action of the lake, the concentration levels off and stabilizes at some saturation level.

b. An overnight express train travels at a constant speed across the Great Plains. The train's distance from its point of origin is a function of time.

c. The population of a small suburb of a Florida city is a function of time. The population began increasing rather slowly, but it has continued to grow at a faster and faster rate.

d. The level of production at a manufacturing plant is a function of capital outlay, i.e. the amount of money invested in the plant. At first, small increases in capital outlay result in large increases in production, but eventually the investors begin to experience diminishing returns on their money, so that although production continues to increase, it is at a disappointingly slow rate.

(1)

x	1	2	3	4	5	6	7	8
y	60	72	86	104	124	149	179	215

(2)

x	1	2	3	4	5	6	7	8
y	60	85	103	120	134	147	159	169

(3)

x	1	2	3	4	5	6	7	8
y	60	120	180	240	300	360	420	480

(4)

x	1	2	3	4	5	6	7	8
y	60	96	118	131	138	143	146	147

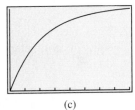

(a)

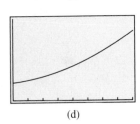

(c)

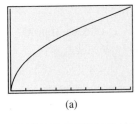

(b)

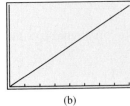

(d)

FIGURE 5.101

16. Four different functions are described below. Match each description with the appropriate table of values and with its graph in Figure 5.102.

a. After a chemical spill that greatly increased the phosphate concentration of Crystal Lake last year, fresh water flowing through the lake has gradually reduced the phosphate concentration to its natural level. The concentration of phosphate since the spill is a function of time.

b. The number of bacteria in a person during the course of an illness is a function of time. It increases rapidly at first, and then decreases slowly as the patient recovers.

c. A squirrel drops a pine cone from the top of a California redwood. The height of the pine cone is a function of time, decreasing ever more rapidly as gravity accelerates its descent.

d. Enrollment in Ginny's Weight Reduction program is a function of time. It began declining last fall. After the holidays enrollment stabilized for a while, but soon began to fall off again.

(1)

x	0	1	2	3	4
y	160	144	96	16	0

(2)

x	0	1	2	3	4
y	20	560	230	90	30

(3)

x	0	1	2	3	4
y	480	340	240	160	120

(4)

x	0	1	2	3	4
y	250	180	170	150	80

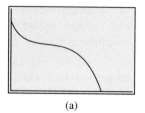
(a)

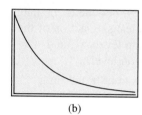
(b)

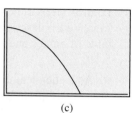
(c)

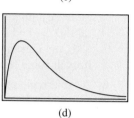
(d)

FIGURE 5.102

17. The table shows the radii, r, of several gold coins in centimeters and their value, v, in dollars.

Radius	0.5	1	1.5	2	2.5
Value	200	800	1800	3200	5000

a. Which graph in Figure 5.103 represents the data?

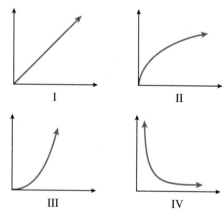

I   II

III   IV

FIGURE 5.103

b. Which equation describes the function?

I. $v = k\sqrt{r}$ III. $v = kr^2$

II. $v = kr$ IV. $v = \dfrac{k}{r}$

18. The table shows how the amount of water, A, flowing past a point on a river is related to the width, W, of the river at that point.

Width (feet)	11	23	34	46
Amount of water (ft³/second)	23	34	41	47

a. Which graph in Figure 5.104 represents the data?

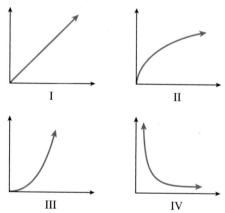

I   II

III   IV

FIGURE 5.104

b. Which equation describes the information?

I. $A = k\sqrt{W}$ III. $A = kW^2$

II. $A = kW$ IV. $A = \dfrac{k}{W}$

In each of Problems 19–22, one quantity varies directly with the square root of the other, that is, $y = k\sqrt{x}$, (a) Find the value of k and write a function relating the variables, (b) Use your function to answer the question, and (c) Graph your function and verify your answer to part (b) on the graph.

19. The stream speed necessary to move a granite particle is a function of the diameter of the particle; faster river currents can move larger particles. Table 5.23 shows the stream speed necessary to move particles of different sizes. What speed is needed to carry a particle with diameter 0.36 centimeters?

TABLE 5.23

Diameter (cm)	Speed (cm/second)
0.01	5
0.04	10
0.09	15
0.16	20

20. The speed at which water comes out of the spigot at the bottom of a water jug is a function of the water level in the jug; the flow slows down as the water level drops. Table 5.24 shows different water levels and the resulting flow speeds. What is the flow speed when the water level is at 16 inches?

TABLE 5.24

Level (in.)	Speed (gal/min)
9	1.50
6.25	1.25
4	1.00
2.25	0.75

21. Table 5.25 gives the distance, d, in miles that you can see from various heights, h, given in feet. How far can you see from an airplane flying at 20,000 feet?

TABLE 5.25

h	100	500	1000	1500
d	12.25	27.39	38.74	47.44

22. When a layer of ice forms on a pond, the thickness of the ice, d, in centimeters is a function of time, t, in minutes. (See Table 5.26.) How thick is the ice after 3 hours?

TABLE 5.26

t	10	30	40	60
d	0.50	0.87	1.01	1.24

Use absolute-value notation to write each expression as an equation or an inequality. (It may be helpful to restate each sentence using the word "distance.")

23. x is six units from the origin.

24. a is seven units from the origin.

25. The distance from p to -3 is five units.

26. The distance from q to -7 is two units.

27. t is within three units of 6.

28. w is no more than one unit from -5.

29. b is at least 0.5 unit from -1.

30. m is more than 0.1 unit from 8.

31. A small pottery is setting up a workshop to produce mugs. Three machines are located on a long table, as shown in Figure 5.105. The potter must use each machine once in the course of producing a mug. Where should the potter stand in order to minimize the distance she must walk to the machines?

Let x represent the coordinate of the potter's station.

a. Write expressions for the distance from the potter's station to each of the machines.

b. Write a function that gives the sum of the distances from the potter's station to the three machines.

c. Graph your function on the domain $[-20, 30]$, and use the graph to answer the question.

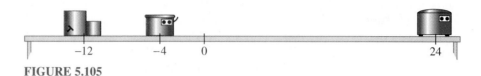

FIGURE 5.105

32. Suppose the pottery in Problem 31 adds a fourth machine to the procedure for producing a mug, located at $x = 16$ in Figure 5.105. Where should the potter stand now to minimize the distance she has to walk while producing each mug?

33. Richard and Marian are moving to Parkville after they graduate to take jobs. The main road through Parkville runs east and west, and crosses a river in the center of town. Richard's job is located 10 miles east of the river on the main road, and Marian's job is 6 miles west of the river. There is a health club they both like located 2 miles east of the river. If they plan to visit the health club every workday, where should Richard and Marian look for an apartment to minimize their total daily driving distance?

34. Romina's Bakery has just signed contracts to provide baked goods for 3 new restaurants located on Route 28 outside of town. The Coffee Stop is 2 miles north of town center, Sneaky Pete's is 8 miles north, and the Sea Shell is 12 miles south. Romina wants to open a branch bakery on Route 28 to handle the new business. Where should she locate the bakery in order to minimize the distance she must drive for deliveries?

35. Graph $y = |x + 3|$. Use your graph to solve the following equations and inequalities.

a. $|x + 3| = 2$ **b.** $|x + 3| \le 4$

c. $|x + 3| > 5$

36. Graph $y = |x - 2|$. Use your graph to solve the following equations and inequalities.

a. $|x - 2| = 5$ **b.** $|x - 2| < 8$

c. $|x - 2| \ge 4$

Use graphs to solve the following equations and inequalities.

37. $|2x - 1| = 4$ **38.** $|3x - 1| = 5$ **39.** $|2x + 6| < 3$ **40.** $|5 - 3x| \le 15$

41. $|3 - 2x| \ge 7$ **42.** $|3x + 2| > 10$

CHAPTER 5 REVIEW

Which of the following tables describe functions? Why or why not?

1.

x	-2	-1	0	1	2	3
y	6	0	1	2	6	8

2.

p	3	-3	2	-2	-2	0
q	2	-1	4	-4	3	0

3.

Student	Score on IQ test	Score on SAT test
(A)	118	649
(B)	98	450
(C)	110	590
(D)	105	520
(E)	98	490
(F)	122	680

4.

Student	Correct answers on math quiz	Quiz grade
(A)	13	85
(B)	15	89
(C)	10	79
(D)	12	82
(E)	16	91
(F)	18	95

5. The total number of barrels of oil pumped by the AQ oil company is given by the formula

$$N(t) = 2000 + 500t$$

where N is the number of barrels of oil t days after a new well is opened. Evaluate $N(10)$ and explain what it means.

6. The number of hours required for a boat to travel upstream between two cities is given by the formula

$$H(v) = \frac{24}{v - 8}, \qquad (v > 8)$$

where v represents the boat's top speed in miles per hour. Evaluate $H(16)$ and explain what it means.

Evaluate each function for the given values.

7. $F(t) = \sqrt{1 + 4t^2}$, $\qquad F(0)$ and $F(-3)$.

8. $G(x) = \sqrt[3]{x - 8}$, $\quad G(0)$ and $G(20)$

9. $h(v) = 6 - |4 - 2v|$, $\quad h(8)$ and $h(-8)$

10. $m(p) = \dfrac{120}{p + 15}$, $\quad m(5)$ and $m(-40)$

11. $P(x) = x^2 - 6x + 5$.
   a. Compute $P(0)$.
   b. Find all values of x for which $P(x) = 0$.

12. $R(x) = \sqrt{4 - x^2}$.
   a. Compute $R(0)$.
   b. Find all values of x for which $R(x) = 0$.

Consider the graphs shown for problems 13 and 14.

13. **a.** Find $f(-2)$ and $f(2)$ in Figure 5.106.
   b. For what value(s) of t is $f(t) = 4$?
   c. Find the t- and $f(t)$-intercepts of the graph.
   d. What is the maximum value of f? For what value(s) of t does f take on its maximum value?

14. **a.** Find $P(-3)$ and $P(3)$ in Figure 5.107
   b. For what value(s) of z is $P(z) = 2$?
   c. Find the z- and $P(z)$-intercepts of the graph.
   d. What is the minimum value of P? For what value(s) of z does P take on its minimum value?

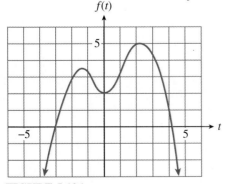

FIGURE 5.106

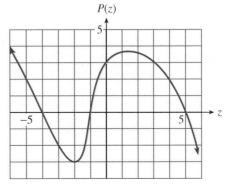

FIGURE 5.107

Use a graphing calculator to graph each function on the given domain. Adjust Ymin and Ymax until you can determine the range of the function using the TRACE **key. Then verify your answer algebraically by evaluating the function. State the domain and corresponding range in interval notation.**

15. $f(t) = -t^2 + 3t$; $-2 \le t \le 4$

16. $g(s) = \sqrt{s - 2}$; $2 \le s \le 6$

17. $F(x) = \dfrac{1}{x + 2}$; $-2 < x \le 4$

18. $H(x) = \dfrac{1}{2 - x}$; $-4 \le x < 2$

Which of the following graphs represent functions?

19.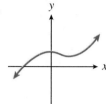

FIGURE 5.108

20.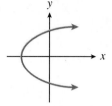

FIGURE 5.109

21.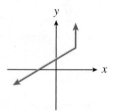

FIGURE 5.110

22.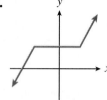

FIGURE 5.111

Graph each function by hand.

23. $f(t) = -2t + 4$

24. $g(s) = -\dfrac{2}{3}s - 2$

25. $p(x) = 9 - x^2$

26. $q(x) = x^2 - 16$

Graph the given function on a graphing calculator. Then use the graph to solve the equations and inequalities. Round your answers to one decimal place if necessary.

27. $y = \sqrt[3]{x}$

    a. Solve $\sqrt[3]{x} = 0.8$

    b. Solve $\sqrt[3]{x} = 1.5$

    c. Solve $\sqrt[3]{x} > 1.7$

    d. Solve $\sqrt[3]{x} \le 1.26$

28. $y = \dfrac{1}{x}$

    a. Solve $\dfrac{1}{x} = 2.5$

    b. Solve $\dfrac{1}{x} = 0.3125$

    c. Solve $\dfrac{1}{x} \ge 0.\overline{2}$

    d. Solve $\dfrac{1}{x} < 5$

29. $y = \dfrac{1}{x^2}$

    a. Solve $\dfrac{1}{x^2} = 0.03$

    b. Solve $\dfrac{1}{x^2} = 6.25$

    c. Solve $\dfrac{1}{x^2} > 0.16$

    d. Solve $\dfrac{1}{x^2} \le 4$

30. $y = \sqrt{x}$

    a. Solve $\sqrt{x} = 0.707$

    b. Solve $\sqrt{x} = 1.7$

    c. Solve $\sqrt{x} < 1.5$

    d. Solve $\sqrt{x} \ge 1.3$

For each table, *y* varies directly or inversely with a power of *x*. Find the power of *x* and the constant of variation, *k*. Write a formula for each function of the form $y = kx^n$ or $y = \dfrac{k}{x^n}$.

31.

x	y
2	4.8
5	30.0
8	76.8
11	145.2

32.

x	y
1.4	75.6
2.3	124.2
5.9	318.6
8.3	448.2

33.

x	y
0.5	40.0
2.0	10.0
4.0	5.0
8.0	2.5

34.

x	y
1.5	320.0
2.5	115.2
4.0	45.0
6.0	20.0

35. The distance, *s*, a pebble falls through a thick liquid varies directly with the square of the length of time, *t*, it falls.

   a. If the pebble falls 28 centimeters in 4 seconds, express the distance it will fall as a function of time.

   b. Find the distance the pebble will fall in 6 seconds.

36. The volume, *V*, of a gas varies directly with the temperature, *T*, and inversely with the pressure, *P*, of the gas.

   a. If $V = 40$ when $T = 300$ and $P = 30$, express the volume of the gas as a function of the temperature and pressure of the gas.

   b. Find the volume when $T = 320$ and $P = 40$.

37. The demand for bottled water is inversely proportional to the price per bottle. If Droplets can sell 600 bottles at $8 each, how many bottles can the company sell at $10 each?

38. The intensity of illumination from a light source varies inversely with the square of the distance from the source. If a reading lamp has an intensity of 100 lumens at a distance of 3 feet, what is its intensity 8 feet away?

39. A person's weight, *w*, varies inversely with the square of his or her distance, *r*, from the center of the earth.

   a. Express *w* as a function of *r*. Let *k* stand for the constant of variation.

   b. Make a rough graph of your function.

   c. How far from the center of the earth must Neil be in order to weigh one-third of his weight on the surface? The radius of the earth is about 3960 miles.

40. The period, *T*, of a pendulum varies directly with the square root of its length, *L*.

   a. Express *T* as a function of *L*. Let *k* stand for the constant of variation.

   b. Make a rough graph of your function.

   c. If a certain pendulum is replaced by a new one four-fifths as long as the old one, what happens to the period?

Sketch graphs to illustrate the following situations.

41. Inga runs hot water into the bathtub until it is about half full. Because the water is too hot, she lets it sit awhile before getting into the tub. After several minutes of bathing, she gets out and drains the tub. Graph the water level in the bathtub as a function of time, from the moment Inga starts filling the tub until it is drained.

42. David turns on the oven and it heats up steadily until the proper baking temperature is reached. The oven maintains that temperature during the time David bakes a pot roast. When he turns the oven off, David leaves the oven door open for a few minutes, and the temperature drops fairly rapidly during that time. After David closes the door, the temperature continues to drop, but at a much slower rate. Graph the temperature of the oven as a function of time, from the moment David first turns on the oven until shortly after David closes the door when the oven is cooling.

Use absolute-value notation to write each expression as an equation or inequality.

43. x is 4 units from the origin.

44. The distance from y to -5 is 3 units.

45. p is within 4 units of 7.

46. q is at least $\frac{3}{10}$ of a unit from -4.

Use a graph to solve each inequality.

47. $|3x - 2| < 4$

48. $|2x + 0.3| \leq 0.5$

49. $|3y + 1.2| \geq 1.5$

50. $\left|3z + \dfrac{1}{2}\right| > \dfrac{1}{3}$

Sketch graphs by hand for each function on the domain $(0, \infty.)$

51. y varies directly with x^2. The constant of variation is $k = 0.25$.

52. y varies directly with x. The constant of variation is $k = 1.5$

53. y varies inversely with x. The constant of variation is $k = 2$.

54. y varies inversely with x^2. The constant of variation is $k = 4$.

For problems 55 and 56, (a) plot the points and sketch a smooth curve through them, (b) use your graph to help you discover the equation that describes the function.

55.

x	$g(x)$
2	12
3	8
4	6
6	4
8	3
12	2

56.

x	$F(x)$
-2	8
-1	1
0	0
1	-1
2	-8
3	-27

In Problems 57–62, (a) use the graph to complete the table of values, (b) by finding a pattern in the table of values, write an equation for the graph.

57.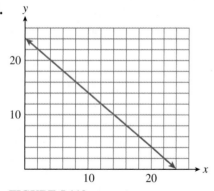

FIGURE 5.112

x	0	4	8		16	
y				10		2

58.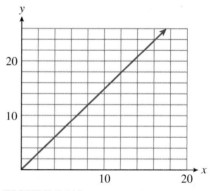

FIGURE 5.113

x	0	4	10		14	
y				18		24

59.

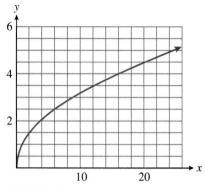

FIGURE 5.114

x	0		4		16	25
y		1		3		

60.

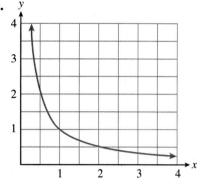

FIGURE 5.115

x		0.5	1	1.5		4
y	4				0.5	

61.

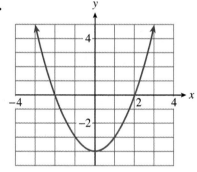

FIGURE 5.116

x	−3	−2		0	1	2
y			−3			

62.

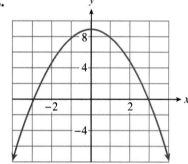

FIGURE 5.117

x	−3	−2		0	1	
y			8			−7

Powers and Roots

In Chapter 5 we studied direct and inverse variation, which are modeled by functions of the form

$$g(x) = kx^n \qquad \text{and} \qquad h(x) = \frac{k}{x^n}$$

respectively. For example, the function

$$F(d) = \frac{k}{d^2}$$

gives the gravitational force, F, exerted by the sun on the planets and other objects in our solar system. This force varies inversely with the square of the distance, d, from the object to the sun. In Section 5.5 we also considered the function

$$f(T) = 20.06 \sqrt{T}$$

which gives the speed of sound as a function of the air temperature, T. In this chapter we'll see that all of these functions are examples of *power functions,* and that by extending our definition of exponent, we can express all of them in the form $f(x) = kx^n$.

 INVESTIGATION 10

Inflating a Balloon

If you blow air into a balloon, what do you think will happen to the air pressure inside the balloon as it expands? Here is what two physics books have to say:

> "The greater the pressure inside, the greater the balloon's volume."
> Boleman, Jay, *Physics, a Window on Our World*

> "Contrary to the process of blowing up a toy balloon, the pressure required to force air into a bubble decreases with bubble size."
> Sears, Francis, *Mechanics, Heat, and Sound*

1. Based on these two quotes and your own intuition, sketch a graph of pressure as a function of the diameter of the balloon. Describe your graph: Is it increasing or decreasing? Is it concave up (bending upward) or concave down (bending downward)?

2. In 1998, two high school students, April Leonardo and Tolu Noah, decided to see for themselves how the pressure inside a balloon changes as the balloon expands. Using a column of water to measure pressure, they collected the following data while blowing up a balloon. Graph their data on the grid in Figure 6.1.

Diameter (cm)	Pressure (cm H$_2$O)
5.7	60.6
7.3	57.2
8.2	47.9
10.7	38.1
12	37.1
14.6	31.9
17.5	28.1
20.5	26.4
23.5	28
25.2	31.4
26.1	34
27.5	37.2
28.4	37.9
29	40.7
30	43.3
30.6	46.6
31.3	50
32.2	61.9

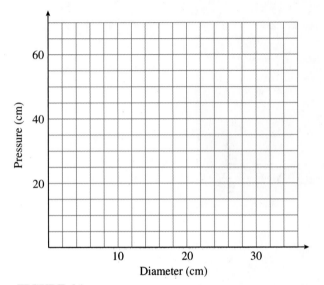

FIGURE 6.1

3. Describe the graph of April and Tolu's data. Does the graph confirm the predictions of the physics books? (We'll return to April and Tolu's experiment later in Chapter 6.)

6.1 Integer Exponents

Recall that a positive integer exponent tells us how many times its base occurs as a factor in an expression. For example,

$$4a^3b^2 \qquad \text{means} \qquad 4aaabb$$

Before studying this chapter, you may want to review Appendix A.5, the Laws of Exponents.

Negative Exponents

Study the list of powers of 2 shown in Table 6.1a, and observe the pattern as we move up the list from bottom to top. Each time the exponent increases by 1 we multiply by another factor of 2. We can continue up the list computing positive exponents as far as we like.

If we move back down the list, we divide by 2 at each step, until we get to the bottom of the list, $2^1 = 2$. What if we continue the list in the same way, dividing by 2 each time we decrease the exponent? The results are shown in Table 6.1b.

TABLE 6.1A Positive Exponents

TABLE 6.1B Integer Exponents

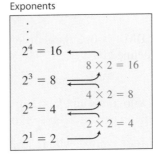

As we continue to divide by 2, we generate fractions whose denominators are powers of 2. In particular, we see that

$$2^{-1} = \frac{1}{2} = \frac{1}{2^1} \qquad \text{and} \qquad 2^{-2} = \frac{1}{4} = \frac{1}{2^2}$$

Based on these observations, we make the following definition.

Definition of Negative Exponent

$$a^{-n} = \frac{1}{a^n} \qquad (a \neq 0)$$

EXAMPLE 1

a. $2^{-3} = \dfrac{1}{2^3} = \dfrac{1}{8}$

b. $9x^{-2} = 9 \cdot \dfrac{1}{x^2} = \dfrac{9}{x^2}$ ∎

COMMON ERRORS

1. A negative exponent does *not* mean that the power is negative! For example,

$$2^{-3} \neq -2^3$$

2. In Example 1b, note that

$$9x^{-2} \neq \frac{1}{9x^2}$$

The exponent, -2, applies *only* to the base x, not to 9. ∎

EXERCISE 1 Write each expression without using negative exponents.

a. 5^{-4} **b.** $5x^{-4}$

EXAMPLE 2

The body mass index, or BMI, is one measure of a person's physical fitness. Your body mass index is defined by

$$BMI = wh^{-2}$$

where w is your weight in kilograms and h is your height in meters. The World Health Organization classifies a person as obese if his or her BMI is 25 or higher.

    a. Calculate the BMI for a woman who is 1.625 meters (64 inches) tall and weighs 54 kilograms (120 pounds).

    b. For a fixed weight, how does BMI vary with height?

    c. Akebono, the heaviest sumo wrestler in history, weighs 501 pounds (227 kilograms). How tall would Akebono have to be to have a BMI under 25?

Solutions

    a. $BMI = 54(1.625^{-2}) = 54\left(\dfrac{1}{1.625^2}\right) = 20.45$

    b. $BMI = \dfrac{w}{h^2}$, so BMI varies inversely with the square of height. That is, BMI decreases as height increases.

    c. To find the height that gives a BMI of 25, we solve the equation $25 = 227h^{-2}$.

$$25 = \frac{227}{h^2} \quad \text{Multiply both sides by } h^2$$
$$25h^2 = 227 \quad \text{Divide both sides by 25.}$$
$$h^2 = 9.08 \quad \text{Extract square roots.}$$
$$h \approx 3.01$$

To have a BMI under 25, Akebono would have to be over 3.01 meters, or 9 feet 11 inches tall. ∎

EXERCISE 2 Solve the equation $0.2x^{-3} = 1.5$

Rewrite without a negative exponent.

Clear the fraction.

Isolate the variable.

Power Functions

A function like

$$f(x) = kx^p$$

where k and p are constants, is called a **power function.** Examples of power functions are

$$V(r) = \frac{4}{3}\pi r^3 \qquad \text{and} \qquad L(T) = 0.8125T^2$$

Functions that describe inverse variation can be written as power functions using negative exponents. In particular, the basic functions

$$f(x) = \frac{1}{x} \qquad \text{and} \qquad g(x) = \frac{1}{x^2}$$

which we studied in Chapter 5, can also be written as

$$f(x) = x^{-1} \qquad \text{and} \qquad g(x) = x^{-2}$$

Their graphs are shown in Figure 6.2. Note that the domains of power functions with negative exponents do not include zero. Most applications are concerned with positive variables only, so many models use only the portion of the graph in the first quadrant.

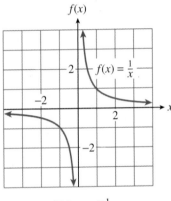

$$f(x) = x^{-1}$$

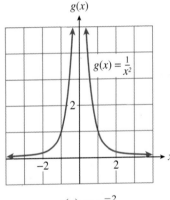

$$g(x) = x^{-2}$$

FIGURE 6.2

EXAMPLE 3

In the middle ages in Europe, castles were built as defensive strongholds. An attacking force would build a huge catapult called a trebuchet to hurl rocks and scrap metal inside the castle walls. The engineers could adjust its range by varying the mass of the projectiles. The mass, m, of the projectile is inversely proportional to the square of the distance, d, to the target.

a. Use a negative exponent to write m as a function of d, $m = f(d)$.

b. The engineers test the trebuchet with a 20-kilogram projectile, which lands 250 meters away. Find the constant of proportionality, and rewrite your formula for m.

c. Graph $m = f(d)$.

d. The trebuchet is 180 meters from the courtyard within the castle. What size projectile will hit the target?

e. The attacking force would like to hurl a 100-kilogram projectile at the castle. How close must they bring their catapult?

Solutions

a. If we use k for the constant of proportionality, then $m = \dfrac{k}{d^2}$. Rewriting this equation with a negative exponent gives $m = kd^{-2}$.

b. Substitute $m = 20$ and $d = 250$ to obtain

$$20 = k(250)^{-2} \quad \text{Multiply both sides by } 250^2.$$
$$1,250,000 = k$$

Thus, $m = 1,250,000d^{-2}$.

c. Evaluate the function for several values of m, or use your calculator to obtain the graph in Figure 6.3.

d. Substitute $d = 180$ into the formula.

$$m = 1,250,000(180)^{-2}$$
$$= \frac{1,250,000}{32,400} \approx 38.58$$

They should use a mass of approximately 38.6 kilograms.

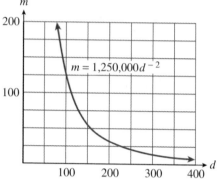

FIGURE 6.3

e. Substitute $m = 100$ into the formula and solve for d.

$$100 = 1,250,000d^{-2} \quad \text{Multiply by } d^2.$$
$$100d^2 = 1,250,000 \quad \text{Divide by 100.}$$
$$d^2 = 12,500 \quad \text{Take square roots.}$$
$$d = \pm\sqrt{12,500}$$

They must locate the catapult $\sqrt{12,500} \approx 111.8$ meters from the castle.   ■

EXERCISE 3   Write each function as a power function in the form $y = kx^p$.

a. $f(x) = \dfrac{12}{x^2}$ **b.** $g(x) = \dfrac{1}{4x}$ **c.** $h(x) = \dfrac{2}{5x^6}$

Transforming Data

In Chapter 1 we used linear regression to fit an equation to a collection of data. When the data points lie roughly on a line, it is fairly easy to estimate a line of best fit by eye, and to find the equation of the regression line. If the data points do not follow a linear pattern, but instead lie near some other curve, it is not so easy to find the equation for an appropriate regression curve. If we know what type of function to try, we can sometimes use a transformation to convert the data to a linear pattern.

EXAMPLE 4

When you press on the plunger of a bicycle pump, the air inside is compressed. The data below show that the pressure inside the pump increases as the volume decreases. Decide whether pressure is inversely proportional to volume, and if so, find the constant of proportionality.

Volume, V (liters)	1.13	0.52	0.26	0.13
Pressure, P (atmospheres)	1.02	2.26	4.51	9.14

Solution

We'd like to know whether the data fit a curve $P = kV^{-1}$, for some value of k. If we let $x = V^{-1}$, the equation $P = kV^{-1}$ becomes $P = kx$. Also, if the original data points (V, P) lie near the curve $P = kV^{-1}$, the transformed data points (x, P) will lie near the line $P = kx$.

We can make a table of (x, P) values by computing $x = \dfrac{1}{V}$ for each of the original data points, as shown. For example, the point $(V, P) = (1.13, 1.02)$ is transformed into the point $(x, P) = \left(\dfrac{1}{1.13}, 1.02\right) \approx (0.885, 1.02)$. When we plot the transformed points, we see that they do show a strong linear trend, as you can see in Figure 6.4.

x	0.885	1.923	3.846	7.692
P	1.02	2.26	4.51	9.14

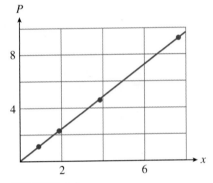

FIGURE 6.4

The slope, k, of the regression line in Figure 6.4 is also the constant of proportionality in the equation $P = kV^{-1}$. By choosing two points on that line, we estimate its slope as 1.2, so the constant of proportionality is 1.2. The graph of $P = 1.2V^{-1}$ and the scatterplot of the original data are shown in Figure 6.5.

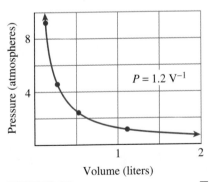

FIGURE 6.5

Because many curves have similar shapes, at least on part of their domains, it is not a good idea to fit a particular type of function to a collection of data without some assurance that the function is appropriate for the situation at hand. Choosing a reasonable type of function requires some knowledge of the variables involved. For Example 4, we can appeal to Boyle's law, which says that pressure and volume are inversely proportional for an ideal gas. Therefore, we might reasonably expect the same relation to hold for the air inside a bicycle pump.

EXERCISE 4

a. Use the transformation $x = \dfrac{1}{r^2}$ to decide whether B is inversely proportional to r^2 for the following data.

r	0.2	0.3	0.4	0.5	0.6	0.7	0.8
B	84	45	26	12	9	8	6

First plot the data on the grid. Is the model $B = \dfrac{k}{r^2}$ reasonable for these data?

Complete the table for the transformed data points, (x, B), by computing $x = \dfrac{1}{r^2}$ for each given data point.

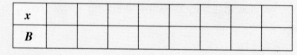

x							
B							

Plot the transformed data on the grid. Is the plot approximately linear?

b. Find a linear equation for the plot of the transformed data. Use that equation to write a power function for B in terms of r. Sketch the graph of the power function on the plot of the original data.

Draw a line of best fit through the data points.

Choose two points on the line and calculate its equation.

Write a formula for B as a function of r, and sketch the graph on the plot of the original data.

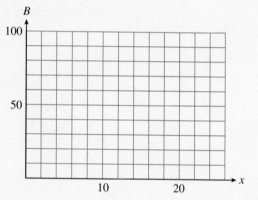

Working with Negative Exponents

A negative exponent denotes the *reciprocal* of a power. In particular, to simplify a fraction with a negative exponent, we compute the positive power of its reciprocal.

EXAMPLE 5

a. $\left(\dfrac{3}{5}\right)^{-2} = \left(\dfrac{5}{3}\right)^2 = \dfrac{25}{9}$

b. $\left(\dfrac{x^3}{4}\right)^{-3} = \left(\dfrac{4}{x^3}\right)^3 = \dfrac{(4)^3}{(x^3)^3} = \dfrac{64}{x^9}$

■

Dividing by a power with a negative exponent is equivalent to multiplying by a positive power.

<table>
<tr><td>

EXAMPLE 6

</td><td>

a. $\dfrac{1}{5^{-3}} = 1 \div 5^{-3}$

$\qquad = 1 \div \dfrac{1}{5^3}$

$\qquad = 1 \times 5^3 = 125$

</td><td>

b. $\dfrac{k^2}{m^{-4}} = k^2 \div m^{-4}$

$\qquad = k^2 \div \dfrac{1}{m^4}$

$\qquad = k^2 m^4$

</td></tr>
</table>

■

EXERCISE 5 Write each expression without using negative exponents.

a. $\left(\dfrac{3}{b^4}\right)^{-2}$ **b.** $\dfrac{12}{x^{-6}}$

Laws of Exponents

The laws of exponents reviewed in Appendix A.5 apply to all integer exponents, positive and negative. When we allow negative exponents, we can simplify the rule for computing quotients of powers.

$$\textbf{II.}\;\; \frac{a^m}{a^n} = a^{m-n} \qquad (a \neq 0)$$

For example, by applying this new version of the law for quotients we find

$$\frac{x^2}{x^5} = x^{2-5} = x^{-3}$$

which is consistent with our previous version of the rule,

$$\frac{x^2}{x^5} = \frac{1}{x^{5-2}} = \frac{1}{x^3}$$

For reference, we restate the laws of exponents below. The laws are valid for all integer exponents m and n, and for $b \neq 0$.

Laws of Exponents

I. $a^m \cdot a^n = a^{m+n}$ **IV.** $(ab)^n = a^n b^n$

II. $\dfrac{a^m}{a^n} = a^{m-n}$ **V.** $\left(\dfrac{a}{b}\right)^n = \dfrac{a^n}{b^n}$

III. $(a^m)^n = a^{mn}$

EXAMPLE 7

a. $x^3 \cdot x^{-5} = x^{3-5} = x^{-2}$ Apply the first law: add exponents.

b. $\dfrac{8x^{-2}}{4x^{-6}} = \dfrac{8}{4}x^{-2-(-6)} = 2x^4$ Apply the second law: subtract exponents.

c. $(5x^{-3})^{-2} = 5^{-2}(x^{-3})^{-2} = \dfrac{x^6}{25}$ Apply laws IV and III.   ∎

You can check that each of the calculations in Example 7 is shorter when we use negative exponents instead of converting the expressions into algebraic fractions.

EXERCISE 6   Simplify by applying the laws of exponents.

a. $(2a^{-4})(-4a^2)$ **b.** $\dfrac{(r^2)^{-3}}{3r^{-4}}$

In Table 6.1b we saw that $2^0 = 1$, and in fact $a^0 = 1$ as long as $a \neq 0$. This is because the quotient of any (nonzero) number divided by itself is 1. So

$$1 = \frac{a^m}{a^m} = a^{m-m} = a^0$$

Thus,

$$a^0 = 1, \qquad a \neq 0$$

For example,

$$3^0 = 1, \qquad (-528)^0 = 1, \qquad \text{and} \qquad (0.024)^0 = 1$$

Scientific Notation

Scientists and engineers regularly encounter very large numbers such as,

$$5{,}980{,}000{,}000{,}000{,}000{,}000{,}000{,}000$$

(the mass of Earth in kilograms) and very small numbers such as

$$0.000\,000\,000\,000\,000\,000\,000\,000\,01\,67$$

(the mass of a hydrogen atom in grams). These numbers can be written in a more compact and useful form by using powers of 10.

In our base 10 number system, multiplying a number by a positive power of 10 has the effect of moving the decimal place k places to the right, where k is the exponent in the power of 10. For example,

$$3.529 \times 10^2 = 352.9 \qquad \text{and} \qquad 25 \times 10^4 = 250{,}000$$

Multiplying a number by a negative power of 10 moves the decimal place to the left. For example,

$$1728 \times 10^{-3} = 1.728 \qquad \text{and} \qquad 4.6 \times 10^{-5} = 0.000046$$

Using this property, we can write any number as the product of a number between 1 and 10 (including 1) and a power of 10. For example, the mass of the earth and the mass of a hydrogen atom can be expressed as

$$5.98 \times 10^{24} \text{ kilograms} \quad \text{and} \quad 1.67 \times 10^{-24} \text{ gram}$$

respectively. A number written in this form is said to be expressed in **scientific notation.**

To Write a Number in Scientific Notation:

1. Locate the decimal point so that there is exactly one nonzero digit to its left.
2. Count the number of places you moved the decimal point; this determines the power of 10.
   a. If the original number is greater than 10, the exponent is positive.
   b. If the original number is less than 1, the exponent is negative.

EXAMPLE 8

Write each number in scientific notation.

a. $478{,}000 = 4.78000 \times 10^5$ b. $0.00032 = 00003.2 \times 10^{-4}$
$\underbrace{}_{\text{5 places}}$ $\underbrace{}_{\text{4 places}}$

$= 4.78 \times 10^5$ $= 3.2 \times 10^{-4}$ ∎

EXERCISE 7   Write each number in scientific notation.

a. 0.063   b. 1480

Your calculator displays numbers in scientific notation if they are too large or too small to fit in the display screen. Try squaring the number 123,456,789 on your calculator. Enter

$$123456789 \quad \boxed{x^2}$$

and the calculator will display the result as

$$1.524157875 \textbf{ E } 16$$

This is how the calculator displays the number $1.524157875 \times 10^{16}$. Notice that the power 10^{16} is displayed as "**E** 16."

To enter a number in scientific form, we use the key labeled $\boxed{\text{EE}}$. For example, to enter 3.26×10^{-18}, use the keying sequence

$$3.26 \; \boxed{\text{EE}} \; \boxed{(-)} \; 18$$

EXAMPLE 9

The average American eats 110 kilograms of meat per year. It takes about 16 kilograms of grain to produce one kilogram of meat, and advanced farming techniques can produce about 6000 kilograms of grain on each hectare of arable land. (The "hectare" is 10,000 square meters, or just under two and a half acres.) Now, the total land area of the earth is about 13 billion hectares, but only about 11% of that land is arable. Is it possible for each of the 5.5 billion people on earth to eat as much meat as Americans do?

Solution

First we'll compute the amount of meat necessary to feed every person on earth 110 kilograms per year. There are 5.5×10^9 people on earth.

$$(5.5 \times 10^9 \text{ people}) \times (110 \text{ kilograms/person}) = 6.05 \times 10^{11} \text{ kilograms of meat}$$

Next we'll compute the amount of grain needed to produce that much meat.

$$(16 \text{ kilograms of grain/kilogram of meat}) \times$$
$$(6.05 \times 10^{11} \text{ kilograms of meat}) = 9.68 \times 10^{12} \text{ kilograms of grain}$$

Next we'll see how many hectares of land are needed to produce that much grain.

$$(9.68 \times 10^{12} \text{ kilograms of grain}) \div (6000 \text{ kilograms/hectare}) = 1.61\overline{3} \times 10^9 \text{ hectares}$$

Finally, we'll compute the amount of arable land available for grain production.

$$0.11 \times (13 \times 10^9 \text{ hectares}) = 1.43 \times 10^9 \text{ hectares}$$

Thus, even if we use every hectare of arable land to produce grain for livestock, we won't have enough to provide every person on earth with 110 kilograms of meat per year. ∎

READING QUESTIONS

1. What does a negative exponent mean?
2. What is a power function?
3. Explain why it makes sense to define $5^0 = 1$.
4. Explain how to write a number in scientific notation.

ANSWERS TO 6.1 EXERCISES

1a. $\dfrac{1}{5^4}$

b. $\dfrac{5}{x^4}$

2. $x = \sqrt[3]{\dfrac{2}{15}} \approx 0.51$

3a. $f(x) = 12x^{-2}$

b. $g(x) = \dfrac{1}{4}x^{-1}$

c. $h(x) = \dfrac{2}{5}x^{-6}$

4a.

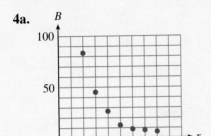

x	25	11.11	6.25	4	2.78	2.04	.56
B	84	45	26	12	9	8	6

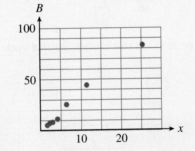

$B = \dfrac{k}{r^2}$ is a reasonable model.

b. The transformed data fall close to the line $B = 3.7x$, so $B \approx \dfrac{3.7}{r^2}$.

5a. $\dfrac{b^8}{9}$ **b.** $12x^6$

6a. $\dfrac{-8}{a^2}$ **b.** $\dfrac{1}{3r^2}$

7a. 6.3×10^{-2} **b.** 1.48×10^3

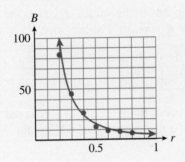

HOMEWORK 6.1

Write without negative exponents and simplify.

1. a. 2^{-1} **b.** $(-5)^{-2}$

   c. $\left(\dfrac{1}{3}\right)^{-3}$ **d.** $\dfrac{1}{(-2)^{-4}}$

2. a. 3^{-2} **b.** $(-2)^{-3}$

   c. $\left(\dfrac{3}{5}\right)^{-2}$ **d.** $\dfrac{1}{(-3)^{-3}}$

3. a. $\dfrac{5}{4^{-3}}$ **b.** $(2q)^{-5}$

   c. $-4x^{-2}$ **d.** $\dfrac{8}{b^{-3}}$

4. a. $\dfrac{3}{2^{-6}}$ **b.** $(4k)^{-3}$

   c. $-7x^{-4}$ **d.** $\dfrac{5}{a^{-5}}$

5. a. $(m - n)^{-2}$ **b.** $y^{-2} + y^{-3}$

   c. $2pq^{-4}$ **d.** $\dfrac{-5y^{-2}}{x^{-5}}$

6. a. $(p + q)^{-3}$ **b.** $z^{-1} - z^{-2}$

   c. $8m^{-2}n^2$ **d.** $\dfrac{-6y^{-3}}{x^{-3}}$

Compute each power.

7. a. 2^3 **b.** $(-2)^3$
   c. 2^{-3} **d.** $(-2)^{-3}$

9. a. $\left(\dfrac{1}{2}\right)^3$ **b.** $\left(-\dfrac{1}{2}\right)^3$
   c. $\left(\dfrac{1}{2}\right)^{-3}$ **d.** $\left(-\dfrac{1}{2}\right)^{-3}$

8. a. 4^2 **b.** $(-4)^2$
   c. 4^{-2} **d.** $(-4)^{-2}$

10. a. $\left(\dfrac{1}{4}\right)^2$ **b.** $\left(-\dfrac{1}{4}\right)^2$
   c. $\left(\dfrac{1}{4}\right)^{-2}$ **d.** $\left(-\dfrac{1}{4}\right)^{-2}$

Use your calculator to fill in the tables. Round your answers to two decimal places.

11. $f(x) = x^{-2}$

a.

x	1	2	4	8	16
x^{-2}					

b. What happens to the values of $f(x)$ as the values of x increase? Explain why.

c.

x	1	0.5	0.25	0.125	0.0625
x^{-2}					

d. What happens to the values of $f(x)$ as the values of x decrease toward 0? Explain why.

12. $g(x) = x^{-3}$

a.

x	1	2	4.5	6.2	9.3
x^{-3}					

b. What happens to the values of $g(x)$ as the values of x increase? Explain why.

c.

x	1.5	0.6	0.1	0.03	0.002
x^{-3}					

d. What happens to the values of $g(x)$ as the values of x decrease toward 0? Explain why.

13. Use your calculator to graph each of the following functions on the window

     Xmin $= -5$ Xmax $= 5$
     Ymin $= -2$ Ymax $= 10$

a. $f(x) = x^2$ **b.** $f(x) = x^{-2}$
c. $f(x) = \dfrac{1}{x^2}$ **d.** $f(x) = \left(\dfrac{1}{x}\right)^2$

e. Which functions have the same graph? Explain your results.

14. Use your calculator to graph each of the following functions on the window

     Xmin $= -3$ Xmax $= 3$
     Ymin $= -5$ Ymax $= 5$

a. $f(x) = x^3$ **b.** $f(x) = x^{-3}$
c. $f(x) = \dfrac{1}{x^3}$ **d.** $f(x) = \left(\dfrac{1}{x}\right)^3$

e. Which functions have the same graph? Explain your results.

Write each expression as a power function using negative exponents.

15. a. $F(r) = \dfrac{3}{r^4}$ **b.** $G(w) = \dfrac{2}{5w^3}$

   c. $H(z) = \dfrac{1}{(3z)^2}$

16. a. $h(s) = \dfrac{9}{s^3}$ **b.** $f(v) = \dfrac{3}{8v^6}$

   c. $g(t) = \dfrac{1}{(5t)^4}$

17. The amount of force or thrust generated by the propeller of a ship is a function of two variables: the diameter of the propeller and its speed, in rotations per minute. The thrust, T, in pounds is proportional to the square of the speed, r, and the fourth power of the diameter, d, in feet.

a. Write a formula for the thrust in terms of the speed and diameter of the propeller.

b. A propeller of diameter 2 feet generates a thrust of 1000 pounds at 100 rotations per minute. Find the constant of variation in the formula for thrust.

18. Refer to Problem 17.

  a. Sketch a graph of the thrust as a function of the diameter of the propeller at a speed of 100 rotations per minute. If the diameter of the propeller is doubled, by what factor does the thrust increase?

  b. Sketch a graph of the thrust as a function of the propeller speed for a propellor of diameter 4 feet. If the speed of the propeller is doubled, by what factor does the thrust increase?

19. In some molecules, called dipoles, electric charge is not uniformly distributed, but collects into positive and negative regions. The electric field, E, at a distance x along the axis of the dipole is inversely proportional to the cube of x.

  a. Write a formula for E as a power function of x.

  b. When $x = 1$, $E = \dfrac{2qL}{4\pi\epsilon}$, where $2qL$ is the dipole moment and ϵ is the charge on one electron. Find the constant (in terms of these parameters) in your power function.

20. The Van der Waals' force F between dipole molecules is responsible for the condensation of a gas into a liquid. The Van der Waals' force is inversely proportional to the seventh power of x, the distance between the molecules.

  a. Write a formula for F as a power function of x.

  b. When $x = 1$, $F = \dfrac{-3\mu^2}{(4\pi\epsilon)^2}$, where μ is the dipole moment and ϵ is the charge on one electron. Find the constant (in terms of these parameters) in your power function.

For Problems 21–24, use the suggested transformation of variables to fit a function to the data.

21. A cyclist works against both air resistance (drag) and rolling resistance. The magnitude of drag is proportional to the square of the cyclist's velocity, and the rolling resistance is a constant. The total force that the cyclist battles at velocity v is therefore given by

$$F = av^2 + b$$

The table shows the average force, F, encountered by a professional cyclist while riding at velocity, v.

Velocity, v (m/s)	5.4	6.6	7.4	7.8	8.7	9.5	9.9
Force, F (Newtons)	12.7	15.6	19.0	22.4	23.4	24.4	30.7

Velocity, v (m/s)	10.0	10.5	10.7	11.4	11.8	11.9
Force, F (Newtons)	27.3	32.2	30.2	36.1	37.6	35.6

  a. Let $x = v^2$ and make a table of (x, F) values, and plot your data.

  b. Draw a regression line, and estimate the values of a and b.

  c. Predict the force the cyclist encounters at a velocity of 10.5 meters per second.

  d. At what velocity will the cyclist meet a force of 25 newtons?

  e. Use your values of a and b to graph $F = av^2 + b$ on a plot of the original data.

Source: Pugh, L.G.C.E., 1974

22. The rate at which photosynthesis occurs in a plant depends on the concentration of inorganic carbon available in the environment. The data show the rate of photosynthesis for the spiked water milfoil, which grows in lakes.

Inorganic carbon (mg/l)	3	4	8	11	21
Rate of photosynthesis (mg/h)		1.3	2	3.1	3	4.2

Inorganic carbon (mg/l)	28	29	30	42.6
Rate of photosynthesis (mg/h)		4.7	5	6	6

The two variables are related by an equation of the form

$$\frac{1}{v} = a \cdot \frac{1}{u} + b$$

where u represents the concentration of inorganic carbon and v gives the rate of photosynthesis.

a. Let $x = \dfrac{1}{u}$ and $y = \dfrac{1}{v}$ and make a table for values for x and y.

b. Plot the transformed data. What is the transformed equation?

c. By drawing a line of best fit, estimate the values of a and b.

d. Use the original equation to explain why the value of $\dfrac{1}{v}$ can be close to b but will always be greater than b.

e. Why is $\dfrac{1}{b}$ called the maximum rate of photosynthesis? Estimate $\dfrac{1}{b}$.

f. Use your values of a and b to graph $v = \left(a \cdot \dfrac{1}{u} + b \right)^{-1}$ on a plot of the original data.

Source: Burton, 1998

23. Delbert conducts an experiment to estimate the value of the gravitational constant, g. He releases a ball from the top of an inclined plane and allows it to roll downward. The table shows how far the ball has rolled at different times after its release.

Time, t (seconds)	1	1.5	2	2.5	3	3.5
Distance, d (meters)	0.95	2.16	3.89	6.12	8.80	11.85

The distance, d, the ball rolls is proportional to the square of the time, t, since its release.

a. Let $x = t^2$ and make a new table of values for (x, d). Then plot the data.

b. Use a regression line for your data to estimate the constant of proportionality.

c. If the slope of the inclined plane is m, the constant of proportionality is $\dfrac{gm}{\sqrt{m^2 + 1}}$, where g is the gravitational constant. The inclined plane in Delbert's experiment has slope 0.1. What value of g does Delbert calculate?

24. Francine also performs an experiment to estimate the value of g. She measures the lengths and the periods of several different pendulums. (The period of a pendulum is the

Length, L (centimeters)	8	15	31	47	58	75
Period, T (seconds)	0.6	0.9	1.2	1.4	1.6	1.8

time required for it to complete one entire back and forth cycle.) Her data are given in the table. The period, T, of a pendulum is proportional to the square root of its length, L.

a. Let $x = \sqrt{L}$ and make a new table of values for (x, T). Then plot the data.

b. Use a regression line to estimate the constant of proportionality.

c. The constant of proportionality is $\dfrac{2\pi}{\sqrt{g}}$, where g is the gravitational constant. What does Francine's experiment predict for the value of g?

25. A hot object such as a star or a lightbulb filament radiates energy over a range of wavelengths, but the wavelength with maximum energy is inversely proportional to the temperature of the object. If temperature is measured in degrees Kelvin, and wavelength in micrometers, the constant of proportionality is 2898. (One micrometer is one thousandth of a millimeter, or $1\,\mu m = 10^{-6}$ meter.)

a. Write a formula for the wavelength of maximum energy, λ_{max}, as a function of temperature, T. This formula, called Wien's law, was discovered in 1894.

b. Our sun's temperature is about $5765°$K. At what wavelength is most of its energy radiated?

c. The color of light depends on its wavelength, as shown in the table. Can you explain why the sun does not appear to be green? Use Wien's law to describe how the color of a star depends on its temperature.

Color	Red	Orange	Yellow
Wavelength (μm)	0.64–0.74	0.59–0.64	0.56–0.59

Color	Green	Blue	Violet
Wavelength (μm)	0.50–0.56	0.44–0.50	0.39–0.44

d. Astronomers cannot measure the temperature of a star directly, but they can determine the color or wavelength of its light. Write a formula for T as a function of λ_{max}.

e. Estimate the temperatures of the following stars, given the approximate value of λ_{max} for each.

Star	λ_{max}	Temperature
R Cygni	1.115	
Betelgeuse	0.966	
Arcturus	0.725	
Polaris	0.414	
Sirius	0.322	
Rigel	0.223	

f. Sketch a graph of T as a function of λ_{max} and locate each star on the graph.

26. Investigation 10 gives data for the pressure inside April and Tolu's balloon as a function of its diameter. Figure 6.6 shows a plot of the first half of these data. As the diameter of the balloon increases from 5 cm to 20 cm, the pressure inside decreases. Can we find a function that describes this portion of the graph?

a. Pressure is the force per unit area exerted by the balloon on the air inside, or $P = \dfrac{F}{A}$.

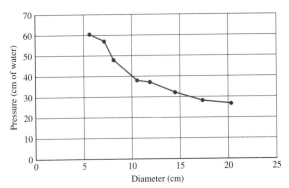

FIGURE 6.6

Because the balloon is spherical, its surface area, A, is given by $A = \pi d^2$. The force exerted by the balloon is not so easy to express in terms of d, but because the force increases as the balloon expands, we'll try an expression of the form $F = kd$, where k is a constant, and see if this fits the data. Use these relationships to express P as a power function of d.

b. Make a table of values for $x = \dfrac{1}{d}$ and P, and plot the points (x, P). Find a line of best fit by eye, or use your calculator's linear regression feature.

c. Use your regression equation to write a formula for P as a function of d. Do your calculations support the hypothesis that P is a power function of d? (We'll return to the balloon experiment in Chapter 10.)

Use the laws of exponents to simplify and write without negative exponents.

27. a. $a^{-3} \cdot a^8$ **b.** $5^{-4} \cdot 5^{-3}$

    c. $\dfrac{p^{-7}}{p^{-4}}$ **d.** $(7^{-2})^5$

29. a. $(4x^{-5})(5x^2)$ **b.** $\dfrac{3u^{-3}}{9u^9}$

    c. $\dfrac{5^6 t^0}{5^{-2} t^{-1}}$

31. a. $(3x^{-2}y^3)^{-2}$ **b.** $\left(\dfrac{6a^{-3}}{b^2}\right)^{-2}$

    c. $\dfrac{5h^{-3}(h^4)^{-2}}{6h^{-5}}$

33. a. Is it true that $(x + y)^{-2} = x^{-2} + y^{-2}$? Explain why or why not.

    b. Give a numerical example to support your answer.

28. a. $b^2 \cdot b^{-6}$ **b.** $4^{-2} \cdot 4^{-6}$

    c. $\dfrac{w^{-9}}{w^2}$ **d.** $(9^{-4})^3$

30. a. $(3y^{-8})(2y^4)$ **b.** $\dfrac{4c^{-4}}{8c^{-8}}$

    c. $\dfrac{3^{10} s^{-1}}{3^{-5} s^0}$

32. a. $(2x^3y^{-4})^{-3}$ **b.** $\left(\dfrac{a^4}{4b^{-5}}\right)^{-3}$

    c. $\dfrac{4v^{-5}(v^{-2})^{-4}}{3v^{-8}}$

34. a. Is it true that $(a - b)^{-1} = a^{-1} - b^{-1}$? Explain why or why not.

    b. Give a numerical example to support your answer.

Write each expression as a sum of terms of the form kx^p.

35. a. $\dfrac{x}{3} + \dfrac{3}{x}$ **b.** $\dfrac{x - 6x^2}{4x^3}$

36. a. $\dfrac{2}{x^2} - \dfrac{x^2}{2}$ **b.** $\dfrac{5x + 1}{(3x)^2}$

37. a. $\dfrac{2}{x^4}\left(\dfrac{x^2}{4} + \dfrac{x}{2} - \dfrac{1}{4}\right)$

    b. $\dfrac{x^2}{3}\left(\dfrac{2}{x^4} - \dfrac{1}{3x^2} + \dfrac{1}{2}\right)$

38. a. $\dfrac{9}{x^3}\left(\dfrac{x^3}{3} - 1 - \dfrac{1}{x^3}\right)$

    b. $\dfrac{x^2}{2}\left(\dfrac{3}{x} - \dfrac{5}{x^3} + \dfrac{7}{x^5}\right)$

Use the distributive law to write each product as a sum of powers.

39. $x^{-1}(x^2 - 3x + 2)$

40. $3x^{-2}(2x^4 + x^2 - 4)$

41. $-3t^{-2}(t^2 - 2 - 4t^{-2})$

42. $-t^{-3}(3t^2 - 1 - t^{-2})$

43. $2u^{-3}(-2u^3 - u^2 + 3u)$

44. $2u^{-1}(-1 - u - 2u^2)$

Factor as indicated, and write the factored form with positive exponents only.

45. $4x^2 + 16x^{-2} = 4x^{-2}(?)$

46. $20y - 15y^{-1} = 5y^{-1}(?)$

47. $3a^{-3} - 3a + a^3 = a^{-3}(?)$

48. $2 - 4q^{-2} - 8q^{-4} = 2q^{-4}(?)$

Write each number in scientific notation.

49. a. 285   **b.** 8,372,000

    c. 0.024   **d.** 0.000523

50. a. 68,742   **b.** 481,000,000,000

    c. 0.421   **d.** 0.000004

Write each number in standard notation.

51. a. 2.4×10^2 **b.** 6.87×10^{15}

    c. 5.0×10^{-3} **d.** 2.02×10^{-4}

52. a. 4.8×10^3 **b.** 8.31×10^{12}

    c. 8.0×10^{-1} **d.** 4.31×10^{-5}

Compute with the aid of a calculator. Write your answers in standard notation.

53. a. $\dfrac{(2.4 \times 10^{-8})(6.5 \times 10^{32})}{5.2 \times 10^{18}}$

    b. $\dfrac{(7.5 \times 10^{-13})(3.6 \times 10^{-9})}{(1.5 \times 10^{-15})(1.6 \times 10^{-11})}$

54. a. $\dfrac{(8.4 \times 10^{-22})(1.6 \times 10^{15})}{3.2 \times 10^{-11}}$

    b. $\dfrac{(9.4 \times 10^{24})(7.2 \times 10^{-18})}{(4.5 \times 10^{26})(6.4 \times 10^{-16})}$

55. In 1999 the public debt of the United States was $5,605,304,000,000.

    a. Express this number in scientific notation.

    b. If the population of the United States in 1999 was 272,712,000, what was the per capita debt (the debt per person) for that year?

56. One light-year is the number of miles traveled by light in one year (365 days). The speed of light is approximately 186,000 miles per second.

    a. Compute the number of miles in one light-year, and express your answer in scientific notation.

    b. The star nearest to the sun is Proxima Centauri, at a distance of 4.3 light-years. How long would it take Pioneer 10 (the first space vehicle to achieve escape velocity from the solar system), traveling at 32,114 miles per hour, to reach Proxima Centauri?

57. The diameter of the galactic disk is about 1.2×10^{18} kilometers, and our sun lies about halfway from the center of the galaxy to the edge of the disk. The sun orbits the galactic center once in 240 million years.

   a. What is the speed of the sun in its orbit, in kilometers per year?

   b. What is its speed in meters per second?

58. Lake Superior has an area of 31,700 square miles and an average depth of 483 feet.

   a. Find the approximate volume of Lake Superior in cubic feet.

   b. If 1 cubic foot of water is equivalent to 7.48 gallons, how many gallons of water are in Lake Superior?

Use rounding to give approximate answers for Problems 59–62, without using a calculator.

59. The average distance from the earth to the sun is 1.5×10^{11} meters. The distance from the sun to Proxima Centauri, the next closest star, is 3.99×10^{16} meters. The most distant stars visible to the unaided eye are 2000 times as far away as Proxima Centauri.

   a. How many times farther is Proxima Centauri from the sun than the sun is from Earth?

   b. How far from the sun are the most distant visible stars?

60. The radius of the earth is 6.37×10^6 meters, and the radius of the sun is 6.96×10^8 meters. The radii of the other stars range from 1% of the solar radius to 1000 times the solar radius.

   a. What fraction of the solar radius is earth's radius?

   b. What is the range of stellar radii, in meters?

61. The earth's crust is composed of a number of separate tectonic plates that float on the mantle and move very slowly relative to each other. For instance, New York lies on the North American plate, and London lies on the Eurasian plate. These plates meet in the Atlantic Ocean, and are separating at a speed of 25 millimeters per year. At that rate, how long will it take for the distance between London and New York to increase by 25 kilometers (about 15.5 miles)?

   Source: Keynes, 1998.

62. (See Problem 61.) Los Angeles, which lies on the Pacific plate, and San Francisco, which lies on the North American plate, are separated by the San Andreas fault. The two plates are sliding past each other at a rate of 20 millimeters per year. Currently, the distance between the cities is 560 kilometers. How long will it be before San Francisco becomes a suburb of Los Angeles?

   Source: Keynes, 1998.

63. Airplanes use radar to detect the distances to other objects. A radar unit transmits a pulse of energy, which bounces off a distant object, and the "echo" of the pulse returns to the sender. The power, P, of the returning echo is inversely proportional to the fourth power of the distance, d, to the object. A radar operator receives an echo of 5×10^{-10} watts from an aircraft 2 nautical miles away.

   a. Express the power of the echo received in picowatts. (1 picowatt $= 10^{-12}$ watts.)

   b. Write a function that expresses P in terms of d using negative exponents. Use picowatts for the units of power.

   c. Complete the table of values for the power of the echo received from objects at various distances.

d (nautical miles)	4	5	7	10
P (picowatts)				

   d. Radar units can typically detect signals as low as 10^{-13} watt. How far away is an aircraft whose echo is 10^{-13} watt? (*Hint:* Convert 10^{-13} watt to picowatts.)

   e. Sketch a graph of P as a function of d. Use units of picowatts on the vertical axis.

64. Halley's comet orbits the sun every 76 years, and was first observed in 240 BCE. Its orbit is highly elliptical, so that its closest approach to the sun (perihelion) is only 0.587 AU, while at its greatest distance (aphelion), the comet is 34.39 AU from the sun. (An AU, or astronomical unit, is the distance from Earth to the sun, 1.5×10^8 kilometers.)

a. Calculate the distances from the sun to Halley's comet in meters at perihelion and aphelion.

b. Halley's comet has a volume of 700 cubic kilometers, and its density is about 0.1 gram per cubic centimeter. Calculate the mass of the comet in kilograms.

c. The gravitational force (in Newtons) exerted by the sun on its satellites is inversely proportional to the square of the

distance to the satellite in meters. The constant of variation is Gm_1m_2, where $m_1 = 1.99 \times 10^{30}$ kilograms is the mass of the sun, m_2 is the mass of the satellite, and $G = 6.67 \times 10^{-11}$ is the gravitational constant. Write a formula for the force, F, exerted by the sun on Halley's comet at a distance of d meters.

d. Calculate the force exerted by the sun on Halley's comet at perihelion and at aphelion.

6.2 Roots and Radicals

In Section 6.1 we saw that inverse variation can be expressed as a power function by using negative exponents. We can also use exponents to denote square roots and other radicals.

n^{th} Roots

Recall that s is a square root of b if $s^2 = b$, and s is a cube root of b if $s^3 = b$. In a similar way we can look for the fourth, fifth, or sixth root of a number. For instance, the fourth root of b is a number s whose fourth power is b. In general, we make the following definition.

> s is called an nth **root of b** if $s^n = b$

We use the symbol $\sqrt[n]{b}$ to denote the nth root of b. An expression of the form $\sqrt[n]{b}$ is called a **radical**, b is called the **radicand**, and n is called the **index** of the radical.

EXAMPLE 1

a. $\sqrt[4]{81} = 3$ because $3^4 = 81$
b. $\sqrt[5]{32} = 2$ because $2^5 = 32$
c. $\sqrt[6]{64} = 2$ because $2^6 = 64$
d. $\sqrt[4]{1} = 1$ because $1^4 = 1$
e. $\sqrt[5]{100,000} = 10$ because $10^5 = 100,000$ ∎

EXERCISE 1   Evaluate each radical.

a. $\sqrt[4]{16}$ **b.** $\sqrt[5]{243}$

Exponential Notation for Radicals

A convenient notation for radicals uses fractional exponents. Consider the expression $9^{1/2}$. What meaning can we attach to an exponent that is a fraction? The third law of exponents says that when we raise a power to a power we multiply the exponents together:

$$(x^a)^b = x^{ab}$$

Therefore, if we square the number $9^{1/2}$, we get

$$(9^{1/2})^2 = 9^{(1/2)(2)} = 9^1 = 9$$

Thus, $9^{1/2}$ is a number whose square is 9. But this means that $9^{1/2}$ is a square root of 9, or

$$9^{1/2} = \sqrt{9} = 3$$

In general, any non-negative number raised to the 1/2 power is equal to the positive square root of the number, or

$$a^{1/2} = \sqrt{a}$$

EXAMPLE 2

  a. $25^{1/2} = 5$ **c.** $(-25)^{1/2}$ is not a real number

  b. $-25^{1/2} = -5$ **d.** $0^{1/2} = 0$ ■

The same reasoning works for roots with any index. For instance, $8^{1/3}$ is the cube root of 8, because

$$(8^{1/3})^3 = 8^{(1/3)(3)} = 8^1 = 8$$

In general, we make the following definition for fractional exponents.

For any natural number $n \geq 2$,

$$a^{1/n} = \sqrt[n]{a}$$

Of course, we can use decimal fractions for exponents as well. For example,

$$\sqrt{x} = x^{1/2} = x^{0.5}$$
$$\sqrt[4]{x} = x^{1/4} = x^{0.25}$$

EXAMPLE 3

  a. $81^{1/4} = \sqrt[4]{81} = 3$ **c.** $100^{0.5} = \sqrt{100} = 10$

  b. $125^{1/3} = \sqrt[3]{125} = 5$ **d.** $16^{0.25} = \sqrt[4]{16} = 2$ ■

COMMON ERROR

Note that

$$25^{1/2} \neq \frac{1}{2}(25) \qquad \text{and} \qquad 125^{1/3} \neq \frac{1}{3}(125)$$

An exponent of $\frac{1}{2}$ denotes the square root of its base, and an exponent of $\frac{1}{3}$ denotes the cube root of its base.   ■

Fractional exponents simplify many calculations involving radicals. You should learn to convert easily between exponential and radical notation. Remember that a negative exponent denotes a reciprocal.

EXAMPLE 4   Convert each radical to exponential notation.

    a. $\sqrt[3]{12} = 12^{1/3}$ **c.** $\dfrac{1}{\sqrt[5]{ab}} = (ab)^{-1/5}$

    b. $\sqrt[4]{2y} = (2y)^{1/4}$ or $(2y)^{0.25}$ **d.** $3\sqrt[6]{w} = 3w^{1/6}$ ■

In Example 4b, the parentheses around $(2y)$ must not be omitted.

EXAMPLE 5   Convert each power to radical notation.

    a. $5^{1/2} = \sqrt{5}$ **c.** $2x^{1/3} = 2\sqrt[3]{x}$

    b. $x^{0.2} = \sqrt[5]{x}$ **d.** $8a^{-1/4} = \dfrac{8}{\sqrt[4]{a}}$ ■

In Example 5d, the exponent $-1/4$ applies only to a, not to $8a$.

Irrational Numbers

What about n^{th} roots such as $\sqrt{23}$ and $\sqrt[3]{5}$, which cannot be evaluated easily? These are examples of *irrational numbers*. We can use a calculator to obtain decimal approximations to irrational numbers. For example, you can verify that

$$\sqrt{23} \approx 4.796 \qquad \text{and} \qquad \sqrt[3]{5} \approx 1.710$$

It is not possible to write down an exact decimal equivalent for an irrational number, but we can find an approximation to as many decimal places as we like. You can read more about irrational numbers in Appendix B.1.

The following keying sequence for evaluating the irrational number $7^{1/5}$ is *incorrect:*

$$7 \;\boxed{\wedge}\; 1 \;\boxed{\div}\; 5 \;\boxed{\text{ENTER}}$$

You can check that this sequence calculates $\dfrac{7^1}{5}$, instead of $7^{1/5}$. Recall that according to the order of operations, powers are computed before multiplications or divisions. We must enclose the exponent $\frac{1}{5}$ in parentheses and enter

$$7 \;\boxed{\wedge}\; \boxed{(}\; 1 \;\boxed{\div}\; 5 \;\boxed{)}\; \boxed{\text{ENTER}}$$

Alternatively, we can enter

$$7 \;\boxed{\wedge}\; 0.2 \;\boxed{\text{ENTER}}$$

because $\frac{1}{5} = 0.2$. ∎

Power Functions

An animal's heart rate is related to its size or mass, with smaller animals generally having faster heart rates. The heart rates of mammals are given approximately by the power function

$$H(m) = km^{-1/4}$$

where m is the animal's mass and k is a constant.

A typical human male weighs about 70 kilograms and has a resting heart rate of 70 beats per minute.

    a. Find the constant of proportionality and write a formula for $H(m)$.

    b. Fill in the table with the heart rates of the mammals whose masses are given.

Animal	Shrew	Rabbit	Cat	Wolf	Horse	Polar bear	Elephant	Whale
Mass (kg)	0.004	2	4	80	300	600	5400	70,000
Heart rate								

    c. Sketch a graph of H for masses up to 6000 kilograms.

Solutions

    a. Substitute $H = 70$ and $m = 70$ into the equation, and solve for k.

$$70 = k \cdot 70^{-1/4}$$

$$k = \frac{70}{70^{-1/4}} \approx 202.5$$

Thus, $H(m) = 202.5m^{-1/4}$.

b. Evaluate the function H for each of the masses given in the table.

Animal	Shrew	Rabbit	Cat	Wolf	Horse	Polar bear	Elephant	Whale
Mass (kg)	0.004	2	4	80	300	600	5400	70,000
Heart rate	805	170	143	68	49	41	24	12

c. Plot the points in the table to obtain the graph shown in Figure 6.7.

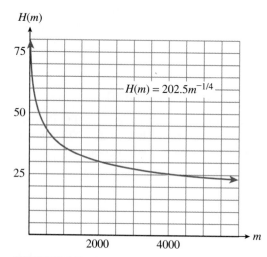

FIGURE 6.7

Many properties relating to the growth of plants and animals can be described by power functions of their mass. The study of the relationship between the growth rates of different parts of an organism, or of organisms of similar type, is called *allometry*. An equation of the form

$$\text{variable} = k(\text{mass})^p$$

used to describe such a relationship is called an *allometric equation*.

Of course, power functions can be expressed using any of the notations we have discussed. For example, the function in Example 6 can be written as

$$H(m) = 202.5m^{-1/4} \quad \text{or} \quad H(m) = 202.5m^{-0.25} \quad \text{or} \quad H(m) = \frac{202.5}{\sqrt[4]{m}}$$

EXERCISE 4

a. Complete the table of values for the power function $f(x) = x^{1/4}$.

x	0	1	5	10	20	50	70	100
$f(x)$								

b. Sketch a graph of $y = f(x)$.

Solving Equations

In Chapter 3 we learned that squaring and taking square roots are *inverse operations,* that is, each operation undoes the effects of the other. We use this fact when solving equations involving x^2 or $\sqrt{x}$.

EXAMPLE 7

When a car brakes suddenly, its speed can be estimated from the length of the skid marks it leaves on the pavement. A formula for the car's speed in miles per hour is $v = f(d) = \sqrt{24d}$, where the length of the skid marks, d, is given in feet.

   a. If a car leaves skid marks 80 feet long, how fast was the car traveling when the driver applied the brakes?

   b. How far will a car skid if its driver applies the brakes while traveling 80 miles per hour?

Solutions

   a. To find the velocity of the car, we evaluate the function for $d = 80$.

$$v = \sqrt{24(80)} \qquad \text{Substitute 80 for } d.$$
$$= \sqrt{1920} \qquad \text{Multiply inside the radical.}$$
$$\approx 43.8178046 \quad \text{Take the square root.}$$

The car was traveling at approximately 44 miles per hour.

   b. We would like to find the value of d when the value of v is known. We substitute $v = 80$ into the equation to obtain

$$80 = \sqrt{24d}$$

Because d appears under a square root in the equation, we first square both sides to get

$$80^2 = \left(\sqrt{24d}\right)^2 \quad \text{Square both sides.}$$
$$6400 = 24d \qquad \text{Divide by 24.}$$
$$266.\overline{6} = d$$

You can check that this value for d works in the original equation. Thus, the car will skid approximately 267 feet. A graph of the function $v = \sqrt{24d}$ is shown in Figure 6.8, along with the points corresponding to the values in parts (a) and (b).

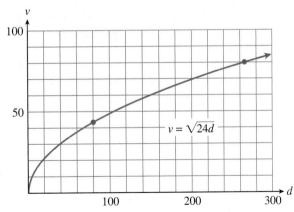

FIGURE 6.8

In general, we can solve an equation in which one side is an nth root of x by raising both sides of the equation to the nth power. To solve an equation involving n^{th} powers of x, we first isolate the power, then take the nth root of both sides.

EXAMPLE 8

A spherical fish tank in the lobby of the Atlantis Hotel holds about 905 cubic feet of water. What is the radius of the fish tank?

Solution

The volume of a sphere is given by the formula $V = \frac{4}{3}\pi r^3$, where r is the radius. Substitute **905** cubic feet for the volume of the sphere, and solve for r.

$$\frac{4}{3}\pi r^3 = 905$$

First, isolate the variable. Divide both sides by $\frac{4}{3}\pi$.

$$\frac{\frac{4}{3}\pi r^3}{\frac{4}{3}\pi} = \frac{905}{\frac{4}{3}\pi}$$

$$r^3 = \frac{905}{\pi} \cdot \frac{3}{4} \approx 216.05$$

Now take the cube root of both sides by raising each side to the power $\frac{1}{3}$.

$$(r^3)^{1/3} \approx 216.05^{1/3}$$

$$r \approx 6$$

The radius of the fish tank is approximately 6 feet. ∎

EXERCISE 5 Solve $5\sqrt[4]{x-1} = 10$.

Isolate the radical.

Raise both sides to the fourth power.

Complete the solution.

Roots of Negative Numbers

You already know that $\sqrt{-9}$ is not a real number, because there is no real number whose square is -9. Similarly, $\sqrt[4]{-16}$ is not a real number, because there is no real number r for which $r^4 = -16$. (Both of these radicals are *complex numbers*. Complex numbers are discussed in Appendix B.2.) In general, we cannot find an even root (square root, fourth root, and so on) of a negative number.

On the other hand, every positive number has *two* roots that are real numbers. For example, both 3 and -3 are square roots of 9. The symbol $\sqrt{9}$ refers only to the positive, or **principal root** of 9. If we want to refer to the negative square root of 9, we must write $-\sqrt{9} = -3$. Similarly, both 2 and -2 are fourth roots of 16, because $2^4 = 16$ and $(-2)^4 = 16$. However, the symbol $\sqrt[4]{16}$ refers to the principal, or positive, fourth root only. Thus,

$$\sqrt[4]{16} = 2 \qquad \text{and} \qquad -\sqrt[4]{16} = -2$$

Things are simpler for odd roots (cube roots, fifth roots, and so on). Every real number, whether positive, negative, or zero, has exactly one real-valued odd root. For example,

$$\sqrt[5]{32} = 2 \qquad \text{and} \qquad \sqrt[5]{-32} = -2$$

Here is a summary of our discussion.

1. Every positive number has two real-valued roots if the index is even, one positive and one negative.
2. A negative number has no real-valued root if the index is even.
3. Every real number, positive or negative, has exactly one real-valued root if the index is odd.

EXAMPLE 9

a. $\sqrt[4]{-625}$ is not a real number. **b.** $-\sqrt[4]{625} = -5$
c. $\sqrt[5]{-1} = -1$ **d.** $\sqrt[4]{-1}$ is not a real number. ■

The same principles apply to powers with fractional exponents. Thus

$$(-32)^{1/5} = -2$$

but $(-64)^{1/6}$ is not a real number. On the other hand,

$$-64^{1/6} = -2$$

because the exponent 1/6 applies only to 64, and the negative sign is applied after the root is computed.

EXERCISE 6 Evaluate each power, if possible.

a. $-81^{1/4}$ **b.** $(-81)^{1/4}$
c. $-64^{1/3}$ **d.** $(-64)^{1/3}$

READING QUESTIONS

1. Define the terms radical, radicand, index, and principal root.
2. What does the notation $x^{1/4}$ mean?
3. What does the notation $x^{0.2}$ mean?
4. What is the inverse operation for raising to the third power (cubing)?

ANSWERS TO 6.2 EXERCISES

1a. 2   **b.** 3

2a. $\sqrt[5]{100{,}000} = 10$ **b.** $\sqrt[4]{625} = 5$

3a. $3(2x)^{-1/4}$ **b.** $-5\sqrt[8]{b}$

4a.

x	0	1	5	10	20	50	70	100
$f(x)$	0	1	1.495	1.778	2.115	2.659	2.893	3.162

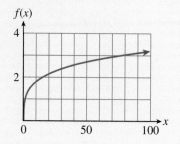

5. 17

6a. -3 **b.** undefined   **c.** -4 **d.** -4

HOMEWORK 6.2

Find the indicated root without using a calculator, then check your answers.

1. a. $\sqrt{121}$ **b.** $\sqrt[3]{27}$ **c.** $\sqrt[4]{625}$ **2. a.** $\sqrt{169}$ **b.** $\sqrt[3]{64}$ **c.** $\sqrt[4]{81}$

3. a. $\sqrt[5]{32}$ **b.** $\sqrt[4]{16}$ **c.** $\sqrt[3]{729}$ **4. a.** $\sqrt[5]{100{,}000}$ **b.** $\sqrt[4]{1296}$ **c.** $\sqrt[3]{343}$

Find the indicated power without using a calculator, then check your answers.

5. a. $9^{1/2}$ **b.** $81^{1/4}$ **c.** $64^{1/6}$ **6. a.** $25^{1/2}$ **b.** $16^{1/4}$ **c.** $27^{1/3}$

7. a. $32^{0.2}$ **b.** $8^{-1/3}$ **c.** $64^{-0.5}$ **8. a.** $625^{0.25}$ **b.** $243^{-1/5}$ **c.** $49^{-0.5}$

Write each expression in radical form.

9. a. $3^{1/2}$ **b.** $4x^{1/3}$ **c.** $(4x)^{0.2}$ **10. a.** $7^{1/2}$ **b.** $3x^{1/4}$ **c.** $(3x)^{0.25}$

11. a. $6^{-1/3}$ **b.** $3(xy)^{-0.125}$ **c.** $(x-2)^{1/4}$ **12. a.** $8^{-1/4}$ **b.** $y(5x)^{-0.5}$ **c.** $(y+2)^{1/3}$

Write each expression in exponential form.

13. a. $\sqrt{7}$ **b.** $\sqrt[3]{2x}$ **c.** $2\sqrt[5]{z}$ **14. a.** $\sqrt{5}$ **b.** $\sqrt[3]{4y}$ **c.** $5\sqrt[3]{x}$

15. a. $\dfrac{-3}{\sqrt[3]{6}}$ **b.** $\sqrt[4]{x-3y}$ **c.** $\dfrac{-1}{\sqrt[5]{1+3b}}$ **16. a.** $\dfrac{2}{\sqrt[5]{3}}$ **b.** $\sqrt[3]{y+2x}$ **c.** $\dfrac{-1}{\sqrt[4]{3a-2b}}$

Use a calculator to approximate each irrational number to the nearest thousandth.

17. a. $2^{1/2}$ **b.** $\sqrt[3]{75}$ **c.** $\sqrt[4]{1.6}$ **18. a.** $3^{1/2}$ **b.** $\sqrt[4]{60}$ **c.** $\sqrt[3]{1.4}$

     d. $365^{-1/3}$ **e.** $0.006^{-0.2}$ **d.** $1058^{-1/4}$ **e.** $1.05^{-0.1}$

19. When the Concorde lands at Heathrow airport in London, the width w of the sonic boom felt on the ground is given in kilometers by the following formula:

$$w = 4\left(\frac{Th}{m}\right)^{1/2}$$

T stands for the temperature on the ground in degrees Kelvin, h is the altitude of the Concorde when it breaks the sound barrier, and m is the drop in temperature for each gain in altitude of one kilometer. Find the width of the sonic boom if the ground temperature is 293° Kelvin, the altitude of the Concorde is 15 kilometers, and the temperature drop is 4° Kelvin per kilometer of altitude.

21. If you walk in the normal way, your speed, v, in meters per second, is limited by the length of your legs, r, according to the formula

$$v \leq \sqrt{gr}$$

where g is the acceleration due to gravity.

a. Complete the table of values for v as a function of r. The constant, g, is approximately 10.

r (meters)	0.2	0.4	0.6	0.8	1.0
v (meters per second)					

b. A typical adult man has legs about 0.9 meters long. How fast can he walk?

c. A typical four-year-old has legs 0.5 meters long. How fast can she walk?

d. Graph maximum walking speed against leg length.

e. Race-walkers can walk as fast as 4.4 meters per second by rotating their hips so that the effective length of their legs is increased. What is that effective length?

f. On the moon the value of g is 1.6. How fast can a typical adult man walk on the moon?

Source: Alexander, 1992

20. The manager of an office supply store must decide how many of each item in stock she should order. The Wilson lot size formula gives the most cost-efficient quantity, Q, in terms of the cost, C, of placing an order, the number of items, N, sold per week, and the weekly inventory cost, I, per item (cost of storage, maintenance, and so on).

$$Q = \left(\frac{2CN}{I}\right)^{1/2}$$

How many reams of computer paper should she order if she sells on average 80 reams per week, the weekly inventory cost for a ream is $0.20, and the cost of ordering, including delivery charges, is $25?

22. When a ship moves through the water, it creates waves that impede its own progress. Because of this resistance, there is an upper limit to the speed at which a ship can travel, given in knots by

$$v_{max} = 1.3\sqrt{L}$$

where L is the length of the vessel in feet.

a. Complete the table of values for v_{max} as a function of L.

L (feet)	200	400	600	800	1000
v_{max} (knots)					

b. Graph maximum speed against vessel length.

c. The world's largest ship, the oil tanker Jahre Viking, is 1054 feet long. What is its top speed?

d. As a ship approaches its maximum speed, the power required increases sharply. Therefore, most merchant ships are designed to cruise at speeds no higher than $v_c = 0.8\sqrt{L}$. Graph v_c on the same axes with v_{max}.

e. What is the cruising speed of the Jahre Viking? What percent of its maximum speed is that?

Source: Gilner, 1972

23. A rough estimate for the radius of the nucleus of an atom is provided by the formula

$$r = kA^{1/3}$$

where A is the mass number of the nucleus, and $k \approx 1.3 \times 10^{-13}$ centimeters.

a. Estimate the radius of the nucleus of an atom of iodine-127, which has mass number 127. If the nucleus is roughly spherical, what is its volume?

b. The nuclear mass of iodine-127 is 2.1×10^{-22} grams. What is the density of the nucleus? (Density is mass per unit volume.)

c. Complete the table of values for the radii of various radioisotopes.

Element	Carbon	Potassium	Cobalt
Mass number, A	14	40	60
Radius, r			

Element	Technetium	Radium
Mass number, A	99	226
Radius, r		

d. Sketch a graph of r as a function of A. (Use units of 10^{-13} centimeters on the vertical axis.)

24. Astronomers cannot measure the mass of a star directly. However, by studying binary stars, they discovered that for large stars (more than three times the mass of our sun), the luminosity or brightness of the star is approximately proportional to the cube of its mass. The mass–luminosity relation allows astronomers to estimate the mass of a star from its luminosity.

a. Write an expression for the mass of a large star as a function of its luminosity.

b. Properties of stars are often measured in solar units, so that, for instance, a star of luminosity 100 is 100 times brighter than the sun. A particular star of luminosity 4000 solar units is found to have a mass 8 times the solar mass. Use these values to calculate the constant of variation in your equation. (Many numerical relationships in astronomy are accurate only to an order of magnitude.)

c. Fill in the table with approximate masses for stars with the given luminosities.

Luminosity	10	100	1000	10,000	50,000
Mass					

d. Sketch a graph of the mass–luminosity relation for stars of luminosity up to 10,000 solar units.

e. The star Formalhaut is part of the southern constellation Piscis Austrini. Its luminosity is 15.2 times the sun's luminosity, and its mass is 14.1 solar masses. How well does the model fit this star?

25. Membership in the County Museum has been increasing slowly since it was built in 1950. The number of members is given by the function

$$M(t) = 72 + 100t^{1/3}$$

where t is the number of years since 1950.

a. How many members were there in 1960? In 1970?

b. In what year did the museum have 400 members? If the membership continues to grow according to the given function, when will the museum have 500 members?

c. Graph the function $M(t)$. How would you describe the growth of the membership over time?

26. Due to improvements in technology, the annual electricity cost of running most major appliances has decreased steadily since 1970. The average annual cost of running a refrigerator is given in dollars by the function

$$C(t) = 148 - 28t^{1/3}$$

where t is the number of years since 1970.

a. How much did it cost to run a refrigerator in 1980? In 1990?

b. When was the cost of running a refrigerator half of the cost in 1970? If the cost continues to decline according to the given function, when will it cost $50 per year to run a refrigerator?

c. Graph the function $C(t)$. Do you think that the cost will continue to decline indefinitely according to the given function? Why or why not?

27. Match each function with the description of its graph in the first quadrant.

$$\text{I. } f(x) = x^2 \quad \text{II. } f(x) = x^{-2} \quad \text{III. } f(x) = x^{1/2}$$

  a. Increasing and concave up

  b. Increasing and concave down

  c. Decreasing and concave up

28. In each pair, match the functions with their graphs in Figure 6.9.

29. a. Graph the functions

$$y_1 = x^{1/2},\ y_2 = x^{1/3},\ y_3 = x^{1/4},\ y_4 = x^{1/5}$$

  in the window

$$\text{Xmin} = 0 \qquad \text{Xmax} = 100$$
$$\text{Ymin} = 0 \qquad \text{Ymax} = 10$$

  What do you observe?

b. Use your graphs to evaluate $100^{1/2}$, $100^{1/3}$, $100^{1/4}$, and $100^{1/5}$.

c. Use your calculator to evaluate $100^{1/n}$ for $n = 10$, $n = 100$, and $n = 1000$. What happens when n gets large?

30. a. Graph the functions

$$y_1 = x^{1/2},\ y_2 = x^{1/3},\ y_3 = x^{1/4},\ y_4 = x^{1/5}$$

  in the window

$$\text{Xmin} = 0 \qquad \text{Xmax} = 1$$
$$\text{Ymin} = 0 \qquad \text{Ymax} = 1$$

  What do you observe?

b. Use your graphs to evaluate $0.5^{1/2}$, $0.5^{1/3}$, $0.5^{1/4}$, and $0.5^{1/5}$.

c. Use your calculator to evaluate $0.5^{1/n}$ for $n = 10$, $n = 100$, and $n = 1000$. What happens when n gets large?

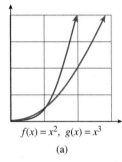

$f(x) = x^2,\ g(x) = x^3$

(a)

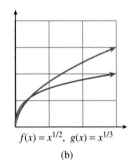

$f(x) = x^{1/2},\ g(x) = x^{1/3}$

(b)

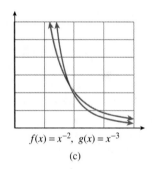

$f(x) = x^{-2},\ g(x) = x^{-3}$

(c)

FIGURE 6.9

Graph each set of functions in the given window. What do you observe?

31. $y_1 = \sqrt{x},\ y_2 = x^2,\ y^3 = x$

$$\text{Xmin} = 0 \qquad \text{Xmax} = 4$$
$$\text{Ymin} = 0 \qquad \text{Ymax} = 4$$

32. $y_1 = \sqrt[3]{x},\ y_2 = x^3,\ y^3 = x$

$$\text{Xmin} = -4 \qquad \text{Xmax} = 4$$
$$\text{Ymin} = -4 \qquad \text{Ymax} = 4$$

33. $y_1 = \sqrt[5]{x},\ y_2 = x^5,\ y^3 = x$

$$\text{Xmin} = -2 \qquad \text{Xmax} = 2$$
$$\text{Ymin} = -2 \qquad \text{Ymax} = 2$$

34. $y_1 = \sqrt[4]{x},\ y_2 = x^4,\ y^3 = x$

$$\text{Xmin} = 0 \qquad \text{Xmax} = 2$$
$$\text{Ymin} = 0 \qquad \text{Ymax} = 2$$

35. a. Write $\sqrt{x}$ with a fractional exponent.

  b. Write $\sqrt{\sqrt{x}}$ with fractional exponents.

  c. Use the laws of exponents to show that $\sqrt{\sqrt{x}} = \sqrt[4]{x}$.

36. a. Write $\sqrt[3]{x}$ with a fractional exponent.

  b. Write $\sqrt{\sqrt[3]{x}}$ with a fractional exponent.

  c. Use the laws of exponents to show that $\sqrt{\sqrt[3]{x}} = \sqrt[6]{x}$.

Simplify.

37. a. $\left(\sqrt[3]{125}\right)^3$ **b.** $\left(\sqrt[4]{2}\right)^4$
  c. $\left(3\sqrt{7}\right)^2$ **d.** $\left(-x^2\sqrt[3]{2x}\right)^3$

38. a. $\left(\sqrt[4]{16}\right)^4$ **b.** $\left(\sqrt[3]{6}\right)^3$
  c. $\left(2\sqrt[3]{12}\right)^3$ **d.** $\left(-a^3\sqrt[4]{a^2}\right)^4$

Solve.

39. $2\sqrt[3]{x} - 5 = -17$ **40.** $3\sqrt[3]{x} + 1 = -11$ **41.** $4(x + 2)^{1/5} = 12$ **42.** $-9(x - 3)^{1/5} = 18$

43. $(2x - 3)^{-1/4} = \dfrac{1}{2}$ **44.** $(5x + 2)^{-1/3} = \dfrac{1}{4}$ **45.** $\sqrt[3]{x^2 - 3} = 3$ **46.** $\sqrt[4]{x^3 - 7} = 2$

47. $\sqrt[3]{2x^2 - 15x} = 5$ **48.** $\sqrt[3]{2x^2 - 11x} = 6$

49. If the radius and height of a right circular cylinder are both equal to a, then its volume is given by the function

$$V = f(a) = \pi a^3$$

a. Complete the table of values, and sketch a graph of f.

a	1	2	4	8
V				

b. Find the height of such a cylinder if its volume is 816.8 cubic meters.

50. The power generated by a windmill is given in watts by the function

$$P = g(v) = 0.015v^3$$

where v is the velocity of the wind.

a. Complete the table of values and sketch a graph of g.

v	10	20	30	40
P				

b. Find the wind velocity needed to generate 500 watts of power.

51. The Stefan-Boltzmann law says that the temperature, T, of the sun can be computed, in degrees Kelvin, from the following formula.

$$sT^4 = \dfrac{L}{4\pi R^2}$$

where $L = 3.9 \times 10^{33}$ is the total luminosity of the sun, $R = 9.96 \times 10^{10}$ centimeters is the radius of the sun, and $s = 5.7 \times 10^{-5}$ is a constant governing radiation. Calculate the temperature of the sun.

52. Poiseuille's law for the flow of liquid through a tube can be used to describe blood flow through an artery. The rate of flow F in liters per minute is given by

$$F = \dfrac{kr^4}{L}$$

where r is the radius of the artery in centimeters, L is its length in centimeters, and $k = 7.8 \times 10^5$ is a constant determined by the blood pressure and viscosity. If a certain artery is 20 centimeters long, what should its radius be in order to allow a blood flow of 5 liters per minute?

Solve each formula for the specified variable.

53. $r = \sqrt[3]{\dfrac{3V}{4\pi}}$ for V **54.** $d = \sqrt[3]{\dfrac{16Mr^2}{m}}$ for M **55.** $R = \sqrt[4]{\dfrac{8Lvf}{\pi p}}$ for p **56.** $T = \sqrt[4]{\dfrac{E}{SA}}$ for A

57. Elasticity is the property of an object that causes it to regain its original shape after being compressed or deformed. One measure of elasticity considers how high the object bounces when dropped onto a hard surface,

$$e = \sqrt{\dfrac{\text{height bounced}}{\text{height dropped}}}$$

a. The table gives the value of e for various types of balls. Calculate the bounce height for each ball when it is dropped from a height of 6 feet onto a wooden floor.

b. Write a formula for e in terms of H, the bounce height, for the data in (a).

Type of ball	Bounce height	e
Baseball		0.50
Basketball		0.75
Golf ball		0.60
Handball		0.80
Softball		0.55
Superball		0.90
Tennisball		0.74
Volleyball		0.75

c. Graph the function from part (b).

d. If Ball A has twice the elasticity of Ball B, how much higher will Ball A bounce than Ball B?

Source: Davis, Kimmet, and Autry, 1986

58. The tone produced by a vibrating string depends on the frequency of the vibration. The frequency in turn depends on the length of the string, its weight, and its tension. In 1636 Marin Mersenne quantified these relationships as follows. The frequency, f, of the vibration is

   i. inversely proportional to the string's length, L

   ii. directly proportional to the square root of the string's tension, T

   iii. inversely proportional to the square root of the string's weight per unit length, w.

a. Write a formula for f that summarizes Mersenne's laws.

b. Sketch a graph of f as a function of L, assuming that T and w are constant. (You do not have enough information to put scales on the axes, but you can show the shape of the graph.)

c. On a piano, the frequency of the highest note is about 4200 Hertz. This frequency is 150 times the frequency of the lowest note, at about 28 Hertz. Ideally, only the lengths of the strings should change, so that all the notes have the same tonal quality. If the string for the highest note is 5 centimeters long, how long should the string for the lowest note be?

d. Sketch a graph of f as a function of T, assuming that L and w are constant.

e. Sketch a graph of f as a function of w, assuming that L and T are constant.

f. The tension of all the strings in a piano should be about the same, to avoid warping the frame. Suggest another way to produce a lower note. (*Hint:* Look at a piano's strings.)

g. The longest string on the piano in part (c) is 133.5 cm long. How much heavier (per unit length) is the longest string than the shortest string?

Source: Berg and Stork, 1982

Write each expression as a sum of terms of the form kx^p.

59. $\dfrac{\sqrt{x}}{4} - \dfrac{2}{\sqrt{x}} + \dfrac{x}{\sqrt{2}}$

60. $\dfrac{\sqrt{3}}{x} + \dfrac{3}{\sqrt{x}} - \dfrac{\sqrt{x}}{3}$

61. $\dfrac{6 - \sqrt[3]{x}}{2\sqrt[3]{x}}$

62. $\dfrac{\sqrt[4]{x} + 2}{2\sqrt[4]{x}}$

63. $x^{-0.5}(x + x^{0.25} - x^{0.5})$

64. $x^{0.5}(x^{-1} + x^{-0.5} + x^{-0.25})$

6.3 Rational Exponents

Powers of the Form $a^{m/n}$

In the last section we considered powers of the form $a^{1/n}$, such as $x^{1/3}$ and $x^{-1/4}$, and saw that $a^{1/n}$ is equivalent to the root $\sqrt[n]{a}$. What about other fractional exponents? What meaning can we attach to a power of the form $a^{m/n}$?

Consider the power $x^{3/2}$. Notice that the exponent $\frac{3}{2} = 3(\frac{1}{2})$, and thus by the third law of exponents we can write

$$\left(x^{1/2}\right)^3 = x^{(1/2)3} = x^{3/2}$$

In other words, we can compute $x^{3/2}$ by first taking the square root of x, and then cubing the result. For example,

$$100^{3/2} = \left(100^{1/2}\right)^3 \qquad \text{Take the square root of 100.}$$
$$= 10^3 = 1000 \qquad \text{Cube the result.}$$

We will define fractional exponents only when the base is a positive number.

$$a^{m/n} = (a^{1/n})^m = (a^m)^{1/n}, \qquad a > 0$$

To compute $a^{m/n}$, we can compute the nth root first, or the mth power, whichever is easier. For example,

$$8^{2/3} = (8^{1/3})^2 = 2^2 = 4$$

or

$$8^{2/3} = (8^2)^{1/3} = 64^{1/3} = 4$$

EXAMPLE 1

a. $81^{3/4} = (81^{1/4})^3$
$= 3^3 = 27$

b. $-27^{5/3} = -(27^{1/3})^5$
$= -3^5 = -243$

c. $27^{-2/3} = \dfrac{1}{(27^{1/3})^2}$
$= \dfrac{1}{3^2} = \dfrac{1}{9}$

d. $5^{3/2} = (5^{1/2})^3$
$\approx (2.236)^3 \approx 11.180$ ∎

You can verify all the calculations in Example 1 on your calculator. For example, to evaluate $81^{3/4}$, key in

81 ∧ (3 ÷ 4) **ENTER**

or simply

81 ∧ 0.75 **ENTER**

EXERCISE 1 Evaluate each power.

a. $32^{-3/5}$

b. $-81^{1.25}$

Power Functions

Perhaps the single most useful piece of information a scientist can have about an animal is its metabolic rate. The metabolic rate is the amount of energy the animal uses per unit of time for its usual activities, including locomotion, thermoregulation, growth, and reproduction. The basal metabolic rate, or BMR, sometimes called the resting metabolic rate, is the minimum amount of energy the animal can expend in order to survive.

EXAMPLE 2

Kleiber's rule states that the basal metabolic rate for many groups of animals is given by

$$B(m) = 68m^{0.75}$$

where m is the mass of the animal in kilograms and the BMR is measured in kilocalories per day.

a. Calculate the BMR for various animals whose masses are given in the table.

Animal	Bat	Squirrel	Raccoon	Lynx	Human	Moose	Rhinoceros
Weight (kg)	0.1	0.6	8	30	70	360	3500
BMR (kcal/day)							

b. Sketch a graph of Kleiber's rule for $0 < m \le 400$.

c. Do larger species eat more or less, relative to their body mass, than smaller ones?

Solutions

a. Evaluate the function for the values of m given. For example, to calculate the BMR of a bat, we compute

$$B(0.1) = 68(0.1)^{0.75} = 12.1$$

A bat expends, and hence must consume, at least 12 kilocalories per day. Evaluate the function to complete the rest of the table.

Animal	Bat	Squirrel	Raccoon	Lynx	Human	Moose	Rhinoceros
Weight (kg)	0.1	0.6	8	30	70	360	3500
BMR (kcal/day)	12	46	323	872	1646	5620	30,943

b. Plot the data from the table to obtain the graph in Figure 6.10.

c. If energy consumption were proportional to body weight, the graph would be a straight line. But because the exponent in Kleiber's rule, $\frac{3}{4}$, is less than 1, the graph is concave down, or bends downward. Therefore, larger species eat less than smaller ones, relative to their body weight. For example, a moose weighs 600 times as much as a squirrel, but its energy requirement is only 122 times the squirrel's.

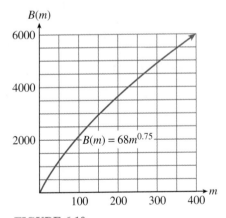

FIGURE 6.10

EXERCISE 2

a. Complete the table of values for the function $f(x) = x^{-3/4}$.

x	0.1	0.2	0.5	1	2	5	8	10
$f(x)$								

b. Sketch the graph of the function.

Radical Notation

Because $a^{1/n} = \sqrt[n]{a}$, we can write any power with a fractional exponent in radical form as follows

$$a^{m/n} = \sqrt[n]{a^m} = (\sqrt[n]{a})^m$$

EXAMPLE 3

a. $125^{4/3} = \sqrt[3]{125^4}$ or $\left(\sqrt[3]{125}\right)^4$

b. $x^{0.4} = x^{2/5} = \sqrt[5]{x^2}$

c. $6w^{-3/4} = \dfrac{6}{\sqrt[4]{w^3}}$ ∎

Usually we will want to convert from radical notation to fractional exponents, because exponential notation is easier to use.

EXAMPLE 4

a. $\sqrt{x^5} = x^{5/2}$

b. $5\sqrt[4]{p^3} = 5p^{3/4}$

c. $\dfrac{3}{\sqrt[5]{t^2}} = 3t^{-2/5}$

d. $\sqrt[3]{2y^2} = (2y^2)^{1/3} = 2^{1/3}y^{2/3}$ ∎

EXERCISE 3 Convert to exponential notation.

a. $\sqrt[3]{6w^2}$

b. $\sqrt[4]{\dfrac{v^3}{s^5}}$

Operations with Rational Exponents

Powers with rational exponents—positive, negative, or zero—obey the laws of exponents discussed in Section 6.1. You may want to review those laws before studying the following examples.

EXAMPLE 5

a. $\dfrac{7^{0.75}}{7^{0.5}} = 7^{0.75-0.5} = 7^{0.25}$ Apply the second law of exponents.

b. $v \cdot v^{-2/3} = v^{1+(-2/3)} = v^{1/3}$ Apply the first law of exponents.

c. $(x^8)^{0.5} = x^{8(0.5)} = x^4$ Apply the third law of exponents.

d. $\dfrac{(5^{1/2}y^2)^2}{(5^{2/3}y)^3} = \dfrac{5y^4}{5^2y^3}$ Apply the fourth law of exponents.

$$= \dfrac{y^{4-3}}{5^{2-1}} = \dfrac{y}{5}$$ Apply the second law of exponents. ∎

EXERCISE 4   Simplify by applying the laws of exponents.

a. $x^{1/3}(x + x^{2/3})$ **b.** $\dfrac{n^{9/4}}{4n^{3/4}}$

Solving Equations

According to the third law of exponents, when we raise a power to another power, we multiply the exponents together. In particular, if the two exponents are reciprocals, then their product is 1. For example,

$$(x^{2/3})^{3/2} = x^{(2/3)(3/2)} = x^1 = x$$

This observation can help us to solve equations involving fractional exponents. For instance, to solve the equation

$$x^{2/3} = 4$$

we raise both sides of the equation to the reciprocal power, 3/2. This gives us

$$(x^{2/3})^{3/2} = 4^{3/2}$$
$$x = 8$$

EXAMPLE 6

Solve $(2x + 1)^{3/4} = 27$.

Solution

Raise both sides of the equation to the reciprocal power, $\frac{4}{3}$.

$$[(2x + 1)^{3/4}]^{4/3} = 27^{4/3}$$ Apply the third law of exponents.
$$2x + 1 = 81$$ Solve as usual.
$$x = 40$$ ∎

EXERCISE 5   Solve the equation $3.2z^{0.6} - 9.7 = 8.7$. Round your answer to two decimal places.

Isolate the power.
Raise both sides to the reciprocal power.

EXAMPLE 7

Small animals such as bats cannot survive for long without eating. The graph in Figure 6.11 shows how the weight, W, of a typical vampire bat decreases over time until its next meal. The curve is the graph of the function

$$W(h) = 130.25h^{-0.126}$$

where h is the number of hours since the bat's most recent meal.

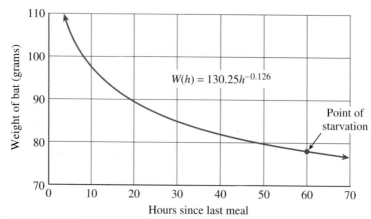

FIGURE 6.11
Source: Wilkinson, 1984

a. How long can the bat survive after eating until its next meal? What is the bat's weight at the point of starvation?

b. When the bat's weight has dropped to 90 grams, how long can it survive before eating again?

c. Fill in the table with the number of hours since eating at various weights.

Weight (grams)	97.5	92.5	85	80
Hours since eating				
Point on graph	A	B	C	D

d. Compute the slope from point A to point B, and from point C to point D.

e. Vampire bats sometimes donate blood (through regurgitation) to other bats that are close to starvation. Suppose a bat at point A on the curve donates 5 grams of blood to a bat at point D. Explain why this strategy is effective for the survival of the bat community.

Solutions

a. By reading the coordinates at the point of starvation, we see that the bat can survive without eating for about 60 hours, and its weight at starvation would have dropped to about 78 grams. We can verify this estimate by evaluating the function at $h = 60$.

$$W(60) = 130.25(60)^{-0.126}$$
$$= 130.25(0.597) \approx 77.8$$

b. The h-coordinate of the point where $W = 90$ is approximately 19, so the bat can survive for another 41 hours. Alternatively, we can solve the equation

$$90 = 130.25h^{-0.126}$$

for h. First divide both sides by 130.25 to get

$$0.69098 = h^{-0.126}$$

Then raise both sides of the equation to the reciprocal of -0.126.

$$0.69098^{-1/0.126} = (h^{-0.126})^{-1/0.126}$$
$$18.8 \approx h$$

Approximately 19 hours have elapsed since the bat last ate, so it can survive for another 41 hours.

c. Read the values from the graph, or solve the equation $W = 130.25h^{-0.126}$ for the given values of W the same way we did in part (b). The points are labeled on the graph in Figure 6.12.

Weight (grams)	97.5	92.5	85	80
Hours since eating	10.0	15.1	29.6	47.9
Point on graph	A	B	C	D

d. The slope from point A to point B is

$$\frac{92.5 - 97.5}{15.1 - 10.0} = \frac{-5 \text{ grams}}{5.1 \text{ hours}}$$

or -0.98 grams per hour; the slope from point C to point D is

$$\frac{80 - 85}{47.9 - 29.6} = \frac{-5 \text{ grams}}{18.3 \text{ hours}}$$

or -0.27 grams per hour.

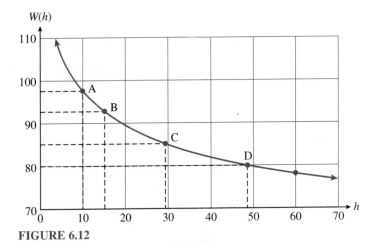

FIGURE 6.12

e. For the same amount of food, the recipient gains 18.3 hours of life, while the donor loses only 5.1 hours. The slope between the points A and B is steeper than the slope between C and D, so a vertical change of 5 grams at point D corresponds to a larger time interval than a change of 5 grams at point A. ■

1. What does the notation $a^{m/n}$ mean?
2. Explain how to evaluate the function $f(x) = x^{-3/4}$ for $x = 625$, without using a calculator.
3. Explain why $x\sqrt{x} = x^{1.5}$.
4. What is the first step in solving the equation $(x - 2)^{-5/2} = 1.8$?

ANSWERS TO 6.3 EXERCISES

1a. $\dfrac{1}{8}$ **b.** -243

2a.

x	0.1	0.2	0.5	1	2	5	8	10
$f(x)$	5.623	3.344	1.682	1	0.595	0.299	0.210	0.178

b.

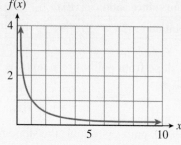

3a. $6^{1/3}w^{2/3}$ **b.** $v^{3/4}s^{-5/4}$

4a. $x^{4/3} + x$ **b.** $\dfrac{n^{3/2}}{4}$ **5.** 18.45

HOMEWORK 6.3

For these problems, assume that all variables represent positive numbers.
Evaluate each power.

1. **a.** $81^{3/4}$ **b.** $125^{2/3}$ **c.** $625^{0.75}$ 2. **a.** $-8^{2/3}$ **b.** $-64^{2/3}$ **c.** $243^{0.4}$
3. **a.** $16^{-3/2}$ **b.** $8^{-4/3}$ **c.** $32^{-1.6}$ 4. **a.** $-125^{-4/3}$ **b.** $-32^{-3/5}$ **c.** $100^{-2.5}$

Write each power in radical form.

5. **a.** $x^{4/5}$ **b.** $b^{-5/6}$ **c.** $(pq)^{-2/3}$ 6. **a.** $y^{3/4}$ **b.** $a^{-2/7}$ **c.** $(st)^{-3/5}$
7. **a.** $3x^{0.4}$ **b.** $4z^{-4/3}$ **c.** $-2x^{0.25}y^{0.75}$ 8. **a.** $5y^{2/3}$ **b.** $6w^{-1.5}$ **c.** $-3x^{0.4}y^{0.6}$

Write each expression with fractional exponents.

9. **a.** $\sqrt[3]{x^2}$ **b.** $2\sqrt[5]{ab^3}$ **c.** $\dfrac{-4m}{\sqrt[6]{p^7}}$ 10. **a.** $\sqrt{y^3}$ **b.** $6\sqrt[5]{(ab)^3}$ **c.** $\dfrac{-2n}{\sqrt[8]{q^{11}}}$

11. **a.** $\sqrt[3]{(ab)^2}$ **b.** $\dfrac{8}{\sqrt[4]{x^3}}$ **c.** $\dfrac{R}{3\sqrt{TK^5}}$ 12. **a.** $\sqrt[3]{ab^2}$ **b.** $\dfrac{5}{\sqrt[3]{y^2}}$ **c.** $\dfrac{S}{4\sqrt{VH^3}}$

Evaluate each root without using a calculator.

13. a. $\sqrt[5]{32^3}$ **b.** $-\sqrt[3]{27^4}$ **c.** $\sqrt[4]{16y^{12}}$

14. a. $\sqrt[4]{16^5}$ **b.** $-\sqrt[3]{125^2}$ **c.** $\sqrt[5]{243x^{10}}$

15. a. $-\sqrt{a^8b^{16}}$ **b.** $\sqrt[3]{8x^9y^{27}}$ **c.** $-\sqrt[4]{81a^8b^{12}}$

16. a. $-\sqrt{a^{10}b^{36}}$ **b.** $\sqrt[3]{64x^6y^{18}}$ **c.** $-\sqrt[5]{32x^{25}y^5}$

Use a calculator to approximate each power or root to the nearest thousandth.

17. a. $12^{5/6}$ **b.** $\sqrt[3]{6^4}$ **c.** $37^{-2/3}$ **d.** $4.7^{2.3}$

18. a. $20^{5/4}$ **b.** $\sqrt[5]{8^3}$ **c.** $128^{-3/4}$ **d.** $16.1^{0.29}$

19. During a flu epidemic in a small town, health officials estimate that the number of people infected t days after the first case was discovered is given by

$$I(t) = 50t^{3/5}$$

a. Complete the table of values.

t	5	10	15	20
$I(t)$				

b. How long will it be before 300 people are ill?

c. Graph the function $I(t)$, and verify your answer to part (b) on your graph.

20. The research division of an advertising firm estimates that the number of people who have seen their ads t days after the campaign begins is given by the function

$$N(t) = 2000t^{5/4}$$

a. Complete the table of values.

t	6	10	14	20
$N(t)$				

b. How long will it be before 75,000 people have seen the ads?

c. Graph the function $N(t)$ and verify your answer to part (b) on your graph.

Graph each set of power functions in the suggested window, and compare the graphs.

21. $y_1 = x$, $y_2 = x^{5/4}$, $y_3 = x^{3/2}$, $y_4 = x^2$, $y_5 = x^{5/2}$

        Xmin $= 0$ Xmax $= 6$
        Ymin $= 0$ Ymax $= 10$

22. $y_1 = x^{2/5}$, $y_2 = x^{1/2}$, $y_3 = x^{2/3}$, $y_4 = x^{3/4}$, $y_5 = x$

        Xmin $= 0$ Xmax $= 6$
        Ymin $= 0$ Ymax $= 4$

23. The "surface to volume ratio" is important in studying how organisms grow, and why animals of different size have different characteristics.

a. Write formulas for the volume, V, and the surface area, A, of a cube in terms of its length, L.

b. Express the length of the cube as a function of its volume. Express the length of the cube as a function of its surface area.

c. Express the surface area of the cube as a function of its volume.

d. Express the surface to volume ratio of a cube in terms of its length. What happens to the surface to volume ratio as L increases?

24. Repeat Problem 23 for the volume and surface area of a sphere in terms of its radius, R.

a. Write formulas for the volume, V, and the surface area, A, of a sphere in terms of its radius, R.

b. Express the radius of the sphere as a function of its volume. Express the radius of the sphere as a function of its surface area.

c. Express the surface area of the sphere as a function of its volume.

d. Express the surface to volume ratio of a sphere in terms of its radius. What happens to the surface to volume ratio as R increases?

25. The drainage basin of a river channel is the area of land that contributes water to the river. Table 6.2 gives the lengths in miles of some of the world's largest rivers and the areas of their drainage basins in square miles.

TABLE 6.2

River	Area of drainage basin	Length
Amazon	2,700,000	4200
Nile	1,400,000	4200
Mississippi	1,300,000	4100
Yangtze	580,000	2900
Volga	480,000	2300
St. Lawrence	460,000	1900
Ganges	440,000	1400
Orinoco	380,000	1400
Indus	360,000	2000
Danube	350,000	1800
Colorado	250,000	1700
Platte	72,000	800
Rhine	63,000	900
Seine	48,000	500
Delaware	12,000	200

a. Plot the data, using units of 100,000 on the horizontal axis and units of 500 on the vertical axis.

b. The length, L, of the channel is related to the area, A, of its drainage basin according to the formula

$$L = 1.05A^{0.58}$$

Graph this function on top of the data points.

c. The drainage basin for the Congo covers about 1,600,000 square miles. Estimate the length of the Congo River.

d. The Rio Grande is 1700 miles long. What is the area of its drainage basin?

Source: Leopold, Wolman, and Miller 1992.

26. In the 1970s Jared Diamond studied the number of bird species on small islands near New Guinea. He found that larger islands support a larger number of different species, according to the formula

$$S = 15.1A^{0.22}$$

where S is the number of species on an island of area A square kilometers.

a. Fill in the table.

A	10	100	1000	5000	10,000
S					

b. Graph the function on the domain $0 < A \leq 10,000$.

c. How many species of birds would you expect to find on Manus Island, with an area of 2100 square kilometers? On Lavongai, whose area is 1140 square kilometers?

d. How large must an island be in order to support 200 different species of bird?

Source: Chapman and Reiss, 1992.

27. The table below shows the exponent, p, in the allometric equation

$$\text{variable} = k \, (\text{body mass})^p$$

for some different variables related to mammals.

Variable	Exponent, p
Home range size	1.26
Lung volume	1.02
Brain mass	0.70
Respiration rate	−0.26

a. Match each equation to one of the graphs shown in Figure 6.13 on page 379.

b. Explain how the value of p in the allometric equation determines the shape of the graph. Consider three cases $p > 1$, $0 < p < 1$, and $p < 0$.

Source: Chapman and Reiss, 1992.

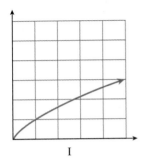
I

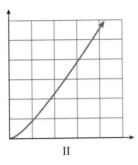
II

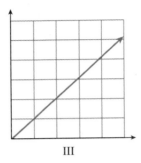
III

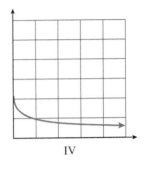
IV

FIGURE 6.13

28. The average body mass of a dolphin is about 150 kilograms, twice the body mass of an average human male.

  a. Using the allometric equations in Problem 27, calculate the ratio of the brain mass of a dolphin to that of a human.

  b. A good-sized brown bear weighs about 300 kilograms, twice the weight of a dolphin. Calculate the ratio of the brain mass of a brown bear to that of a dolphin.

  c. Use a ratio to compare the heartbeat frequencies of a dolphin and a human, and those of a brown bear and a dolphin. (See Example 6 of Section 6.2.)

29. The gourd species *Tricosanthes* grows according to the formula $L = ad^{2.2}$, where L is its length and d is its width. The species *Lagenaria* has the growth law $L = ad^{0.81}$.

  a. By comparing the exponents, which gourd grows into a long, thin shape, and which is relatively fatter? Which species is called the snake gourd, and which is the bottle gourd?

  b. The snake gourd reaches a length of 2 meters (200 cm) with a diameter of only 4 cm. Find the value of a in its growth law.

  c. The bottle gourd is 10 cm long and 7 cm in diameter at maturity. Find the value of a in its growth law.

  d. The giant bottle gourd grows to a length of 23 cm with a diameter of 20 cm. Does it grow according to the same law as standard bottle gourds?

Source: Burton, 1998.

30. As a fiddler crab grows, one claw (called the chela) grows much faster than the rest of the body. The table shows the mass of the chela, C, versus the mass of the rest of the body, b, for a number of fiddler crabs.

b	65	110	170	205	300	360	615
C	6	15	30	40	68	110	240

  a. Plot the data.

  b. On the same axes, graph the function $C = 0.007b^{1.63}$. Describe how the function fits the data.

  c. Using the function in part (b), predict the chela mass of a fiddler crab if the rest of its body weighs 400 mg.

  d. The chela from a fiddler crab weighs 250 mg. How much does the rest of its body weigh?

  e. As the body mass of a fiddler crab doubles from 100 mg to 200 mg, by what factor does the mass of its chela increase? How about as the body mass doubles from 200 mg to 400 mg?

Source: Burton, 1998.

31. Are world record times for track events proportional to the length of the race? Table 6.3 gives the men's and women's world records in 1999 for races from 1 kilometer to 30 kilometers in length.

TABLE 6.3

Distance (km)	Men's record (min)	Women's record (min)
1	2.203	2.483
1.5	3.433	3.841
2	4.798	5.423
3	7.345	8.102
5	12.656	14.468
10	26.379	29.530
20	56.927	66.813
25	73.93	89.487
30	89.313	107.093

a. On separate graphs, plot the men's and women's times against distance. Does time appear to be proportional to distance?

b. Use slopes to decide whether the graphs of time versus distance are in fact linear.

c. Both sets of data can be modeled by power functions of the form $t = kx^b$, where b is called the "fatigue index." Graph the function $M(x) = 2.22x^{1.085}$ over the men's data points, and $W(x) = 2.45x^{1.106}$ over the women's data. Describe how the graphs of the two functions differ. Explain why b is called the fatigue index.

32. Fell running is a popular sport in the hills or "fells" of the British Isles. Fell running records depend on the altitude gain over the course of the race as well as its length. The equivalent horizontal distance for a race of length x kilometers with an ascent of y kilometers is given by $x + Ny$, where N is Naismith's number (see Problem 41 of Section 1.3.) The record times for women's races are approximated in minutes by $t = 2.43(x + 9.5y)^{1.15}$, and men's times by $t = 2.18(x + 8.0y)^{1.14}$.

a. Whose times show a greater fatigue index, men or women? (See Problem 31.)

b. Whose times are more strongly affected by ascents?

c. Predict the winning times for both men and women in a 56-kilometer race with an ascent of 2750 meters.

Source: Scarf, 1998.

33. The climate of a region has a great influence on the types of animals that can survive there. Extremes of temperature create difficult living conditions, so the diversity of wildlife decreases as the annual variability in temperature increases. Along the west coast of North America, the number of species of mammals, M, is approximately related to the temperature range, R (in degrees Celsius), by the function $M = f(R) = 433.8R^{-0.742}$.

a. Use your calculator to sketch the graph of this function for temperature ranges up to 30°C.

b. How many species would you expect to find in a region where the temperature range is 10°C? Label the corresponding point on your graph.

c. If 50 different species are found in a certain region, what temperature range would you expect the region to experience? Label the corresponding point on your graph.

d. Evaluate the function to find $f(9)$, $f(10)$, $f(19)$, and $f(20)$. What do these values represent? Calculate the change in the number of species as the temperature range increases from 9°C to 10°C, and from 19°C to 20°C. Which 1° increase results in a greater decrease in diversity? Explain your answer in terms of slopes on your graph.

Source: Chapman and Reiss, 1992.

34. A bicycle ergometer is used to measure the amount of power generated by a cyclist. The scatterplot in Figure 6.14 shows how long an athlete was able to sustain various levels of power output. The curve is the graph of $y = 500x^{-0.29}$, which approximately models the data.

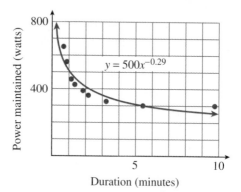

FIGURE 6.14

a. As represented in the graph, which variable is independent and which is dependent?

b. The athlete maintained 650 watts of power for 40 seconds. What power output does the equation predict for 40 seconds?

c. The athlete maintained 300 watts of power for 10 minutes. How long does the equation predict that power output can be maintained?

d. In 1979 a remarkable pedal-powered aircraft called the Gossamer Albatross was successfully flown across the English Channel. The flight took 3 hours. According to the equation, what level of power can be maintained for 3 hours?

e. The Gossamer Albatross needed 250 watts of power to keep it airborne. For how long can 250 watts be maintained, according to the given equation?

Source: Alexander, 1992.

35. In 1981 John Damuth collected data on the average body mass, m, and the average population density, D, for 307 species of herbivores. He found that, very roughly,

$$D = km^{-0.75}$$

Although the data are fairly scattered around this curve, it is a reasonable description of the relationship between population density and body mass.

a. Explain why you might expect an animal's rate of food consumption to be proportional to its metabolic rate. (See Example 2 for an explanation of metabolic rate.)

b. Explain why you might expect the population density of a species to be inversely proportional to the rate of food consumption of an individual animal.

c. Use Kleiber's rule from Example 2 and your answers to parts (a) and (b) to explain why Damuth's proposed formula for population density is reasonable.

d. Sketch a graph of the function D. You do not have enough information to put scales on the axes, but you can show the shape of the graph. (*Hint:* Graph the function for $k = 1$.)

Source: Burton, 1998.

36. The seeds of a given species of plant are often sown close together, either naturally or by design in a nursery or cultivated field. At first each plant grows unhindered, but as it gets larger, its roots and branches need more space, and the individual plants begin to compete with each other. Some plants die, and the density of plants decreases. Studies on pine plantations in the 1930s showed that, after competition begins, the average mass of an individual tree is a power function of the density of the trees per unit area, given by

$$M(d) = kd^{-1.5}$$

Subsequent studies have shown that the exponent varies somewhat from species to species, but the value -1.5 is common enough that the formula is known as the "$-\frac{3}{2}$ self-thinning law." To simplify the following argument, suppose that a pine tree is shaped like a tall circular cone, and that as it grows, its height is always a constant multiple of its base radius, r.

a. Explain why the area taken up by a single tree is inversely proportional to the plant density, d.

b. Explain why the base radius of a tree is proportional to the square root of the area the tree fills. Write r as a power function of d.

c. Write a formula for the volume of the tree in terms of its base radius, r. Use part (b) to write the volume as a power function of d.

d. Explain why the mass (or weight) of a pine tree is roughly proportional to its volume, and hence justify the self-thinning law.

e. Sketch a graph of the function M. You do not have enough information to put scales on the axes, but you can show the shape of the graph. (*Hint:* Graph the function for $k = 1$.)

Source: Chapman and Reiss, 1992.

Simplify by applying the laws of exponents. Write your answers with positive exponents only.

37. a. $4a^{6/5}a^{4/5}$ **b.** $9b^{4/3}b^{1/3}$

38. a. $(-2m^{2/3})^4$ **b.** $(-5n^{3/4})^3$

39. a. $\dfrac{8w^{9/4}}{2w^{3/4}}$ **b.** $\dfrac{12z^{11/3}}{4z^{5/3}}$

40. a. $(-3u^{5/3})(5u^{-2/3})$ **b.** $(-2v^{7/8})(-3v^{-3/8})$

41. a. $\dfrac{k^{3/4}}{2k}$ **b.** $\dfrac{4h^{2/3}}{3h}$

42. a. $c^{-2/3}\left(\dfrac{2}{3}c^2\right)$ **b.** $\dfrac{r^3}{4}(r^{-5/2})$

43. Birds' eggs typically lose 10–20% of their mass during incubation. The embryo metabolizes lipid during growth, and this process releases water vapor through the porous shell. The incubation time for birds' eggs is a function of the mass of the egg, and has been experimentally determined as

$$I(m) = 12.0m^{0.217}$$

where m is measured in grams and I in days.

a. Calculate the incubation time (to the nearest day) for the wren, whose eggs weigh about 2.5 grams, and the greylag goose, whose eggs weigh 46 grams.

b. The rate of water loss from the egg is also a function of its mass, and appears to follow the rule

$$W(m) = 0.015m^{0.742}$$

Combine the functions $I(m)$ and $W(m)$ to calculate the fraction of the initial egg mass that is lost during the entire incubation period.

c. Explain why your result shows that most eggs lose about 18% of their mass during incubation.

Source: Burton, 1998

44. The incubation time for birds' eggs is given by

$$I(m) = 12.0m^{0.217}$$

where m is measured in grams and I in days.

(See Problem 43.) Before hatching, the eggs take in oxygen at the rate of

$$O(m) = 22.2m^{0.77}$$

in milliliters per day.

a. Combine the functions $I(m)$ and $O(m)$ to calculate the total amount of oxygen taken in by the egg during its incubation.

b. Use your result from part (a) to explain why total oxygen consumption per unit mass is approximately inversely proportional to incubation time. (Oxygen consumption is a reliable indicator of metabolic rate, and it is reasonable that incubation time should be inversely proportional to metabolic rate.)

c. Predict the oxygen consumption per gram of a herring gull's eggs, given that their incubation time is 26 days. (The actual value is 11 milliliters per day.)

Source: Burton, 1998

Solve. Round your answers to the nearest thousandth if necessary.

45. $x^{2/3} - 1 = 15$

46. $x^{3/4} + 3 = 11$

47. $x^{-2/5} = 9$

48. $x^{-3/2} = 8$

49. $2(5.2 - x^{5/3}) = 1.4$

50. $3(8.6 - x^{5/2}) = 6.5$

51. Kepler's law gives a relation between the period, p, of a planet's revolution, in years, and its average distance, a, from the sun:

$$p^2 = Ka^3$$

where $K = 1.243 \times 10^{-24}$, a is measured in miles, and p in years.

a. Solve Kepler's law for p in terms of a.

Use the distributive law to find the product.

57. $2x^{1/2}(x - x^{1/2})$

58. $x^{1/3}(2x^{2/3} - x^{1/3})$

59. $\dfrac{1}{2}y^{-1/3}(y^{2/3} + 3y^{-5/6})$

60. $3y^{-3/8}\left(\dfrac{1}{4}y^{-1/4} + y^{3/4}\right)$

61. $(2x^{1/4} + 1)(x^{1/4} - 1)$

62. $(2x^{1/3} - 1)(x^{1/3} + 1)$

63. $(a^{3/4} - 2)^2$

64. $(a^{2/3} + 3)^2$

b. Find the period of Mars if its average distance from the sun is 1.417×10^8 miles.

52. Refer to Kepler's law, $p^2 = Ka^3$, in Problem 51.

a. Solve Kepler's law for a in terms of p.

b. Find the distance from Venus to the sun if its period is 0.615 years.

53. If $f(x) = (3x - 4)^{3/2}$, find x so that $f(x) = 27$.

54. If $g(x) = (6x - 2)^{5/3}$, find x so that $g(x) = 32$.

55. If $S(x) = 12x^{-5/4}$, find x so that $S(x) = 20$.

56. If $T(x) = 9x^{-6/5}$, find x so that $T(x) = 15$.

Factor out the smallest power from each expression. Write your answers with positive exponents only.

65. $x^{3/2} + x = x(?)$

66. $y - y^{2/3} = y^{2/3}(?)$

67. $y^{3/4} - y^{-1/4} = y^{-1/4}(?)$

68. $x^{-3/2} + x^{-1/2} = x^{-3/2}(?)$

69. $a^{1/3} + 3 - a^{-1/3} = a^{-1/3}(?)$

70. $3b - b^{3/4} + 4b^{-3/4} = a^{-3/4}(?)$

Write each expression without using negative exponents.

1. $4 \cdot 10^{-3}$

2. $7a^{-5}$

Solve each equation.

3. $6t^{-3} = \dfrac{3}{500}$

4. $3.5 - 2.4p^{-2} = -6.1$

Write each function in the form $y = kx^p$.

5. $f(x) = \dfrac{2}{3x^4}$

6. $g(x) = \dfrac{8x^7}{29}$

7. a. Plot the data. Does it appear that v is approximately proportional to u^2?

u	0.2	0.3	0.4	0.5	0.6	0.7	0.8
v	0.148	0.333	0.592	0.925	1.332	1.813	2.368

  b. Use the transformation $x = u^2$ to complete the table of values.

x							
v							

  c. Plot the transformed data (x, v). Do the data appear approximately linear? Fit a linear equation to the data points.

  d. Use your equation in x and v to write a formula for v as a function of u. Graph this function on the plot of the original data, and evaluate the fit.

8. Plot the data and describe the shape of the graph.

a	0.2	0.3	0.4	0.5	0.6	0.7	0.8
b	0.422	0.570	0.692	0.793	0.880	0.953	1.018

  b. Use the transformations $x = \dfrac{1}{a}$ and $y = \dfrac{1}{b}$ to complete the table of values.

x							
y							

  c. Plot the transformed data (x, y). Do the data appear approximately linear? Fit a linear equation to the data points.

  d. Use your equation in x and v to write an equation of the form $\dfrac{1}{b} = m \cdot \dfrac{1}{a} + k$. Graph this function on the plot of the original data, and evaluate the fit.

Write each expression without using negative exponents.

9. $\dfrac{3w^{-1}}{t^{-7}w^2}$

10. $\left(\dfrac{5}{n}\right)^{-2}$

Write each number in scientific notation.

11. 1,234,000

12. 0.0000567

Evaluate each radical.

13. $\sqrt[4]{0.0081}$

14. $\sqrt[5]{-7776}$

Write each power with radical notation and evaluate if possible.

15. $4096^{1/6}$

16. $0.0049^{0.5}$

17. $0.00243^{0.4}$

18. $512^{2/3}$

19. $-256^{1/4}$

20. $(-256)^{1/4}$

21. $(-1728)^{1/3}$

22. $-1728^{1/3}$

Convert to exponential notation.

23. $-2\sqrt[3]{5m}$

24. $\dfrac{73}{\sqrt[5]{22q}}$

25. $\dfrac{9}{\sqrt[4]{(2p)^3}}$

26. $13\sqrt[7]{\dfrac{6q^2}{p^3}}$

Convert to radical notation.

27. $11h^{1/5}$

28. $10r^{0.25}$

29. $\dfrac{2}{3}c^{0.8}$

30. $-8d^{2/3}$

Simplify by applying the laws of exponents. Write your answers with positive exponents only.

31. $(7t)^3(7t)^{-1}$

32. $\dfrac{36r^{-2}s}{9r^{-3}s^4}$

33. $\dfrac{(2k^{-1})^{-4}}{4k^{-3}}$

34. $(2w^{-3})^{-2}(2w^{-3})^5(-5w^2)$

35. $\dfrac{8a^{-3/4}}{a^{-11/4}}$

36. $b^{2/3}(4b^{-2/3} - b^{1/3})$

Solve each equation. If the solution is irrational, round your answer to two decimal places.

37. $8\sqrt[4]{x + 6} = 24$

38. $9.8 = 7\sqrt[3]{z - 4}$

39. $\dfrac{2}{3}(2y + 1)^{0.2} = 6$

40. $1.3w^{0.3} + 4.7 = 5.2$

Complete the table of values and graph the given function.

41. $f(x) = x^{0.3}$

x	0	1	5	10	20	50	70	100
$f(x)$								

42. $f(x) = x^{-0.7}$

x	0.1	0.2	0.5	1	2	5	8	10
$f(x)$								

6.4 The Distance and Midpoint Formulas

Computing Distances in the Coordinate Plane

Many airplanes carry an inertial navigational system (INS), which computes the airplane's current position relative to its starting point. Suppose that Francine left Oldfield Airport in her light plane and is now 253 miles east and 124 miles north of the airport. She knows that Preston Airport is 187 miles east and 201 miles south of Oldfield Airport. Which airport is closer to Francine's present location?

We can sketch the situation on the xy-plane. If we place Oldfield Airport at the origin, then Francine is located at the point $F(253, 124)$, and Preston Airport is at $P(187, -201)$, as shown in Figure 6.15a.

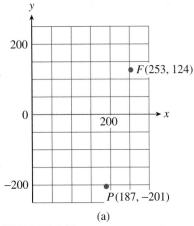

(a)

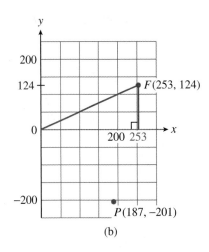

(b)

FIGURE 6.15

The distance *OF* from Francine to Oldfield Airport is the hypotenuse of a right triangle whose base is 253 miles and whose height is 124 miles. (See Figure 6.15b.) If we apply the Pythagorean theorem to this triangle, we find

$$(OF)^2 = 253^2 + 124^2$$
$$= 79{,}385$$
$$OF = \sqrt{79{,}385} \approx 281.7$$

Francine is approximately 282 miles from Oldfield Airport.

The distance *FP* from Francine to Preston Airport is the hypotenuse of another right triangle, as shown in Figure 6.16. To find the base of this triangle, we compute the difference in the *x*-coordinates of points *P* and *F*:

$$253 - 187 = 66$$

so the base is 66 miles long. To find the height of the triangle, we compute the difference in the *y*-coordinates:

$$124 - (-201) = 325$$

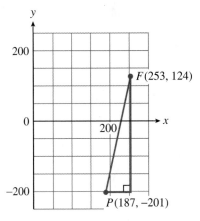

FIGURE 6.16

so the height is 325 miles. Applying the Pythagorean theorem again gives us

$$(FP)^2 = 66^2 + 325^2$$
$$= 109{,}981$$
$$FP = \sqrt{109{,}981} \approx 331.6$$

Francine is about 332 miles from Preston Airport. At the moment, she is closer to Oldfield Airport.

The Distance Formula

By using the Pythagorean theorem, we can derive a formula for the distance between two points, P_1 and P_2, in terms of their coordinates. We first label a right triangle, as we did in the example above. Draw a horizontal line through P_1 and a vertical line through P_2. These lines meet at a point P_3, as shown in Figure 6.17. The x-coordinate of P_3 is the same as the x-coordinate of P_2, and the y-coordinate of P_3 is the same as the y-coordinate of P_1. Thus, the coordinates of P_3 are (x_2, y_1).

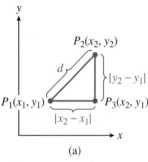

(a)   (b)

FIGURE 6.17

The distance between P_1 and P_3 is $|x_2 - x_1|$, and the distance between P_2 and P_3 is $|y_2 - y_1|$. These two numbers are the lengths of the legs of the right triangle. The length of the hypotenuse is the distance between P_1 and P_2, which we'll call d. By the Pythagorean theorem,

$$d^2 = (x_2 - x_1)^2 + (y_2 - y_1)^2$$

Taking the (positive) square root of each side of this equation gives us the **distance formula.**

The **distance** d between points $P_1(x_1, y_1)$ and $P_2(x_2, y_2)$ is

$$d = \sqrt{(x_2 - x_1)^2 + (y_2 - y_1)^2}$$

EXAMPLE 1 Find the distance between $(2, -1)$ and $(4, 3)$.

Solution

Substitute $(2, -1)$ for (x_1, y_1) and $(4, 3)$ for (x_2, y_2) in the distance formula to obtain

$$
\begin{aligned}
d &= \sqrt{(x_2 - x_1)^2 + (y_2 - y_1)^2} \\
&= \sqrt{(4 - 2)^2 + [3 - (-1)]^2} \\
&= \sqrt{4 + 16} \\
&= \sqrt{20} \approx 4.47
\end{aligned}
$$

In Example 1 we obtain the same answer if we use $(4, 3)$ for P_1 and $(2, -1)$ for P_2.

$$d = \sqrt{(2 - 4)^2 + [(-1) - 3]^2}$$
$$= \sqrt{4 + 16} = \sqrt{20}$$

EXERCISE 1

a. Find the distance between the points $(-5, 3)$ and $(3, -9)$.

b. Plot the points, and illustrate how the Pythagorean theorem is used in calculating the distance.

Finding the Midpoint

Suppose that Francine flies from her present position at $(253, 124)$ toward Preston Airport at $(187, -201)$. What are her coordinates when she is halfway to Preston?

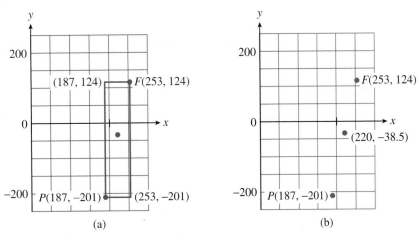

FIGURE 6.18

When she is halfway to Preston, Francine will be at the center of the rectangle shown in Figure 6.18a. The x-coordinate of the center is the average of the x-coordinates at the left and right sides, or

$$x_m = \frac{187 + 253}{2} = 220$$

Similarly, the y-coordinate of the center is the average of the y-coordinates at the bottom and the top.

$$y_m = \frac{-201 + 124}{2} = -38.5$$

Francine will be 220 miles east and 38.5 miles south of Oldfield when she is halfway to Preston.

The Midpoint Formula

If we know the coordinates of two points, we can calculate the coordinates of the point halfway between them using the **midpoint** formula. Each coordinate of the midpoint is the average of the corresponding coordinates of the two points.

> The **midpoint** of the line segment joining the points $P_1(x_1, y_1)$ and $P_2(x_2, y_2)$ is the point $M(\bar{x}, \bar{y})$, where
>
> $$\bar{x} = \frac{x_1 + x_2}{2} \qquad \text{and} \qquad \bar{y} = \frac{y_1 + y_2}{2}$$

The proof of the formula uses similar triangles and is left as an exercise.

EXAMPLE 2

Find the midpoint of the line segment joining the points $(-2, 1)$ and $(4, 3)$.

Solution

Substitute $(-2, 1)$ for (x_1, y_1) and $(4, 3)$ for (x_2, y_2) in the midpoint formula to obtain

$$\bar{x} = \frac{x_1 + x_2}{2} = \frac{-2 + 4}{2} = 1 \qquad \text{and} \qquad \bar{y} = \frac{y_1 + y_2}{2} = \frac{1 + 3}{2} = 2$$

The midpoint of the segment is the point $(\bar{x}, \bar{y}) = (1, 2)$. ∎

EXERCISE 2

a. Find the midpoint of the line segment joining the points $(-5, 3)$ and $(3, -9)$.

b. Plot both points and draw a rectangle with the points as opposite vertices. Illustrate that the midpoint is the center of the rectangle.

Circles

A **circle** is the set of all points in a plane that lie at a given distance, called the **radius,** from a fixed point called the **center.** We can use the distance formula to find an equation for a circle. First consider the circle in Figure 6.19a, whose center is the origin, $(0, 0)$.

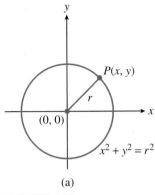

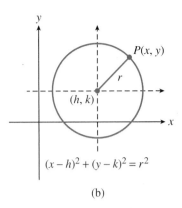

(a)

(b)

FIGURE 6.19

The distance from the origin to any point $P(x, y)$ on the circle is r. Therefore,

$$\sqrt{(x - 0)^2 + (y - 0)^2} = r$$

or, squaring both sides,

$$(x - 0)^2 + (y - 0)^2 = r^2$$

Thus, the equation for a circle of radius t centered at the origin is

$$x^2 + y^2 = r^2$$

Now consider the circle in Figure 6.19b, whose center is the point (h, k). Every point $P(x, y)$ on the circle lies a distance r from (h, k), so the equation of the circle is given by the following formula.

The equation for a circle of radius r centered at the point is (h, k) is

$$(x - h)^2 + (y - k)^2 = r^2$$

This equation is the *standard form* for a circle of radius r with center at (h, k). It is easy to graph a circle if its equation is given in standard form.

EXAMPLE 3

Graph the circles.

   a. $(x - 2)^2 + (y + 3)^2 = 16$ **b.** $x^2 + (y - 4)^2 = 7$

Solutions

   a. The graph of $(x - 2)^2 + (y + 3)^2 = 16$ is a circle with radius 4 and center at $(2, -3)$. To sketch the graph, first locate the center of the circle. (The center is not part of the graph of the circle.) From the center, move a distance of 4 units (the radius of the circle) in each of four directions:

up, down, left, and right. This locates four points that lie on the circle: $(2, 1)$, $(2, -7)$, $(-2, -3)$, and $(6, -3)$. Sketch the circle through these four points. (See Figure 6.20.)

b. The graph of $x^2 + (y - 4)^2 = 7$ is a circle with radius $\sqrt{7}$ and center at $(0, 4)$. From the center, move approximately $\sqrt{7}$, or 2.6 units in each of the four coordinate directions to obtain the points $(0, 6.6)$, $(0, 1.4)$, $(-2.6, 4)$, and $(2.6, 4)$. Sketch the circle through these four points. (See Figure 6.21.)

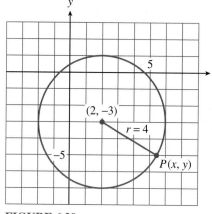

FIGURE 6.20

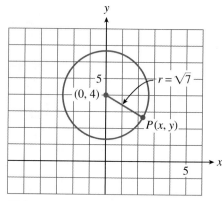

FIGURE 6.21 ■

EXERCISE 3

a. State the center and radius of the circle

$$(x + 3)^2 + (y + 2)^2 = 16$$

b. Graph the circle.

General Form for Circles

The equations of circles often appear in a general quadratic form, rather than the standard form described above. For example, we can expand the squares of binomials in Example 3a,

$$(x - 2)^2 + (y + 3)^2 = 16$$

to obtain

$$x^2 - 4x + 4 + y^2 + 6y + 9 = 16$$

or

$$x^2 + y^2 - 4x + 6y - 3 = 0$$

This is a quadratic equation in two variables. Such an equation describes a circle if the coefficients of the quadratic, or squared, terms are equal. We'll learn more about quadratic equations in Chapter 11.

Conversely, an equation of the form

$$x^2 + y^2 + ax + by + c = 0$$

can be converted to standard form by completing the square in both variables. Once this is done, the center and radius of the circle can be determined directly from the equation.

EXAMPLE 4

Write the equation of the circle

$$x^2 + y^2 + 8x - 2y + 6 = 0$$

in standard form, and graph the equation.

Solution

Prepare to complete the square in both variables by writing the equation as

$$(x^2 + 8x + \underline{}) + (y^2 - 2y + \underline{}) = -6$$

Complete the square in x by adding 16 to each side of the equation, and complete the square in y by adding 1 to each side, to get

$$(x^2 + 8x + 16) + (y^2 - 2y + 1) = -6 + 16 + 1$$

from which we obtain the standard form,

$$(x + 4)^2 + (y - 1)^2 = 11$$

Thus, the circle has its center at $(-4, 1)$, and its radius is $\sqrt{11}$, or approximately 3.3. The graph is shown in Figure 6.22.

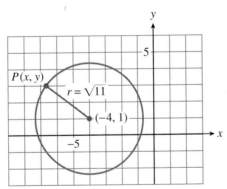

FIGURE 6.22

EXERCISE 4 Write the equation of the circle

$$x^2 + y^2 - 14x + 4y + 25 = 0$$

in standard form.

We can write an equation for any circle if we can find its center and radius.

EXAMPLE 5

Find an equation for the circle whose diameter has endpoints $(7, 5)$ and $(1, -1)$.

Solution

The center of the circle is the midpoint of its diameter. (See Figure 6.23.) Use the midpoint formula to find the center.

$$h = \bar{x} = \frac{7 + 1}{2} = 4$$

$$k = \bar{y} = \frac{5 - 1}{2} = 2$$

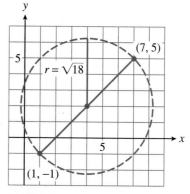

Thus, the center is the point $(h, k) = (4, 2)$. The radius is the distance from the center to either of the endpoints of the diameter, say the point $(7, 5)$. Use the distance formula with the points $(7, 5)$ and $(4, 2)$ to find the radius.

FIGURE 6.23

$$r = \sqrt{(7 - 4)^2 + (5 - 2)^2}$$

$$= \sqrt{3^2 + 3^2} = \sqrt{18}$$

Finally, substitute 4 for h and 2 for k (the coordinates of the center) and $\sqrt{18}$ for r (the radius) into the standard form

$$(x - h)^2 + (y - k)^2 = r^2$$

to obtain

$$(x - 4)^2 + (y - 2)^2 = 18$$ ∎

READING QUESTIONS

1. State the distance formula. The distance formula is based on which familiar theorem?
2. State the midpoint formula. The coordinates of the midpoint are the _____ of the coordinates of the endpoints.
3. State the standard form for the equation of a circle.
4. What algebraic technique is used to write a quadratic equation in standard form?

1a. $\sqrt{208}$ **b.**

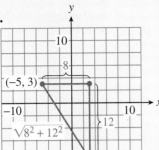

2a. $(-1, -3)$ **b.**

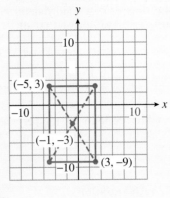

3a. center $(-3, -2)$, radius 4 **b.**
4. $(x - 7)^2 + (y + 2)^2 = 28$

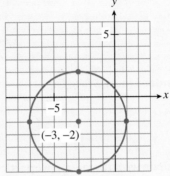

HOMEWORK 6.4

Find the distance between each of the given pairs of points, and find the midpoint of the segment joining them.

1. $(1, 1), (4, 5)$
2. $(-1, 1), (5, 9)$
3. $(2, -3), (-2, -1)$
4. $(5, -4), (-1, 1)$
5. $(3, 5), (-2, 5)$
6. $(-2, -5), (-2, 3)$

7. Leanne is sailing 3 miles west and 5 miles south of the harbor. She heads directly toward an island that is 8 miles west and 7 miles north of the harbor.

   a. How far is Leanne from the island?

   b. How far will Leanne be from the harbor when she is halfway to the island?

8. Dominic is 100 meters east and 250 meters north of Kristy. He is walking directly toward a tree that is 220 meters east and 90 meters north of Kristy.

   a. How far is Dominic from the tree?

   b. How far will Dominic be from Kristy when he is halfway to the tree?

Sketch a diagram for each problem on graph paper, then solve the problem.

9. Find the perimeter of the triangle with vertices $(10, 1), (3, 1), (5, 9)$.

10. Find the perimeter of the triangle with vertices $(-1, 5), (8, -7), (4, 1)$.

11. Show that the rectangle with vertices $(-4, 1)$, $(2, 6), (7, 0)$, and $(1, -5)$ is a square.

12. Show that the triangle with vertices $(0, 0)$, $(6, 0)$ and $(3, 3)$ is an isosceles right triangle—that is, a right triangle with two sides of the same length.

13. The points $(1, 6)$, $(5, 2)$, $(-2, 3)$, $(2, -1)$ are the vertices of a quadrilateral. Show that its diagonals are of equal length.

14. Two opposite vertices of a square are $A(-9, -5)$ and $C(3, 3)$.
   a. Find the length of a diagonal of the square.
   b. Find the length of the side of the square.

15. Show that the point $C(\sqrt{5}, 2 + \sqrt{5})$ is the same distance from $A(2, 0)$ and $B(-2, 4)$.

16. Show that the points $(-2, 1)$, $(0, -1)$, and $(\sqrt{3} - 1, \sqrt{3})$ are the vertices of an equilateral triangle.

17. Find the equation of the perpendicular bisector of the line segment joining $A(2, 1)$ and $B(1, 3)$.

18. The point $A(3, 5)$ is one vertex of a square. The diagonals of the square intersect at the point $(7, 2)$. Find the coordinates of the other three vertices.

Graph each equation.

19. $x^2 + y^2 = 25$ **20.** $x^2 + y^2 = 16$

21. $4x^2 + 4y^2 = 16$ **22.** $2x^2 + 2y^2 = 18$

23. $(x - 4)^2 + (y + 2)^2 = 9$

24. $(x - 1)^2 + (y - 3)^2 = 16$

25. $(x + 3)^2 + y^2 = 10$

26. $x^2 + (y + 4)^2 = 12$

Write each equation in standard form. State the center and radius of the circle.

27. $x^2 + y^2 + 2x - 4y - 6 = 0$

28. $x^2 + y^2 - 6x + 2y - 4 = 0$

29. $x^2 + y^2 + 8x = 4$

30. $x^2 + y^2 - 10y = 2$

Write an equation for the circle with the given properties.

31. Center at $(-2, 5)$, radius $2\sqrt{3}$.

32. Center at $(4, -3)$, radius $2\sqrt{6}$.

33. Center at $\left(\frac{3}{2}, -4\right)$, one point on the circle $(4, -3)$.

34. Center at $\left(\frac{-3}{2}, \frac{-1}{2}\right)$, one point on the circle $(-4, -2)$.

35. Endpoints of a diameter at $(1, 5)$ and $(3, -1)$.

36. Endpoints of a diameter at $(3, 6)$ and $(-5, 2)$.

37. Center at $(-3, -1)$, the x-axis is tangent to the circle.

38. Center at $(1, 7)$, the y-axis is tangent to the circle.

39. Find an equation for the circle that passes through the points $(2, 3)$, $(3, 2)$, and $(-4, -5)$. (*Hint:* Find values for a, b, and c so that the three points lie on the graph of $x^2 + y^2 + ax + by + c = 0$.)

40. Find an equation for the circle that passes through the points $(0, 0)$, $(6, 0)$, and $(0, 8)$. (See the hint for Problem 39.)

41. In Figure 6.24, $M(\bar{x}, \bar{y})$ is the midpoint of the line segment $\overline{AB}$. Using similar triangles, show that

$$AC = 2AE \qquad \text{and} \qquad BC = 2BD$$

(You may assume that $\angle ACB$, $\angle MDB$, and $\angle AEM$ are right angles.)

42. Use the result of Problem 41 and the distance formula to show that

$$\bar{x} = \frac{x_1 + x_2}{2} \qquad \text{and} \qquad \bar{y} = \frac{y_1 + y_2}{2}$$

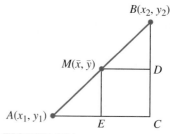

FIGURE 6.24

6.5 Working with Radicals

In some situations radical notation is more convenient to use than exponents. In these cases we often simplify radical expressions algebraically before using a calculator to obtain decimal approximations.

Properties of Radicals

Because $\sqrt[n]{a} = a^{1/n}$, we can use the laws of exponents to derive two important properties that are useful in simplifying radicals.

Property 1. $\quad \sqrt[n]{ab} = \sqrt[n]{a}\,\sqrt[n]{b} \qquad$ for $a, b \geq 0$

Property 2. $\quad \sqrt[n]{\dfrac{a}{b}} = \dfrac{\sqrt[n]{a}}{\sqrt[n]{b}} \qquad$ for $a \geq 0, b > 0$

As examples, you can verify that
$$\sqrt{36} = \sqrt{4}\,\sqrt{9} \qquad \text{and} \qquad \sqrt[3]{\frac{1}{8}} = \frac{\sqrt[3]{1}}{\sqrt[3]{8}}$$

Simplifying Radicals

We use Property (1) to simplify radical expressions by factoring the radicand. For example, to simplify $\sqrt[3]{108}$, we look for perfect cubes that divide evenly into 108. The easiest way to do this is to try the perfect cubes in order: 1, 8, 27, 64, 125, . . . and so on, until we find one that is a factor. For this example, we find that $108 = 27 \cdot 4$. Using Property (1), we write
$$\sqrt[3]{108} = \sqrt[3]{27}\,\sqrt[3]{4}$$

Simplify the first factor to find
$$\sqrt[3]{108} = 3\sqrt[3]{4}$$

This expression is considered simpler than the original radical because the new radicand, 4, is smaller than the original, 108.

We can also simplify radicals containing variables. If the exponent on the variable is a multiple of the index, we can extract the variable from the radical. For instance,
$$\sqrt[3]{x^{12}} = x^{12/3} = x^4$$

(You can verify this by noting that $(x^4)^3 = x^{12}$.) If the exponent on the variable is not a multiple of the index, we factor out the highest power that is a multiple. For example,

$$\sqrt[3]{x^{11}} = \sqrt[3]{x^9 \cdot x^2} \qquad \text{Apply Property (1).}$$
$$= \sqrt[3]{x^9} \cdot \sqrt[3]{x^2} \qquad \text{Simplify } \sqrt[3]{x^9} = x^{9/3}.$$
$$= x^3\sqrt[3]{x^2}$$

EXAMPLE 1

Simplify each radical.

  a. $\sqrt{18x^5}$ **b.** $\sqrt[3]{24x^6y^8}$

Solutions

a. The index of the radical is 2, so we look for perfect square factors of $18x^5$. The factor 9 is a perfect square, and x^4 has an exponent divisible by 2. Thus

$$\sqrt{18x^5} = \sqrt{9x^4 \cdot 2x} \quad \text{Apply Property (1).}$$
$$= \sqrt{9x^4}\sqrt{2x} \quad \text{Take square roots.}$$
$$= 3x^2\sqrt{2x}$$

b. The index of the radical is 3, so we look for perfect cube factors of $24x^6y^8$. The factor 8 is a perfect cube, and x^6 and y^6 have exponents divisible by 3. Thus

$$\sqrt[3]{24x^6y^8} = \sqrt[3]{8x^6y^6 \cdot 3y^2} \quad \text{Apply Property (1).}$$
$$= \sqrt[3]{8x^6y^6}\sqrt[3]{3y^2} \quad \text{Take cube roots.}$$
$$= 2x^2y^2\sqrt[3]{3y^2}$$ ∎

EXERCISE 1 Simplify $\sqrt[3]{250b^7}$.

Look for perfect cube factors of $250b^7$.

Apply Property (1).

Take cube roots.

COMMON ERROR

Property (1) applies only to *products* under the radical, not to sums or differences. Thus, for example,

$$\sqrt{4 \cdot 9} = \sqrt{4}\sqrt{9} = 2 \cdot 3 \qquad \text{but} \qquad \sqrt{4 + 9} \neq \sqrt{4} + \sqrt{9}$$

and

$$\sqrt[3]{x^3y^6} = \sqrt[3]{x^3}\sqrt[3]{y^6} = xy^2 \qquad \text{but} \qquad \sqrt[3]{x^3 - y^6} \neq \sqrt[3]{x^3} - \sqrt[3]{y^6}$$ ∎

To simplify roots of fractions, we use Property (2), which allows us to write the expression as a quotient of two radicals.

EXAMPLE 2

  a. $\sqrt{\dfrac{3}{4}} = \dfrac{\sqrt{3}}{\sqrt{4}} = \dfrac{\sqrt{3}}{2}$ **b.** $\sqrt[3]{\dfrac{5}{8}} = \dfrac{\sqrt[3]{5}}{\sqrt[3]{8}} = \dfrac{\sqrt[3]{5}}{2}$ ∎

We can also use Properties (1) and (2) to simplify products and quotients of radicals.

EXAMPLE 3

Simplify.

  a. $\sqrt[4]{6x^2}\,\sqrt[4]{8x^3}$ **b.** $\dfrac{\sqrt[3]{16y^5}}{\sqrt[3]{y^2}}$

Solutions

a. First apply Property (1) to write the product as a single radical, then simplify.

$$\sqrt[4]{6x^2}\,\sqrt[4]{8x^3} = \sqrt[4]{48x^5} \qquad \text{Factor out perfect fourth powers.}$$
$$= \sqrt[4]{16x^4}\,\sqrt[4]{3x} \quad \text{Simplify.}$$
$$= 2x\sqrt[4]{3x}$$

b. Apply Property (2) to write the quotient as a single radical.

$$\frac{\sqrt[3]{16y^5}}{\sqrt[3]{y^2}} = \sqrt[3]{\frac{16y^5}{y^2}} \qquad \text{Reduce.}$$
$$= \sqrt[3]{16y^3} \qquad \text{Simplify: factor out perfect cubes.}$$
$$= \sqrt[3]{8y^3}\,\sqrt[3]{2}$$
$$= 2y\sqrt[3]{2} \qquad\blacksquare$$

EXERCISE 2 Simplify.

a. $\sqrt{\dfrac{18x^5}{25y^4}}$

b. $\sqrt[3]{2a^2}\,\sqrt[3]{6a^2}$

Sums and Differences of Radicals

You know that sums or differences of like terms can be combined by adding or subtracting their coefficients.

$$3xy + 5xy = (3 + 5)xy = 8xy$$

"Like radicals," that is, radicals of the same index and radicand, can be combined in the same way.

EXAMPLE 4

a. $3\sqrt{3} + 4\sqrt{3} = (3 + 4)\sqrt{3}$ b. $4\sqrt[3]{y} - 6\sqrt[3]{y} = (4 - 6)\sqrt[3]{y}$
$\qquad\qquad\quad = 7\sqrt{3}$ $\qquad\qquad\qquad\qquad\qquad = -2\sqrt[3]{y}$ $\blacksquare$

COMMON ERRORS

1. In Example 4a, $3\sqrt{3} + 4\sqrt{3} \ne 7\sqrt{6}$. Only the coefficients are added; the radicand does not change.

2. Sums of radicals with different radicands or different indices *cannot* be combined. Thus,

$$\sqrt{11} + \sqrt{5} \ne \sqrt{16}$$
$$\sqrt[3]{10x} - \sqrt[3]{2x} \ne \sqrt[3]{8x}$$

and

$$\sqrt[3]{7} + \sqrt{7} \ne \sqrt[5]{7}$$

None of the expressions above can be simplified. $\qquad\blacksquare$

Products

According to Property (1) on page 395, radicals of the same index can be multiplied together.

$$\sqrt[n]{a}\,\sqrt[n]{b} = \sqrt[n]{ab} \qquad (a, b \geq 0)$$

Thus, for example,

$$\sqrt{2}\sqrt{18} = \sqrt{36} = 6 \qquad \text{and} \qquad \sqrt[3]{2x}\,\sqrt[3]{4x^2} = \sqrt[3]{8x^3} = 2x$$

For products involving binomials, we can apply the distributive law.

EXAMPLE 5

a. $\sqrt{3}\left(\sqrt{2x} + \sqrt{6}\right) = \sqrt{3 \cdot 2x} + \sqrt{3 \cdot 6}$
$$= \sqrt{6x} + \sqrt{18} = \sqrt{6x} + 3\sqrt{2}$$

b. $\left(\sqrt{x} - \sqrt{y}\right)\left(\sqrt{x} + \sqrt{y}\right) = \sqrt{x^2} - \sqrt{xy} + \sqrt{xy} - \sqrt{y^2}$
$$= x - y \qquad\blacksquare$$

EXERCISE 4 Expand $\left(\sqrt{5} - 2\sqrt{3}\right)^2$.

Rationalizing the Denominator

It is easier to work with radicals if there are no roots in the denominators of fractions. We can use the fundamental principle of fractions to remove radicals from the denominator. This process is called **rationalizing the denominator.** For square roots, we multiply the numerator and denominator of the fraction by the radical in the denominator.

EXAMPLE 6 Rationalize the denominator of each fraction.

a. $\sqrt{\dfrac{1}{3}}$
b. $\dfrac{\sqrt{2}}{\sqrt{50x}}$

Solutions

a. Apply Property (2) to write the radical as a quotient.

$$\sqrt{\frac{1}{3}} = \frac{\sqrt{1}}{\sqrt{3}} = \frac{1}{\sqrt{3}} \qquad \text{Multiply numerator and denominator by } \sqrt{3}.$$

$$= \frac{1 \cdot \sqrt{3}}{\sqrt{3} \cdot \sqrt{3}} = \frac{\sqrt{3}}{3}$$

b. It is always best to simplify the denominator before rationalizing.

$$\frac{\sqrt{2}}{\sqrt{50x}} = \frac{\sqrt{2}}{5\sqrt{2x}} \qquad \text{Multiply numerator and denominator by } \sqrt{2x}.$$

$$= \frac{\sqrt{2} \cdot \sqrt{2x}}{5\sqrt{2x} \cdot \sqrt{2x}} \qquad \text{Simplify.}$$

$$= \frac{\sqrt{4x}}{5(2x)} = \frac{2\sqrt{x}}{10x} = \frac{\sqrt{x}}{5x} \qquad \blacksquare$$

EXERCISE 5 Rationalize the denominator of $\dfrac{-\sqrt{3}}{\sqrt{7}}$.

If the denominator of a fraction is a *binomial* in which one or both terms are radicals, we can use a special building factor to rationalize it. First, recall that

$$(p - q)(p + q) = p^2 - q^2$$

where the product consists of perfect squares only. Each of the two factors $p - q$ and $p + q$ is said to be the **conjugate** of the other.

Now consider a fraction of the form

$$\frac{a}{b + \sqrt{c}}$$

If we multiply the numerator and denominator of this fraction by the conjugate of the denominator, we get

$$\frac{a(b - \sqrt{c})}{(b + \sqrt{c})(b - \sqrt{c})} = \frac{ab - a\sqrt{c}}{b^2 - (\sqrt{c})^2} = \frac{ab - a\sqrt{c}}{b^2 - c}$$

The denominator of the fraction no longer contains any radicals—it has been rationalized.

Multiplying numerator and denominator by the conjugate of the denominator also works on fractions of the form

$$\frac{a}{\sqrt{b} + c} \qquad \text{and} \qquad \frac{a}{\sqrt{b} + \sqrt{c}}$$

We leave the verification of these cases as exercises.

EXAMPLE 7

Rationalize the denominator: $\dfrac{x}{\sqrt{2} + \sqrt{x}}$.

Solution

Multiply numerator and denominator by the conjugate of the denominator, $\sqrt{2} - \sqrt{x}$.

$$\frac{x(\sqrt{2} - \sqrt{x})}{(\sqrt{2} + \sqrt{x})(\sqrt{2} - \sqrt{x})} = \frac{x\sqrt{2} - x\sqrt{x}}{2 - x} \qquad \blacksquare$$

Simplifying $\sqrt[n]{a^n}$

In Section 6.2 we saw that raising to a power is the "inverse" operation for extracting roots; that is

$$\left(\sqrt[n]{a}\right)^n = a$$

For example, if $a = 16$ and $n = 4$, then

$$\left(\sqrt[4]{16}\right)^4 = 2^4 = 16$$

Now consider the power and root operations in the opposite order; that is, consider

$$\sqrt[n]{a^n}$$

If the index n is an odd number, then

$$\sqrt[n]{a^n} = a \qquad (\textbf{n odd})$$

For example,

$$\sqrt[3]{2^3} = \sqrt[3]{8} = 2 \qquad \text{and} \qquad \sqrt[3]{(-2)^3} = \sqrt[3]{-8} = -2$$

However, if n is even, $\sqrt[n]{a^n}$ is always positive, regardless of whether a itself is positive or negative. For example if $a = -3$, then

$$\sqrt{(-3)^2} = \sqrt{9} = 3$$

Thus, we have the following special relationship for even roots.

$$\sqrt[n]{a^n} = |a|$$

We summarize our results as follows.

1. If n is odd, $\sqrt[n]{a^n} = a$
2. If n is even, $\sqrt[n]{a^n} = |a|$
    In particular, $\sqrt{a^2} = |a|$

EXAMPLE 8

  a. $\sqrt{16x^2} = 4|x|$ **b.** $\sqrt{(x-1)^2} = |x-1|$ ■

1. State two properties of radicals.
2. What are "like radicals," and when are they significant?
3. Explain how to rationalize a monomial denominator.
4. Explain how to rationalize a binomial denominator.
5. Explain how to simplify $\sqrt[n]{a^n}$. Consider two cases.

ANSWERS TO 6.5 EXERCISES

1. $5b^2\sqrt[3]{2b}$ **2a.** $\dfrac{3x^2\sqrt{2x}}{5y^2}$ **2b.** $a\sqrt[3]{12a}$

3. $2\sqrt[3]{5x^2} - 3\sqrt[3]{2x^2}$ **4.** $17 - 4\sqrt{15}$ **5.** $\dfrac{-\sqrt{21}}{7}$

6. $3 + \sqrt{6}$

HOMEWORK 6.5

Simplify. Assume that all variables represent positive numbers.

1. a. $\sqrt{18}$ **b.** $\sqrt[3]{24}$ **c.** $-\sqrt[4]{64}$ **2. a.** $\sqrt{50}$ **b.** $\sqrt[3]{54}$ **c.** $-\sqrt[4]{162}$

3. a. $\sqrt{60,000}$ **b.** $\sqrt[3]{900,000}$ **c.** $\sqrt[3]{\dfrac{-40}{27}}$ **4. a.** $\sqrt{800,000}$ **b.** $\sqrt[3]{24,000}$ **c.** $\sqrt[4]{\dfrac{80}{625}}$

5. a. $\sqrt[3]{x^{10}}$ **b.** $\sqrt{27z^3}$ **c.** $\sqrt[4]{48a^9}$ **6. a.** $\sqrt[3]{y^{16}}$ **b.** $\sqrt{12t^5}$ **c.** $\sqrt[3]{81b^8}$

Simplify.

7. a. $-\sqrt{18s}\,\sqrt{2s^3}$ **b.** $\sqrt[3]{7h^2}\,\sqrt[3]{-49h}$ **8. a.** $\sqrt{3w^3}\,\sqrt{27w^3}$ **b.** $-\sqrt[4]{2m^3}\,\sqrt[4]{8m}$

   c. $\sqrt{16 - 4x^2}$ **c.** $\sqrt{9Y^2 + 18}$

9. a. $\sqrt[3]{8A^3 + A^6}$ **b.** $\dfrac{\sqrt{45x^3y^3}}{\sqrt{5y}}$ **10. a.** $\sqrt[3]{b^9 - 27b^3}$ **b.** $\dfrac{\sqrt{98x^2y^3}}{\sqrt{xy}}$

   c. $\dfrac{\sqrt[3]{8b^7}}{\sqrt[3]{a^6b^2}}$ **c.** $\dfrac{\sqrt[3]{16r^4}}{\sqrt[3]{4t^3}}$

Simplify and combine like terms.

11. $3\sqrt{7} + 2\sqrt{7}$ **12.** $5\sqrt{2} - 3\sqrt{2}$ **13.** $4\sqrt{3} - \sqrt{27}$ **14.** $\sqrt{75} + 2\sqrt{3}$

15. $\sqrt{50x} + \sqrt{32x}$ **16.** $\sqrt{8y} - \sqrt{18y}$

17. $3\sqrt[3]{16} - \sqrt[3]{2} - 2\sqrt[3]{54}$ **18.** $\sqrt[3]{81} + 2\sqrt[3]{24} - 3\sqrt[3]{3}$

Multiply.

19. $2(3 - \sqrt{5})$ **20.** $5(2 - \sqrt{7})$ **21.** $\sqrt{2}(\sqrt{6} + \sqrt{10})$ **22.** $\sqrt{3}(\sqrt{12} - \sqrt{15})$

23. $\sqrt[3]{2}(\sqrt[3]{20} - 2\sqrt[3]{12})$ **24.** $\sqrt[3]{3}(2\sqrt[3]{18} + \sqrt[3]{36})$ **25.** $(\sqrt{x} - 3)(\sqrt{x} + 3)$ **26.** $(2 + \sqrt{x})(2 - \sqrt{x})$

27. $(\sqrt{2} - \sqrt{3})(\sqrt{2} + 2\sqrt{3})$ **28.** $(\sqrt{3} - \sqrt{5})(2\sqrt{3} + \sqrt{5})$

29. $\left(\sqrt{5} - \sqrt{2}\right)^2$

30. $\left(\sqrt{2} - 2\sqrt{3}\right)^2$

31. $\left(\sqrt{a} - 2\sqrt{b}\right)^2$

32. $\left(\sqrt{2a} - 2\sqrt{b}\right)\left(\sqrt{2a} + 2\sqrt{b}\right)$

Verify by substitution that the number is a solution of the quadratic equation.

33. $x^2 - 2x - 3 = 0$; $1 + \sqrt{3}$

34. $x^2 + 4x - 1 = 0$; $-2 + \sqrt{5}$

35. $x^2 + 6x - 9 = 0$; $-3 + 3\sqrt{2}$

36. $4x^2 - 20x + 22 = 0$; $\dfrac{5 - \sqrt{3}}{2}$

Rationalize the denominator.

37. $\dfrac{6}{\sqrt{3}}$

38. $\dfrac{10}{\sqrt{5}}$

39. $\sqrt{\dfrac{7x}{18}}$

40. $\sqrt{\dfrac{27x}{20}}$

41. $\sqrt{\dfrac{2a}{b}}$

42. $\sqrt{\dfrac{5p}{q}}$

43. $\dfrac{2\sqrt{3}}{\sqrt{2k}}$

44. $\dfrac{6\sqrt{2}}{\sqrt{3v}}$

45. $\dfrac{4}{1 + \sqrt{3}}$

46. $\dfrac{3}{7 - \sqrt{2}}$

47. $\dfrac{x}{x - \sqrt{3}}$

48. $\dfrac{y}{\sqrt{5} - y}$

49. $\dfrac{\sqrt{6} - 3}{2 - \sqrt{6}}$

50. $\dfrac{\sqrt{x} + \sqrt{y}}{\sqrt{x} - \sqrt{y}}$

51. Use your calculator to graph each function, and explain the result.

   a. $y = \sqrt{x^2}$ **b.** $y = \sqrt[3]{x^3}$

52. Use your calculator to graph each function, and explain the result.

   a. $y = (x^4)^{1/4}$ **b.** $y = (x^5)^{1/5}$

For these problems, do not assume that variables represent positive numbers. Use absolute value bars as necessary to simplify the radicals.

53. a. $\sqrt{4x^2}$ **b.** $\sqrt{(x - 5)^2}$ **c.** $\sqrt{x^2 - 6x + 9}$ **54. a.** $\sqrt{9x^2y^4}$ **b.** $\sqrt{(2x - 1)^2}$ **c.** $\sqrt{9x^2 - 6x + 1}$

In these problems we rationalize denominators with index greater than 2. Follow the method described in Problems 55 and 56.

55. Rationalize the denominator of $\dfrac{1}{\sqrt[3]{2x}}$.

We need a third power, $(2x)^3$, under the radical in order to extract the root. Therefore we must multiply the denominator $\sqrt[3]{2x}$ by *two* additional factors of $\sqrt[3]{2x}$. Complete the solution:

$$\dfrac{1}{\sqrt[3]{2x}} = \dfrac{1 \cdot \sqrt[3]{2x}\,\sqrt[3]{2x}}{\sqrt[3]{2x} \cdot \sqrt[3]{2x}\,\sqrt[3]{2x}} = ?$$

56. Rationalize the denominator of $\sqrt[5]{\dfrac{6}{16x^3}}$.

First note that

$$\sqrt[5]{\dfrac{6}{16x^3}} = \sqrt[5]{\dfrac{6}{2^4x^3}}$$

Because we want a fifth power, $(2x)^5$, in the denominator, we multiply numerator and denominator by $\sqrt[5]{2x^2}$. Complete the solution:

$$\sqrt[5]{\dfrac{6}{16x^3}} = \dfrac{\sqrt[5]{6} \cdot \sqrt[5]{2x^2}}{\sqrt[5]{2^4x^3} \cdot \sqrt[5]{2x^2}} = ?$$

57. $\dfrac{1}{\sqrt[3]{x^2}}$

58. $\dfrac{1}{\sqrt[4]{y^3}}$

59. $\sqrt[3]{\dfrac{2}{3y}}$

60. $\sqrt[4]{\dfrac{2}{3x}}$

61. $\sqrt[4]{\dfrac{x}{8y^3}}$

62. $\sqrt[3]{\dfrac{x}{4y^2}}$

63. $\dfrac{9x^3}{\sqrt[4]{27x}}$

64. $\dfrac{15x^4}{\sqrt[3]{5x}}$

6.6   Radical Equations

A **radical equation** is one in which the variable appears under a square root or other radical. For example,

$$\sqrt{x + 3} = 4$$

is a radical equation. We solved simple radical equations like this one in Section 6.2. Because squaring and taking square roots are inverse operations, we can solve this equation by squaring both sides in order to "undo" the radical.

$$\left(\sqrt{x + 3}\right)^2 = 4^2$$
$$x + 3 = 16$$
$$x = 13$$

You can check that $x = 13$ is the solution for this equation.

Similarly, the operations of cubing and taking cube roots are inverse operations. This relationship enables us to solve equations involving $\sqrt[3]{x}$.

EXAMPLE 1

Solve $4\sqrt[3]{x - 9} = 12$.

Solution

First isolate the radical by dividing both sides of the equation by 4.

$$\sqrt[3]{x - 9} = 3$$

Next, cube both sides of the equation.

$$\left(\sqrt[3]{x - 9}\right)^3 = 3^3$$
$$x - 9 = 27$$
$$x = 36$$

The solution is 36. We can also solve the equation by graphing $y = 4\sqrt[3]{x - 9}$, as shown in Figure 6.25. The point $(36, 12)$ lies on the graph, so $x = 36$ is the solution of the equation $4\sqrt[3]{x - 9} = 12$.

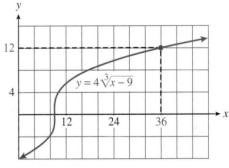

FIGURE 6.25

EXERCISE 1   Solve $6 + 2\sqrt[4]{12 - v} = 10$.

Isolate the radical.

Raise each side to the fourth power.

Complete the solution.

Extraneous Solutions

It is important to check the solution to a radical equation, because it is possible to introduce false or **extraneous** solutions when we square both sides of the equation. For example, the equation

$$\sqrt{x} = -5$$

has no solution, because $\sqrt{x}$ is never a negative number. However, if we try to solve the equation by squaring both sides, we find

$$\left(\sqrt{x}\right)^2 = (-5)^2$$
$$x = 25$$

You can check that 25 is *not* a solution to the original equation, $\sqrt{x} = -5$, because $\sqrt{25}$ does not equal -5.

If each side of an equation is raised to an odd power, extraneous solutions will not be introduced. However, if we raise both sides to an even power, we should check each solution in the original equation.

EXAMPLE 2   Solve the equation $\sqrt{x + 2} + 4 = x$.

Solution

First, isolate the radical expression on one side of the equation. (This will make it easier to square both sides.)

$$\sqrt{x + 2} = x - 4 \qquad \text{Square both sides of the equation.}$$
$$\left(\sqrt{x + 2}\right)^2 = (x - 4)^2$$
$$x + 2 = x^2 - 8x + 16 \qquad \text{Subtract } x + 2 \text{ from both sides.}$$
$$x^2 - 9x + 14 = 0 \qquad \text{Factor the left side.}$$
$$(x - 2)(x - 7) = 0 \qquad \text{Set each factor equal to zero.}$$
$$x = 2 \qquad \text{or} \qquad x = 7$$

Check   Does $\sqrt{2 + 2} + 4 = 2$?   No; 2 is not a solution.

Does $\sqrt{7 + 2} + 4 = 7$?   Yes; 7 is a solution.

The apparent solution 2 is extraneous. The only solution to the original equation is 7. We can verify the solution by graphing the equations

$$y_1 = \sqrt{x + 2} \qquad \text{and} \qquad y_2 = x - 4$$

as shown in Figure 6.26. The graphs intersect in only one point, $(7, 3)$, so there is only one solution, $x = 7$.

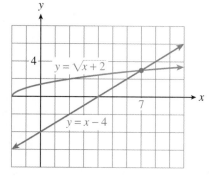

FIGURE 6.26

When we square both sides of an equation, it is *not* correct to square each term of the equation separately. Thus, in Example 2, the original equation is *not* equivalent to

$$\left(\sqrt{x+2}\right)^2 + 4^2 = x^2$$

This is because $(a + b)^2 \neq a^2 + b^2$. Instead, we must square the *entire* left side of the equation as a binomial, like this:

$$\left(\sqrt{x+2} + 4\right)^2 = x^2$$

or we may proceed as shown in Example 2. ∎

EXERCISE 2   Solve $2x - 5 = \sqrt{40 - 3x}$.

Square both sides.
Solve the quadratic equation.
Check for extraneous roots.

Solving Formulas

We can also solve formulas involving radicals for one variable in terms of the others.

EXAMPLE 3

Solve the formula $t = \sqrt{1 + s^2}$ for s.

Solution

The variable we want is under a radical, so we square both sides of the equation.

$$t^2 = 1 + s^2 \qquad \text{Subtract 1 from both sides.}$$
$$t^2 - 1 = s^2 \qquad \text{Take square roots.}$$
$$s = \pm\sqrt{t^2 - 1}$$

∎

EXERCISE 3   Solve the formula $r - 2 = \sqrt[3]{V - Bh}$ for h.

Equations with More than One Radical

Sometimes it is necessary to square both sides of an equation more than once in order to eliminate all the radicals.

EXAMPLE 4

Solve $\sqrt{x - 7} + \sqrt{x} = 7$.

Solution

First isolate the more complicated radical on one side of the equation. (This will make it easier to square both sides.) We will subtract $\sqrt{x}$ from both sides.

$$\sqrt{x-7} = 7 - \sqrt{x}$$

Now square each side to remove one radical. Be careful when squaring the binomial $7 - \sqrt{x}$.

$$\left(\sqrt{x-7}\right)^2 = \left(7 - \sqrt{x}\right)^2$$

$$x - 7 = 49 - 14\sqrt{x} + x$$

Collect like terms, and isolate the radical on one side of the equation.

$$-56 = -14\sqrt{x} \quad \text{Divide both sides by } -14.$$

$$4 = \sqrt{x}$$

Now square again to obtain

$$(4)^2 = \left(\sqrt{x}\right)^2$$

$$16 = x$$

Check   Does $\sqrt{16-7} + \sqrt{16} = 7$?   Yes. The solution is 16.   ■

COMMON ERROR

Recall that we cannot solve a radical equation by squaring each term separately. In other words, it is *incorrect* to begin Example 4 by writing

$$\left(\sqrt{x-7}\right)^2 + \left(\sqrt{x}\right)^2 = 7^2$$

We must square the *entire expression* on each side of the equal sign as one piece.   ■

EXERCISE 4   Solve $\sqrt{3x+1} = 6 - \sqrt{9-x}$.

Square both sides.
Isolate the radical term.
Divide both sides by 4.
Square both sides again.
Solve the quadratic equation.
Check for extraneous roots.

Applications

Equations involving radicals arise in a variety of applications.

EXAMPLE 5

Two oil derricks are located the same distance offshore from a straight section of coast. A supply boat returns from the derrick at point A to the harbor at C (see

Figure 6.27), reloads, and sails to the derrick at *B*, traveling a total distance of 102 miles. How far are the derricks from the coast?

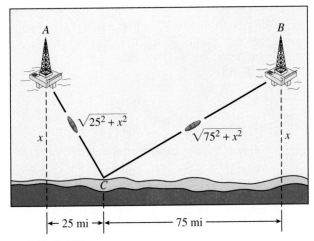

FIGURE 6.27

Solution

Let *x* represent the distance from each derrick to the coast. Use the Pythagorean theorem to write an expression for the distance traveled by the supply boat on each leg of its journey. The sum of these distances is 102 miles.

$$\sqrt{x^2 + 25^2} + \sqrt{x^2 + 75^2} = 102$$

Solve the equation. First, rewrite the equation with one radical on each side, and square both sides.

$$\left(\sqrt{x^2 + 25^2}\right)^2 = \left(102 - \sqrt{x^2 + 75^2}\right)^2$$
$$x^2 + 625 = 10{,}404 - 204\sqrt{x^2 + 5625} + x^2 + 5625$$

Collect like terms and square again.

$$204\sqrt{x^2 + 5625} = 15{,}404 \qquad \text{Divide both sides by 204.}$$
$$\sqrt{x^2 + 5625} \approx 75.5 \qquad \text{Square both sides.}$$
$$x^2 + 5625 \approx 5701.73 \qquad \text{Solve by subtracting 5625 from each side}$$
$$x \approx \pm 8.8 \qquad\qquad\;\; \text{and extracting roots.}$$

Because *x* must be greater than 0, we find that $x \approx 8.8$. The derricks are about 8.8 miles offshore. ∎

We can also solve the equation in Example 5 graphically. Graph the functions

$$Y_1 = \sqrt{x^2 + 25^2} + \sqrt{x^2 + 75^2}$$
$$Y_2 = 102$$

in the window

$$\text{Xmin} = 0 \qquad \text{Xmax} = 20$$
$$\text{Ymin} = 95 \qquad \text{Ymax} = 110$$

to obtain the graph shown in Figure 6.28. The solution of the equation

$$\sqrt{x^2 + 25^2} + \sqrt{x^2 + 75^2} = 102$$

is the x-coordinate of the intersection point of the two graphs. Using the [TRACE] or the **intersect** feature, you can verify that $x \approx 8.8$ miles.

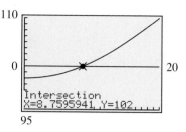

FIGURE 6.28

READING QUESTIONS

1. What is a radical equation?
2. What is the opposite operation for taking a fourth root?
3. What is an extraneous solution?
4. How are extraneous solutions introduced?
5. Explain why the following first step for solving the radical equation is incorrect:

$$\sqrt{x - 5} + \sqrt{2x - 1} = 8$$
$$(x - 5) + (2x - 1) = 64$$

ANSWERS TO 6.6 EXERCISES

1. $v = -4$ **2.** $x = 5$ **3.** $h = \dfrac{V - r^3 + 6r^2 - 12r + 8}{B}$

4. $x = 5$ or $x = 8$

HOMEWORK 6.6

Solve each radical equation.

1. $\sqrt{x} - 5 = 3$ **2.** $\sqrt{x} - 4 = 1$

3. $\sqrt{y + 6} = 2$ **4.** $\sqrt{y - 3} = 5$

5. $4\sqrt{z} - 8 = -2$ **6.** $-3\sqrt{z} + 14 = 8$

7. $5 + 2\sqrt{6 - 2w} = 13$

8. $8 - 3\sqrt{9 + 2w} = -7$

9. The period of a pendulum is the time it takes for the pendulum to complete one entire swing, from left to right and back again. The greater the length, L, of the pendulum, the longer its period, T. In fact, if L is measured in feet, then the period is given in seconds by

$$T = 2\pi \sqrt{\frac{L}{32}}$$

Suppose you are standing in the Convention Center in Portland, Oregon, and you time the period of its Foucault pendulum (the longest in the world). Its period is approximately 10.54 seconds. Approximately how long is the pendulum?

10. If you are flying in an airplane at an altitude of h miles, on a clear day you can see a distance of d miles to the horizon, where $d = 89.4\sqrt{h}$. At what altitude will you be able to see for a distance of 100 miles? How high is that in feet?

Solve each formula for the indicated variable.

11. $T = 2\pi\sqrt{\dfrac{L}{g}}$ for L

12. $T = 2\pi\sqrt{\dfrac{m}{k}}$ for m

13. $r = \sqrt{t^2 - s^2}$ for s

14. $c = \sqrt{a^2 - b^2}$ for b

Solve.

15. $3z + 4 = \sqrt{3z + 10}$

16. $2x - 3 = \sqrt{7x - 3}$

17. $2x + 1 = \sqrt{10x + 5}$

18. $4x + 5 = \sqrt{3x + 4}$

19. $\sqrt{y + 4} = y - 8$

20. $4\sqrt{x - 4} = x$

21. $\sqrt{2y - 1} = \sqrt{3y - 6}$

22. $\sqrt{4y + 1} = \sqrt{6y - 3}$

23. $\sqrt{x - 3}\sqrt{x} = 2$

24. $\sqrt{x}\sqrt{x - 5} = 6$

25. $\sqrt{y + 4} = \sqrt{y + 20} - 2$

26. $4\sqrt{y} + \sqrt{1 + 16y} = 5$

27. $\sqrt{x} + \sqrt{2} = \sqrt{x + 2}$

28. $\sqrt{4x + 17} = 4 - \sqrt{x + 1}$

29. $\sqrt{5 + x} + \sqrt{x} = 5$

30. $\sqrt{y + 7} + \sqrt{y + 4} = 3$

31. The orthocenter of a triangle is the point that is equidistant from the vertices of the triangle. Find the orthocenter of the triangle ABC with vertices $A(2, -1)$, $B(-2, 3)$, and $C(2, 5)$. (*Hint:* Let (x, y) be the coordinates of the orthocenter. Write two equations in x and y about the distances from the orthocenter to the vertices of the triangle.)

32. The point (a, b) is equidistant from the origin, the point $(6, 8)$ and the point $(6, 0)$. Find the values of a and b.

33. Grandview College is located 8 miles from a long straight stretch of the state highway. Station KGVC broadcasts from the college and has a signal range of 15 miles. Francine leaves the college and travels along the highway at a speed of 60 miles per hour (or 1 mile per minute). (See Figure 6.29)

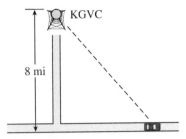

FIGURE 6.29

a. How far has Francine traveled from the junction t minutes after turning onto the highway?

b. Express Francine's distance from the college as a function of t.

c. Graph your function in the window

$$\text{Xmin} = 0 \qquad \text{Xmax} = 23.5$$
$$\text{Ymin} = 0 \qquad \text{Ymax} = 25$$

d. How long will it be until Francine is out of radio range for station KGVC?

34. Brenda is flying at an altitude of 1500 feet on a heading that will take her directly above Van Nuys airport. Her speed is 70 knots (1 knot equals 1.15 miles per hour), and she passes over the airport at precisely 10 a.m. The airport has a radio range of 4 miles. *Hint:* Convert all units to miles. (See Figure 6.30.)

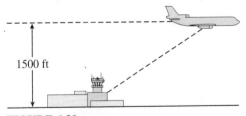

FIGURE 6.30

a. What is Brenda's horizontal distance from the airport at t minutes after 10 a.m.?

b. Express Brenda's distance from the airport as a function of t.

c. Graph your function in the window

$$Xmin = 0 \qquad Xmax = 0.1$$
$$Ymin = 0 \qquad Ymax = 5$$

d. How long will it be before Brenda is out of radio range for Van Nuys airport?

35. A UFO is hovering directly over your head at an altitude of 700 feet. It begins descending at a rate of 10 feet per second. At the same time, you start to run at a rate of 15 feet per second.

a. What is the altitude of the UFO t seconds later?

b. How far have you run t seconds later?

c. Sketch a diagram of your position and the UFO at time t seconds. Label the sketch with your answers to parts (a) and (b).

d. Express the distance D between you and the UFO as a function of t. (Use your answers to parts (a) and (b).)

e. What is the distance between you and the UFO when $t = 10$?

f. Graph your function D in the window

$$Xmin = 0 \qquad Xmax = 94$$
$$Ymin = 0 \qquad Ymax = 1240$$

Verify your answer to part (e) on the graph. Use the graph to find the distance between you and the UFO when $t = 45$.

g. When is the distance between you and the UFO the smallest? (Use the graph to estimate to the nearest second.)

h. Actually, the window we used in part (f) is too big. Find the domain of the function D and change Xmax to a more realistic value. (*Hint*: When does the UFO reach the ground?)

36. A private plane flying north at 200 miles per hour leaves St. Louis at noon. A small jet plane flying east at 200 miles per hour is 600 miles west of St. Louis at noon.

a. How far is the jet from St. Louis t hours after noon (in hundreds of miles)?

b. How far is the private plane from St. Louis t hours after noon (in hundreds of miles)?

c. Sketch a diagram showing the positions of the two planes t hours after noon. Label the sketch with your answers to parts (a) and (b).

d. Express the distance d between the two planes as a function of t. (Use your answers to parts (a) and (b).)

e. What is the distance between the two planes at 1 p.m.?

f. Graph your function in the window

$$Xmin = 0 \qquad Xmax = 4.7$$
$$Ymin = 0 \qquad Ymax = 10$$

Verify your answer to part (e) on the graph. Use your graph to find the distance between the planes at 12:30.

g. When is the distance between the two planes the smallest? (Use the graph to estimate to the nearest tenth of an hour.)

h. Actually, the window we used in part (f) is too big. Find the domain of the function d and change Xmax to a more realistic value. (*Hint*: When does the jet reach St. Louis?)

37. Delbert drives past a service station on the highway and 5 miles later turns right onto a country road. After he has traveled for 1 mile the car's engine throws a rod, and he decides to walk back to the service station. Delbert can walk 4 miles per hour along the road, but he estimates that he can only make 3 miles per hour through the fields. He would like to reach the service station in the shortest time possible. (See Figure 6.31.)

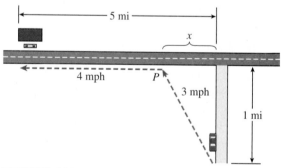

FIGURE 6.31

a. How long will it take him if he retraces his route, walking along the road and then along the highway?

b. How long will it take him if he heads out across the fields, straight for the service station?

c. What if walks through the fields to some intermediate point P on the highway, as shown in the figure, and then continues along the highway? Let the distance from the junction to point P be x miles, and express Delbert's walking time, t, as a function of x. Use the following steps:

    i. Express the distance Delbert walks along the road and the distance he walks through the fields in terms of x.

    ii. Express the time Delbert takes on each part of his walk in terms of x.

    iii. Add the two times you found above.

d. Graph your function in the window

$$\text{Xmin} = -5 \qquad \text{Xmax} = 10$$
$$\text{Ymin} = 0 \qquad \text{Ymax} = 3$$

Locate the points corresponding to your answers to parts (a) and (b).

e. Choose an appropriate domain and range for this problem. (What are the smallest and largest values of x that make sense?)

f. Graph your function on the domain from part (e). Use your graph to find the value of x that results in the shortest walking time for Delbert to reach the service station. What is that time?

38. The telephone company is laying a cable from an island 12 miles offshore to a relay station 18 miles down the coast, as shown in Figure 6.32. It costs $8,000 per mile to lay the cable underwater and $3,000 per mile to lay the cable underground. The telephone company would like to find the cheapest route to run their cable.

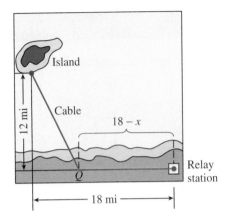

FIGURE 6.32

a. How much would it cost to run the cable directly to the shore (12 miles) and then 18 miles along the shore to the station?

b. How much would it cost to run the cable directly to the station, entirely underwater?

c. What if the phone company runs the cable from the island to some intermediate point Q on the shore, and then continues laying the cable along the shore? Let the distance from the relay station to the point Q be $18 - x$ miles, and express the cost of laying the cable (in thousands of dollars) as a function of x. Use the following steps:

    i. Express the length of the cable laid underwater and the length of the cable laid underground in terms of x.

    ii. Express the cost of laying each portion of the cable in terms of x.

    iii. Add the two costs you found above.

d. Graph your function in the window

$$\text{Xmin} = -10 \qquad \text{Xmax} = 20$$
$$\text{Ymin} = 0 \qquad \text{Ymax} = 200$$

Locate the points corresponding to your answers to parts (a) and (b).

e. Choose an appropriate domain and range for this problem. (What are the smallest and largest values of x that make sense?)

f. Graph your function on the domain from part (e). Use your graph to find the value of x that results in the least expensive route for the cable. What is the cost for that route?

Write without negative exponents and simplify.

1. **a.** $(-3)^{-4}$ **b.** 4^{-3}

2. **a.** $\left(\dfrac{1}{3}\right)^{-2}$ **b.** $\dfrac{3}{5^{-2}}$

3. **a.** $(3m)^{-5}$ **b.** $-7y^{-8}$

4. **a.** $a^{-1} + a^{-2}$ **b.** $\dfrac{3q^{-9}}{r^{-2}}$

5. **a.** $6c^{-7} \cdot 3^{-1}c^4$ **b.** $\dfrac{11z^{-7}}{3^{-2}z^{-5}}$

6. **a.** $(2d^{-2}k^3)^{-4}$ **b.** $\dfrac{2w^3(w^{-2})^{-3}}{5w^{-5}}$

7. The speed of light is approximately 186,000 miles per second.

   a. How long will it take light to travel a distance of 1 foot? (1 mile = 5280 feet) Express your answer in both scientific and standard notation.

   b. How long does it take sunlight to reach the earth, a distance of 92,956,000 miles?

8. The national debt is approaching 6×10^{12} dollars. How many hours would it take you to earn an amount equal to the national debt if you were paid $20 per hour? Express your answer in standard notation, both in terms of hours and in terms of years.

9. In the twenty-first century, spacecraft may be able to travel at speeds of 3×10^7 meters per second, 1000 times their current speed. (At that speed you could circumnavigate the earth in 1.3 seconds.)

   a. How long would it take to reach the sun at this speed? (See Problem 59 in Section 6.1.)

   b. How long would it take to reach Proxima Centauri?

   c. What fraction of the speed of light $(3 \times 10^8$ meters per second) is this speed?

10. **a.** Use the data in the table to calculate the density of each of the planets. (Assume the planets are spherical.)

Planet	Radius (km)	Mass (10^{20} kg)	Density (kg/m^3)
Mercury	2440	3302	
Venus	6052	48,690	
Earth	6378	59,740	
Mars	3397	6419	
Jupiter	71,490	18,990,000	
Saturn	60,270	5,685,000	
Uranus	25,560	866,200	
Neptune	24,765	1,028,000	
Pluto	1150	150	

   b. The planets are composed of three broad categories of materials: rocky materials, "icy" materials (including water), and the materials that dominate the sun, namely hydrogen and helium. The density of rock varies from 3000 to 8000 kg/m^3. Which of the planets could be composed mainly of rock?

   Source: Open University, 1998

11. The faster a cyclist travels, the more energy he expends. Studies indicate that the net energy expenditure is approximately proportional to the square of cycling speed. The data in the table show the energy expenditure of a competitive cyclist.

Cycling speed (m/s), v	5.6	7.5	8.8	10.4	11.8
Energy (cal/kg sec), E	0.9	1.8	2.6	3.9	5

   a. Make a table of values for (v^2, E), and plot the data.

   b. Draw a regression line, and find an equation for energy expenditure as a function of cycling speed.

   c. Is energy expenditure exactly proportional to speed squared? Why or why not?

d. Using your formula from part (b), estimate the net energy expenditure for a speed of 9.5 meters per second.

Source: Pugh, 1974

12. The time needed for a microwave oven to heat up an item is inversely proportional to the power output of the oven. The table shows power output and heating times for a Christmas pudding.

Power output (watts), P	450	500	650	750
Heating time (minutes), T	14.5	13	10	9

a. Make a table for values of $\left(\dfrac{1}{P}, T\right)$, and plot the data.

b. Use a regression line to estimate the value of the constant of proportionality, k.

c. How long would it take to heat the pudding in an 800-watt microwave oven?

d. What power output would be required to heat the pudding in 12 minutes?

Write each power in radical form.

13. a. $25m^{1/2}$ **b.** $8n^{-1/3}$

14. a. $(13d)^{2/3}$ **b.** $6x^{2/5}y^{3/5}$

15. a. $(3q)^{-3/4}$ **b.** $7(uv)^{3/2}$

16. a. $(a^2 + b^2)^{0.5}$ **b.** $(16 - x^2)^{0.25}$

Write each radical as a power with a fractional exponent.

17. a. $2\sqrt[3]{x^2}$ **b.** $\dfrac{1}{4}\sqrt[4]{x}$

18. a. $z^2\sqrt{z}$ **b.** $z\sqrt[3]{z}$

19. a. $\dfrac{6}{\sqrt[4]{b^3}}$ **b.** $\dfrac{-1}{3\sqrt[3]{b}}$

20. a. $\dfrac{-4}{\left(\sqrt[4]{a}\right)^2}$ **b.** $\dfrac{2}{\left(\sqrt{a}\right)^3}$

21. According to the theory of relativity, the mass of an object traveling at velocity, v, is given by the function

$$m = \frac{M}{\sqrt{1 - \dfrac{v^2}{c^2}}}$$

where M is the mass of the object at rest and c is the speed of light. Find the mass of a man traveling at a velocity of $0.7c$ if his rest mass is 80 kilograms.

22. The cylinder of smallest surface area for a given volume has a radius and height both equal to $\sqrt[3]{\dfrac{V}{\pi}}$. Find the dimensions of the tin can of smallest surface area with volume 60 cubic inches.

23. Two businesswomen start a small company to produce saddle bags for bicycles. The number q of saddle bags they can produce depends on the amount of money, m, they invest and the number of hours of labor, w, they employ, according to the formula

$$q = 0.6m^{1/4}w^{3/4}$$

where m is measured in thousands of dollars.

a. If the businesswomen invest \$100,000 and employ 1600 hours of labor in their first month of production, how many saddle bags can they expect to produce?

b. With the same initial investment, how many hours of labor would they need in order to produce 200 saddle bags?

24. A child who weighs w pounds and is h inches tall has a surface area (in square inches) given approximately by

$$S = 8.5h^{0.35}w^{0.55}$$

a. What is the surface area of a child who weighs 60 pounds and is 40 inches tall?

b. What is the weight of a child who is 50 inches tall and whose surface area is 397 square inches?

25. Membership in the Wildlife Society has grown according to the function

$$M(t) = 30t^{3/4}$$

where t is the number of years since its founding in 1970.

   a. Sketch a graph of the function $M(t)$.

   b. What was the society's membership in 1990?

   c. In what year will the membership be 810 people?

26. The heron population in Saltmarsh Refuge is estimated by conservationists at

$$P(t) = 360t^{-2/3}$$

where t is the number of years since the refuge was established in 1990.

   a. Sketch a graph of the function $P(t)$.

   b. How many heron were there in 1995?

   c. In what year will there be only 40 heron left?

27. A brewery wants to replace its old vats with larger ones. To estimate the cost of the new equipment, the accountant uses the 0.6 rule for industrial costs, which states that the cost of a new container is approximately $N = Cr^{0.6}$, where C is the cost of the old container and r is the ratio of the capacity of the new container to the old one.

   a. If an old vat cost $5000, sketch a graph of N as a function of r.

   b. How much should the accountant budget for a new vat that holds 1.8 times as much as the old one?

28. If a quantity of air expands without changing temperature, its pressure in pounds per square inch is given by $P = kV^{-1.4}$, where V is the volume of the air in cubic inches and $k = 2.79 \times 10^4$.

   a. Sketch a graph of P as a function of V.

   b. Find the air pressure of an air sample when its volume is 50 cubic inches.

29. Manufacturers of ships (and other complex products) find that the average cost of producing a ship decreases as more of those ships are produced. This relationship is called the *experience curve*, and is given by the equation

$$C = ax^{-b}$$

where C is the average cost per ship in millions of dollars and x is the number of ships produced. The value of the constant b depends on the complexity of the ship.

   a. What is the significance of the constant of proportionality a? (*Hint:* What is the value of C if only one ship is built?)

   b. For one kind of ship, $b = \frac{1}{8}$ and the cost of producing the first ship is $12 million. Write the equation for C as a function of x using radical notation.

   c. Compute the cost per ship when two ships have been built. By what percent does the cost per ship decrease? By what percent does the cost per ship decrease from building 2 ships to building 4 ships?

   d. By what percent does the average cost decrease from building n ships to building $2n$ ships? (In the shipbuilding industry, the average cost per ship usually decreases by 5 to 10% each time the number of ships doubles.)

   Source: Storch, Hammon, and Bunch, 1988

30. A population is in a period of "supergrowth" if its rate of growth, R, at any time is proportional to P^k, where P is the population at that time and k is a constant greater than 1. Suppose R is given by

$$R = 0.015P^{1.2}$$

where P is measured in thousands and R is measured in thousands per year.

   a. Find R when $P = 20,000$, when $P = 40,000$, and when $P = 60,000$.

   b. What will the population be when its rate of growth is 5000 per year?

   c. Graph R and use your graph to verify your answers to parts (a) and (b).

Evaluate each function for the given values.

31. $Q(x) = 4x^{5/2}$

   a. $Q(16)$ **b.** $Q\left(\dfrac{1}{4}\right)$

   c. $Q(3)$ **d.** $Q(100)$

32. $T(w) = -3w^{2/3}$

   a. $T(27)$ **b.** $T\left(\dfrac{1}{8}\right)$

   c. $T(20)$ **d.** $T(1000)$

Solve.

33. $2\sqrt{w} - 5 = 21$

34. $16 - 3\sqrt{w} = -5$

35. $12 - \sqrt{5v + 1} = 3$

36. $3\sqrt{17 - 4v} - 8 = 19$

37. $x - 3\sqrt{x + 2} = 0$

38. $\sqrt{x + 1} + \sqrt{x + 8} = 7$

39. $(x + 7)^{1/2} + x^{1/2} = 7$

40. $(y - 3)^{1/2} + (y + 4)^{1/2} = 7$

41. $\sqrt[3]{x + 1} = 2$

42. $x^{2/3} + 2 = 6$

43. $(x - 1)^{-3/2} = \dfrac{1}{8}$

44. $(2x + 1)^{-1/2} = \dfrac{1}{3}$

Solve each formula for the indicated variable.

45. $t = \sqrt{\dfrac{2v}{g}}$, for g

46. $q - 1 = 2\sqrt{\dfrac{r^2 - 1}{3}}$, for r

47. $R = \dfrac{1 + \sqrt{p^2 + 1}}{2}$, for p

48. $q = \sqrt[3]{\dfrac{1 + r^2}{2}}$, for r

49. Two highways intersect at right angles as shown in Figure 6.33. At the instant when a car heading east at 50 miles per hour passes the intersection a car traveling north at 40 miles per hour is already 5 mile north of the intersection. When will the cars be 200 miles apart?

50. Two radio antennae are 75 feet apart. They are supported by a guy wire attached to the first antenna at a height of 20 feet, anchored to the ground between the antennae, and attached to the second antenna at a height of 25 feet. If the wire is 90 feet long, how far from the base of the second antenna should it be anchored to the ground? (See Figure 6.34.)

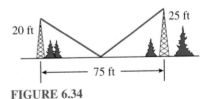

FIGURE 6.34

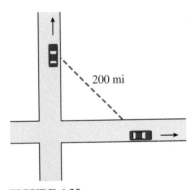

FIGURE 6.33

51. Find the perimeter of the triangle with vertices $A(-1, 2)$, $B(5, 4)$, $C(1, -4)$. Determine if $\triangle ABC$ is a right triangle.

52. Find the perimeter of the triangle obtained by joining the midpoints of the sides of $\triangle ABC$ from Problem 51.

Graph each equation.

53. $x^2 + y^2 = 81$

54. $3x^2 + 3y^2 = 12$

55. $(x - 2)^2 + (y + 4)^2 = 9$

56. $x^2 + y^2 + 6y = 0$

Write the equation for the circle with the given properties.

57. Center at $(5, -2)$, radius $4\sqrt{2}$

58. Center at $(7, -1)$, one point of the circle $(2, 3)$

59. Endpoints of diameter at $(-2, 3)$ and $(4, 5)$

60. Passing through $(6, 2)$, $(-1, 1)$, and $(0, -6)$.

Write each radical or radical expression in simplest form.

61. a. $\sqrt{\dfrac{125p^9}{a^4}}$ **b.** $\sqrt[3]{\dfrac{24v^2}{w^6}}$

62. a. $\dfrac{\sqrt{a^5 b^3}}{\sqrt{ab}}$ **b.** $\dfrac{\sqrt{x}\sqrt{xy^3}}{\sqrt{y}}$

63. a. $\sqrt[3]{8a^3 - 16b^6}$ **b.** $\sqrt[3]{8a^3}\,\sqrt[3]{-16b^6}$

64. a. $\sqrt{4t^2 + 24t^6}$ **b.** $\sqrt{4t^2}\,\sqrt{24t^6}$

65. a. $\left(x - 2\sqrt{x}\right)^2$

66. a. $2v\sqrt{v} + 3\sqrt{v^3} - \dfrac{v^2}{\sqrt{v}}$

   b. $\left(x - 2\sqrt{x}\right)\left(x + 2\sqrt{x}\right)$

   b. $12\sqrt[3]{w^2} - \sqrt[3]{w}\left(-2\sqrt[3]{w}\right) + \dfrac{8w}{\sqrt[3]{w}}$

Rationalize the denominator.

67. a. $\dfrac{7}{\sqrt{5y}}$ **b.** $\dfrac{6d}{\sqrt{2d}}$

68. a. $\sqrt{\dfrac{3r}{11s}}$ **b.** $\sqrt{\dfrac{26}{2m}}$

69. a. $\dfrac{-3}{\sqrt{a} + 2}$ **b.** $\dfrac{-3}{\sqrt{z} - 4}$

70. a. $\dfrac{2x - \sqrt{3}}{x - \sqrt{3}}$ **b.** $\dfrac{m - \sqrt{3}}{5m + 2\sqrt{3}}$

Exponential and Logarithmic Functions

7.1 Exponential Growth and Decay

We first consider two examples of population growth.

INVESTIGATION 11

Population Growth

A. In a laboratory experiment, researchers establish a colony of 100 bacteria and monitor its growth. The experimenters discover that the colony triples in population every day.

1. Fill in Table 7.1 showing the population, $P(t)$, of bacteria t days later.

TABLE 7.1

t	$P(t)$
0	100
1	
2	
3	
4	
5	

$P(0) = 100$

$P(1) = 100 \cdot 3 =$

$P(2) = [100 \cdot 3] \cdot 3 =$

$P(3) =$

$P(4) =$

$P(5) =$

2. Plot the data points from Table 7.1 on Figure 7.1 and connect them with a smooth curve.

3. Write a function that gives the population of the colony at any time t in days. (Express the values you calculated in part (1) using powers of 3. Do you see a connection between the value of t and the exponent on 3?)

4. Graph your function from part (3) using a calculator. (Use Table 7.1 to choose an appropriate domain and range.) The graph should resemble your hand-drawn graph from part (2).

5. Evaluate your function to find the number of bacteria present after 8 days. How many bacteria are present after 36 hours?

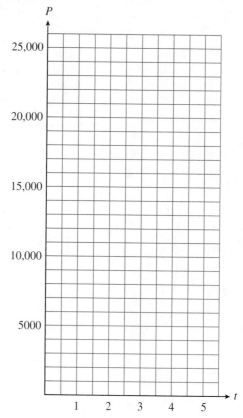

FIGURE 7.1

B. Under ideal conditions the number of rabbits in a certain area can double every 3 months. A rancher estimates that 60 rabbits live on his land.

1. Fill in Table 7.2 showing the population, $P(t)$, of rabbits t months later.

TABLE 7.2

t	$P(t)$
0	60
3	
6	
9	
12	
15	

$P(0) = 60$

$P(3) = 60 \cdot 2 =$

$P(6) = [60 \cdot 2] \cdot 2 =$

$P(9) =$

$P(12) =$

$P(15) =$

2. Plot the data points from Table 7.2 on Figure 7.2 and connect them with a smooth curve.

3. Write a function that gives the population of rabbits at any time t in months. (Express the values you calculated in part (a) using powers

of 2. Note that the population of rabbits is multiplied by 2 every 3 months. If you know the value of t, how do you find the corresponding exponent in $P(t)$?

4. Graph your function from part (3) using a calculator. (Use Table 7.2 to choose an appropriate domain and range.) The graph should resemble your hand-drawn graph from part (2).

5. Evaluate your function to find the number of rabbits present after 2 years. How many rabbits are present after 8 months?

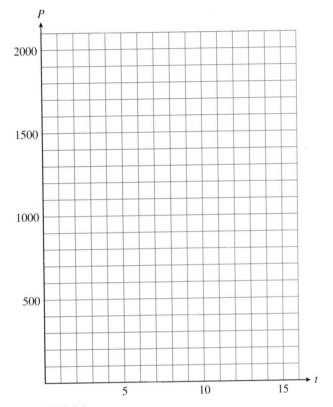

FIGURE 7.2

Growth Factors

The functions in Investigation 11 describe **exponential growth.** During each time interval of a fixed length the population is *multiplied* by a certain constant amount. In Part A the bacteria population grows by a factor of 3 every day. For this reason we say that 3 is the **growth factor** for the function. Functions that describe exponential growth can be expressed in the standard form

$$P(t) = P_0 a^t$$

where $P_0 = P(0)$ is the initial value of the function and a is the growth factor. For the bacteria population we have

$$P(t) = 100 \cdot 3^t$$

so $P_0 = 100$ and $a = 3$.

In Part B the rabbit population grew by a factor of 2 every 3 months. We can write the function as

$$P(t) = 60 \cdot 2^{t/3}$$

To see the growth factor for this function, we use the third law of exponents to write $2^{t/3}$ in another form. Recall that to raise a power to a power we multiply exponents, so

$$(2^{1/3})^t = 2^{t(1/3)} = 2^{t/3}$$

The growth law for the rabbit population is thus

$$P(t) = 60 \cdot (2^{1/3})^t$$

so that the initial value of the function is $P_0 = 60$ and the growth factor is $a = 2^{1/3}$, or approximately 1.26. The rabbit population grows by a factor of about 1.26 every month.

If the units are the same, a population with a larger growth factor grows faster than one with a smaller growth factor.

EXAMPLE 1

A lab technician compares the growth of two species of bacteria. She starts two colonies of 50 bacteria each. Species A doubles in population every 2 days, and species B triples every 3 days. Find the growth factor for each species.

Solution

A function describing the growth of species A is

$$P(t) = 50 \cdot 2^{t/2} = 50 \cdot (2^{1/2})^t$$

so the growth factor for species A is $2^{1/2}$, or approximately 1.41. For species B,

$$P(t) = 50 \cdot 3^{t/3} = 50 \cdot (3^{1/3})^t$$

so the growth factor for species B is $3^{1/3}$, or approximately 1.44. Species B grows faster than species A.   ■

> **EXERCISE 1** In 1999, analysts expected the number of Internet service providers to double in five years.
>
> **a.** What was the annual growth factor for the number of Internet service providers?
>
> **b.** If there were 5078 Internet service providers in April of 1999, estimate the number of providers in April of 2000 and in April of 2001.
>
> **c.** Write a formula for $I(t)$, the number of Internet service providers t years after 1999.
>
> Source: LA Times, Sept 6, 1999

Percent Increase

Exponential growth occurs in other circumstances, too. For example, if the interest on a savings account is compounded annually, the amount of money in the account grows exponentially.

Consider a principal of $100 invested at 5% interest compounded annually. At the end of 1 year the amount is

$$\text{Amount} = \text{Principal} + \text{Interest}$$

$$A = P + Prt$$
$$= 100 + 100(0.05)(1) = 105$$

It will be more useful to write the formula for the amount in factored form.

$$A = P + Prt \qquad \text{Factor out } P.$$
$$= P(1 + rt)$$

With this version of the formula, the calculation for the amount at the end of 1 year looks like this:

$$A = P(1 + rt)$$
$$= 100[1 + 0.05(1)]$$
$$= 100(1.05) = \mathbf{105}$$

The amount, $105, becomes the new principal for the second year. To find the amount at the end of the second year, we apply the formula again, with $P = 105$.

$$A = P(1 + rt)$$
$$= \mathbf{105}[1 + 0.05(1)]$$
$$= 105(1.05) = 110.25$$

Observe that to find the amount at the end of each year we multiply the principal by a factor of $1 + r = 1.05$. Thus, we can express the amount at the end of the second year as

$$A = [100(1.05)](1.05)$$
$$= 100(1.05)^2$$

and at the end of the third year as

$$A = [100(1.05)^2](1.05)$$
$$= 100(1.05)^3$$

We organize our results into Table 7.3, where $A(t)$ represents the amount of money in the account after t years.

TABLE 7.3

t	$A(t)$
0	100
1	$100(1.05) = 105$
2	$100(1.05)^2 = 110.25$
3	$100(1.05)^3 = 115.76$

For this example, a formula for the amount after t years is

$$A(t) = 100(1.05)^t$$

In general, for an initial investment of P dollars at an interest rate, r, compounded annually, the amount accumulated after t years is

$$A(t) = P(1 + r)^t$$

This function describes exponential growth with an initial value of P and a growth factor of $a = 1 + r$. The interest rate, r, which indicates the *percent increase* in the account each year, corresponds to a *growth factor* of $1 + r$. The notion of percent increase is often used to describe the growth factor for quantities that grow exponentially.

EXAMPLE 2

During a period of rapid inflation prices rose by 12% over 6 months. At the beginning of the inflationary period a pound of butter cost $2.

  a. Make a table of values showing the rise in the cost of butter over the next two years.

  b. Write a function that gives the price of a pound of butter t years after inflation began.

  c. How much did a pound of butter cost after 3 years? After 15 months?

  d. Graph the function found in (b).

Solutions

  a. The *percent increase* in the cost of butter is 12% every 6 months. Therefore, the *growth factor* for the cost of butter is $1 + 0.12 = 1.12$ every half year. If $P(t)$ represents the price of butter after t years, then $P(0) = 2$ and we multiply the price by 1.12 every half year, as shown in Table 7.4.

TABLE 7.4

t	$P(t)$	
0	2	$P(0) = 2.00$
$\frac{1}{2}$	$2(1.12)$	$P\left(\frac{1}{2}\right) = 2.24$
1	$2(1.12)^2$	$P(1) = 2.51$
$\frac{3}{2}$	$2(1.12)^3$	$P\left(\frac{3}{2}\right) = 2.81$
2	$2(1.12)^4$	$P(2) = 3.15$

  b. Look closely at the second column of Table 7.4. After t years of inflation the original price of $2 has been multiplied $2t$ times by a factor of 1.12. Thus,

$$P = 2(1.12)^{2t}$$

  c. To find the price of a pound of butter at any time after inflation began, we evaluate the function at the appropriate value of t.

$$P(3) = 2(1.12)^{2(3)}$$
$$= 2(1.12)^6 \approx 3.95$$

After 3 years the price was \$3.95. Fifteen months is 1.25 years, so we evaluate $P(1.25)$.

$$P(1.25) = 2(1.12)^{2(1.25)}$$
$$= 2(1.12)^{2.5} \approx 2.66$$

After 15 months the price was \$2.66.

d. To graph the function

$$P(t) = 2(1.12)^{2t}$$

we evaluate it for several values, as shown in Table 7.5. Plot the points and connect them with a smooth curve to obtain the graph shown in Figure 7.3.

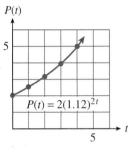

FIGURE 7.3 ∎

TABLE 7.5

t	$P(t)$
0	2.00
1	2.51
2	3.15
3	3.95
4	4.95

In Example 2 we can rewrite the formula for $P(t)$ as follows.

$$P(t) = 2(1.12)^{2t}$$
$$= 2[(1.12)^2]^t$$
$$= 2(1.2544)^t$$

Thus, the annual growth factor for the price of butter is 1.2544, and the annual percent growth rate is 25.44%.

EXERCISE 2 In 1998, the average annual cost of a public college was \$10,069, and costs were climbing by 6% per year.

a. Complete the table and sketch a graph of $C(t)$.

t	0	5	10	15	20	25
$C(t)$						

b. Write a formula for $C(t)$, the cost of one year of college t years after 1998.

c. If the percent growth rate remains steady, how much would a year of college cost in 2001?

d. How much would a year of college cost in 2020?

Exponential Decay

In the examples above, exponential growth was modeled by increasing functions of the form

$$P(t) = P_0 a^t$$

where $a > 1$. The function $P(t) = P_0 a^t$ is a *decreasing* function if $0 < a < 1$. In this case the function is said to describe exponential decay, and the constant a is called the **decay factor.** In Investigation 12 we consider two examples of exponential decay.

Exponential Decay

A. A small coal-mining town has been losing population since 1940, when 5000 people lived there. At each census thereafter (taken at 10-year intervals) the population has declined to approximately 0.90 of its earlier figure.

1. Fill in Table 7.6 showing the population $P(t)$ of the town t years after 1940.

   TABLE 7.6

t	P(t)
0	5000
10	
20	
30	
40	
50	

   $P(0) = 5000$

   $P(10) = 5000 \cdot 0.90 =$

   $P(20) = [5000 \cdot 0.90] \cdot 0.90 =$

   $P(30) =$

   $P(40) =$

   $P(50) =$

2. Plot the data points from Table 7.6 on Figure 7.4 and connect them with a smooth curve.

3. Write a function that gives the population of the town at any time t in years after 1940. (Express the values you calculated in part (1) using powers of 0.90. Do you see a connection between the value of t and the exponent on 0.90?)

4. Graph your function from part (3) using a calculator. (Use Table 7.6 to choose an appropriate domain and range.) The graph should resemble your hand-drawn graph from part (2).

5. Evaluate your function to find the population of the town in 1995. What was the population in 2000?

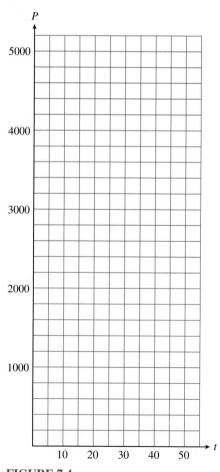

FIGURE 7.4

B. A plastic window coating 1 millimeter thick decreases the light coming through the window by 25%. This means that 75% of the original amount of light comes through 1 millimeter of the coating. Each additional millimeter of coating reduces the light by another 25%.

1. Fill in Table 7.7 showing the percent of the light $P(x)$ that shines through x millimeters of the window coating.

   TABLE 7.7

x	$P(x)$
0	100
1	
2	
3	
4	
5	

   $P(0) = 100$

   $P(1) = 100 \cdot 0.75 =$

   $P(2) = [100 \cdot 0.75] \cdot 0.75 =$

   $P(3) =$

   $P(4) =$

   $P(5) =$

2. Plot the data points from Table 7.7 on Figure 7.5 and connect them with a smooth curve.

3. Write a function that gives the percent of the light that shines through x millimeters of the coating. (Express the values you calculated in part (1) using powers of 0.75. Do you see a connection between the value of x and the exponent on 0.75?)

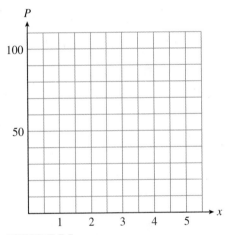

FIGURE 7.5

4. Graph your function from part (3) using a calculator. (Use your table of values to choose an appropriate domain and range.) The graph should resemble your hand-drawn graph from part (2).

5. Evaluate your function to find the percent of the light that comes through 6 millimeters of plastic coating. What percent comes through $\frac{1}{2}$ millimeter?

Decay Factor

On page 422 we noted that a percent increase of r (in decimal form) corresponds to a growth factor of $a = 1 + r$. A percent decrease of r corresponds to a decay factor of $a = 1 - r$. In Part B of Investigation 12, each millimeter of plastic reduced the amount of light by 25%, so $r = 0.25$, and the decay factor for the function $P(x)$ is

$$a = 1 - r$$
$$= 1 - 0.25 = 0.75$$

EXERCISE 3 The number of butterflies visiting a nature station is declining by 18% per year. In 1998, 3600 butterflies visited the nature station.

a. What is the decay factor in the annual butterfly count?

b. Write a formula for $B(t)$, the number of butterflies t years after 1998.

c. Complete the table and sketch a graph of $B(t)$.

t	0	2	4	6	8	10
$B(t)$						

We summarize our observations about exponential growth and decay functions as follows.

Exponential Growth and Decay Functions

The function

$$P(t) = P_0 a^t$$

models exponential growth and decay.

$$P_0 = P(0) \text{ is the } \textbf{initial value} \text{ of } P;$$

$$a \text{ is the } \textbf{growth} \text{ or } \textbf{decay factor.}$$

1. If $a > 1$, $P(t)$ is increasing, and $a = 1 + r$, where r represents percent increase.

2. If $0 < a < 1$, $P(t)$ is decreasing, and $a = 1 - r$, where r represents percent decrease.

Comparing Linear Growth and Exponential Growth

It may be helpful to compare the notions of linear growth and exponential growth. Consider the two functions

$$L(t) = 5 + 2t \qquad \text{and} \qquad E(t) = 5 \cdot 2^t \qquad (t \geq 0)$$

whose graphs are shown in Figure 7.6.

L is a linear function with y-intercept 5 and slope 2; E is an exponential function with initial value 5 and growth factor 2. The growth factor of an exponential function is in a sense analogous to the slope of a linear function: Each measures how quickly the function is increasing (or decreasing). However, for each unit increase in t, 2 units are *added* to the value of $L(t)$, whereas the value of $E(t)$ is *multiplied* by 2. An exponential function with growth factor 2 grows much more rapidly than a linear function with slope 2. You can see the rapid growth of the exponential function by comparing the graphs in Figure 7.6 or the function values in Table 7.8.

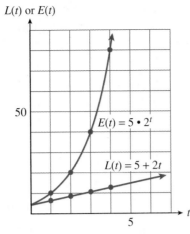

$L(t)$ or $E(t)$

50

$E(t) = 5 \cdot 2^t$

$L(t) = 5 + 2t$

5   t

FIGURE 7.6

TABLE 7.8

t	$L(t)$	$E(t)$
0	5	5
1	7	10
2	9	20
3	11	40
4	13	80

EXAMPLE 3

A solar energy company sold $80,000 worth of solar collectors last year, its first year of operation. This year its sales rose to $88,000, an increase of 10%. The marketing department must estimate its projected sales for the next 3 years.

a. If the marketing department predicts that sales will grow linearly, what should it expect the sales total to be next year? Graph the projected sales figures over the next 3 years, assuming that sales will grow linearly.

b. If the marketing department predicts that sales will grow exponentially, what should it expect the sales total to be next year? Graph the projected sales figures over the next 3 years, assuming that sales will grow exponentially.

TABLE 7.9

t	$L(t)$	$E(t)$
0	80,000	80,000
1	88,000	88,000
2	96,000	96,800
3	104,000	106,480
4	112,000	117,128

Solutions

a. Let $L(t)$ represent the company's total sales t years after starting business, where $t = 0$ is considered the first year of operation. Assuming that sales grow linearly, L is a linear function of the form $L(t) = mt + b$. $L(0) = 80,000$, so the intercept b is 80,000. The slope m of the graph is

$$\frac{\Delta S}{\Delta t} = \frac{8000 \text{ dollars}}{1 \text{ year}} = 8000$$

where $\Delta S = 8000$ is the increase in sales during the first year. Thus, $L(t) = 8000t + 80,000$, and the expected sales total for the next year is

$$L(2) = 8000\,(2) + 80,000 = 96,000$$

b. Let $E(t)$ represent the company's sales under the assumption that sales will grow exponentially. Then E is a function of the form $E(t) = E_0 a^t$. The percent increase in sales over the first year was $r = 0.10$, so the growth rate is $a = 1 + r = 1.10$. The initial value E_0 is 80,000. Thus, $E(t) = 80,000(1.10)^t$, and the expected sales total for the next year is

$$E(2) = 80,000(1.10)^2$$
$$= 96,800$$

Evaluate each function at several points (as in Table 7.9) to obtain the graphs shown in Figure 7.7.   ∎

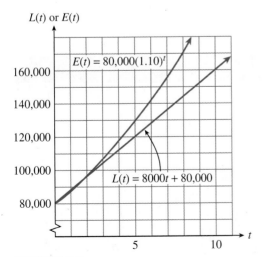

$L(t)$ or $E(t)$

$E(t) = 80,000(1.10)^t$

160,000

140,000

120,000

100,000

$L(t) = 8000t + 80,000$

80,000

5   10 t

FIGURE 7.7

1. State a formula for exponential growth.
2. How is the formula for exponential decay different from the formula for exponential growth?
3. What is the growth factor for a population that grows 4% annually?
4. What is the decay factor for a population that declines by 4% annually?
5. Explain the difference between linear growth and exponential growth.

ANSWERS TO 7.1 EXERCISES

1a. $2^{1/5}$ **b.** 5733 and 6700   **c.** $I(t) = 5078 \cdot 2^{t/5}$

2a.

t	0	5	10	15	20	25
$C(t)$	10,069	13,475	18,032	24,131	32,293	43,215

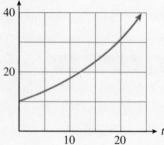

b. $C(t) = 10{,}069 \cdot 1.06^t$ **c.** \$11,992 per year   **d.** \$36,284

3a. 0.82   **b.** $B(t) = 3600 \cdot 0.82^t$

c.

t	0	2	4	6	8	10
$B(t)$	3600	2421	1628	1094	736	495

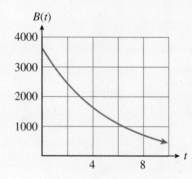

HOMEWORK 7.1

For Problems 1–8,

a. Complete the table of values.

b. Write a function that describes the exponential growth.

c. Graph the function.

d. Evaluate the function at the given values.

1. A colony of bacteria starts with 300 organisms and doubles every week. How many bacteria will there be after 8 weeks? After 5 days?

Weeks	0	1	2	3	4
Bacteria					

2. A population of 24 fruit flies triples every month. How many fruit flies will there be after 6 months? After 3 weeks? (Assume that a month equals 4 weeks.)

Months	0	1	2	3	4
Fruit flies					

3. A typical beehive contains 20,000 insects. The population can increase in size by a factor of 2.5 every 6 weeks. How many bees could there be after 4 weeks? After 20 weeks?

Weeks	0	6	12	18	24
Bees					

4. A rancher who started with 800 head of cattle finds that his herd increases by a factor of 1.8 every 3 years. How many head of cattle will he have after 1 year? After 10 years?

Years	0	3	6	9	12
Cattle					

5. A sum of $4000 is invested in an account that pays 8% interest compounded annually. How much is in the account after 2 years? After 10 years?

Years	0	1	2	3	4
Account balance					

6. Otto invests $600 in an account that pays 7.3% interest compounded annually. How much is in Otto's account after 3 years? After 6 years?

Years	0	1	2	3	4
Account balance					

7. Since 1963 housing prices have risen an average of 5% per year. Paul bought a house for $20,000 in 1963. How much was the house worth in 1975? In 1990?

Years since 1963	0	1	2	3	4
Value of house					

8. Sales of Windsurfers have increased 12% per year since 1990. If Sunsails sold 1500 Windsurfers in 1990, how many did it sell in 1995? How many should it sell in 2002?

Years since 1990	0	1	2	3	4
Windsurfers					

For Problems 9–14,

a. Complete the table of values.

b. Write a function that describes exponential decay.

c. Graph the function.

d. Evaluate the function at the given values.

9. During a vigorous spraying program the mosquito population was reduced to $\frac{3}{4}$ of its previous size every 2 weeks. If the mosquito population was originally estimated at 250,000, how many mosquitoes remained after 3 weeks of spraying? After 8 weeks?

Weeks	0	2	4	6	8
Mosquitoes					

10. The number of perch in Hidden Lake has declined to half of its previous level every 5 years since 1960, when the perch population was estimated at 8000. How many perch were there in 1970? In 1988?

Years	0	5	10	15	20
Perch					

11. Scuba divers find that the water in Emerald Lake filters out 15% of the sunlight for each 4 feet that they descend. How much sunlight penetrates to a depth of 20 feet? To a depth of 45 feet?

Feet	0	4	8	12	16
% of light					

12. Arch's motorboat cost $15,000 in 1990 and has depreciated by 10% every 3 years. How much was the boat worth in 1999? In 2000?

Years	0	3	6	9	12
Value of boat					

13. Plutonium-238 is a radioactive element that decays over time into a less harmful element at a rate of 0.8% per year. A power plant has 50 pounds of plutonium-238 to dispose of. How much plutonium-238 will be left after 10 years? After 100 years?

Years	0	1	2	3	4
Pounds of plutonium-238					

14. Iodine-131 is a radioactive element that decays at a rate of 8.3% per day. How much of a 12-gram sample will be left after 1 week? After 15 days?

Days	0	1	2	3	4
Grams of iodine-131					

15. a. Explain why $P(t) = 2 \cdot 3^t$ and $Q(t) = 6^t$ are not the same function.

   b. Complete the table of values for P and Q, showing that their values are not the same.

t	0	1	2
$P(t)$			
$Q(t)$			

16. a. Explain why $P(t) = 4 \cdot \left(\frac{1}{2}\right)^t$ and $Q(t) = 2^t$ are not the same function.

   b. Complete the table of values for P and Q, showing that their values are not the same.

t	0	1	2
$P(t)$			
$Q(t)$			

17. In the 1940s David Lack undertook a study of the European robin. He tagged 130 one-year-old robins and found that on average 35.6% of the birds survived each year.

   a. According to the data, how many robins would have originally hatched to produce 130 one-year-olds?

   b. Write a formula for the number of the original robins still alive after t years.

   c. Graph your function.

   d. One of the original robins actually survived for 9 years. How many robins does the model predict will survive for 9 years?

   Source: Burton, 1998.

18. Many insects grow by discrete amounts each time they shed their exoskeletons. Dyar's rule says that the size of the insect increases by a constant ratio at each stage.

   a. Dyar measured the width of the head of a caterpillar of a swallowtail butterfly at each stage. The caterpillar's head was initially approximately 42 millimeters wide, and 63.84 millimeters wide after its first stage. Find the growth ratio.

   b. Write an equation for the width of the caterpillar's head at the nth stage.

   c. Graph your equation.

   d. What head width does the model predict after 5 stages?

   Source: Burton, 1998.

19. In 1772, the astronomer Johann Bode promoted a formula for the orbital radii of the six planets known at the time. This formula calculated the orbital radius, r, as a function of the planet's position, n, in line from the sun.

a. Evaluate Bode's law,

$$r(n) = 0.4 + 0.3(2^{n-1})$$

for the values in the table. (Use a large negative number, such as $n = -100$, to approximate $r(-\infty)$.)

n	$-\infty$	1	2	3	4	5	6
$r(n)$							

b. How do the values of $r(n)$ compare with the actual orbital radii of the planets shown in Table 7.10? (The radii are given in astronomical units (AU). One AU is the distance from the earth to the sun, about 149.6×10^6 kilometers.) Assign values of n to each of the planets so that they conform to Bode's law.

TABLE 7.10

Planet	Mercury	Venus	Earth
Orbital radius (AU)	0.39	0.72	1.00

Planet	Mars	Jupiter	Saturn
Orbital radius (AU)	1.52	5.20	9.54

c. In 1781, William Herschel discovered the planet Uranus at a distance of 19.18 AU from the sun. If $n = 7$ for Uranus, what does Bode's law predict for the orbital radius of Uranus?

d. None of the planets' orbital radii corresponds to $n = 2$ in Bode's law. However, in 1801 the first of a group of asteroids between the orbits of Mars and Jupiter was discovered. The asteroids have orbital radii between 2.5 and 3.0 AU. If we consider the asteroids as one "planet" at $n = 2$, what orbital radius does Bode's law predict?

e. In 1846 Neptune was discovered 30.6 AU from the sun, and in 1930 Pluto was discovered 39.4 AU from the sun. What orbital radii does Bode's law predict for these planets?
Source: Bolton, 1974.

20. In 1665 there was an outbreak of the plague in London. Table 7.11 shows the number of people who died of plague during each week of the summer that year.

TABLE 7.11

Week	Deaths	Week	Deaths
0, May 9	9	12, August 1	2010
1, May 16	3	13, August 8	2817
2, May 23	14	14, August 15	3880
3, May 30	17	15, August 22	4237
4, June 6	43	16, August 29	6102
5, June 13	112	17, September 5	6988
6, June 20	168	18, September 12	6544
7, June 27	267	19, September 19	7165
8, July 4	470	20, September 26	5533
9, July 11	725	21, October 3	4929
10, July 18	1089	22, October 10	4327
11, July 25	1843		

a. Scale horizontal and vertical axes for the entire data set, but plot only the data for the first 8 weeks of the epidemic, from May 9 through July 4. On the same axes graph the function $f(x) = 3.98(1.83)^x$.

b. At what rate was the number of victims increasing during the first 8 weeks?

c. Add data points for July 11 through October 10 to your graph. Describe the progress of the epidemic relative to the function f, and offer an explanation.

d. Make a table showing the total number of plague victims at the end of each week, and plot the data. Describe the graph.
Source: Bolton, 1974.

Complete each table describing exponential growth or decay. Round values to two decimal places if necessary. (Hint: First find the growth or decay factor.)

21.

t	P
0	8
1	12
2	18
3	
4	

22.

t	P
0	4
1	5
2	6.25
3	
4	

23.

x	Q
0	20
1	24
2	
3	
4	

24.

x	Q
0	100
1	105
2	
3	
4	

25.

w	N
0	120
1	96
2	
3	
4	

26.

w	N
0	640
1	480
2	
3	
4	

Solve each equation. (See Section 6.2 to review solving equations involving a power of the variable.) Round your answer to two places if necessary.

27. $768 = 12a^3$

28. $1875 = 3a^4$

29. $14{,}929.92 = 5000a^6$

30. $151{,}875 = 20{,}000a^5$

31. $1253 = 260(1 + r)^{12}$

32. $116{,}473 = 48{,}600(1 + r)^{15}$

33. $56.27 = 78(1 - r)^8$

34. $10.56 = 12.4(1 - r)^{20}$

Calculate the growth factor for each population to determine which one grows faster.

35. A researcher starts 2 populations of fruit flies of different species, each with 30 flies. Species A increases by 30% in 6 days and species B increases by 20% in 4 days.

    a. What was the population of species A after 6 days? Find the daily growth factor for species A.

    b. What was the population of species B after 4 days? Find the daily growth factor for species B.

    c. Which species multiplies more rapidly?

36. A biologist isolates two strains of a particular virus and monitors the growth of each, starting with samples of 0.01 gram. Strain A increases by 10% in 8 hours and strain B increases by 12% in 9 hours.

    a. How much did the sample of strain A weigh after 8 hours? What was its hourly growth factor?

    b. How much did the sample of strain B weigh after 9 hours? What was its hourly growth factor?

    c. Which strain of virus grows more rapidly?

In Problems 37–40, assume that each population grows exponentially with constant annual percent increase, _r_.

37. **a.** The population of the state of Texas was 9,579,700 in 1960. Write a formula in terms of r for the population of Texas t years later.

    b. In 1970, the population was 11,196,700. Write an equation and solve for r. What was the annual percent rate of increase to the nearest hundredth of a percent?

38. **a.** The population of the state of Florida was 4,951,600 in 1960. Write a formula in terms of r for the population of Florida t years later.

    b. In 1970 the population was 6,789,400. Write an equation and solve for r. What was the annual percent rate of increase to the nearest hundredth of a percent?

39. a. The population of Rainville was 10,000 in 1970 and doubled in 20 years. What was the annual percent rate of increase to the nearest hundredth percent?

b. The population of Elmira was 350,000 in 1970 and doubled in 20 years. What was the annual percent rate of increase to the nearest hundredth of a percent?

c. If a population doubles in 20 years, does the percent rate of increase depend on the size of the original population?

d. The population of Grayling doubled in 20 years. What was the annual percent rate of increase to the nearest hundredth of a percent?

40. a. The population of Boomtown was 300 in 1908 and tripled in 7 years. What was the annual percent rate of increase to the nearest hundredth of a percent?

b. The population of Fairview was 15,000 in 1962 and tripled in 7 years. What was the annual percent rate of increase to the nearest hundredth of a percent?

c. If a population triples in 7 years, does the percent rate of increase depend on the size of the original population?

d. The population of Pleasant Lake tripled in 7 years. What was the annual percent rate of increase to the nearest hundredth of a percent?

In Problems 41–44, compare linear and exponential growth.

41. At a large university three students start a rumor that final exams have been canceled. After two hours six students (including the first three) have heard the rumor.

a. Assuming that the rumor grows linearly, complete the table below for $L(t)$, the number of students who have heard the rumor after t hours. Then write a formula for the function $L(t)$. Graph the function.

t	0	2	4	6	8
$L(t)$					

b. Complete the table below assuming that the rumor grows exponentially. Write a formula for the function $E(t)$, and graph it on the same set of axes with $L(t)$.

t	0	2	4	6	8
$E(t)$					

42. Over the weekend the Midland Infirmary identifies four cases of Asian flu. Three days later it has treated a total of ten cases.

a. Assuming that the number of flu cases grows linearly, complete the table below for $L(t)$, the number of people infected after t days. Then write a formula for the function $L(t)$. Graph the function.

t	0	3	6	9	12
$L(t)$					

b. Complete the table below assuming that the flu grows exponentially. Write a formula for the function $E(t)$, and graph it on the same set of axes with $L(t)$.

t	0	3	6	9	12
$E(t)$					

43. A population of geese grew from 400 to 720 in 5 years.

a. If the population grew linearly, what was its annual rate of growth?

b. If the population grew exponentially, what was its annual growth factor? What was its annual percent increase?

44. The world's population of tigers declined from 10,400 in 1980 to 6000 in 1998.

a. If the population declined linearly, what was its annual rate of decay?

b. If the population declined exponentially, what was its annual decay factor? What was its annual percent rate of decrease?

7.2 Exponential Functions

In Section 7.1 we studied functions that describe exponential growth or decay. More formally, we define an **exponential function** to be one of the form

$$f(x) = ab^x, \quad \text{where} \quad b > 0 \quad \text{and} \quad b \neq 1, \quad a \neq 0$$

Some examples of exponential functions are

$$f(x) = 5^x, \quad P(t) = 250(1.7)^t, \quad \text{and} \quad g(n) = 2.4(0.3)^n$$

The constant a is the y-intercept of the graph because

$$f(0) = a \cdot b^0 = a \cdot 1 = a$$

For the examples above, we find that the y-intercepts are

$$f(0) = 5^0 = 1$$
$$P(0) = 250(1.7)^0 = 250, \quad \text{and}$$
$$g(0) = 2.4(0.3)^0 = 2.4$$

The positive constant b is called the **base** of the exponential function. There are several constraints on the bases. We do not allow $b < 0$ as a base because if b is negative, then b^x is not a real number for some values of x. (For example, if $b = -4$ and $f(x) = (-4)^x$, then $f\left(\frac{1}{2}\right) = (-4)^{1/2}$ is an imaginary number.) We also exclude $b = 1$ as a base because $1^x = 1$ for all values of x; hence the function $f(x) = 1^x$ is actually the constant function $f(x) = 1$.

Graphs of Exponential Functions

The graphs of exponential functions have two characteristic shapes, depending on whether the base b is greater than 1 or less than 1. As typical examples, consider the graphs of $f(x) = 2^x$ and $g(x) = \left(\frac{1}{2}\right)^x$ shown in Figure 7.8. Some values for f and g are recorded in Tables 7.12 and 7.13.

TABLE 7.12

x	$f(x)$
-3	$\frac{1}{8}$
-2	$\frac{1}{4}$
-1	$\frac{1}{2}$
0	1
1	2
2	4
3	8

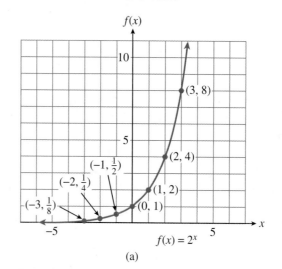

(a)

FIGURE 7.8

TABLE 7.13

x	$g(x)$
-3	8
-2	4
-1	2
0	1
1	$\frac{1}{2}$
2	$\frac{1}{4}$
3	$\frac{1}{8}$

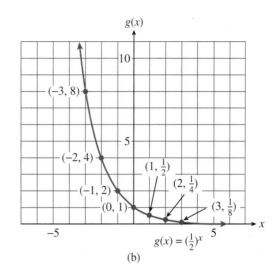

(b)

Notice that $f(x) = 2^x$ is an increasing function and $g(x) = \left(\dfrac{1}{2}\right)^x$ is a decreasing function. In general, exponential functions have the following properties.

Properties of Exponential Functions, $f(x) = ab^x$, $a > 0$

1. Domain: all real numbers.
2. Range: all positive numbers.
3. If $b > 1$, the function is increasing; if $0 < b < 1$, the function is decreasing.

In Table 7.12 you can see that as the x-values decrease toward negative infinity, the corresponding y-values decrease toward zero. As a result, the graph of f decreases toward the x-axis as we move to the left. Thus, the negative x-axis is a horizontal asymptote for exponential functions with $b > 1$. For exponential functions with $0 < b < 1$ the positive x-axis is an asymptote, as illustrated in Figure 7.8b. (See Section 5.3 to review asymptotes.)

In Example 1 we compare two increasing exponential functions. The larger the value of the base, b, the faster the function grows. In this example, both functions have $a = 1$.

EXAMPLE 1

Compare the graphs of $f(x) = 3^x$ and $g(x) = 4^x$.

Solution

Evaluate each function for several convenient values, as shown in Table 7.14.

TABLE 7.14

x	$f(x)$	$g(x)$
-2	$\frac{1}{9}$	$\frac{1}{16}$
-1	$\frac{1}{3}$	$\frac{1}{4}$
0	1	1
1	3	4
2	9	16

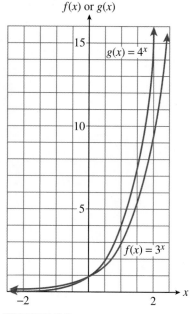

FIGURE 7.9

   Plot the points for each function and connect them with smooth curves. For positive x-values, $g(x)$ is always larger than $f(x)$, and is increasing more rapidly. In Figure 7.9, $g(x) = 4^x$ climbs more rapidly than $f(x) = 3^x$. Both graphs cross the y-axis at $(0, 1)$. ■

   For decreasing exponential functions, those with bases between 0 and 1, the smaller the base the more steeply the graph decreases. For example, compare the graphs of $p(x) = 0.8^x$ and $q(x) = 0.5^x$ shown in Figure 7.10.

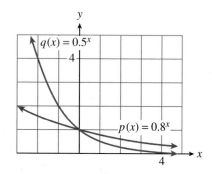

FIGURE 7.10

EXERCISE 1

a. State the ranges of the functions f and g from Example 1 on the domain $[-2, 2]$.

b. State the ranges of the functions p and q shown in Figure 7.10 on the domain $[-2, 2]$. Round your answers to two decimal places.

You may recall from Chapter 4 that the graphs of the functions $y = x^2 - 4$ and $y = (x - 4)^2$ have the same shape as the basic parabola, $y = x^2$. In a similar way, we can shift or stretch the graph of an exponential function while the basic shape is preserved. (We'll learn more about transformations of graphs in Chapter 10.)

EXAMPLE 2

Use a graphing calculator to graph the following functions. Describe how these graphs compare with the graph of $y = 2^x$.

a. $f(x) = 2^x + 3$ **b.** $g(x) = 2^{x+3}$

Solutions

The graphs of $y = 2^x$, $f(x) = 2^x + 3$, and $g(x) = 2^{x+3}$ are shown in Figure 7.11 using the standard window.

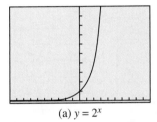

FIGURE 7.11

(a) $y = 2^x$ (b) $y = 2^x + 3$ (c) $y = 2^{x+3}$

a. The graph of $f(x) = 2^x + 3$ shown in Figure 7.11(b) has the same basic shape as that of $y = 2^x$, but it has a horizontal asymptote at $y = 3$ instead of at $y = 0$ (the x-axis). If every point on the graph of $y = 2^x$ were to move 3 units upward, the result would be the graph of $y = 2^x + 3$. If we graph $f(x) = 2^x + 3$ in the window

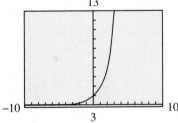

$$\text{Xmin} = -10 \quad \text{Xmax} = 10$$
$$\text{Ymin} = 3 \quad \text{Ymax} = 13$$

as shown in Figure 7.12, we can see that it has the same shape as $y = 2^x$.

FIGURE 7.12

b. The graph of $g(x) = 2^{x+3}$ shown in Figure 7.11(c) also has the same basic shape as $y = 2^x$, but this time the graph has been translated horizontally. If every point on the graph of $y = 2^x$ were to be moved 3 units to the left, the result would be the graph of $g(x) = 2^{x+3}$. ∎

EXERCISE 2 Which of the functions below have the same graph? Explain why.

a. $f(x) = \left(\dfrac{1}{2}\right)^x$ **b.** $g(x) = -2^x$ **c.** $h(x) = 2^{-x}$

Comparing Exponential and Power Functions

Exponential functions are not the same as the power functions we studied in Chapter 6. A power function, $h(x) = kx^p$, has a variable base and a constant exponent, as in $h(x) = 2x^3$. An exponential function has a constant base and a variable exponent, as in $f(x) = 2(3^x)$.

Compare the values for these two functions in Table 7.15. We can tell that f is exponential because its values increase by a factor of 3 for each unit increase in x. In general, we can identify the base of an exponential function in the following way: If its table shows successive integer values of x, we divide any function value by the previous one. For this function, we see that $54 \div 18 = 3$, or $18 \div 6 = 3$.

TABLE 7.15

x	$h(x) = 2x^3$
-3	-54
-2	-16
-1	-2
0	0
1	2
2	16
3	54

x	$f(x) = 2(3^x)$
-3	$\frac{2}{27}$
-2	$\frac{1}{4}$
-1	$\frac{2}{3}$
0	2
1	6
2	18
3	54

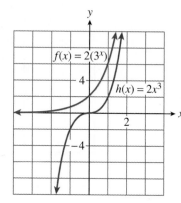

FIGURE 7.13

As you would expect, the graphs of the two functions are also quite different. (See Figure 7.13.)

EXERCISE 3 Which of the following functions are exponential functions, and which are power functions?

a. $F(x) = 1.5^x$ **b.** $G(x) = 3x^{1.5}$
c. $H(x) = 3^{1.5x}$ **d.** $K(x) = (3x)^{1.5}$

Exponential Equations

An **exponential equation** is one in which the variable is part of an exponent. For example, the equation

$$3^x = 81 \tag{1}$$

is exponential. Many exponential equations can be solved by writing both sides of the equation as powers with the same base. To solve Equation (1), we write

$$3^x = 3^4$$

which is true if and only if $x = 4$. In general, if two equivalent powers have the same base, then their exponents must be equal also.

Sometimes the laws of exponents can be used to express both sides of an equation as single powers of a common base.

EXAMPLE 3

Solve the following equations.

a. $3^{x-2} = 9^3$ **b.** $27 \cdot 3^{-2x} = 9^{x+1}$

Solutions

a. Using the fact that $9 = 3^2$, write each side of the equation as a power of 3.

$$3^{x-2} = (3^2)^3$$
$$3^{x-2} = 3^6$$

Now equate the exponents to obtain

$$x - 2 = 6$$
$$x = 8$$

b. Write each factor as a power of 3.

$$3^3 \cdot 3^{-2x} = (3^2)^{x+1}$$

Use the laws of exponents to simplify each side.

$$3^{3-2x} = 3^{2x+2}$$

Now equate the exponents to obtain

$$3 - 2x = 2x + 2$$
$$-4x = -1$$
$$x = \frac{1}{4}$$

     ■

EXERCISE 4   Solve the equation $2^{x+2} = 128$.

Write each side as a power of 2.
Equate exponents.

Exponential equations arise frequently in the study of exponential growth.

EXAMPLE 4

During the summer a population of fleas doubles in number every 5 days. If a population starts with 10 fleas, how long will it be before there are 10,240 fleas?

Solution

Let P represent the number of fleas present after t days. The original population of 10 is multiplied by a factor of 2 every 5 days, or

$$P(t) = 10 \cdot 2^{t/5}$$

Set $P = 10,240$ and solve for t.

$$10,240 = 10 \cdot 2^{t/5} \quad \text{Divide both sides by 10.}$$
$$1024 = 2^{t/5} \quad \text{Write 1024 as a power of 2.}$$
$$2^{10} = 2^{t/5}$$

Equate the exponents to get $10 = \frac{t}{5}$ or $t = 50$. The population will grow to 10,240 fleas in 50 days. ■

EXERCISE 5 During an advertising campaign in a large city, the makers of Chip-O's corn chips estimate that the number of people who have heard of Chip-O's increases by a factor of 8 every 4 days.

a. If 100 people are given trial bags of Chip-O's to start the campaign, write a function $N(t)$ for the number of people who have heard of Chip-O's after t days of advertising.

b. Use your calculator to graph the function $N(t)$ on the domain $0 \le t \le 15$.

c. How many days should they run the campaign in order for Chip-O's to be familiar to 51,200 people? Use algebraic methods to find your answer, and verify on your graph.

Graphical Solution of Exponential Equations

It is not always so easy to express both sides of the equation as powers of the same base. In the following sections we will develop more general methods for finding exact solutions to exponential equations. But we can use a graphing calculator to obtain approximate solutions.

EXAMPLE 5

Use the graph of $y = 2^x$ to find an approximate solution to the equation $2^x = 5$ accurate to the nearest hundredth.

Solution

Enter $Y_1 = 2$ [∧] X and use the standard graphing window ([ZOOM] [6]) to obtain the graph shown in Figure 7.14(a). We are looking for a point on this graph with y-coordinate 5. Using the **Trace** feature, we see that the y-coordinates are too small when $x < 2.1$ and too large when $x > 2.4$. The solution we want lies somewhere between $x = 2.1$ and $x = 2.4$, but this approximation is not accurate enough.

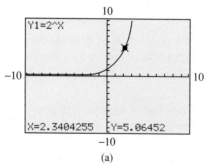
(a)

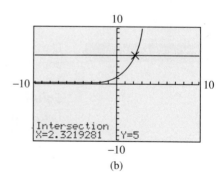
(b)

FIGURE 7.14

To improve our approximation, we'll use the intersect feature. Set $Y_2 = 5$ and press $\boxed{\text{GRAPH}}$. The x-coordinate of the intersection point of the two graphs is the solution of the equation $2^x = 5$. Activating the **intersect** command results in Figure 7.14(b), and we see that, to the nearest hundredth, the solution is 2.32. We can verify that our estimate is reasonable by substituting into the equation.

$$2^{2.32} \stackrel{?}{=} 5$$

We enter $2 \boxed{\wedge} 2.32 \boxed{\text{ENTER}}$ to get 4.993322196. This number is not equal to 5, but it is close, so we believe that $x = 2.32$ is a reasonable approximation to the solution of the equation $2^x = 5$. ■

EXERCISE 6 Use the graph of $y = 5^x$ to find an approximate solution to $5^x = 285$, accurate to two decimal places.

Exponential Regression

If we suspect that a collection of data can be described by an exponential model, we can use a calculator to find a function of the form $y = ab^x$ to fit the data. For example, Table 7.16 shows the amount of nitrogen used annually as a fertilizer in the United Kingdom since 1913.

TABLE 7.16

Year	1913	1929	1950	1960	1970	1980
Nitrogen (kilotonnes)	38	54	215	413	800	1261

First press $\boxed{\text{STAT}}$ $\boxed{\text{ENTER}}$ and enter the data, just as you would for linear regression. (We let $t = 0$ correspond to the year 1900, so that 1913 is $t = 13$, and so on.) Then press $\boxed{\text{STAT}}$ $\boxed{\triangleright}$ $\boxed{0}$ for **ExpReg,** the exponential regression equation, and press $\boxed{\text{ENTER}}$. The regression equation should be approximately $y = 14.44(1.057)^x$.

To see a graph of the data and the exponential function, turn on **Plot 1** and select the scatterplot icon. Press $\boxed{\text{Y=}}$ and clear out any old function definitions. Position the cursor after Y_1 and press $\boxed{\text{VARS}}$ $\boxed{5}$ $\boxed{\triangleright}$ $\boxed{\triangleright}$ $\boxed{\text{ENTER}}$ to store the regression equation in Y_1. Finally, press $\boxed{\text{ZOOM}}$ $\boxed{9}$ to see the graph in Figure 7.15.

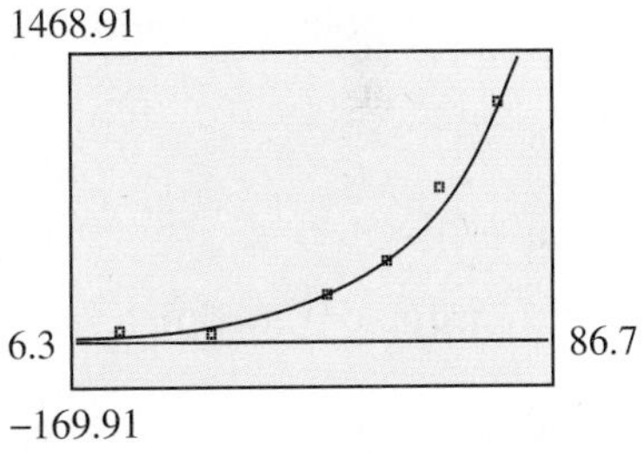

FIGURE 7.15

READING QUESTIONS

1. What is an exponential function?
2. How can you tell whether an exponential function is increasing or decreasing?
3. State the domain and range of an exponential function.
4. How is an exponential function different from a power function?
5. Explain the algebraic technique for solving exponential equations described in this section.

ANSWERS TO 7.2 EXERCISES

1a. $f: \left[\frac{1}{9}, 9\right]$; $g: \left[\frac{1}{16}, 16\right]$ **b.** p: [0.64, 1.56]; q [0.25, 4]

2. (a) and (c)

3. exponential; (a), (c); power: (b), (d)   **4.** $x = 5$

5a. $N(t) = 100 \cdot 8^{t/4}$ **b.** 250,000

   c. 12 days

6. $x \approx 3.51$

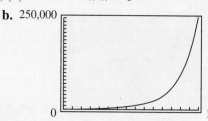

HOMEWORK 7.2

Find the y-intercept of each exponential function and decide whether the graph is increasing or decreasing.

1. a. $f(x) = 26(1.4)^x$ **c.** $h(x) = 75\left(\frac{4}{5}\right)^x$ **b.** $g(x) = 1.2(0.84)^x$ **d.** $k(x) = \frac{2}{3}\left(\frac{9}{8}\right)^x$

2. a. $M(x) = 1.5(0.05)^x$ **c.** $P(x) = \left(\frac{5}{8}\right)^x$ **b.** $N(x) = 0.05(1.05)^x$ **d.** $Q(x) = \left(\frac{4}{3}\right)^x$

Decide whether each function is an exponential function, a power function, or neither.

3. a. $g(t) = 3t^{0.4}$ **c.** $D(x) = 6x^{1/2}$ **b.** $h(t) = 4(0.3)^t$ **d.** $E(x) = 4x + x^4$

4. a. $R(w) = 5(5)^{w-1}$ **c.** $M(z) = 0.2z^{1.3}$ **b.** $Q(w) = 2^w - w^2$ **d.** $N(z) = z^{-3}$

Graph each pair of functions on the same axes by making a table of values and plotting points by hand. Choose appropriate scales for the axes.

5. a. $f(x) = 3^x$ **b.** $g(x) = \left(\frac{1}{3}\right)^x$ **6. a.** $F(x) = \left(\frac{1}{10}\right)^x$ **b.** $G(x) = 10^x$

7. a. $h(t) = 4^{-t}$ **b.** $q(t) = -4^t$ **8. a.** $P(t) = -5^t$ **b.** $R(t) = 5^{-t}$

In each group of functions, which have identical graphs? Explain why.

9. a. $h(x) = 6^x$ **c.** $m(x) = 6^{-x}$ **b.** $k(x) = \left(\frac{1}{6}\right)^x$ **d.** $n(x) = \frac{1}{6^x}$

10. a. $Q(t) = 5^t$ **c.** $F(t) = \left(\frac{1}{5}\right)^{-t}$ **b.** $R(t) = \left(\frac{1}{5}\right)^t$ **d.** $G(t) = \frac{1}{5^{-t}}$

For Problems 11–14, (a) use a graphing calculator to obtain the graphs on the domain [−5, 5], (b) give the range of the function on that domain, accurate to hundredths.

11. $g(t) = 1.3^t$ **12.** $h(t) = 2.4^t$ **13.** $N(x) = 0.8^x$ **14.** $P = 0.7^x$

Solve each equation algebraically.

15. $5^{x+2} = 25^{4/3}$

16. $3^{x-1} = 27^{1/2}$

17. $3^{2x-1} = \dfrac{\sqrt{3}}{9}$

18. $2^{3x-1} = \dfrac{\sqrt{2}}{16}$

19. $4 \cdot 2^{x-3} = 8^{-2x}$

20. $9 \cdot 3^{x+2} = 81^{-x}$

21. $27^{4x+2} = 81^{x-1}$

22. $16^{2-3x} = 64^{x+5}$

23. $10^{x^2-1} = 1000$

24. $5^{x^2-x-4} = 25$

25. Before the advent of antibiotics an outbreak of cholera might spread through a city so that the number of cases doubled every 6 days.

   a. Twenty-six cases were discovered on July 5. Write a function for the number of cases of cholera t days later.

   b. Use your calculator to graph your function on the interval $0 \le t \le 90$.

   c. When should hospitals expect to be treating 106,496 cases? Use algebraic methods to find your answer, and verify on your graph.

26. An outbreak of ungulate fever can sweep through the livestock in a region so that the number of animals affected triples every 4 days.

   a. A rancher discovers 4 cases of ungulate fever among his herd. Write a function for the number of cases of ungulate fever t days later.

   b. Use your calculator to graph your function on the interval $0 \le t \le 20$.

   c. If the rancher does not act quickly, how long will it be until 324 head are affected? Use algebraic methods to find your answer, and verify on your graph.

27. A color television set loses 30% of its value every 2 years.

   a. Write a function for the value of a television set t years after it was purchased if it cost $700 originally.

   b. Use your calculator to graph your function on the interval $0 \le t \le 20$.

   c. How long will it be before a $700 television set depreciates to $343? Use algebraic methods to find your answer, and verify on your graph.

28. A mobile home loses 20% of its value every 3 years.

   a. A certain mobile home costs $20,000. Write a function for its value after t years.

   b. Use your calculator to graph your function on the interval $0 \le t \le 30$.

   c. How long will it be before a $20,000 mobile home depreciates to $12,800? Use algebraic methods to find your answer, and verify on your graph.

Use a graphing calculator to find an approximate solution accurate to the nearest hundredth.

29. $3^{x-1} = 4$

30. $2^{x+3} = 5$

31. $4^{-x} = 7$

32. $6^{-x} = 3$

Which of the following tables could describe exponential functions? Explain why or why not. If the function is exponential, find its growth or decay factor.

33. a.

x	y
0	3
1	6
2	12
3	24
4	48

b.

t	P
0	6
1	7
2	10
3	15
4	22

c.

x	N
0	2
1	6
2	34
3	110
4	258

d.

p	R
0	405
1	135
2	45
3	15
4	5

34. a.

t	y
1	100
2	50
3	$33\frac{1}{3}$
4	25
5	20

b.

x	P
1	$\frac{1}{2}$
2	1
3	2
4	4
5	8

c.

h	a
0	70
1	7
2	0.7
3	0.07
4	0.007

d.

t	Q
0	0
1	$\frac{1}{4}$
2	1
3	$\frac{9}{4}$
4	4

35. The graph of $f(x) = P_0 a^x$ is shown in Figure 7.16.

a. Read the value of P_0 from the graph.

b. Make a short table of values for the function by reading values from the graph. Does your table confirm that the function is exponential?

c. Use your table to calculate the growth factor a.

d. Using your answers to parts (a) and (c), write a formula for $f(x)$.

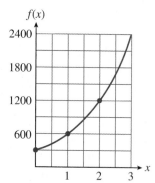

FIGURE 7.16

36. The graph of $g(x) = P_0 a^x$ is shown in Figure 7.17.

a. Read the value of P_0 from the graph.

b. Make a short table of values for the function by reading values from the graph. Does your table confirm that the function is exponential?

c. Use your table to calculate the growth factor a.

d. Using your answers to parts (a) and (c), write a formula for $g(x)$.

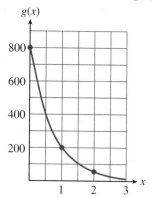

FIGURE 7.17

37. For several days after the Northridge earthquake on January 17, 1994, the area received a number of significant aftershocks. The color graph in Figure 7.18 shows that the number of aftershocks decreased exponentially over time. The graph of the function $S(d) = S_0 a^d$, shown in black, approximates the data.

a. Read the value of S_0 from the graph.

b. Find an approximation for the decay factor a by comparing two points on the graph. (Some of the points on the graph of $S(d)$ are approximately $(1, 82)$, $(2, 45)$, $(3, 25)$, and $(4, 14)$.)

c. Using your answers to (a) and (b), write a formula for $S(d)$.

Source: Los Angeles Times, June 27, 1995

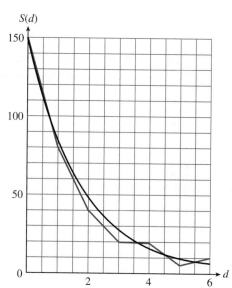

FIGURE 7.18

38. The frequency of a musical note depends on its pitch. The graph in Figure 7.19 shows that the frequency increases exponentially. The function $F(p) = F_0 a^p$ gives the frequency as a function of the number of half-tones, p, above the starting point on the scale.

   a. Read the frequency F_0 from the graph. (This is the frequency of the note A above middle C.)

   b. Find an approximation for the growth factor a by comparing two points on the graph. (Some of the points on the graph of $F(p)$ are approximately $(1, 466)$, $(2, 494)$, $(3, 523)$, and $(4, 554)$.)

   c. Using your answers to (a) and (b), write a formula for $F(p)$.

   d. The frequency doubles when you raise a note by one octave, which is equivalent to 12 half-tones. Use this information to find an *exact* value for a.

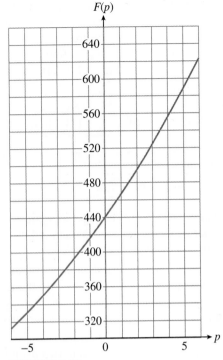

FIGURE 7.19

39. The order of a stream or river is a measure of its relative size. A first-order stream is the smallest, one that has no tributaries. Second-order streams have only first-order streams as tributaries. Third-order streams may have first- and second-order streams as tributaries, and so on. The Mississippi River is an example of a tenth-order stream, and the Columbia River is ninth order. Both the number of streams of a given order and their average length are exponential functions of their order.

   a. Using the given values, find a function $N(x) = ab^{x-1}$ for the number of streams of a given order.

   b. Complete the column for the number of streams of a given order. (Round to the nearest whole number of streams for each order.)

   c. Find a function $L(x) = ab^{x-1}$ for the average length of streams of a given order, and complete that column.

   d. Find the total length of all streams of each order, and hence estimate the total length of all stream channels in the United States.

   Source: Leopold, Wolman, and Miller, 1992.

Order	Number	Average length	Total length
1	1,600,000	1	
2	339,200	2.3	
3			
4			
5			
6			
7			
8			
9			
10			

40. Related species living in the same area often evolve in different sizes to minimize competition for food and habitat. Here are the masses of eight species of fruit pigeon found in New Guinea, ranked from smallest to largest.

Size rank	1	2	3	4	5
Mass (grams)	49	76	123	163	245

Size rank	6	7	8
Mass (grams)	414	592	802

a. Plot the masses of the pigeons against their order of increasing size. What kind of function might fit the data?

b. Compute the ratios of the masses of successive sizes of fruit pigeons. Are the ratios approximately constant? What does this information tell you about your answer to part (a)?

c. Compute the average ratio to two decimal places. Using this ratio, estimate the mass of a hypothetical fruit pigeon of size rank zero.

d. Using your answers to part (c), write an exponential function that approximates the data. Graph this function on top of the data and evaluate the fit.

Source: Burton, 1998.

41. Dilute hydrochloric acid reacts with sodium thiosulphate solution to give a precipitate of sulfur. The time it takes for the reaction to occur is a function of the concentration of sodium thiosulphate, as shown in the table.

Concentration (moles/liter)	0.15	0.12	0.09	0.06	0.03
Reaction time (seconds)	43	55	66	105	243

a. Plot reaction time, t, against concentration of sodium thiosulphate, c. What sort of function does the graph appear to be?

b. Make a table of values for the reaction rate, $r = \dfrac{1}{t}$, as a function of concentration, c, and plot the data.

c. Find the regression line for the data in part (b).

d. Another student suggests that "reaction rate is proportional to concentration." Do your calculations support this claim? Explain.

e. Write a formula for reaction time, t, as a function of concentration, c, Does this function provide a reasonable fit for your graph in part (a)?

Source: From *Chemistry*, by J. A. Hunt and A. Sykes, Copyright © 1984 Longman Group Ltd. Reprinted by permission.

42. (continuation of Problem 41) We'll repeat the experiment, but this time we'll hold the concentration of sodium thiosulphate constant and vary the temperature of the solution.

Temperature (°C)	20	30	40	50	60
Reaction time (seconds)	280	132	59	31	17

a. Plot reaction time, t, against the temperature, T. What sort of function does the graph appear to be?

b. Make a table of values for the reaction rate, $r = \dfrac{1}{t}$, as a function of temperature, T, and plot the data.

c. Use your calculator's exponential regression feature to find a regression equation for the data in part (b).

d. For many chemical reactions, the reaction rate doubles for each increase in temperature of 10°C. Write a formula for reaction rate that describes this function. Does your formula agree with the regression equation from part (c)?

e. Write a formula for reaction time, t, as a function of concentration, C. Does this function provide a reasonable fit for your graph in part (a)?

Source: From *Chemistry*, by J. A. Hunt and A. Sykes, Copyright © 1984 Longman Group Ltd. Reprinted by permission.

43. The Koch snowflake is an example of a fractal. It is named in honor of the Swiss mathematician Niels Fabian Helge von Koch (1870–1924). To construct a Koch snowflake, draw an equilateral triangle with sides of length 1 unit. This is stage $n = 0$. Divide each side into 3 equal segments and draw a smaller equilateral triangle on each middle segment, as shown in Figure 7.20. The new figure (stage $n = 1$) is a 6-pointed star with 12 sides. Repeat the process to obtain stage $n = 2$: Trisect each of the 12 sides and draw an equilateral triangle on each middle third. If you continue this process forever, the resulting figure is the Koch snowflake.

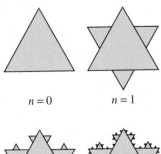
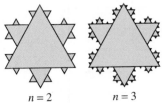

$n = 0$ $n = 1$

$n = 2$ $n = 3$

FIGURE 7.20

a. We will consider several functions related to the Koch snowflake.

$S(n)$ is the length of each side in stage n

$N(n)$ is the number of sides in stage n

$P(n)$ is the perimeter of the snowflake at stage n

Fill in the table describing the snowflake at each stage.

Stage n	$S(n)$	$N(n)$	$P(n)$
0			
1			
2			
3			

b. Write an expression for $S(n)$.

c. Write an expression for $N(n)$.

d. Write an expression for $P(n)$.

e. What happens to the perimeter as n gets larger and larger?

f. As n increases, the area of the snowflake increases also. Is the area of the completed Koch snowflake finite or infinite?

44. The Sierpinski carpet is another fractal. It is named for the Polish mathematician Waclaw Sierpinski (1882–1969). To build a Sierpinski carpet, start with a unit square (sides of length 1 unit.) For stage $n = 1$, trisect each side and partition the square into 9 smaller squares. Remove the center square, leaving a hole surrounded by 8 squares, as shown in Figure 7.21. For stage $n = 2$, repeat the process on each of the remaining 8 squares. If you continue this process forever, the resulting figure is the Sierpinski carpet.

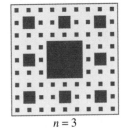

$n = 0$ $n = 2$

$n = 1$ $n = 3$

FIGURE 7.21

a. We will consider several functions related to the Sierpinski carpet.

$S(n)$ is the side of a new square at stage n

$A(n)$ is the area of a new square at stage n

$N(n)$ is the number of new squares removed at stage n

$R(n)$ is the total area removed at stage n

$T(n)$ is the total area remaining at stage n

Fill in the table describing the carpet at each stage.

Stage n	$S(n)$	$A(n)$	$N(n)$	$R(n)$	$T(n)$
1					
2					
3					

b. Write an expression for $S(n)$.

c. Write an expression for $A(n)$.

d. Write an expression for $N(n)$.

e. Write an expression for $R(n)$.

f. Write an expression for $T(n)$.

g. What happens to the area remaining as n approaches infinity?

Fill in the tables. Graph each pair of functions on the same set of axes. Then answer the questions below for Problems 45–46.

45.

x	$f(x) = x^2$	$g(x) = 2^x$
-2		
-1		
0		
1		
2		
3		
4		
5		
6		

46.

x	$f(x) = x^3$	$g(x) = 3^x$
-2		
-1		
0		
1		
2		
3		
4		
5		
6		

a. Give the range of f and the range of g.

b. For how many values of x does $f(x) = g(x)$?

c. Estimate the value(s) of x for which $f(x) = g(x)$.

d. For what values of x is $f(x) < g(x)$?

e. Which function grows more rapidly for large values of x?

For Problems 47–50, (a) use the order of operations to explain why the two functions are different. (b) Complete the table of values comparing the values of the two functions.

47. $f(x) = 2^{x-1}$, $g(x) = 2^x - 1$

x	$y = 2^x$	$f(x)$	$g(x)$
-2			
-1			
0			
1			
2			

48. $f(x) = 3^x + 2$, $g(x) = 3^{x+2}$

x	$y = 3^x$	$f(x)$	$g(x)$
-2			
-1			
0			
1			
2			

49. $f(x) = -3^x$, $g(x) = 3^{-x}$

x	$y = 3^x$	$f(x)$	$g(x)$
-2			
-1			
0			
1			
2			

50. $f(x) = 2^{-x}$, $g(x) = -2^x$

x	$y = 2^x$	$f(x)$	$g(x)$
-2			
-1			
0			
1			
2			

Use a calculator to make tables of values and to graph each pair of functions in the same window. Describe how the second graph differs from the first.

51. $y_1 = 3^x$
$y_2 = 3^x - 5$

52. $y_1 = 4^x$
$y_2 = 4^x + 2$

53. $y_1 = 4^t$
$y_2 = 4^{t-5}$

54. $y_1 = 6^t$
$y_2 = 6^{t+2}$

55. $f(h) = 2^{-h}$
$g(h) = -2^{-h}$

56. $g(p) = 5^{-p}$
$h(p) = -5^{-p}$

57. $N(t) = 10^t$
$M(t) = 20 + 10^{t+2}$

58. $P(t) = 10^t$
$Q(t) = 10^{t-3} - 50$

7.3 Logarithms

In this section we introduce a new mathematical tool called a *logarithm*, which will help us solve exponential equations.

Suppose that a colony of bacteria doubles in size every day. If the colony starts with 50 bacteria, how long will it be before there are 800 bacteria? We answered questions of this type in Section 7.2 by writing and solving an approximate exponential equation. The function

$$P(t) = 50 \cdot 2^t$$

gives the number of bacteria present on day t, so we must solve the equation

$$800 = 50 \cdot 2^t$$

Dividing both sides by 50 yields

$$16 = 2^t$$

The solution to this equation is the answer to the question "To what power must we raise 2 in order to get 16?" The value of t that solves the equation is called the base 2 **logarithm** of 16. Because $2^4 = 16$, the base 2 logarithm of 16 is **4**. We write this as

$$\log_2 16 = 4$$

In other words, we solve an exponential equation by computing a logarithm. You can check that $t = 4$ solves the problem stated above.

$$P(4) = 50 \cdot 2^4 = 800$$

In this example, the unknown exponent was called a logarithm. In general, we make the following definition.

> The **base b logarithm of x,** written **$\log_b x$,** is the exponent to which b must be raised in order to yield x.

Some logarithms, like some square roots, are easy to evaluate, while others require a calculator. We'll start with the easy ones.

EXAMPLE 1

a. $\log_3 9 = 2$ because $3^2 = 9$.

b. $\log_5 125 = 3$ because $5^3 = 125$.

c. $\log_4 \dfrac{1}{16} = -2$ because $4^{-2} = \dfrac{1}{16}$.

d. $\log_5 \sqrt{5} = \dfrac{1}{2}$ because $5^{1/2} = \sqrt{5}$.

∎

EXERCISE 1   Find each logarithm.

a. $\log_3 81$

b. $\log_{10} \dfrac{1}{1000}$

From the definition of a logarithm and the examples above we see that the statements

$$y = \log_b x \qquad \text{and} \qquad x = b^y \tag{1}$$

are equivalent. The logarithm, y, is the same as the *exponent* in $x = b^y$. Thus, *a logarithm is an exponent;* it is the exponent to which b must be raised to yield x.

The equations in (1) allow us to convert from logarithmic to exponential form, or vice versa. You should memorize these conversion equations, as we will use them frequently.

As special cases of the equivalence in (1) we can compute the following useful logarithms.

For any base $b > 0$,

$\log_b b = 1$	because	$b^1 = b$
$\log_b 1 = 0$	because	$b^0 = 1$
$\log_b b^x = x$	because	$b^x = b^x$

EXAMPLE 2

a. $\log_2 2 = 1$ **b.** $\log_5 1 = 0$ **c.** $\log_3 3^4 = 4$ ∎

Solving Exponential Equations

We use logarithms to solve exponential equations, just as we use square roots to solve quadratic equations. Consider the two equations

$$x^2 = 25 \quad \text{and} \quad 2^x = 8$$

We solve the first equation by taking a square root, and we solve the second equation by computing a logarithm.

$$x = \pm\sqrt{25} = \pm 5 \quad \text{and} \quad x = \log_2 8 = 3$$

The operation of taking a base b logarithm is the *inverse* of raising the base b to a power, just as extracting square roots is the inverse of squaring a number.

Every exponential equation can be rewritten in logarithmic form by using the conversion equations in (1). Thus

$$3 = \log_2 8 \quad \text{and} \quad 8 = 2^3$$

are equivalent statements, just as

$$5 = \sqrt{25} \quad \text{and} \quad 25 = 5^2$$

are equivalent statements. Rewriting an exponential equation in logarithmic form is a basic strategy for finding its solution.

EXAMPLE 3

Rewrite each equation in logarithmic form.

a. $2^{-1} = \dfrac{1}{2}$ **b.** $a^{1/5} = 2.8$

c. $6^{1.5} = T$ **d.** $M^v = 3K$

Solutions

First identify the base b, then the exponent or logarithm y. Use the conversion in (1) to rewrite the equation $b^y = x$ in the form $\log_b x = y$.

a. $\log_2 \dfrac{1}{2} = -1$ **b.** $\log_a 2.8 = \dfrac{1}{5}$

c. $\log_6 T = 1.5$ **d.** $\log_M 3K = v$ ∎

EXERCISE 2 Rewrite each equation in logarithmic form.

a. $8^{-1/3} = \dfrac{1}{2}$ **b.** $5^x = 46$

Approximating Logarithms

Suppose we would like to solve the equation

$$2^x = 26$$

The solution of this equation is $x = \log_2 26$, but can we find a decimal approximation for this value? There is no integer power of 2 that equals 26, because

$$2^4 = 16 \quad \text{and} \quad 2^5 = 32$$

Thus, $\log_2 26$ must be between 4 and 5. We can use trial and error to find the value of $\log_2 26$ to the nearest tenth. Use your calculator to make a table of values for $y = 2^x$, starting with $x = 4$ and using increments of 0.1.

TABLE 7.17

x	2^x		x	2^x
4	$2^4 = 16$		4.5	$2^{4.5} = 22.627$
4.1	$2^{4.1} = 17.148$		4.6	$2^{4.6} = 24.251$
4.2	$2^{4.2} = 18.379$		4.7	$2^{4.7} = 25.992$
4.3	$2^{4.3} = 19.698$		4.8	$2^{4.8} = 27.858$
4.4	$2^{4.4} = 21.112$		4.9	$2^{4.9} = 29.857$

From Table 7.17 we see that 26 is between $2^{4.7}$ and $2^{4.8}$, and is closer to $2^{4.7}$. To the nearest tenth, $\log_2 26 \approx 4.7$.

Trial and error can be a time-consuming process. In Example 4 we illustrate a graphical method for estimating the value of a logarithm.

EXAMPLE 4 Approximate $\log_3 7$ to the nearest hundredth.

Solution

If $\log_3 7 = x$, then $3^x = 7$. We will use the graph of $y = 3^x$ to approximate a solution to $3^x = 7$. Graph $Y_1 = 3 \boxed{\wedge} X$ and $Y_2 = 7$ in the standard window

(ZOOM 6) to obtain the graph shown in Figure 7.22. Activate the **intersect** feature to find that the two graphs intersect at the point (1.7712437, 7). Because this point lies on the graph of $y = 3^x$, we know that

$$3^{1.7712437} \approx 7 \quad \text{or} \quad \log_3 7 \approx 1.7712437$$

To the nearest hundredth, $\log_3 7 \approx 1.77$.

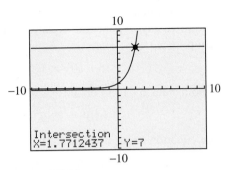

FIGURE 7.22

EXERCISE 3

a. Rewrite the equation $3^x = 90$ in logarithmic form.

b. Use a graph to approximate the solution to the equation in part (a). Round your answer to three decimal places.

Base 10 Logarithms

Some logarithms are used so frequently in applications that their values are programmed into scientific and graphing calculators. These are the base 10 logarithms, such as

$$\log_{10} 1000 = 3 \quad \text{and} \quad \log_{10} 0.01 = -2$$

Base 10 logarithms are called **common logarithms,** and the subscript 10 is often omitted, so that "log x" is understood to mean "$\log_{10} x$." To evaluate a base 10 logarithm, we use the LOG key on a calculator. Many logarithms are irrational numbers, and the calculator gives as many digits as its display allows. We can then round off to the desired accuracy.

EXAMPLE 5

Approximate the following logarithms to 2 decimal places.

a. log 6.5 **b.** log 256

Solutions

a. The keying sequence LOG 6.5 ENTER yields 0.8129133566, so log 6.5 ≈ 0.81.

b. The keying sequence LOG 256 ENTER yields 2.408239965, so log 256 ≈ 2.41.

We can check the approximations found in Example 5 with our conversion equations. Remember that a logarithm is an exponent, and in this example the base is 10. We find that

$$10^{0.81} \approx 6.45654229 \qquad \text{and} \qquad 10^{2.41} \approx 257.0395783$$

so our approximations are reasonable, although you can see that rounding a logarithm to 2 decimal places does lose some accuracy. Rounding logarithms to 4 decimal places is customary.

We can now solve any exponential equation with base 10. For instance, to solve the equation

$$16 \cdot 10^t = 360$$

we first divide both sides by 16 to obtain

$$10^t = 22.5$$

then convert the equation to logarithmic form and evaluate.

$$t = \log_{10} 22.5 \approx 1.352182518$$

To 4 decimal places, the solution is 1.3522.

EXERCISE 4 Solve $20 + 10^x = 220$.

Isolate the power of 10.
Rewrite in logarithmic form.
Evaluate.

In general, to solve exponential equations involving powers of 10, we can use the following steps.

Steps for Solving Exponential Equations

1. Isolate the power on one side of the equation.
2. Rewrite the equation in logarithmic form.
3. Use a calculator, if necessary, to evaluate the logarithm.
4. Solve for the variable.

EXAMPLE 6

Solve the equation $38 = 95 - 15 \cdot 10^{0.4x}$

Solution

First, isolate the power of 10: Subtract 95 from both sides of the equation and divide by -15 to obtain

$$-57 = -15 \cdot 10^{0.4x} \qquad \text{Divide by } -15.$$
$$3.8 = 10^{0.4x}$$

Convert the equation to logarithmic form as

$$\log_{10} 3.8 = 0.4x$$

Solving for x yields

$$\frac{\log_{10} 3.8}{0.4} = x$$

We can evaluate this expression on the calculator by entering

[LOG] 3.8 [ )] [÷] 0.4 [ENTER]

which yields 1.449458992. Thus, to four decimal places, $x \approx 1.4495$. ∎

Application to Exponential Functions

We've seen that exponential functions are used to describe some applications of growth and decay, $P(t) = P_0 a^t$. Two common questions that arise in connection with exponential models are

   1. Given a value of t, find the corresponding value of $P(t)$.

   2. Given a value of $P(t)$, find the corresponding value of t.

To answer the first question, we evaluate the function $P(t)$ at the appropriate value. To answer the second question, we must solve an exponential equation, and this usually involves logarithms.

EXAMPLE 7

The value of a large tractor originally worth $30,000 depreciates exponentially according to the formula $V(t) = 30{,}000 \, (10)^{-0.04t}$, where t is in years. When will the tractor be worth half its original value?

Solution

We want to find the value of t for which $V(t) = 15{,}000$. That is, we want to solve the equation

$$15{,}000 = 30{,}000 \, (10)^{-0.04t}$$

Divide both sides by 30,000 to obtain

$$0.5 = 10^{-0.04t}$$

Convert the equation to logarithmic form as

$$\log_{10} 0.5 = -0.04t$$

and divide by -0.04 to obtain

$$\frac{\log_{10} 0.5}{-0.04} = t$$

To evaluate this expression, key in

[LOG] 0.5 [ )] [÷] [(−)] 0.04 [ENTER]

to find $t \approx 7.525749892$. The tractor will be worth $15,000 in approximately $7\frac{1}{2}$ years. ∎

At this stage it seems we will be able to solve only exponential equations in which the base is 10. However, in Section 7.5 we will see how the properties of logarithms enable us to solve exponential equations with any base.

Logarithmic Equations

We can also use the conversion

$$y = \log_b x \qquad \text{if and only if} \qquad x = b^y$$

to convert expressions involving logarithms to exponential form.

EXAMPLE 8

Convert each equation to exponential form.

a. $\log_{10} 0.001 = z$ **b.** $\log_3 20 = t$

c. $\log_b (3x + 1) = 3$ **d.** $\log_q p = w$

Solutions

First identify the base b, and then the exponent or logarithm y. Use the conversion to write the equation $\log_b x = y$ in the form $b^y = x$.

a. $10^z = 0.001$ **b.** $3^t = 20$

c. $b^3 = 3x + 1$ **d.** $q^w = p$ ∎

Simple equations involving logarithms can sometimes be solved by rewriting them in exponential form.

EXAMPLE 9

Solve for the unknown value in each equation.

a. $\log_2 x = 3$ **b.** $\log_b 2 = \dfrac{1}{2}$

c. $\log_3 (2x - 1) = 4$ **d.** $2(\log_3 x) - 1 = 4$

Solutions

Write each equation in exponential form and solve for the variable.

a. $2^3 = x$

$x = 8$

b. $b^{1/2} = 2$ Square both sides.

$(b^{1/2})^2 = 2^2$

$b = 4$

c. $2x - 1 = 3^4$

$2x = 82$

$x = 41$

d. $2(\log_3 x) = 5$ Isolate $\log_3 x$

$\log_3 x = \dfrac{5}{2}$

$x = 3^{5/2}$ ∎

EXERCISE 6   Solve for x: $\log_4 x = 3$

READING QUESTIONS

1. What is the logarithm base b of a positive number?
2. Explain how to convert a logarithmic equation to exponential form.
3. What is a common logarithm?
4. How can you use logarithms to solve exponential equations?

ANSWERS TO 7.3 EXERCISES

1a. 4

  b. -3

2a. $\log_8\left(\dfrac{1}{2}\right) = \dfrac{-1}{3}$

  b. $\log_5 46 = x$

3a. $\log_3 90 = x$

  b. $x \approx 4.096$

4. $x = \log_{10} 200 \approx 2.3010$

5a. 25.85%

  b. $t \approx 10.4$ (2004)

  c. No, the percent of homes with computers cannot exceed 100%.

6. $x = 64$

HOMEWORK 7.3

Find each logarithm without using a calculator.

1. a. $\log_7 49$ **b.** $\log_2 32$ **2. a.** $\log_4 64$ **b.** $\log_3 27$

3. a. $\log_3 \sqrt{3}$ **b.** $\log_3 \dfrac{1}{3}$ **4. a.** $\log_5 \dfrac{1}{5}$ **b.** $\log_5 \sqrt{5}$

5. a. $\log_4 4$ **b.** $\log_6 1$ **6. a.** $\log_{10} 1$ **b.** $\log_{10} 10$

7. a. $\log_8 8^5$ **b.** $\log_7 7^6$ **8. a.** $\log_{10} 10^{-4}$ **b.** $\log_{10} 10^{-6}$

9. a. $\log_{10} 0.1$ **b.** $\log_{10} 0.001$ **10. a.** $\log_{10} 10{,}000$ **b.** $\log_{10} 1000$

Rewrite each equation in logarithmic form.

11. $t^{3/2} = 16$ **12.** $v^{5/3} = 12$ **13.** $0.8^{1.2} = M$ **14.** $3.7^{2.5} = Q$

15. $x^{5t} = W - 3$ **16.** $z^{-3t} = 2P + 5$ **17.** $3^{-0.2t} = 2N_0$ **18.** $10^{1.3t} = 3M_0$

For Problems 19–22, (a) Solve each equation, writing your answer as a logarithm, (b) use trial and error to approximate the logarithm to one decimal place.

19. $4^x = 2.5$ **20.** $2^x = 0.2$ **21.** $10^x = 0.003$ **22.** $10^x = 4500$

Use a graph to approximate each logarithm to the nearest hundredth. (*Hint*: Graph an appropriate function $y = b^x$.)

23. $\log_{10} 7$ **24.** $\log_{10} 50$ **25.** $\log_3 67.9$ **26.** $\log_5 86.3$

Use a calculator to approximate each logarithm to four decimal places.

27. a. $\log_{10} 54.3$ **b.** $\log_{10} 2344$ **28. a.** $\log_{10} 27.9$ **b.** $\log_{10} 1476$

     c. $\log_{10} 0.073$ **d.** $\log_{10} 0.00614$ **c.** $\log_{10} 0.6942$ **d.** $\log_{10} 0.0104$

Solve for *x*. Round your answers to hundredths.

29. $10^{-3x} = 5$ **30.** $10^{-5x} = 76$ **31.** $25 \cdot 10^{0.2x} = 80$

32. $8 \cdot 10^{1.6x} = 312$ **33.** $12.2 = 2(10^{1.4x}) - 11.6$ **34.** $163 = 3(10^{0.7x}) - 49.3$

35. $3(10^{-1.5x}) - 14.7 = 17.1$ **36.** $4(10^{-0.6x}) + 16.1 + 28.2$ **37.** $80(1 - 10^{-0.2x}) = 65$

38. $250(1 - 10^{-0.3x}) = 100$

The atmospheric pressure decreases with altitude above the surface of the earth. For Problems 39–44, use the relationship

$$P(a) = 30(10)^{-0.09a}$$

between altitude *a* in miles and atmospheric pressure *P* in inches of mercury. Graph this function in the window

$$\text{Xmin} = 0 \qquad \text{Xmax} = 9.4$$
$$\text{Ymin} = 0 \qquad \text{Ymax} = 30$$

Solve the equations below algebraically, and verify with your graph.

39. The elevation of Mount Everest, the highest mountain in the world, is 29,028 feet. What is the atmospheric pressure at the top? (*Hint*: 1 mile = 5280 feet)

40. The elevation of Mount McKinley, the highest mountain in the United Sates, is 20,320 feet. What is the atmospheric pressure at the top?

41. How high above sea level is the atmospheric pressure 20.2 inches of mercury?

42. How high above sea level is the atmospheric pressure 16.1 inches of mercury?

43. Find the height above sea level at which the atmospheric pressure is equal to one-half the pressure at sea level.

44. Find the height above sea level at which the atmospheric pressure is equal to one-fourth the pressure at sea level.

45. The population of the state of California increased during the years 1960 to 1970 according to the formula

$$P(t) = 15{,}717{,}000(10)^{0.0104t}$$

where t is measured in years since 1960.

a. What was the population in 1970?

b. Assuming the same rate of growth, estimate the population of California in the years 1980, 1990, and 2000.

c. When did the population of California reach 20,000,000?

d. When did the population reach 30,000,000?

e. Graph the function P with a suitable domain and range, and verify your answers to parts (a) through (d).

46. The population of the state of New York increased during the years 1960 to 1970 according to the formula

$$P(t) = 16{,}782{,}000(10)^{0.0036t}$$

where t is measured in years since 1960.

a. What was the population in 1970?

b. Assuming the same rate of growth, estimate the population of New York in the years 1980, 1990, and 2000.

c. When did the population of New York reach 20,000,000?

d. When will the population reach 30,000,000?

e. Graph the function P with a suitable domain and range, and verify your answers to parts (a) through (d).

Convert each logarithmic equation to exponential form.

47. $\log_{16} 256 = w$ **48.** $\log_9 729 = y$ **49.** $\log_b 9 = -2$ **50.** $\log_b 8 = -3$

51. $\log_{10} A = -2.3$ **52.** $\log_{10} C = -4.5$ **53.** $\log_4 36 = 2q - 1$ **54.** $\log_5 3 = 6 - 2p$

55. $\log_u v = w$ **56.** $\log_m n = p$

Solve for the unknown value.

57. $\log_b 8 = 3$ **58.** $\log_b 625 = 4$ **59.** $\log_2 y = -1$ **60.** $\log_5 y = -2$

61. $\log_b 10 = \dfrac{1}{2}$ **62.** $\log_b 0.1 = -1$ **63.** $\log_2 (3x - 1) = 5$ **64.** $\log_5 (9 - 4x) = 3$

65. $3(\log_7 x) + 5 = 7$ **66.** $5(\log_2 x) + 6 = -14$

Simplify each expression.

67. $\log_2 (\log_4 16)$ **68.** $\log_5 (\log_5 5)$ **69.** $\log_{10} [\log_3 (\log_5 125)]$ **70.** $\log_{10} [\log_2 (\log_3 9)]$

71. $\log_2 [\log_2 (\log_3 81)]$ **72.** $\log_4 [\log_2 (\log_3 81)]$ **73.** $\log_b (\log_b b)$ **74.** $\log_b (\log_a a^b)$

MIDCHAPTER REVIEW

1. The population of Poplar Heights was 5000 in 1990 and doubled every four years.

a. What was the annual growth factor for the population?

b. Write a formula for $P(t)$, the population of Poplar Heights t years after 1990.

c. Sketch a graph of $P(t)$ for $t = 0$ to $t = 20$.

2. Travis began raising alpaca on his ranch in 1985. He started with 8 animals and his flock tripled in six years.

a. What was the annual growth factor for the population of alpaca?

b. Write a formula for $N(t)$, the number of alpaca t years after 1985.

c. Sketch a graph of $N(t)$ for $t = 0$ to $t = 24$.

For Problems 3–6, (a) complete the table of values, (b) write a function that describes the exponential growth or decay, (c) use your calculator to graph the function, (d) answer the question in the problem.

3. Housing prices in Lorton have risen an average of 10% per year since 1992. If Marlene bought a house for $135,000 in 1992, how much was the house worth in 1999?

Years since 1992	0	1	2	3	4
Value of house					

4. Sales of personal computers have increased 2.3% per year since 1990. If Compucalc sold 500 personal computers in 1990, how many did they sell in 1998?

Years since 1990	0	1	2	3	4
Computers sales					

5. The number of agriculture majors at Valley State College has been declining by $\frac{1}{5}$ each year. If there were 200 agriculture majors in 1995, how many were there in 2000?

Years since 1995	0	1	2	3	4
Agriculture majors					

6. Angel bought a sport utility vehicle in 1998 for $24,000. It depreciates by 8% each year. How much will it be worth in 2008?

Years since 1998	0	1	2	3	4
Value of SUV					

Graph the function, and give its domain and range.

7. $f(x) = 1.8^x$

8. $g(x) = (0.65)^x$

Solve.

9. $5^{x-3} = 625$

10. $16(2^{2x-5}) = 64$

Use a graph to find the solution, accurate to two decimal places.

11. $2^x = 3$

12. $(0.3)^x = 0.1$

Evaluate each logarithm.

13. $\log_6 \dfrac{1}{36}$

14. $\log_2 1024$

15. Rewrite in logarithmic form: $64^{-1/6} = \dfrac{1}{2}$

16. Rewrite in exponential form: $2.4 = \log_m 5H$

Solve. Round your answers to four decimal places.

17. $5(10^x) = 30$

18. $12 - 30(10^{-0.2x}) = 11.25$

19. A nationwide association of cosmetologists finds that news of a new product will spread among its members so that the number of cosmetologists who have tried the product is increased by a factor of 9 every 5 weeks.

a. If 20 cosmetologists try a new product at its launching, write a function $N(t)$ for the number of cosmetologists who have tried the product after t weeks.

b. Use your calculator to graph the function $N(t)$ on the domain $0 \le t \le 20$.

c. How long will it be before 14,580 cosmetologists have tried a new product? Use algebraic methods to find your answer, and verify on your graph.

20. Java House had 12 coffee shops in 1990, when it started a nationwide franchise. The number of shops t years later is given by $F(t) = 12(10^{0.25t})$.

a. Use your calculator to graph $F(t)$ for $0 \le t \le 20$.

b. When will there be 12,000 Java House shops? When will there be 50,000 shops?

Solve.

21. $\log_{1/2} x = 5$

22. $\log_3 (3x - 9) + 6 = 10$

7.4 Logarithmic Functions

In this section we will study **logarithmic functions.** For example, the function

$$f(x) = \log_2 x$$

defined for positive values of x, is a logarithmic function. In order to understand logarithmic functions better, we first investigate how they are related to some more familiar functions, the exponential functions.

Inverse of a Function

In Chapter 6 we saw that raising to the nth power and taking n^{th} roots are inverse operations. For example, if we first cube a number and then take the cube root of the result, we return to the original number.

$$x = 5 \quad \rightarrow \quad x^3 = 125 \quad \rightarrow \quad \sqrt[3]{x^3} = \sqrt[3]{125} = 5$$

| Cube the number | Take the cube root of the result | Original number |

If we graph the functions $f(x) = x^3$ and $g(x) = \sqrt[3]{x}$ on the same set of axes, we see that the graphs are related in an interesting way, as shown in Figure 7.23. We say that the graphs are "symmetric about the line $y = x$," which means that if we were to place a mirror along the line $y = x$, each graph would be the reflection of the other.

TABLE 7.18

x	$f(x) = x^3$
-2	-8
-1	-1
$-\frac{1}{2}$	$-\frac{1}{8}$
0	0
$\frac{1}{2}$	$\frac{1}{8}$
1	1
2	8

TABLE 7.19

x	$g(x) = \sqrt[3]{x}$
-8	-2
-1	-1
$-\frac{1}{8}$	$-\frac{1}{2}$
0	0
$\frac{1}{8}$	$\frac{1}{2}$
1	1
8	2

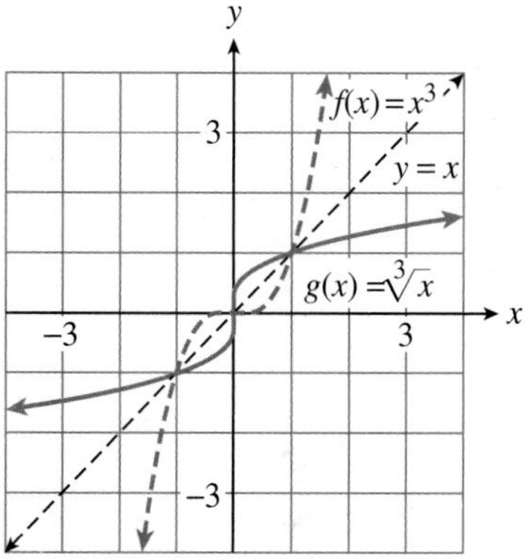

FIGURE 7.23

We say that the two functions $f(x) = x^3$ and $g(x) = \sqrt[3]{x}$ are **inverse functions.** We will study inverse functions in more detail in Chapter 10, but for now we will just observe some basic facts about their relationship.

The table of values for $g(x) = \sqrt[3]{x}$ (Table 7.19) can be obtained from the table for $f(x) = x^3$ (Table 7.18) by interchanging the values of x and y in each ordered pair. This makes sense when we recall that each function undoes the effect of the other. In fact, we can state the following rule to define the cube root function.

$$y = \sqrt[3]{x} \quad \text{if and only if} \quad x = y^3$$

A similar rule relates the operations of raising a base b to a power and taking a base b logarithm. That is,

$$y = \log_b x \quad \text{if and only if} \quad x = b^y$$

This relationship means that the function $g(x) = \log_b x$ is the inverse of the function $f(x) = b^x$. In particular, the function $g(x) = \log_2 x$ is the inverse of $f(x) = 2^x$. Each function "undoes" the effects of the other. For example, start with $x = 3$, apply f and then apply g to the result.

$$x = 3 \quad \rightarrow \quad f(3) = 2^3 = 8 \quad \rightarrow \quad g(8) = \log_2 8 = 3$$

Apply the
exponential function   Apply the
log function

We return to our original number, 3. Thus,

$$\log_2 (2^3) = 3$$

This equation holds for any value of x and for any base $b > 0$. In other words, applying first the exponential function and then the log function will return the original input value. We can also apply the two functions in the opposite order. For example,

$$2^{\log_2 8} = 8$$

In general, we have the following rules.

$$\log_b b^x = x \qquad \text{and} \qquad b^{\log_b x} = x$$

These two equations tell us that the exponential function $y = b^x$ and the logarithmic function $y = \log_b x$ "undo" each other, and hence are inverse functions.

EXERCISE 1 Simplify each expression.

a. $\log_{10} 10^6$

b. $10^{\log_{10} 1000}$

c. $\log_4 4^6$

d. $4^{\log_4 64}$

Graphs of Logarithmic Functions

We can obtain a table of values for $g(x) = \log_2 x$ by making a table for $f(x) = 2^x$ and then interchanging the columns, as shown in Tables 7.20 and 7.21. You can see that the graphs of $f(x) = 2^x$ and $g(x) = \log_2 x$, shown in Figure 7.24, are symmetric about the line $y = x$.

TABLE 7.20

x	$f(x) = 2^x$
-2	$\frac{1}{4}$
-1	$\frac{1}{2}$
0	1
1	2
2	4

TABLE 7.21

x	$g(x) = \log_2 x$
$\frac{1}{4}$	-2
$\frac{1}{2}$	-1
1	0
2	1
4	2

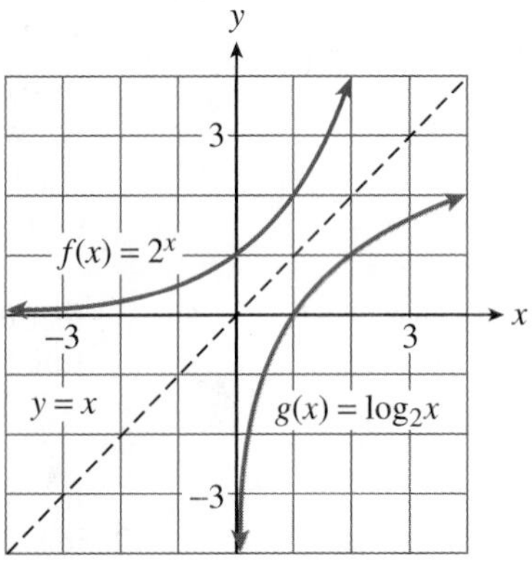

FIGURE 7.24

The same procedure works for graphing log functions with any base: If we want to find values for the function $y = \log_b x$, we can find values for the exponential function $y = b^x$, and then interchange the x and y values in each ordered pair.

EXAMPLE 1

Graph the function $f(x) = 10^x$ and its inverse $g(x) = \log_{10} x$ on the same axes.

Solution

Make a table of values for the function $f(x) = 10^x$. A table of values for the inverse function, $g(x) = \log_{10} x$, can be obtained by interchanging the components of each ordered pair in the table for f. (See Tables 7.22 and 7.23.)

TABLE 7.22

x	$f(x)$
-2	0.01
-1	0.1
0	1
1	10
2	100

TABLE 7.23

x	$g(x)$
0.01	-2
0.1	-1
1	0
10	1
100	2

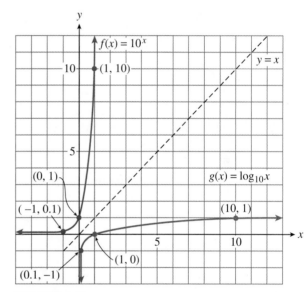

FIGURE 7.25

Plot each set of points and connect them with smooth curves to obtain the graphs shown in Figure 7.25. ∎

While the exponential growth function increases very rapidly for positive values, its inverse, the logarithmic function, grows extremely slowly, as you can see in Figure 7.25. In addition, the logarithmic function $y = \log_b x$ for any base $b > 0$, $b \neq 1$, has the following properties.

Properties of Logarithmic Functions $y = \log_b x$

1. Domain: all positive real numbers.
2. Range: all real numbers.
3. The graphs of $y = \log_b x$ and $y = b^x$ are symmetric about the line $y = x$.

Because the domain of a logarithmic function includes only the *positive* real numbers, *the logarithm of a negative number or zero is undefined.*

Evaluating Logarithmic Functions

We can use the $\boxed{\text{LOG}}$ key on a calculator to evaluate the function $f(x) = \log_{10} x$.

EXAMPLE 2

Let $f(x) = \log_{10} x$. Evaluate the following.

a. $f(35)$ **b.** $f(-8)$ **c.** $2f(16) + 1$

Solutions

a. $f(35) = \log_{10} 35 \approx 1.544$

b. Because -8 is not in the domain of f, $f(-8)$, or $\log_{10}(-8)$ is undefined.

c. $2f(16) + 1 = 2(\log_{10} 16) + 1$
$\approx 2(1.204) + 1 = 3.408$ ∎

EXAMPLE 3

Evaluate the expression $T = \dfrac{\log_{10}\left(\dfrac{M_f}{M_0} + 1\right)}{k}$ for $k = 0.028$, $M_f = 1832$, and $M_0 = 15.3$.

Solution

Follow the order of operations and calculate

$$T = \frac{\log_{10}\left(\dfrac{1832}{15.3} + 1\right)}{0.028} = \frac{\log_{10}(120.739)}{0.028}$$

$$\approx \frac{2.082}{0.028} \approx 74.35$$

A graphing calculator keying sequence for this computation is

$\boxed{\text{LOG}}$ 1832 $\boxed{\div}$ 15.3 $\boxed{+}$ 1 $\boxed{)}$ $\boxed{\div}$ 0.028 $\boxed{\text{ENTER}}$ ∎

EXERCISE 2 Evaluate $t = \dfrac{1}{k}\log_{10}\dfrac{C_H}{C_L}$ for $k = 0.05$, $C_H = 2$, and $C_L = 0.5$.

EXAMPLE 4

If $f(x) = \log_{10} x$, find x so that $f(x) = -3.2$.

Solution

We must solve the equation $\log_{10} x = -3.2$. Rewriting the equation in exponential form yields

$$x = 10^{-3.2} \approx 0.00063$$ ∎

In Example 4 the expression $10^{-3.2}$ can be evaluated in two different ways with a calculator. We can use the $\boxed{\wedge}$ key and press

$$10 \; \boxed{\wedge} \; \boxed{(-)} \; 3.2 \; \boxed{\text{ENTER}}$$

which gives 6.30957344 **E**−4, or approximately 0.00063. Alternatively, we can use the calculator's 10^x function and press

$$\boxed{\text{2nd}} \; \boxed{\text{LOG}} \; \boxed{(-)} \; 3.2 \; \boxed{\text{ENTER}}$$

which gives the same answer as before.

Logarithmic Models

Because logarithmic functions grow very slowly, they are useful for modeling phenomena that take on a very wide range of values. Loudness, or intensity, of sound is one example. Table 7.24 shows the intensity of some common sounds, measured in watts per square meter.

TABLE 7.24

Sound	Intensity	Decibels
Whisper	10^{-10}	20
Background music	10^{-8}	40
Loud conversation	10^{-6}	60
Heavy traffic	10^{-4}	80
Jet airplane	10^{-2}	100
Thunder	10^{-1}	110

Thunder is 10^9, or one billion times, more intense than a whisper. It would be impossible to show such a wide range of values on a graph and still maintain reasonable precision. The decibel is a logarithmic scale used for measuring sound intensity. Notice that there is a variation of only 90 decibels between a whisper and thunder. We will consider the decibel scale shortly.

Another quantity with a wide variation in values is the acidity of a chemical substance, measured by its concentration of hydrogen ions, denoted by $[H^+]$.

EXAMPLE 5

Chemists use a logarithmic scale called pH to measure acidity, where

$$pH = -\log_{10} [H^+] \tag{1}$$

pH values fall between 0 and 14, with 7 indicating a neutral solution. The lower the pH value, the more acidic the substance.

a. Calculate the pH of a solution with a hydrogen ion concentration of 3.98×10^{-5}.

b. The water in a swimming pool should be maintained at a pH of 7.5. What is the hydrogen ion concentration of the water?

Solutions

a. Use a calculator to evaluate the function in (1) with $[H^+] = 3.98 \times 10^{-5}$.

$$pH = -\log_{10}(3.98 \times 10^{-5}) \approx 4.4$$

b. Solve the equation

$$7.5 = -\log_{10}[H^+]$$

for $[H^+]$. First, write

$$-7.5 = \log_{10}[H^+]$$

then rewrite the equation in exponential form to get

$$[H^+] = 10^{-7.5} \approx 3.2 \times 10^{-8}$$

The hydrogen ion concentration of the water is 3.2×10^{-8}. ∎

> **EXERCISE 3** The pH of the water in a tide pool is 8.3. What is the hydrogen ion concentration of the water?

Two more logarithmic models describe the loudness of a sound and the strength of an earthquake. Both these models use expressions of the form $\log\left(\dfrac{a}{b}\right)$. Be careful to follow the order of operations when using these models. We must compute the quotient $\dfrac{a}{b}$ before taking a logarithm. In particular, it is *not* true that $\log\left(\dfrac{a}{b}\right)$ can be simplified to $\dfrac{\log a}{\log b}$.

EXAMPLE 6

The loudness of a sound is measured in decibels D by

$$D = 10\log_{10}\left(\frac{I}{10^{-12}}\right) \tag{2}$$

where I is the intensity of its sound waves (in watts per square meter).

a. A whisper generates about 10^{-10} watts per square meter at a distance of 3 feet. Find the number of decibels for a whisper 3 feet away.

b. Normal conversation registers at about 40 decibels. How many times more intense than a whisper is normal conversation?

Solutions

a. Evaluate the function in (2) with $I = 10^{-10}$ to find

$$D = 10\log_{10}\left(\frac{10^{-10}}{10^{-12}}\right) = 10\log_{10}10^2$$

$$= 10(2) = 20 \text{ decibels}$$

b. Let I_w stand for the intensity of a whisper and let I_c stand for the intensity of normal conversation. We are looking for the ratio $\dfrac{I_c}{I_w}$. From part (a) we know that

$$I_w = 10^{-10}$$

and from Equation (2) we have

$$40 = 10 \log_{10}\left(\frac{I_c}{10^{-12}}\right)$$

which we can solve for I_c. Dividing both members of the equation by 10 and rewriting in exponential form, we have

$$\frac{I_c}{10^{-12}} = 10^4 \qquad\qquad \text{Multiply both sides by } 10^{-12}.$$

$$I_c = 10^4(10^{-12}) = 10^{-8}$$

Finally, compute the ratio $\dfrac{I_c}{I_w}$.

$$\frac{I_c}{I_w} = \frac{10^{-8}}{10^{-10}} = 10^2$$

Normal conversation is 100 times more intense than a whisper. ∎

EXERCISE 4   The noise of city traffic registers at about 70 decibels.

a. What is the intensity of traffic noise, in watts per square meter?

b. How many times more intense is traffic noise than conversation?

EXAMPLE 7

One method for measuring the magnitude of an earthquake compares the amplitude A of its seismographic trace with the amplitude A_0 of the smallest detectable earthquake. The log of their ratio is the Richter magnitude, M. Thus,

$$M = \log_{10}\left(\frac{A}{A_0}\right) \tag{3}$$

a. The Northridge earthquake of January 1994 registered 6.9 on the Richter scale. What would be the magnitude of an earthquake 100 times as powerful as the Northridge quake?

b. How many times more powerful than the Northridge quake was the San Francisco earthquake of 1989, which registered 7.1 on the Richter scale?

Solutions

a. Let A_N represent the amplitude of the Northridge quake and let A_H represent the amplitude of a quake 100 times more powerful. From Equation (3) we have

$$6.9 = \log_{10}\left(\frac{A_N}{A_0}\right)$$

or, rewriting in exponential form,

$$\frac{A_N}{A_0} = 10^{6.9}$$

Now, $A_H = 100A_N$, so

$$\frac{A_H}{A_0} = \frac{100A_N}{A_0}$$
$$= 100\left(\frac{A_N}{A_0}\right) = 10^2(10^{6.9})$$
$$= 10^{8.9}$$

Thus, from Equation (3) the magnitude of the more powerful quake is

$$\log_{10}\left(\frac{A_H}{A_0}\right) = \log_{10} 10^{8.9}$$
$$= 8.9$$

b. Let A_S stand for the amplitude of the San Francisco earthquake. Then we are looking for the ratio A_S/A_N. First we'll use Equation (3) to compute values for A_S and A_N.

$$6.9 = \log_{10}\left(\frac{A_N}{A_0}\right) \qquad \text{and} \qquad 7.1 = \log_{10}\left(\frac{A_S}{A_0}\right)$$

Rewriting each equation in exponential form, we have

$$\frac{A_N}{A_0} = 10^{6.9} \qquad \text{and} \qquad \frac{A_S}{A_0} = 10^{7.1}$$

or

$$A_N = 10^{6.9}A_0 \qquad \text{and} \qquad A_S = 10^{7.1}A_0$$

Now we can compute the ratio we want.

$$\frac{A_S}{A_N} = \frac{10^{7.1}A_0}{10^{6.9}A_0} = 10^{0.2}$$

The San Francisco earthquake was $10^{0.2}$, or approximately 1.58, times as powerful as the Northridge quake. ■

An earthquake 100, or 10^2, times as strong is only two units greater in magnitude on the Richter scale. In general, a *difference* of K units on the Richter scale (or any logarithmic scale) corresponds to a *factor* of 10^K units in the intensity of the quake.

READING QUESTIONS

1. How are the graphs of inverse functions related?
2. Give the domain and range of a logarithmic function.
3. Sketch a typical logarithmic function.
4. Explain why you can't take the log of a negative number.

ANSWERS TO 7.4 EXERCISES

1a. 6   **b.** 1000   **c.** 6   **d.** 64

2. $20 \log_{10} 4 \approx 12.04$ **3.** $[H]^+ = 10^{-8.3} \approx 5.01 \times 10^{-9}$

4a. $I = 10^{-5}$ **b.** 1000

In Problems 1–4, (a) make tables of values for each exponential function and its inverse logarithmic function, and (b) graph both functions on the same set of axes.

1. $f(x) = 2^x$

2. $f(x) = 3^x$

3. $f(x) = \left(\dfrac{1}{3}\right)^x$

4. $f(x) = \left(\dfrac{1}{2}\right)^x$

In Problems 5–10, $f(x) = \log_{10} x$. Evaluate.

5. a. $f(487)$ **b.** $f(2.16)$

6. a. $f(93)$ **b.** $f(6.95)$

7. a. $f(-7)$ **b.** $6f(28)$

8. a. $f(0)$ **b.** $3f(41)$

9. a. $18 - 5f(3)$ **b.** $\dfrac{2}{5 + f(0.6)}$

10. a. $15 - 4f(7)$ **b.** $\dfrac{3}{2 + f(0.2)}$

Evaluate each expression for the given values.

11. $R = \dfrac{1}{L}\log_{10}\left(\dfrac{P}{L - P}\right);$ for $L = 8500$ and $P = 3600$

12. $T = \dfrac{H\log_{10}\dfrac{N}{N_0}}{\log_{10}\dfrac{1}{2}};$ for $H = 5730$, $N = 180$, and $N_0 = 920$

13. $M = \sqrt{\dfrac{\log_{10} H}{k \log_{10} H_0}};$ for $H = 0.93$, $H_0 = 0.02$, and $k = 0.006$

14. $h = a - \sqrt{\dfrac{\log_{10} B}{t}};$ for $a = 56.2$, $B = 78$, and $t = 0.3$

In Problems 15–20, $f(x) = \log_{10} x$. Solve for x.

15. $f(x) = 1.41$ **16.** $f(x) = 2.3$

17. $f(x) = 0.52$ **18.** $f(x) = 0.8$

19. $f(x) = -1.3$ **20.** $f(x) = -1.69$

In Problems 21–34, see Examples 5–7 for the appropriate formulas.

21. The hydrogen ion concentration of vinegar is about 6.3×10^{-4}. Calculate the pH of vinegar.

22. The hydrogen ion concentration of spinach is about 3.2×10^{-6}. Calculate the pH of spinach.

23. The pH of lime juice is 1.9. Calculate its hydrogen ion concentration.

24. The pH of ammonia is 9.8. Calculate its hydrogen ion concentration.

25. A lawn mower generates a noise of intensity 10^{-2} watts per square meter. Find the decibel level of the sound of a lawn mower.

26. A jet airplane generates a noise of intensity 100 watts per square meter at a distance of 100 feet. Find the decibel level for a jet airplane.

27. The loudest sound emitted by any living source is made by the blue whale. Its whistles have been measured at 188 decibels and are detectable 500 miles away. Find the intensity of the blue whale's whistle in watts per square meter.

28. The loudest sound created in a laboratory registered at 210 decibels. The energy from such a sound is sufficient to bore holes in solid material. Find the intensity of a 210-decibel sound.

29. At a concert by The Who in 1976, the sound level 50 meters from the stage registered 120 decibels. How many times more intense was this than a 90-decibel sound (the threshold of pain for the human ear)?

30. The loudest scientifically measured shouting by a human being registered 123.2 decibels. How many times more intense was this than normal conversation at 40 decibels?

31. In 1964 an earthquake in Alaska measured 8.4 on the Richter scale. An earthquake measuring 4.0 is considered small and causes little damage. How many times stronger was the Alaska quake than one measuring 4.0?

32. On April 30, 1986, an earthquake in Mexico City measured 7.0 on the Richter scale. On September 21 a second earthquake, measuring 8.1, hit Mexico City. How many times stronger was the September quake than the one in April?

33. A small earthquake measured 4.2 on the Richter scale. What would be the magnitude of an earthquake three times as strong?

34. Earthquakes measuring 3.0 on the Richter scale often go unnoticed. What would be the magnitude of a quake 200 times as strong as a 3.0 quake?

35. The water velocity at any point in a stream or river is related to the logarithm of the depth at that point. For the Hoback River near Bondurant, Wyoming, $v = 2.63 + 1.03 \log d$, where v is the velocity of the water, in feet per second, and d is the vertical distance from the stream bed, in feet, at that point. For Pole Creek near Pinedale, Wyoming, $v = 1.96 + 0.65 \log d$. Both streams are 1.2 feet deep at the locations mentioned.

   a. Complete the table of values for each stream.

Distance from bed (feet)	0.2	0.4	0.6	0.8	1.0	1.2
Velocity, Hoback River (ft/sec)						
Velocity, Pole Creek (ft/sec)						

   b. If you double the distance from the bed, by how much does the velocity increase in each stream?

   c. Plot both functions on the same graph.

   d. The average velocity of the entire stream can be closely approximated as follows: Measure the velocity at 20% of the total depth of the stream from the surface and at 80% of the total depth, then average these 2 values. Find the average velocity for the Hoback River and for Pole Creek.

   Source: Leopold, Wolman, and Miller, 1992.

36. In a psychology experiment, volunteers were asked to memorize a list of nonsense words and 24 hours later were tested to see how many of the words they recalled. On average, the subjects had forgotten 20% of the words. The researchers found that the more lists their volunteers memorized, the larger the fraction of words they were unable to recall.

Number of lists, n	1	4	8	12	16	20
Percent forgotten, F	20	40	55	66	74	80

   a. Plot the data. What sort of function seems to fit the data points?

   b. Psychologists often describe rates of forgetting by logarithmic functions. Graph the function

   $$f(n) = 16.6 + 46.3 \log x$$

   on the same graph with your data. Comment on the fit.

   c. What happens to the function $f(n)$ as n grows increasingly large? Does this behavior accurately reflect the situation being modeled?

   Source: Underwood, 1964.

37. Let $f(x) = 3^x$ and $g(x) = \log_3 x$.

   a. Compute $f(4)$.

   b. Compute $g[f(4)]$.

   c. Explain why $\log_3 3^x = x$ for any x.

   d. Compute $\log_3 3^{1.8}$.

   e. Simplify $\log_3 3^a$.

38. Let $f(x) = \log_2 x$ and $g(x) = 2^x$.

   a. Compute $f(32)$.

   b. Compute $g[f(32)]$.

   c. Explain why $2^{\log_2 x} = x$ for $x > 0$.

   d. Compute $2^{\log_2 6}$.

   e. Simplify $2^{\log_2 Q}$.

39. Each part of Figure 7.26 shows a portion of the graph of one of the following functions. Match each function with its graph.

a. $f(x) = 2^x$ **b.** $f(x) = x^2$

c. $f(x) = \dfrac{2}{x}$ **d.** $f(x) = \sqrt{x}$

e. $f(x) = \log_2 x$ **f.** $f(x) = \left(\dfrac{1}{2}\right)^x$

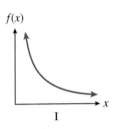
$f(x)$

I

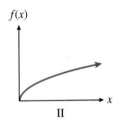
$f(x)$

II

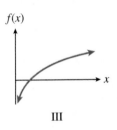
$f(x)$

III

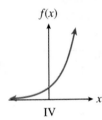
$f(x)$

IV

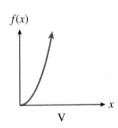
$f(x)$

V

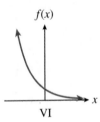
$f(x)$

VI

FIGURE 7.26

40. Choose the graph in Figure 7.27 for each function described below.

a. The area, A, of a pentagon is a quadratic function of the length, l, of its side.

b. The strength, F, of a hurricane varies inversely with its speed, s.

c. The price of food has increased by 3% every year for a decade.

d. The magnitude, M, of a star is a logarithmic function of its brightness, I.

e. The speed of the train increased at a constant rate.

f. If you don't practice a foreign language, you lose $\frac{1}{8}$ of the words in your working vocabulary, V, each year.

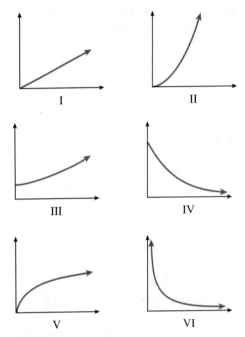

I II

III IV

V VI

FIGURE 7.27

41. a. What is the inverse operation for "raise 6 to the power x"?

b. What is the inverse operation for "take the log base 5 of x"?

43. a. How large must x be before the graph of $y = \log_{10} x$ reaches a height of 4?

b. How large must x be before the graph of $y = \log_{10} x$ reaches a height of 8?

45. What is the domain of the function $f(x) = \log_b (x - 9)$?

42. a. The log base 3 of a number is 5. What is the number?

b. Four raised to a certain power is 32. What is the exponent?

44. a. How large must x be before the graph of $y = \log_2 x$ reaches a height of 5?

b. How large must x be before the graph of $y = \log_2 x$ reaches a height of 10?

46. What is the domain of the function $f(x) = \log_b (16 - 3x)$?

47. a. Complete the following table.

x	x^2	$\log_{10} x$	$\log_{10} x^2$
1			
2			
3			
4			
5			
6			

b. Do you notice a relationship between $\log_{10} x$ and $\log_{10} x^2$? State the relationship as an equation.

48. a. Complete the following table.

x	$\dfrac{1}{x}$	$\log_{10} x$	$\log_{10} \dfrac{1}{x}$
1			
2			
3			
4			
5			
6			

b. Do you notice a relationship between $\log_{10} x$ and $\log_{10} \dfrac{1}{x}$? State the relationship as an equation.

Assuming that the relationships you found in Problems 47 and 48 hold for logarithms to any base, complete the following tables and use them to graph the given functions.

49.

x	$y = \log_e x$
1	0
2	0.693
4	
16	
$\frac{1}{2}$	
$\frac{1}{4}$	
$\frac{1}{16}$	

50.

x	$y = \log_f x$
1	0
2	0.431
4	
16	
$\frac{1}{2}$	
$\frac{1}{4}$	
$\frac{1}{16}$	

7.5   Properties of Logarithms

Because logarithms are actually exponents, they have several properties that can be derived from the laws of exponents. (Refer to Appendix A.5 to review the laws of exponents.) Here are the laws we'll need at present.

1. To multiply two powers with the same base, add the exponents and leave the base unchanged.

$$a^m \cdot a^n = a^{m+n}$$

2. To divide two powers with the same base, subtract the exponents and leave the base unchanged.

$$\frac{a^m}{a^n} = a^{m-n}$$

3. To raise a power to a power, keep the same base and multiply the exponents.

$$(a^m)^n = a^{mn}$$

Each of these laws corresponds to one of three properties of logarithms.

Properties of Logarithms

If $x, y > 0$, then

1. $\log_b (xy) = \log_b x + \log_b y$

2. $\log_b \dfrac{x}{y} = \log_b x - \log_b y$

3. $\log_b x^m = m \log_b x$

We'll consider proofs of the three properties of logarithms in the Homework Problems. For now, study the examples below, keeping in mind that a logarithm is an exponent.

1. Property (1):

$$\underset{5}{\log_2 32} = \log_2 (4 \cdot 8) = \underset{= \quad 2 \quad + \quad 3}{\log_2 4 + \log_2 8} \qquad \text{because} \qquad \begin{array}{l} 2^5 = 2^2 \cdot 2^3 \\ 32 = 4 \cdot 8 \end{array}$$

2. Property (2):

$$\underset{3}{\log_2 8} = \log_2 \frac{16}{2} = \underset{= \quad 4 \quad - \quad 1}{\log_2 16 - \log_2 2} \qquad \text{because} \qquad \begin{array}{l} 2^3 = \dfrac{2^4}{2^1} \\ 8 = \dfrac{16}{2} \end{array}$$

3. Property (3):

$$\underset{6}{\log_2 64} = \log_2 (4)^3 = \underset{= \quad 3 \cdot 2}{3 \log_2 4} \qquad \text{because} \qquad \begin{array}{l} (2^2)^3 = 2^6 \\ (4)^3 = 64 \end{array}$$

Of course, the properties are not primarily useful for computing logs of constants, but for simplifying expressions that contain variables. We can use them to write such expressions in more convenient forms for solving exponential and logarithmic equations.

EXAMPLE 1

Simplify $\log_b \sqrt{xy}$.

Solution

First, express $\sqrt{xy}$ using a fractional exponent.

$$\log_b \sqrt{xy} = \log_b (xy)^{1/2}$$

Then apply Property (3) to write

$$\log_b (xy)^{1/2} = \frac{1}{2} \log_b (xy)$$

Finally by Property (1) we have

$$\frac{1}{2} (\log_b xy) = \frac{1}{2} (\log_b x + \log_b y)$$

Thus,

$$\log_b \sqrt{xy} = \frac{1}{2}(\log_b x + \log_b y)$$ ■

EXERCISE 1 Simplify $\log_b \dfrac{x}{y^2}$.

We can also use the three properties of logarithms to write sums and differences of logarithms as a single logarithm.

EXAMPLE 2 Express $3(\log_b x - \log_b y)$ as a single logarithm with a coefficient of 1.

Solution

Use the properties of logarithms. Begin by applying Property (2).

$$3(\log_b x - \log_b y) = 3 \log_b \left(\frac{x}{y}\right) \quad \text{Apply Property (3).}$$

$$= \log_b \left(\frac{x}{y}\right)^3$$

Therefore,

$$3(\log_b x - \log_b y) = \log_b \left(\frac{x}{y}\right)^3$$ ■

COMMON ERRORS Be careful when using the properties of logarithms! Compare the statements below:

1.
$$\log_b (xy) = \log_b x + \log_b y \qquad \text{(Property 1)}$$
but
$$\log_b (x + y) \neq \log_b x + \log_b y$$

2.
$$\log_b \left(\frac{x}{y}\right) = \log_b x - \log_b y \qquad \text{(Property 2)}$$
but
$$\log_b \frac{x}{y} \neq \frac{\log_b x}{\log_b y}$$ ■

Solving Logarithmic Equations

The properties of logarithms are useful in solving both exponential and logarithmic equations. To solve logarithmic equations, we must first combine any expressions involving logs into a single logarithm.

EXAMPLE 3

Solve $\log_{10}(x + 1) + \log_{10}(x - 2) = 1$.

Solution

Use Property (1) of logarithms to rewrite the left-hand side as a single logarithm.

$$\log_{10}(x + 1)(x - 2) = 1$$

Once the left-hand side is expressed as a *single* logarithm, we can rewrite the equation in exponential form as

$$(x + 1)(x - 2) = 10^1$$

from which

$$x^2 - x - 2 = 10 \quad \text{Subtract 10 from both sides.}$$
$$x^2 - x - 12 = 0 \quad \text{Factor the left side.}$$
$$(x - 4)(x + 3) = 0 \quad \text{Apply the zero-factor principle.}$$

Thus,

$$x = 4 \quad \text{or} \quad x = -3$$

The number -3 is not a solution of the original equation because $\log_{10}(x + 1)$ and $\log_{10}(x - 2)$ are not defined for $x = -3$. The solution of the original equation is 4. ∎

EXERCISE 2 Solve $\log_{10} x + \log_{10} 2 = 3$.
Rewrite the left side as a single logarithm.
Rewrite the equation in exponential form.
Solve for x.

Solving Exponential Equations

By using Property (3), we can now solve exponential equations in which the base is not 10. For example, to solve the equation

$$5^x = 7 \tag{1}$$

we could rewrite the equation in logarithmic form to obtain the exact solution

$$x = \log_5 7$$

However, if we want a decimal approximation for the solution, we begin by taking the base 10 logarithm of both sides of Equation (1) to get

$$\log_{10}(5^x) = \log_{10} 7$$

Then use Property (3) to rewrite the left-hand side as

$$x \log_{10} 5 = \log_{10} 7$$

and divide both sides by $\log_{10} 5$ to get

$$x = \frac{\log_{10} 7}{\log_{10} 5}$$

On your calculator enter the sequence

$$\boxed{\text{LOG}}\ 7\ \boxed{)}\ \boxed{\div}\ \boxed{\text{LOG}}\ 5\ \boxed{\text{ENTER}}$$

to find that

$$x \approx 1.2091$$

COMMON ERROR Do not confuse the expression $\dfrac{\log_{10} 7}{\log_{10} 5}$ with $\log_{10}\left(\dfrac{7}{5}\right)$; they are not the same! We can rewrite

$$\log_{10}\left(\frac{7}{5}\right) = \log_{10} 7 - \log_{10} 5$$

using Property (2), but we cannot rewrite $\dfrac{\log_{10} 7}{\log_{10} 5}$. You can check on your calculator that

$$\frac{\log_{10} 7}{\log_{10} 5} \neq \log_{10} 7 - \log_{10} 5$$

EXAMPLE 4 Solve $1640 = 80 \cdot 6^{0.03x}$.

Solution
Divide both sides by 80 to obtain

$$20.5 = 6^{0.03x}$$

Take the base 10 logarithm of both sides of the equation and use Property (3) of logarithms to get

$$\log_{10} 20.5 = \log_{10} 6^{0.03x}$$
$$= 0.03x \log_{10} 6$$

On the right side of the equation, x is multiplied by the two constants 0.03 and $\log_{10} 6$. Therefore, to solve for x, we must divide both sides of the equation by $0.03 \log_{10} 6$. We use a calculator to evaluate the answer.

$$x = \frac{\log_{10} 20.5}{0.03 \log_{10} 6} \approx 56.19$$

COMMON ERROR In Example 4, do not try to simplify

$$80 \cdot 6^{0.03x} \rightarrow 480^{0.03x} \quad \text{Incorrect!}$$

Remember that the order of operations tells us to compute the power $6^{0.03x}$ *before* multiplying by 80.

EXERCISE 3 Solve $5(1.2)^{2.5x} = 77$.

Divide both sides by 5.
Take the log of both sides.
Apply Property (3) to simplify the left side.
Solve for x.

EXAMPLE 5

The population of Silicon City was 6500 in 1970 and has been tripling every 12 years. When did the population reach 75,000?

Solution

The population of Silicon City grows according to the formula

$$P(t) = 6500 \cdot 3^{t/12}$$

where t is the number of years after 1970. (See Section 7.1 to review exponential growth.) We want to find the value of t for which $P(t) = 75,000$; that is, we want to solve the equation

$$6500 \cdot 3^{t/12} = 75,000$$

Divide both sides by 6500 to get

$$3^{t/12} = \frac{150}{13}$$

then take the base 10 logarithm of both sides and solve for t.

$$\log_{10}(3^{t/12}) = \log_{10}\left(\frac{150}{13}\right) \qquad \text{Apply Property (3).}$$

$$\frac{t}{12}\log_{10} 3 = \log_{10}\left(\frac{150}{13}\right) \qquad \text{Divide by } \log_{10} 3 \text{; multiply by 12.}$$

$$t = \frac{12 \ \log_{10}\dfrac{150}{13}}{\log_{10} 3}$$

Use a calculator to evaluate the answer.

$$t \approx 26.71$$

The population of Silicon City reached 75,000 about 27 years after 1970, or in 1997. ■

EXERCISE 4 Traffic on U.S. highways is growing by 2.7% per year. How long will it take the volume of traffic to double?

Source: *Time*, Jan 25, 1999

Solving Formulas

The techniques discussed above can also be used to solve formulas involving exponential or logarithmic expressions for one variable in terms of the others.

EXAMPLE 6

Solve $P = Cb^{kt}$ for t $(C, k \neq 0)$.

Solution

First express the power b^{kt} in terms of the other variables.

$$b^{kt} = \frac{P}{C}$$

Write the exponential equation in logarithmic form.

$$kt = \log_b \frac{P}{C}$$

Multiply each member by $1/k$.

$$t = \frac{1}{k} \log_b \frac{P}{C}$$

■

EXERCISE 5 Solve $N = N_0 \log_b (ks)$ for s.

Divide both sides by N_0.
Rewrite in exponential form.
Solve for k.

READING QUESTIONS

1. State the three properties of logarithms.
2. Explain how to solve an exponential equation whose base is not 10.
3. Explain why $12 \cdot 10^{3x}$ is not the same as 120^{3x}.
4. Explain why $\log\left(\dfrac{3}{4}\right)$ is not the same as $\dfrac{\log 3}{\log 4}$.
5. Explain why $\log (x + y)$ is not equal to $\log x + \log y$, in general.

ANSWERS TO 7.5 EXERCISES

1. $\log_b x - 2 \log_b y$ **2.** $x = 500$ **3.** $x = \dfrac{\log_{10} 15.4}{2.5 \log_{10} 1.2} \approx 5.999$

4. about 26 years   **5.** $s = \dfrac{b^{N/N_0}}{k}$

HOMEWORK 7.5

Use Properties (1), (2), and (3) on page 473 to write each expression in terms of simpler logarithms. Assume that all variables denote positive numbers.

1. a. $\log_b 2x$ **b.** $\log_b \dfrac{x}{y}$ **c.** $\log_b \dfrac{xy}{z}$ **d.** $\log_b x^3$

2. a. $\log_b xy$ **b.** $\log_b \dfrac{y}{x}$ **c.** $\log_b \dfrac{x}{yz}$ **d.** $\log_b x^{1/3}$

3. a. $\log_b \sqrt{x}$ **b.** $\log_b \sqrt[3]{x^2}$ **c.** $\log_b x^2 y^3$ **d.** $\log_b \dfrac{x^{1/2} y}{z^2}$

4. a. $\log_b \sqrt[5]{y}$ **b.** $\log_b \sqrt{x^3}$ **c.** $\log_b x^{1/3} y^2$ **d.** $\log_b \dfrac{xy^2}{z^{1/2}}$

5. a. $\log_{10} \sqrt[3]{\dfrac{xy^2}{z}}$ **b.** $\log_{10} \sqrt{\dfrac{2L}{R^2}}$ **c.** $\log_{10} 2\pi \sqrt{\dfrac{l}{g}}$ **d.** $\log_{10} 2y \sqrt[3]{\dfrac{x}{y}}$

6. a. $\log_{10} \sqrt{(s - a)(s - b)}$ **b.** $\log_{10} \sqrt{s^2(s - a)^3}$

Given that $\log_b 2 = 0.6931$, $\log_b 3 = 1.0986$, and $\log_b 5 = 1.6094$, find the value of each expression. (Hint: For example, $\log_b 15 = \log_b 3 + \log_b 5$.)

7. a. $\log_b 6$ **b.** $\log_b \dfrac{2}{5}$ **8. a.** $\log_b 10$ **b.** $\log_b \dfrac{3}{2}$

9. a. $\log_b 9$ **b.** $\log_b \sqrt{50}$ **10. a.** $\log_b 25$ **b.** $\log_b 75$

Solve each equation by using logarithms base 10.

11. $2^x = 7$ **12.** $3^x = 4$ **13.** $3^{x+1} = 8$ **14.** $2^{x-1} = 9$

15. $4^{x^2} = 15$ **16.** $3^{x^2} = 21$ **17.** $4.26^{-x} = 10.3$ **18.** $2.13^{-x} = 8.1$

19. $25 \cdot 3^{2.1x} = 47$ **20.** $12 \cdot 5^{1.5x} = 85$ **21.** $3600 = 20 \cdot 8^{-0.2x}$ **22.** $0.06 = 50 \cdot 4^{-0.6x}$

Solve algebraically, and check with a graph.

23. A culture of *Salmonella* bacteria is started with 0.01 gram and triples in weight every 16 hours.

     a. Write a function for the weight of the culture as a function of time.

     b. How long will it take for the culture to weigh 0.5 gram?

24. In 1975 Summit City used 4.2×10^6 kilowatt-hours of electricity, and the demand for electricity has increased by a factor of 1.5 every 10 years since then.

     a. Write a function for the demand for electricity as a function of time.

     b. When will Summit City need 10 million kilowatt-hours annually?

25. The concentration of a certain drug injected into the bloodstream decreases by 20% each hour as the drug is eliminated from the body. If the initial dose creates a concentration of 0.7 milligrams per milliliter,

     a. Write a function for the concentration of the drug in the bloodstream as a function of time.

     b. The minimum effective concentration of the drug is 0.4 milligrams per milliliter. When should the second dose be administered?

     c. Verify your answer with a graph.

26. A small pond is tested for pollution and the concentration of toxic chemicals is found to be 80 parts per million. Clean water enters the pond from a stream, mixes with the polluted water, then leaves the pond so that the pollution level is reduced by 10% each month.

     a. Write a function for the concentration of toxic chemicals as a function of time.

     b. How long will it be before the concentration of toxic chemicals reaches a safe level of 25 parts per million?

     c. Verify your answer with a graph.

Decide whether the two expressions are equal or not.

27. a. $\log_2 (4 \cdot 8)$, $(\log_2 4)(\log_2 8)$

     b. $\log_2 (16 + 16)$, $\log_2 16 + \log_2 16$

28. a. $\log_3 (3 \cdot 27)$, $\log_3 3 + \log_3 27$

     b. $\log_3 (9^2)$, $(\log_3 9)^2$

29. a. $\log_{10} \left(\dfrac{240}{10} \right)$, $\dfrac{\log_{10} 240}{\log_{10} 10}$

     b. $\log_{10} \left(\dfrac{1}{2} \cdot 80 \right)$, $\dfrac{1}{2} \log_{10} 80$

30. a. $\log_{10} (75 - 15)$, $\log_{10} 75 - \log_{10} 15$

     b. $\log_{10} \left(\dfrac{75}{15} \right)$, $\log_{10} 75 - \log_{10} 15$

Graph each pair of functions on your calculator. Explain the result.

31. $f(x) = \log (2x)$, $g(x) = \log 2 + \log x$

32. $f(x) = \log \left(\dfrac{x}{3} \right)$, $g(x) = \log x - \log 3$

33. $f(x) = \log \left(\dfrac{1}{x} \right)$, $g(x) = -\log x$

34. $f(x) = \log (x^3)$, $g(x) = 3 \log x$

35. The annual rate of growth of Hickory Corners is 3.7%.

   a. Write a function for the population of Hickory Corners as a function of time. Let the initial population be P_0.

   b. How long will it take for the population to double?

   c. Suppose $P_0 = 500$. Graph your function in the window Xmin = 0, Xmax = 94, Ymin = 0, Ymax = 10,000.

   d. Use **intersect** to verify that the population doubles from 500 to 1000 people, from 1000 to 2000 people, and from 2000 to 4000 people in equal periods of time. (This time period is called the doubling time.)

36. In 1986 the inflation rate in Bolivia was 8000% annually. The unit of currency in Bolivia is the boliviano.

   a. Write a function for the price of an item as a function of time. Let P_0 be its initial price.

   b. How long did it take for prices to double?

   c. Suppose $P_0 = 5$. Graph your function in the window Xmin = 0, Xmax = 0.94, Ymin = 0, Ymax = 100.

   d. Use **intersect** to verify that the price of an item doubles from 5 to 10 bolivianos, from 10 to 20, and from 20 to 40 in equal periods of time.

37. Radioactive potassium-42, which is used by cardiologists as a tracer, decays at a rate of 5.4% per hour.

   a. Suppose you have 400 milligrams of potassium-42. How long will it take for half of the sample to decay? This time is called the half-life of potassium-42.

   b. How long will it take for three-fourths of the sample to decay? For seven-eighths of the sample?

38. Radium-226 decays at a rate of 0.4% per year.

   a. Find the half-life of radium-226. (See Problem 37.)

   b. How long will it take for three-fourths of a sample of radium-226 to decay? For seven-eighths of the sample?

39. In 1619 Johannes Kepler formulated a relationship between the radius of a planet's orbit and its orbital period, the time it takes for the planet to revolve about the sun.

   a. Use the data from Table 7.25 to plot the orbital times of the planets against their orbital radii. What sort of function does the graph appear to be?

TABLE 7.25

Planet	Orbital radius (AU)	Orbital time (years)
Mercury	0.387	0.241
Venus	0.723	0.615
Earth	1.000	1.000
Mars	1.524	1.881
Jupiter	5.203	11.862
Saturn	9.539	29.46
Uranus	19.182	84.01
Neptune	30.058	164.79
Pluto	39.349	248.43

   b. Suppose that orbital time, t, is a power function of orbital radius, r, so that $t = r^k$, where k is a constant. Let $X = \log r$ and $Y = \log t$, and show that $Y = kX$.

   c. Use Table 7.25 to complete the table below for the nine planets.

Planet	Mercury	Venus	Earth	Mars	Jupiter
$X = \log r$					
$Y = \log t$					

Planet	Saturn	Uranus	Neptune	Pluto
$X = \log r$				
$Y = \log t$				

   d. Plot the points (X, Y) and find the equation of the regression line for the data.

   e. Use your regression equation to find the value of k, then write a formula for t as a function of r. Does your function fit the data you plotted in part (a)?

   f. Find the orbital time for an asteroid whose orbit has radius 2.8 AU.

40. Inside your ear, the basilar membrane contains nerve endings that convert sound waves to electrical impulses, which enable your brain to interpret what you hear. The basilar membrane is about 3.5 centimeters long, starting at the oval window, which is the opening to the inner ear. The position along the basilar membrane at which maximum nerve response occurs is a function of the frequency of the sound wave. Sounds with the lowest frequency, around 20 hertz, are registered at the far end from the oval window. Two tones separated by one octave (a factor of two in frequency) register at positions about 3.5 millimeters apart on the basilar membrane.

a. Complete the table for the entire length of the basilar membrane, showing the frequencies of various tones and the distance from the oval window where they register. Each tone is one octave higher than the previous one.

Octave	0	1	2	$\cdots$
Frequency (hertz)	20	40	80	$\cdots$
Distance (mm)	35	31.5	28	$\cdots$

b. What is the highest frequency most people can hear? How many octaves does the audible range cover?

c. Plot frequency, f, as a function of distance, d. What sort of function does the graph appear to be?

d. Write an equation for frequency, f, in terms of number of octaves, n.

e. Write an equation for distance, d, in terms of number of octaves, n. Solve your equation for n in terms of d.

f. Use your expression for n from part (e) to write f as a function of d, and simplify. Does your function fit the graph you plotted in part (c)?

Source: Berg and Stork, 1982.

Express as a single logarithm with a coefficient of 1.

41. a. $\log_b 8 - \log_b 2$ **b.** $2\log_b x + 3\log_b y$ **c.** $-2\log_b x$

42. a. $\log_b 5 + \log_b 2$ **b.** $\dfrac{1}{4}\log_b x - \dfrac{3}{4}\log_b y$ **c.** $-\log_b x$

43. a. $\dfrac{1}{2}(\log_{10} y + \log_{10} x - 3\log_{10} z)$ **b.** $\dfrac{1}{2}\log_b 16 + 2(\log_b 2 - \log_b 8)$

44. a. $\dfrac{1}{3}(\log_{10} x - 2\log_{10} y - \log_{10} z)$ **b.** $\dfrac{1}{2}(\log_b 6 + 2\log_b 4) - \log_b 2$

Solve each logarithmic equation.

45. $\log_{10} x + \log_{10}(x + 21) = 2$

46. $\log_{10}(x + 3) + \log_{10} x = 1$

47. $\log_8(x + 5) - \log_8 2 = 1$

48. $\log_{10}(x - 1) - \log_{10} 4 = 2$

49. $\log_{10}(x + 2) + \log_{10}(x - 1) = 1$

50. $\log_4(x + 8) + \log_4(x + 2) = 2$

51. $\log_3(x - 2) - \log_3(x + 1) = 3$

52. $\log_{10}(x + 3) - \log_{10}(x - 1) = 1$

For Problems 53–58, use the following formula for compound interest. If P dollars is invested at an annual interest rate, r (expressed as a decimal), compounded n times yearly, the amount, A, after t years is given by

$$A = P\left(1 + \frac{r}{n}\right)^{nt}$$

53. a. Suppose that $5000 is invested at 12% annual interest. Write formulas for the amount Y_1 in the account (as a function of t) if the interest is compounded annually, and the amount Y_2 in the account if the interest is compounded quarterly. Graph Y_1 and Y_2 together in a suitable window.

    b. Find the compounded amount of $5000 invested at 12% for 10 years when compounded annually and when compounded quarterly. Verify with your graphs.

    c. Are the graphs of Y_1 and Y_2 the same? Use your calculator to study tables for the annual amounts. When do the 2 amounts first differ by 5% of the original investment, that is, by $250?

55. What rate of interest is required so that $1000 will yield $1900 after 5 years if the interest rate is compounded monthly?

57. How long will it take a sum of money to triple if it is invested at 10% compounded daily?

54. a. Suppose that $800 is invested at 7% annual interest. Write formulas for the amount Y_1 in the account (as a function of t) if the interest is compounded semiannually, and the amount Y_2 in the account if the interest is compounded monthly. Graph Y_1 and Y_2 together in a suitable window.

    b. Find the compounded amount of $800 invested at 7% for 15 years when compounded semiannually and when compounded monthly. Verify with your graphs.

    c. Are the graphs of Y_1 and Y_2 the same? Use your calculator to study tables for the annual amounts. When do the 2 amounts first differ by 1% of the original investment, that is, by $8?

56. What rate of interest is required so that $400 will yield $600 after 3 years if the interest rate is compounded quarterly?

58. How long will it take a sum of money to increase fivefold if it is invested at 10% compounded quarterly?

Solve each formula for the specified variable.

59. $A = A_0(10^{kt} - 1)$,   for t

61. $w = pv^q$,   for q

63. $t = T\log_{10}\left(1 + \frac{A}{k}\right)$,   for A

60. $B = B_0(1 - 10^{-kt})$,   for t

62. $l = p^a q^b$,   for b

64. $\log_{10} R = \log_{10} R_0 + kt$,   for R

In Problems 65–67 we will use the laws of exponents to prove the corresponding properties of logarithms.

65. We will use the first law of exponents, $a^p \cdot a^q = a^{p+q}$, to prove the first property of logarithms.

    a. Let $m = \log_b x$ and $n = \log_b y$. Rewrite these relationships in exponential form.

$$x = \underline{\quad} \quad \text{and} \quad y = \underline{\quad}$$

    b. Now consider the expression $\log_b (xy)$. Replace x and y by your answers to part (a).

    c. Apply the first law of exponents to your expression in part (b).

    d. Use the definition of logarithm to simplify your answer to part (c).

    e. Refer to the definitions of m and n in part (a) to finish the proof.

66. We will use the second law of exponents, $\dfrac{a^p}{a^q} = a^{p-q}$, to prove the second property of logarithms.

    a. Let $m = \log_b x$ and $n = \log_b y$. Rewrite these relationships in exponential form.

$$x = \underline{\quad} \quad \text{and} \quad y = \underline{\quad}$$

b. Now consider the expression $\log_b\left(\dfrac{x}{y}\right)$.
Replace x and y by your answers to part (a).

c. Apply the second law of exponents to your expression in part (b).

d. Use the definition of logarithm to simplify your answer to part (c).

e. Refer to the definitions of m and n in part (a) to finish the proof.

67. We will use the third law of exponents, $(a^p)^q = a^{pq}$, to prove the third property of logarithms.

a. Let $m = \log_b x$. Rewrite this relationship in exponential form.

$$x = \underline{\quad\quad}$$

b. Now consider the expression $\log_b (x^k)$. Replace x by your answer to part (a).

c. Apply the third law of exponents to your expression in part (b).

d. Use the definition of logarithm to simplify your answer to part (c).

e. Refer to the definition of m in part (a) to finish the proof.

68. a. Explain why $\log_b 1 = 0$.

b. Explain why $\log_b b^x = x$.

c. Explain why $b^{\log_b x} = x$.

7.6 The Natural Base

There is another base for logarithms and exponential functions that is often used in applications. This base is an irrational number called e, where

$$e \approx 2.71828182846$$

The number e is essential for numerous advanced topics and is often called the **natural base.**

The Natural Exponential Function

The **natural exponential function** is the function $f(x) = e^x$. Values for e^x can be obtained with a calculator using the [e^x] key ([2nd] [LN] on most calculators). For example, you can evaluate e^1 by pressing

[2nd] [LN] 1

TABLE 7.26

x	$y = 2^x$	$y = e^x$	$y = 3^x$
-3	0.125	0.050	0.037
-2	0.250	0.135	0.111
-1	0.500	0.368	0.333
0	1	1	1
1	2	2.718	3
2	4	7.389	9
3	8	20.086	27

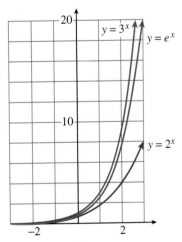

FIGURE 7.28

to confirm the value of e given above.

Because e is a number between 2 and 3, the graph of $y = e^x$ lies between the graphs of $y = 2^x$ and $y = 3^x$. Compare the tables of values (Table 7.26) and the graphs of the three functions in Figure 7.28. You should verify the table and graphs on your calculator.

The Natural Logarithmic Function

The base e logarithm of a number x, or $\log_e x$, is called the **natural logarithm** of x and is denoted by **ln x.** Thus,

$$\ln x = \log_e x$$

The natural logarithm of x is the exponent to which e must be raised to produce x. For example, the natural logarithm of 10, or ln 10, is the solution of the equation

$$e^y = 10$$

You can verify on your calculator that

$$e^{2.3} \approx 10 \qquad \text{or} \qquad \ln 10 \approx 2.3$$

In general,

$$y = \ln x \qquad \text{if and only if} \qquad e^y = x$$

In particular, we see that

$$\ln e = 1 \qquad \text{because} \qquad e^1 = e$$
$$\ln 1 = 0 \qquad \text{because} \qquad e^0 = 1$$

> **EXERCISE 1** Use your calculator to evaluate.
>
> **a.** e^2 **b.** $e^{3.5}$ **c.** $e^{-0.5}$
>
> **d.** ln 100   **e.** ln 0.01   **f.** $\ln e^3$

We use natural logarithms in the same way that we use logs to other bases. The properties of logarithms we studied in Section 7.5 also apply to logarithms base e.

Properties of Natural Logarithms

If $x, y > 0$, then

1. $\ln(xy) = \ln x + \ln y$

2. $\ln \dfrac{x}{y} = \ln x - \ln y$

3. $\ln x^m = m \ln x$

The following properties are also particularly useful.

$$\ln e^x = x \qquad e^{\ln x} = x$$

These two properties tell us that the functions $y = e^x$ and $y = \ln x$ "undo" each other, so they are inverse functions, as you can see in Table 7.27. The graphs of both functions are shown in Figure 7.29.

TABLE 7.27

x	$y = e^x$
-2	0.135
-1	0.368
0	1
1	2.718
2	7.389

x	$y = \ln x$
0.135	-2
0.368	-1
1	0
2.718	1
7.389	2

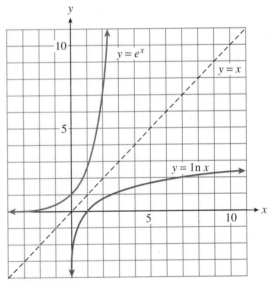

FIGURE 7.29

You can see from its graph that the domain of the natural log function is the set of all positive numbers. Consequently, the natural logs of negative numbers and zero are undefined.

Solving Equations

We use the natural logarithm to solve exponential equations with base e. The techniques we've learned for solving other exponential equations also apply to equations with base e.

EXAMPLE 1

Solve each equation for x.

    a. $e^x = 0.24$ **b.** $\ln x = 3.5$

Solutions

    a. Convert the equation to logarithmic form and evaluate using a calculator.

$$x = \ln 0.24 \approx -1.427$$

    b. Convert the equation to exponential form and evaluate.

$$x = e^{3.5} \approx 33.115$$ ■

To solve more complicated exponential equations, we isolate the power on one side of the equation before converting to logarithmic form.

EXAMPLE 2

Solve $140 = 20e^{0.4x}$.

Solution

First, divide each side by 20 to obtain

$$7 = e^{0.4x}$$

and then convert the equation to logarithmic form as

$$0.4x = \ln 7 \quad \text{Divide both sides by 0.4.}$$
$$x = \frac{\ln 7}{0.4}$$

Rounded to four decimal places, $x \approx 4.8648$.   ■

We can solve the equation in Example 2,

$$7 = e^{0.4x}$$

by "taking the natural logarithm of both sides." This gives us

$$\ln 7 = \ln e^{0.4x} \quad \text{Simplify the right side.}$$
$$\ln 7 = 0.4x$$

because $\ln e^a = a$ for any number a. We then proceed with the solution as before.

EXERCISE 2   Solve $80 - 16e^{-0.2x} = 70.3$.

Subtract 80 from both sides and divide by -16.

Take the natural log of both sides.

Divide by -0.2.

EXAMPLE 3

Solve $P = \dfrac{a}{1 + be^{-kt}}$ for t.

Solution

Multiply both sides of the equation by the denominator, $1 + be^{-kt}$, to get

$$P(1 + be^{-kt}) = a$$

then isolate the power, e^{-kt}:

$$1 + be^{-kt} = \frac{a}{P} \qquad \text{Subtract 1 from both sides and simplify.}$$

$$be^{-kt} = \frac{a}{P} - 1 = \frac{a - P}{P} \qquad \text{Divide both sides by } b.$$

$$e^{-kt} = \frac{a - P}{bP}$$

Take the natural logarithm of both sides to get

$$\ln e^{-kt} = \ln \frac{a - P}{bP}$$

and recall that $\ln e^x = x$ to simplify the left side.

$$-kt = \ln \frac{a - P}{bP}$$

Finally, divide both sides by $-k$ to solve for t.

$$t = \frac{-1}{k} \ln \frac{a - P}{bP} \qquad\qquad \blacksquare$$

EXERCISE 3   Solve $N = Ae^{-kt}$ for k.

Divide both sides by A.
Take the natural log of both sides.
Divide both sides by $-t$.

Exponential Growth and Decay

In Section 7.1 we considered functions of the form

$$P(t) = P_0 \cdot a^t \tag{1}$$

which describe exponential growth when $a > 1$ and exponential decay when $0 < a < 1$. Exponential growth and decay can also be modeled by functions of the form

$$P(t) = P_0 \cdot e^{kt} \tag{2}$$

where we have substituted e^k for the growth factor a in Equation (1) so that

$$P(t) = P_0 \cdot a^t$$
$$= P_0 \cdot (e^k)^t = P_0 \cdot e^{kt}$$

We can find the value of k by solving the equation $a = e^k$ for k to get $k = \ln a$.

For instance, in Investigation 11 on page 419 we found that a colony of bacteria grew according to the formula

$$P(t) = 100 \cdot 3^t$$

We can express this function in the form $P(t) = 100 \cdot e^{kt}$ if we set

$$3 = e^k \qquad \text{or} \qquad k = \ln 3 \approx 1.0986$$

Thus the growth law for the colony of bacteria can be written

$$P(t) \approx 100 \cdot e^{1.0986t}$$

The function

$$P(t) = P_0 e^{kt}$$

describes exponential growth if the value of k is positive, as in the formula for the colony of bacteria. If the value of k is negative, the function describes exponential decay.

EXAMPLE 4

Express the decay law $N(t) = 60(0.8)^t$ in the form $N(t) = N_0 e^{kt}$.

Solution

For this decay law, $N_0 = 60$ and $a = 0.8$. We'd like to find a value for k so that $e^k = a = 0.8$; that is, we must solve the equation

$$e^k = 0.8 \qquad\qquad \text{Take natural log of both sides.}$$
$$\ln e^k = \ln 0.8 \qquad\qquad \text{Simplify.}$$
$$k = \ln 0.8 \approx -0.2231$$

Replacing a by e^k, we find that the decay law is

$$N(t) \approx 60e^{-0.2231t}$$

■

EXERCISE 4 From 1994 to 1998, the number of personal computers using the Internet grew according to the formula $N(t) = 2.8e^{0.85t}$, where $t = 0$ in 1994 and N is in millions.

a. Evaluate $N(1)$. By what percent did the number of Internet users grow in one year?

b. Express the growth law in the form $N(t) = N_0(1 + r)^t$.
(*Hint:* $e^k = 1 + r$.)

Source: *Los Angeles Times*, September 6, 1999

Applications

Exponential functions using base e are used to model a number of interesting phenomena. For example, some savings institutions offer accounts on which the interest is **compounded continuously.** The amount accumulated in such an account after t years at interest rate r is given by the function

$$A(t) = P e^{rt}$$

where P is the principal invested.

EXAMPLE 5

Suppose you invest $500 in an account that pays 8% interest compounded continuously. You leave the money in the account without making any additional deposits or withdrawals.

    a. Write a formula that gives the value of your account, $A(t)$, after t years.

    b. Make a table of values showing $A(t)$ for the first 5 years.

    c. How much will the account be worth after 10 years?

    d. Graph the function $A(t)$.

    e. How long will it be before the account is worth $1000?

Solutions

    a. Substitute 500 for P and 0.08 for r to find

$$A(t) = 500e^{0.08t}$$

    b. Evaluate the formula for $A(t)$ to obtain Table 7.28.

TABLE 7.28

t	$A(t)$
0	500
1	541.64
2	586.76
3	635.62
4	688.56
5	745.91

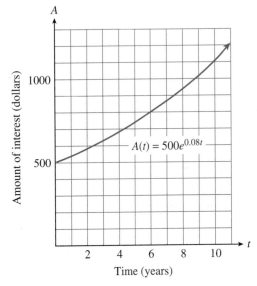

FIGURE 7.30

    c. Evaluate $A(t)$ for $t = 10$.

$$A(10) = 500e^{0.08(10)}$$
$$= 500e^{0.8}$$
$$\approx 500(2.2255) = 1112.77$$

The account will be worth $1112.77 after 10 years.

d. The graph of $A(t)$ is shown in Figure 7.30.

e. Substitute 1000 for $A(t)$ and solve the equation.

$$1000 = 500e^{0.08t} \qquad \text{Divide both sides by 500.}$$
$$2 = e^{0.08t} \qquad \text{Take natural log of both sides.}$$
$$\ln 2 = \ln e^{0.08t} = 0.08t \qquad \text{Divide both sides by 0.08.}$$
$$t = \frac{\ln 2}{0.08} \approx 8.6643$$

The account will be worth $1000 after approximately 8.7 years. ■

Radioactive decay is another common application of exponential functions. Radioactive elements decay over time into more stable isotopes of the same element. The amount of radioactive material left at any time t is given by a decreasing exponential function.

EXERCISE 5 A scientist isolates 25 grams of krypton-91, which decays according to the formula $N(t) = 25e^{-0.07t}$, where t is in seconds.

a. Complete the table of values showing the amount of krypton-91 left at ten-second intervals over the first minute.

t	0	10	20	30	40	50	60
$N(t)$							

b. Use the table to choose a suitable window and graph the function $N(t)$.

c. Write and solve an equation to answer the question: How long does it take for 60% of the krypton-91 to decay? (*Hint:* If 60% of the krypton-91 has decayed, 40% of the original 25 grams remains.)

1. What is the natural base?
2. Explain why $\ln e^x = x$.
3. State a formula for exponential growth using base e.
4. How is the formula for exponential decay in base e different from the formula for exponential growth?

ANSWERS TO 7.6 EXERCISES

1a. $e^2 \approx 7.389$ **b.** $e^{3.5} \approx 33.115$ **c.** $e^{-0.5} \approx 0.6065$

d. $\ln 100 \approx 4.605$ **e.** $\ln 0.01 \approx -4.605$ **f.** $\ln e^3 = 3$

2. $x = -5 \ln\left(\dfrac{9.7}{16}\right) \approx 2.5$ **3.** $k = \dfrac{-\ln(N/A)}{t}$

4a. $N(1) \approx 6.55$, 134%   **b.** $N(t) \approx 2.8(1 + 1.3396)^t$

5a.

t	0	10	20	30	40	50	60
N(t)	2.5	12.42	6.16	3.06	1.52	0.75	0.37

c. $25e^{-0.07t} = 0.40(25)$; $t = \dfrac{\ln(0.4)}{-0.07} \approx 13.09$ seconds

b.

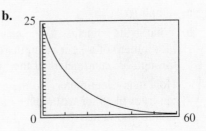

HOMEWORK 7.6

Use your calculator to complete the table for each function Then choose a suitable window and graph the function.

t	−10	−5	0	5	10	15	20
f(x)							

1. $f(x) = e^{0.2x}$ **2.** $f(x) = e^{0.6x}$ **3.** $f(x) = e^{-0.3x}$ **4.** $f(x) = e^{-0.1x}$

Solve for x. Round your answers to four decimal places.

5. a. $e^x = 1.9$ **b.** $e^x = 45$ **c.** $e^x = 0.3$ **6. a.** $e^x = 2.1$ **b.** $e^x = 60$ **c.** $e^x = 0.9$

7. a. $\ln x = 1.42$ **b.** $\ln x = 0.63$ **c.** $\ln x = -2.6$ **8. a.** $\ln x = 2.03$ **b.** $\ln x = 0.59$ **c.** $\ln x = -3.4$

9. The number of bacteria in a culture grows according to the function

$$N(t) = N_0 e^{0.04t}$$

where N_0 is the number of bacteria present at time $t = 0$ and t is the time in hours.

a. Write a growth law for a sample in which 6000 bacteria were present initially.

b. Make a table of values for $N(t)$ in 5-hour intervals over the first 30 hours.

c. Graph $N(t)$.

d. How many bacteria were present at $t = 24$ hours?

e. How much time must elapse (to the nearest tenth of an hour) for the original 6000 bacteria to increase to 100,000?

10. Hope invests $2000 in a savings account that pays $5\frac{1}{2}\%$ annual interest compounded continuously.

a. Write a formula that gives the amount of money, $A(t)$, in Hope's account after t years.

b. Make a table of values for $A(t)$ in 2-year intervals over the first 10 years.

c. Graph $A(t)$.

d. How much will Hope's account be worth after 7 years?

e. How long will it take for the account to grow to $5000?

11. The intensity I (in lumens) of a light beam after passing through t centimeters of a filter having an absorption coefficient of 0.1 is given by the function

$$I(t) = 1000e^{-0.1t}$$

  a. Graph $I(t)$.

  b. What is the intensity (to the nearest tenth of a lumen) of a light beam that has passed through 0.6 centimeters of the filter?

  c. How many centimeters (to the nearest tenth) of the filter will reduce the illumination to 800 lumens?

12. X-rays can be absorbed by a lead plate so that

$$I(t) = I_0 e^{-1.88t}$$

where I_0 is the X-ray count at the source and $I(t)$ is the X-ray count behind a lead plate of thickness t inches.

  a. Graph $I(t)$.

  b. What percent of an X-ray beam will penetrate a lead plate $\frac{1}{2}$ inch thick?

  c. How thick should the lead plate be in order to screen out 70% of the X-rays?

Express each exponential function in the form $P(t) = P_0 a^t$. Is the function increasing or decreasing?

13. $P(t) = 20e^{0.4t}$ 14. $P(t) = 0.8e^{1.3t}$ 15. $P(t) = 6500e^{-2.5t}$ 16. $P(t) = 1.7e^{-0.02t}$

17. **a.** Fill in the table, rounding your answers to four decimal places.

x	0	0.5	1	1.5	2	2.5
e^x						

  b. Compute the ratio of each function value to the previous one. Explain the result.

18. **a.** Fill in the table, rounding your answers to four decimal places.

x	0	2	4	6	8	10
e^x						

  b. Compute the ratio of each function value to the previous one. Explain the result.

19. **a.** Fill in the table, rounding your answers to the nearest integer.

x	0	0.6931	1.3863	2.0794	2.7726	3.4657	4.1589
e^x							

  b. Subtract each x-value from the next one. Explain the result.

20. **a.** Fill in the table, rounding your answers to the nearest integer.

x	0	1.0986	2.1972	3.2958	4.3944	5.4931	6.5917
e^x							

  b. Subtract each x-value from the next one. Explain the result.

Solve. Round your answers to two decimal places.

21. $6.21 = 2.3e^{1.2x}$ 22. $22.26 = 5.3e^{0.4x}$ 23. $6.4 = 20e^{0.3x} - 1.8$ 24. $4.5 = 4e^{2.1x} + 3.3$

25. $46.52 = 3.1e^{1.2x} + 24.2$ 26. $1.23 = 1.3e^{2.1x} - 17.1$

27. $16.24 = 0.7e^{-1.3x} - 21.7$ 28. $55.68 = 0.6e^{-0.7x} + 23.1$

Solve each equation for the specified variable.

29. $y = e^{kt}$,   for t

30. $\dfrac{T}{R} = e^{t/2}$,   for t

31. $y = k(1 - e^{-t})$,   for t

32. $B - 2 = (A + 3)e^{-t/3}$,   for t

33. $T = T_0 \ln(k + 10)$,   for k

34. $P = P_0 + \ln 10k$,   for k

35. a. Fill in the table, rounding your answers to three decimal places.

n	0.39	3.9	39	390
$\ln n$				

b. Subtract each natural logarithm in your table from the next one. (For example, compute $\ln 3.9 - \ln 0.39$.) Explain the result.

37. a. Fill in the table, rounding your answers to three decimal places.

n	2	4	8	16
$\ln n$				

b. Divide each natural logarithm in your table by $\ln 2$. Explain the result.

36. a. Fill in the table, rounding your answers to three decimal places.

n	0.64	6.4	64	640
$\ln n$				

b. Subtract each natural logarithm in your table from the next one. (For example, compute $\ln 6.4 - \ln 0.64$.) Explain the result.

38. a. Fill in the table, rounding your answers to three decimal places.

n	5	25	125	625
$\ln n$				

b. Divide each natural logarithm in your table by $\ln 5$. Explain the result.

For Problems 39–44, (a) express each growth or decay law in the form $N(t) = N_0 e^{kt}$ (b) check your answer by graphing both forms of the function on the same axes. Do they have the same graph?

39. $N(t) = 100 \cdot 2^t$

40. $N(t) = 50 \cdot 3^t$

41. $N(t) = 1200(0.6)^t$

42. $N(t) = 300(0.8)^t$

43. $N(t) = 10(1.15)^t$

44. $N(t) = 1000(1.04)^t$

45. The population of Citrus Valley was 20,000 in 1980. In 1990 it was 35,000.

a. What is P_0 if $t = 0$ in 1980?

b. Use the population in 1990 to find the growth factor e^k.

c. Write a growth law of the form $P(t) = P_0 e^{kt}$ for the population of Citrus Valley.

d. If it continues at the same rate of growth, what will the population be in 2010?

47. Cobalt-60 is a radioactive isotope used in the treatment of cancer. A 500-milligram sample of cobalt-60 decays to 385 milligrams after 2 years.

a. Using $P_0 = 500$, find the decay factor e^k for cobalt-60.

b. Write a decay law $N(t) = N_0 e^{kt}$ for cobalt-60.

c. How much of the original sample will be left after 10 years?

46. In 1981 a copy of *Time* magazine cost $1.50. In 1988 the cover price was $2.00.

a. What is P_0 if $t = 0$ in 1981?

b. Use the price in 1988 to find the growth factor e^k.

c. Find a growth law of the form $P(t) = P_0 e^{kt}$ for the price of *Time*.

d. In 1999, a copy of *Time* cost $3.50. Did the price of the magazine continue to grow at the same rate from 1981 to 1999?

48. Weed seeds can survive for a number of years in the soil. An experiment on cultivated land found 155 million weed seeds per acre, and in the following years the experimenters prevented the seeds from coming to maturity and producing new weeds. Four years later there were 13.6 million seeds per acre.

a. Find the annual decay factor e^k for the number of weed seeds in the soil.

b. Write an exponential formula with base e for the number of weed seeds that survived after t years.

Source: Burton, 1998.

49. Delbert invests $500 in an account that pays 9.5% interest compounded continuously.

a. Write a formula for $A(t)$ that gives the amount of money in Delbert's account after t years.

b. How long will it take Delbert's investment to double to $1000?

c. How long will it take Delbert's money to double again, to $2000?

d. Graph $A(t)$ and illustrate the doubling time on your graph.

e. Choose any point (t_1, A_1) on the graph, then find the point on the graph with vertical coordinate $2A_1$. Verify that the difference in the t-coordinates of the two points is the doubling time.

50. Technetium-99m (Tc-99m) is an artificially produced radionuclide used as a tracer for producing images of internal organs such as the heart, liver, and thyroid. A solution of Tc-99m with initial radioactivity of 10,000 becquerels (Bq) decays according to the formula

$$N(t) = 10,000e^{-0.1155t}$$

where t is in hours.

a. How long will it take the radioactivity to fall to half its initial value, or 5000 Bq?

b. How long will it take the radioactivity to be halved again?

c. Graph $N(t)$ and illustrate the half-life on your graph.

d. Choose any point (t_1, N_1) on the graph, then find the point on the graph with vertical coordinate $0.5N_1$. Verify that the difference in the t-coordinates of the two points is the half-life.

51. All living things contain a certain amount of the isotope carbon-14. When an organism dies, the carbon-14 decays according to the formula

$$N(t) = N_0e^{-0.000124t}$$

where t is measured in years. Scientists can estimate the age of an organic object by measuring the amount of carbon-14 remaining.

a. When the Dead Sea scrolls were discovered in 1947, they had 78.8% of their original carbon-14. How old were the Dead Sea scrolls then?

b. What is the half-life of carbon-14; that is, how long does it take for half of an object's carbon-14 to decay?

52. The growth of plant populations can be measured by the amount of pollen they produce. The pollen from a population of pine trees that lived more than 9500 years ago in Norfolk, England, was deposited in the layers of sediment in a lake basin and dated with radiocarbon techniques. Figure 7.31 shows the rate of pollen accumulation plotted against time, and the fitted curve $P(t) = 650e^{0.00932t}$.

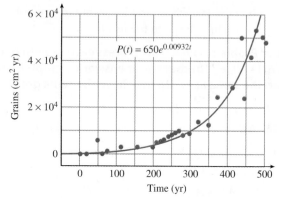

FIGURE 7.31

Data from "Postglacial population expansion of forest trees in Norfolk, U.K.," by K. D. Bennett, 1983, *Nature, 303,* 164–167. Copyright © 1983 Macmillan Magazines, Ltd. Found in *Biology by Numbers,* by R. F. Burton. Copyright © 1998, Cambridge University Press. Reprinted with permission of Cambridge University Press, Macmillan Magazines, Ltd., and K. D. Bennett.

a. What was the annual rate of growth in pollen accumulation?

b. Find the doubling time for the pollen accumulation, that is, the time it took for the amount of pollen to double.

c. By what factor did the pollen increase over a period of 500 years?

53. The half-life (the time it takes for half of a sample of radioactive material to decay) of iodine-131 is approximately 8 days.

a. If a sample initially contains N_0 grams of iodine-131, how much will it contain after 8 days? How much will it contain after 16 days? After 32 days?

b. Use your answers to part (a) to sketch a graph of $N(t)$, the amount of iodine-131 remaining, versus time. (Choose an arbitrary height for N_0 on the vertical axis.)

c. Calculate k, and hence find a decay law of the form $N(t) = N_0 e^{kt}$, where $k < 0$, for iodine-131.

54. The half-life of hydrogen-3 is 12.5 years.

   a. If a sample initially contains N_0 grams of hydrogen-3, how much will it contain after 12.5 years? How much will it contain after 25 years?

   b. Use your answers to part (a) to sketch a graph of $N(t)$, the amount of hydrogen-3 remaining, versus time. (Choose an arbitrary height for N_0 on the vertical axis.)

   c. Calculate k, and hence find a decay law of the form $N(t) = N_0 e^{kt}$, where $k < 0$, for hydrogen-3.

55. A Geiger counter measures the amount of radioactive material present in a substance. The table shows the count rate for a sample of iodine-128 as a function of time.

Time (min)	0	10	20	30	40
Counts/sec	120	90	69	54	42

Time (min)	50	60	70	80	90
Counts/sec	33	25	19	15	13

   a. Graph the data and use your calculator's exponential regression feature to fit a curve to them.

   b. Write your equation in the form $G(t) = G_0 e^{kt}$.

   c. Calculate the half-life of iodine-128.

   Source: © Andrew Hunt and Alan Sykes, *Chemistry*, 1984. Reprinted by permission of Pearson Education Limited.

56. The table shows the count rate for sodium-24 registered by a Geiger counter as a function of time.

Time (min)	0	10	20	30	40
Counts/sec	180	112	71	45	28

Time (min)	50	60	70	80	90
Counts/sec	18	11	7	4	3

   a. Graph the data and use your calculator's exponential regression feature to fit a curve to them.

b. Write your equation in the form $G(t) = G_0 e^{kt}$.

c. Calculate the half-life of sodium-24.

Source: © Andrew Hunt and Alan Sykes, *Chemistry*, 1984. Reprinted by permission of Pearson Education Limited.

57. Do hedgerows planted at the boundaries of a field have a good or bad effect on crop yields? Hedges provide some shelter for the crops, and retain moisture, but they may compete for nutrients or create too much shade. Results of studies on the microclimates produced by hedges are summarized in Figure 7.32, which shows how crop yields increase or decrease as a function of distance from the hedgerow.

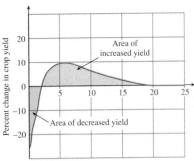

FIGURE 7.32
Source: Briggs and Courtney, 1985.

   a. We'll use trial-and-improvement to fit a curve to the graph. First graph $y_1 = xe^{-x}$ in the window $Xmin = -2$, $Xmax = 5$, $Ymin = -1$, $Ymax = 1$ to see that it has the right shape.

   b. Graph $y_2 = (x - 2)e^{-(x-2)}$ on the same axes. How is the graph of y_2 different from the graph of y_1?

   c. Next we'll find the correct scale by trying functions of the form $y = a(x - 2)e^{-(x-2)/b}$. Experiment with different whole-number values of a and b. How do the values of a and b affect the curve?

   d. Graph $y = 5(x - 2)e^{-(x-2)/4}$ in the window $Xmin = -5$, $Xmax = 25$, $Ymin = -20$, $Ymax = 25$. This function is a reasonable approximation for the curve in Figure 7.32. Compare the area of decreased yield (below the x-axis) with the area of increased yield (above the x-axis). Which area is larger? Is the overall effect of hedgerows on crop yield good or bad?

e. About how far from the hedgerow do the beneficial effects extend? If the average hedgerow is about 2.5 meters tall, how large should the field be to exploit their advantages?

58. Organic matter in the ground decomposes over time, and if the soil is cultivated properly, the fraction of its original organic carbon content is given by

$$C(t) = \frac{a}{b} - \frac{a - b}{b} e^{-bt}$$

where t is in years, and a and b are constants.

a. Write and simplify the formula for $C(t)$ if $a = 0.01$, $b = 0.028$.

b. Graph $C(t)$ in the window Xmin = 0, Xmax = 200, Ymin = 0, Ymax = 1.5.

c. What value does $C(t)$ approach as t increases? Compare this value to $\frac{a}{b}$.

d. The half-life of this function is the amount of time until $C(t)$ declines halfway to its limiting value, $\frac{a}{b}$. What is the half-life?

Source: Briggs and Courtney, 1985.

Problems 59–64 deal with the "change of base" formula.

59. Follow the steps below to calculate $\log_8 20$.

a. Let $x = \log_8 20$. Write the equation in exponential form.

b. Take the logarithm base 10 of both sides of your new equation.

c. Simplify and solve for x.

60. Follow the steps in Problem 59 to calculate $\log_8 5$.

61. Use Problem 59 to find a formula for calculating $\log_8 Q$, where Q is any positive number.

62. Find a formula for calculating $\log_b Q$, where $b > 1$ and Q is any positive number.

63. Find a formula for calculating $\ln Q$ in terms of $\log_{10} Q$.

64. Find a formula for calculating $\log_{10} Q$ in terms of $\ln Q$.

CHAPTER 7 REVIEW

For Problems 1–4, do the following.

a. Make a table of values for the situation described.

b. Write a function that describes the exponential growth or decay.

c. Graph the function.

d. Evaluate the function at the given values.

1. The number of computer science degrees awarded by Monroe College has increased by a factor of $\frac{3}{2}$ every 5 years since 1974. If the college granted 8 degrees in 1974, how many did it award in 1984? In 1995?

2. The price of public transportation has been rising by 10% per year since 1975. If it cost $0.25 to ride the bus in 1975, how much did it cost in 1985? How much will it cost in the year 2010 if the current trend continues?

3. A certain medication is eliminated from the body at a rate of 15% per hour. If an initial dose of 100 milligrams is taken at 8 a.m., how much is left at 12 noon? At 6 p.m.?

4. After the World Series, sales of T-shirts and other memorabilia declines 30% per week. If $200,000 worth of souvenirs were sold during the Series, how much will be sold 4 weeks later? Six weeks after the Series?

Graph each function.

5. $f(t) = 1.2^t$

6. $g(t) = 0.6^{-t}$

7. $P(x) = 2^x - 3$

8. $R(x) = 2^{x + 3}$

Solve each equation.

9. $3^{x+2} = 9^{1/3}$ **10.** $2^{x-1} = 8^{-2x}$

11. $4^{2x+1} = 8^{x-3}$ **12.** $3^{x^2-4} = 27$

13. "Within belts of uniform moisture conditions and comparable vegetation, the organic matter content of soil decreases exponentially with increasing temperature." Data indicate that the organic content doubles with each 10°C decrease in temperature. Write a formula for this function, stating clearly what each variable represents.

Source: Leopol, Wolman, and Miller, 1992.

14. In 1951 a study of barley yields under diverse soil conditions led to the formula

$$Y = cV^a G^b$$

where V is a soil texture rating, G is a drainage rating, and a, b, and c are constants. In fields with similar drainage systems, the formula gives yields Y as a function of V, the soil texture. What type of function is it? If it is an increasing function, what can you say about a?

Source: Briggs and Courtney, 1985.

Find each logarithm.

15. $\log_2 16$ **16.** $\log_4 2$ **17.** $\log_3 \dfrac{1}{3}$ **18.** $\log_7 7$

19. $\log_{10} 10^{-3}$ **20.** $\log_{10} 0.0001$

Write each equation in exponential form.

21. $\log_2 3 = x - 2$ **22.** $\log_n q = p - 1$

Write each equation in logarithmic form.

23. $0.3^{-2} = x + 1$ **24.** $4^{0.3t} = 3N_0$

Solve for the unknown value.

25. $\log_3 \dfrac{1}{3} = y$ **26.** $\log_3 x = 4$ **27.** $\log_b 16 = 2$ **28.** $\log_2 (3x - 1) = 3$

29. $4 \cdot 10^{1.3x} = 20.4$ **30.** $127 = 2(10^{0.5x}) - 17.3$ **31.** $3(10^{-0.7x}) + 6.1 = 9$ **32.** $40(1 - 10^{-1.2x}) = 30$

Evaluate each expression.

33. $k = \dfrac{1}{t}(\log_{10} N - \log_{10} N_0);$ for $t = 2.3$, $N = 12{,}000$, and $N_0 = 9{,}000$

34. $P = \dfrac{1}{k}\sqrt{\dfrac{\log_{10} N}{t}};$ for $k = 0.4$, $N = 48$, and $t = 1.2$

35. $h = k \log_{10}\left(\dfrac{N}{N - N_0}\right);$ for $k = 1.2$, $N = 6400$, and $N_0 = 2000$

36. $Q = \dfrac{1}{t}\left(\dfrac{\log_{10} M}{\log_{10} N}\right);$ for $t = 0.3$, $M = 180$, and $N = 460$

Write each expression in terms of simpler logarithms. (Assume that all variables and variable expressions denote positive real numbers.)

37. $\log_b\left(\dfrac{xy^{1/3}}{z^2}\right)$ **38.** $\log_b \sqrt{\dfrac{L^2}{2R}}$

39. $\log_{10}\left(x\sqrt[3]{\dfrac{x}{y}}\right)$ **40.** $\log_{10} \sqrt{(s - a)(s - g)^2}$

Write each expression as a single logarithm with coefficient 1.

41. $\frac{1}{3}(\log_{10} x - 2 \log_{10} y$

42. $\frac{1}{2} \log_{10} (3x) - \frac{2}{3} \log_{10} y$

43. $\frac{1}{3} \log_{10} 8 - 2(\log_{10} 8 - \log_{10} 2)$

44. $\frac{1}{2}(\log_{10} 9 + 2 \log_{10} 4) + 2 \log_{10} 5$

Solve each logarithmic equation.

45. $\log_3 x + \log_3 4 = 2$

46. $\log_2 (x + 2) - \log_2 3 = 6$

47. $\log_{10} (x - 1) + \log_{10} (x + 2) = 3$

48. $\log_{10} (x + 2) - \log_{10} (x - 3) = 1$

Solve each equation by using base 10 logarithms.

49. $3^{x - 2} = 7$ **50.** $4 \cdot 2^{1.2x} = 64$

51. $1200 = 24 \cdot 6^{-0.3x}$ **52.** $0.08 = 12 \cdot 3^{-1.5x}$

53. Solve $N = N_0(10^{kt})$ for t

54. Solve $Q = R_0 + R \log_{10} kt$ for t

55. The population of Dry Gulch has been declining according to the function

$$P(t) = 3800 \cdot 2^{-t/20}$$

where t is the number of years since the town's heyday in 1910.

a. What was the population of Dry Gulch in 1990?

b. In what year will the population dip below 120 people?

56. The number of compact discs produced each year by Delta Discs is given by the function

$$N(t) = 8000 \cdot 3^{t/4}$$

where t is the number of years since discs were introduced in 1980.

a. How many discs did Delta produce in 1989?

b. In what year will Delta first produce over 2 million discs?

57. a. Write a formula for the cost of a camera t years from now if it costs $90 now and the inflation rate is 6% annually.

b. How much will the camera cost 10 months from now?

c. How long will it be before the camera costs $120?

58. a. Write a formula for the cost of a sofa t years from now if it costs $1200 now and the inflation rate is 8% annually.

b. How much will the sofa cost 20 months from now?

c. How long will it be before the sofa costs $1500?

Solve each equation.

59. $e^x = 4.7$ **60.** $e^x = 0.5$

61. $\ln x = 6.02$ **62.** $\ln x = -1.4$

63. $4.73 = 1.2e^{0.6x}$ **64.** $1.75 = 0.3e^{-1.2x}$

65. The voltage V across a capacitor in a certain circuit is given by the function

$$V(t) = 100(1 - e^{-0.5t})$$

where t is the time in seconds.

a. Make a table of values and graph $V(t)$ for $t = 0$ to $t = 10$.

b. Describe the graph. What happens to the voltage in the long run?

c. How much time must elapse (to the nearest hundredth of a second) for the voltage to reach 75 volts?

66. If the population of a particular animal is very small, inbreeding will cause a loss of genetic diversity. In a population of N individuals, the percent of the species' original genetic variation that remains after t generations is given by

$$V = V_0\left(1 - \frac{1}{2N}\right)^t$$

a. Assuming $V_0 = 100$, graph V as a function of t for three different values of N: $N = 1000, 100,$ and 10.

b. Fill in the table to compare the values of V after 5 generations, after 50 generations, and after 100 generations.

Population size	Number of generations		
	5	50	100
1000			
100			
10			

c. Studies of the cheetah have revealed variation at only 3.2% of its genes. (Other species show variation at 10 to 43% of their genes.) The population of cheetah may be less than 5000. Use trial and error to discover how many generations it will take before the cheetah's genetic variation is reduced to 1%, assuming the population can be maintained at its current level.

Source: Chapman and Reiss, 1992.

67. Solve $y = 12e^{-kt} + 6$ for t.

68. Solve $N = N_0 + 4 \ln (k + 10)$ for k.

69. Express $N(t) = 600(0.4)^t$ in the form $N(t) = N_0 e^{kt}$

70. Express $N(t) = 100(1.06)^t$ in the form $N(t) = N_0 e^{kt}$

71. An eccentric millionaire offers you a summer job for the month of June. She will pay you 2 cents for your first day of work and will double your wages every day thereafter. (Assume that you work every day, including weekends.)

a. Make a table showing your wages on each day. Do you see a pattern?

b. Write a function that gives your wages in terms of the number of days you have worked.

c. How much will you make on June 15? On June 30?

72. The king of Persia offered one of his subjects anything he desired in return for services rendered. The subject requested that the king give him an amount of grain calculated as follows: Place one grain of wheat on the first square of a chessboard, two grains on the second square, four grains on the third square, and so on, until the entire chessboard is covered.

a. Make a table showing the number of grains of wheat on each square of the chessboard.

b. Write a function for the amount of wheat on each square.

c. How many grains of wheat should be placed on the last (64th) square?

Polynomial and Rational Functions

8.1 Polynomial Functions

We have already encountered some examples of polynomial functions. Linear functions

$$f(x) = ax + b$$

and quadratic functions

$$f(x) = ax^2 + bx + c$$

are special cases of polynomial functions. In general, a **polynomial function** has the form

$$f(x) = a_n x^n + a_{n-1} x^{n-1} + a_{n-2} x^{n-2} + \cdots + a_2 x^2 + a_1 x + a_0$$

where $a_0, a_1, a_2, \ldots a_n$ are constants. The coefficient of the highest power term, the constant a_n, is called the **lead coefficient.**

Before studying this chapter, you may want to review Appendix A.6, Products and Factoring.

 INVESTIGATION 13

Polynomial Models

In this Investigation we consider some examples of polynomial functions.

A. In order to detect an airplane at a range of R nautical miles, a radar antenna must transmit a signal whose power, P, in watts is given by

$$P(R) = 0.1R^4 \qquad (1)$$

TABLE 8.1

R	$P(R)$
0	
2	
4	
6	
8	
10	

1. Complete Table 8.1 for the function P.

2. Graph the function on the grid in Figure 8.1.

3. Use the graph to approximate the range at which a plane can be detected when the transmitted power is 500 watts.

4. Verify your answer to (3) by solving an algebraic equation.

5. What happens to the value of P when R is doubled, say from 2 miles to 4 miles?

6. Describe your graph: Give the intercepts and discuss the concavity. Will the graph continue to increase as R increases?

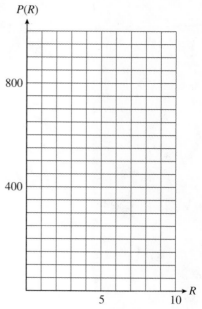

FIGURE 8.1

B. An accounting firm is designing a large work space divided into square cubicles with 8-foot ceilings. The walls and ceiling of each cubicle will be covered with a sound-proofing material, except for the door, which is 7 feet tall by 3 feet wide.

1. How much material will be needed for each cubicle if it measures 10 feet by 10 feet? Follow the steps below.

    Step 1 Compute the areas:

    Area of ceiling = _____

    Area of each wall = _____

    Area of door = _____

    Step 2 Add the areas of the ceiling and each wall, and then subtract the area of the door.

    Area of ceiling + 4 (Area of each wall) − Area of door

    Total = _____

2. Now follow the same steps for a cubicle of arbitrary size. Let d represent the dimensions (length and width) of the cubicle, and write a function for the amount of sound-proofing material, A, needed for each cubicle in terms of d.

    Step 1 Compute the areas:

    Area of ceiling = _____

    Area of each wall = _____

    Area of door = _____

Step 2 Add the areas of the ceiling and each wall, and then subtract the area of the door.

Area of ceiling + 4 (Area of each wall) − Area of door

$$A(d) = $$

TABLE 8.2

d	A(d)
5	
6	
8	
12	
15	

3. Complete Table 8.2 for the function $A(d)$.

4. Plot the values from your table on the grid in Figure 8.2. The graph of $A(d)$ is part of what familiar shape?

5. How big can the cubicles be if the sound-proofing material costs $1.20 per square foot, and the designers can budget at most $300 per cubicle for the material? Follow the steps below.

   Step 1 Divide the amount budgeted by the cost per square foot to find the largest area the designers can afford for each cubicle.

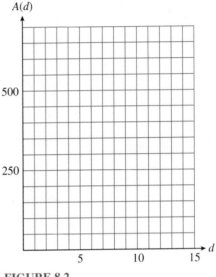

FIGURE 8.2

$$\frac{300 \text{ dollars}}{1.20 \text{ dollars per square foot}} = \underline{\hspace{1cm}} \text{ square feet}$$

   Step 2 Set your formula for $A(d)$ equal to your value from Step 1.

   Step 3 Solve the equation for d. (What kind of equation is it?)

6. Locate the point on your graph that corresponds to the cubicle in part (5).

C. Leon is flying his plane to Au Gres, Michigan. He maintains a constant altitude until he passes over a marker just outside the neighboring town of Omer, when he begins his descent for landing. During the descent, his altitude in feet is given by

$$A(x) = 128x^3 - 960x^2 + 8000$$

where x is the number of miles Leon has traveled since passing over the marker in Omer.

TABLE 8.3

x	A(x)
0.5	
1.0	
1.5	
2.0	
2.5	

1. What is Leon's altitude when he begins his descent?

2. Complete Table 8.3 for the function $A(x)$.

3. Use a graphing calculator to graph $A(x)$ in the window

$$\text{Xmin} = 0 \qquad \text{Xmax} = 5$$
$$\text{Ymin} = 0 \qquad \text{Ymax} = 8000$$

   Use the ⌐TRACE⌐ feature to verify the values in Table 8.3.

4. Use the ⌐TRACE⌐ feature to discover how far from Omer Leon will travel before landing. (In other words, how far is Au Gres from Omer?)

5. Verify your answer to part (4) algebraically.

Products of Polynomials

When we multiply two or more polynomials together, we get another polynomial of higher degree.

EXAMPLE 1

Compute the products.

   a. $(x + 2)(5x^3 - 3x^2 + 4)$ **b.** $(x - 3)(x + 2)(x - 4)$

Solutions

   a. $(x + 2)(5x^3 - 3x^2 + 4)$ Apply the distributive law.

$$= x(5x^3 - 3x^2 + 4) + 2(5x^3 - 3x^2 + 4) \quad \text{Apply the distributive law again.}$$
$$= 5x^4 - 3x^3 + 4x + 10x^3 - 6x^2 + 8 \quad \text{Combine like terms.}$$
$$= 5x^4 + 7x^3 - 6x^2 + 4x + 8$$

   b. $(x - 3)\underline{(x + 2)(x - 4)}$ Multiply two of the factors first.

$$= (x - 3)(x^2 - 2x - 8) \qquad \text{Apply the distributive law.}$$
$$= x(x^2 - 2x - 8) - 3(x^2 - 2x - 8) \qquad \text{Apply the distributive law again.}$$
$$= x^3 - 2x^2 - 8x - 3x^2 + 6x + 24 \qquad \text{Combine like terms.}$$
$$= x^3 - 5x^2 - 2x + 24$$

EXERCISE 1   Multiply $(y + 2)(y^2 - 2y + 3)$

In Example 1a we multiplied a polynomial of degree one by a polynomial of degree three, and the product was a polynomial of degree four. In Example 1b the product of three first-degree polynomials is a third-degree polynomial. In general,

The degree of a product of polynomials is the sum of the degrees of the factors.

This fact will be useful to us in Section 8.2 when we graph polynomial functions.

There are a few special products resulting in cubic polynomials that are useful enough to merit specific mention. In the Homework problems you will be asked to verify the following products.

Cube of a Binomial

1. $(x + y)^3 = x^3 + 3x^2y + 3xy^2 + y^3$
2. $(x - y)^3 = x^3 - 3x^2y + 3xy^2 - y^3$

If you become familiar with these general forms, you can use them as patterns to find specific examples of such products.

EXAMPLE 2

Write $(2w - 3)^3$ as a polynomial.

Solution
Use product (2) above, with x replaced by $2w$ and y replaced by 3.

$$(x - y)^3 = x^3 - 3x^2y + 3xy^2 - y^3$$
$$(2w - 3)^3 = (2w)^3 - 3(2w)^2(3) + 3(2w)(3)^2 - 3^3 \quad \text{Simplify.}$$
$$= 8w^3 - 36w^2 + 54w - 27 \qquad \blacksquare$$

Of course, we can also expand the product in Example 2 simply by polynomial multiplication and arrive at the same answer.

EXERCISE 2 Write $(5 + x^2)^3$ as a polynomial.

Factoring Cubics

Another pair of products is useful for factoring cubic polynomials. In the Homework problems you will be asked to verify the following products.

$$(x + y)(x^2 - xy + y^2) = x^3 + y^3$$
$$(x - y)(x^2 + xy + y^2) = x^3 - y^3$$

Viewing these products from right to left, we have the following special factorizations for the sum and difference of two cubes.

Factoring the Sum or Difference of Two Cubes

3. $x^3 + y^3 = (x + y)(x^2 - xy + y^2)$
4. $x^3 - y^3 = (x - y)(x^2 + xy + y^2)$

When we recognize a polynomial as a sum or difference of two perfect cubes, we then identify the two cubed expressions and apply the formula.

EXAMPLE 3

Factor each polynomial.

a. $8a^3 + b^3$

b. $1 - 27h^6$

Solutions

a. This polynomial is a sum of two cubes. The cubed expressions are $2a$ (because $(2a)^3 = 8a^3$) and b. Use formula (3) above as a pattern, replacing x by $2a$ and y by b.

$$x^3 \ + y^3 = (x + y) \ \ (x^2 \ - \ \ xy + y^2)$$
$$(2a)^3 + b^3 = (2a + b)((2a)^2 - (2a)b + b^2) \quad \text{Simplify.}$$
$$= (2a + b) \ (4a^2 - 2ab \ + b^2)$$

b. This polynomial is a difference of two cubes. The cubed expressions are 1 (because $1^3 = 1$) and $3h^2$ (because $(3h^2)^3 = 27h^6$). Use formula (4) above as a pattern, replacing x by 1 and y by $3h^2$.

$$x^3 - \ \ \ y^3 = (x - \ \ y)(x^2 + \ \ xy \ + \ y^2)$$
$$1^3 - (3h^2)^3 = (1 - 3h^2)(1^2 + 1(3h^2) + (3h^2)^2) \quad \text{Simplify.}$$
$$= (1 - 3h^2)(1 \ + 3h^2 \ \ + 9h^4) \qquad\blacksquare$$

EXERCISE 3 Factor $125n^3 - p^3$.

READING QUESTIONS

1. What is a polynomial function?
2. Describe the product of a polynomial of degree 3 and a polynomial of degree 4.
3. State two formulas for the cube of a binomial.
4. State formulas for factoring the sum and difference of two cubes.

ANSWERS TO 8.1 EXERCISES

1. $y^3 - y + 6$

2. $125 + 75x^2 + 15x^4 + x^6$

3. $(5n - p)(25n^2 + 5np + p^2)$

Multiply.

1. $(3x - 2)(4x^2 + x - 2)$

2. $(2x + 3)(3x^2 - 4x + 2)$

3. $(x - 2)(x - 1)(x - 3)$

4. $(z - 5)(z + 6)(z - 1)$

5. $(2a^2 - 3a + 1)(3a^2 + 2a - 1)$

6. $(b^2 - 3b + 5)(2b^2 - b + 1)$

7. $(y - 2)(y + 2)(y + 4)(y + 1)$

8. $(z + 3)(z + 2)(z - 1)(z + 1)$

Find the first three terms, in ascending powers, of the product. (Do not compute the entire product!)

9. $(2 - x + 3x^2)(3 + 2x - x^2 + 2x^4)$

10. $(1 + x - 2x^2)(-3 + 2x - 4x^3)$

11. $(1 - 2x^2 - x^4)(4 + x^2 - 2x^4)$

12. $(3 + 2x)(5 - 2x^2 - 3x^3 - x^5 + 2x^6)$

Find the indicated term in each product. (Do not compute the entire product!)

13. $(4 + 2x - x^2)(2 - 3x + 2x^2);\quad x^2$

14. $(1 - 2x + 3x^2)(6 - x - x^3);\quad x^3$

15. $(3x + x^3 - 7x^5)(1 + 4x - 3x^2);\quad x^3$

16. $(2 + 3x^2 + 2x^4)(2 - x - x^2 - x^4);\quad x^4$

Without performing the multiplication, give the degree of each product.

17. **a.** $(x^2 - 4)(3x^2 - 6x + 2)$

  b. $(x - 3)(2x - 5)(x^3 - x + 2)$

  c. $(3x^2 + 2x)(x^3 + 1)(-2x^2 + 8)$

18. **a.** $(6x^2 - 1)(4x^2 - 9)$

  b. $(3x + 4)(3x + 1)(2x^3 + x^2 - 7)$

  c. $(x^2 - 3)(2x^3 - 5x^2 + 2)(-x^3 - 5x)$

Every polynomial factors into a product of a constant and linear factors of the form $(x - a)$, where a can be either real or complex. How many linear factors are in the factored form of the given polynomial?

19. **a.** $x^4 - 2x^3 + 4x^2 + 8x - 6$

  b. $2x^5 - x^3 + 6x - 4$

20. **a.** $x^6 - 6x$

  b. $x^3 + 3x^2 - 2x + 1$

Verify the following products discussed in the text.

21. $(x + y)^3 = x^3 + 3x^2y + 3xy^2 + y^3$

22. $(x - y)^3 = x^3 - 3x^2y + 3xy^2 - y^3$

23. $(x + y)(x^2 - xy + y^2) = x^3 + y^3$

24. $(x - y)(x^2 + xy + y^2) = x^3 - y^3$

25. **a.** As if you were addressing a classmate, explain how to remember the formula for expanding $(x + y)^3$. In particular, mention the exponents on each term and the numerical coefficients.

  b. Explain how to remember the formula for expanding $(x - y)^3$, assuming your listener already knows the formula for $(x + y)^3$.

26. **a.** As if you were addressing a classmate, explain how to remember the formula for factoring a sum of two cubes. Pay particular attention to the placement of the variables and the signs of the terms.

  b. Explain how to remember the formula for factoring a difference of two cubes, assuming your listener already knows how to factor a sum of two cubes.

Use the formulas for the cube of a binomial to expand the products.

27. $(1 + 2z)^3$

28. $(1 - x^2)^3$

29. $\left(1 - 5\sqrt{t}\,\right)^3$

30. $\left(1 - \dfrac{3}{a}\right)^3$

Write each product as a polynomial and simplify.

31. $(x - 1)(x^2 + x + 1)$

32. $(x + 2)(x^2 - 2x + 4)$

33. $(2x + 1)(4x^2 - 2x + 1)$

34. $(3x - 1)(9x^2 + 3x + 1)$

35. $(3a - 2b)(9a^2 + 6ab + 4b^2)$

36. $(2a + 3b)(4a^2 - 6ab + 9b^2)$

Factor completely.

37. $x^3 + 27$

38. $y^3 - 1$

39. $a^3 - 8b^3$

40. $27a^3 + b^3$

41. $x^3 y^3 - 1$

42. $8 + x^3 y^3$

43. $27a^3 + 64b^3$

44. $8a^3 - 125b^3$

45. $125a^3 b^3 - 1$

46. $64a^3 b^3 + 1$

47. a. A closed box has a square base of length and width x inches and a height of 8 inches, as shown in Figure 8.3. Write a polynomial function $S(x)$ that gives the surface area of the box in terms of the dimensions of the base.

b. What is the surface area of a box of length and width 18 inches?

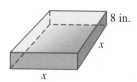

FIGURE 8.3

48. a. An empty reflecting pool is twice as long as it is wide and is three feet deep, as illustrated in Figure 8.4. Write a polynomial function $S(x)$ that gives the surface area of the empty pool.

b. What is the surface area of the pool if it is 12 feet wide?

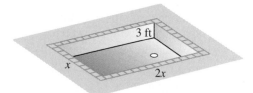

FIGURE 8.4

49. a. Write a polynomial function $A(x)$ that gives the area of the front face of the speaker frame (the region in color) in Figure 8.5.

b. If $x = 8$ inches, find the area of the front face of the frame.

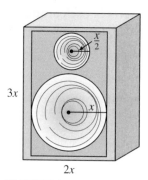

FIGURE 8.5

50. a. A Norman window is shaped like a rectangle whose length is twice its width, surmounted by a semicircle. (See Figure 8.6.) Write a polynomial function $A(x)$ that gives its area.

b. If $x = 3$ feet, find the area of the window.

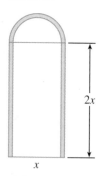

FIGURE 8.6

51. a. A grain silo is built in the shape of a cylinder with a hemisphere on top. (See Figure 8.7.) Write an expression for the volume of the silo in terms of the radius and height of the cylindrical portion of the silo.

b. If the total height of the silo is five times its radius, write a polynomial function $V(x)$ in one variable for its volume.

FIGURE 8.7

52. a. A cold medication capsule is shaped like a cylinder with a hemispherical cap on each end. (See Figure 8.8.) Write an expression for the volume of the capsule in terms of the radius and length of the cylindrical portion.

b. If the radius of the capsule is one-fourth of its overall length, write a polynomial function $V(x)$ in one variable for its volume.

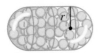

FIGURE 8.8

53. Jack invests \$500 in an account bearing interest rate, r, compounded annually. This means that each year his account balance is increased by a factor of $1 + r$.

a. Write expressions for the amount of money in Jack's account after 2 years, after 3 years, and after 4 years.

b. Expand the expressions you found in part (a) as polynomials.

c. How much money will be in Jack's account at the end of 2 years, 3 years, and 4 years if the interest rate is 8%?

54. A small company borrows \$800 for start-up costs and agrees to repay the loan at interest rate, r, compounded annually. This means that each year the debt is increased by a factor of $1 + r$.

a. Write expressions for the amount of money the company will owe if it repays the loan after 2 years, after 3 years, or after 4 years.

b. Expand the expressions you found in part (a) as polynomials.

c. How much money will the company owe after 2 years, after 3 years, or after 4 years at an interest rate of 12%?

55. A paper company plans to make boxes without tops from sheets of cardboard 12 inches wide and 16 inches long. They will cut out four squares of side x inches from the corners of the sheet and fold up the edges as shown in Figure 8.9.

a. Write expressions in terms of x for the length, width, and height of the resulting box.

b. Write a formula for the volume, V, of the box as a function of x.

c. What is the domain of the function V? (What are the largest and smallest reasonable values for x?)

d. Make a table of values for $V(x)$ on its domain

e. Graph your function V in a suitable window.

f. Use your graph to find the value of x that will yield a box with maximum possible volume. What is the maximum possible volume?

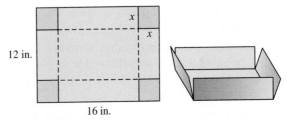

FIGURE 8.9

56. The paper company also plans to make boxes with tops from 12-inch by 16-inch sheets of cardboard by cutting out the shaded areas shown in Figure 8.10 on page 510 and folding along the dashed lines.

a. Write expressions in terms of x for the length, width, and height of the resulting box.

b. Write a formula for the volume, V, of the box as a function of x.

c. What is the domain of the function V? (What are the largest and smallest reasonable values for x?)

d. Make a table of values for $V(x)$ on its domain

e. Graph your function V in a suitable window.

f. Use your graph to find the value of x that will yield a box with maximum possible volume. What is the maximum possible volume?

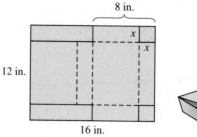

FIGURE 8.10

Use your graphing calculator to help you answer the questions in each problem. Then verify your answers algebraically.

57. A doctor who is treating a heart patient wants to prescribe medication to lower the patient's blood pressure. The body's reaction to this medication is a function of the dose administered. If the patient takes x milliliters of the medication, his blood pressure should decrease by $R = f(x)$ points, where

$$f(x) = 3x^2 - \frac{1}{3}x^3$$

a. For what values of x is $R = 0$?

b. Find a suitable domain for the function, and explain why you chose this domain.

c. Graph the function f on its domain.

d. How much should the patient's blood pressure drop if he takes 2 milliliters of medication?

e. What is the maximum drop in blood pressure that can be achieved with this medication?

f. There may be risks associated with a large change in blood pressure. How many milliliters of the medication should be administered to produce half the maximum possible drop in blood pressure?

58. A soup bowl has the shape of a hemisphere of radius 6 centimeters. The volume of the soup in the bowl, $V = f(x)$, is a function of the depth, x, of the soup. (See Figure 8.11.)

a. What is the domain of f? Why?

b. The function f is given by

$$f(x) = 6\pi x^2 - \frac{\pi}{3}x^3$$

Graph the function on its domain.

c. What is the volume of the soup if it is 3 centimeters deep?

d. What is the maximum volume of soup that the bowl can hold?

e. Find the depth of the soup (to within 2 decimal places of accuracy) when the bowl is filled to half its capacity.

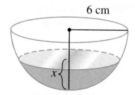

FIGURE 8.11

59. The population $P(t)$ of Cyberville has been growing according to the formula

$$P(t) = t^3 - 63t^2 + 1403t + 900$$

where t is the number of years since 1960.

a. Graph $P(t)$ in the window

$$\text{Xmin} = 0 \qquad \text{Xmax} = 47$$
$$\text{Ymin} = 0 \qquad \text{Ymax} = 20000$$

b. What was the population in 1960? 1975? 1994?

c. By how much did the population grow from 1960 to 1961? From 1975 to 1976? From 1994 to 1995?

d. Approximately when was the population growing at the slowest rate; that is, when is the graph the least steep?

60. The annual profit $P(t)$ (in thousands of dollars) of the Enviro Company is given by

$$P(t) = 2t^3 - 152t^2 + 3400t + 30$$

where t is the number of years since 1950, the first year that the company showed a profit.

a. Graph $P(t)$ on the window

$$\text{Xmin} = 0 \qquad \text{Xmax} = 94$$
$$\text{Ymin} = 0 \qquad \text{Ymax} = 50000$$

b. What was the profit in 1950? 1970? 1990?

c. How did the profit change from 1950 to 1951? From 1970 to 1971? From 1990 to 1991?

d. During which years did the profit decrease from one year to the next?

61. During an earthquake Nordhoff Way split in two, and one section shifted up several centimeters. Engineers created a ramp from the lower section to the upper section. In the coordinate system shown in Figure 8.12, the ramp is part of the graph of

$$y = f(x) = 0.00004x^3 - 0.006x^2 + 20$$

a. By how much did the upper section of the street shift during the earthquake?

b. What is the horizontal distance from the bottom of the ramp to the raised part of the street?

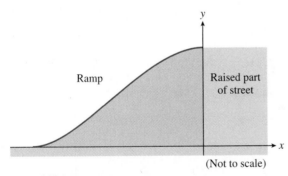

FIGURE 8.12

62. The off ramp from a highway connects to a parallel one-way road. Figure 8.13 shows the highway, the off ramp, and the road. The road lies on the x-axis, and the off ramp begins at a point on the y-axis, with distances in meters. The off ramp is part of the graph of the polynomial

$$y = f(x) = 0.00006x^3 - 0.009x^2 + 30$$

a. How far east of the exit does the off ramp meet the one-way road?

b. How far apart are the highway and the road?

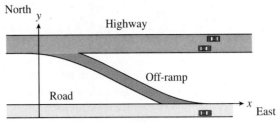

FIGURE 8.13

63. The number of minutes of daylight per day in Chicago is approximated by the polynomial

$$H(t) = 0.000\,000\,525t^4 - 0.0213t^2 + 864$$

where t is the number of days since the summer solstice. The approximation is valid for $-74 < t < 74$. (A negative value of t corresponds to a number of days before the summer solstice.)

a. Use a table of values with increments of 10 days to estimate the range of the function on its domain.

b. Graph the polynomial on its domain.

c. How many minutes of daylight are there on the summer solstice?

d. How much daylight is there two weeks before the solstice?

e. When are the days more than 14 hours long?

f. When are the days less than 13 hours long?

64. The water level (in feet) at a harbor is approximated by the polynomial

$$W(t) = 0.00733t^4 - 0.332t^2 + 9.1$$

where t is the number of hours since the high tide. The approximation is valid for $-4 \le t \le 4$. (A negative value of t corresponds to a number of hours before the high tide.)

a. Use a table of values to estimate the range of the function on its domain.

b. Graph the polynomial on its domain.

c. What is the water level at high tide?

d. What is the water level 3 hours before high tide?

e. When is the water level below 8 feet?

f. When is the water level above 7 feet?

8.2 Graphing Polynomial Functions

In Section 8.1 we considered several applications of polynomial functions. Although most applications use only a portion of the graph of a particular polynomial, we can learn a lot about these functions by taking a more global view of their behavior.

Classifying Polynomials by Degree

The graph of a polynomial function depends first of all on its degree. We have already studied the graphs of polynomials of degrees zero, one, and two. A polynomial of degree zero is a constant, and its graph is a horizontal line. An example of such a polynomial function is $f(x) = 3$. (See Figure 8.14a.) A polynomial of degree one is a linear function, and its graph is a straight line. The function $f(x) = 2x - 3$ is an example of a polynomial of degree one. (See Figure 8.14b.)

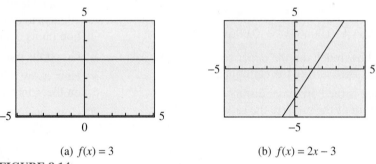

(a) $f(x) = 3$ (b) $f(x) = 2x - 3$

FIGURE 8.14

Quadratic functions, such as $f(x) = -2x^2 + 6x + 8$, are polynomials of degree two. The graph of every quadratic function is a parabola, with the same basic shape as the standard parabola, $y = x^2$. (See Figure 8.15.)

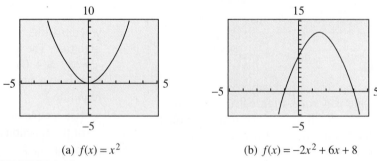

(a) $f(x) = x^2$ (b) $f(x) = -2x^2 + 6x + 8$

FIGURE 8.15

Cubic Polynomials

Do the graphs of all *cubic*, or third-degree, polynomials have a basic shape in common? We can graph a few examples and find out.

EXAMPLE 1

Graph the cubic polynomial $P(x) = x^3 - 4x$, and compare its graph with that of the standard cubic, $y = x^3$.

Solution

The graph of the standard cubic is shown in Figure 8.16a. To help us understand the graph of the polynomial $P(x) = x^3 - 4x$, we will make a table of values by evaluating the function. We can do this by hand or use the table feature on the graphing calculator.

x	-3	-2	-1	0	1	2	3
$P(x)$	-15	0	3	0	-3	0	15

The graph of $P(x) = x^3 - 4x$ is shown in Figure 8.16b. It is not exactly the same shape as the basic cubic, but it is similar, especially at the edges of the graphs.

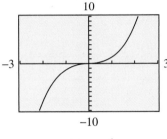

(a) $y = x^3$

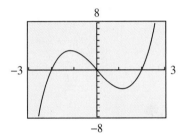

(b) $P(x) = x^3 - 4x$

FIGURE 8.16

Despite the differences in the central portions of the two graphs, they exhibit similar "end" behavior. For very large and very small values of x, both graphs look like the power function $y = x^3$. The y-values increase from $-\infty$ toward zero in the third quadrant, and increase from zero toward $+\infty$ in the first quadrant. In simpler terms, we might say that the graphs start at the lower left and extend to the upper right. All cubic polynomials display this behavior when their lead coefficients (the coefficient of the x^3 term) are positive.

Both of the graphs in Example 1 are smooth curves without any breaks or holes. This smoothness is a feature of the graphs of all polynomial functions. The domain of any polynomial function is the entire set of real numbers.

EXERCISE 1

a. Complete the table of values for $C(x) = -x^3 - 2x^2 + 4x + 4$.

x	-4	-3	-2	-1	0	1	2	3	4
y									

b. Graph $y = C(x)$ in the standard window. Compare the graph to the graphs in Example 1: What similarities do you notice? What differences?

Quartic Polynomials

Now let's compare the end behavior of two quartic, or fourth-degree, polynomials.

EXAMPLE 2

Graph the polynomials $f(x) = x^4 - 10x^2 + 9$ and $g(x) = x^4 + 2x^3$, and compare.

Solution

For each function we make a table of values, as in Table 8.4.

TABLE 8.4

x	-4	-3	-2	-1	0	1	2	3	4
$f(x)$	105	0	-15	0	9	0	-15	0	105

x	-3	-2	-1	0	1	2	3
$g(x)$	27	0	-1	0	3	32	135

The graphs are shown in Figure 8.17.

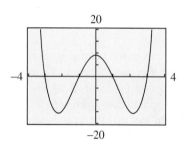

(a) $f(x) = x^4 - 10x^2 + 9$

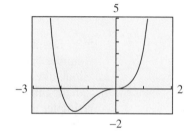

(b) $g(x) = x^4 + 2x^3$

FIGURE 8.17

All the essential features of the graphs are shown in these viewing windows. The graphs continue forever in the directions indicated, without any additional twists or turns.

As in Example 1, both graphs have similar end behavior. The y-values decrease from $+\infty$ toward zero as x increases from $-\infty$, and the y-values increase toward $+\infty$ as x increases to $+\infty$. This end behavior is similar to that of the power function $y = x^4$. Its graph also starts at the upper left and extends to the upper right. ■

In Examples 1 and 2 we have seen polynomials of degree three, whose graphs have a characteristic shape illustrated in Figure 8.16, and polynomials of degree four, whose graphs are illustrated by Figure 8.17. In the problems for this section you will consider more graphs of polynomials of various degrees to help you verify the following observations.

1. *A polynomial of **odd** degree (with positive lead coefficient) has negative y-values for large negative x, and positive y-values for large positive x. (See Figure 8.18)*

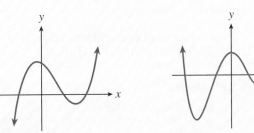

FIGURE 8.18 **FIGURE 8.19**

2. *A polynomial of **even** degree (with positive lead coefficient) has positive y-values for both large positive and large negative x. (See Figure 8.19.)*

X-Intercepts and the Factor Theorem

In Chapter 4 we saw that the *x*-intercepts of a quadratic polynomial $f(x) = ax^2 + bx + c$ occur at values of *x* for which $f(x) = 0$, that is, at the real-valued solutions of the equation $ax^2 + bx + c = 0$. The same holds true for polynomials of higher degree.

Solutions of the equation $P(x) = 0$ are called **zeros** of the polynomial P. In Example 1 we graphed the cubic polynomial $P(x) = x^3 - 4x$. Its x-intercepts are the solutions of the equation $x^3 - 4x = 0$, which we can solve by factoring the polynomial $P(x)$.

$$x^3 - 4x = 0$$
$$x(x - 2)(x + 2) = 0$$

The zeros of P are 0, 2, and -2. Each zero of P corresponds to a factor of $P(x)$. This result suggests the following theorem, which holds for any polynomial P.

Factor Theorem

Let $P(x)$ be a polynomial with real-number coefficients. Then $(x - a)$ is a factor of $P(x)$ if and only if $P(a) = 0$.

Because a polynomial function of degree n can have at most n linear factors of the form $(x - a)$, it follows that P can have at most n distinct zeros. Another way of saying this is that if $P(x)$ is a polynomial of nth degree, the equation $P(x) = 0$ can have at most n distinct solutions, some of which may be complex. Because only real-valued solutions appear on the graph as x-intercepts,

A polynomial of degree n can have at most n x-intercepts.

If some of the zeros of P are complex numbers, they will not appear on the graph, so a polynomial of degree n *may* have *fewer* than n x-intercepts.

EXAMPLE 3

Find the zeros of each polynomial, and list the x-intercepts of its graph.

    a. $f(x) = x^3 + 6x^2 + 9x$ **b.** $g(x) = x^4 - 3x^2 - 4$

Solutions

    a. Factor the polynomial to obtain

$$f(x) = x(x^2 + 6x + 9)$$
$$= x(x + 3)(x + 3)$$

By the Factor Theorem, the zeros of f are 0, -3, and -3. (We say that f has a zero *of multiplicity two* at -3.) Because all of these are real numbers, all will appear as x-intercepts on the graph. Thus, the x-intercepts occur at $(0, 0)$ and at $(-3, 0)$.

    b. Factor the polynomial to obtain

$$g(x) = (x^2 - 4)(x^2 + 1)$$
$$= (x - 2)(x + 2)(x^2 + 1)$$

By extracting roots, we find that the solutions of

$$x^2 + 1 = 0$$

are

$$x = \pm\sqrt{-1} = \pm i$$

Thus, the zeros of g are $-2, 2, -i,$ and i. (Refer to Section 3.6 for the definitions of i and $-i$.) Only the first two of these are real numbers, so the graph has only two x-intercepts, at $(-2, 0)$ and $(2, 0)$. The graphs of both polynomials are shown in Figure 8.20.

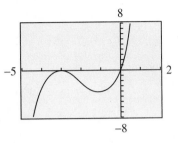

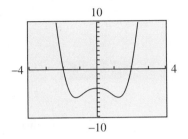

(a) $f(x) = x^3 + 6x^2 + 9x$ (b) $g(x) = x^4 - 3x^2 - 4$

FIGURE 8.20

EXERCISE 3

a. Find the zeros of $P(x) = -x^4 + x^3 + 2x^2$ by factoring.

b. Sketch a rough graph of $y = P(x)$ by hand.

Zeros of Multiplicity Two or Three

The appearance of the graph near an x-intercept is determined by the multiplicity of the zero there. Both real zeros of the polynomial $g(x) = x^4 - 3x^2 - 4$ in Example 3b are of multiplicity one, and the graph *crosses* the x-axis at each intercept. However, the polynomial $f(x) = x^3 + 6x^2 + 9x$ in Example 3a has a zero of multiplicity two at $x = -3$. The graph of f just *touches* the x-axis and then reverses direction without crossing the axis.

To understand what happens in general, compare the graphs of the three polynomials in Figure 8.21. In Figure 8.21a, $L(x) = x - 2$ has a zero of multiplicity one at $x = 2$, and its graph crosses the x-axis there. In Figure 8.21b, $Q(x) = (x - 2)^2$ has a zero of multiplicity two at $x = 2$, and its graph touches the x-axis there but changes direction without crossing. In Figure 8.21c, $C(x) = (x - 2)^3$ has a zero of multiplicity three at $x = 2$. In this case the graph makes an **S**-shaped curve at the intercept, like the graph of $y = x^3$.

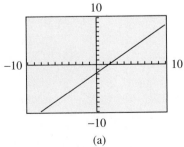

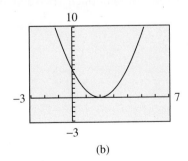

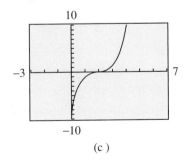

(a)   (b)   (c)

FIGURE 8.21

Near its x-intercepts the graph of a polynomial takes one of the characteristic shapes illustrated in Figure 8.21. (Although we will not consider zeros of multiplicity greater than 3, they correspond to similar behavior in the graph: At a zero of odd multiplicity the graph has an **S**-shaped curve at the intercept, and at a zero of even multiplicity the graph changes direction without crossing the x-axis.)

EXAMPLE 4

Graph the polynomial

$$f(x) = (x + 2)^3(x - 1)(x - 3)^2$$

Solution

The polynomial has degree six, an even number, so its graph starts at the upper left and extends to the upper right. Its y-intercept is

$$f(0) = (2)^3(-1)(-3)^2 = -72$$

f has a zero of multiplicity three at $x = -2$, a zero of multiplicity one at $x = 1$, and a zero of multiplicity two at $x = 3$. The graph has an **S**-shaped curve at $x = -2$, crosses the x-axis at $x = 1$, touches the x-axis and then changes direction at $x = 3$, as shown in Figure 8.22.

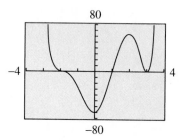

FIGURE 8.22

EXERCISE 4   Sketch a rough graph of $f(x) = (x + 3)(x - 1)^2$ by hand. Label the x- and y-intercepts.

READING QUESTIONS

1. Describe the graphs of polynomials of degrees zero, one, and two.
2. What is a zero of a polynomial?
3. How are zeros related to the factors of a polynomial?
4. What do the zeros tell you about the graph of a polynomial?
5. Discuss zeros of multiplicity two and three.

ANSWERS TO 8.2 EXERCISES

1a.

x	-4	-3	-2	-1	0	1	2	3	4
y	20	1	-4	-1	4	5	-4	-29	-76

b.

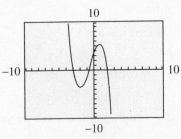

Both graphs have three x-intercepts, but the function in Example 1 has end behavior like $y = x^3$, and this function has end behavior like $y = -x^3$.

2a.

x	-4	-3	-2	-1	0	1	2	3	4
y	-286	-106	-30	-4	2	-6	-46	-160	-414

b.

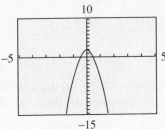

The graphs all have end behavior like a fourth degree power function $y = ax^4$. The end behavior of the graphs in Example 2 is the same as that of $y = x^4$, but the graph here has end behavior like $y = -x^4$.

3a. $-1, 0, 2$ **b.**

4.

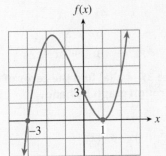

HOMEWORK 8.2

Use your calculator to graph each cubic (third-degree) polynomial. Write a sentence or two describing how the graphs are similar and how they differ.

1. $y = x^3 + 4$

2. $y = x^3 - 8$

3. $y = -2 - 0.05x^3$

4. $y = 5 - 0.02x^3$

5. $y = x^3 - 3x$

6. $y = 9x - x^3$

7. $y = x^3 + 5x^2 - 4x - 20$

8. $y = -x^3 - 2x^2 + 5x + 6$

Use a calculator to graph each cubic polynomial. Which graphs are the same?

9. a. $y = x^3 - 2$
   b. $y = (x - 2)^3$
   c. $y = x^3 - 6x^2 + 12x - 8$

10. a. $y = x^3 + 3$
   b. $y = (x + 3)^3$
   c. $y = x^3 + 9x^2 + 27x + 27$

Use your calculator to graph each quartic (fourth-degree) polynomial. Write a sentence or two describing how the graphs are similar and how they differ.

11. $y = 0.5x^4 - 4$

12. $y = 0.3x^4 + 1$

13. $y = -x^4 + 6x^2 - 10$

14. $y = x^4 - 8x^2 - 8$

15. $y = x^4 - 3x^3$

16. $y = -x^4 - 4x^3$

17. $y = -x^4 - x^3 - 2$

18. $y = x^4 + 2x^3 + 4x^2 + 10$

Sketch a rough graph of each polynomial function by hand.

19. $q(x) = (x + 4)(x + 1)(x - 1)$

20. $p(x) = x(x + 2)(x + 4)$

21. $G(x) = (x - 2)^2(x + 2)^2$

22. $F(x) = (x - 1)^2(x - 3)^2$

23. $h(x) = x^3(x + 2)(x - 2)$

24. $H(x) = (x + 1)^3(x - 2)^2$

25. $P(x) = (x + 4)^2(x + 1)^2(x - 1)^2$

26. $Q(x) = x^2(x - 5)(x - 1)^2(x + 2)$

For Problems 27–34, (a) use your calculator to graph each polynomial and locate the x-intercepts, (b) write the polynomial in factored form.

27. $P(x) = x^3 - 7x - 6$

28. $Q(x) = x^3 + 3x^2 - x - 3$

29. $R(x) = x^4 - x^3 - 4x^2 + 4x$

30. $S(x) = x^4 + 3x^3 - x^2 - 3x$

31. $p(x) = x^3 - 3x^2 - 6x + 8$

32. $q(x) = x^3 + 6x^2 - x - 30$

33. $r(x) = x^4 - x^3 - 10x^2 + 4x + 24$

34. $s(x) = x^4 + x^3 - 8x^2 - 2x + 12$

For Problems 35–44, (a) find the zeros of each polynomial by factoring, (b) sketch a rough graph by hand.

35. $P(x) = x^4 + 4x^2$

36. $P(x) = x^3 + 3x$

37. $f(x) = x^4 + 4x^3 + 4x^2$

38. $g(x) = x^4 + 4x^3 + 3x^2$

39. $g(x) = 4x - x^3$

40. $f(x) = 8x - x^4$

41. $k(x) = x^4 - 10x^2 + 16$

42. $m(x) = x^4 - 15x^2 + 36$

43. $r(x) = (x^2 - 1)(x + 3)^2$

44. $s(x) = (x^2 - 9)(x - 1)^2$

Find the equation in factored form of each polynomial graphed below.

45.

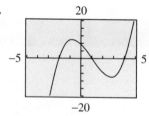

FIGURE 8.23

46.

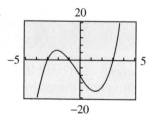

FIGURE 8.24

47.

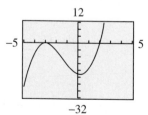

FIGURE 8.25

48.

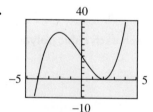

FIGURE 8.26

49.

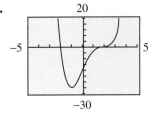

FIGURE 8.27

50.

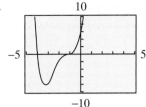

FIGURE 8.28

For Problems 51–54, (a) write the formula for each function, and graph with a calculator, (b) describe how the graph of each function differs from the graph of $y = f(x)$.

51. $f(x) = x^3 - 4x$
  a. $y = f(x) + 3$
  b. $y = f(x) - 5$
  c. $y = f(x - 2)$
  d. $y = f(x + 3)$

52. $f(x) = x^3 - x^2 + x - 1$
  a. $y = f(x) + 4$
  b. $y = f(x) - 4$
  c. $y = f(x - 3)$
  d. $y = f(x + 5)$

53. $f(x) = x^4 - 4x^2$
  a. $y = f(x) + 6$
  b. $y = f(x) - 2$
  c. $y = f(x - 4)$
  d. $y = f(x + 2)$

54. $f(x) = x^4 + 3x^3$
  a. $y = f(x) + 5$
  b. $y = f(x) - 3$
  c. $y = f(x - 2)$
  d. $y = f(x + 4)$

8.3   Rational Functions

A **rational function** is one of the form

$$f(x) = \frac{P(x)}{Q(x)}$$

where $P(x)$ and $Q(x)$ are polynomials. The graphs of rational functions can be quite different from the graphs of polynomials.

EXAMPLE 1

Francine is planning a 60-mile training flight through the desert on her cycle-plane, a pedal-driven aircraft. If there is no wind, she can pedal at an average speed of 15 miles per hour, so she can complete the flight in 4 hours.

  a. If there is a headwind of x miles per hour, it will take Francine longer to fly 60 miles. Express the time it will take for Francine to complete the training flight as a function of x.

  b. Make a table of values for the function.

  c. Graph the function and explain what it tells you about the time Francine should allow for the ride.

Solutions

a. If there is a headwind of x miles per hour, Francine's ground speed will be $15 - x$ miles per hour. Using the fact that $\text{time} = \dfrac{\text{distance}}{\text{rate}}$, we find that the time needed for the flight will be

$$t = f(x) = \frac{60}{15 - x}$$

b. Evaluate the function for several values of x, as shown in Table 8.5.

TABLE 8.5

x	0	3	5	7	9	10
t	4	5	6	7.5	10	12

For example, if the headwind is **5** miles per hour, then

$$t = \frac{60}{15 - 5} = \frac{60}{10} = 6$$

Francine's effective speed is only 10 miles per hour, and it will take her 6 hours to fly the 60 miles. The table shows that as the speed of the headwind increases, the time required for the flight increases also.

c. The graph of the function is shown in Figure 8.29. You can use your calculator with the window

$$\text{Xmin} = -8.5 \qquad \text{Xmax} = 15$$
$$\text{Ymin} = 0 \qquad \text{Ymax} = 30$$

to verify the graph. In particular, the point $(0, 4)$ lies on the graph. This point tells us that if there is no wind, Francine can fly the 60 miles in 4 hours, as we calculated earlier.

The graph is increasing, as indicated by the table of values. In fact, as the speed of the wind gets close to 15 miles per hour, Francine's flying time becomes extremely large. In theory, if the wind speed were exactly 15 miles per hour Francine would never complete her flight. On the graph, the time becomes infinite at $x = 15$.

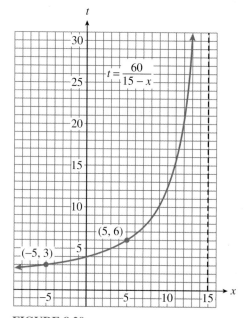

FIGURE 8.29

What about negative values for x? If we interpret a negative headwind as a tailwind, Francine's flying time should decrease for negative x-values. For example, if $x = -5$ there is a tailwind of 5 miles per hour, so Francine's effective speed is 20 miles per hour, and she can complete the flight in 3 hours. As the tailwind gets stronger (that is, as we move farther to the left in the x-direction), Francine's flying time continues to decrease, and the graph approaches the x-axis. ■

The vertical dashed line at $x = 15$ on the graph of $t = \dfrac{60}{15 - x}$ is a *vertical asymptote* for the graph. We first encountered asymptotes in Section 5.3 when we studied the graph of $y = \dfrac{1}{x}$. Locating the vertical asymptotes of a rational function is an important part of determining the shape of the graph.

EXAMPLE 2

EarthCare decides to sell T-shirts to raise money. They make an initial investment of $100 to pay for the design of the T-shirt and to set up the printing process. After that the T-shirts cost $5 each for labor and materials.

a. Express EarthCare's average cost per T-shirt as a function of the number of T-shirts they produce.

b. Make a table of values for the function.

c. Graph the function and explain what it tells you about the cost of the T-shirts.

Solutions

a. If EarthCare produces x T-shirts, their total costs will be $100 + 5x$ dollars. To find the average cost per T-shirt, we divide the total cost by the number of T-shirts produced, to get

$$C = g(x) = \frac{100 + 5x}{x}$$

b. Evaluate the function for several values of x, as shown in Table 8.6.

TABLE 8.6

x	1	2	4	5	10	20
C	105	55	40	25	15	10

If EarthCare makes only one T-shirt, its cost is $105. But if they make more T-shirts, the cost of the original $100 investment is distributed among them. For example, the average cost per T-shirt for 2 T-shirts is

$$\frac{100 + 5(2)}{2} = 55$$

and the average cost for 5 T-shirts is

$$\frac{100 + 5(5)}{5} = 25$$

c. The graph is shown in Figure 8.30. You can use your calculator with the window

$$\begin{array}{ll} \text{Xmin} = 0 & \text{Xmax} = 470 \\ \text{Ymin} = 0 & \text{Ymax} = 30 \end{array}$$

to verify the graph. Use the TRACE key to locate on the graph several points from the table of values. For example, the point $(5, 25)$ indicates that if EarthCare makes 5 T-shirts, the cost per shirt is $25.

The graph shows that as the number of T-shirts increases, the average cost per shirt continues to decline, but not as rapidly as at first. Eventually the average cost levels off and approaches $5 per T-shirt. For example, if EarthCare produces 400 T-shirts, the average cost per shirt is

$$\frac{100 + 5(400)}{400} = 5.25$$

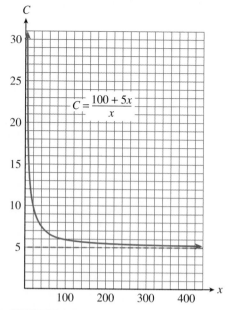

FIGURE 8.30

The horizontal line $C = 5$ on the graph of $C = \dfrac{100 + 5x}{x}$ is a *horizontal asymptote*. As x increases, the graph approaches the line $C = 5$ but never actually meets it. The average price per T-shirt will always be slightly more than $5. Horizontal asymptotes are also important in sketching the graphs of rational functions.

Graphing Rational Functions

Most applications of rational functions have restricted domains; that is, they make sense for only a subset of the real numbers on the *x*-axis. Consequently, only a portion of the graph is useful for analyzing the application. However, a knowledge of the general shape and properties of the whole graph can be very helpful in understanding a rational function.

As we stated earlier, a rational function is a quotient of two polynomials. Some examples of rational functions are shown below.

$$f(x) = \frac{2}{(x - 3)^2} \qquad g(x) = \frac{x}{x + 1}$$

$$h(x) = \frac{2x^2}{x^2 + 4} \qquad k(x) = \frac{x^2 - 1}{x^2 - 9}$$

Because we cannot divide by zero, a rational function $f(x) = \dfrac{P(x)}{Q(x)}$ is undefined for any value $x = a$ where $Q(a) = 0$. These *x*-values are not in the domain of the function.

EXAMPLE 3 Find the domains of the rational functions *f, g, h,* and *k* defined above.

Solution

The domain of *f* is the set of all real numbers except 3, because the denominator, $(x - 3)^2$, equals 0 when $x = 3$. The domain of *g* is the set of all real numbers except -1, because $x + 1$ equals 0 when $x = -1$. The denominator of the function *h*, $x^2 + 4$, is never equal to zero, so the domain of *k* is all the real numbers. The domain of *k* is the set of all real numbers except 3 and -3, because $x^2 - 9$ equals 0 when $x = 3$ or $x = -3$. ∎

We only need to exclude the zeros of the *denominator* from the domain of a rational function. We do not exclude the zeros of the numerator. In fact, the zeros of the numerator include the zeros of the rational function itself, because a fraction is equal to zero when its numerator is zero but its denominator is not zero.

EXERCISE 1 Find the domain of $F(x) = \dfrac{x - 2}{x + 4}$.

Vertical Asymptotes

As we saw in Section 8.2, a polynomial function is defined for all values of *x*, and its graph is a smooth curve without any breaks or holes. The graph of a rational function, on the other hand, will have breaks or holes at those *x*-values where it is undefined.

For example, consider the function

$$f(x) = \frac{2}{(x - 3)^2}$$

given above. This function is undefined for $x = 3$, so there is no point on the graph with *x*-coordinate 3. However, we can make a table of values for other values of *x*. Plotting the ordered pairs in Table 8.7 results in the points shown in Figure 8.31.

TABLE 8.7

x	*y*
0	$\frac{2}{9}$
1	$\frac{1}{2}$
2	2
3	undefined
4	2
5	$\frac{1}{2}$
6	$\frac{2}{9}$

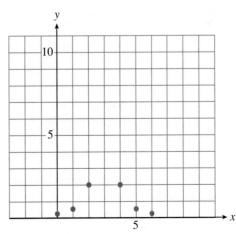

FIGURE 8.31

To fill in the rest of the graph, first use your calculator to make tables for x-values greater than 6 or less than 0, as shown in Figure 8.32. (Set the **Table Setup** menu for **Ask** on the Independent variable and **Auto** on the Dependent variable.) These tables show that as x gets very large in absolute value, $\dfrac{2}{(x-3)^2}$ gets closer to zero. Consequently, the graph approaches the x-axis as we move away from the origin, as shown in Figure 8.33.

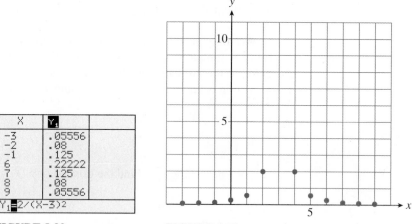

FIGURE 8.32   **FIGURE 8.33**

Next, make tables by choosing x-values close to 3, as shown in Figure 8.34. As we choose x-values closer and closer to 3, $(x-3)^2$ gets closer to zero, so the fraction $\dfrac{2}{(x-3)^2}$ gets very large in absolute value. The graph has a vertical asymptote at $x = 3$. This means that the graph approaches, but never touches, the vertical line $x = 3$. We indicate the vertical asymptote on the completed graph by a dashed line, as shown in Figure 8.35.

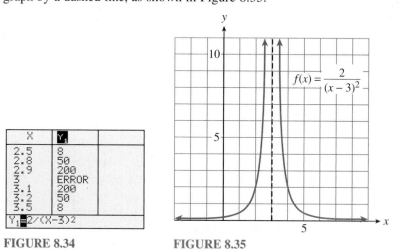

FIGURE 8.34   **FIGURE 8.35**

In general, we have the following result.

Vertical Asymptotes

If $Q(a) = 0$ but $P(a) \neq 0$, then the graph of the rational function

$f(x) = \dfrac{P(x)}{Q(x)}$ has a vertical asymptote at $x = a$.

EXERCISE 2 Find the vertical asymptotes of $G(x) = \dfrac{4x^2}{x^2 - 4}$.

If $P(a)$ and $Q(a)$ are both zero, then the graph of the rational function $\dfrac{P(x)}{Q(x)}$ may have a "hole" at $x = a$ rather than an asymptote. (This possibility is considered in Problems 65–68 of Homework 8.4.)

Near a vertical asymptote the graph of a rational function has one of the four characteristic shapes illustrated in Figure 8.36. Locating the vertical asymptotes can help us make a quick sketch of a rational function.

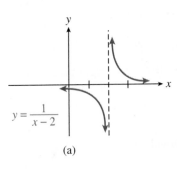

(a)

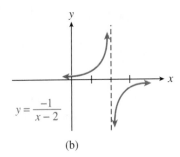

(b)

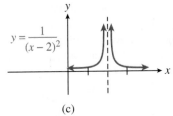

(c)

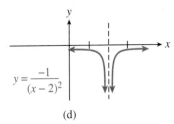

(d)

FIGURE 8.36

EXAMPLE 4 Locate the vertical asymptotes and sketch the graph of $g(x) = \dfrac{x}{x + 1}$.

Solution

The denominator, $x + 1$, equals zero when $x = -1$. Because the numerator does not equal zero when $x = -1$, there is a vertical asymptote at $x = -1$. The asymptote separates the graph into two pieces. Use the TABLE feature of your calculator to evaluate $g(x)$ for several values of x on either side of the asymptote, as shown in Figure 8.37. Plot the points found in this way, and connect the points on either side of the asymptote to obtain the graph shown in Figure 8.38.

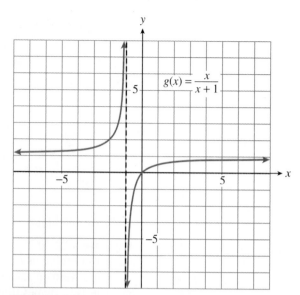

X	Y₁
-8	1.1429
-2	2
-1.2	6
-.9	-9
0	0
4	.8
8	.88889

Y₁=X/(X+1)

FIGURE 8.37   **FIGURE 8.38**   ■

Horizontal Asymptotes

Look again at the graph of $g(x) = \dfrac{x}{x + 1}$ in Example 4. As $|x|$ gets large—that is, as we move away from the origin along the x-axis in either direction—the corresponding y-values get closer and closer to 1. The graph approaches, but never coincides with, the line $y = 1$. We say that the graph has a **horizontal asymptote** at $y = 1$.

When does a rational function $f(x) = \dfrac{P(x)}{Q(x)}$ have a horizontal asymptote? It depends on the degrees of the two polynomials $P(x)$ and $Q(x)$. The degree of the numerator of $g(x)$ in Example 4 is equal to the degree of the denominator. Equivalently, the highest power of x in the numerator (one, in this case) is the same as the highest power in the denominator.

Consider the three rational functions whose graphs are shown in Figure 8.39.

$$r(x) = \frac{x + 1}{x^2} \qquad\qquad s(x) = \frac{x^2 - 4}{2x^2} \qquad\qquad t(x) = \frac{x^2 + 1}{x - 1}$$

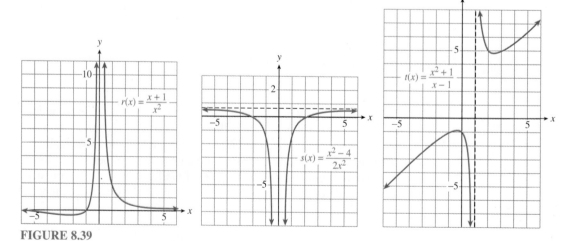

FIGURE 8.39

The graph of $r(x) = \dfrac{x+1}{x^2}$ in Figure 8.39a has a horizontal asymptote at $y = 0$, the x-axis, because the degree of the denominator is larger than the degree of the numerator. Higher powers of x grow much more rapidly than smaller powers. Thus, for large values of $|x|$ the denominator is much bigger in absolute value than the numerator of $r(x)$, and consequently the function values approach 0.

The graph of $s(x) = \dfrac{x^2 - 4}{2x^2}$ in Figure 8.39b has a horizontal asymptote at $y = \dfrac{1}{2}$, because the numerator and denominator of the fraction have the same degree. For large values of $|x|$, the terms of lower degree are negligible compared to the squared terms. As x increases, $s(x)$ is approximately equal to $\dfrac{x^2}{2x^2}$, or $\dfrac{1}{2}$. Thus, the function values approach a constant value of $\dfrac{1}{2}$.

The graph of $t(x) = \dfrac{x^2 + 1}{x - 1}$ in Figure 8.39c does not have a horizontal asymptote, because the degree of the numerator is larger than the degree of the denominator. As $|x|$ increases, $x^2 + 1$ grows much faster than $x - 1$, so their ratio does not approach a constant value. The function values increase without bound.

We summarize our discussion as follows.

Horizontal Asymptotes

Suppose $f(x) = \dfrac{P(x)}{Q(x)}$ is a rational function, where the degree of $P(x)$ is m and the degree of $Q(x)$ is n.

1. If $m < n$, the graph of f has a horizontal asymptote at $y = 0$.
2. If $m = n$, the graph of f has a horizontal asymptote at $y = \dfrac{a}{b}$, where a is the lead coefficient of $P(x)$ and b is the lead coefficient of $Q(x)$.
3. If $m > n$, the graph of f does not have a horizontal asymptote.

EXERCISE 3 Find the horizontal asymptote of $G(x) = \dfrac{4x^2}{x^2 - 4}$.

EXAMPLE 5 Locate the horizontal asymptotes and sketch the graph of $h(x) = \dfrac{2x^2}{x^2 + 4}$.

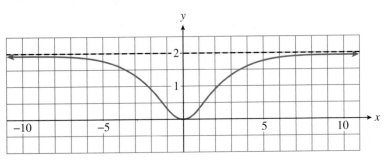

FIGURE 8.40

Solution

The numerator and denominator of the fraction are both second-degree polynomials, so the graph does have a horizontal asymptote. The lead coefficients of $P(x)$ and $Q(x)$ are 2 and 1, respectively, so the horizontal asymptote is $y = \dfrac{2}{1}$, or $y = 2$.

The function h does not have a vertical asymptote because the denominator, $x^2 + 4$, is never equal to zero. The y-intercept of the graph is the point $(0, 0)$. We can plot several points by evaluating the function at convenient x-values, and use the asymptote to help us sketch the graph, as shown in Figure 8.40. ■

E X E R C I S E 4 Locate the horizontal and vertical asymptotes and sketch the graph of

$$k(x) = \frac{x^2 - 1}{x^2 - 9}$$

Vertical asymptotes:

Horizontal asymptote:

x-intercepts:

y-intercept:

x	-4	-1	1	4
$k(x)$				

1. What is a rational function?
2. How are the graphs of rational functions different from the graphs of polynomials?
3. Describe how to locate the vertical asymptotes of the graph of a rational function.
4. Describe how to locate the horizontal asymptotes of a rational function.

A N S W E R S   T O   8 . 3   E X E R C I S E S

1. $x \neq -4$ **2.** $x = -2$ and $x = 2$ **3.** $y = 4$
4. $y = 1; x = -3, x = 3$

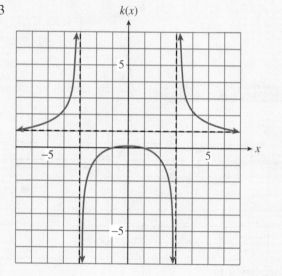

1. The eider duck, one of the world's fastest flying birds, can exceed an airspeed of 65 miles per hour. A flock of eider ducks is migrating south at an average airspeed of 50 miles per hour against a moderate headwind. Their next feeding grounds are 150 miles away.

   a. Express the ducks' travel time, t, as a function of the wind speed, v.

   b. Complete the table showing the travel time for various windspeeds. What happens to the travel time as the headwind increases?

v	0	5	10	15	20	25	30	35	40	45	50
t											

   c. Use the table to choose an appropriate window and graph your function $t(v)$. Give the equations of any horizontal or vertical asymptotes. What does the vertical asymptote signify in the context of the problem?

2. The fastest fish in the sea may be the bluefin tuna, which has been clocked at 43 miles per hour in short sprints. A school of tuna is migrating a distance of 200 miles at an average speed of 36 miles per hour in still water, but they have run into a current flowing against their direction of travel.

   a. Express the tunas' travel time, t, as a function of the current speed, v.

   b. Complete the table showing the travel time for various current speeds. What happens to the travel time as the current increases?

v	0	4	8	12	16	20	24	28	32	36
t										

   c. Use the table to choose an appropriate window and graph your function $t(v)$. Give the equations of any horizontal or vertical asymptotes. What does the vertical asymptote signify in the context of the problem?

3. The cost in thousands of dollars for immunizing $p\%$ of the residents of Emporia against a dangerous new disease is given by the function

   $$C(p) = \frac{72p}{100 - p}$$

   a. What is the domain of C?

   b. Complete the table showing the cost of immunizing various percentages of the population.

p	0	15	25	40	50	75	80	90	100
C									

   c. Graph the function C. (Use Xmin = 6, Xmax = 100, and appropriate values of Ymin and Ymax.) What percentage of the population can be immunized if the city is able to spend $108,000?

   d. For what values of p is the total cost more than $1,728,000?

   e. The graph has a vertical asymptote. What is it? What is its significance in the context of this problem?

4. The cost in thousands of dollars for extracting $p\%$ of a precious ore from a mine is given by the equation

   $$C(p) = \frac{360p}{100 - p}$$

   a. What is the domain of C?

   b. Complete the table showing the cost of extracting various percentages of the ore.

p	0	15	25	40	50	75	80	90	100
C									

   c. Graph the function C. (Use Xmin = 6, Xmax = 100, and appropriate values of Ymin and Ymax.) What percentage of the ore can be extracted if $540,000 can be spent on the extraction?

   d. For what values of p is the total cost less than $1,440,000?

(*continued on page 532*)

e. The graph has a vertical asymptote. What is it? What is its significance in the context of this problem?

5. The total cost in dollars of producing n calculators is approximately $20{,}000 + 8n$.

   a. Express the cost per calculator, C, as a function of the number, n, of calculators produced.

   b. Complete the table showing the cost per calculator for various production levels.

n	100	200	400	500	1000	2000	4000	5000	8000
C									

   c. Graph the function $C(n)$ for the cost per calculator. Use the window

   Xmin $= 0$ Xmax $= 9400$
   Ymin $= 0$ Ymax $= 50$

   d. How many calculators should be produced so that the cost per calculator is $18?

   e. For what values of n is the cost less than $12 per calculator?

   f. Find the horizontal asymptote of the graph. What does it represent in this context?

6. The number of loaves of Mom's Bread sold each day is approximated by the demand function

$$D(p) = \frac{100}{1 + (p - 1.10)^4}$$

where p is the price per loaf in dollars.

   a. Complete the table showing the demand for Mom's Bread at various prices per loaf. Round the values of $D(p)$ to the nearest whole number.

p	0.25	0.50	1.00	1.25	1.50	1.75
Demand						

p	2.00	2.25	2.50	2.75	3.00
Demand					

   b. Graph the demand function in the window

   Xmin $= 0$ Xmax $= 3.74$
   Ymin $= 0$ Ymax $= 1.70$

What happens to the demand for Mom's Bread as the price increases?

   c. Add a row to your table to show the daily revenue from Mom's Bread at various prices.

p	0.25	0.50	1.00	1.25	1.50	1.75
Demand						
Revenue						

p	2.00	2.25	2.50	2.75	3.00
Demand					
Revenue					

   d. Using the formula for $D(p)$, write an expression $R(p)$ that approximates the total daily revenue as a function of the price, p.

   e. Graph the revenue function $R(p)$ in the same window with $D(p)$. Estimate the maximum possible revenue. Does the maximum for $D(p)$ occur at the same value of p as the maximum for $R(p)$?

   f. Find the horizontal asymptote of the graphs. What does it represent in this context?

7. A computer store sells approximately 300 of its most popular model per year. The manager would like to minimize her annual inventory cost by ordering the optimal number of computers, x, at regular intervals. If she orders x computers in each shipment, the cost of storage will be $6x$ dollars, and the cost of reordering will be $\frac{300}{x}(15x + 10)$ dollars. The inventory cost is the sum of the storage cost and the reordering cost.

   a. Use the distributive law to simplify the expression for the reordering cost. Then express the inventory cost, C, as a function of x.

   b. Complete the table of values for the inventory cost for various reorder sizes.

x	10	20	30	40	50	60	70	80	90	100
C										

   c. Graph your function C in the window

$$\text{Xmin} = 0 \qquad \text{Xmax} = 150$$
$$\text{Ymin} = 4500 \qquad \text{Ymax} = 5500$$

Estimate the minimum possible value for C.

d. How many computers should the manager order in each shipment so as to minimize the inventory cost? How many orders will she make during the year?

e. Graph the function $y = 6x + 4500$ in the same window with the function C. What do you observe?

8. A chain of electronics stores sells approximately 500 portable phones every year. The owner would like to minimize his annual inventory cost by ordering the optimal number of phones, x, at regular intervals. The cost of storing the phones will then be $2x$ dollars, and the cost of reordering will be $\dfrac{500}{x}(4x + 10)$. The total annual inventory cost is the sum of the storage cost and the reordering cost.

a. Use the distributive law to simplify the expression for the reordering cost. Then express the inventory cost, C, as a function of x.

b. Complete the table of values for the inventory cost for various reorder sizes.

x	10	20	30	40	50	60	70	80	90	100
C										

c. Graph your function C in the window

$$\text{Xmin} = 0 \qquad \text{Xmax} = 150$$
$$\text{Ymin} = 2000 \qquad \text{Ymax} = 2500$$

Estimate the minimum possible value for C.

d. How many portable phones should the owner order in each shipment so as to minimize the inventory cost? How many orders will he make during the year?

e. Graph the function $y = 2x + 2000$ in the same window with the function C. What do you observe?

9. Francine wants to make a rectangular box. In order to simplify construction and keep her costs down, she plans for the box to have a square base and a total surface area of 96

square centimeters. She would like to know the largest volume that such a box can have.

a. If the square base has length x centimeters, show that the height of the box is $h = \dfrac{24}{x} - \dfrac{x}{2}$ centimeters (*Hint:* The surface area of the box is the sum of the areas of the six sides of the box.)

b. Write an expression for the volume, V, of the box as a function of the length, x, of its base.

c. Complete the table showing the heights and volumes of the box for various base lengths.

x	1	2	3	4	5	6	7
h							
V							

Explain why the values of h and V are negative when $x = 7$.

d. Graph your expression for volume, $V(x)$, in an appropriate window. Approximate the maximum possible volume for a box of surface area 96 square centimeters.

e. What value of x gives the maximum volume?

f. Graph the height, $h(x)$, in the same window with $V(x)$. What is the height of the box with greatest volume? (Find the height directly from your graph and verify by using the formula given for $h(x)$.)

10. Delbert wants to make a box with a square base and a volume of 64 cubic centimeters. He would like to know the smallest surface area that such a box can have.

a. If the square base has length x centimeters, show that the height of the box must be $h = \dfrac{64}{x^2}$ centimeters.

b. Write an expression for the surface area, S, of the box as a function of the length, x, of its base. (*Hint:* The surface area of the box is the sum of the areas of the six sides of the box.)

c. Complete the table, on page 534 showing the heights and surface areas of the box for various base lengths.

x	1	2	3	4	5	6	7	8
h								
S								

d. Graph your expression for surface area, $S(x)$, in an appropriate window. Approximate the minimum possible surface area for Delbert's box.

e. What value of x gives the minimum surface area?

f. Graph the height, $h(x)$, in the same window with $S(x)$. What is the height of the box with the smallest surface area? (Find the height directly from your graph and verify by using the formula given for $h(x)$.)

11. A train whistle sounds higher when the train is approaching you than when it is moving away from you. This phenomenon is known as the Doppler effect. If the actual pitch of the whistle is 440 hertz (this is the A note below middle C), then the note you hear will have the pitch

$$P(v) = \frac{440(332)}{332 - v}$$

where the velocity, v, in meters per second is positive as the train approaches and negative when the train is moving away. (The number 332 that appears in this expression is the speed of sound in meters per second.)

a. Complete the table of values showing the pitch of the whistle at various train velocities.

v	-100	-75	-50	-25	0	25	50	75	100
P									

b. Graph the function P. (Use the window Xmin $= -94$, Xmax $= 94$, and appropriate values of Ymin and Ymax.)

c. What is the velocity of the train if the note you hear has a pitch of 415 hertz (corre-

sponding to the note A-flat)? A pitch of $553.\overline{3}$ hertz (C-sharp)?

d. For what velocities will the pitch you hear be greater than 456.5 hertz?

e. The graph has a vertical asymptote (although it is not visible in the suggested window). Where is it and what is its significance in this context?

12. The maximum altitude (in meters) attained by a projectile shot from the surface of the earth is

$$h(v) = \frac{6.4 \times 10^6 v^2}{(19.6)(6.4 \times 10^6) - v^2}$$

where v is the speed (in meters per second) at which the projectile was launched. (The radius of the earth is 6.4×10^6 meters, and the constant 19.6 is related to the earth's gravitational constant.)

a. Complete the table of values showing the maximum altitude for various launch velocities.

v	100	200	300	400	500
h					

v	600	700	800	900	1000
h					

b. Graph the function h. (Use the window Xmin $= 0$, Xmax $= 940$, and appropriate values of Ymin and Ymax.)

c. Approximately what speed is needed to attain an altitude of 4000 meters? An altitude of 16 kilometers?

d. For what velocities will the projectile attain an altitude exceeding 32 kilometers?

e. The graph has a vertical asymptote (although it is not visible in the suggested window). Where is it and what is its significance in this context?

For Problems 13–30, (a) sketch the horizontal and vertical asymptotes for each function, (b) use the asymptotes to help you sketch the rest of the graph.

13. $y = \dfrac{1}{x + 3}$

14. $y = \dfrac{1}{x - 3}$

15. $y = \dfrac{2}{x^2 - 5x + 4}$

16. $y = \dfrac{4}{x^2 - x - 6}$

17. $y = \dfrac{x}{x + 3}$

18. $y = \dfrac{x}{x - 2}$

19. $y = \dfrac{x + 1}{x + 2}$

20. $y = \dfrac{x - 1}{x - 3}$

21. $y = \dfrac{2x}{x^2 - 4}$ **22.** $y = \dfrac{x}{x^2 - 9}$ **23.** $y = \dfrac{x - 2}{x^2 + 5x + 4}$ **24.** $y = \dfrac{x + 1}{x^2 - x - 6}$

25. $y = \dfrac{x^2 - 1}{x^2 - 4}$ **26.** $y = \dfrac{2x^2}{x^2 - 1}$ **27.** $y = \dfrac{x + 1}{(x - 1)^2}$ **28.** $y = \dfrac{2(x^2 - 1)}{x^2 + 4}$

29. $y = \dfrac{x}{x^2 + 3}$ **30.** $y = \dfrac{x^2 + 2}{x^2 + 4}$

31. Graph the curve known as Newton's Serpentine: $y = \dfrac{4x}{x^2 + 1}$.

32. Graph the curve known as the Witch of Agnesi: $y = \dfrac{8}{x^2 + 4}$.

33. a. Show that the equation $\dfrac{1}{y} - \dfrac{1}{x} = \dfrac{1}{k}$ is equivalent to $y = \dfrac{kx}{x + k}$ on their common domain.

  b. Graph the function $y = \dfrac{kx}{x + k}$ for $k = 1$, 2, and 3 in the window

    Xmin = 0   Xmax = 30
    Ymin = 0   Ymax = 4

  Describe the graphs.

34. Consider the graph of $y = \dfrac{ax}{x + k}$, where a and k are positive constants.

  a. What is the horizontal asymptote of the graph?

  b. Show that for $x = k$, $y = \dfrac{a}{2}$.

  c. Sketch the graph of $y = \dfrac{ax}{x + k}$ for $a = 4$ and $k = 10$ in the window

    Xmin = 0   Xmax = 60
    Ymin = 0   Ymax = 5

  Illustrate your answers to parts (a) and (b) on the graph.

For Problems 35–36, (a) use your answers to Problem 34 to find equations of the form $y = \dfrac{ax}{x + k}$ for the graphs shown, (b) check your answer with a graphing calculator.

35.

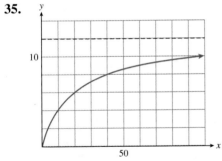

FIGURE 8.41

36.

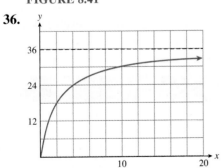

FIGURE 8.42

37. The Michaelis-Menten equation is the rate equation for chemical reactions catalyzed by enzymes. The speed of the reaction, v, is a function of the initial concentration of the reactant, s, and is given by

$$v = \frac{Vs}{s + K}$$

where V is the maximum possible reaction rate, and K is called the Michaelis constant.

  a. What value does v approach as s increases?

  b. What is the value of v when $s = K$?

  c. The table gives data from reactions of the enzyme D-amino acid oxidase.

s	0.33	0.66	1.00	1.66	2.50	3.33	6.66
v	0.08	0.14	0.20	0.30	0.39	0.46	0.58

  Plot the data and estimate the values of V and K from your graph.

  d. Graph the function $v = \dfrac{0.88s}{s + 3.34}$ on top of your data points.

  Souce: Holme and Peck, 1993.

38. a. Refer to the Michaelis-Menten equation in Problem 37. Solve for $\dfrac{1}{v}$, and write your new equation in the form $\dfrac{1}{v} = a \cdot \dfrac{1}{s} + b$. Express a and b in terms of V and K.

b. Use the data from part (c) of Problem 37 to make a table of values for $\left(\dfrac{1}{s}, \dfrac{1}{v}\right)$.

c. Plot the points $\left(\dfrac{1}{s}, \dfrac{1}{v}\right)$, and use linear regression to find the line of best fit.

d. Use your values for a and b to solve for V and K.

39. a. Refer to your equation in part (a) of Problem 38. Write an equation for $\dfrac{s}{v}$ of the form $\dfrac{s}{v} = as + b$. Express a and b in terms of V and K.

b. Use the data from part (c) of Problem 37 to make a table of values for $\left(s, \dfrac{s}{v}\right)$.

c. Plot the points $\left(s, \dfrac{s}{v}\right)$, and use linear regression to find the line of best fit.

d. Use your values for a and b to solve for V and K.

40. Animals spend most of their time hunting or foraging for food to keep themselves alive. Knowing the rate at which an animal (or population of animals) eats can help us determine its metabolic rate, or its impact on its habitat. The rate of eating is proportional to the availability of food in the area, but has an upper limit imposed by mechanical considerations, such as how long it takes the animal to capture and ingest its prey.

a. Sketch a graph of eating rate as a function of quantity of available food. This will be a qualitative graph only; you do not have enough information to put scales on the axes.

b. Suppose that the rate at which an animal catches its prey is proportional to the number of prey available, or $r_c = ax$, where a is a constant and x is the number of available prey. The rate at which it handles and eats the prey is constant, $r_h = b$. Write expressions for T_c and T_h, the times for catching and handling N prey.

c. Show that the rate of food consumption is given by

$$y = \frac{abx}{b + ax} = \frac{bx}{b/a + x}$$

Hint: $y = \dfrac{N}{T}$, where N is the number of prey consumed in a time interval, T, where $T = T_c + T_h$.

d. In a study of ladybirds it was discovered that larvae in their second stage of development consumed aphids at a rate of y_2 aphids per day, given by

$$y_2 = \frac{20x}{x + 16}$$

where x is the number of aphids available. Larvae in the third stage ate at rate y_3 given by

$$y_3 = \frac{90x}{x + 79}$$

Graph both of these functions on the domain $0 \le x \le 140$.

e. What is the maximum rate at which ladybird larvae in each stage of development can consume aphids?

Source: Burton, 1998.

MIDCHAPTER REVIEW

Multiply

1. $(t + 4)(t^2 - t - 1)$

2. $(b + 3)(2b - 1)(2b + 5)$

Write as a polynomial.

3. $(v - 10)^3$

4. $(a + 2b^2)^3$

Factor.

5. $y^3 + 27x^3$

6. $x^9 - 8$

Graph each polynomial in the standard window.

7. $y = x^3 - 3x + 2$

8. $y = -0.1(x^4 - 6x^3 + x^2 + 24x + 16)$

Find the zeros of the polynomial, and sketch the graph by hand.

9. $P(x) = x^3 - 8x$

10. $Q(x) = x^4 - 12x^2$

Sketch the graph by hand. Label the x- and y-intercepts.

11. $g(x) = (x + 4)^2(x - 2)$

12. $f(x) = x^3(x + 3)^2$

State the domain of each function.

13. $h(x) = \dfrac{x^2 - 9}{x(x^2 - 4)}$

14. $j(x) = \dfrac{x^2 - 3x + 10}{x^2(x^2 + 1)}$

For Problems 15 and 16, (a) sketch the horizontal and vertical asymptotes for each function, (b) use the asymptotes to help you sketch the graph.

15. $F(x) = \dfrac{2x}{x^2 - 1}$

16. $G(x) = \dfrac{2}{x^2 - 1}$

8.4　Operations on Algebraic Fractions

The algebraic expression that defines a rational function is called a **rational expression** or an **algebraic fraction**. Operations on algebraic fractions follow the same rules as operations on common fractions.

Reducing Fractions

When we reduce an ordinary fraction such as $\frac{24}{36}$, we are using the fundamental principle of fractions.

Fundamental Principle of Fractions

If we multiply or divide the numerator and denominator of a fraction by the same (non-zero) number, the new fraction is equivalent to the old one. In symbols,

$$\frac{ac}{bc} = \frac{a}{b}, \qquad (b, c \neq 0)$$

Thus,

$$\frac{24}{36} = \frac{2 \cdot 12}{3 \cdot 12} = \frac{2}{3}$$

We can use the same procedure to reduce algebraic fractions: We look for common factors in the numerator and denominator and then apply the fundamental principle.

EXAMPLE 1

Reduce each algebraic fraction.

a. $\dfrac{8x^3y}{6x^2y^3}$

b. $\dfrac{6x - 3}{3}$

Solutions

Factor out any common factors from the numerator and denominator. Then divide numerator and denominator by the common factors.

a. $\dfrac{8x^3y}{6x^2y^3} = \dfrac{4x \cdot 2x^2y}{3y^2 \cdot 2x^2y}$

$= \dfrac{4x}{3y^2}$

b. $\dfrac{6x + 3}{3} = \dfrac{3(2x + 1)}{3}$

$= 2x + 1$ ■

EXERCISE 1   Reduce $\dfrac{2a + 8b}{4b}$.

If the numerator or denominator of the fraction contains more than one term, it is especially important to *factor* before attempting to apply the fundamental principle. We can divide out common *factors* from the numerator and denominator of a fraction, but the fundamental principle does *not* apply to common *terms*. For example,

$$\frac{2xy}{3y} = \frac{2x}{3}$$

because y is a common factor in the numerator and denominator. However,

$$\frac{2x + y}{3 + y} \neq \frac{2x}{3}$$

because y is a common term, but is *not a common factor* of the numerator and denominator. Furthermore,

$$\frac{5x + 3}{5y} \neq \frac{x + 3}{y}$$

because 5 is not a factor of the *entire* numerator.

EXAMPLE 2

Reduce each fraction.

a. $\dfrac{4x + 2}{4}$

b. $\dfrac{9x^2 + 3}{6x + 3}$

Solutions

Factor the numerator and denominator. Then divide numerator and denominator by the common factors.

a. $\dfrac{4x + 2}{4} = \dfrac{2(2x + 1)}{2(2)}$

$= \dfrac{2x + 1}{2}$

b. $\dfrac{9x^2 + 3}{6x + 3} = \dfrac{3(3x^2 + 1)}{3(2x + 1)}$

$= \dfrac{3x^2 + 1}{2x + 1}$ ■

Note that in Example 2a above,

$$\frac{4x + 2}{4} \neq x + 2$$

and in Example 2b,

$$\frac{9x^2 + 3}{6x + 3} \neq \frac{9x^2}{6x}$$

EXERCISE 2 Reduce $\dfrac{16t^2 + 4}{4t + 4}$.

We summarize the procedure for reducing algebraic fractions as follows.

To Reduce an Algebraic Fraction:

1. Factor the numerator and denominator.
2. Divide the numerator and denominator by any common factors.

EXAMPLE 3 Reduce each fraction.

a. $\dfrac{x^2 - 7x + 6}{36 - x^2}$
b. $\dfrac{27x^3 - 1}{9x^2 - 1}$

Solutions

a. Factor numerator and denominator to obtain

$$\frac{(x - 6)(x - 1)}{(6 - x)(6 + x)}$$

The factor $x - 6$ in the numerator is the opposite of the factor $6 - x$ in the denominator. That is, $x - 6 = -1(6 - x)$. Thus,

$$\frac{-1(6 - x)(x - 1)}{(6 - x)(6 + x)} = \frac{-1(x - 1)}{6 + x} = \frac{1 - x}{6 + x}$$

b. The numerator of the fraction is a difference of two cubes, and the denominator is a difference of two squares. Factor each to obtain

$$\frac{(3x - 1)(9x^2 + 3x + 1)}{(3x - 1)(3x + 1)} = \frac{9x^2 + 3x + 1}{3x + 1}$$

EXERCISE 3 Reduce $\dfrac{m^2 - 9}{m^2 - 3m - 18}$.

Products of Fractions

To multiply two or more common fractions together, we multiply their numerators together and multiply their denominators together. The same is true for a product of algebraic fractions. For example,

$$\frac{6x^2}{y} \cdot \frac{xy}{2} = \frac{6x^2 \cdot xy}{y \cdot 2} = \frac{6x^3 y}{2y} \quad \text{Reduce.}$$

$$= \frac{3x^3 \cancel{(2y)}}{\cancel{2y}} = 3x^3$$

We can simplify the process by first factoring each numerator and denominator and dividing out any common factors.

$$\frac{6x^2}{y} \cdot \frac{xy}{2} = \frac{\cancel{2} \cdot 3x^2}{\cancel{y}} \cdot \frac{x\cancel{y}}{\cancel{2}} = 3x^3$$

In general, we have the following procedure for finding the product of two algebraic fractions.

To Multiply Algebraic Fractions:

1. Factor each numerator and denominator.
2. Divide out any factors that appear in both a numerator and a denominator.
3. Multiply together the numerators; multiply together the denominators.

EXAMPLE 4   Find each product.

a. $\dfrac{5}{x^2 - 1} \cdot \dfrac{x + 2}{x}$ **b.** $\dfrac{4y^2 - 1}{4 - y^2} \cdot \dfrac{y^2 - 2y}{4y + 2}$

Solutions

a. The denominator of the first fraction factors into $(x + 1)(x - 1)$. There are no common factors to divide out, so we multiply the numerators together and multiply the denominators together.

$$\frac{5}{x^2 - 1} \cdot \frac{x + 2}{x} = \frac{5(x + 2)}{x(x^2 - 1)} = \frac{5x + 10}{x^3 - x}$$

b. Factor each numerator and each denominator. Look for common factors.

$$\frac{4y^2 - 1}{4 - y^2} \cdot \frac{y^2 - 2y}{4y + 2} = \frac{(2y - 1)\cancel{(2y + 1)}}{\cancel{(2 - y)}(2 + y)} \cdot \frac{y\overset{-1}{\cancel{(y - 2)}}}{2\cancel{(2y + 1)}} \quad \begin{array}{l}\text{Divide out common factors.} \\ \text{Note: } y - 2 = -(2 - y)\end{array}$$

$$= \frac{-y(2y - 1)}{2(y + 2)}$$

∎

Quotients of Fractions

To divide two algebraic fractions, we multiply the first fraction by the reciprocal of the second fraction. For example,

$$\frac{2x^3}{3y} \div \frac{4x}{5y^2} = \frac{2x^3}{3y} \cdot \frac{5y^2}{4x}$$

$$= \frac{\cancel{2x} \cdot x^2}{3\cancel{y}} \cdot \frac{5y \cdot \cancel{y}}{2 \cdot \cancel{2x}} = \frac{5x^2y}{6}$$

If the fractions involve polynomials of more than one term, we may need to factor each numerator and denominator in order to recognize any common factors. This suggests the following procedure for dividing algebraic fractions.

To Divide Algebraic Fractions:

1. Multiply the first fraction by the reciprocal of the second fraction.
2. Factor each numerator and denominator.
3. Divide out any factors that appear in both a numerator and a denominator.
4. Multiply together the numerators; multiply together the denominators.

EXAMPLE 5

Find each quotient.

a. $\dfrac{x^2 - 1}{x + 3} \div \dfrac{x^2 - x - 2}{x^2 + 5x + 6}$ **b.** $\dfrac{6ab}{2a + b} \div 4a^2b$

Solutions

a. Multiply the first fraction by the reciprocal of the second fraction.

$$\frac{x^2 - 1}{x + 3} \div \frac{x^2 - x - 2}{x^2 + 5x + 6} = \frac{x^2 - 1}{x + 3} \cdot \frac{x^2 + 5x + 6}{x^2 - x - 2} \qquad \text{Factor.}$$

$$= \frac{(x - 1)\cancel{(x + 1)}}{\cancel{x + 3}} \cdot \frac{\cancel{(x + 3)}(x + 2)}{\cancel{(x + 1)}(x - 2)}$$

$$= \frac{(x - 1)(x + 2)}{x - 2}$$

b. Multiply the first fraction by the reciprocal of the second fraction.

$$\frac{6ab}{2a + b} \div 4a^2b = \frac{3 \cdot \cancel{2ab}}{2a + b} \cdot \frac{1}{2a \cdot \cancel{2ab}} \qquad \text{Divide out common factors.}$$

$$= \frac{3}{2a(2a + b)} = \frac{3}{4a^2 + 2ab} \qquad \blacksquare$$

Polynomial Division

If a quotient of two polynomials cannot be reduced, we can sometimes simplify the expression by treating it as a division. When the degree of the numerator is greater than the degree of the denominator, the quotient will be the sum of a polynomial and a simpler algebraic fraction.

If the divisor is a monomial, we can simply divide the monomial into each term of the numerator.

EXAMPLE 6

Divide $\dfrac{9x^3 - 6x^2 + 4}{3x}$.

Solution

Divide $3x$ into each term of the numerator.

$$\frac{9x^3 - 6x^2 + 4}{3x} = \frac{9x^3}{3x} - \frac{6x^2}{3x} + \frac{4}{3x}$$

$$= 3x^2 - 2x + \frac{4}{3x}$$

The quotient is the sum of a polynomial, $3x^2 - 2x$, and an algebraic fraction, $\dfrac{4}{3x}$. ■

If the denominator is not a monomial, we can use a method similar to the long division algorithm used in arithmetic.

EXAMPLE 7

Divide $\dfrac{2x^2 + x - 7}{x + 3}$.

Solution

First write

$$x + 3\overline{)2x^2 + x - 7}$$

and divide $2x^2$ (the first term of the numerator) by x (the first term of the denominator) to obtain $2x$. (It may be helpful to write down the division: $\dfrac{2x^2}{x} = 2x$.)

Write $2x$ above the quotient bar as the first term of the quotient, as shown below.

Next, multiply $x + 3$ by $2x$ to obtain $2x^2 + 6x$, and subtract this product from $2x^2 + x - 7$.

$$
\begin{array}{r}
2x \\
x + 3 \overline{)\,2x^2 + x - 7} \\
-(2x^2 + 6x) \\
\hline
-5x - 7
\end{array}
$$

Repeating the process, divide $-5x$ by x to obtain -5. Write -5 as the second term of the quotient. Then multiply $x + 3$ by -5 to obtain $-5x - 15$, and subtract.

$$
\begin{array}{r}
2x - 5 \\
x + 3 \overline{)\,2x^2 + x - 7} \\
-(2x^2 + 6x) \\
\hline
-5x - 7 \\
-(-5x - 15) \\
\hline
8
\end{array}
$$

Because the degree of 8 is less than the degree of $x + 3$, the division is finished. The quotient is $2x - 5$, with a remainder of 8. We write the remainder as a fraction to obtain

$$\frac{2x^2 + x - 7}{x + 3} = 2x - 5 + \frac{8}{x + 3} \qquad \blacksquare$$

When using polynomial division it helps to write the polynomials in descending powers of the variable. If the numerator is "missing" any terms, we can insert terms with zero coefficients so that like powers will be aligned. For example, to perform the division

$$\frac{3x - 1 + 4x^3}{2x - 1}$$

we first write the numerator in descending powers as $4x^3 + 3x - 1$. We insert $0x^2$ between $4x^3$ and $3x$ and set up the quotient as

$$2x - 1 \overline{)\,4x^3 + 0x^2 + 3x - 1}$$

We then proceed as in Example 7. You can check that the quotient is

$$2x^2 + x + 2 + \frac{1}{2x - 1}$$

EXERCISE 7   Divide $\dfrac{4 + 8y^2 - 3y^3}{3y + 1}$.

READING QUESTIONS

1. Explain how to reduce an algebraic fraction.
2. When multiplying algebraic fractions, what should you do before attempting to "cancel"?
3. Explain how to divide one fraction by another.

4. What will you get if you divide a polynomial into another polynomial of higher degree?

ANSWERS TO 8.4 EXERCISES

1. $\dfrac{a + 4b}{2b}$

2. $\dfrac{4t^2 + 1}{t + 1}$

3. $\dfrac{m - 3}{m - 6}$

4. $\dfrac{5p(3n - 1)}{9n^2 + 1}$

5. $\dfrac{3x(x + 2y)}{4y(x - y)}$

6. $3a + 1 - \dfrac{1}{2a}$

7. $-y^2 + 3y - 1 + \dfrac{5}{3y + 1}$

HOMEWORK 8.4

Reduce each algebraic fraction.

1. $\dfrac{14c^2d}{-7c^2d^3}$

2. $\dfrac{-12r^2st}{-6rst^2}$

3. $\dfrac{4x + 6}{6}$

4. $\dfrac{2y - 8}{8}$

5. $\dfrac{6a^3 - 4a^2}{4a}$

6. $\dfrac{3x^3 - 6x^2}{6x^2}$

7. $\dfrac{6 - 6t^2}{(t - 1)^2}$

8. $\dfrac{4 - 4x^2}{(x + 1)^2}$

9. $\dfrac{2y^2 - 8}{2y + 4}$

10. $\dfrac{5y^2 - 20}{2y - 4}$

11. $\dfrac{6 - 2v}{v^3 - 27}$

12. $\dfrac{4 - 2u}{u^3 - 8}$

13. $\dfrac{4x^3 - 36x}{6x^2 + 18x}$

14. $\dfrac{5x^2 + 10x}{5x^3 + 20x}$

15. $\dfrac{y^2 - 9x^2}{(3x - y)^2}$

16. $\dfrac{(2x - y)^2}{y^2 - 4x^2}$

17. $\dfrac{2x^2 + x - 6}{x^2 + x - 2}$

18. $\dfrac{6x^2 - x - 1}{2x^2 + 9x - 5}$

19. $\dfrac{8z^3 - 27}{4z^2 - 9}$

20. $\dfrac{8z^3 - 1}{4z^2 - 1}$

21. Which of the following fractions are equivalent to $2a$ (on their common domain)?

   a. $\dfrac{2a + 4}{4}$

   b. $\dfrac{4a^2 - 2a}{2a - 1}$

   c. $\dfrac{4a^2 - 2a}{2a}$

   d. $\dfrac{a + 3}{2a^2 + 6a}$

22. Which of the following fractions are equivalent to $3b$ (on their common domain)?

   a. $\dfrac{9b^2 - 3b}{3b}$

   b. $\dfrac{b + 2}{3b^2 + 6b}$

   c. $\dfrac{3b - 9}{9}$

   d. $\dfrac{9b^2 - 3b}{3b - 1}$

23. Which of the following fractions are equivalent to -1 (where they are defined)?

   a. $\dfrac{2a + b}{2a - b}$

   b. $\dfrac{-(a + b)}{b - a}$

   c. $\dfrac{2a^2 - 1}{2a^2}$

   d. $\dfrac{-a^2 + 3}{a^2 + 3}$

24. Which of the following fractions are equivalent to -1 (where they are defined)?

   a. $\dfrac{2a - b}{b - 2a}$

   b. $\dfrac{-b^2 - 2}{b^2 + 2}$

   c. $\dfrac{3b^2 - 1}{3b^2 + 1}$

   d. $\dfrac{b - 1}{b}$

Write each product as a single fraction in lowest terms.

25. $\dfrac{-4}{3np} \cdot \dfrac{6n^2p^3}{16}$

26. $\dfrac{14a^3b}{3b} \cdot \dfrac{-6}{7a^2}$

27. $5a^2b^2 \cdot \dfrac{1}{a^3b^3}$

28. $15x^2y \cdot \dfrac{3}{35xy^2}$

29. $\dfrac{5x + 25}{2x} \cdot \dfrac{4x}{2x + 10}$

30. $\dfrac{3y}{4xy - 6y^2} \cdot \dfrac{2x - 3y}{12x}$

31. $\dfrac{4a^2 - 1}{a^2 - 16} \cdot \dfrac{a^2 - 4a}{2a + 1}$

32. $\dfrac{9x^2 - 25}{2x - 2} \cdot \dfrac{x^2 - 1}{6x - 10}$

33. $\dfrac{2x^2 - x - 6}{3x^2 + 4x + 1} \cdot \dfrac{3x^2 + 7x + 2}{2x^2 + 7x + 6}$

34. $\dfrac{3x^2 - 7x - 6}{2x^2 - x - 1} \cdot \dfrac{2x^2 - 9x - 5}{3x^2 - 13x - 10}$

35. $\dfrac{3x^4 - 48}{x^4 - 4x^2 - 32} \cdot \dfrac{4x^4 - 8x^3 + 4x^2}{2x^4 + 16x}$

36. $\dfrac{x^4 - 3x^3}{x^4 + 6x^2 - 27} \cdot \dfrac{x^4 - 81}{3x^4 - 81x}$

Write each quotient as a single fraction in lowest terms.

37. $\dfrac{4x - 8}{3y} \div \dfrac{6x - 12}{y}$

38. $\dfrac{6y - 27}{5x} \div \dfrac{4y - 18}{x}$

39. $\dfrac{a^2 - a - 6}{a^2 + 2a - 15} \div \dfrac{a^2 - 4}{a^2 + 6a + 5}$

40. $\dfrac{a^2 + 2a - 15}{a^2 + 3a - 10} \div \dfrac{a^2 - 9}{a^2 - 9a + 14}$

41. $\dfrac{x^3 + y^3}{x} \div \dfrac{x + y}{3x}$

42. $\dfrac{8x^3 - y^3}{x + y} \div \dfrac{2x - y}{x^2 - y^2}$

43. $1 \div \dfrac{x^2 - 1}{x + 2}$

44. $1 \div \dfrac{x^2 + 3x + 1}{x - 2}$

45. $(x^2 - 5x + 4) \div \dfrac{x^2 - 1}{x^2}$

46. $(x^2 - 9) \div \dfrac{x^2 - 6x + 9}{3x}$

47. $\dfrac{x^2 + 3x}{2y} \div 3x$

48. $\dfrac{2y^2 + y}{3x} \div 2y$

Divide.

49. $\dfrac{18r^2s^2 - 15rs + 6}{3rs}$

50. $\dfrac{8a^2x^2 - 4ax^2 + ax}{2ax}$

51. $\dfrac{15s^{10} - 21s^5 + 6}{-3s^2}$

52. $\dfrac{25m^6 - 15m^4 + 7}{-5m^3}$

53. $\dfrac{4y^2 + 12y + 7}{2y + 1}$

54. $\dfrac{4t^2 - 4t - 5}{2t - 1}$

55. $\dfrac{x^3 + 2x^2 + x + 1}{x - 2}$

56. $\dfrac{2x^3 - 3x^2 - 2x + 4}{x + 1}$

57. $\dfrac{4z^2 + 5z + 8z^4 + 3}{2z + 1}$

58. $\dfrac{7 - 3t^3 - 23t^2 + 10t^4}{2t + 3}$

59. $\dfrac{x^4 - 1}{x - 2}$

60. $\dfrac{y^5 + 1}{y - 1}$

Verify that the given value is a zero of the polynomial. Find the other zeros. (Hint: Use polynomial division to write $P(x) = (x - a)Q(x)$, then factor $Q(x)$.)

61. $P(x) = x^3 - 2x^2 + 1; \quad a = 1$

62. $P(x) = x^3 + 2x^2 - 1; \quad a = -1$

63. $P(x) = x^4 - 3x^3 - 10x^2 + 24x; \quad a = -3$

64. $P(x) = x^4 + 5x^3 - x^2 - 5x; \quad a = -5$

The following are examples of functions whose graphs have "holes."

a. Find the domain of the function.

b. Reduce the fraction to lowest terms.

c. Graph the function. (Hint: The graph of the original function is identical to the graph of the function in part (b) except that certain points are excluded from the domain.) Indicate a "hole" in the graph by an open circle.

65. $y = \dfrac{x^2 - 4}{x - 2}$ **66.** $y = \dfrac{x^2 - 1}{x + 1}$ **67.** $y = \dfrac{x + 1}{x^2 - 1}$ **68.** $y = \dfrac{x - 3}{x^2 - 9}$

8.5   More Operations on Algebraic Fractions

Sums and Differences of Like Fractions

Algebraic fractions with the same denominator are called **like fractions.** To add or subtract like fractions, we combine their numerators and keep the same denominator for the sum or difference. This method is an application of the distributive law.

EXAMPLE 1

Find each sum or difference.

a. $\dfrac{2x}{9z^2} + \dfrac{5x}{9z^2}$ b. $\dfrac{2x - 1}{x + 3} - \dfrac{5x - 3}{x + 3}$

Solutions

a.   Because these are like fractions, we add their numerators and keep the same denominator.

$$\frac{2x}{9z^2} + \frac{5x}{9z^2} = \frac{2x + 5x}{9z^2} = \frac{7x}{9z^2}$$

b.   Be careful to subtract the *entire* numerator of the second fraction: use parentheses to show that the subtraction applies to both terms of $5x - 3$.

$$\frac{2x - 1}{x + 3} - \frac{5x - 3}{x + 3} = \frac{2x - 1 - (5x - 3)}{x + 3}$$

$$= \frac{2x - 1 - 5x + 3}{x + 3} = \frac{-3x + 2}{x + 3}$$ ∎

EXERCISE 1   Subtract $\dfrac{a + 1}{a^2 - a} - \dfrac{5 - 3a}{a^2 - a}$.

Lowest Common Denominator

To add or subtract fractions with different denominators, we must first find a *common denominator*. For arithmetic fractions we use the smallest natural number that is exactly divisible by each of the given denominators. For example, to add the fractions $\frac{1}{6}$ and $\frac{3}{8}$, we use 24 as the common denominator, because 24 is the smallest natural number that both 6 and 8 divide into evenly.

We define the **lowest common denominator (LCD)** of two or more algebraic fractions as the polynomial of least degree that is exactly divisible by each of the given denominators.

EXAMPLE 2

Find the LCD for the fractions $\dfrac{3x}{x+2}$ and $\dfrac{2x}{x-3}$.

Solution

The LCD is a polynomial that has as factors both $x + 2$ and $x - 3$. The simplest such polynomial is $(x + 2)(x - 3)$, or $x^2 - x + 6$. For our purposes it will be more convenient to leave the LCD in factored form, $(x + 2)(x - 3)$. ∎

The LCD in Example 2 was easy to find because each original denominator consisted of a single factor; that is, neither denominator could be factored. In that case the LCD is just the product of the original denominators. We can always find a common denominator by multiplying together all the denominators in the given fractions, but this may not give us the *simplest* or *lowest* common denominator. Using anything other than the simplest possible common denominator will complicate our work needlessly.

If any of the denominators in the given fractions can be factored, we factor them before looking for the LCD.

To Find the LCD of Algebraic Fractions:

1. Factor each denominator completely.

2. Include each different factor in the LCD as many times as it occurs in any *one* of the given denominators.

EXAMPLE 3

Find the LCD for the fractions $\dfrac{2x}{x^2-1}$ and $\dfrac{x+3}{x^2+x}$.

Solution

Factor the denominators of each of the given fractions.

$$x^2 - 1 = (x - 1)(x + 1) \qquad \text{and} \qquad x^2 + x = x(x + 1)$$

The factor $(x - 1)$ occurs once in the first denominator, the factor x occurs once in the second denominator, and the factor $(x + 1)$ occurs once in each denominator. Therefore we include in our LCD one copy of each of these factors. The LCD is $x(x + 1)(x - 1)$. ∎

In Example 3, we do not include two factors of $(x + 1)$ in the LCD. We need only one factor of $(x + 1)$ because $(x + 1)$ occurs only once in either denominator. You should check that each original denominator divides evenly into our LCD, $x(x + 1)(x - 1)$. ∎

EXERCISE 2 Find the LCD for the fractions $\dfrac{1}{2x^3(x - 1)^2}$ and $\dfrac{3}{4x^2 - 4x}$.

Building Fractions

After finding the LCD we *build* each fraction to an equivalent one with the LCD as its denominator. The new fractions will be like fractions, and we can combine them as explained above.

Building a fraction is the opposite of reducing a fraction, in the sense that we multiply, rather than divide, the numerator and denominator by an appropriate factor. To find the **building factor**, we compare the factors of the original denominator with those of the desired common denominator.

EXAMPLE 4 Build each of the fractions $\dfrac{3x}{x + 2}$ and $\dfrac{2x}{x - 3}$ to equivalent fractions with the LCD $(x + 2)(x - 3)$ as denominator.

Solution

Compare the denominator of the given fraction to the LCD. We see that the fraction $\dfrac{3x}{x + 2}$ needs a factor of $(x - 3)$ in its denominator, so $(x - 3)$ is the building factor for the first fraction. We multiply the numerator and denominator of the first fraction by $(x - 3)$ to obtain an equivalent fraction.

$$\frac{3x}{x + 2} = \frac{3x(x - 3)}{(x + 2)(x - 3)} = \frac{3x^2 - 9x}{x^2 - x - 6}$$

The fraction $\dfrac{2x}{x - 3}$ needs a factor of $(x + 2)$ in the denominator, so we multiply numerator and denominator by $(x + 2)$.

$$\frac{2x}{x - 3} = \frac{2x(x + 2)}{(x - 3)(x + 2)} = \frac{2x^2 + 4x}{x^2 - x - 6}$$ ∎

The two new fractions we obtained in Example 4 are like fractions; they have the same denominator.

Sums and Differences of Unlike Fractions

We are now ready to add or subtract algebraic fractions with unlike denominators. We will do this in four steps.

To Add or Subtract Fractions with Unlike Denominators:

1. Find the LCD for the given fractions.
2. Build each fraction to an equivalent fraction with the LCD as its denominator.
3. Add or subtract the numerators of the resulting like fractions. Use the LCD as the denominator of the sum or difference.
4. Reduce the sum or difference, if possible.

EXAMPLE 5

Subtract $\dfrac{3x}{x + 2} - \dfrac{2x}{x - 3}$.

Solution

Step 1 The LCD for these fractions is $(x + 2)(x - 3)$.

Step 2 We build each fraction to an equivalent one with the LCD, as we did in Example 4.

$$\frac{3x}{x + 2} = \frac{3x^2 - 9x}{x^2 - x - 6} \qquad \text{and} \qquad \frac{2x}{x - 3} = \frac{2x^2 + 4x}{x^2 - x - 6}$$

Step 3 Combine the numerators over the same denominator.

$$\frac{3x}{x + 2} - \frac{2x}{x - 3} = \frac{3x^2 - 9x}{x^2 - x - 6} - \frac{2x^2 + 4x}{x^2 - x - 6} \qquad \text{Subtract the numerators.}$$
$$= \frac{(3x^2 - 9x) - (2x^2 + 4x)}{x^2 - x - 6} = \frac{x^2 - 13x}{x^2 - x - 6}$$

Step 4 Reduce the result, if possible. If we factor both numerator and denominator, we find

$$\frac{x(x - 13)}{(x - 3)(x + 2)}$$

The fraction cannot be reduced. ■

EXERCISE 3 Add $\dfrac{2}{x^2 - x - 2} + \dfrac{2}{x^2 + 2x + 1}$.

EXAMPLE 6

Write as a single fraction: $1 + \dfrac{2}{a} - \dfrac{a^2 + 2}{a^2 + a}$.

Solution
Step 1 To find the LCD, factor each denominator:

$$a = a$$
$$a^2 + a = a(a + 1)$$

The LCD is $a(a + 1)$.

Step 2 Build each term to an equivalent fraction with the LCD as denominator. (The building factors for each fraction are shown in color.) The third fraction already has the LCD for its denominator.

$$1 = \frac{1 \cdot a(a + 1)}{1 \cdot a(a + 1)} = \frac{a^2 + a}{a(a + 1)}$$

$$\frac{2}{a} = \frac{2 \cdot (a + 1)}{a \cdot (a + 1)} = \frac{2a + 2}{a(a + 1)}$$

$$\frac{a^2 + 2}{a^2 + a} = \frac{a^2 + 2}{a(a + 1)}$$

Step 3 Combine the numerators over the LCD.

$$1 + \frac{2}{a} - \frac{a^2 + 2}{a^2 + a} = \frac{a^2 + a}{a(a + 1)} + \frac{2a + 2}{a(a + 1)} - \frac{a^2 + 2}{a(a + 1)}$$

$$= \frac{a^2 + a + (2a + 2) - (a^2 + 2)}{a(a + 1)}$$

$$= \frac{3a}{a(a + 1)}$$

Step 4 Reduce the fraction to find

$$\frac{3\cancel{a}}{\cancel{a}(a + 1)} = \frac{3}{a + 1}$$ ∎

Applications

As we saw in Section 8.3, the formulas for rational functions are algebraic fractions. It is often useful to simplify the formula for a function before using it.

EXAMPLE 7

When estimating their travel time, pilots must take into account the prevailing winds. A tailwind adds to the plane's ground speed, while a headwind decreases the ground speed. Skyhigh Airlines is setting up a shuttle service from Dallas to Phoenix, a distance of 800 miles.

   a. Express the time needed for a one-way trip, without wind, as a function of the speed of the plane.

   b. Suppose there is a prevailing wind of 30 miles per hour blowing from the west. Write expressions for the flying time from Dallas to Phoenix and from Phoenix to Dallas.

   c. Write an expression for the round trip flying time, excluding stops, with a 30-mile-per-hour wind from the west, as a function of the plane's speed. Simplify your expression.

Solutions

a. Recall that time $= \dfrac{\text{distance}}{\text{rate}}$. If we let r represent the speed of the plane in still air, then the time required for a one-way trip is

$$f(r) = \frac{800}{r}$$

b. On the trip from Dallas to Phoenix the plane encounters a headwind of 30 miles per hour, so its actual ground speed is $r - 30$. On the return trip the plan enjoys a tailwind of 30 miles per hour, so its actual ground speed is $r + 30$. Therefore, the flying times are

$$\text{Dallas to Phoenix: } \frac{800}{r - 30}$$

and

$$\text{Phoenix to Dallas: } \frac{800}{r + 30}$$

c. The round trip flying time from Dallas to Phoenix and back is

$$F(r) = \frac{800}{r - 30} + \frac{800}{r + 30}$$

The LCD for these fractions is $(r - 30)(r + 30)$. Thus,

$$\frac{800}{r - 30} + \frac{800}{r + 30} = \frac{800(r + 30)}{(r - 30)(r + 30)} + \frac{800(r - 30)}{(r + 30)(r - 30)}$$

$$= \frac{(800r + 24000) + (800r - 24000)}{(r + 30)(r - 30)} = \frac{1600r}{r^2 - 900} \quad \blacksquare$$

Complex Fractions

A fraction that contains one or more fractions in either its numerator or its denominator or both is called a **complex fraction.** For example,

$$\frac{\dfrac{2}{3}}{\dfrac{5}{6}} \qquad \text{and} \qquad \frac{x + \dfrac{3}{4}}{x - \dfrac{1}{2}}$$

are complex fractions. Like simple fractions, complex fractions represent quotients. For the examples above,

$$\frac{\dfrac{2}{3}}{\dfrac{5}{6}} = \frac{2}{3} \div \frac{5}{6}$$

and

$$\frac{x + \dfrac{3}{4}}{x - \dfrac{1}{2}} = \left(x + \frac{3}{4} \right) \div \left(x - \frac{1}{2} \right)$$

We can always simplify a complex fraction into a standard algebraic fraction. If the denominator of the complex fraction is a single term, we can treat the fraction as a division problem and multiply the numerator by the reciprocal of the denominator. Thus,

$$\frac{\dfrac{2}{3}}{\dfrac{5}{6}} = \frac{2}{3} \div \frac{5}{6} = \frac{2}{3} \cdot \frac{6}{5} = \frac{4}{5}$$

If the numerator or denominator of the complex fraction contains more than one term, it is easier to use the fundamental principle of fractions to simplify the expression.

EXAMPLE 8   Simplify $\dfrac{x + \dfrac{3}{4}}{x - \dfrac{1}{2}}$.

Solution

Consider all of the simple fractions that appear in the complex fraction; in this example $\frac{1}{2}$ and $\frac{3}{4}$. The LCD of these fractions is 4. If we multiply the numerator and denominator of the complex fraction by 4, we will eliminate the fractions within the fraction. Be sure to multiply *each* term of the numerator and *each* term of the denominator by 4.

$$\frac{4\left(x + \dfrac{3}{4}\right)}{4\left(x - \dfrac{1}{2}\right)} = \frac{4(x) + 4\left(\dfrac{3}{4}\right)}{4(x) - 4\left(\dfrac{1}{2}\right)} = \frac{4x + 3}{4x - 2}$$

Thus, the original complex fraction is equivalent to the simple fraction $\dfrac{4x + 3}{4x - 2}$. ∎

We summarize the method for simplifying complex fractions as follows.

To Simplify a Complex Fraction:

1. Find the LCD of all the fractions contained in the complex fraction.
2. Multiply the numerator and the denominator of the complex fraction by the LCD.
3. Reduce the resulting simple fraction, if possible.

EXERCISE 4   Simplify $\dfrac{x - 2}{x - \dfrac{4}{x}}$.

Negative Exponents

Algebraic fractions are sometimes written using negative exponents. (You can review negative exponents in Section 6.1.)

EXAMPLE 9

Write each expression as a single algebraic fraction.

a. $x^{-1} - y^{-1}$ **b.** $(x^{-2} + y^{-2})^{-1}$

Solutions

a. $x^{-1} - y^{-1} = \dfrac{1}{x} - \dfrac{1}{y}$ or $\dfrac{y - x}{xy}$

b. $(x^{-2} + y^{-2})^{-1} = \left(\dfrac{1}{x^2} + \dfrac{1}{y^2}\right)^{-1} = \left(\dfrac{y^2 + x^2}{x^2 y^2}\right)^{-1} = \left(\dfrac{x^2 y^2}{y^2 + x^2}\right)$ ∎

When working with fractions and exponents, it is important to avoid some tempting but *incorrect* algebraic operations.

COMMON ERRORS

1. In Example 9a, note that

$$\frac{1}{x} - \frac{1}{y} \neq \frac{1}{x - y}$$

For example, you can check that, for $x = 2$ and $y = 3$,

$$\frac{1}{2} - \frac{1}{3} \neq \frac{1}{2 - 3} = -1$$

2. In Example 9b, note that

$$(x^{-2} + y^{-2})^{-1} \neq x^2 + y^2$$

In general, the fourth law of exponents does *not* apply to sums and differences; that is,

$$(a + b)^n \neq a^n + b^n$$ ∎

EXERCISE 5   Simplify $(1 + x^{-1})^{-1}$.

READING QUESTIONS

1. Explain how to add or subtract fractions with the same denominator.
2. What is the first step in adding fractions with different denominators?
3. What does it mean to build a fraction, and when would you do so?
4. How can you use an LCD to simplify a complex fraction?

HOMEWORK 8.5

Write each sum or difference as a single fraction in lowest terms.

1. $\dfrac{x}{2} - \dfrac{3}{2}$

2. $\dfrac{y}{7} - \dfrac{5}{7}$

3. $\dfrac{1}{6}a + \dfrac{1}{6}b - \dfrac{5}{6}c$

4. $\dfrac{1}{3}x - \dfrac{2}{3}y + \dfrac{1}{3}z$

5. $\dfrac{x-1}{2y} + \dfrac{x}{2y}$

6. $\dfrac{y+1}{b} - \dfrac{y-1}{b}$

7. $\dfrac{3}{x+2y} - \dfrac{x-3}{x+2y} - \dfrac{x-1}{x+2y}$

8. $\dfrac{2}{a-3b} - \dfrac{b-2}{a-3b} + \dfrac{b}{a-3b}$

Find the LCD for each pair of fractions.

9. $\dfrac{5}{6(x+y)^2}$, $\dfrac{3}{4xy^2}$

10. $\dfrac{1}{8(a-b)^2}$, $\dfrac{5}{12a^2b^2}$

11. $\dfrac{2a}{a^2+5a+4}$, $\dfrac{2}{(a+1)^2}$

12. $\dfrac{3x}{x^2-3x+2}$, $\dfrac{3}{(x-1)^2}$

13. $\dfrac{x+2}{x^2-x}$, $\dfrac{x+1}{(x-1)^3}$

14. $\dfrac{y-1}{y^2+2y}$, $\dfrac{y-3}{(y+2)^2}$

Write each sum or difference as a single fraction in lowest terms.

15. $\dfrac{x}{2} + \dfrac{2x}{3}$

16. $\dfrac{3y}{4} + \dfrac{y}{3}$

17. $\dfrac{5}{6}y - \dfrac{3}{4}y$

18. $\dfrac{3}{4}x - \dfrac{1}{6}x$

19. $\dfrac{x+1}{2x} + \dfrac{2x-1}{3x}$

20. $\dfrac{y-2}{4y} + \dfrac{2y-3}{3y}$

21. $\dfrac{5}{x} + \dfrac{3}{x-1}$

22. $\dfrac{2}{y+2} + \dfrac{3}{y}$

23. $\dfrac{y}{2y-1} - \dfrac{2y}{y+1}$

24. $\dfrac{2x}{3x+1} - \dfrac{x}{x-2}$

25. $\dfrac{y-1}{y+1} - \dfrac{y-2}{2y-3}$

26. $\dfrac{x-2}{2x+1} - \dfrac{x+1}{x-1}$

27. $\dfrac{7}{5x-10} - \dfrac{5}{3x-6}$

28. $\dfrac{2}{3y+6} - \dfrac{3}{2y+4}$

29. $\dfrac{y-1}{y^2-3y} - \dfrac{y+1}{y^2+2y}$

30. $\dfrac{x+1}{x^2+2x} - \dfrac{x-1}{x^2-3x}$

31. $x - \dfrac{1}{x}$

32. $1 + \dfrac{1}{y}$

33. $x + \dfrac{1}{x-1} - \dfrac{1}{(x-1)^2}$

34. $y - \dfrac{2}{y^2-1} + \dfrac{3}{y+1}$

35. River Queen Tours offers a 50-mile round trip excursion on the Mississippi River on a paddle wheel boat. The current in the Mississippi is 8 miles per hour.

   a. Express the time required for the downstream journey as a function of the speed of the paddle wheel boat in still water.

   b. Write a function for the time required for the return trip upstream.

   c. Write and simplify an expression for the time needed for the round trip as a function of the boat's speed.

37. Two pilots for the Flying Express parcel service receive packages simultaneously. Orville leaves Boston for Chicago at the same time Wilbur leaves Chicago for Boston. Each selects an air speed of 400 miles per hour for the 900-mile trip. The prevailing winds blow from east to west.

   a. Express Orville's flying time as a function of the wind speed.

   b. Write a function for Wilbur's flying time.

   c. Who reaches his destination first? By how much time (in terms of wind speed)?

36. A rowing team can maintain a speed of 15 miles per hour in still water. The team's daily training session includes a 5-mile run up the Red Cedar River and the return downstream.

   a. Express the team's time on the upstream leg as a function of the speed of the current.

   b. Write a function for the team's time on the downstream leg.

   c. Write and simplify an expression for the total time for the training run as a function of the current's speed.

38. On New Year's Day a blimp leaves its berth in Carson, California, and heads north for the Rose Bowl, 23 miles away. There is a breeze from the north at 6 miles per hour.

   a. Express the time required for the trip as a function of the blimp's air speed.

   b. Write a function for the time needed for the return trip.

   c. Which trip takes longer? By how much time (in terms of the blimp's air speed)?

Write each complex fraction as a simple fraction in lowest terms.

39. $\dfrac{\dfrac{2}{a} + \dfrac{3}{2a}}{5 + \dfrac{1}{a}}$
   40. $\dfrac{\dfrac{2}{y} + \dfrac{1}{2y}}{y + \dfrac{y}{2}}$
   41. $\dfrac{1 + \dfrac{2}{a}}{1 - \dfrac{4}{a^2}}$
   42. $\dfrac{9 - \dfrac{1}{x^2}}{3 - \dfrac{1}{x}}$
   43. $\dfrac{h + \dfrac{h}{m}}{1 + \dfrac{1}{m}}$
   44. $\dfrac{1 + \dfrac{1}{p}}{1 - \dfrac{1}{p}}$

45. $\dfrac{1}{1 - \dfrac{1}{q}}$
   46. $\dfrac{4}{\dfrac{2}{v} + 2}$
   47. $\dfrac{L + C}{\dfrac{1}{L} + \dfrac{1}{C}}$
   48. $\dfrac{H - T}{\dfrac{H}{T} - \dfrac{T}{H}}$
   49. $\dfrac{\dfrac{4}{x^2} - \dfrac{4}{z^2}}{\dfrac{2}{z} - \dfrac{2}{x}}$
   50. $\dfrac{\dfrac{6}{b} - \dfrac{6}{a}}{\dfrac{3}{a^2} - \dfrac{3}{b^2}}$

51. The focal length of a lens is given by the formula

$$\frac{1}{f} = \frac{1}{p} + \frac{1}{q}$$

where f stands for the focal length, p is the distance from the object viewed to the lens, and q is the distance from the image to the lens. Suppose you estimate that the distance from your cat (the object viewed) to your camera lens is 60 inches greater than the distance from the lens to the film inside the camera, where the image forms.

   a. Express $1/f$ as a function of q.

   b. Write and simplify a function for f in terms of q.

52. If two resistors R_1 and R_2 in an electrical circuit are connected in parallel, the total resistance R in the circuit is given by

$$\frac{1}{R} = \frac{1}{R_1} + \frac{1}{R_2}$$

   a. Suppose that the second resistor, R_2, is 10 ohms greater than the first. Express $1/R$ as a function of the first resistor.

   b. Write and simplify a function for R in terms of the first resistor.

53. Andy drives 300 miles to Lake Tahoe at 70 miles per hour and returns home at 50 miles per hour. What is his average speed for the round trip? (It is not 60 miles per hour!)

a. Write expressions for the time it takes for each leg of the trip if Andy drives a distance d at speed r_1 and returns at speed r_2.

b. Write expressions for the total distance and total time for the trip.

c. Write an expression for the average speed for the entire trip.

d. Write your answer to part (c) as a simple fraction.

e. Use your formula to answer the question stated in the problem.

54. The owner of a print shop volunteers to produce flyers for his candidate's campaign. His large printing press can complete the job in 4 hours, and the smaller model can finish the flyers in 6 hours. How long will it take to print the flyers if he runs both presses simultaneously?

a. Suppose that the large press can complete a job in t_1 hours and the smaller press takes t_2 hours. Write expressions for the fraction of a job that each press can complete in 1 hour.

b. Write an expression for the fraction of a job that can be completed in 1 hour with both presses running simultaneously.

c. Write an expression for the amount of time needed to complete the job with both presses running.

d. Write your answer to part (c) as a simple fraction.

e. Use your formula to answer the question stated in the problem.

Write each expression as a single algebraic fraction.

55. $x^{-2} + y^{-2}$ **56.** $x^{-2} - y^{-2}$ **57.** $2w^{-1} - (2w)^{-2}$ **58.** $3w^{-3} + (3w)^{-1}$

59. $a^{-1}b - ab^{-1}$ **60.** $a - b^{-1}a - b^{-1}$ **61.** $(x^{-1} + y^{-1})^{-1}$ **62.** $(1 - xy^{-1})^{-1}$

63. $\dfrac{x + x^{-2}}{x}$ **64.** $\dfrac{x^{-1} - y}{x^{-1}}$ **65.** $\dfrac{a^{-1} + b^{-1}}{(ab)^{-1}}$ **66.** $\dfrac{x}{x^{-2} - y^{-2}}$

Write each complex fraction as a simple fraction in lowest terms, and rationalize the denominator.

67. $\dfrac{\dfrac{2}{\sqrt{7}}}{1 - \dfrac{\sqrt{3}}{\sqrt{7}}}$

68. $\dfrac{\dfrac{1}{4}}{\dfrac{\sqrt{5}}{2\sqrt{2}} - \dfrac{\sqrt{3}}{2\sqrt{2}}}$

69. $\dfrac{\dfrac{\sqrt{3}}{2} + \dfrac{1}{\sqrt{2}}}{1 - \dfrac{\sqrt{3}}{2} \cdot \dfrac{1}{\sqrt{2}}}$

70. $\dfrac{\dfrac{1}{\sqrt{3}} - \dfrac{\sqrt{5}}{3}}{1 + \dfrac{1}{\sqrt{3}} \cdot \dfrac{\sqrt{5}}{3}}$

8.6 Equations that Include Algebraic Fractions

When working with rational functions, we often need to solve equations that involve algebraic fractions. In Example 1 of Section 8.3 we wrote a function that gave the time Francine needs for a 60-mile training run on her cycle-plane, in terms of the wind speed, x.

$$t = f(x) = \frac{60}{15 - x}$$

If it takes Francine 9 hours to cover 60 miles, what is the speed of the wind? We can answer this question by reading values from the graph of f, as shown in Figure 8.43. When $t = 9$, the value of x is between 8 and 9, so the wind speed is between 8 and 9 miles per hour.

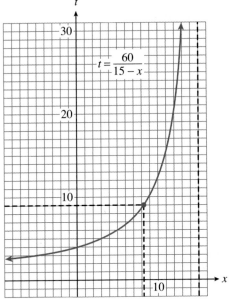

FIGURE 8.43

Solving Equations with Fractions Algebraically

If we need a more accurate value for the wind speed, we can solve the equation

$$\frac{60}{15 - x} = 9$$

To solve an equation involving an algebraic fraction, we multiply each side of the equation by the denominator of the fraction. This has the effect of "clearing" the fraction, and gives us an equivalent equation without fractions.

EXAMPLE 1

Solve the equation $\dfrac{60}{15 - x} = 9$

Solution

Multiply both sides of the equation by $15 - x$ to obtain

$$(15 - x)\frac{60}{15 - x} = 9(15 - x)$$
$$60 = 9(15 - x) \quad \text{Apply the distributive law.}$$

From here we can proceed as usual.

$$60 = 135 - 9x \quad \text{Subtract 135 from both sides.}$$
$$-75 = -9x \quad \quad \text{Divide by } -9.$$
$$8.\overline{3} = x$$

The wind speed was $8.\overline{3}$ or $8\frac{1}{3}$ miles per hour. ∎

If the equation contains more than one fraction, we can clear all the denominators at once by multiplying both sides by the LCD of the fractions.

EXAMPLE 2   Rani times herself as she kayaks 30 miles down the Derwent River with the help of the current. Returning upstream against the current she manages only 18 miles in the same amount of time. Rani knows that she can kayak at a rate of 12 miles per hour in still water. What is the speed of the current?

Solution

If we let x represent the speed of the current, we can fill in the following table.

	Distance	Rate	Time
Downstream	30	$12 + x$	$\dfrac{30}{12+x}$
Upstream	18	$12 - x$	$\dfrac{18}{12-x}$

Because Rani paddled for equal amounts of time upstream and downstream, we have the equation

$$\frac{30}{12+x} = \frac{18}{12-x}$$

The LCD for the fractions in this equation is $(12 + x)(12 - x)$. We multiply both sides of the equation by the LCD to obtain

$$(12+x)(12-x)\frac{30}{12+x} = \frac{18}{12-x}(12+x)(12-x)$$

$$30(12 - x) = 18(12 + x)$$

Solving this equation, we find

$$360 - 30x = 216 + 18x$$
$$144 = 48x$$
$$3 = x$$

The speed of the current is 3 miles per hour. ∎

EXERCISE 2   Solve $\dfrac{x}{6-x} = \dfrac{1}{2}$.

Solving Equations with Fractions Graphically

We can solve the equation in Example 2 graphically by considering two functions, one for each side of the equation. Graph the two functions

$$Y_1 = \frac{30}{12 + x} \qquad \text{and} \qquad Y_2 = \frac{18}{12 - x}$$

in the window

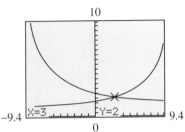

$$\text{Xmin} = -9.4 \qquad \text{Xmax} = 9.4$$
$$\text{Ymin} = 0 \qquad \text{Ymax} = 10$$

to obtain the graph shown in Figure 8.44.

FIGURE 8.44

The function Y_1 gives the time it takes Rani to paddle 30 miles downstream, and Y_2 gives the time it takes her to paddle 18 miles upstream. Both of these times depend on the speed of the current, x. We are looking for a value of x that makes Y_1 and Y_2 equal. This occurs at the intersection point of the two graphs, $(3, 2)$. Thus, the speed of the current is 3 miles per hour, as we found in Example 2. The y-coordinate of the intersection point gives the time Rani paddled on each part of her trip: 2 hours each way.

Extraneous Solutions

A rational function is undefined for any values of x that make its denominator equal zero. These values are not in the domain of the function, and therefore cannot be solutions to equations involving the function. Consider the equation

$$\frac{x}{x - 3} = \frac{3}{x - 3} + 2 \tag{1}$$

When we multiply both sides by the LCD, $x - 3$, we obtain

$$(x - 3)\frac{x}{x - 3} = (x - 3)\frac{3}{x - 3} + (x - 3)2$$

or

$$x = 3 + 2x - 6$$

whose solution is

$$x = 3$$

However, $x = 3$ is *not* a solution of the original equation (1). Both sides of the equation are undefined at $x = 3$. If you graph the two functions

$$Y_1 = \frac{x}{x - 3} \qquad \text{and} \qquad Y_2 = \frac{3}{x - 3} + 2$$

you will find that the graphs never intersect, which means that there is no solution to equation (1).

What went wrong with our method of solution? We multiplied both sides of the equation by $x - 3$, which is zero when $x = 3$, so we really multiplied both sides of the equation by zero. Multiplying by zero does not produce an equivalent equation, and false solutions may be introduced. An apparent solution that does not satisfy the original equation is called an **extraneous solution.** Whenever we multiply an equation by an expression containing the variable, we should check that the solution obtained is not excluded from the domain of the rational functions involved.

When solving an equation with fractions algebraically we must be careful to multiply *each* term of the equation by the LCD, no matter whether each term involves fractions.

EXAMPLE 3

a. Solve the equation $\dfrac{6}{x} + 1 = \dfrac{1}{x + 2}$ algebraically.

b. Solve the same equation graphically.

Solutions

a. To solve the equation algebraically, we multiply both sides by the LCD, $x(x + 2)$. Notice that we multiply *each term* on the left side by the LCD, to get

$$x(x + 2)\left(\frac{6}{x} + 1\right) = x(x + 2)\left(\frac{1}{x + 2}\right)$$

or

$$6(x + 2) + x(x + 2) = x$$

Use the distributive law to remove the parentheses and write the result in standard form.

$$6x + 12 + x^2 + 2x = x$$
$$x^2 + 7x + 12 = 0$$

This is a quadratic equation that we can solve by factoring.

$$(x + 3)(x + 4) = 0$$

so the solutions are $x = -3$ and $x = -4$. Neither of these values causes either denominator to equal zero, so they are not extraneous solutions.

b. To solve the equation graphically, graph the two functions

$$Y_1 = \frac{6}{x} + 1 \qquad \text{and} \qquad Y_2 = \frac{1}{x + 2}$$

in the window

$$\text{Xmin} = -4.7 \qquad \text{Xmax} = 4.7$$
$$\text{Ymin} = -10 \qquad \text{Ymax} = 10$$

as shown in Figure 8.45a.

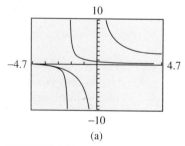

(a)

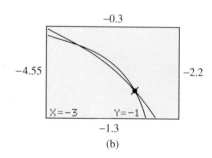

(b)

FIGURE 8.45

We see that the first graph has an asymptote at $x = 0$, and the second graph has one at $x = -2$. It appears that the two graphs may intersect around $x = -3$. To investigate further, we change the window settings to

$$\text{Xmin} = -4.55 \qquad \text{Xmax} = -2.2$$
$$\text{Ymin} = -1.3 \qquad \text{Ymax} = -0.3$$

to obtain the close-up view shown in Figure 8.45b. In this window we can see that the graphs intersect in two distinct points, and by using the $\boxed{\text{TRACE}}$ we find that their x-coordinates are $x = -3$ and $x = -4$. ∎

EXERCISE 3 Solve $\dfrac{9}{x^2 + x - 2} + \dfrac{1}{x^2 - x} = \dfrac{4}{x - 1}$.

Formulas

Algebraic fractions may appear in formulas that relate several variables. If we want to solve for one variable in terms of the others, we may need to clear the fractions.

EXAMPLE 4

Solve the formula $p = \dfrac{v}{q + v}$ for v.

Solution

Because the variable we want appears in the denominator, we must first multiply both sides of the equation by that denominator, $q + v$.

$$(q + v)p = \dfrac{v}{q + v}(q + v)$$

or

$$(q + v)p = v$$

Apply the distributive law on the left side, and collect all terms that involve v on one side of the equation.

$$qp + vp = v \qquad \text{Subtract } vp \text{ from both sides.}$$
$$qp = v - vp$$

We cannot combine the two terms containing v because they are not like terms. However, we can *factor out* v, so that the right side is written as a single term containing the variable v. We can then complete the solution.

$$qp = v(1 - p) \qquad \text{Divide both sides by } 1 - p.$$
$$\dfrac{qp}{1 - p} = v$$

∎

EXERCISE 4 Solve for a: $\dfrac{2ab}{a + b} = H$.

1. What is the first step in solving an equation that includes algebraic fractions?

2. If the equation also contains terms without fractions, should you multiply those terms by the LCD?

3. What are extraneous solutions, and when might they arise?

4. If you are solving a formula and two or more terms contain the variable you are solving for, what should you do?

ANSWERS TO 8.6 EXERCISES

1. $x = -2, x = 4$ **2.** $x = 2$ **3.** $x = \dfrac{-1}{2}$

4. $a = \dfrac{bH}{2b - H}$

HOMEWORK 8.6

Solve each equation algebraically.

1. $\dfrac{6}{w + 2} = 4$ **2.** $\dfrac{12}{r - 7} = 3$ **3.** $9 = \dfrac{h - 5}{h - 2}$ **4.** $-3 = \dfrac{v + 1}{v - 6}$

5. $\dfrac{15}{s^2} = 8$ **6.** $\dfrac{3}{m^2} = 5$ **7.** $4.3 = \sqrt{\dfrac{18}{y}}$ **8.** $6.5 = \dfrac{52}{\sqrt{z}}$

9. The total weight, S, that a beam can support is given in pounds by

$$S = \frac{182.6wh^2}{l}$$

where w is the width of the beam in inches, h is its height in inches, and l is the length of the beam in feet. A beam over the doorway in an interior wall of a house must support 1600 pounds. If the beam is 4 inches wide and 9 inches tall, how long can it be?

10. If two appliances are connected in parallel in an electrical circuit, the total resistance R in the circuit is given by

$$R = \frac{ab}{a + b}$$

where a and b are the resistances of the two appliances. If one appliance has a resistance of 18 ohms, and the total resistance in the circuit is measured at 12 ohms, what is the resistance of the second appliance?

11. A flock of eider ducks is making a 150-mile flight at an average air speed of 50 miles per hour against a moderate headwind.

    a. Express the ducks' travel time, t, as a function of the wind speed, v, and graph the function in the window

        Xmin = 0   Xmax = 50
        Ymin = 0   Ymax = 20

    b. Write and solve an equation to find the wind speed if the flock makes its trip in 4 hours. Label the corresponding point on your graph.

12. A school of bluefin tuna is making a 200-mile trip at an average speed of 36 miles per hour in still water, but is swimming into a current.

    a. Express the tuna's travel time, t, as a function of the current speed, v, and graph the function on the window

        Xmin = 0   Xmax = 36
        Ymin = 0   Ymax = 50

    b. Write and solve an equation to find the current speed if the school makes its trip in 8 hours. Label the corresponding point on your graph.

13. The cost in thousands of dollars for immunizing $p\%$ of the residents of Emporia against a dangerous new disease is given by the function

$$C(p) = \frac{72p}{100 - p}$$

Write and solve an equation to determine what percent of the population can be immunized for $168,000.

14. The cost in thousands of dollars for extracting $p\%$ of a precious ore from a mine is given by the equation

$$C(p) = \frac{360p}{100 - p}$$

Write and solve an equation to determine what percentage of the ore can be extracted for $390,000.

For Problems 15–18, (a) solve the equation graphically by graphing two functions, one for each side of the equation, (b) solve the equation algebraically.

15. $\dfrac{2x}{x + 1} = \dfrac{x + 1}{2}$

16. $\dfrac{3}{2x + 1} = \dfrac{2x - 3}{x}$

17. $\dfrac{2}{x + 1} = \dfrac{x}{x + 1} + 1$

18. $\dfrac{5}{x - 3} = \dfrac{x + 2}{x - 3} + 3$

19. The manager of Joe's Burgers discovers that he will sell $\dfrac{160}{x}$ burgers per day if the price of a burger is x dollars. On the other hand, he can afford to make $6x + 49$ burgers if he charges x dollars apiece for them.

 a. Graph the *demand function,* $D(x) = \dfrac{160}{x}$,

 and the *supply function,* $S(x) = 6x + 49$, in the same window. At what price x does the demand for burgers equal the number that Joe can afford to supply? This value for x is called the *equilibrium price.*

 b. Write and solve an equation to verify your equilibrium price.

20. A florist finds that she will sell $\dfrac{300}{x}$ dozen roses per week if she charges x dollars for a dozen. Her suppliers will sell her $5x - 55$ dozen roses if she sells them at x dollars per dozen.

 a. Graph the demand function, $D(x) = \dfrac{300}{x}$,

 and the supply function, $S(x) = 5x - 55$, in the same window. At what equilibrium price x will the florist sell all the roses she purchases?

 b. Write and solve an equation to verify your equilibrium price.

21. Francine wants to fence a rectangular area of 3200 square feet to grow vegetables for her family of three.

 a. Express the length of the garden as a function of its width.

 b. Express the perimeter, P, of the garden as a function of its width.

 c. Graph your function for perimeter and find the coordinates of the lowest point on the graph. Interpret those coordinates in the context of the problem.

 d. Francine has 240 feet of chain link to make a fence for the garden, and she would like to know what the width of the garden should be. Write an equation that describes this situation.

 e. Solve your equation and find the dimensions of the garden.

22. The cost of wire fencing is $7.50 per foot. A rancher wants to enclose a rectangular pasture of 1000 square feet with this fencing.

 a. Express the length of the pasture as a function of its width.

 b. Express the cost of the fence as a function of its width.

 c. Graph your function for the cost and find the coordinates of the lowest point on the graph. Interpret those coordinates in the context of the problem.

(continued on page 563)

d. The rancher has $1050 to spend on the fence, and she would like to know what the width of the pasture should be. Write an equation to describe this situation.

e. Solve your equation and find the dimensions of the pasture.

23. A proportion is an equation in which each side is a ratio: $\dfrac{a}{b} = \dfrac{c}{d}$. Show that this equation may be rewritten as $ad = bc$.

24. Suppose that y varies directly with x, and (a, b) and (c, d) are two points on the graph of y in terms of x. Show that $\dfrac{b}{a} = \dfrac{d}{c}$.

Solve each proportion using your result from Problem 23.

25. $\dfrac{3}{4} = \dfrac{y+2}{12-y}$

26. $\dfrac{-3}{4} = \dfrac{y-7}{y+14}$

27. $\dfrac{50}{r} = \dfrac{75}{r+20}$

28. $\dfrac{30}{r} = \dfrac{20}{r-10}$

Use your result from Problem 24 to write and solve a proportion for the problem.

29. Property taxes on a house vary directly with the value of the house. If the taxes on a house worth $120,000 are $2700, what would the taxes be on a house assessed at $275,000?

30. The cost of electricity varies directly with the number of units (BTU's) consumed. If a typical household in the Midwest uses 83 million BTU's of electricity annually and pays $1236, how much will a household that uses 70 million BTU's annually spend for energy?

31. Distances on a map vary directly with actual distances. The scale on a map of Michigan uses $\frac{3}{8}$ inch to represent 10 miles. If Isle Royale is $1\frac{11}{16}$ inches long on the map, what is the actual length of the island?

32. The dimensions of an enlargement vary directly with the dimensions of the original. A photographer plans to enlarge a photograph that measures 8.3 centimeters by 11.2 centimeters to produce a poster that is 36 centimeters wide. How long will the poster be?

33. The Forest Service tags 200 perch and releases them into Spirit Lake. One month later it captures 80 perch and finds that 18 of them are tagged. What is the Forest Service's estimate of the original perch population of the lake?

34. The Wildlife Commission tags 30 Canada Geese at one of its migratory feeding grounds. When the geese return, the commission captures 45 geese, of which 4 are tagged. What is the commission's estimate of the number of geese that use the feeding ground?

35. The highest point on earth is Mount Everest in Tibet, with an elevation of 8848 meters. The deepest part of the ocean is the Challenger Deep in the Mariana Trench, near Indonesia, 11,034 meters below sea level.

a. What is the total height variation in the surface of the earth?

b. What percentage of the earth's radius, 6400 kilometers, is this variation?

c. If the earth were shrunk to the size of a basketball, with a radius of 4.75 inches, what would be the corresponding height of Mount Everest?

36. Shortly after the arrival of human beings at the Hawaiian islands around 400 AD, many species of birds became extinct. Fossils of 29 different species have been found, but some species may have left no fossils for us to find. We can estimate the total number of extinct species using a proportion. Of 9 species that are still alive, biologists have found fossil evidence of 7.

a. Assume that the same fraction of extinct species have left fossil records, and calculate the total number of extinct species.

b. Give 2 reasons why this estimate may not be completely accurate.

Source: Burton, 1998.

37. In Figure 8.46, the rectangle ABCD is divided into a square and a smaller rectangle, CDEF. The two rectangles ABCD and CDEF are similar (their corresponding sides are proportional.) A rectangle ABCD with this property is called a "golden rectangle," and the ratio of its length to its width is called the golden ratio. The golden ratio appears frequently in art and nature, and is considered to give the most pleasing proportions to many figures.

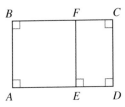

FIGURE 8.46

We'll compute the golden ratio as follows.

a. In Figure 8.46, let $AB = 1$ and $AD = x$. What are the lengths of AE, ED, and CD?

b. Write a proportion in terms of x for the similarity of rectangles $ABCD$ and $CDEF$. Be careful to match up the corresponding sides.

c. Solve your proportion for x. Find the golden ratio, $\dfrac{AD}{AB} = \dfrac{x}{1}$

38. Figure 8.47 shows the graphs of two equations, $y = x$ and $y = \dfrac{1}{x} + 1$.

a. Find the x-coordinate of the intersection point of the two graphs.

b. Compare your answer to the golden ratio you computed in Problem 37.

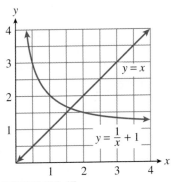

FIGURE 8.47

Solve each formula for the specified variable.

39. $S = \dfrac{a}{1 - r}$, for r **40.** $I = \dfrac{E}{r + R}$, for R **41.** $H = \dfrac{2xy}{x + y}$, for x **42.** $M = \dfrac{ab}{a + b}$, for b

43. $F = \dfrac{Gm_1 m_2}{d^2}$, for d **44.** $F = \dfrac{kq_1 q_2}{r^2}$, for r **45.** $\dfrac{1}{Q} + \dfrac{1}{I} = \dfrac{2}{r}$, for r **46.** $\dfrac{1}{R} = \dfrac{1}{A} + \dfrac{1}{B}$, for B

47. The time, T, it takes for the moon to eclipse the sun totally is given in minutes by the formula

$$T = \dfrac{1}{v}\left(\dfrac{rD}{R} - d\right)$$

where d is the diameter of the moon, D is the diameter of the sun, r is the distance from the earth to the moon, R is the distance from the earth to the sun, and v is the speed of the moon.

a. Solve the formula for v in terms of the other variables.

b. It takes 2.68 minutes for the moon to eclipse the sun. Calculate the speed of the moon, given the following values:

$d = 3.48 \times 10^3$ km $D = 1.41 \times 10^6$ km
$r = 3.82 \times 10^5$ km $R = 1.48 \times 10^8$ km

48. The stopping distance, s, for a car traveling at speed v meters per second is given in meters by

$$s = vT + \dfrac{v^2}{2a}$$

where T is the reaction time of the driver and a is the average deceleration as the car brakes. Suppose that all the cars on a crowded motorway maintain the appropriate spacing determined by the stopping distance for their speed. What speed allows the maximum flow of cars along the road per unit time? Using the formula $time = \dfrac{distance}{speed}$, we see that the time interval t between cars is

$$t = \dfrac{s}{v} + \dfrac{L}{v}$$

where L is the length of the car. To achieve the maximum flow of cars, we would like t to be as small as possible.

a. Substitute the expression for s into the formula for t, and simplify.

b. A typical reaction time is $T = 0.7$ seconds, a typical car length is $L = 5$ meters, and $a = 7.5$ meters per second squared. With

(continued on page 566)

these values, graph t as a function of v in the window

$$\text{Xmin} = 0 \qquad \text{Xmax} = 20$$
$$\text{Ymin} = 0 \qquad \text{Ymax} = 3$$

c. To 1 decimal place, what value of v gives the minimum value of y? Convert your answer to miles per hour.

Source: Bolton, 1974.

49. Many endangered species have fewer than 1000 individuals left. To preserve the species, captive breeding programs must maintain a certain effective population, N, given by

$$N = \frac{4FM}{F + M}$$

where F is the number of breeding females and M the number of breeding males.

a. What is the effective population if there are equal numbers of breeding males and females?

b. In 1972 a breeding program for Speke's gazelle was established with just 3 female gazelle. Graph the effective population, N, as a function of the number of males.

c. What is the largest effective population that can be created with 3 females? How many males are needed to achieve the maximum value?

d. With 3 females, for what value of M is $N = M$?

e. The breeding program for Speke's gazelle began with only 1 male. What was the effective population?

Source: Chapman and Reiss, 1992.

50. When a drug or chemical is injected into a patient, biological processes begin removing that substance. If no more of the substance is introduced, the body removes a fixed fraction of the substance each hour. The amount of substance remaining in the body at time, t, is an exponential decay function, so there is a *biological half-life* to the substance denoted by T_b. If the substance is a radioisotope, it undergoes radioactive decay and so has a physical half-life as well, denoted T_p. The *effective half-life*, denoted by T_e, is related to the biological and physical half-lives by the equation

$$\frac{1}{T_e} = \frac{1}{T_b} + \frac{1}{T_p}$$

The radioisotope ^{131}I is used as a "label" for the human serum albumin. The physical half-life of ^{131}I is 8 days.

a. If ^{131}I is cleared from the body with a biological half-life of 21 days, what is the effective half-life of ^{131}I?

b. The biological half-life of a substance varies considerably from person to person. If the biological half-life of ^{131}I is x days, what is the effective half-life?

c. Let $f(x)$ represent the effective half-life of ^{131}I when the biological half-life is x days. Graph $y = f(x)$.

d. What would the biological half-life of ^{131}I need to be to produce an effective half-life of 6 days? Label the corresponding point on your graph.

e. For what possible biological half-lives of ^{131}I will the effective half-life be less than 4 days?

Source: Pope, 1989.

Solve the equation algebraically.

51. $\dfrac{3}{x - 2} = \dfrac{1}{2} + \dfrac{2x - 7}{2x - 4}$

52. $\dfrac{2}{x + 1} + \dfrac{1}{3x + 3} = \dfrac{1}{6}$

53. $\dfrac{4}{x + 2} - \dfrac{1}{x} = \dfrac{2x - 1}{x^2 + 2x}$

54. $\dfrac{1}{x - 1} + \dfrac{2}{x + 1} = \dfrac{x - 2}{x^2 - 1}$

55. $\dfrac{x}{x + 2} - \dfrac{3}{x - 2} = \dfrac{x^2 + 8}{x^2 - 4}$

56. $\dfrac{4}{2x - 3} + \dfrac{4x}{4x^2 - 9} = \dfrac{1}{2x + 3}$

57. $\dfrac{4}{3x} + \dfrac{3}{3x + 1} + 2 = 0$

58. $-3 = \dfrac{-10}{x + 2} + \dfrac{10}{x + 5}$

59. A chartered sightseeing flight over the Grand Canyon is scheduled to return to its departure point in 3 hours. The pilot would like to cover a distance of 144 miles before turning around, and he hears on the weather service that there will be a headwind of 20 miles per hour on the outward journey.

    a. Express the time it takes for the outward journey as a function of the airspeed of the plane.

    b. Express the time it takes for the return journey as a function of the speed of the plane.

    c. Graph the sum of the 2 functions and find the point on the graph with y-coordinate 3. Interpret the coordinates of the point in the context of the problem.

    d. The pilot would like to know what airspeed to maintain in order to complete the tour in 3 hours. Write an equation to describe this situation.

    e. Solve your equation to find the appropriate airspeed.

60. Two student pilots leave the airport at the same time. They both fly at an airspeed of 180 miles per hour, but the first flies with the wind and the second flies against the wind.

    a. Express the time it takes the first pilot to travel 500 miles as a function of the wind speed.

    b. Express the time it takes the second pilot to travel 400 miles as a function of the wind speed.

    c. Graph the 2 functions in the same window, and find the coordinates of the intersection point. Interpret those coordinates in the context of the problem.

    d. Both pilots check in with their instructors at the same time, and the first pilot has traveled 500 miles while the second pilot has gone 400 miles. Write an equation to describe this situation.

    e. Solve your equation to find the speed of the wind.

CHAPTER 8 REVIEW

Multiply.

1. $(2x - 5)(x^2 - 3x + 2)$

2. $(b^2 - 2b - 3)(2b^2 + b - 5)$

Factor.

3. $8x^3 - 27z^3$

4. $1 + 125a^3b^3$

5. The expression $\dfrac{n}{6}(n - 1)(n - 2)$ gives the number of different 3-item pizzas that can be created from a list of n toppings.

    a. Write the expression as a polynomial.

    b. If Mitch's Pizza offers 12 different toppings, how many different combinations for 3-item pizzas can be made?

    c. Use a table or graph to determine how many different toppings are needed in order to be able to have more than 1000 possible combinations for 3-item pizzas.

6. The expression $n(n - 1)(n - 2)$ gives the number of different triple-scoop ice cream cones that can be created from a list of n flavors.

    a. Write the expression as a polynomial.

    b. If Zanner's Ice Cream Parlor offers 21 flavors, how many different triple-scoop ice cream cones can be made?

    c. Use a table or graph to determine how many different flavors are needed in order to have more than 10,000 possible triple-scoop ice cream cones.

Find the zeros of each polynomial function.

7. $Q(x) = x^5 - 4x^3$

8. $R(x) = 2x^3 + 3x^2 - 2x$

For Problems 9–18, (a) find the x-intercepts, (b) sketch the graph.

9. $f(x) = (x - 2)(x + 1)^2$

10. $g(x) = (x - 3)^2(x + 2)$

11. $G(x) = x^2(x - 1)(x + 3)$

12. $F(x) = (x + 1)^2(x - 2)^2$

13. $V(x) = x^3 - x^5$

14. $H(x) = x^4 - 9x^2$

15. $P(x) = x^3 + x^2 - x - 1$

16. $y = x^3 + x^2 - 2x$

17. $y = x^4 - 7x^2 + 6$

18. $y = x^4 + x^3 - 3x^2 - 3x$

19. The radius, r, of a cylindrical can should be one-half its height, h.

   a. Express the volume, V, of the can as a function of its height.

   b. What is the volume of the can if its height is 2 centimeters? 4 centimeters?

   c. Graph the volume as a function of the height and verify your results of part (b) graphically. What is the approximate height of the can if its volume is 100 cubic centimeters?

20. The Twisty-Freez machine dispenses soft ice cream in a cone-shaped peak with a height 3 times the radius of its base. The ice cream comes in a round bowl with base diameter, d.

   a. Express the volume, V, of Twisty-Freez in the bowl as a function of d.

   b. How much Twisty-Freez comes in a 3-inch-diameter dish? 4-inch?

   c. Graph the volume as a function of the diameter and verify your results of part (b) graphically. What is the approximate diameter of a Twisty-Freez if its volume is 5 cubic inches?

21. Our galaxy is shaped like a disk with a diameter of 1.2×10^{21} meters and thickness of 2×10^{19} meters. How does its shape compare to a compact disk, with a ratio of diameter to thickness of 100 to 1?

22. A person will float in fresh water if his or her density is less than or equal to 1 kilogram per liter, the density of water. $\left(\text{Density is given by the formula } density = \dfrac{weight}{volume}.\right)$ Suppose a swimmer weighs $50 + F$ kilograms, where F is the amount of fat her body contains.

   a. Calculate the volume of her nonfat body mass if its density is 1.1 kilograms per liter.

   b. Calculate the volume of the fat if its density is 0.901 kilograms per liter.

   c. The swimmer's lungs hold 2.6 liters of air. Write an expression for the total volume of her body, including the air in her lungs.

   d. Write an expression for the density of the swimmer's body.

   e. Write an equation for the amount of fat needed for the swimmer to float in fresh water.

   f. Solve your equation. What percent of the swimmer's weight is fat?

   g. Suppose the swimmer's lungs can hold 4.6 liters of air. What percent body fat does she need to be buoyant?

   Source: Burton, 1998.

23. A new health club opened up, and the manager kept track of the number of active members over its first few months of operation. The equation below gives the number, N, of active members, in hundreds, t months after the club opened.

$$N = \frac{44t}{40 + t^2}$$

   a. Use your calculator to graph the function N on a suitable domain.

   b. How many active members did the club have after 8 months?

   c. In which months did the club have 200 active members?

   d. When does the health club have the largest number of active members? What happens to the number of active members as time goes on?

24. A small lake in a state park has become polluted by runoff from a factory upstream. The cost for removing p percent of the pollution from the lake is given, in thousands of dollars, by

$$C = \frac{25p}{100 - p}$$

   a. Use your calculator to graph the function

C on a suitable domain.

b. How much will it cost to remove 40% of the pollution?

c. How much of the pollution can be removed for $100,000?

d. What happens to the cost as the amount of pollution to be removed increases? How much will it cost to remove all the pollution?

For Problems 25–30, (a) identify all asymptotes and intercepts, (b) sketch the graph.

25. $y = \dfrac{1}{x - 4}$

26. $y = \dfrac{2}{x^2 - 3x - 10}$

27. $y = \dfrac{x - 2}{x + 3}$

28. $y = \dfrac{x - 1}{x^2 - 2x - 3}$

29. $y = \dfrac{3x^2}{x^2 - 4}$

30. $y = \dfrac{2x^2 - 2}{x^2 - 9}$

31. The Explorer's Club is planning a canoe trip to travel 90 miles up the Lazy River and return in 4 days. Club members plan to paddle for 6 hours each day, and they know that the current in the Lazy River is 2 miles per hour.

a. Express the time it will take for the upstream journey as a function of their paddling speed in still water.

b. Express the time it will take for the downstream journey as a function of their paddling speed in still water.

c. Graph the sum of the 2 functions and find the point on the graph with *y*-coordinate 24. Interpret the coordinates of the point in the context of the problem.

d. The Explorer's Club would like to know what average paddling speed they must maintain in order to complete their trip in 4 days. Write an equation to describe this situation.

e. Solve your equation to find the required paddling speed.

32. Pam lives on the banks of the Cedar River and makes frequent trips in her outboard motorboat. The boat travels at 20 miles per hour in still water.

a. Express the time it takes Pam to travel 8 miles upstream to the gas station as a function of the speed of the current.

b. Express the time it takes Pam to travel 12 miles downstream to Marie's house as a function of the speed of the current.

c. Graph the 2 functions in the same window, and find the coordinates of the intersection point. Interpret those coordinates in the context of the problem.

d. Pam traveled to the gas station in the same time it took her to travel to Marie's house. Write an equation to describe this situation.

e. Solve your equation to find the speed of the current in the Cedar River.

Reduce each fraction to lowest terms.

33. $\dfrac{2a^2(a - 1)^2}{4a(a - 1)^3}$

34. $\dfrac{4y - 6}{6}$

35. $\dfrac{2x^2y^3 - 4x^3y}{4x^2y}$

36. $\dfrac{(x - 2y)^2}{4y^2 - x^2}$

37. $\dfrac{a^2 - 6a + 9}{2a^2 - 18}$

38. $\dfrac{4x^2y^2 + 4xy + 1}{4x^2y^2 - 1}$

Write each expression as a single fraction in lowest terms.

39. $\dfrac{2a^2}{3b} \cdot \dfrac{15b^2}{a}$

40. $-\dfrac{1}{3}ab^2 \cdot \dfrac{3}{4}a^3b$

41. $\dfrac{4x + 6}{2x} \cdot \dfrac{6x^2}{(2x + 3)^2}$

42. $\dfrac{4x^2 - 9}{3x - 3} \cdot \dfrac{x^2 - 1}{4x - 6}$

43. $\dfrac{a^2 - a - 2}{a^2 - 4} \div \dfrac{a^2 + 2a + 1}{a^2 - 2a}$

44. $\dfrac{a^3 - 8b^3}{a^2b} \div \dfrac{a^2 - 4ab + 4b^2}{ab^2}$

45. $1 \div \dfrac{4x^2 - 1}{2x + 1}$

46. $\dfrac{y^2 + 2y}{3x} \div 4y$

Divide.

47. $\dfrac{36x^6 - 28x^4 + 16x^2 - 4}{4x^4}$

48. $\dfrac{y^3 + 3y^2 - 2y - 4}{y + 1}$

49. $\dfrac{x^3 - 4x^2 + 2x + 3}{x - 2}$

50. $\dfrac{x^2 + 2x^3 - 1}{2x - 1}$

Write each expression as a single fraction in lowest terms.

51. $\dfrac{x + 2}{3x} - \dfrac{x - 4}{3x}$

52. $\dfrac{5}{6}b - \dfrac{1}{3}b + \dfrac{3}{4}b$

53. $\dfrac{3}{2x - 6} - \dfrac{4}{x^2 - 9}$

54. $\dfrac{1}{y^2 + 4y + 4} + \dfrac{3}{y^2 - 4}$

55. $\dfrac{2a + 1}{a - 3} - \dfrac{-2}{a^2 - 4a + 3}$

56. $a - \dfrac{1}{a^2 + 2a + 1} + \dfrac{3}{a^2 - 1}$

Write each complex fraction as a simple fraction in lowest terms.

57. $\dfrac{\dfrac{3}{4} - \dfrac{1}{2}}{\dfrac{3}{4} + \dfrac{1}{2}}$

58. $\dfrac{y - \dfrac{2y}{x}}{1 + \dfrac{2}{x}}$

59. $\dfrac{x - 4}{x - \dfrac{16}{x}}$

60. $\dfrac{\dfrac{1}{x - 1}}{1 - \dfrac{1}{x^2}}$

Solve.

61. $\dfrac{y + 3}{y + 5} = \dfrac{1}{3}$

62. $\dfrac{z^2 + 2}{z^2 - 2} = 3$

63. $\dfrac{x}{x - 2} = \dfrac{2}{x - 2} + 7$

64. $\dfrac{3x}{x + 1} - \dfrac{2}{x^2 + x} = \dfrac{4}{x}$

Solve for the indicated variable.

65. $V = C\left(1 - \dfrac{t}{n}\right)$ for n

66. $r = \dfrac{dc}{1 - ec}$ for c

67. $\dfrac{p}{q} = \dfrac{r}{q + r}$ for q

68. $I = \dfrac{E}{R + \dfrac{r}{n}}$ for R

Write each expression as a single fraction involving positive exponents.

69. $x^{-3} + y^{-1}$

70. $\dfrac{x^{-1}}{y} - \dfrac{x}{y^{-1}}$

71. $\dfrac{x^{-1} - y}{x - y^{-1}}$

72. $\dfrac{x^{-1} + y^{-1}}{x^{-1}}$

73. $\dfrac{x^{-1} - y^{-1}}{(x - y)^{-1}}$

74. $\dfrac{(xy)^{-1}}{x^{-1} - y^{-1}}$

CHAPTER

9

Sequences and Series

In this chapter we study functions with domain a subset of the natural numbers, rather than an interval of the real line. We will also learn techniques for summing a sequence and for finding powers of a binomial. These skills are used in discrete mathematics, which has applications in finance, computer science, and many other fields.

Scheduling a Soccer League

Francine is the Program Director for the City Park Summer Program. This week she is planning the soccer league for seven-to-nine-year-olds. In this league, each team plays every other team once before the play-offs. The number of games Francine must schedule (not including the play-off games) depends on how many teams she allows in the league. To get a feel for the numbers involved, Francine does some calculations.

1. How many games will be needed if there is only one team in the league? (This event is not very likely, but Francine is looking for a pattern, so she considers all possibilities.) How many games will be needed if there are two teams? Start recording your findings in Table 9.1.

2. The number of games gets harder to calculate as the number of teams increases. Francine discovers that she can model the problem by drawing one dot for each team in the league, and then connecting each dot to

every other dot with a straight line. The number of lines represents the number of games needed. The diagrams for three teams and for four teams are shown in Figure 9.1. Use similar diagrams to fill in Table 9.1 up to eight teams. (Be careful that no three of your dots "line up.")

TABLE 9.1

Teams	Games
1	
2	
3	
4	
5	
6	
7	
8	

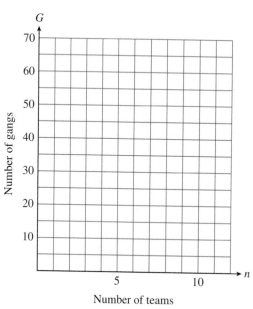

FIGURE 9.1

3. Plot the data points on the graph in Figure 9.2, using the number of teams (n) on the horizontal axis, and the number of games (G) on the vertical axis. For this problem, we cannot really connect the points because n cannot have fractional values. (Remember that n is the number of teams.) However, we will draw a faint curve through the points anyway to help us see a pattern. Do the points lie on a straight line?

G

Number of gangs

70

60

50

40

30

20

10

5   10 n

Number of teams

FIGURE 9.2

TABLE 9.2

n	G	ΔG
1	0	—
2	1	1
3	3	2
4		
5		
6		
7		
8		
9		
10		
11		
12		

4. Can you see a quick way to obtain more entries in the table? Here is a hint: Add another column to your table, as shown in Table 9.2. We will call this column ΔG, which means "change in G." As you move down the table, how much does G increase at each step? The first few values of ΔG are filled in for you. Do you see a pattern? Use this pattern to fill in the table up to $n = 12$.

5. Explain how you would compute the number of games needed for 21 teams if k games are needed for 20 teams.

6. If $G = f(n)$ represents the total number of games needed for n teams, write an equation that shows how to compute $f(n + 1)$ in terms of $f(n)$.

7. Summarize your work: In your own words, describe the relationship between the number of teams in the league and the number of games Francine must schedule. Include in your discussion a description of your graph.

9.1 Sequences

Definitions and Notation

Consider the following function: Gwynn would like to compete in a triathlon, but she needs to improve her swimming. She begins a training schedule, swimming twenty laps a day for the first week and increasing that number by six laps each week. Table 9.3 gives the first few values of the function.

TABLE 9.3

Week	1	2	3	4	5	6	7	8	n
Number of laps	20	26	32	38	44	50	56	62	$f(n)$

The function f makes sense only for domain values that are positive integers. We would not ask how many laps Gwynn swims in week 4.63, or in week -6. A function whose domain is a set of successive positive integers is called a **sequence.** Most people think of a *sequence* as a list of objects in which the order is important, and we often present a mathematical sequence in just that way. The information in Table 9.3 can be displayed more simply by listing only the range values, in order.

$$20, 26, 32, 38, 44, 50, 56, 62$$

When we list them in this way, the range values are called the **terms** of the sequence. The domain values are indicated implicitly by the position of the term. For example, the third term of the sequence, 32, is the value of $f(3)$.

We often use the notation a_n instead of $f(n)$ to refer to the terms of a sequence. Thus, $a_1 = f(1)$, $a_2 = f(2)$, and so on.

EXAMPLE 1

Let a_n be the number of seats in the nth row of a theater. Write an equation using subscript notation for each of the following statements.

a. The first row has 30 seats.

b. The twentieth row has 68 seats.

c. Row 18 has twice as many seats as row 2.

d. Row n has $2n + 28$ seats.

e. Row $n + 1$ has 2 more seats than the previous row.

Solutions

a. a_1 represents the number of seats in row 1, so $a_1 = 30$.

b. a_{20} represents the number of seats in row 20, so $a_{20} = 68$.

c. $a_{18} = 2a_2$

d. $a_n = 2n + 28$

e. $a_{n+1} = a_n + 2$ (Row n is the row before row $n + 1$.) ∎

In Example 1, $a_n = 2n + 28$ is a formula for the number of seats in the nth row of the theater. The expression $2n + 28$ is called the **general term** for the sequence.

EXERCISE 1 Let A_n be the number of dollars in a bank account at the end of the nth year since the account began. Write an equation using subscript notation for each of the following statements.

a. The account had $1149.90 at the end of the second year.

b. At the end of year 4 the account had $85.75 more than at the end of year 3.

c. At the end of year 5 the account had 1.07 times as much as at the end of year 4.

EXAMPLE 2

Find the first four terms in each sequence with the given general term.

a. $a_n = \dfrac{n(n + 1)}{2}$ **b.** $a_n = (-1)^n 2^n$

Solutions

Evaluate each general term for successive values of n.

a. $a_1 = \dfrac{1(1 + 1)}{2} = 1;$ $\qquad a_2 = \dfrac{2(2 + 1)}{2} = 3$

$a_3 = \dfrac{3(3 + 1)}{2} = 6;$ $\qquad a_4 = \dfrac{4(4 + 1)}{2} = 10$

The first four terms are 1, 3, 6, 10.

b. $a_1 = (-1)^1 2^1 = -2;$ $\qquad a_2 = (-1)^2 2^2 = 4$

$a_3 = (-1)^3 2^3 = -8;$ $\qquad a_4 = (-1)^4 2^4 = 16$

The first four terms are $-2, 4, -8, 16$. ∎

EXERCISE 2 Find the first four terms in the sequence with the general term $b_n = (n + 1)^2 - n^2$.

Applications of Sequences

Sequences are useful for describing situations that have discrete values, rather than continuous ones. For example, some savings accounts accrue interest only once a year.

EXAMPLE 3

Suppose you deposit $8000 in a savings account that pays 5% interest compounded annually. How much money will be in the account at the end of each of the next four years?

Solution

During the first year the account will earn 5% of 8000 or $0.05(8000) = 400$. Thus, at the end of the first year the account will contain the original $8000 plus the $400 interest, for a total of $8400.

At the end of the second year the account will have the $8400 from the previous year, plus 5% of 8400 or $0.05(8400) = 420$ in interest. We can write this sum as

$$8400 + 8400(0.05) = 8400(1 + 0.05) = 8400(1.05) \text{ dollars}$$

At the end of the second year there will be $8400(1.05) = $8820 in the account. In fact, each new balance is found by multiplying the previous balance by 1.05. (See Section 7.1 to review compound interest.) The balances at the ends of the third and fourth years are

$$\$8820(1.05) = \$9261, \text{ and } \$9261(1.05) = \$9724.05$$

The annual balances form the sequence 8400, 8820, 9261, 9724.05, as shown in Table 9.4. ■

TABLE 9.4

n	a_n
1	8400
2	8820
3	9261
4	9724.05

EXERCISE 3 You just finished your last cup of coffee and have 100 mg of caffeine in your system. For each hour that passes, the amount of caffeine in your system decreases by 14%. How much caffeine is in your system at the start of each of the next four hours?

EXAMPLE 4

Suppose you deposit $8000 in a savings account that pays 5% *simple* annual interest. In this case, the interest is earned only on the initial $8000 and not compounded. How much money will be in the account at the end of each of the next 4 years?

Solution

As in Example 3, the account earns 5% of $8000 or $0.05(8000) = 400$ during the first year. At the end of the first year the account will contain the original $8000 plus $400 interest for a total of $8400.

During the second year interest will be earned only on the initial $8000 deposited, for another $400. At the end of the second year the account will have the $8400 from the previous year plus $400 interest for a total of

$$\$8400 + \$400 = \$8800$$

TABLE 9.5

n	b_n
1	8400
2	8800
3	9200
4	9600

In fact, the balance at the end of each succeeding year is found by adding $400 simple interest to the previous balance. The annual balances form the sequence 8400, 8800, 9200, 9600, as shown in Table 9.5. ■

EXERCISE 4 The financing agreement on your $12,000 car requires you to pay $1200 now and $208.87 a month for five years. What is the total amount you have paid after each of the first 4 payments?

EXAMPLE 5

A regular polygon is a geometric figure in which all the sides are equal in length. For example, an equilateral triangle and a square are regular polygons. All the interior angles in a regular polygon are equal also. Find the general term a_n for the sequence that gives the size of an interior angle in a regular polygon of n sides, as illustrated in Figure 9.3.

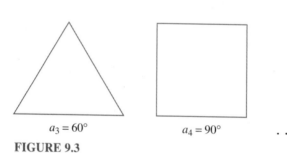

$a_3 = 60°$ $a_4 = 90°$ . . . $a_n = ?°$

FIGURE 9.3

Solution

This sequence starts with a_3 because we cannot have a polygon with fewer than three sides. We already know the values of a_3 and a_4: the angles in an equilateral triangle are each 60°, and in a square the angles are each 90°. We would like to find a formula for the size of the angles in any regular polygon.

If we can find the sum of *all* the angles in a regular polygon, we can divide by n to find the size of each. (You can check that this idea works for the equilateral triangle and the square.) To find the sum of the angles, notice that any polygon can be partitioned into triangles, as shown in Figure 9.4.

By sketching some examples, convince yourself that every polygon of n sides (for $n \geq 3$) can be partitioned into $n - 2$ triangles. Because the angles in every triangle add up to 180°, the sum of the angles in an n-sided polygon is $(n - 2)$ times 180°. To find the size of just one of the angles, we divide the sum by n. This gives us the general term of the sequence.

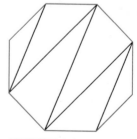

FIGURE 9.4

$$a_n = \frac{(n - 2)180}{n}$$

■

Recursively Defined Sequences

A sequence is defined **recursively** if each term of the sequence is defined in terms of its predecessors. For example, the sequence defined by

$$a_1 = 2, \qquad a_{n+1} = 3a_n - 2$$

is a recursive sequence. Its first four terms are

$$a_1 = 2$$
$$a_2 = 3a_1 - 2 = 3(2) - 2 = 4$$
$$a_3 = 3a_2 - 2 = 3(4) - 2 = 10$$
$$a_4 = 3a_3 - 2 = 3(10) - 2 = 28$$

EXAMPLE 6

Make a table showing the first five terms of the recursive sequence

$$a_1 = -1 \qquad a_{n+1} = (a_n)^2 - 4$$

Solution

The first term is given. We find each subsequent term by using the recursive formula.

$$a_2 = (a_1)^2 - 4 = (-1)^2 - 4 = -3$$
$$a_3 = (a_2)^2 - 4 = (-3)^2 - 4 = 5$$
$$a_4 = (a_3)^2 - 4 = 5^2 - 4 = 21$$
$$a_5 = (a_4)^2 - 4 = (21)^2 - 4 = 437$$

The first five terms are $-1, -3, 5, 21, 437$, as shown in Table 9.6

TABLE 9.6

n	1	2	3	4	5
a_n	-1	-3	5	21	437

■

The definition of a recursive sequence must include a starting point, usually the first term of the sequence, and a recursive formula for calculating the next term of the sequence in terms of the previous term (or terms).

EXAMPLE 7

Find a recursive definition for the sequence in Example 3 if the money is kept in the account for n years.

Solution

If we use the letter a for that sequence, we have $a_1 = 8400$. Each successive term was obtained by multiplying the previous term by 1.05, which means that

$$a_n = 1.05a_{n-1}$$

The sequence is determined by the recursive definition

$$a_1 = 8400, \qquad a_n = 1.05a_{n-1}$$

■

EXAMPLE 8

Find a recursive definition for the sequence in Example 4 if the money is kept in the account for n years.

Solution

If we use the letter b for that sequence, we have $b_1 = 8400$. Each successive term was obtained by adding 400 to the previous term, which means that

$$b_n = b_{n-1} + 400$$

The sequence is determined by the recursive definition

$$b_1 = 8400, \qquad b_n = b_{n-1} + 400$$

∎

EXAMPLE 9

Karen joins a savings plan in which she deposits \$200 per month and receives 12% annual interest compounded monthly.

   a. Find a recursively defined sequence that gives the amount of money in Karen's account n months later.

   b. Find the first 4 terms of the sequence.

Solutions

   a. In the first month Karen deposits \$200, so $a_1 = 200$. Each month thereafter Karen receives 1% interest (one-twelfth of 12% annual interest) on the previous month's balance, then adds \$200. For example, before she makes her deposit in the second month the account has

$$200 + 0.01(200) = 1.01(200)$$

She then adds \$200 to this amount for a total of

$$a_2 = 1.01(200) + 200$$

In general, after the nth deposit, Karen's account contains a_n dollars. In the next month she earns 1% interest on that balance, giving her

$$a_n + 0.01a_n = 1.01a_n$$

Then she deposits another \$200 for a total of

$$a_{n+1} = 1.01a_n + 200$$

Thus, the recursive sequence is defined by

$$a_1 = 200, \qquad a_{n+1} = 1.01a_n + 200$$

   b. Evaluate the general formula found in part (a) for $n = 1, 2, 3, 4$.

$$a_1 = 200$$
$$a_2 = 1.01a_1 + 200$$
$$= 1.01(200) + 200 = 402$$

$$a_3 = 1.01a_2 + 200$$
$$= 1.01(402) + 200 = 606.02$$
$$a_4 = 1.01a_3 + 200$$
$$= 1.01(606.02) + 200 = 812.08 \qquad \blacksquare$$

READING QUESTIONS

1. What distinguishes a sequence from an ordinary function?
2. What are the range values of a sequence called?
3. Give an example of a situation in which a sequence is more appropriate than a function whose domain is an interval of real numbers.
4. What is a recursively defined sequence? How is such a sequence defined?

ANSWERS TO 9.1 EXERCISES

1a. $A_2 = 1149.90$ b. $A_4 = 85.75 + A_3$ c. $A_5 = 1.07A_4$

2. 3, 5, 7, 9   3. 100 mg, 86 mg, 74 mg, 63.6 mg (approximately)

4. \$1200, \$1408.87, \$1614.74, \$1826.61

5. $C_1 = 100,\ C_{n+1} = .86C_n$

HOMEWORK 9.1

Find the first four terms in the sequence whose general term is given.

1. $a_n = n - 5$

2. $b_n = 2n - 3$

3. $c_n = \dfrac{n^2 - 2}{2}$

4. $d_n = \dfrac{3}{n^2 + 1}$

5. $s_n = 1 + \dfrac{1}{n}$

6. $t_n = \dfrac{n}{2n - 1}$

7. $u_n = \dfrac{n(n - 1)}{2}$

8. $v_n = \dfrac{5}{n(n + 1)}$

9. $w_n = (-1)^n$

10. $A_n = (-1)^{n+1}$

11. $B_n = \dfrac{(-1)^n(n - 2)}{n}$

12. $C_n = (-1)^{n-1}3^{n+1}$

13. $D_n = 1$

14. $E_n = -1$

For Problems 15–20 suppose that $a_1, a_2, a_3, \ldots$ is a sequence. Write an equation using subscript notation for each sentence.

15. The first term of the sequence is $\frac{4}{3}$.

16. The third term of the sequence is 8 more than the second term.

17. The nth term is 3 times the previous term.

18. The nth term is one-third of the term that follows it.

19. The $(n + 1)$st term is one-third of the term that follows it.

20. The $(n + 1)$st term is 3 times the previous term.

Find the indicated term for each sequence.

21. $D_n = 2^n - n$; $\quad D_6$

22. $E_n = \sqrt{n + 1}$; $\quad E_{11}$

23. $x_n = \log n$; $\quad x_{26}$

24. $y_n = \log(n + 1)$; $\quad y_9$

25. $z_n = 2\sqrt{n}$; $\quad z_{20}$

26. $U_n = \dfrac{n + 1}{n - 1}$; $\quad U_{17}$

Make a table showing the first five terms of the recursively defined sequence.

27. $s_1 = 3$, $\quad s_n = S_{n-1} + 2$

28. $c_1 = 6$, $\quad c_n = c_{n-1} - 4$

29. $d_1 = 24$, $\quad d_{n+1} = \dfrac{-1}{2}d_n$

30. $r_1 = 27$, $\quad r_{n+1} = \dfrac{2}{3}r_n$

31. $t_1 = 1$, $\quad t_{n+1} = (n + 1)t_n$

32. $x_1 = 1$, $\quad x_n = \left(\dfrac{n}{n - 1}\right)x_{n-1}$

33. $w_1 = 100$, $\quad w_n = 1.10w_{n-1} + 100$

34. $q_1 = 100$, $\quad q_{n+1} = 0.9x_n + 100$

For Problems 35–42, (a) make a table showing the first four terms of each sequence, (b) write an equation to define the sequence *recursively*.

35. A new car costs $14,000 and depreciates in value by 15% each year. How much is the car worth after n years?

36. Krishna takes a job as an executive secretary for $21,000 per year with a guaranteed 5% raise each year. What will his salary be after n years?

37. A long distance phone call costs $1.10 to make the connection and an additional $0.45 for each minute. What is the cost of a call that lasts n minutes?

38. Bettina earns $1000 per month plus $57 for each satellite dish that she sells. What is her monthly income when she sells n satellite dishes?

39. Geraldo inherits an annuity of $50,000 that earns 12% annual interest compounded monthly. If he withdraws $500 at the end of each month, what is the value of the annuity after n months?

40. Eve borrowed $18,000 for a new car at 6% annual interest compounded monthly. If she pays $400 per month toward the loan, how much does she owe after n months?

41. Majel must take 10 milliliters of a medication directly into her bloodstream at constant intervals. During each time interval her kidneys filter out 20% of the drug present just after the most recent dose. How much of the drug will be in her bloodstream after n doses?

42. A forest contains 64,000 trees. According to a new logging plan, each year 5% of the trees will be cut down and 16,000 new trees will be planted. How many trees will be in the forest after n years?

43. a. Draw three non-collinear points in the plane. (The points should not lie on the same line.) How many distinct lines are determined by the points? (In other words, how many different lines can you draw by choosing two of the points and joining them?)

  b. Add a fourth point to your diagram. Now how many lines are determined?

  c. Let L_n stand for the number of distinct lines determined by n non-collinear points. Make a table showing the first five terms of the sequence.

  d. Find a recursive formula for the sequence L_n.

44. a. Draw two distinct non-parallel lines in the plane. In how many points do the lines intersect?

  b. Add a third line to your diagram that is not parallel to either of the first two lines. (The new line should not pass through the intersection point of the previous lines.) How many intersection points are there?

  c. Let P_n stand for the number of intersection points determined by n lines in the plane, no two of which are parallel. Make a table showing the first five terms of the sequence.

d. Find a recursive formula for the sequence P_n.

45. a. The Fibonacci sequence is found throughout nature. For example, the number of spirals in a sunflower or on a pineapple is a term of the Fibonacci sequence. It is named after the Italian mathematician Fibonacci, who used it to model the growth of a population of rabbits. The Fibonacci sequence is defined recursively by

$$f_1 = 1, \quad f_2 = 1, \quad f_{n+2} = f_n + f_{n+1}$$

Make a table showing the first 16 terms of the Fibonacci sequence.

b. Calculate the quotients $\dfrac{f_{n+1}}{f_n}$ for $n = 1$ to $n = 15$. What do you observe? Now find a decimal approximation for the "golden ratio," $\dfrac{1 + \sqrt{5}}{2}$.

46. The Lucas sequence is defined recursively by

$$L_1 = 2, \quad L_2 = 1, \quad L_{n+2} = L_n + L_{n+1}$$

a. Find the first ten terms of the Lucas sequence.

b. Calculate $(L_{n+1})^2 - L_n(L_{n+2})$ for $n = 1$ to $n = 8$. What do you notice?

Use a calculator to evaluate a large number of terms for each recursive sequence. What happens to the terms as n gets larger?

47. $a_1 = 1, \quad a_n = \dfrac{1}{1 + a_{n-1}} + 1$

48. $b_1 = 1, \quad b_n = \dfrac{2}{1 + b_{n-1}} + 1$

49. $c_1 = 3, \quad c_n = \dfrac{\sqrt{1 + c_{n-1}}}{2}$

50. $d_1 = 8, \quad d_n = \dfrac{\sqrt{1 + d_{n-1}}}{2}$

51. $s_1 = 1, \quad s_n = \dfrac{1}{2}\left(s_{n-1} + \dfrac{4}{s_{n-1}}\right)$

52. $t_1 = 1, \quad t_n = \dfrac{1}{2}\left(t_{n-1} + \dfrac{9}{t_{n-1}}\right)$

53. Here is a simplified model for the growth of a population of fish. The size of the stock is measured by the combined weight of the population, called the biomass. The population begins with two fish, a male and a female, each weighing one unit. The fish grow one unit of mass each year, and die after their third year. Every pair of fish gives birth to a new pair of unit-mass fish each year.

a. Let A_n, B_n, and C_n represent the number of one-unit, two-unit, and three-unit fish in the population in year n, and let S_n be the total biomass of the stock in year n. Complete the table.

b. Plot the biomass S_n as a function of the year, and connect the data points with a smooth curve.

c. Write recursive formulas for A_n, B_n, and C_n, starting with $A_0 = 2$, $B_0 = 0$, and $C_0 = 0$.

n	A_n	B_n	C_n	S_n
0	2	0	0	2
1	2	2	0	6
2	4	2	2	14
3				
4				
5				
6				
7				
8				

d. Write a recursive formula for S_n in terms of A_n, B_n, and C_n. Can you express S_n using only terms of the sequence A_n?

Source: Hayward, 1992.

54. The model in Problem 53 is only useful for a limited time, because the fish population cannot continue to grow indefinitely. The model described below is called a logistic model. Suppose a fishery starts with a stock of 20 fish, so $S_0 = 20$. The growth G_n of the population in year n is given by

$$G_n = 0.001S_n(400 - S_n)$$

and

$$S_n = S_{n-1} + G_{n-1}$$

a. Make a table showing values of S_n and G_n up to $n = 20$, rounding values of G_n to whole numbers.

b. Plot the values of G_n. What is the shape of the graph? In what year does the fish population grow the most?

c. Plot the values of S_n. What is the limiting value of the fish population? Describe the growth of the population.

Source: Hayward, 1992.

9.2 Arithmetic and Geometric Sequences

Arithmetic Sequences

Suppose that a charter tour bus service charges $50 plus $15 for each passenger. The cost of a tour is then a function of the number of passengers. Because the number of passengers, n, can be only a natural number, the domain of the function is the set of natural numbers 1, 2, 3, . . . , up to the capacity of the tour bus.

If C_n represents the cost of a tour for n passengers, then

$$C_1 = 50 + 15(1) = 65$$
$$C_2 = 50 + 15(2) = 80$$
$$C_3 = 50 + 15(3) = 95$$

and in general

$$C_n = 50 + 15n$$

Each term of this sequence can be obtained from the previous one by adding 15. A sequence in which each term can be obtained from the previous term by adding a fixed amount is called an **arithmetic sequence.**

The fixed amount we add to each term, or the difference between two successive terms, is called the **common difference.** In the example above the common difference is 15. If we denote the first term of an arithmetic sequence by a and the common difference by d, then the sequence can be defined recursively by

$$a_1 = a$$
$$a_{n+1} = a_n + d$$

EXAMPLE 1

Find the first 4 terms of an arithmetic sequence with first term 6 and common difference 3.

Solution

The first term is 6, so we have $a_1 = 6$. To find each subsequent term, add 3 to the previous term.

$$a_2 = a_1 + 3 = 6 + 3 = 9$$
$$a_3 = a_2 + 3 = 9 + 3 = 12$$
$$a_4 = a_3 + 3 = 12 + 3 = 15$$

The first four terms are 6, 9, 12, 15. ■

EXERCISE 1   Find the first four terms of an arithmetic sequence with first term 100 and common difference -30.

An arithmetic sequence defines a linear function of n. In Figure 9.5 compare the graph of the linear function $f(x) = 2x + 3$, whose domain is the set of all real numbers, and the graph of the arithmetic sequence $a_n = 2n + 3$, whose domain is the set of natural numbers. The common difference, 2, of the sequence corresponds to the slope of the linear function.

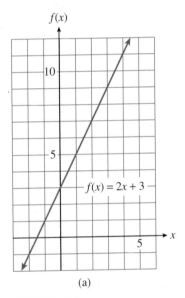

(a)

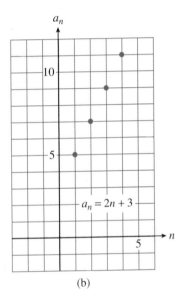

(b)

FIGURE 9.5

The General Term of an Arithmetic Sequence

We have found a recursive definition for an arithmetic sequence, but we can also find a non-recursive definition. That is, we can find a formula for the general term. Consider an arithmetic sequence with first term a and common difference d. The

first term is a

second term is $a + d$

third term is $a + d + d = a + 2d$

fourth term is $a + d + d + d = a + 3d$

.

.

.

nth term is $a + d + d + \cdots + d = a + (n - 1)d$

Thus, we have the following property.

The nth term of an arithmetic sequence is

$$a_n = a + (n - 1)d \qquad (1)$$

We can use Equation (1) to find a particular term of an arithmetic sequence if we know the first term and the common difference.

EXAMPLE 2   Find the fourteenth term of the arithmetic sequence $-6, -1, 4, \ldots$.

Solution
First, find the common difference by subtracting any term from its successor.

$$d = -1 - (-6) = 5$$

Then use Equation (1) with $n = 14$.

$$a_{14} = -6 + (14 - 1)5 = 59 \qquad \blacksquare$$

EXERCISE 2   Find the hundredth term of the arithmetic sequence $1, 3, 5, 7, \ldots$.

EXAMPLE 3   Suppose you deposit $8000 in an account that pays 5% *simple* annual interest. (See Example 4 of Section 9.1.)

   a.   Find a formula for the amount of money in the account after n years.
   b.   How much money will be in the account after 20 years?

Solutions
   a.   Each year, 0.05(8000) or $400 interest is added to the account balance. Thus, the annual balances, b_n, form an arithmetic sequence whose first term is 8400 and whose common difference is 400. Hence,

$$b_n = 8400 + (n - 1)400$$

   b.   Substitute $n = 20$ into the formula for the general term found in part (a).

$$b_{20} = 8400 + (20 - 1)400 = 16{,}000$$

After 20 years there will be $16,000 in the account. $\blacksquare$

Geometric Sequences

Suppose a national junior chess tournament starts with 1024 invited contestants. At the end of each round, the winners move on to the next level. Thus, after each round there are half as many contestants as before, and the number of remaining contestants is a function of the number of rounds completed. Because the number of rounds n is a natural number, the domain of the function is a set of natural numbers.

If C_n represents the number of contestants after n rounds of competition, then

$$C_1 = 1024\left(\frac{1}{2}\right) = 512$$

$$C_2 = 512\left(\frac{1}{2}\right) = 1024\left(\frac{1}{2}\right)^2 = 256$$

$$C_3 = 256\left(\frac{1}{2}\right) = 1024\left(\frac{1}{2}\right)^3 = 128$$

In general,

$$C_n = 1024\left(\frac{1}{2}\right)^n$$

Each term of the sequence defined above can be obtained from the previous one by multiplying by $\frac{1}{2}$. A sequence in which each term can be obtained from the previous term by multiplying by a fixed amount is called a **geometric sequence.**

The fixed amount we multiply each term by is the ratio of two successive terms and is called the **common ratio.** In the example above the common ratio is $\frac{1}{2}$. If we denote the first term of a geometric sequence by a and the common ratio by r, then the sequence can be defined recursively by

$$a_1 = a$$
$$a_{n+1} = ra_n$$

EXAMPLE 4   Find the first four terms of a geometric sequence whose first term is 64 and whose common ratio is $\frac{5}{4}$.

Solution

The first term is 64, so we have $a_1 = 64$. To find each subsequent term, multiply the previous term by $\frac{5}{4}$.

$$a_2 = \frac{5}{4}a_1 = \frac{5}{4}(64) = 80$$

$$a_3 = \frac{5}{4}a_2 = \frac{5}{4}(80) = 100$$

$$a_4 = \frac{5}{4}a_3 = \frac{5}{4}(100) = 125$$

The first four terms are 64, 80, 100, 125.   ■

> **EXERCISE 3**   Find the first four terms of a geometric sequence whose first term is 1000 and whose common ratio is 1.06.

A geometric sequence defines an exponential function of n. In Figure 9.6 compare the graphs of the exponential function $f(x) = 100(2)^{x-1}$, whose domain is the set of real numbers, and the geometric sequence $a_n = 100(2)^{n-1}$, whose domain is the set of natural numbers. The common ratio of the geometric sequence corresponds to the base of the exponential function.

Recall that an arithmetic sequence defines a linear function whose slope corresponds to the common difference. Just as an exponential function grows much faster in the long run than a linear function, so does a geometric sequence grow much faster than an arithmetic sequence.

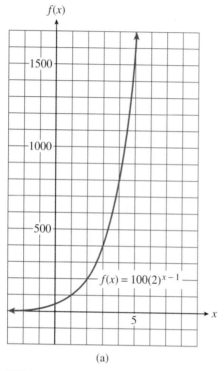

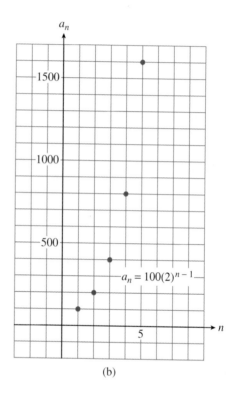

(a)   (b)

FIGURE 9.6

EXAMPLE 5   Identify the following sequences as arithmetic, geometric, or neither.

   a. $3, 5, 7, \ldots$ **b.** $3, -6, 12, \ldots$

   c. $3, 6, 10, \ldots$ **d.** $3, 1, \dfrac{1}{3}, \ldots$

Solutions

   a. This sequence is arithmetic. Each term is obtained from the previous term by adding 2.

   b. This sequence is geometric. Each term is obtained from the previous term by multiplying by -2.

   c. This sequence is neither arithmetic or geometric.

   d. This sequence is geometric. Each term is obtained from the previous term by multiplying by $\frac{1}{3}$.   ■

The General Term of a Geometric Sequence

We can now find a nonrecursive formula for the general term of a geometric sequence. If we denote the first term of the geometric sequence by a, then the

second term is	ar
third term is	$ar \cdot r + ar^2$
fourth term is	$ar^2 \cdot r = ar^3$
$\vdots$	
n^{th} term is	ar^{n-1}

In general we have the following property.

The nth term of a geometric sequence is

$$a_n = ar^{n-1} \tag{2}$$

We can use Equation (2) to find a particular term of a geometric sequence if we know the first term and the common ratio.

EXAMPLE 6

Find the ninth term of the geometric sequence $-24, 12, -6, \ldots$

Solution

First, find the common ratio by dividing any term by its predecessor.

$$r = \frac{12}{-24} = \frac{-1}{2}$$

Then use Equation (2) with $a = -24$, $r = -\frac{1}{2}$, and $n = 9$.

$$a_9 = -24\left(-\frac{1}{2}\right)^8 = \frac{-3}{32}$$ ■

EXAMPLE 7

Suppose you deposit $8000 in an account that pays 5% interest compounded annually. (See Example 3 of Section 9.1.)

a. Find a formula for the amount of money in the account after n years.

b. How much money will be in the account after 20 years?

Solutions

a. Each year, the account balance is multiplied by 1.05. Thus, the annual balances c_n form a geometric sequence whose first term is 8400 and whose common ratio is 1.05. Hence,

$$c_n = 8400(1.05)^{n-1}$$

b. Substitute $n = 20$ into the formula for the general term found in part (a).

$$c_{20} = 8400(1.05)^{19} \approx 21,226.38164$$

After 20 years, there will be $21,226.38 in the account. ∎

EXAMPLE 8

Find a nonrecursive definition for each sequence.

a. $a_1 = 2, \quad a_n = a_{n-1} + 3$ **b.** $\quad b_1 = 2, \quad b_{n-1} = 3b_n$

Solutions

a. From the definition we see that the first few terms of the sequence are 2, 5, 8, 11. Because each new term is found by adding 3 to the previous term, we see that this is an arithmetic sequence with a common difference of 3. This means that the sequence has the form

$$a_n = 2 + (n - 1)3$$

or

$$a_n = 3n - 1$$

b. The first four terms of this sequence are 2, 6, 18, 54. Because each new term is found by multiplying the previous term by 3, we see that this is a geometric sequence with a common ratio of 3. This means that the sequence has the form

$$b_n = 2 \cdot 3^{n-1}$$ ∎

EXERCISE 6   Find a nonrecursive definition for each sequence.

a. $c_1 = 100, \quad c_{k+1} = 1.03c_k$

b. $d_1 = 2, \quad d_k = d_{k-1} - 5$

1. What is an arithmetic sequence? What is the common difference?
2. State a formula for the general term of an arithmetic sequence.

3. An arithmetic sequence defines what kind of function of n?

4. What is a geometric sequence? What is the common ratio?

5. State a formula for the general term of a geometric sequence.

6. A geometric sequence defines what kind of function of n?

HOMEWORK 9.2

Identify each sequence as arithmetic, geometric, or neither.

1. $-2, -6, -18, -54, \ldots$ **2.** $-2, -6, -10, -14, \ldots$ **3.** $16, 8, 0, -8, \ldots$ **4.** $16, 8, 4, 2, \ldots$

5. $-1, 1, -1, 1, \ldots$ **6.** $1, 3, 6, 10, \ldots$ **7.** $1, 4, 9, 16, \ldots$ **8.** $5, -5, 5, -5, \ldots$

9. $27, 9, 3, 1, \ldots$ **10.** $\dfrac{-1}{3}, \dfrac{1}{3}, 1, \dfrac{5}{3}, \ldots$ **11.** $\dfrac{2}{3}, -1, \dfrac{3}{2}, -\dfrac{9}{4}, \ldots$ **12.** $2, -8, 32, -128, \ldots$

Find the first four terms of each sequence.

13. $a = 2, d = 4$

14. $a = 7, d = 3$

15. $a = \dfrac{1}{2}, d = \dfrac{1}{4}$

16. $a = \dfrac{2}{3}, d = \dfrac{1}{3}$

17. $a = 2.7, d = -0.8$

18. $a = 5.9, d = -1.3$

19. $a = 5, r = -2$

20. $a = -4, r = 3$

21. $a = 9, r = \dfrac{2}{3}$

22. $a = 25, r = \dfrac{4}{5}$

23. $a = 60, r = 0.4$

24. $a = 10, r = 0.3$

Find the next three terms, and an expression for the general term.

25. $3, 7, 11, \ldots$ **26.** $-10, -20, -30, \ldots$ **27.** $-1, -5, -9, \ldots$ **28.** $-6, -1, 4, \ldots$

29. $\dfrac{2}{3}, \dfrac{4}{3}, \dfrac{8}{3}, \ldots$ **30.** $6, 3, \dfrac{3}{2}, \ldots$ **31.** $4, -2, 1, \ldots$ **32.** $\dfrac{1}{2}, \dfrac{-3}{2}, \dfrac{9}{2}, \ldots$

33. Find the twelfth term in the arithmetic sequence $2, \frac{5}{2}, 3, \ldots$

34. Find the tenth term in the arithmetic sequence $\frac{3}{4}, 2, \frac{13}{4}, \ldots$

35. Find the eighth term in the geometric sequence $-3, \frac{3}{2}, \frac{-3}{4}, \ldots$

36. Find the sixth term in the geometric sequence $-5, -1, \frac{-1}{5}, \ldots$

37. Find the first term of a geometric sequence with fifth term 48 and common ratio 2.

38. Find the first term of a geometric sequence with fifth term 1 and common ratio $\frac{1}{2}$.

39. How many terms are in the sequence $\frac{1}{8}, \frac{1}{4}, \frac{1}{2}, \ldots, 512$?

40. How many terms are in the sequence $\frac{27}{64}, \frac{9}{16}, \frac{3}{4}, \ldots, \frac{64}{27}$?

Find a non-recursive definition for each sequence.

41. $s_1 = 3, \quad s_n = s_{n-1} + 2$

42. $c_1 = 6, \quad c_n = c_{n-1} - 4$

43. $x_1 = 0, \quad x_{n+1} = x_n - 3$

44. $y_1 = -1, \quad y_{n+1} = y_n + 5$

45. $d_1 = 24, \quad d_{n+1} = \dfrac{-1}{2} d_n$

46. $r_1 = 27, \quad r_{n+1} = \dfrac{2}{3} r_n$

47. $w_1 = 1, \quad w_n = 2w_{n-1}$

48. $q_1 = 7, \quad q_n = 3q_{n-1}$

49. An outdoor theater has 30 seats in the first row, 32 seats in the second row, 34 seats in the third row, and so on, with 2 more seats in any one row than in the previous. Let s_n be the number of seats in the nth row of the theater.

    a. Graph the first 5 terms of the sequence.

    b. How many seats are there in the fiftieth row?

50. Gwynn is training for a triathlon. In her first week of training, she bicycled a total of 20 miles, then increased to 24 miles in her second week, 28 miles in her third week, and so on, increasing her mileage by 4 miles per week. Let m_n be Gwynn's mileage in her n^{th} week of training.

    a. Graph the first 5 terms of the sequence.

    b. How many miles will she bicycle in her eighteenth week of training?

51. The cost of drilling a well increases the deeper you drill. Mel's Wells charges $50 for the first 5 feet, $55 for the second 5 feet, $60 for the third 5 feet, and so on. Let d_n be the cost of drilling the nth 5 feet.

    a. Graph the first 5 terms of the sequence.

    b. How much does Mel charge to drill from a depth of 65 feet to a depth of 70 feet?

52. The cost of having the windows washed in a high-rise office building depends on the height of the window. We Do Windows charges $0.50 per square yard of glass below 10 feet high, $0.60 per square yard for windows be-tween 10 and 20 feet above the ground, $0.70 per square yard for windows between 20 and 30 feet above the ground, and so on. Let w_n be the cost of washing windows on the nth floor if each floor is 10 feet high.

    a. Graph the first 5 terms of the sequence.

    b. How much does We Do Windows charge to wash 100 square yards of windows on the twenty-third floor?

53. Valerie's grandmother deposited $500 into a college fund for Valerie on the day she was born. The fund earns 5% interest compounded annually. Let V_n be the value of the deposit, including interest, n years later.

    a. Graph the first 5 terms of the sequence.

    b. How much will the deposit be worth on Valerie's eighteenth birthday?

54. When he was 25 years old, Bruce won $2000 on the lottery and deposited the money into a retirement fund. The fund earns 6% interest compounded annually. Let R_n be the value of the deposit, including interest, n years later.

    a. Graph the first 5 terms of the sequence.

    b. How much will the deposit be worth when Bruce turns 65 years old?

55. One hundred kilograms of a toxic chemical was dumped illegally into a clean reservoir. A filter can remove 20% of the chemical still present each week (so that 80% of the previous amount remains.) Let c_n be the amount of the chemical remaining after n weeks.

    a. Graph the first 5 terms of the sequence.

    b. How much of the chemical will remain in the water after 20 weeks?

56. A heart patient is given 40 milliliters of a medication by injection. Each hour, 15% of the medicine still present is eliminated from the body (so that 85% of the previous amount remains.) Let m_n be the amount of medicine remaining in the patient's body after n hours.

    a. Graph the first 5 terms of the sequence.

    b. How much of the medication is left in the patient's body after 10 hours?

9.3 Series

Often we are more interested in the *sum* of a sequence than in the sequence itself. For example, suppose that for five years you have made payments of $100 a month into a savings account that pays 12% annual interest compounded monthly. Then the sequence

$$p_n = 100(1.01)^n$$

gives the current value of the payment you made n months ago (plus interest). However, you are probably more interested in the *total* amount in your savings account, which is the *sum* of the terms p_n from $n = 1$ to $n = 60$.

The sum of the terms of a sequence is called a **series.** Although the words "sequence" and "series" are often used interchangeably in everyday English, they have different meanings in mathematics. A sequence is a *list* of numbers, whereas a series is a *single* number obtained by computing a sum. For example, the list of numbers

$$2, 4, 8, 16, 32$$

is a sequence. The *sum* of those five terms is the series

$$2 + 4 + 8 + 16 + 32 = 62$$

We use the symbol S_n to denote the sum of the first n terms of a sequence. Thus, for the sequence above, $S_5 = 62$.

EXAMPLE 1 Find the series S_6 for the sequence with general term $a_n = 3n + 1$.

Solution
The first 6 terms of the sequence are 4, 7, 10, 13, 16, and 19. Thus,

$$S_6 = 4 + 7 + 10 + 13 + 16 + 19 = 69 \qquad \blacksquare$$

> **EXERCISE 1** Find the series S_5 for the sequence with general term $b_n = n^2$.

Arithmetic Series

Obtaining a formula for the sum of the first n terms of a sequence is usually very difficult. But in the special case of an arithmetic sequence, it's not hard. As an example, consider the sum of the first 12 terms of the sequence with general term $a_n = 4n + 1$. That is,

$$S_{12} = 5 + 9 + 13 + \cdots + 41 + 45 + 49 \tag{1}$$

In mathematics, as in other fields, making discoveries often depends on looking at familiar objects in a new way. To find our formula for arithmetic series, we first write the terms of the series in the opposite order.

$$S_{12} = 49 + 45 + 41 + \cdots + 13 + 9 + 5 \qquad (2)$$

By adding Equations (1) and (2) term by term, we find

$$\begin{aligned} S_{12} &= 5 + 9 + 13 + \cdots 41 + 45 + 49 \\ \underline{S_{12}} &= \underline{49 + 45 + 41 + \cdots 13 + 9 + 5} \\ 2S_{12} &= 54 + 54 + 54 + \cdots 54 + 54 + 54 \end{aligned}$$

Each term of the sum is the same, namely, 54. The term 54 occurs 12 times (because we are adding 12 terms of the original sequence), so

$$2S_{12} = 12(54)$$

or

$$S_{12} = \frac{12(54)}{2} = 324$$

Notice that the number 54 is the sum of the first term of the arithmetic series, 5, and the last term, 49. In other words, $54 = a_1 + a_n$. This is the key observation we need to produce our formula. In general, this sum $a_1 + a_n$ occurs n times and results in twice the series we want. Thus,

$$2S_n = n(a_1 + a_n)$$

and dividing both sides by 2 gives us the following formula for computing an arithmetic series.

$$S_n = \frac{n}{2}(a_1 + a_n)$$

EXAMPLE 2 Find the sum of the first 15 odd integers.

Solution

The odd integers form an arithmetic sequence with first term $a_1 = 1$ and common difference 2. Thus, the general term of the sequence is $a_n = 1 + (n - 1)2$, and we would like to find the series S_{15}. We begin by finding the last term of the series, a_{15}.

$$a_{15} = 1 + (15 - 1) = 29$$

Next we use the formula for S_n with $n = 15$, $a_1 = 1$, and $a_{15} = 29$.

$$S_{15} = \frac{15}{2}(1 + 29) = 225 \qquad \blacksquare$$

EXERCISE 2 Find the sum of the first 100 terms in the arithmetic sequence 2, 5, 8, 11, ...

EXAMPLE 3

Arlene starts a new job in a print shop at a salary of $800 per month. If she keeps up with the training program her salary will increase by $35 per month. How much will Arlene have earned at the end of the 18-month training program?

Solution

Arlene's monthly salary is an arithmetic sequence with $a_1 = 800$ and $d = 35$. Thus, the general term of the sequence is $a_n = 800 + (n - 1)35$, and we would like to find the series S_{18}. We begin by finding the last term of the series, a_{18}.

$$a_{18} = 800 + (18 - 1)35 = 1395$$

Next we use the formula for S_n with $n = 18$, $a_1 = 800$, and $a_{18} = 1395$.

$$S_{18} = \frac{18}{2}(800 + 1395) = 19{,}755$$

Arlene's total earnings for the 18 months will be $19,755. ∎

Geometric Series

We can also find a formula for the sum of the first n terms of a geometric sequence. (Recall that the general term of a geometric sequence is $a_n = ar^{n-1}$.) As an example, we'll compute the sum of the first 9 terms of the sequence $a_n = 3^n$. (Observe that this is a geometric sequence with $a = 3$ and $r = 3$. Its general term is thus $a_n = 3(3)^{n-1}$, which is the same as 3^n.) Then

$$S_9 = 3 + 3^2 + 3^3 + \cdots + 3^8 + 3^9 \tag{3}$$

Instead of writing the terms in the opposite order, as we did for the arithmetic series, this time we will multiply each term of S_9 by the common ratio, 3.

$$3S_9 = 3(3) + 3(3)^2 + 3(3)^3 + \cdots + 3(3)^8 + 3(3)^9$$

or

$$3S_9 = 3^2 + 3^3 + 3^4 \cdots + 3^8 + 3^9 + 3^{10} \tag{4}$$

Next we subtract Equation (3) from Equation (4). Most of the terms will cancel out in the subtraction.

$$
\begin{array}{l}
3S_9 = \quad\quad \cancel{3^2} + \cancel{3^3} + \cdots + \cancel{3^8} + \cancel{3^9} + 3^{10} \\
\underline{-[S_9 = \quad 3 + \cancel{3^2} + \cancel{3^3} + \cdots + \cancel{3^8} + \cancel{3^9} \quad\quad]} \\
2S_9 = -3 \quad\quad\quad\quad\quad\quad\quad\quad\quad + 3^{10}
\end{array}
$$

Thus,

$$2S_9 = 3^{10} - 3$$

or

$$S_9 = \frac{3^{10} - 3}{2}$$

which simplifies to 29,523. The second term of the numerator is a_1 (3 in this example) and the first term is a_{10} ($3 \cdot 3^9$ or 3^{10}). The denominator of the expression, 2, is equal to $r - 1$. In general, we can derive the following formula for computing a geometric series.

$$S_n = \frac{a_{n+1} - a_1}{r - 1}$$

This formula is valid as long as the common ratio, r, is not equal to 1. In the Homework problems we will derive this formula and the formula for arithmetic series.

EXAMPLE 4

Find the sum of the first five terms of the sequence $5, \frac{10}{3}, \frac{20}{9}, \frac{40}{27}, \ldots$

Solution

The sequence is geometric with $a_1 = 5$. We find r by dividing $\frac{10}{3}$ by 5.

$$\frac{10}{3} \div 5 = \frac{2}{3}$$

Thus, the general term of the series is $a_n = 5\left(\frac{2}{3}\right)^{n-1}$. We use the formula for a geometric series with $n = 5$.

$$S_5 = \frac{a_6 - a_1}{r - 1}$$

where $a_1 = 5$, $a_6 = 5\left(\frac{2}{3}\right)^5$, and $r = \frac{2}{3}$. Thus

$$S_5 = \frac{5\left(\frac{2}{3}\right)^5 - 5}{\frac{2}{3} - 1} = \frac{5\left(\frac{32}{243} - 1\right)}{\frac{-1}{3}}$$

$$= 5\left(\frac{32 - 243}{243}\right)\left(\frac{-3}{1}\right) \approx 13.02 \qquad \blacksquare$$

EXERCISE 3   Find the sum of the first 10 terms in the geometric sequence $100, 110, 121, \ldots$.

EXAMPLE 5

Payam's starting salary as an engineer is $20,000 with a 5% annual raise for each of the next 5 years, depending on suitable progress. If Payam receives each salary increase, how much will he make over the next 6 years?

Solution

Payam's salary is multiplied each year by a factor of 1.05, so its values form a geometric sequence with a common ratio of $r = 1.05$. The general term of the sequence is $a_n = 20{,}000(1.05)^{n-1}$. His total income over the 6 years will be the sum of the first 6 terms of the sequence. Thus,

$$S_6 = \frac{a_7 - a_1}{r - 1} = \frac{20{,}000(1.05)^6 - 20{,}000}{1.05 - 1}$$
$$\approx 136{,}038.26$$

Payam will earn \$136,038.26 over the next 6 years. ■

1. What is the difference between a sequence and a series?
2. Describe a method for finding the sum of an arithmetic sequence.
3. Describe a method for finding the sum of a geometric sequence.
4. How can you decide whether a given sequence is arithmetic or geometric (or neither)?

ANSWERS TO 9.3 EXERCISES

1. 55   **2.** 15050   **3.** $1000(1.1^{10} - 1) \approx 1593.74$

HOMEWORK 9.3

Evaluate each arithmetic series.

1. The sum of the first 9 terms of the sequence $a_n = -4 + 3n$

2. The sum of the first 10 terms of the sequence $a_n = 5 - 2n$

3. The sum of the first 16 terms of the sequence $a_n = 18 - \dfrac{4}{3}n$

4. The sum of the first 13 terms of the sequence $a_n = -6 - \dfrac{1}{2}n$

5. The sum of the first 30 terms of the sequence $a_n = 1.6 + 0.2n$

6. The sum of the first 25 terms of the sequence $a_n = 2.5 + 0.3n$

Evaluate each geometric series.

7. The sum of the first five terms of $a_n = 2(-4)^{n-1}$

8. The sum of the first eight terms of $a_n = 12(3)^{n-1}$

9. The sum of the first nine terms of $a_n = -48\left(\dfrac{1}{2}\right)^{n-1}$

10. The sum of the first six terms of $a_n = 81\left(\dfrac{2}{3}\right)^{n-1}$

11. The sum of the first four terms of $a_n = 18(1.15)^{n-1}$

12. The sum of the first four terms of $a_n = 512(0.72)^{n-1}$

Identify each series as arithmetic, geometric, or neither, and then evaluate it.

13. $2 + 4 + 6 + \cdots + 96 + 98 + 100$

14. $1 + 3 + 5 + \cdots + 95 + 97 + 99$

15. $2 + 4 + 8 + 16 + \cdots + 256 + 512 + 1024$

16. $1 + 3 + 9 + 27 + \cdots + 6561 + 19{,}683$

17. $1 + 8 + 27 + 64 + 125 + 216 + 343$

18. $1 + 11 + 111 + 1111 + 11{,}111 + 111{,}111$

19. $87 + 84 + 81 + 78 + \cdots + 45 + 42 + 39$

20. $1 + (-2) + (-5) + \cdots + (-41) + (-44)$

21. $6 + 2 + \dfrac{2}{3} + \cdots + \dfrac{2}{81} + \dfrac{2}{243}$

22. $12 + 3 + \dfrac{3}{4} + \cdots + \dfrac{3}{64} + \dfrac{3}{256}$

Write a series to describe each problem, and then evaluate it.

23. Find the sum of all the even integers from 14 to 88.

24. Find the sum of all multiples of 7 from 14 to 105.

25. A clock strikes once at one o'clock, twice at two o'clock, and so on. How many times will the clock strike in a twelve-hour period?

26. Jessica puts one candle on the cake at her daughter's first birthday, two candles at her second birthday, and so on. How many candles will Jessica have used after her daughter's sixteenth birthday?

27. A rubber ball is dropped from a height of 24 feet and returns to three-fourths of its previous height on each bounce.

    a. How high does the ball bounce after hitting the floor for the third time?

    b. How far has the ball traveled vertically when it hits the floor for the fourth time?

28. A Yorkshire terrier can jump 3 feet into the air on his first bounce and five-sixths the height of his previous jump on each successive bounce.

    a. How high can the terrier go on his fourth bounce?

    b. How far has the terrier traveled vertically when he returns to the ground after his fourth bounce?

29. Sales of Brussels Sprouts dolls peaked at $920,000 in 1991 and began to decline at a steady rate of $40,000 per year. What total revenue should the manufacturer expect to gain from sale of the dolls from 1991 to 2000?

30. It takes Alida 20 minutes to type the first page of her term paper, but each subsequent page takes her 40 seconds less than the previous one. How long will it take her to type her 30-page paper?

31. A computer takes 0.1 second to perform the first iteration of a certain loop, and each subsequent iteration takes 0.05 second longer than the previous one. How long will it take the computer to perform 50 iterations?

32. Richard's water bill was $63.50 last month. If his bill increases by $2.30 per month, how much should he expect to pay for water during the next 10 months?

33. Sales of Energy Ranger dolls peaked at $920,000 in 1991 and began to decline by 8% per year. What total revenue should the manufacturer expect to gain from sale of the dolls from 1991 to 2000?

34. It takes Emily 20 minutes to type the first page of her term paper, but each subsequent page takes only 95% as long as the previous one. How long will it take her to type her 30-page paper?

35. A computer takes 0.1 second to perform the first iteration of a certain loop, and each subsequent iteration requires 20% longer than the previous one. How long will it take the computer to perform 50 iterations?

36. Megan's water bill was $63.50 last month. If her bill increases by 2% per month, how much should she expect to pay for water during the next 10 months?

37. Jim and Nora begin a college fund for their son David by depositing $500 into an account each year, beginning on the day David was born. If the account earns an interest rate of 5% compounded annually, how much will be in the account on David's eighteenth birthday?

38. Ben begins an Individual Retirement Account when he turns 25 years old, depositing $2000 into the account each year. If the account earns 6% interest compounded annually, how much will he have in the account when he turns 65 years old?

39. Suppose that you are given 1¢ on the first day of the month, 2¢ on the second day, 4¢ on the third day, and so on, each day's payment being twice the previous day's. What would be your total income on the thirtieth day?

40. According to legend, a man who had pleased the Persian king asked for the following reward. The man was to receive a single grain of wheat for the first square of a chessboard, two grains for the second square, four grains for the third square, and so on, doubling the amount for each square up to the sixty-fourth square. How many grains would he receive in all?

Derive formulas for arithmetic and geometric series by following the indicated steps.

41. In this problem we will find a formula for the sum of the arithmetic series

$$1 + 2 + 3 + \cdots + (N - 2) + (N - 1) + N$$

the sum of the first N positive integers.

a. How many terms are in the series?

b. If we call this sum S, then

$$S = 1 + 2 + 3 + \cdots + (N - 2) + (N - 1) + N$$

Rewrite the sum by reversing the order of the terms on the right side of the equation and add the result "columnwise" to the equation above. What is the sum of the two numbers in any *one* column on the right side?

c. From part (a) you know how many columns there are. Based on that number and your answer to part (b), write an expression for $2S$ and solve for S.

d. Use your answer from part (c) to write a formula for the sum of the arithmetic series

$$1 + 2 + 3 + \cdots + (N - 2) + (N - 1) + N$$

42. In this problem we will find a formula for the sum of an arithmetic series when we are given the first term F, the last term L, the common difference d, and the number of terms N.

a. If we call the sum S, then

$$S = F + (F + d) + (F + 2d) + \cdots + (L - 2d) + (L - d) + L$$

Rewrite the sum by reversing the order of the terms on the right side of the equation and add the result "columnwise" to the equation above. What is the sum of the two numbers in any *one* column on the right side?

b. You know that the original series has N terms. Based on your answer to part (b), write an expression for $2S$ and solve for S.

c. Use your answer from part (c) to write a formula for the sum of an arithmetic series when we are given the first term F, the last term L, the common difference d, and the number of terms N.

43. In this problem we will find a formula for the sum of a geometric series

$$1 + r + r^2 + \cdots + r^{N-1} + r^N$$

where r is an arbitrary constant.

a. What is the common ratio of consecutive terms?

b. If we call the sum A, then

$$A = 1 + r + r^2 + \cdots + r^{N-1} + r^N$$

Multiply both sides of the equation by the common ratio. What is the result?

c. Subtract your last equation from the equation for A given above and simplify. What is the result?

d. Starting with your answer to part (c), solve for A. (*Hint:* Start by factoring out A on the left side of the equation.)

e. Use your answer from part (d) to write a formula for

$$1 + r + r^2 + \cdots + r^{N-1} + r^N$$

44. In this problem we will find a formula for the sum of a geometric series

$$a + ar + ar^2 + \cdots + ar^{N-1} + ar^N$$

where a and r are arbitrary constants.

a. What is the common ratio of consecutive terms?

b. If we call this sum A, then

$$A = a + ar + ar^2 + \cdots + ar^{N-1} + ar^N$$

Multiply both sides of the equation by the common ratio. What is the result?

c. Subtract your last equation from the equation for A given above, simplify, and factor out A on the left side of the equation and a on the right. What is the result?

d. Starting with your answer to part (c), solve for A.

e. Use your answer from part (d) to write a formula for

$$a + ar + ar^2 + \cdots + ar^{N-1} + ar^N.$$

45. In this problem we will find a formula for the sum of an arithmetic series when we are given the first term F, the common difference d, and the number of terms N, but we are *not* given the last term.

a. Because the common difference is d, when we add d to a term we get the next term. The first term is F, so the second term is $F + d$, and the third term is

$$(F + d) + d = F + 2d$$

How can you express the third term? How can you express the ninth term?

b. How can you express the Nth (or last) term L of the series in terms of F, d, and N?

c. Now you know the first term F, the last term L, the common difference d, and the number of terms N. Use your answer to part (16) and the formula from Problem 42 in order to find the value of the series.

46. In this problem we will find a formula for the sum of a geometric series

$$ar^M + ar^{M+1} + ar^{M+2} + \cdots + ar^{N-1} + ar^N$$

where a and r are arbitrary constants. M and N are positive integers with $M < N$.

a. What is the common ratio of consecutive terms?

b. If we call this sum A, then

$$A = ar^M + ar^{M+1} + ar^{M+2}$$
$$+ \cdots + ar^{N-1} + ar^N$$

Multiply both sides of the equation by the common ratio. What is the result?

c. Subtract your last equation from the equation for A given above, simplify, and factor out an A on the left side of the equation and an a on the right. What is the result?

d. Starting with your answer to part (c), solve for A.

e. Use your answer from part (d) to write a formula for the value of the geometric series

$$ar^M + ar^{M+1} + ar^{M+2} + \cdots + ar^{N-1} + ar^N$$

MIDCHAPTER REVIEW

Write an equation using subscript notation for each statement.

1. The balance owed at month n was $200 less than 1.0025 times the balance owed at month $n - 1$. (Let B_n be the balance owed at month n.)

2. Each night's profit was $850 less than the profit on the previous night.

Find the first four terms in the sequence with the given general term.

3. $c_n = \dfrac{n(n + 1)(2n + 1)}{6}$

4. $d_k = (-1)^k k$

5. $r_i = 3$

6. $a_n = \dfrac{n - 1}{n + 1}$

For Problems 7–10,
a. Find the first 4 terms of the sequence.
b. Find a recursive definition of the sequence.
c. Find a formula for the general term.

7. The amount of caffeine in a pregnant woman's system decreases by 7% each hour. Suppose a pregnant woman has 100 mg of caffeine in her system at the start of the first hour. The sequence c_n gives the amount of caffeine in her system each hour.

8. The velocity of an object decreases by 32 feet/second each second that it is under the influence of gravity. Suppose that the object begins with an initial velocity of 100 feet/second. The sequence v_n gives the velocity, in feet/second, of the object each second.

9. a_n is an arithmetic sequence with first term -7 and common difference -6.

10. g_n is a geometric sequence with first term 12 and common ratio $\frac{1}{2}$.

For Problems 11–14, identify the sequence as arithmetic, geometric, or neither. State the common difference or common ratio when appropriate.

11. 58.4, 53.1, 47.8, 42.5, . . .

12. 250, 300, 360, 432

13. 1.1, 10.1, 100.1, 1000.1

14. $\frac{1}{2}, \frac{1}{3}, \frac{1}{4}, \frac{1}{5}$

Find the specified term.

15. Find the 8th term of the geometric sequence with first term 72 and common ratio $\frac{1}{2}$.

16. Find the 200th term of the arithmetic sequence with first term 8 and common difference 13.

Find the value of the given series.

17. Find the series S_5 for the sequence with general term $p_n = 1000(1.1)^n$.

18. Find the series S_{10} for the sequence with general term $b_n = 8$.

19. Find the series S_{15} for the sequence with general term $q_n = 3.42 + 2.16n$.

20. Find the series S_6 for the sequence with general term $s_n = 3n^2$.

9.4 Sigma Notation and Infinite Geometric Series

Summation Notation

There is a convenient notation for representing a series when we know an expression for the general term. For example, suppose we would like to sum the first fifteen terms of the sequence

$$4, 7, 10, \ldots, 3n + 1, \ldots$$

Because the terms of the series are obtained by replacing n in the general term $3n + 1$ by the numbers 1 through 15, we might express the sum as

The sum, as n runs from 1 to 15, of $3n + 1$

Thus, instead of writing out all the terms of the series, we merely indicate which terms are to be included.

To abbreviate the expression further we use the Greek letter Σ (called "sigma") to stand for "the sum." We indicate the first and last values of n below and above the summation symbol Σ as shown below.

$$S_{15} = \sum_{n=1}^{15} (3n + 1)$$

The letter n is called the **index of summation;** it is like a variable because it represents numerical values. Any letter can be used for the index of summation; i, j, and k are other common choices. Of course, the letter used for the index of summation does not affect the sum.

EXAMPLE 1

Use sigma notation to represent the sum of the first 20 terms of the sequence

$$-1, 2, 7, \ldots, k^2 - 2, \ldots$$

Solution

The general term of the sequence is $k^2 - 2$, and the first term is -1, which is obtained by letting $k = 1$ in the formula for the general term. Thus,

$$S_{20} = \sum_{k=1}^{20} (k^2 - 2)$$

∎

EXERCISE 1   Use sigma notation to represent the sum of the first 50 terms of the sequence

$$5, 8, 11 \ldots, 3k + 2, \ldots$$

The **expanded form** of a series written in sigma notation is obtained by writing out all the terms of the series. For example,

$$\sum_{m=4}^{8} \frac{3}{m} = \frac{3}{4} + \frac{3}{5} + \frac{3}{6} + \frac{3}{7} + \frac{3}{8}$$

The series above has five terms, which is one more than the difference of the upper and lower limits of summation.

Recall that the general term of an arithmetic series is a linear function of the index, and the general term of a geometric series is an exponential function. If we recognize a given series as one of these two types, we can use the formulas developed in the last section to evaluate the sum.

EXAMPLE 2

Compute the value of each series.

a. $\displaystyle\sum_{i=1}^{13} (90 - 5i)$ **b.** $\displaystyle\sum_{k=0}^{9} 2^k$

Solutions

   a.   The general term $90 - 5i$ is linear, so this is an arithmetic series. By writing out the first few terms of the series,

$$85 + 80 + 75 + 70 + \cdots$$

   we can verify that the first term of the series is $a_1 = 85$ and the common difference is $d = -5$. In order to use the formula for arithmetic series, we also need to know the last term of the series. We substitute $n = 13$ in the general term to find $a_{13} = 90 - 5(13) = 25$. Thus,

$$\sum_{i=1}^{13} (90 - 5i) = \frac{13}{2}(85 + 25) = 715$$

b. The general term of this series is exponential, so the series is geometric. By writing out a few terms of the series,

$$1 + 2 + 4 + 8 + \cdots$$

we confirm that the first term of the series is $a = 1$ and the common ratio is $r = 2$. The series has 10 terms, from $k = 0$ to $k = 9$. Finally, we substitute these values into the formula for geometric series to obtain

$$\sum_{k=0}^{9} 2^k = \frac{2^{10} - 1}{2 - 1} = 1023 \qquad \blacksquare$$

EXERCISE 2   Compute the value of each series.

a. $\displaystyle\sum_{k=1}^{50} (3k + 2)$ b. $\displaystyle\sum_{n=1}^{8} 10^n$

If a given series is not arithmetic or geometric, we must use direct methods to compute the sum.

EXAMPLE 3

Compute the value of each series.

a. $\displaystyle\sum_{m=1}^{5} m^2$

b. $\displaystyle\sum_{p=1}^{800} 5$

Solutions

a. Because the general term m^2 is neither linear nor exponential, we know that the series is not arithmetic or geometric. However, we can expand the series and evaluate it directly.

$$\sum_{m=1}^{5} m^2 = 1^2 + 2^2 + 3^2 + 4^2 + 5^2$$
$$= 1 + 4 + 9 + 16 + 25 = 55$$

b. The general term is the number 5. Because the index p runs from 1 to 800, we are adding 800 terms, each of which is 5. Thus

$$\sum_{p=1}^{800} 5 = 5 + 5 + 5 + \cdots + 5 + 5 + 5$$
$$= 800(5) = 4000 \qquad \blacksquare$$

EXERCISE 3   Compute the value of each series.

a. $\displaystyle\sum_{k=0}^{20} \frac{1}{3}$

b. $\displaystyle\sum_{k=0}^{4} \frac{k}{k + 1}$

Infinite Series

A series with infinitely many terms is called an **infinite series.** Is it possible to add infinitely many terms and arrive at a finite sum? In some cases, if the terms added are small enough, the answer is yes. Consider the infinite geometric series

$$\frac{1}{2} + \frac{1}{4} + \frac{1}{8} + \frac{1}{16} + \cdots$$

The nth **partial sum** of the series is the sum of its first n terms, and is denoted by S_n. Thus,

$$S_1 = \frac{1}{2}$$

$$S_2 = \frac{1}{2} + \frac{1}{4} = \frac{3}{4}$$

$$S_3 = \frac{1}{2} + \frac{1}{4} + \frac{1}{8} = \frac{7}{8}$$

$$S_4 = \frac{1}{2} + \frac{1}{4} + \frac{1}{8} + \frac{1}{16} = \frac{15}{16}$$

$$\vdots$$

The new term added for each partial sum gets smaller and smaller: $\frac{1}{2}, \frac{1}{4}, \frac{1}{8}, \frac{1}{16}, \cdots$. As n increases—as we add more and more terms of the series—the partial sums appear to be "approaching" 1. That is, as n becomes very large S_n gets very close to 1. It seems reasonable that the sum of *all* the terms of the series is 1.

We can make this conjecture more plausible by examining the formula for the nth partial sum of a geometric series.

$$S_n = \frac{a_{n+1} - a_1}{r - 1}$$

$$= \frac{ar^n - a}{r - 1} = \frac{a - ar^n}{1 - r} \qquad (r \neq 1)$$

Concentrate on the second term of the numerator, ar^n. This term is the only part of the formula that depends on n. What happens to ar^n as n increases? Consider two examples: If $r = \frac{1}{2}$, then

$$r^2 = \left(\frac{1}{2}\right)^2 = \frac{1}{4}, \qquad r^3 = \left(\frac{1}{2}\right)^3 = \frac{1}{8}, \qquad r^4 = \left(\frac{1}{2}\right)^4 = \frac{1}{16}$$

and so on, with $\left(\frac{1}{2}\right)^n$ becoming smaller and smaller for larger values of n. Each time we multiply by another factor of $r = \frac{1}{2}$, the product gets smaller, because $\frac{1}{2} < 1$. On the other hand, if $r > 1$, multiplying by r makes the product larger. If $r = \frac{3}{2}$ for example, then

$$r^2 = \left(\frac{3}{2}\right)^2 = \frac{9}{4}, \qquad r^3 = \left(\frac{3}{2}\right)^3 = \frac{27}{8}, \qquad r^4 = \left(\frac{3}{2}\right)^4 = \frac{81}{16}$$

and so on. In this case the powers of r are increasing. In general, we have the following result.

1. If $0 < |r| < 1$, then r^n gets closer to zero as n increases.
2. If $|r| > 1$, then r^n does not approach a finite number as n increases.

Now return to the formula for S_n and write it in the form

$$S_n = \frac{a}{1 - r}(1 - r^n)$$

where we have factored a from the numerator. If $|r| < 1$, then r^n approaches 0 and the factor $(1 - r^n)$ gets closer and closer to 1 as n grows larger. Consequently, if we compute S_n for larger and larger values of n, the sum approaches the value

$$\frac{a}{1 - r}$$

This analysis motivates us to define the sum of an infinite geometric series as follows.

$$S_\infty = \frac{a}{1 - r} \quad \text{if} \quad -1 < r < 1$$

If $|r| \geq 1$, as in the infinite series

$$3 + 6 + 12 + 24 + \cdots$$

where $r = 2$, the terms become larger as n increases and the sum of the series is not a finite number. In this case the series does not have a sum.

EXAMPLE 4

Make a table showing the first five partial sums of each series. Then use the formula to find the sum, if it exists.

a. $\displaystyle\sum_{j=0}^{\infty} 30(0.8)^j$ **b.** $\displaystyle\sum_{m=0}^{\infty} 3\left(\frac{4}{3}\right)^m$

Solutions

a. Evaluate the general term $30(0.8)^j$ for $j = 0, 1, 2, 3, 4$ and compute the partial sums. For example,

$$S_1 = 30(0.8)^0 = 30$$
$$S_2 = 30(0.8)^0 + 30(0.8)^1 = 54$$

and so on. The first five partial sums, rounded to hundredths, are shown in the table.

n	1	2	3	4	5
S_n	30	54	73.2	88.56	100.85

To compute the sum algebraically, note that this is an infinite geometric series with $a = 30$ and $r = 0.8$. The series has a sum because $|r| < 1$. Thus,

$$S_\infty = \frac{a}{1 - r} = \frac{30}{1 - 0.8} = 150$$

b. Evaluate the general term $3\left(\frac{4}{3}\right)^m$ for $m = 0, 1, 2, 3, 4$ and compute the partial sums. For example,

$$S_1 = 3\left(\frac{4}{3}\right)^0 = 3$$

$$S_2 = 3\left(\frac{4}{3}\right)^0 + 3\left(\frac{4}{3}\right)^1 = 7$$

and so on. The first five partial sums, rounded to hundredths, are shown in the table.

n	1	2	3	4	5
S_n	3	7	12.33	19.44	28.93

This is a geometric series with $a = 3$ and $r = \frac{4}{3}$. The infinite series does not have a sum because $|r| > 1$. ∎

EXERCISE 4   Find the sum, if it exists, or state that the series does not have a sum.

a. $\displaystyle\sum_{j=0}^{\infty} 13\left(\frac{7}{6}\right)^j$

b. $\displaystyle\sum_{m=0}^{\infty} 5.9(0.9)^m$

Repeating Decimals

An interesting application of geometric series involves repeating decimals. Recall that the decimal representation of a rational number either terminates, as does 0.75, or repeats a pattern of digits. For example, you probably recognize 0.3333 as the decimal representation of $\frac{1}{3}$. It is easy to find the decimal form of a fraction: Just divide the denominator into the numerator. Is there a way to find the fractional form of a repeating decimal?

Consider the repeating decimal

$$0.21\overline{21}$$

We will first write this number as an infinite geometric series.

$$0.21 + 0.0021 + 0.000021 + \cdots$$

The first term of this series is $a = 0.21$, or $\frac{21}{100}$, and its common ratio is $r = 0.01$, or $\frac{1}{100}$. Because $|r| < 1$, the series has a sum given by

$$S_\infty = \frac{a}{1-r} = \frac{\frac{21}{100}}{1 - \frac{1}{100}}$$

$$= \frac{\frac{21}{100}}{\frac{99}{100}} = \frac{21}{99} = \frac{7}{33}$$

Thus, the decimal number $0.2121\overline{21}$ is equal to the fraction $\frac{7}{33}$.

EXAMPLE 5   Find a fraction equivalent to $0.3\overline{7}$.

Solution

The decimal can be written as $0.3 + 0.0\overline{7}$. We will find a fraction equivalent to the repeating decimal $0.0\overline{7}$ and add that to 0.3, or $\frac{3}{10}$. Write $0.0\overline{7}$ as a series.

$$0.0\overline{7} = \frac{7}{100} + \frac{7}{1000} + \frac{7}{10,000} + \cdots$$

This is an infinite geometric series with first term $a = \frac{7}{100}$ and common ratio $r = \frac{1}{10}$. The sum of the series is given by

$$S_\infty = \frac{a}{1-r} = \frac{\frac{7}{100}}{1 - \frac{1}{10}}$$

$$= \frac{\frac{7}{100}}{\frac{9}{10}} = \frac{7}{100} \cdot \frac{10}{9} = \frac{7}{90}$$

Thus,

$$0.3\overline{7} = \frac{3}{10} + \frac{7}{90} = \frac{17}{45} \qquad\blacksquare$$

EXERCISE 5   Find a fraction equivalent to $0.\overline{8}$.

READING QUESTIONS

1. Explain how to use sigma notation to define a series.
2. What is an infinite series?
3. What is a partial sum of an infinite series?
4. State a formula for evaluating an infinite geometric series. Under what conditions is the formula valid?
5. A repeating decimal can be rewritten as what kind of series?

ANSWERS TO 9.4 EXERCISES

1. $\displaystyle\sum_{k=1}^{50} (3k + 2)$ **2a.** 3925   **c.** 111,111,110

3a. 7   **b.** $\dfrac{23}{12}$

4a. no sum   **b.** 59   **5.** $\dfrac{8}{9}$

Write each sum in expanded form.

1. $\displaystyle\sum_{i=1}^{4} i^2$

2. $\displaystyle\sum_{i=1}^{3} (3i - 2)$

3. $\displaystyle\sum_{j=5}^{7} (j - 2)$

4. $\displaystyle\sum_{j=2}^{6} (j^2 + 1)$

5. $\displaystyle\sum_{k=1}^{4} k(k + 1)$

6. $\displaystyle\sum_{k=2}^{6} \frac{k}{2}(k + 1)$

7. $\displaystyle\sum_{m=1}^{4} \frac{(-1)^m}{2^m}$

8. $\displaystyle\sum_{m=3}^{5} \frac{(-1)^{m+1}}{m - 2}$

Write each series using sigma notation.

9. $1 + 3 + 5 + 7$

10. $2 + 4 + 6 + 8$

11. $5 + 5^3 + 5^5 + 5^7$

12. $4^3 + 4^5 + 4^7 + 4^9 + 4^{11}$

13. $1 + 4 + 9 + 16 + 25$

14. $1 + 8 + 27 + 64 + 125$

15. $\dfrac{1}{2} + \dfrac{2}{3} + \dfrac{3}{4} + \dfrac{4}{5} + \dfrac{5}{6}$

16. $\dfrac{2}{1} + \dfrac{3}{2} + \dfrac{4}{3} + \dfrac{5}{4} + \dfrac{6}{5}$

17. $\dfrac{1}{1} + \dfrac{2}{3} + \dfrac{3}{5} + \dfrac{4}{7} + \dfrac{5}{9} + \dfrac{6}{11}$

18. $\dfrac{3}{1} + \dfrac{5}{3} + \dfrac{7}{5} + \dfrac{9}{7} + \dfrac{11}{9}$

19. $\dfrac{1}{1} + \dfrac{2}{2} + \dfrac{4}{3} + \dfrac{8}{4} + \cdots$

20. $\dfrac{1}{2} + \dfrac{3}{4} + \dfrac{9}{6} + \dfrac{27}{8} + \cdots$

Identify each series as arithmetic, geometric, or neither and then evaluate.

21. $\displaystyle\sum_{i=1}^{6} (i^2 + 1)$

22. $\displaystyle\sum_{i=1}^{5} 3i^2$

23. $\displaystyle\sum_{j=1}^{4} \frac{1}{j}$

24. $\displaystyle\sum_{j=0}^{4} \frac{2}{j + 1}$

25. $\displaystyle\sum_{k=1}^{100} 1$

26. $\displaystyle\sum_{k=13}^{300}$

27. $\displaystyle\sum_{q=1}^{20} 3q$

28. $\displaystyle\sum_{p=1}^{30} 2p$

29. $\displaystyle\sum_{k=1}^{200} k$

30. $\displaystyle\sum_{k=1}^{150} k$

31. $\displaystyle\sum_{n=1}^{6} n^3$

32. $\displaystyle\sum_{n=1}^{7} n^2$

33. $\displaystyle\sum_{n=0}^{30} (3n - 1)$

34. $\displaystyle\sum_{k=0}^{20} (5k + 2)$

35. $\displaystyle\sum_{k=0}^{25} (5 - 2k)$

36. $\displaystyle\sum_{p=0}^{15} (2 - 3p)$

37. $\displaystyle\sum_{j=0}^{10} 2 \cdot 5^j$

38. $\displaystyle\sum_{j=0}^{10} 3 \cdot 2^j$

39. $\displaystyle\sum_{m=0}^{12} 50(1.08)^m$

40. $\displaystyle\sum_{m=0}^{18} 300(1.12)^m$

For Problems 41–48, (a) make a table of values showing the first five partial sums of the series, (b) evaluate the series algebraically.

41. $\displaystyle\sum_{n=1}^{\infty} \left(\frac{1}{2}\right)^n$

42. $\displaystyle\sum_{n=1}^{\infty} \left(\frac{1}{3}\right)^n$

43. $\displaystyle\sum_{k=1}^{\infty} 12(0.15)^{k-1}$

44. $\displaystyle\sum_{k=1}^{\infty} 25(0.08)^{k-1}$

45. $\displaystyle\sum_{j=0}^{\infty} 4\left(\frac{-3}{5}\right)^j$

46. $\displaystyle\sum_{j=0}^{\infty} 6\left(\frac{-2}{5}\right)^j$

47. $\displaystyle\sum_{n=4}^{\infty} 3\left(\frac{1}{2}\right)^n$

48. $\displaystyle\sum_{n=3}^{\infty} 2\left(\frac{1}{3}\right)^n$

In Problems 49–56, find a fraction equivalent to each of the given repeating decimals.

49. $0.\overline{4}$

50. $0.\overline{6}$

51. $0.3\overline{1}$

52. $0.\overline{45}$

53. $2.4\overline{10}$

54. $3.0\overline{27}$

55. $0.12\overline{8}$

56. $0.8\overline{3}$

57. The arc length through which the bob of a pendulum moves is nine-tenths of its preceding arc length. Approximately how far will the bob move before coming to rest if the first arc length is 12 inches?

58. A force is applied to a particle moving in a straight line in such a fashion that each second it moves only one-half of the distance it moved the preceding second. If the particle moves 10 centimeters the first second, approximately how far will it move before coming to rest?

59. A ball returns to two-thirds of its preceding height on each bounce. If the ball is dropped from a height of 6 feet, approximately what is the total distance the ball travels before coming to rest? (*Hint:* Do separate computations for the total distance the ball falls and the total distance it moves upward.)

60. If a ball is dropped from a height of 10 feet and returns to three-fifths of its preceding height on each bounce, approximately what is the total distance the ball travels before coming to rest? (See the hint for Problem 59.)

9.5 The Binomial Expansion

In Chapter 8 we studied products of polynomials, and in particular we found expanded forms for powers of binomials such as $(a + b)^2$ and $(a + b)^3$. The amount of work involved in expanding such powers increases as the exponent gets larger. In this section we will learn how to raise a binomial to any positive integer power, without having to perform the polynomial products.

INVESTIGATION 15

Powers of Binomials

In this Investigation we will look for patterns in the expansion of $(a + b)^n$. We begin by computing a number of such powers.

Expand each power and fill in the blanks. Arrange the terms in each expansion in descending powers of a.

1. $(a + b)^0 = $ _____.

2. $(a + b)^1 = $ _____.

3. $(a + b)^2 = $ _____.

4. $(a + b)^3 = $ _____.

(*Hint:* Start by writing

$$(a + b)^3 = (a + b)(a + b)^2$$

and use your answer to Step 3.)

5. $(a + b)^4 = $ _____.

(*Hint:* Start by writing

$$(a + b)^4 = (a + b)(a + b)^3$$

and use your answer to Step 4.)

6. $(a + b)^5 = $ _____.

7. Do you see a relationship between the exponent n and the number of terms in the expansion of $(a + b)^n$? (Notice that for $n = 0$ we have $(a + b)^0 = 1$, which has one term.) Fill in Table 9.7.

TABLE 9.7

n	Number of terms in $(a + b)^n$
0	
1	
2	
3	
4	
5	

8. **First observation:** In general, the expansion of $(a + b)^n$ has _____ terms.

9. Next we'll consider the exponents on a and b in each term of the expansions. Refer to your expanded powers in parts 1–5, and fill in Table 9.8.

TABLE 9.8

n	First term of $(a + b)^n$	Last term of $(a + b)^n$	Sum of exponents on a and b in each term
0			
1			
2			
3			
4			
5			

10. **Second observation:** In any term of the expansion of $(a + b)^n$, the sum of the exponents on a and on b is _____ .

11. In fact, we can be more specific in describing the exponents in the expansions. We will use k to label the *terms* in the expansion of $(a + b)^n$, starting with $k = 0$. For example, for $n = 2$ we label the terms as follows:

$$(a + b)^2 = \underset{k = 0}{a^2} + \underset{k = 1}{2ab} + \underset{k = 2}{b^2}$$

We can make a table showing the exponents on a and b in each term of $(a + b)^2$. (See Table 9.9.)

Case $n = 2$:   TABLE 9.9

k	0	1	2
Exponent on a	2	1	0
Exponent on b	0	1	2

Complete the following tables for the cases $n = 3$, $n = 4$, and $n = 5$.

Case $n = 3$:   TABLE 9.10

k	0	1	2	3
Exponent on a	3			
Exponent on b	0			

Case $n = 4$:   TABLE 9.11

k	0	1	2	3	4
Exponent on a	4				
Exponent on b	0				

Case $n = 5$: **TABLE 9.12**

k	0	1	2	3	4	5
Exponent on a	5					
Exponent on b	0					

12. **Third observation:** The variable factors of the kth term in the expansion of $(a + b)^n$ may be expressed as _____ . (Fill in the correct powers in terms of n and k for a and b.)

In the next Investigation we will look for patterns in the *coefficients* of the terms of the expansions.

Powers of Other Binomials

We can use what we learned from Investigation 15 to raise other binomials to powers.

EXAMPLE 1

Expand each of the following.

    a. $(x + 1)^3$ **b.** $(2m - n)^4$

Solutions

    a. We know from Investigation 15 that

$$(a + b)^3 = a^3 + 3a^2b + 3ab^2 + b^3$$

Replace the a with x and the b with 1 to get

$$(x + 1)^3 = (x)^3 + 3(x)^2(1) + 3(x)(1)^2 + (1)^3$$
$$= x^3 + 3x^2 + 3x + 1$$

    b. This time we take the expansion

$$(a + b)^4 = a^4 + 4a^3b + 6a^2b^2 + 4ab^3 + b^4$$

from Investigation 15, and let $a = 2m$ and $b = -n$.

$$(2m - n)^4 = (2m)^4 + 4(2m)^3(-n) + 6(2m)^2(-n)^2 + 4(2m)(-n)^3 + (-n)^4$$
$$= 16m^4 + 4(8m^3)(-n) + 6(4m^2)(n^2) + 4(2m)(-n^3) + n^4$$
$$= 16m^4 - 32m^3n + 24m^2n^2 - 8mn^3 + n^4 \qquad \blacksquare$$

EXERCISE 1 Expand $(r + 2s)^3$.

The Binomial Coefficient

Even without carrying out the multiplication, we know from our investigation that the expansion of $(a + b)^6$ should have 7 terms, with the exponents on a ranging from 6 down to 0 and the exponents on b ranging from 0 up to 6. The expansion should have the form

$$(a + b)^6 = \_\,a^6 + \_\,a^5b + \_\,a^4b^2 + \_\,a^3b^3 + \_\,a^2b^4 \_\,ab^5 + \_\,b^6$$

We use the notation $_nC_k$ for the coefficient of the kth term in the expansion of $(a + b)^n$. For example, the coefficient of a^6 in the expansion above is denoted by $_6C_0$, the coefficient of a^5b^1 is $_6C_1$, the coefficient of a^4b^2 is $_6C_2$, and so on. Note that the 6 in front of the C indicates that $n = 6$, and the number following C corresponds to the exponent on b. The symbol $_nC_k$ is called the **binomial coefficient.**

> The **binomial coefficient** $_nC_k$ is the coefficient of the term containing b^k in the expansion of $(a + b)^n$.

EXAMPLE 2

Use your expansions from Investigation 15 to evaluate the following binomial coefficients.

    a. $_4C_3$ **b.** $_5C_2$

Solutions

  a. $_4C_3$ is the coefficient of the term containing b^3 in the expansion of $(a + b)^4$. Referring to Step 5 in Investigation 15, we see that the coefficient of the term $4ab^3$ is 4. Thus, $_4C_3 = 4$.

  b. $_5C_2$ is the coefficient of the term containing b^2 in the expansion of $(a + b)^5$. Referring to Step 6 in Investigation 15, we see that the coefficient of the term $10a^3b^2$ is 10. Thus, $_5C_2 = 10$. ∎

> **EXERCISE 2** Evaluate the binomial coefficients.
>
> **a.** $_5C_4$ **b.** $_5C_0$

Pascal's Triangle

To get a clearer picture of the binomial coefficients, consider again the expansions of $(a + b)^n$ you calculated in Investigation 15, but this time look only at the numerical coefficients of each term.

$n = 0$				1			
$n = 1$				1 1			
$n = 2$			1 2 1				
$n = 3$		1 3 3 1					
$n = 4$	1 4 6 4 1						
$n = 5$	1 5 10 10 5 1						

This triangular array of numbers is known as **Pascal's triangle.** It has many interesting and surprising properties that have been extensively studied. We might first make the following observations.

1. Each row of Pascal's triangle begins with the number ___ and ends with the number ___.

2. The second number and the next-to-last number in the nth row are ___.

There is an interesting pattern for the rest of the numbers in each row. Pick any number in the row $n = 4$, and look at the two closest numbers in the previous row. (For example, if you picked 6, the two closest numbers in the previous row are 3 and 3.) Do you see a relationship between the numbers? Try the same thing with several numbers in row $n = 5$ to test your theory.

3. Starting with the row $n = 2$, any number in the triangle (except the first and last 1s in each row) can be found by _____.

4. Using your answer to Step 3, continue Pascal's triangle to include the row for $n = 6$.

The numbers in Pascal's triangle are the binomial coefficients we have been looking for. Specifically, the number in the kth position (starting with $k = 0$) of the nth row of the triangle is $_nC_k$. Thus, we can use the numbers in Pascal's triangle to expand $(a + b)^n$.

5. Use Pascal's triangle to find the binomial coefficient $_6C_4$.

6. Expand: $(a + b)^6 = $ _____

7. Expand: $(x - 2)^6 = $ _____
   (*Hint:* Use your answer to Step 6, replacing a by x and b by -2.)

8. Continue Pascal's triangle to include the row for $n = 7$.

Using Pascal's Triangle

Combining what we learned in Investigation 15 about the form of an expansion with our method for finding the coefficients from Pascal's triangle, we can now write the expanded form of any binomial power without having to perform the multiplication.

EXAMPLE 3

Find the expanded form of $(3r - q)^6$ without performing the multiplication.

Solution

Start by writing down the expansion of $(a + b)^6$, using your knowledge from Investigation 14.

$$(a + b)^6 = a^6 + \underline{\quad} a^5b + \underline{\quad} a^4b^2 + \underline{\quad} a^3b^3 + \underline{\quad} a^2b^4 + \underline{\quad} ab^5 + b^6$$

Next, fill in the blanks with the binomial coefficients from Pascal's triangle. Consult Step 4 of Investigation 16 to obtain

$$(a + b)^6 = a^6 + 6a^5b + 15a^4b^2 + 20a^3b^3 + 15a^2b^4 + 6ab^5 + b^6$$

Finally, replace a by $3r$ and replace b by $-q$ and simplify to find

$$(3r - q)^6 = (3r)^6 + 6(3r)^5(-q) + 15(3r)^4(-q)^2 + 20(3r)^3(-q)^3 + 15(3r)^2(-q)^4 + 6(3r)(-q)^5 + (-q)^6$$
$$= 729r^6 - 14584^5q + 1215r^4q^2 - 540r^3q^3 + 135r^2q^4 - 18rq^5 + q^6 \quad \blacksquare$$

EXERCISE 3 Expand $(w + 10z)^6$.

Factorial Notation

Although we can find the binomial coefficients in the expansion of $(a + b)^n$ for any exponent n by extending Pascal's triangle far enough, this process is tedious for large values of n. In such cases it would be more convenient to have a formula for calculating the binomial coefficients directly. For this formula we need a special symbol ***n*!** (read "***n* factorial**"), which is defined for nonnegative integers n as follows.

$$n! = n \cdot (n - 1) \cdot (n - 2) \cdot \cdots \cdot 3 \cdot 2 \cdot 1$$

For example,

$$6! = 6 \cdot 5 \cdot 4 \cdot 3 \cdot 2 \cdot 1 = 720 \qquad \text{and} \qquad 4! = 4 \cdot 3 \cdot 2 \cdot 1 = 24$$

The factorial symbol applies only to the variable or number it follows; for example, $3 \cdot 4!$ is not equal to $(3 \cdot 4)!$

EXAMPLE 4

Write each expression in expanded form and simplify.

   a. $2n!$ for $n = 4$ **b.** $(2n - 1)!$ for $n = 4$

Solutions

   a. $2n! = 2 \cdot 4! = 2(4 \cdot 3 \cdot 2 \cdot 1) = 48$

   b. $(2n - 1)! = 7! = 7 \cdot 6 \cdot 5 \cdot 4 \cdot 3 \cdot 2 \cdot 1 = 5040$ ■

EXAMPLE 5

Write $(3n + 1)!$ in factored form, showing the first three factors and the last three factors.

Solution

$$(3n + 1)! = (3n + 1)(3n)(3n - 1) \cdot \cdots \cdot 3 \cdot 2 \cdot 1$$ ■

   Note that

$$7! = 7 \cdot (6 \cdot 5 \cdot 4 \cdot 3 \cdot 2 \cdot 1) = 7 \cdot 6!$$

and

$$5! = 5 \cdot (4 \cdot 3 \cdot 2 \cdot 1) = 5 \cdot 4!$$

In general, we have the following relationship.

$$n! = n(n - 1)! \tag{1}$$

This relationship can be helpful in simplifying expressions involving factorials.

EXAMPLE 6

Write each expression in expanded form and simplify.

   a. $\dfrac{6!}{3!}$ **b.** $\dfrac{4!6!}{8!}$

Solutions

a. $\dfrac{6!}{3!} = \dfrac{6 \cdot 5 \cdot 4 \cdot 3!}{3!} = 6 \cdot 5 \cdot 4 = 120$

b. $\dfrac{4!6!}{8!} = \dfrac{4 \cdot 3 \cdot 2 \cdot 1 \cdot 6!}{8 \cdot 7 \cdot 6!} = \dfrac{3}{7}$ ■

COMMON ERRORS In Example 6a, notice that $\dfrac{6!}{3!}$ is *not* equal to 2!. ■

> **EXERCISE 4** Write $\dfrac{5!}{3!2!}$ in expanded form and simplify.

The Binomial Coefficient in Factorial Notation

Our definition for $n!$ only makes sense when n is a positive integer. We will also define $n!$ for $n = 0$. In order to be consistent with Equation (1) we must have

$$1! = 1 \cdot (1 - 1)!$$

or

$$1! = 1 \cdot 0!$$

This leads us to make the following definition.

$$0! = 1$$

Note that both 1! and 0! are equal to 1.

Earlier we introduced the notation $_nC_k$ for the numerical coefficients in the expansion of $(a + b)^n$. These binomial coefficients, which are given by Pascal's triangle, can also be expressed using factorial notation as follows.

$$_nC_k = \dfrac{n!}{(n - k)!k!}$$

EXAMPLE 7 Evaluate each binomial coefficient.

a. $_6C_2$ **b.** $_9C_8$

Solutions

a. $_6C_2 = \dfrac{6!}{(6 - 2)! \, 2!}$

$= \dfrac{6!}{4! \, 2!} = \dfrac{6 \cdot 5 \cdot 4!}{(4!)(2 \cdot 1)} = 15$

b. $\displaystyle {}_9C_8 = \frac{9!}{(9-8)!\,8!}$

$$= \frac{9!}{1!\,8!} = \frac{9 \cdot 8!}{1 \cdot 8!} = 9 \qquad \blacksquare$$

EXERCISE 5 Evaluate the binomial coefficient ${}_8C_6$.

You might want to check that the formula works for $k = 0$ when we use $0! = 1$. You should find that

$$_nC_0 = \frac{n!}{(n-0)!(0)!} = \frac{n!}{n!(1)} = 1$$

In other words, the formula correctly gives us the coefficient of a^n in the expansion of $(a + b)^n$.

We now have two methods for computing the binomial coefficient $_nC_k$: Pascal's triangle and our formula using factorials. When n is small, especially if all the coefficients of the binomial expansion are needed, Pascal's triangle is often the easier choice. But when n is large or if only one coefficient is required, the factorial formula for $_nC_k$ is probably quicker than Pascal's triangle.

EXAMPLE 8

Find the coefficient of $m^{11}n^3$ in the expansion of $(m + n)^{14}$.

Solution

Because we are simply replacing a with m and b with n in the expansion of $(a + b)^{14}$, the term containing $m^{11}n^3$ has $n = 14$ and $k = 3$. Thus, the coefficient of $m^{11}n^3$ is $_{14}C_3$.

$$_{14}C_3 = \frac{14!}{(14-3)!3!} = \frac{14 \cdot 13 \cdot 12 \cdot 11!}{(11)!\,3!} = \frac{14 \cdot 13 \cdot 12}{3 \cdot 2 \cdot 1} = 364$$

The coefficient of $m^{11}n^3$ is 364. $\qquad \blacksquare$

EXERCISE 6 Find the coefficient of s^3t^4 in the expansion of $(s + t)^7$.

The Binomial Theorem

Consider the expansion of $(a + b)^6$ written with the $_nC_k$ notation[1] for the coefficients.

$$(a + b)^6 = {}_6C_0a^6b^0 + {}_6C_1a^5b^1 + {}_6C_2a^4b^2 + {}_6C_3a^3b^3 + {}_6C_4a^2b^4 + {}_6C_5ab^5 + {}_6C_6b^6$$
$$\quad\;\; k=0 \qquad k=1 \qquad\; k=2 \qquad\;\; k=3 \qquad\;\; k=4 \qquad\;\; k=5 \qquad\;\; k=6$$

Check that each term can be written in the form

$$_6C_ka^{6-k}b^k$$

---

[1]There are other common notations for binomial coefficients. Instead of $_nC_k$ you may see $\binom{n}{k}$ or $C_{n,k}$ or $C(n, k)$ or C_k^n.

for $k = 0$ to $k = 6$. This means that we can use sigma notation to write the sum of the terms in the expansion.

$$(a + b)^6 = \sum_{k=0}^{6} {}_6C_k a^{6-k}b^k \tag{1}$$

Notice that the sigma notation indicates terms for $k = 0$ to $k = 6$, so there are 7 terms in the expansion, as there should be. Also note that the sum of the exponents on a and b is $(6 - k) + k = 6$.

Equation (1) is a special case of the **binomial theorem,** which uses everything we have learned to write the expanded form of a power of a binomial in the most compact form possible. We can write the general form of the theorem, for positive integers n, as follows.

The Binomial Theorem

$$(a + b)^n = \sum_{k=0}^{n} {}_nC_k\, a^{n-k}b^k$$

EXAMPLE 9

a. Use sigma notation to write the expanded form for $(x - 2y)^{10}$.

b. Find the term containing y^7, and simplify.

Solutions

a. Apply the binomial theorem with $n = 10$, replacing a by x and b by $-2y$, to get

$$(x - 2y)^{10} = \sum_{k=0}^{10} {}_{10}C_k x^{10-k}(-2y)^k$$

b. The term containing y^7 corresponds to $k = 7$. The $k = 7$ term looks like

$${}_{10}C_7 x^{10-7}(-2y)^7$$

or

$${}_{10}C_7 x^3(-2y)^7$$

The value of ${}_{10}C_7$ is

$${}_{10}C_7 = \frac{10!}{(10 - 7)!\, 7!} = \frac{10 \cdot 9 \cdot 8 \cdot 7!}{(3)!\, 7!} = \frac{10 \cdot 9 \cdot 8}{3 \cdot 2 \cdot 1} = 120$$

and $(-2y)^7 = -128y^7$. Thus, the term we want is

$${}_{10}C_7(x)^3(-2y)^7 = 120x^3(-128y^7) = -15{,}360x^3y^7 \qquad \blacksquare$$

EXERCISE 7

a. Use sigma notation to write the expanded form for $(2m + 3n)^7$.

b. Find the term containing m^3, and simplify.

Using a Graphing Calculator to Compute Factorials and Binomial Coefficients

Both factorials and binomial coefficients can be accessed from the probability sub-menu of the [MATH] menu on the TI-83 calculator. (Press [MATH] and then use the arrow keys to highlight **PRB.**) Note that your calculator uses $_nC_r$ for the binomial coefficient, instead of $_nC_k$. (See Figure 9.7)

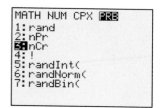

FIGURE 9.7

EXAMPLE 10

Use your calculator to evaluate each expression.

    a. $10!$ **b.** $_{15}C_7$ **c.** $\dfrac{13!}{5!\,8!}$

Solutions

  a. To access the factorial command, !, we press [MATH] ◁ 4. Thus, to evaluate $10!$, key in

$$10 \;[\text{ MATH }]\; ◁ \;4\; [\text{ ENTER }]$$

and the calculator returns the result: 3,628,800.

  b. To evaluate a binomial coefficient, we first enter n and then access the $_nC_r$ command by pressing [MATH] ◁ 3. Thus, to evaluate $_{15}C_7$, we first key in

$$15 \;[\text{ MATH }]\; ◁ \;3$$

The display will show "15nCr," and the calculator is ready for the value of r. Now key in

$$7 \;[\text{ ENTER }]$$

and the calculator shows 6435.

  c. To evaluate $\dfrac{13!}{5!\,8!}$, we key in the expression

$$13! \div (5!\,8!)$$

accessing the factorial command each time as shown in part (a). Or, recognize that this fraction is the factorial equivalent of $_{13}C_8$. Access the binomial coefficient command as in part (b) and key in

$$13nCr8 \;[\text{ ENTER }]$$

to obtain 1287. ∎

READING QUESTIONS

1. What does the symbol $_nC_k$ denote?
2. Explain how to use Pascal's triangle when expanding a power of a binomial.
3. What is factorial notation?
4. State a formula for the binomial coefficients in factorial notation.
5. Explain how to use the binomial theorem to expand $(a + b)^7$.

HOMEWORK 9.5

Answer the questions in Problems 1–4 without expanding the power.

1. How many terms are in the expansion of $(a + b)^{50}$? In the expansion of $(2x + 3y)^{100}$?

2. How many terms are in the expansion of $(a + b)^{75}$? In the expansion of $(5x - 7y)^{200}$?

3. What is the sum of the exponents on x and y in each term of the expansion of $(x + y)^{100}$? In the expansion of $(8x - 7y)^{50}$?

4. What is the sum of the exponents on x and y in each term of the expansion of $(x + y)^{200}$? In the expansion of $(9x - 4y)^{75}$?

5. Write down the portion of Pascal's triangle corresponding to rows from $n = 0$ to $n = 10$. How many rows are involved?

6. Write down the portion of Pascal's triangle corresponding to rows from $n = 0$ to $n = 12$. How many rows are involved?

Write each power in expanded form.

7. $(x + 3)^5$

8. $(2x + y)^4$

9. $(z - 3)^4$

10. $(2w - 1)^5$

11. $\left(2x - \dfrac{y}{2}\right)^3$

12. $\left(\dfrac{x}{3} + 3\right)^6$

13. $(x^2 - 3)^7$

14. $(1 - y^2)^5$

15. $(x + y)^5$

16. $(x - y)^6$

17. $(p - 2q)^4$

18. $(m + 3n)^8$

19. Simplify $(1 + 5t)^3 + (1 - 5t)^3$.

20. Simplify $(3 + 2a)^4 + (3 - 2a)^4$.

21. Write $\left(z - \dfrac{1}{z}\right)^5$ in descending powers of z.

22. Write $\left(v + \dfrac{1}{v}\right)^4$ in descending powers of v.

Write in expanded form and simplify.

23. a. $5!$ b. $\dfrac{9!}{7!}$

    c. $\dfrac{5!\,7!}{12!}$ d. $\dfrac{8!}{2!\,(8-2)!}$

24. a. $7!$ b. $\dfrac{12!}{11!}$

    c. $\dfrac{12!\,4!}{16!}$ d. $\dfrac{10!}{4!\,(10-4)!}$

Evaluate each binomial coefficient.

25. a. ${}_9C_6$ b. ${}_{12}C_3$

    c. ${}_{20}C_{18}$ d. ${}_{14}C_9$

26. a. ${}_8C_5$ b. ${}_{13}C_4$

    c. ${}_{18}C_{16}$ d. ${}_{16}C_7$

27. Find, in ascending powers of x, the first three terms in the expansions of
  a. $(1 - 2x)^7$ **b.** $(2 - x)^6$

28. Find, in descending powers of y, the first three terms of
  a. $(5y - 2)^4$ **b.** $(3y + 1)^5$

29. Find, in descending powers of u, the first three nonzero terms of $(4u - 1)^5 - (4u + 1)^5$.

30. Find, in ascending powers of t, the first three nonzero terms of $(1 - 3t)^6 - (1 + 3t)^6$.

31. Find the first four terms of
  a. $(1 - 5c)^6$ **b.** $(1 - 4c)(1 - 5c)^6$

32. Find the first four terms of
  a. $(1 + 2a)^9$ **b.** $(1 + 3a)(1 + 2a)^9$

33. Simplify $(1 + \sqrt{3})^4 + (1 - \sqrt{3})^4$.

34. Simplify $(\sqrt{2} + \sqrt{3})^4 + (\sqrt{2} - \sqrt{3})^4$.

Find the coefficient of the indicated term.

35. $(x + y)^{20}$; $x^{13}y^7$

36. $(x - y)^{15}$; $x^{12}y^3$

37. $(a - 2b)^{12}$; a^5b^7

38. $(2a - b)^{12}$; a^8b^4

39. $(x - \sqrt{2})^{10}$; x^4

40. $\left(x - \dfrac{1}{2}\right)^8$; x^5

41. $(a^3 - b^3)^9$; $a^{18}b^9$

42. $(x^2 - y^2)^7$; $x^{10}y^4$

Refer to Pascal's triangle and Investigation 16 to answer the questions.

43. Write out the terms of the sequence 11^n for $n = 0, 1, 2, 3, 4$. How are these related to the rows of Pascal's triangle? Can you explain why? (*Hint:* Rewrite 11 as $10 + 1$.)

44. Write out the terms of the sequence 1.1^n for $n = 0, 1\ 2, 3, 4$. How are these related to the rows of Pascal's triangle? Can you explain why? (*Hint:* Rewrite 1.1 as $1 + 0.1$.)

 **INVESTIGATION 17**

Other Properties of Pascal's Triangle

Pascal's triangle has other interesting properties besides providing the binomial coefficients.

1. Add up the numbers in each row of your Pascal's triangle and fill in Table 9.13.

2. The sum of the entries in row n of Pascal's triangle is _____ .

Now add entries of Pascal's triangle along the diagonal lines indicated below. Fill in two more row of Pascal's triangle and find two more diagonal sums.

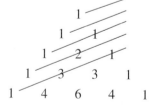

TABLE 9.13

Row n	Sum of entries
0	
1	
2	
3	
4	
5	
6	

3. Write the sequence of numbers obtained by taking sums along the indicated diagonal lines. _____ , _____ , _____ , _____ , _____ , _____ , _____ , . . .

4. What is the name of this sequence? (*Hint:* See Problem 45 of Section 9.1.)

The binomial coefficient ${}_nC_k$ also gives the number of different ways you can choose k distinct items from a set of n items. Finding these numbers is important in determining the probability that certain events will occur.

5. Find the number of different choices you can make for a four-flavor ice cream sundae if there are six flavors to choose from. (*Caution:* When you are deciding which entry in the row corresponds to $k = 4$, remember that the leftmost entry corresponds to $k = 0$.)

6. How many ways can you choose a debate team of three members from a debate club of five members?

7. Evaluate the series

$$\sum_{r=0}^{10} {}_{10}C_r$$

(*Hint:* This is the sum of all the binomial coefficients in the expansion of $(a + b)^n$. What did you learn in Investigation 16 about the sum of the terms in a row of Pascal's triangle?)

8. Evaluate the series

$$\sum_{r=0}^{12} {}_{12}C_r$$

(*Hint:* See Step 7.)

9. How many ways can you choose 6 objects from a set of 49? (This tells the total number of possible choices in a lottery that requires a participant to pick 6 numbers from 1 to 49.)

10. How many ways can you choose 5 cards out of a deck of 52 distinct cards? (This tells the total number of different poker hands.)

CHAPTER 9 REVIEW

Find the first four terms in the sequence whose general term is given.

1. $a_n = \dfrac{n}{n^2 + 1}$

2. $b_n = \dfrac{(-1)^{n-1}}{n}$

Find the first five terms in the recursively defined sequence.

3. $c_1 = 5; \quad c_{n+1} = c_n - 3$

4. $d_1 = 1; \quad d_{n+1} = -\dfrac{3}{4} d_n$

For Problems 5–8, (a) find the first four terms of each sequence, (b) determine equations to define the sequence *recursively*.

5. Rick purchased a sailboat for $1800. How much is it worth after n years if it depreciates in value by 12% each year?

6. Sally earns $24,000 per year. If she receives a 6% raise each year, what will her salary be after n years?

7. To fight off an infection, Garrison receives a 30 milliliter dose of an antibiotic followed by doses of 15 milliliters at regular intervals. Between doses, Garrison's kidneys remove 25% of the antibiotic that was present after the previous dose. How much of the antibiotic is present after n doses?

8. Opal joined the Weight Losers Club. At the end of the first meeting she weighed 187 pounds, and she lost 2 pounds from one meeting to the next. How much does Opal weigh after n meetings?

Find the indicated term for each sequence.

9. $x_n = (-1)^n(n - 2)^2; \quad x_7$

10. $y_n = \sqrt{n^3 - 2}; \quad y_3$

11. The tenth term in the arithmetic sequence that begins $-4, 0, \ldots$

12. The sixth term in the arithmetic sequence that begins $x - a, x + a, \ldots$

13. The eighth term in the geometric sequence that begins $\frac{16}{27}, \frac{-8}{9}, \frac{4}{3}, \ldots$

14. The fifth term in the geometric sequence with third term $\frac{-2}{3}$ and sixth term $\frac{16}{81}$.

15. The twenty-third term of the arithmetic sequence $-84, -74, -66, \ldots$

16. The ninth term of the arithmetic sequence $\frac{-1}{2}, 1, \frac{5}{2}, \ldots$

17. The first term of an arithmetic sequence is 8 and the twenty-eighth term is 89. Find the twenty-first term.

18. What term in the arithmetic sequence $5, 2, -1, \ldots$ is -37?

Identify each sequence as arithmetic, geometric, or neither. For each arithmetic and geometric sequence, find the next four terms of the sequence and a non-recursive expression for the general term.

19. $-1, \frac{1}{2}, \frac{-1}{4}, \frac{1}{8}, \ldots$

20. $12, 9, 3, 1, \ldots$

21. $6, 1, -4, -9, \ldots$

22. $1, -4, 16, -64, \ldots$

23. $-1, 2, -4, 8, \ldots$

24. $\frac{2}{3}, \frac{1}{2}, \frac{3}{8}, \frac{9}{32}, \ldots$

25. first term 3, common difference -4

26. first term $\frac{1}{4}$, common difference $\frac{1}{2}$

27. first term 12, common ratio -4

28. first term 6, common ratio $\frac{1}{3}$

Write the sum in expanded form.

29. $\displaystyle\sum_{k=2}^{5} k(k - 1)$

30. $\displaystyle\sum_{j=2}^{\infty} \frac{j}{2j - 1}$

Write the sum using sigma notation.

31. The first 12 terms of $1, 3, 7, \ldots, 2^k - 1, \ldots$

32. The fourth through fifteenth terms of $x, 4x^2, 9x^3, \ldots, k^2x^k, \ldots$

Identify each series as arithmetic, geometric, or neither, and then evaluate.

33. The sum of the first 12 terms of the sequence $a_n = 3n - 2$

34. The sum of the first 20 terms of the sequence $b_n = 1.4 + 0.1n$

35. $\displaystyle\sum_{i=1}^{6} (3i - 1)$

36. $\displaystyle\sum_{k=1}^{12} \left(\frac{2}{3}k - 1\right)$

37. $\displaystyle\sum_{j=1}^{5} \left(\frac{1}{3}\right)^j$

38. $\displaystyle\sum_{k=1}^{6} 2^{k-1}$

39. $\displaystyle\sum_{n=1}^{5} (-1)^n(n + 1)$

40. $\displaystyle\sum_{n=1}^{4} \frac{n}{n + 1}$

41. $-3 + 2 + \left(\dfrac{-4}{3}\right) + \left(\dfrac{8}{9}\right) + \cdots$

42. $\displaystyle\sum_{i=1}^{\infty} 3\left(\dfrac{1}{3}\right)^{i-1}$

43. A rubber ball is dropped from a height of 12 feet and returns to two-thirds of its previous height on each bounce. How high does the ball bounce after hitting the floor for the fourth time?

44. The property taxes on the Hardestys' family home were $840 in 1994. If the taxes increase by 2% each year, what will the taxes be in 2000?

45. a. Find the sum of all integral multiples of 6 between 10 and 100.

 b. Write the sum in (a) using sigma notation.

46. Kathy planted a 7-foot silver maple tree in 1994. If the tree grows 1.3 feet each year, in what year will it be 20 feet tall?

47. A rubber ball is dropped from a height of 12 feet and returns to three-fourths of its previous height on each bounce. Approximately what is the total distance the ball travels before coming to rest?

48. Suppose you will be paid 1¢ on the first day of June, 3¢ on the second, 9¢ on the third, etc., so that each new day you are paid three times what you received the previous day. What would be the total amount you will receive in the month of June?

Find a fraction equivalent to each of the given repeating decimals.

49. $3.222\overline{2}$

50. $0.4181\overline{818}$

Write each power in expanded form.

51. $(x - 2)^5$

52. $\left(\dfrac{x}{2} - y\right)^4$

Evaluate.

53. $\dfrac{6!}{3!\,(6 - 3)!}$

54. $\dfrac{9!}{5!\,(9 - 5)!}$

55. $_7C_2$

56. $_{16}C_{14}$

57. $\displaystyle\sum_{k=0}^{5} {}_5C_k$

58. $\displaystyle\sum_{k=0}^{6} {}_6C_k(1.4)^{6-k}(0.6)^k$

Find the coefficient of the indicated term.

59. $(x - 2y)^9;\ x^6y^3$

60. $\left(\dfrac{x}{2} - 3\right)^8;\ x$

More About Functions

In this chapter we consider some additional topics about functions. Investigation 18 uses a graphing calculator to explore transformations of functions.

 **INVESTIGATION 18**

Transformations of Graphs

A. How is the graph of a function affected if we add a constant to the formula? In other words, how is the graph of $y = f(x) + k$ different from the graph of $y = f(x)$?

    1. **a.** Graph the absolute value function, $y = |x|$, in the standard graphing window. The TI-83 calculator uses the notation $y = \text{abs}(x)$ instead of $y = |x|$ for the absolute value function, and you will find abs (x) in the **Catalog.** After the "$Y_1 =$" prompt, press **2nd** 0 **ENTER** to choose **abs(**, then enter **X**).

        b. Use your calculator to create a table of values for $y = |x|$. Use **TblStart** $= -3$ and **ΔTbl** $= 1$.

    2. Graph $Y_2 = |x| - 3$ in the same window with Y_1.

        a. Does the graph of Y_2 have the same shape as the graph of Y_1? How is the graph of Y_2 different from the graph of Y_1?

        b. Use the **Table** to compare the function values for Y_2 with the function values for Y_1. (Locate a specific x-value in the table and compare the Y_1 and Y_2 values for that input. How are they related? Try the same comparison for several other x-values.)

3. Now change your definition of Y_2 to $Y_2 = |x| + 3$.

    a. Does the graph of this new Y_2 have the same shape as the graph of Y_1? How is the graph of Y_2 different from the graph of Y_1?

    b. Use the **Table** to compare the function values for Y_2 with the function values for Y_1. How are they related?

4. Define $Y_2 = |x| + k$ for various positive and negative values for k and compare the graph and **Table** values of Y_2 with those of Y_1. Describe how the value of k alters the graph of Y_1.

5. Now change the basic function to the square root function, $Y_1 = \sqrt{x}$, and repeat part (4) with $Y_2 = \sqrt{x} + k$. Do your answers to part (4) still hold for the new functions?

6. Summarize your observations for part A: How does the graph of $y = f(x) + k$ compare with the graph of $y = f(x)$?

B. How is the graph of a function affected if we add a constant to the independent variable? In other words, how is the graph of $y = f(x + h)$ different from the graph of $y = f(x)$?

1. Define $Y_1 = |x|$ and $Y_2 = |x - 2|$.

    a. Does the graph of Y_2 have the same shape as the graph of Y_1? How is the graph of Y_2 different from the graph of Y_1?

    b. Use the **Table** to compare the function values for Y_2 with the function values for Y_1. (Locate a specific output value in the Y_1 column, and note its corresponding x-value. Now find the same output value in the Y_2 column, and note its x-value. How are the two x-values related? Try the same comparison for several other output values.)

2. Now define $Y_2 = |x + 4|$.

    a. Does the graph of this new Y_2 have the same shape as the graph of Y_1? How is the graph of Y_2 different from the graph of Y_1?

    b. Use the **Table** to compare the function values for Y_2 with the function values for Y_1. How are they related?

3. Define $Y_2 = |x + h|$ for various positive and negative values for h and compare the graph and **Table** values of Y_2 with those of Y_1. Describe how the value of h affects the graph of Y_2.

4. Now change the basic function to the square root function, $Y_1 = \sqrt{x}$, and repeat part (3) with $Y_2 = \sqrt{x + h}$. Do your answers to part (3) still hold for the new functions?

5. Summarize your observations for part B: How does the graph of $y = f(x + h)$ compare with the graph of $y = f(x)$?

C. How is the graph of a function affected if we multiply the formula by a constant? In other words, how is the graph of $y = af(x)$ different from the graph of $y = f(x)$?

1. Define $Y_1 = |x|$ and $Y_2 = 2|x|$.

    a. Does the graph of Y_2 have the same shape as the graph of Y_1? How is the graph of Y_2 different from the graph of Y_1?

b. Use the **Table** to compare the function values for Y_2 with the function values for Y_1. (Locate a specific x-value in the table and compare the Y_1 and Y_2 values for that input. How are they related? Try the same comparison for several other x-values.)

2. Now define $Y_2 = \frac{1}{2}|x|$.

    a. Does the graph of this new Y_2 have the same shape as the graph of Y_1? How is the graph of Y_2 different from the graph of Y_1?

    b. Use the **Table** to compare the function values for Y_2 with the function values for Y_1. How are they related?

3. Define $Y_2 = a|x|$ for various values of a, including $a > 1$ and $0 < a < 1$. Compare the graph and **Table** values of Y_2 with those of Y_1. Describe how the value of a affects the graph of Y_2.

4. Now change the basic function to the square root function, $Y_1 = \sqrt{x}$, and repeat part (3) with $Y_2 = a\sqrt{x}$. Do your answers to part (3) still hold for the new functions?

5. Now define $Y_2 = -|x|$.

    a. Does the graph of Y_2 have the same shape as the graph of Y_1? How is the graph of Y_2 different from the graph of Y_1?

    b. Use the **Table** to compare the function values for Y_2 with the function values for Y_1. How are they related?

6. Define $Y_2 = a|x|$ for various *negative* values of a, including $a < -1$ and $-1 < a < 0$. Compare the graph and **Table** values of Y_2 with those of Y_1. Describe how the value of a affects the graph of Y_2 when a is negative.

7. Now change the basic function to the square root function, $Y_1 = \sqrt{x}$, and repeat part (6) with $Y_2 = a\sqrt{x}$, for negative values of a. Do your answers to part (6) still hold for the new functions?

8. Summarize your observations for part C: How does the graph of $y = a f(x)$ compare with the graph of $y = f(x)$? Include in your discussion the cases $a > 1$, $0 < a < 1$, and $a < 0$.

10.1   Transformations of Functions

Models for real situations are often variations of the basic functions introduced in Section 5.3. In this section we explore how certain changes in the formula for a function affect its graph. In particular, we will compare the graph of $y = f(x)$ with the graphs of $y = f(x) + k$, $y = f(x + h)$, and $y = a f(x)$ for different values of the constants k, h, and a. Such variations are called **transformations** of the graph.

Vertical Translations

We have already seen examples of transformations in our study of quadratic functions. For instance, the graphs of $f(x) = x^2 + 4$ and $g(x) = x^2 - 4$ are variations of the basic parabola, as shown in Figure 10.1.

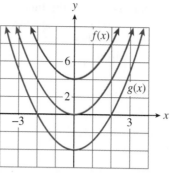

FIGURE 10.1

TABLE 10.1

x	-2	-1	0	1	2
$y = x^2$	4	1	0	1	4
$f(x) = x^2 + 4$	8	5	4	5	8

TABLE 10.2

x	-2	-1	0	1	2
$y = x^2$	4	1	0	1	4
$g(x) = x^2 - 4$	0	-3	-4	-3	0

Each y-value in Table 10.1 for $f(x)$ is 4 units greater than the corresponding y-value for the basic parabola. Consequently, each point on the graph of $f(x)$ is 4 units higher than the corresponding point on the basic parabola. Similarly, each point on the graph of $g(x)$ is 4 units lower than the corresponding point on the basic parabola.

The graphs of $y = f(x)$ and $y = g(x)$ are said to be **translations** of the graph of $y = x^2$. They are shifted to a different location in the plane but retain the same size and shape as the original graph. In general we have the following principles.

Vertical Translations

Compared with the graph of $y = f(x)$,

1. the graph of $y = f(x) + k$ ($k > 0$) is shifted *upward k* units.
2. the graph of $y = f(x) - k$ ($k > 0$) is shifted *downward k* units.

EXAMPLE 1

Graph the following functions.

a. $g(x) = |x| + 3$
b. $h(x) = \dfrac{1}{x} - 2$

Solutions

a. Table 10.3 shows that the y-values for $g(x)$ are each three units greater than the corresponding y-values for the absolute value function. The graph of $g(x) = |x| + 3$ is a translation of the basic graph $y = |x|$, shifted upward three units, as shown in Figure 10.2.

TABLE 10.3

x	-2	-1	0	1	2
$y = |x|$	2	1	0	1	2
$g(x) = |x| + 3$	5	4	3	4	5

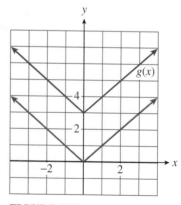

FIGURE 10.2

b. Table 10.4 shows that the y-values for $h(x)$ are each two units smaller than the corresponding y-values for $y = \dfrac{1}{x}$. The graph of $h(x) = \dfrac{1}{x} - 2$ is a translation of the basic graph of $y = \dfrac{1}{x}$, shifted downward two units, as shown in Figure 10.3.

TABLE 10.4

x	-2	-1	$\frac{1}{2}$	1	2
$y = \dfrac{1}{x}$	$\frac{-1}{2}$	-1	2	1	$\frac{1}{2}$
$h(x) = \dfrac{1}{x} - 2$	$\frac{-5}{2}$	-3	0	-1	$\frac{-3}{2}$

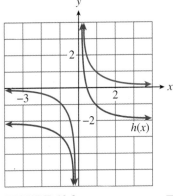

FIGURE 10.3

EXERCISE 1

a. Graph the function $f(x) = |x| + 1$.

b. How is the graph different from the graph of $y = |x|$?

Horizontal Translations

Now consider the graphs of $f(x) = (x + 2)^2$ and $g(x) = (x - 2)^2$ shown in Figure 10.4. Compared with the graph of the basic function $y = x^2$, the graph of $f(x) = (x + 2)^2$ is shifted two units to the *left*. You can see why this happens by studying the function values in Table 10.5. Locate a particular y-value for $y = x^2$, say, $y = 1$. You must move two units to the left in the table to find the same y-value for $f(x)$, as shown by the arrow. In fact, each y-value for $f(x)$ occurs two units to the left when compared to the same y-value for $y = x^2$.

TABLE 10.5

x	-3	-2	-1	0	1	2	3
$y = x^2$	9	4	1	0	1	4	9
$f(x) = (x + 2)^2$	1	0	1	4	9	16	25

TABLE 10.6

x	-3	-2	-1	0	1	2	3
$y = x^2$	9	4	1	0	1	4	9
$g(x) = (x - 2)^2$	25	16	9	4	1	0	1

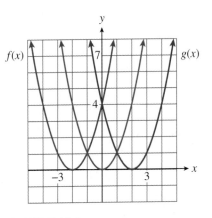

FIGURE 10.4

Similarly, the graph of $g(x) = (x - 2)^2$ is shifted two units to the *right* compared to the graph of $y = x^2$. In Table 10.6, each y-value for $g(x)$ occurs two units to the right of the same y-value for $y = x^2$. In general, we have the following principle.

Horizontal Translations

Compared with the graph of $y = f(x)$,

1. the graph of $y = f(x + h)$ $(h > 0)$ is shifted h units to the *left*.
2. the graph of $y = f(x - h)$ $(h > 0)$ is shifted h units to the *right*.

EXAMPLE 2   Graph the following functions.

   a. $g(x) = \sqrt{x + 1}$ **b.** $h(x) = \dfrac{1}{(x - 3)^2}$

Solutions

   a. Table 10.7 shows that each y-value for $g(x)$ occurs one unit to the left of the same y-value for the graph of $y = \sqrt{x}$. Consequently, each point on the graph of $y = g(x)$ is shifted one unit to the left of $y = \sqrt{x}$, as shown in Figure 10.5.

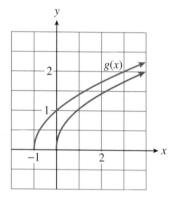

FIGURE 10.5

TABLE 10.7

x	-1	0	1	2	3
$y = \sqrt{x}$	undefined	0	1	1.414	1.732
$g(x) = \sqrt{x + 1}$	0	1	1.414	1.732	2

   b. Table 10.8 shows that each y-value for $h(x)$ occurs three units to the right of the same y-value for the graph of $y = \dfrac{1}{x^2}$. Consequently, each point on the graph of $y = h(x)$ is shifted three units to the right of $y = \dfrac{1}{x^2}$, as shown in Figure 10.6.

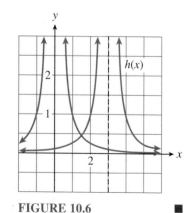

FIGURE 10.6

TABLE 10.8

x	-1	0	1	2	3	4
$y = \dfrac{1}{x^2}$	1	undefined	1	$\frac{1}{4}$	$\frac{1}{9}$	$\frac{1}{16}$
$h(x) = \dfrac{1}{(x - 3)^2}$	$\frac{1}{16}$	$\frac{1}{9}$	$\frac{1}{4}$	1	undefined	1

The graphs of some functions involve both horizontal and vertical translations.

EXAMPLE 3

Graph $f(x) = (x + 4)^3 + 2$.

Solution

We identify the basic graph from the structure of the formula for $f(x)$. In this case the basic graph is $y = x^3$, so we begin by locating a few points on that graph, as shown in Figure 10.7. We'll perform the translations separately, following the order of operations. First, we sketch a graph of $y = (x + 4)^3$ by shifting each point on the basic graph four units to the left. Then we move each point up two units to obtain the graph of $f(x) = (x + 4)^3 + 2$. All three graphs are shown in Figure 10.7.

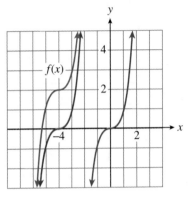

FIGURE 10.7 ∎

Scale Factors

We have seen that *adding* a constant to the expression defining a function results in a translation of its graph. What happens if we *multiply* the expression by a constant? Consider the graphs of the functions

$$f(x) = 2x^2, \qquad g(x) = \frac{1}{2}x^2, \qquad \text{and} \qquad h(x) = -x^2$$

shown in Figures 10.8a, 10.8b, and 10.8c, and compare each to the graph of $y = x^2$.

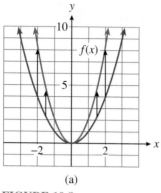

(a)

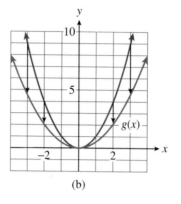

(b)

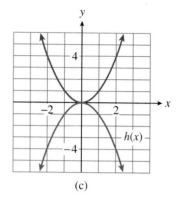

(c)

FIGURE 10.8

x	$y = x^2$	$f(x) = 2x^2$
-2	4	8
-1	1	2
0	0	0
1	1	2
2	4	8

x	$y = x^2$	$g(x) = \frac{1}{2}x^2$
-2	4	2
-1	1	$\frac{1}{2}$
0	0	0
1	1	$\frac{1}{2}$
2	4	2

x	$y = x^2$	$h(x) = -x^2$
-2	4	-4
-1	1	-1
0	0	0
1	1	-1
2	4	-4

Compared to the graph of $y = x^2$, the graph of $f(x) = 2x^2$ is expanded, or stretched, vertically by a factor of 2. The y-coordinate of each point on the graph has been doubled, as you can see in the table of values, so each point on the graph of f is twice as far from the x-axis as its counterpart on the basic graph $y = x^2$. The graph of $g(x) = \frac{1}{2}x^2$ is compressed vertically by a factor of $\frac{1}{2}$; each point is half as far from the x-axis as its counterpart on the graph of $y = x^2$. The graph of $h(x) = -x^2$ is "flipped" or reflected about the x-axis; the y-coordinate of each point on the graph of $y = x^2$ is replaced by its opposite.

In general we have the following principles.

Scale Factors

Compared with the graph of $y = f(x)$, the graph of $y = a\,f(x)$, where $a \neq 0$, is

1. expanded vertically by a factor of $|a|$ if $|a| > 1$.
2. compressed vertically by a factor of $|a|$ if $0 < |a| < 1$.
3. reflected about the x-axis if $a < 0$.

The constant a is called the **scale factor** for the graph.

EXAMPLE 4

Graph the following functions.

a. $g(x) = 3\sqrt[3]{x}$ **b.** $h(x) = \dfrac{-1}{2}|x|$

Solutions

a. The graph of $g(x) = 3\sqrt[3]{x}$ is an expansion of the basic graph $y = \sqrt[3]{x}$ by a factor of 3, as shown in Figure 10.9. Each point on the basic graph has its y-coordinate tripled.

b. The graph of $h(x) = \dfrac{-1}{2}|x|$ is a compression of the basic graph $y = |x|$ by a factor of $\frac{1}{2}$, combined with a reflection about the x-axis. You may find it helpful to graph the function in two steps, as shown in Figure 10.10.

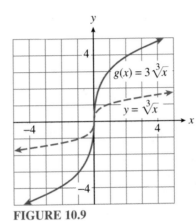

FIGURE 10.9

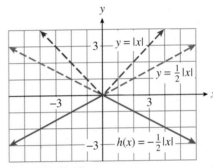

FIGURE 10.10 ∎

EXERCISE 4

a. Graph the function $f(x) = 2|x|$.

b. How is the graph different from the graph of $y = |x|$?

---

1. How does a vertical translation affect the formula for a function? Give an example.
2. How does a horizontal translation affect the formula for a function? Give an example.
3. How does a scale factor affect the formula for a function? Give an example.
4. How is the graph of $y = -f(x)$ different from the graph of $y = f(x)$?

1a.

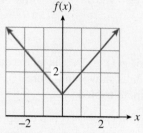

b. The graph is shifted 1 unit upward.

2a.

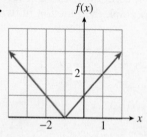

b. The graph is shifted 1 unit to the left.

3a.

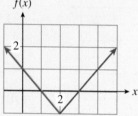

b. The graph is shifted 2 units to the right and 1 unit down.

4a.

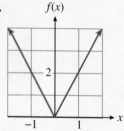

b. The graph is stretched vertically by a factor of 2.

HOMEWORK 10.1

For Problems 1–12, (a) describe how to transform one of the basic graphs to obtain the graph of the given function, (b) sketch the basic graph and the graph of the given function by hand on the same axes. Label the coordinates of three points on the graph of the given function.

1. $f(x) = |x| - 2$

2. $g(x) = (x + 1)^3$

3. $g(s) = \sqrt[3]{s - 4}$

4. $f(s) = s^2 + 3$

5. $F(t) = \dfrac{1}{t^2} + 1$

6. $G(t) = \sqrt{t - 2}$

7. $G(r) = (r + 2)^2$

8. $F(r) = \dfrac{1}{r - 4}$

9. $H(d) = \sqrt{d} - 3$

10. $h(d) = \sqrt[3]{d} + 5$

11. $h(v) = \dfrac{1}{v + 6}$

12. $H(v) = \dfrac{1}{v^2} - 2$

For Problems 13–24, (a) the graph of each function can be obtained from one of the basic graphs by two translations. Describe the translations. (b) Sketch the basic graph and the graph of the given function by hand on the same axes. Label the coordinates of three points on the graph of the given function.

13. $f(x) = 2 + (x - 3)^2$

14. $f(x) = (x + 4)^2 + 1$

15. $g(z) = \dfrac{1}{z + 2} - 3$

16. $g(z) = \dfrac{1}{z - 1} + 1$

17. $F(u) = \sqrt{u + 4} + 4$

18. $F(u) = \sqrt{u - 3} - 5$

19. $G(t) = |t - 5| - 1$

20. $G(t) = |t + 4| + 2$

21. $H(w) = \dfrac{1}{(w - 1)^2} + 6$

22. $H(w) = \dfrac{1}{(w + 3)^2} - 3$

23. $f(t) = \sqrt[3]{t - 8} - 1$

24. $f(t) = \sqrt[3]{t + 1} + 8$

For Problems 25–34, (a) identify the scale factor for each function and describe how it affects the graph of the corresponding basic function, (b) sketch the basic graph and the graph of the given function by hand on the same axes. Label the coordinates of three points on the graph of the given function.

25. $f(x) = \dfrac{1}{3}|x|$

26. $H(x) = -3|x|$

27. $h(z) = \dfrac{-2}{z^2}$

28. $g(z) = \dfrac{2}{z}$

29. $G(v) = -3\sqrt{v}$

30. $F(v) = -4\sqrt[3]{v}$

31. $g(s) = \dfrac{-1}{2}s^3$

32. $f(s) = \dfrac{1}{8}s^3$

33. $H(x) = \dfrac{1}{3x}$

34. $h(x) = \dfrac{-1}{4x^2}$

For Problems 35–44, (a) describe the graph as a transformation of a basic function, or of $y = 2^x$, (b) give an equation for the function graphed.

35.

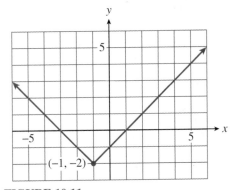

FIGURE 10.11

36.

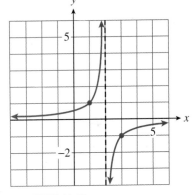

FIGURE 10.12

37.

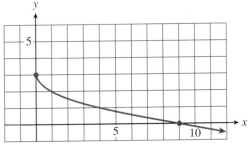

FIGURE 10.13

38.

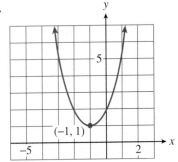

FIGURE 10.14

39.

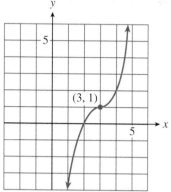

FIGURE 10.15

40.

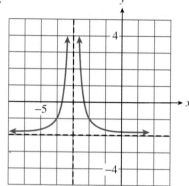

FIGURE 10.16

41.

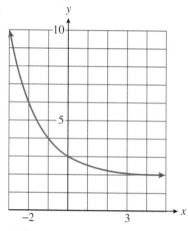

FIGURE 10.17

42.

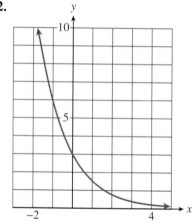

FIGURE 10.18

43.

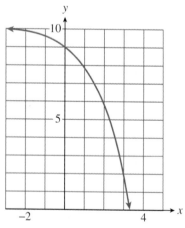

FIGURE 10.19

44.

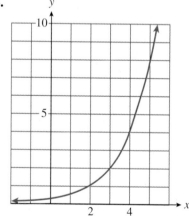

FIGURE 10.20

For Problems 45–48, (a) write each equation in the form $y = (x - p)^2 + q$ by completing the square, (b) using horizontal and vertical translations, sketch the graph by hand.

45. $y = x^2 - 4x + 7$ **46.** $y = x^2 - 2x - 1$ **47.** $y = x^2 + 2x - 3$ **48.** $y = x^2 + 4x + 5$

10.2 Using Function Notation

Operations with Function Notation

Recall that the elements of the domain are the "input" values for a function and the elements of the range are the corresponding "output" values. Thus, if we choose $x = 0$ as an input value for the function $f(x) = \sqrt{x} + 4$, the output value is $f(0) = \sqrt{0} + 4 = 2$.

EXAMPLE 1

Find the range element associated with each domain element.

    a. $g(x) = \dfrac{2}{2x - 1}$, $x = -3$ **b.** $H(x) = 4 - |x + 2|$, $x = -6$

Solutions

    a. Evaluate the function g for $x = -3$.

$$g(-3) = \frac{2}{2(-3) - 1} = \frac{-2}{7}$$

    b. Evaluate the function H for $x = -6$.

$$H(-6) = 4 - |-6 + 2| = 4 - 4 = 0$$ ∎

EXERCISE 1 Find $F(-1)$ if $F(x) = -x^2 - x + 2$.

Sometimes we need to evaluate a function at an algebraic expression, rather than at a specific number.

EXAMPLE 2

TrailGear manufactures backpacks at a cost of

$$C(x) = 3000 + 20x$$

for x backpacks. The company finds that the monthly demand for backpacks increases by 50% during the summer. Their backpacks are produced at several small co-ops in different states.

    a. If each co-op usually produces b backpacks per month, how many should they produce during the summer months?

    b. What costs for producing backpacks should they expect during the summer?

Solutions

a. An increase of 50% means an additional 50% of the current production level, b. Therefore, a co-op that produced b backpacks per month during the winter should increase production to $b + 0.5b$, or $1.5b$ backpacks per month in the summer.

b. The cost of producing $1.5b$ backpacks will be

$$C(\mathbf{1.5b}) = 3000 + 20(\mathbf{1.5b}) = 3000 + 30b \qquad \blacksquare$$

EXAMPLE 3

Evaluate the function $f(x) = 4x^2 - x + 5$ for the following expressions.

a. $x = 2h$ **b.** $x = a + 3$

Solutions

a. $f(\mathbf{2h}) = 4(\mathbf{2h})^2 - (\mathbf{2h}) + 5$

$= 4(4h^2) - 2h + 5 = 16h^2 - 2h + 5$

b. $f(\mathbf{a+3}) = 4(\mathbf{a+3})^2 - (\mathbf{a+3}) + 5$

$= 4(a^2 + 6a + 9) - a - 3 + 5$

$= 4a^2 + 24a + 36 - a + 2$

$= 4a^2 + 23a + 38 \qquad \blacksquare$

COMMON ERROR

In Example 3, notice that

$$f(2h) \neq 2f(h)$$

and

$$f(a + 3) \neq f(a) + f(3)$$

To compute $f(a) + f(3)$, we must first compute $f(a)$ and $f(3)$, then add them.

$$f(a) + f(3) = (4a^2 - a + 5) + (4 \cdot 3^2 - 3 + 5)$$
$$= 4a^2 - a + 43$$

In general, it is *not* true that $f(a + b) = f(a) + f(b)$. Remember that the parentheses in the expression $f(x)$ do not indicate multiplication, so the distributive law does not apply to the expression $f(a + b)$. $\blacksquare$

EXERCISE 2   Evaluate the function $G(t) = 1 - t^2$ for the following expressions.

a. $t = 3w$ **b.** $t = s + 1$

EXAMPLE 4

Let $f(x) = x^3 - 1$, and evaluate each expression.

a. $f(2) + f(3)$ **b.** $f(2 + 3)$ **c.** $2f(x) + 3$

Solutions

a. $f(2) + f(3) = (2^3 - 1) + (3^3 - 1)$
$$= 7 + 26 = 33$$

b. $f(2 + 3) = f(5) = 5^3 - 1 = 124$

c. $2f(x) + 3 = 2(x^3 - 1) + 3$
$$= 2x^3 - 2 + 3 = 2x^3 + 1 \qquad \blacksquare$$

EXERCISE 3 Let $F(r) = \dfrac{3}{r} + 1$, and evaluate each expression.

a. $F(-1) + F(3)$ **b.** $F(-1 + 3)$ **c.** $-1F(r) + 3$

Functions Defined Piecewise

A function may be defined by different formulas on different portions of the x-axis. Such a function is said to be defined "piecewise." To graph a function defined piecewise, we consider each piece of the x-axis separately.

EXAMPLE 5

Graph the function defined by

$$f(x) = \begin{cases} x + 1 & \text{if } x \le 1 \\ 3 & \text{if } x > 1 \end{cases}$$

Solution

Think of the plane as divided into two regions by the vertical line $x = 1$, as shown in Figure 10.21. In the left-hand region ($x \le 1$), graph the line $y = x + 1$. (The fastest way to graph the line is to plot its intercepts, $(-1, 0)$ and $(0, 1)$.) Notice that the value $x = 1$ is included in the first region, so $f(1) = 2$, and the point $(1, 2)$ is included on the graph. We indicate this with a solid dot at the point $(1, 2)$. In the right-hand region ($x > 1$), graph the horizontal line $y = 3$. The value $x = 1$ is *not* included in the second region, so the point $(1, 3)$ is *not* part of the graph. We indicate this with an open circle at the point $(1, 3)$.

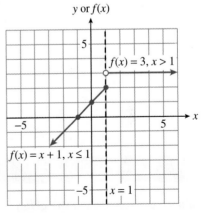

FIGURE 10.21

$\blacksquare$

The absolute value function $f(x) = |x|$ is another example of a function that is defined piecewise.

$$f(x) = |x| = \begin{cases} x & \text{if } x \ge 0 \\ -x & \text{if } x < 0 \end{cases}$$

To sketch the absolute value function, we can graph the line $y = -x$ in the second quadrant, and the line $y = x$ in the first quadrant.

READING QUESTIONS

1. What is the set of input values for a function called? The set of output values?

2. Explain why $f(a + b) \neq f(a) + f(b)$ in general.

3. What is a piecewise-defined function?

ANSWERS TO 10.2 EXERCISES

1. 2   **2a.** $1 - 9w^2$ **b.** $-s^2 - 2s$

3a. 0   **b.** $\dfrac{5}{2}$ **c.** $\dfrac{-3}{r} + 2$

4.

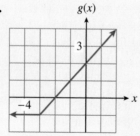

HOMEWORK 10.2

Evaluate the function at the given algebraic expressions.

1. $G(s) = 3s^2 - 6s$

    a. $G(3a)$ **b.** $G(a + 2)$

    c. $G(a) + 2$ **d.** $G(-a)$

2. $h(x) = 2x^2 + 6x - 3$

    a. $h(2a)$ **b.** $h(a + 3)$

    c. $h(a) + 3$ **d.** $h(-a)$

3. $g(x) = 8$

  a. $g(2)$ **b.** $g(8)$

  c. $g(a + 1)$ **d.** $g(-x)$

4. $f(t) = -3$

  a. $f(4)$ **b.** $f(-3)$

  c. $f(b - 2)$ **d.** $f(-t)$

5. $P(x) = x^3 - 1$

  a. $P(2x)$ **b.** $2P(x)$

  c. $P(x^2)$ **d.** $[P(x)]^2$

6. $Q(t) = 5t^3$

  a. $Q(2t)$ **b.** $2Q(t)$

  c. $Q(t^2)$ **d.** $[Q(t)]^2$

For the functions in Problems 7–14, compute the following.

a. $f(2) + f(3)$ **c.** $f(a) + f(b)$

b. $f(2 + 3)$ **d.** $f(a + b)$

For which functions does $f(a + b) = f(a) + f(b)$?

7. $f(x) = 3x - 2$ **8.** $f(x) = 1 - 4x$ **9.** $f(x) = x^2 + 3$ **10.** $f(x) = x^2 - 1$

11. $f(x) = \sqrt{x + 1}$ **12.** $f(x) = \sqrt{6 - x}$ **13.** $f(x) = \dfrac{-2}{x}$ **14.** $f(x) = \dfrac{3}{x}$

For Problems 15–16, find the coordinates of the indicated points and write an algebraic expression using function notation for the indicated quantity in each figure.

15. a. The length of the vertical line segment on the y-axis.

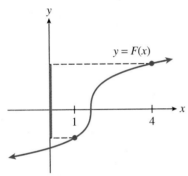

FIGURE 10.22

  b. The increase in y as x increases from a to b.

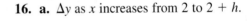

FIGURE 10.23

16. a. Δy as x increases from 2 to $2 + h$.

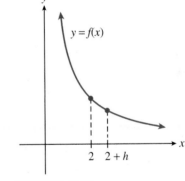

FIGURE 10.24

  b. The shaded area.

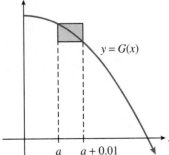

FIGURE 10.25

For Problems 17–20, (a) find the coordinates of the indicated points on the graph of $y = f(x)$, (b) write an algebraic expression using function notation for the slope of the line segment joining points P and Q.

17.

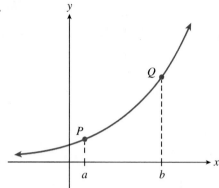

FIGURE 10.26

18.

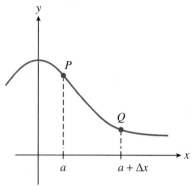

FIGURE 10.27

19.

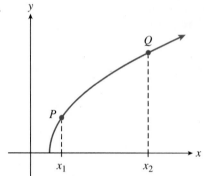

FIGURE 10.28

20.

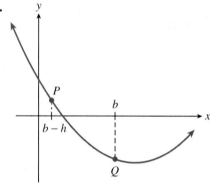

FIGURE 10.29

Graph by hand the following piecewise-defined functions.

21. $f(x) = \begin{cases} -2 & \text{if } x \le 1 \\ x - 3 & \text{if } x > 1 \end{cases}$

22. $h(x) = \begin{cases} -x + 2 & \text{if } x \le -1 \\ 3 & \text{if } x > -1 \end{cases}$

23. $G(t) = \begin{cases} 3t + 9 & \text{if } t < -2 \\ -3 - \dfrac{1}{2}t & \text{if } t \ge -2 \end{cases}$

24. $F(s) = \begin{cases} \dfrac{1}{3}s + 3 & \text{if } s < 3 \\ 2s - 3 & \text{if } s \ge 3 \end{cases}$

25. $H(t) = \begin{cases} t^2 & \text{if } t \le 1 \\ \dfrac{1}{2}t + \dfrac{1}{2} & \text{if } t > 1 \end{cases}$

26. $g(t) = \begin{cases} \dfrac{3}{2}t + 7 & \text{if } t \le -2 \\ t^2 & \text{if } t > -2 \end{cases}$

27. $k(x) = \begin{cases} |x| & \text{if } x \le 2 \\ \sqrt{x} & \text{if } x > 2 \end{cases}$

28. $S(x) = \begin{cases} \dfrac{1}{x} & \text{if } x < 1 \\ |x| & \text{if } x \ge 1 \end{cases}$

29. $D(x) = \begin{cases} |x| & \text{if } x < -1 \\ x^3 & \text{if } x \ge -1 \end{cases}$

30. $m(x) = \begin{cases} x^2 & \text{if } x \le \dfrac{1}{2} \\ |x| & \text{if } x > \dfrac{1}{2} \end{cases}$

31. $P(t) = \begin{cases} t^3 & \text{if } t \le 1 \\ \dfrac{1}{t^2} & \text{if } t > 1 \end{cases}$

32. $Q(t) = \begin{cases} t^2 & \text{if } t \le -1 \\ \sqrt[3]{t} & \text{if } t > -1 \end{cases}$

33. Lead nitrate and potassium iodide react in solution to produce lead iodide, which settles out, or "precipitates," as a yellow compound at the bottom of the container. As you add more lead nitrate to the solution, more lead iodide is produced until all the potassium iodide is used up. The table shows the height of the precipitate in the container as a function of the amount of lead nitrate added.

Lead nitrate solution (cc)	0.5	1.0	1.5	2.0	2.5
Height of precipitate (mm)	2.8	4.8	6.2	7.4	9.5

Lead nitrate solution (cc)	3.0	3.5	4.0
Height of precipitate (mm)	9.6	9.6	9.6

a. Plot the data. By eye, sketch a piecewise linear function with two parts to fit the data points.

b. Calculate the slope of the increasing part of the graph, including units. What is the significance of the slope?

c. Write a formula for your piecewise function.

d. Interpret your graph in the context of the problem.

Source: © Andrew Hunt and Alan Sykes, *Chemistry*, 1984. Reprinted by permission of Pearson Education Limited.

34. The graph in Figure 10.30 shows the temperature of one gram of water as a function of the amount of heat applied, in calories.

a. How much heat is required to raise the temperature of one gram of water by one degree?

b. How much heat is required to convert one gram of ice to water?

c. How much heat is required to convert one gram of water to steam?

d. Write a piecewise function to describe the graph.

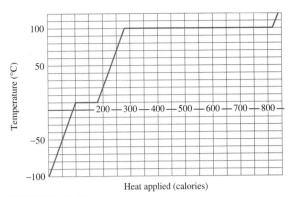

FIGURE 10.30

35. The function $f(t)$ describes a volunteer's heart rate during a treadmill test.

$$f(t) = \begin{cases} 100 & 0 \le t < 3 \\ 56t - 68 & 3 \le t < 4 \\ 186 - 500(0.5)^t & 4 \le t < 9 \\ 100 + 6.6(0.6)^{t-14} & 9 \le t < 20 \end{cases}$$

The heart rate is given in beats per minute and t is in minutes.

a. Sketch the graph of the function.

b. The treadmill test began with walking at 5.5 kilometers per hour, then jogging, starting at 12 kilometers per hour and increasing to 14 kilometers per hour, and finished with a cool-down walking period. Identify each of these activities on the graph, and describe the volunteer's heart rate during each phase.

Source: Davis and Autry, 1986

36. In Chapter 6 we considered the data from April and Tolu's experiment, which measured the pressure inside a spherical balloon as a function of its diameter. Refer to the graph of the data you made in Investigation 10, and recall that the pressure is given by the formula

$$P = \frac{F}{A}$$

where F is the force exerted by the balloon and A is the surface area of the balloon.

a. Hooke's law says that the force exerted by an elastic band is given by $F = kx$, where the band is stretched to length x. A spherical balloon is not precisely like an elastic band, and for $d \leq 21$, the data suggest that the force is not precisely proportional to d, but can be modeled by the power function $F(d) = 662d^{1.3}$. Write a formula for the pressure as a function of d for $d \leq 21$.

b. As the balloon expands, we approach the limits of its elasticity, and Hooke's law no longer applies. For $d > 21$, the function $F(d) = 1308.7(1.165)^d$ can be used to model the force of the balloon. Write a

formula for the pressure as a function of d for $d > 21$.

c. Write a piecewise formula for the pressure inside the balloon as a function of its diameter. Graph the function on top of the data from April and Tolu's experiment. How well does the function fit the data? Comment specifically on the data just before the balloon pops.

10.3 Inverse Functions

We often think of a function as a machine or process that acts on the elements of the domain (the "input" values) to produce the elements of the range (the "output" values). (See Figure 10.31.) For example, the function $f(x) = 2x$ doubles each input value, as shown in Table 10.9.

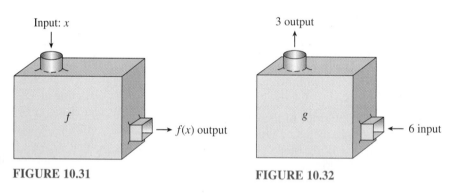

FIGURE 10.31   **FIGURE 10.32**

TABLE 10.9

x	$f(x)$
1	2
2	4
3	6
4	8

TABLE 10.10

x	$g(x)$
2	1
4	2
6	3
8	4

The **inverse** of a function f is obtained by running the function machine in reverse: The elements of the *range* of f are used as the input values, and the output values produced by the inverse function machine are the corresponding elements of the domain of f. (See Figure 10.32.) Consequently, we can make a table of values for the inverse function, g, by interchanging the columns of a table for f. (See Table 10.10.) Thus $g(2) = 1$ because $f(1) = 2$, $g(4) = 2$ because $f(2) = 4$, and so on.

We can make the following observation in general about inverse functions.

Suppose g is the inverse function for f. Then

$$g(b) = a \quad \text{if and only if} \quad f(a) = b$$

EXAMPLE 1 Suppose g is the inverse function for f, and we know the following function values for f:

$$f(-3) = 5, \quad f(2) = 1, \quad f(5) = 0$$

Find $g(5)$ and $g(0)$.

Solution

$g(5) = -3$ because $f(-3) = 5$, and $g(0) = 5$ because $f(5) = 0$. Tables may be helpful in visualizing the two functions, as shown in Tables 10.11a and 10.11b.

TABLE 10.11a

x	$f(x)$
-3	5
2	1
5	0

$\rightarrow$ Interchange the columns $\rightarrow$

TABLE 10.11b

x	$g(x)$
5	-3
1	2
0	5

■

> **EXERCISE 1** Suppose g is the inverse function for f, and we know the following function values for f:
>
> $$f(-1) = 0, \quad f(0) = 1, \quad f(1) = 2$$
>
> Find $g(0)$ and $g(1)$.

Finding a Formula for the Inverse Function

When we swap the columns in a table of values to find the inverse function, we are really interchanging the input and output variables. If a function is defined by an equation, we can find a formula for its inverse function in the same way: Interchange the variables in the equation. For the function $f(x) = 2x$ (or $y = 2x$), we interchange the variables x and y to obtain $x = 2y$. Solving this new equation for y, we find $y = \dfrac{x}{2}$. The inverse function for $f(x) = 2x$ is given by the formula $g(x) = \dfrac{x}{2}$.

EXAMPLE 2 **a.** Find the inverse of the function $f(x) = 4x - 3$.

  b. Make a table of values for $f(x)$ and a table for its inverse function.

Solutions

a. Write the equation for f in the form

$$y = 4x - 3$$

Interchange the variables to obtain

$$x = 4y - 3$$

and solve for y.

$$y = \frac{x}{4} + \frac{3}{4}$$

The inverse function is $g(x) = \dfrac{x}{4} + \dfrac{3}{4}$.

b. Choose values for x and evaluate $4x - 3$ for those x-values, as shown in Table 10.12a.

TABLE 10.12a

x	$f(x)$
0	-3
1	1
2	5
3	9

TABLE 10.12b

x	$g(x)$
-3	0
1	1
5	2
9	3

To make a table for g, we could choose values for x and evaluate $\dfrac{x}{4} + \dfrac{3}{4}$, but because g is the inverse function for f, we can simply interchange the columns in our table for f, as shown in Table 10.12b. ∎

EXERCISE 2 Find a formula for $g(x)$, the inverse of the function $f(x) = \dfrac{x}{3} - 5$.

The inverse function g "undoes" the effect of the function f. For example, the function $f(x) = 2x$ doubles the elements of its domain, and its inverse function, $g(x) = \dfrac{x}{2}$, halves them. If we apply the function f to a given input value and then apply the function g to the output from f, the end result will be the original input value. This is like feeding a number into the function machine and then feeding the result back in with the machine running in reverse. For example, consider the function $f(x) = 2x$ and its inverse, $g(x) = \dfrac{x}{2}$ again. If we choose $x = 5$ as an input value, we find that

$$f(5) = 2 \cdot 5 = 10, \qquad \text{and} \qquad g(10) = \frac{10}{2} = 5$$

Example 3 illustrates the fact that if g is the inverse function for f, then f is also the inverse function for g.

EXAMPLE 3

a. Find the inverse of the function $f(x) = x^3 + 2$.

b. Show that the inverse function "undoes" the effect of f on $x = 2$.

c. Show that f "undoes" the effect of the inverse function on $x = -25$.

Solutions

a. Write the equation for f in the form

$$y = x^3 + 2$$

Interchange the variables to obtain

$$x = y^3 + 2$$

and solve for y.

$$y^3 = x - 2 \qquad \text{Subtract 2 from both sides.}$$
$$y = \sqrt[3]{x - 2} \qquad \text{Take the cube root of both sides.}$$

The inverse function is $g(x) = \sqrt[3]{x - 2}$.

b. First evaluate the function f for $x = 2$.

$$f(2) = 2^3 + 2 = \mathbf{10}$$

Then evaluate the inverse function g at $x = 10$.

$$g(\mathbf{10}) = \sqrt[3]{10 - 2} = \sqrt[3]{8} = \mathbf{2}$$

We started and ended with 2.

c. First evaluate the function g for $x = -25$.

$$g(\mathbf{-25}) = \sqrt[3]{-25 - 2} = \mathbf{-3}$$

Then evaluate the inverse function f for $x = -3$.

$$f(\mathbf{-3}) = (-3)^3 + 2 = \mathbf{-25}$$

We started and ended with -25. ∎

EXERCISE 3

a. Find $g(x)$, the inverse of the function $f(x) = \sqrt[5]{x} + 2$.

b. Show that g "undoes" the effect of f on $x = -3$.

c. Show that f "undoes" the effect of g on $x = 2$.

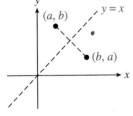

FIGURE 10.33

Graph of the Inverse Function

The graphs of a function and its inverse are related in an interesting way. To see this relationship, we first observe in Figure 10.33 that the points (a, b) and (b, a) are always located symmetrically with respect to the graph of $y = x$. Now, for every point (a, b) on the graph of f, the point (b, a) is on the graph of the inverse of f. Thus the graphs of $y = f(x)$ and its inverse, $y = g(x)$, are reflections of each other about the line $y = x$. Figure 10.34 shows the graphs of $y = 4x - 3$ and its inverse $y = \dfrac{x}{4} + \dfrac{3}{4}$, from Example 1, along with the graph of $y = x$.

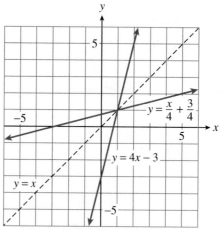

FIGURE 10.34

EXAMPLE 4

Graph the function

$$f(x) = x^3 + 2$$

and its inverse,

$$g(x) = \sqrt[3]{x - 2}$$

which we found in Example 2, on the same set of axes.

Solution

The graph of f is the graph of $y = x^3$ translated two units upward. The graph of $g(x)$ is the graph of $y = \sqrt[3]{x}$ translated two units to the right. The two graphs are symmetric about the line $y = x$. (See Figure 10.35.)

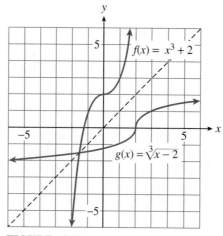

FIGURE 10.35

Horizontal Line Test

In some cases the inverse of a function is not itself a function. For example, to find the inverse of $f(x) = x^2$, first interchange x and y to obtain $x = y^2$ and then solve for y to get $y = \pm\sqrt{x}$. The graphs of f and its inverse are shown in Figure 10.36. Because the graph of the inverse does not pass the vertical line test, it is *not* a function. (See Section 5.2 to review the vertical line test.)

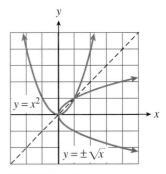

FIGURE 10.36

For many applications it is important to know whether or not the inverse of f is a function. This can be determined from the graph of f. When we interchange x and y to find a formula for the inverse, horizontal lines of the form $y = k$ become vertical lines of the form $x = k$. Thus if the graph of the *inverse* is to pass the vertical line test, the graph of the *original function* must pass the horizontal line test, namely, that no horizontal line should intersect the graph in more than one point. Notice that the graph of $f(x) = x^2$ does not pass the horizontal line test, so we would not expect its inverse to be a function.

Horizontal Line Test

If no horizontal line intersects the graph of a function more than once, then the inverse is also a function.

EXAMPLE 5   Which of the functions in Figure 10.37 have inverses that are also functions?

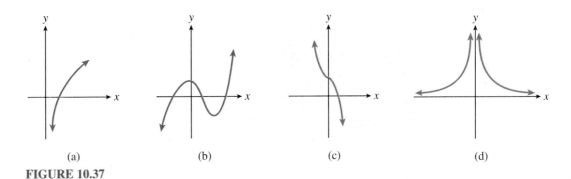

(a)   (b)   (c)   (d)

FIGURE 10.37

Solution

In each case, apply the horizontal line test to determine whether the inverse is a function. Because no horizontal line intersects their graphs more than once, the functions pictured in Figure 10.37(a) and (c) have inverses that are also functions. The functions in (b) and (d) do not have inverses that are functions. ■

EXERCISE 5 Which of the functions shown below have inverses that are also functions?

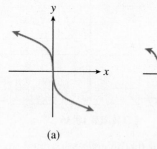

(a)

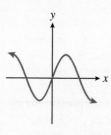

(b)

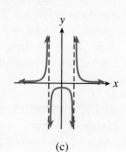

(c)

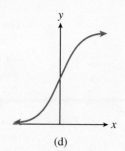

(d)

Inverse Function Notation

If the inverse of a function f is also a function, then the inverse is often denoted by the symbol f^{-1}, read "f inverse." For example, the function $f(x) = x^3 + 2$ passes the horizontal line test (see Example 4), so its inverse is a function and can be denoted by $f^{-1}(x) = \sqrt[3]{x - 2}$.

EXAMPLE 6

If $h(x) = 2x - 6$, find $h^{-1}(10)$.

Solution

First find the inverse function for $y = 2x - 6$. Interchange x and y to get

$$x = 2y - 6$$

and solve for y.

$$2y = x + 6$$

$$y = \frac{x}{2} + 3$$

The inverse function is

$$h^{-1}(x) = \frac{x}{2} + 3$$

Now evaluate the inverse function at $x = 10$.

$$h^{-1}(10) = \frac{10}{2} + 3 = 8$$

■

Although the same symbol, $^{-1}$, is used for both reciprocals and inverse functions, the two notions are not equivalent. That is, the inverse of a given function is usually not the same as the reciprocal of that function. In Example 6 note that $h^{-1}(x)$ is not the same as the reciprocal of $h(x)$, because $\dfrac{1}{h(x)} = \dfrac{1}{2x - 6}$. To avoid confusion, we use the notation $\dfrac{1}{h}$ to refer to the reciprocal of the function h.

EXERCISE 6 If $f(x) = \dfrac{1}{x + 1}$, find $f^{-1}(1)$.

READING QUESTIONS

1. Explain how the inverse of a function is related to the original function.
2. Explain how to find a formula for the inverse of a function.
3. How is the graph of the inverse function related to the graph of the original function?
4. What test can we apply to a function to decide whether it has an inverse?

ANSWERS TO 10.3 EXERCISES

1. $g(0) = -1$, $g(1) = 0$ 2. $g(x) = 3x + 15$
3a. $g(x) = x^5 - 2$ b. $f(-3) = -1$, and $g(-1) = -3$
c. $g(2) = 30$, and $f(30) = 2$
4. 5. (a) and (d)   6. $f^{-1}(1) = 0$

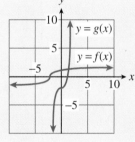

HOMEWORK 10.3

For Problems 1–12, (a) make a table of values for the function, then make a table for its inverse, (b) find a formula for the inverse of the function, (c) graph the function and its inverse on the same set of axes, along with the graph of $y = x$.

1. $f(x) = 2x - 6$
2. $f(x) = 3x - 1$
3. $f(x) = x^3 + 1$
4. $f(x) = \sqrt[3]{x + 1}$
5. $f(x) = \log x$
6. $f(x) = e^x$
7. $f(x) = \dfrac{1}{x - 1}$
8. $f(x) = \dfrac{1}{x} - 3$
9. $f(x) = \log(x) - 1$
10. $f(x) = \dfrac{1}{x}$
11. $f(x) = 7(10^x)$
12. $f(x) = \ln(x - 3)$

13. a. Find the inverse g of the function $f(x) = (x - 2)^3$.

  b. Show that g "undoes" the effect of f on $x = 4$.

  c. Show that f "undoes" the effect of g on $x = -8$.

  d. Graph the function and its inverse on the same set of axes, along with the graph of $y = x$.

14. a. Find the inverse g of the function $f(x) = \dfrac{2}{x + 1}$.

  b. Show that g "undoes" the effect of f on $x = 3$.

  c. Show that f "undoes" the effect of g on $x = -1$.

  d. Graph the function and its inverse on the same set of axes, along with the graph of $y = x$.

15. a. Find the inverse g of the function $f(x) = 10^{2x}$.

  b. Show that g "undoes" the effect of f on $x = -1$.

  c. Show that f "undoes" the effect of g on $x = 100$.

  d. Graph the function and its inverse on the same set of axes, along with the graph of $y = x$.

16. a. Find the inverse g of the function $f(x) = 5 + 10^{-x}$.

  b. Show that g "undoes" the effect of f on $x = 1$.

  c. Show that f "undoes" the effect of g on $x = 15$.

  d. Graph the function and its inverse on the same set of axes, along with the graph of $y = x$.

Which of the functions in Problems 17–24 have inverses that are also functions?

17. a. 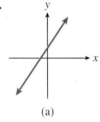 **b.** 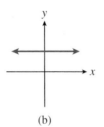 **c.** 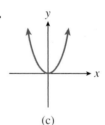 **d.**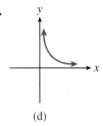

(a)   (b)   (c)   (d)

FIGURE 10.38

18. a. 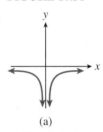 **b.** 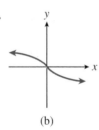 **c.** 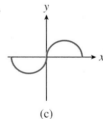 **d.**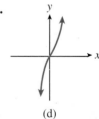

(a)   (b)   (c)   (d)

FIGURE 10.39

19. a. $f(x) = x$ **b.** $f(x) = x^2$
20. a. $f(x) = x^3$ **b.** $f(x) = x^4$

21. a. $f(x) = \dfrac{1}{x}$ **b.** $f(x) = \dfrac{1}{x^2}$
22. a. $f(x) = \sqrt{x}$ **b.** $f(x) = \sqrt[3]{x}$

23. a. $f(x) = 30 - 10^{-x}$ **b.** $f(x) = 30 - 10^{-x^2}$
24. a. $f(x) = \log(2x - 1)$
    b. $f(x) = \log(x^2 + 1)$

25. For each of the functions listed below, select the graph of its *inverse* function, if possible, from the figures labeled I–VI. (The inverse of one of the functions is not shown.)

a. $f(x) = 2^x$

b. $f(x) = x^2$

c. $f(x) = \dfrac{2}{x}$

d. $f(x) = \sqrt{x}$

e. $f(x) = \log_2 x$

f. $f(x) = \left(\dfrac{1}{2}\right)^x$

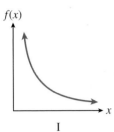

I

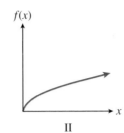

II

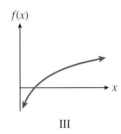

III

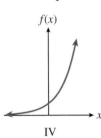

IV

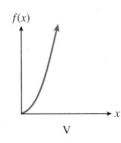

V

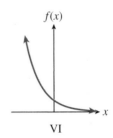

VI

FIGURE 10.40

26. Match each function shown in (a)–(d) with its inverse from I–IV. Find a formula for each function and its inverse.

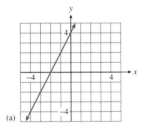

(a)

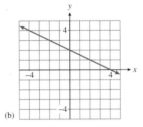

(b)

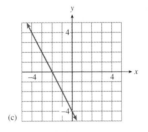

(c)

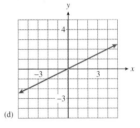

(d)

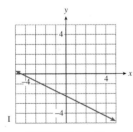

I

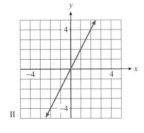

II

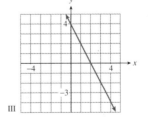

III

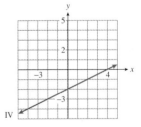

IV

FIGURE 10.41

27. If $F(t) = \frac{2}{3}t + 1$, find $F^{-1}(5)$.

28. If $G(s) = \frac{s - 3}{4}$, find $G^{-1}(-2)$.

29. If $m(v) = 6 - \frac{2}{v}$, find $m^{-1}(-3)$.

30. If $p(z) = 1 - 2z^3$, find $p^{-1}(7)$.

31. If $f(x) = \frac{x + 2}{x - 1}$, find $f^{-1}(2)$.

32. If $g(n) = \frac{3n + 1}{n - 3}$, find $g^{-1}(-2)$.

33. If $h(r) = \log_2 r$, find $h^{-1}(8)$.

34. If $H(w) = 3^w$, find $H^{-1}\left(\frac{1}{9}\right)$.

35. Let $f(-1) = 0, f(0) = 1, f(1) = -2$, and $f(2) = -1$.

   a. Make a table of values for $f(x)$ and another table for its inverse function.

   b. Find $f^{-1}(1)$.

   c. Find $f^{-1}(-1)$

36. Let $f(-2) = 1, f(-1) = -2, f(0) = 0$, and $f(1) = -1$.

   a. Make a table of values for $f(x)$ and another table for its inverse function.

   b. Find $f^{-1}(-1)$.

   c. Find $f^{-1}(1)$.

37. $f(x) = x^3 + x + 1$.

   a. Make a table of values for $f(x)$ and another table for its inverse function.

   b. Find $f^{-1}(1)$.

   c. Find $f^{-1}(3)$.

38. $f(x) = x^5 + x^3 + 7$.

   a. Make a table of values for $f(x)$ and another table for its inverse function.

   b. Find $f^{-1}(7)$.

   c. Find $f^{-1}(5)$.

39. If $f(x) = x^{2/3}$, is the inverse of f a function? Explain why or why not.

40. If $f(x) = x^{4/3}$, is the inverse of f a function? Explain why or why not.

41. The formula $C = \frac{5}{9}(F - 32)$ shows how to compute the Celsius temperature, C, in terms of the Fahrenheit temperature, F.

   a. Find the Celsius temperature of 104° Fahrenheit.

   b. Find the Fahrenheit temperature of 37° Celsius by substituting $C = 37$ in the given formula and solving for F.

   c. Find a formula that shows how to compute F in terms of C.

   d. Find the Fahrenheit temperature of 37°C by using your formula from part (c).

42. The amount, A, in a savings account paying 5% interest compounded continuously is given by the formula $A = 1000e^{0.05t}$, where t is the number of years since the initial principal of $1000 was deposited.

   a. Find the amount in the account after 7 years.

   b. Determine how long is required for the original principal to double by substituting $A = 2000$ and solving for t.

   c. Find a formula that shows the time, t, in terms of A.

   d. Use your formula from part (c) to find the time needed before the principal is $2000.

Graph each function.

1. $f(x) = x^2 - 1$ **2.** $g(x) = \sqrt{x} + 1$ **3.** $h(x) = \sqrt{x+1}$ **4.** $k(x) = (x-1)^2$

5. $F(x) = \dfrac{1}{x+2} - 3$ **6.** $G(x) = \sqrt[3]{x-3} + 1$ **7.** $H(x) = \dfrac{-1}{x}$ **8.** $K(x) = \dfrac{-1}{x^2}$

Evaluate the function for the specified domain values.

9. $g(p) = \dfrac{12}{1-p}$ for $p = -5$

10. $f(w) = -w^2 + 6$ for $w = -3$

11. $K(s) = \sqrt[3]{4-s}$ for $s = -b^2$

12. $h(t) = \dfrac{t+3}{t-5}$ for $t = x + 5$

13. Let $r(t) = (t+1)^3$, and evaluate each expression.

   a. $r(-3+3)$ **b.** $r(-3) + r(3)$

   c. $-3r(x) + 3$

14. Let $W(m) = \sqrt{25-m}$, and evaluate each expression.

   a. $5W(y^2) - 16$ **b.** $W(5-16)$

   c. $W(5) - W(16)$

Graph the piecewise-defined function.

15. $f(x) = \begin{cases} \dfrac{x}{3} + 4 & \text{if } x \le 0 \\ 4 - x & \text{if } x > 0 \end{cases}$

16. $f(x) = \begin{cases} 3 - x & \text{if } x \le 1 \\ 4 - 2x & \text{if } x > 1 \end{cases}$

Suppose that f has an inverse function f^{-1}. Use the given values of f to find the indicated values of f^{-1}.

17. $f(2) = 4, f(4) = 6, f(6) = 12$. Find $f^{-1}(4)$ and $f^{-1}(6)$.

18. $f(0) = 10, f(5) = 0, f(10) = -22$. Find $f^{-1}(0)$ and $f^{-1}(-22)$.

Find a formula for the inverse function.

19. $f(x) = \sqrt[3]{2x+1}$

20. $f(x) = \dfrac{5}{x+1}$

21. a. Show that the function f of Problem 19 "undoes" the effect of f^{-1} on $x = -3$.

   b. Show that f^{-1} "undoes" the effect of f on $x = 13$.

23. Graph the functions f and f^{-1} of Problem 19 on the same set of axes.

22. a. Show that the function f of Problem 20 "undoes" the effect of f^{-1} on $x = 1$.

   b. Show that f^{-1} "undoes" the effect of f on $x = -2$.

24. Graph the functions f and f^{-1} of Problem 20 on the same set of axes.

25. Which of the functions in Figure 10.42 have inverses that are also functions?

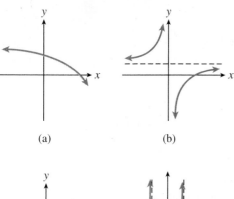

(a)   (b)

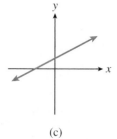
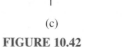

(c)

(d)

FIGURE 10.42

26. Which of the functions in Figure 10.43 have inverses that are also functions?

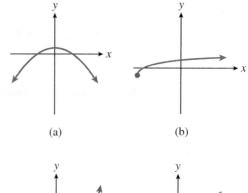

(a)   (b)

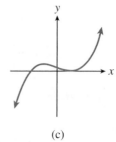
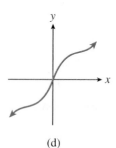

(c)   (d)

FIGURE 10.43

10.4   More about Modeling

We have used a variety of functions to model different physical phenomena or processes. In the next investigation we consider a simple example of a periodic function. A **periodic function** is one that repeats its range values at evenly spaced intervals, or periods, of the domain. Periodic functions are used to model phenomena that exhibit cyclical behavior, such as growth patterns in plants and animals, radio waves, and planetary motion.

INVESTIGATION 19

Periodic Functions

Delbert is standing at the point $(5, 0)$ on the square shown in Figure 10.44. If he walks around the square in the counter-clockwise direction, his position at any time depends on the distance d that he has walked. In other words, Delbert's y-coordinate is a function of the distance he has walked.

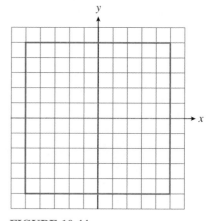

FIGURE 10.44

1. What is the y-coordinate of Delbert's position when he has walked 2 units? 5 units? 8 units? (Start filling in Table 10.13 in part (6).)

2. What is the y-coordinate of Delbert's position when he has walked between 0 and 5 units?

3. After Delbert reaches the upper right corner of the square, he will turn left and start walking along the top of the square. What is his y-coordinate along the top of the square?

4. What is the y-coordinate of Delbert's position when he has walked between 5 units and 15 units from the start?

5. Delbert will turn left again after he reaches the upper left corner and start walking down the left side of the square. What will be the y-coordinate of his position after he has walked a total of 18 units? 20 units? 22 units?

6. Fill in the rest of Table 10.13 with the y-coordinates of Delbert's position when he has walked a distance d. (Assume that he continues to go around the square more than once.)

TABLE 10.13

d	0	2	5	8	10	12	15	18	20	22	25	28	30	32	35	38	40	42	45	48	50
y																					

7. On the grid in Figure 10.45, plot a graph of Delbert's y-coordinate versus distance.

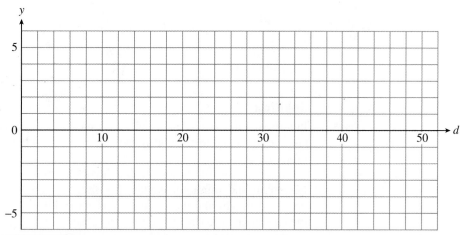

FIGURE 10.45

8. How far does Delbert walk before he starts his second time around the square? Before he starts his third time around?

9. Suppose you know where Delbert is on the square at some given time. Can you predict where he will be after he has walked another 40 units?

10. Let $y = f(d)$. What does your answer to part (9) say about $f(d)$ and $f(d + 40)$ for any positive value of d?

11. Find a formula for $f(d)$ when $0 \le d \le 5$.

12. Find a formula for $f(d)$ when $5 \le d \le 15$.

13. Find a formula for $f(d)$ when $15 \le d \le 25$.

14. How does the graph of f for $0 \le d \le 40$ compare with the graph for $40 \le d \le 80$?

15. Describe how the graph of f would continue for $0 \le d \le 400$.

16. Choose any positive value for d. How does the value of $f(d)$ compare with the value of $f(d + 20)$?

Periodic Functions

Here is another example about periodic functions.

EXAMPLE 1

Imagine a grandfather clock. As the minute hand sweeps around, the height of its tip changes with time. Which of the graphs in Figure 10.46 best represents the height of the tip of the minute hand as a function of time?

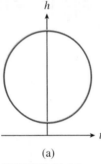
(a)

(b)

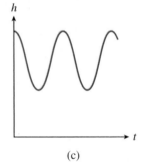
(c)

FIGURE 10.46

Solution

Figure 10.46(a) is not the graph of a function at all: Some values of t, such as $t = 0$, correspond to more than one value of h, which is not possible in the graph of a function.

Figure 10.46(b) shows the height of the minute hand varying between a maximum and minimum value. The height decreases at a constant rate (the graph is straight and the slope is constant) until the minimum is reached, and then increases at a constant rate. But notice in Figure 10.47 that during the 10 minutes from 12:10 to 12:20 the height of the minute hand decreases about half the diameter of the clock, while from 12:20 to 12:30 the height decreases only about a quarter of the diameter of the clock. Thus the height of the minute hand does *not* decrease at a constant rate.

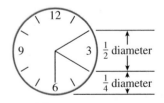
FIGURE 10.47

Figure 10.46(c) is the best choice. The graph is curved because the slopes are not constant. The graph is steep when the height is changing rapidly, and the graph is nearly horizontal when the height is changing slowly. The height changes slowly near the hour and the half-hour, and more rapidly near the quarter-hours. ■

EXERCISE 1 Consider the swinging pendulum of a grandfather clock. Which of the graphs in Figure 10.46 best represents the height of the pendulum as a function of time?

Some Geometric Models

We often need knowledge of geometry in order to find a formula for a function.

EXAMPLE 2

Consider several different lines that pass through the point (2, 1) and form right triangles in the first quadrant, as shown in Figure 10.48 The shape of the triangle depends on the slope of the line.

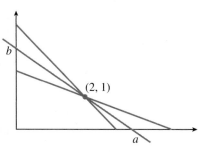

FIGURE 10.48

  a. Find the area of the triangle as a function of the slope, m, of the line through (2, 1).

  b. What is the domain of this function?

  c. Graph your function for values of m between -3 and 0.

  d. For what value of m does the triangle have the smallest area?

Solutions

  a. First, draw several different triangles that satisfy the conditions of the problem. Let a and b stand for the x- and y-intercepts of the line through (2, 1). The values of a and b depend on the slope of the line, and the area of the triangle is $A = \frac{1}{2} \cdot$ base $\cdot$ height, or

$$A = \frac{1}{2} \cdot a \cdot b$$

  Thus, we will be able to express the area A as a function of slope if we can express both a and b in terms of the slope.

  The point-slope equation of the line is

$$y - 1 = m(x - 2)$$

  To find the x-intercept, we substitute the point $(a, 0)$ into the equation of the line and solve for a.

$$0 - 1 = m(a - 2) \quad \text{Divide by } m.$$

$$\frac{-1}{m} = a - 2 \quad \text{Add 2.}$$

$$2 - \frac{1}{m} = a$$

This tells us that the x-intercept is $a = 2 - \dfrac{1}{m}$.

To find the y-intercept, we substitute the point $(0, b)$ into the equation of the line and solve for b.

$$b - 1 = m(0 - 2) \quad \text{Add 1.}$$
$$b = 1 - 2m$$

The y-intercept is $b = 1 - 2m$.

Because the area of the triangle is $A = \frac{1}{2} \cdot a \cdot b$, we obtain

$$A(m) = \frac{1}{2}\left(2 - \frac{1}{m}\right)(1 - 2m)$$

b. The domain of the function is the set of all values of the slope that will give a triangle in the first quadrant. We see that we have such a triangle exactly when the slope is negative, so the domain is the set of all negative numbers.

c. Use your calculator to graph the function $A(m)$ on the domain $[-3, 0)$. The graph is shown in Figure 10.49. Each point on the graph represents a different triangle. The x-coordinate gives the slope of the line that forms the triangle, and the y-coordinate gives the area of the triangle. For example, the point $(-1.5, 5.3)$ represents the triangle formed by the line with slope -1.5. The area of that triangle is $5\frac{1}{3}$ square units.

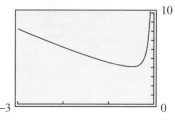

FIGURE 10.49

d. Use the $\boxed{\text{TRACE}}$ feature to find the coordinates of the lowest point on the graph. This point represents the triangle with the smallest area. Its x-coordinate is approximately -0.5. By zooming in we can verify that the coordinates of the minimum point are $(-0.5, 4)$, so the smallest triangle is formed by a line of slope -0.5, and its area is 4. ∎

EXERCISE 2 Draw a rectangle in the first quadrant as follows: Two sides lie on the coordinate axes, and the upper right vertex is a point P on the graph of $y = 12 - 2x$, as shown in the figure. The shape of the rectangle depends on which point on the line $y = 12 - 2x$ is used as a vertex.

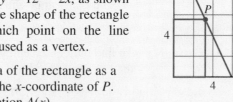

a. Find the area of the rectangle as a function of the x-coordinate of P. Call the function $A(x)$.

b. What is the domain of the function $A(x)$?

c. Graph the function $A(x)$ on its domain.

d. What is the maximum value of $A(x)$?

1. What is a periodic function?
2. What is true about the graph of a periodic function?
3. Describe one way to find the minimum value of a function.

ANSWERS TO 10.4 EXERCISES

1. (c) **2a.** $A = 12x - 2x^2$ **b.** $0 < x < 6$

c. **d.** 18

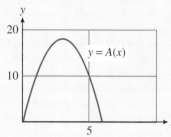

$y = A(x)$

HOMEWORK 10.4

1. Refer to Investigation 19. Sketch a graph of the x-coordinate of Delbert's position as a function of time.

2. In Example 1 on page 656 we considered the tip of the minute hand on a grandfather clock. Suppose that the face of the clock is one foot in diameter, and the bottom of the clock is five feet from the ground. Sketch a graph of the x-coordinate of the tip as a function of time.

3. At a ski slope, the lift chairs take 5 minutes to travel from the bottom, at an elevation of 3000 feet, to the top, at elevation 4000 feet. The cable supporting the ski lift chairs is a loop turning on pulleys at a constant speed. At the top and bottom, the chairs are at a constant elevation for a few seconds to allow skiers to get on and off.

  a. Sketch a graph of $h(t)$, the height of one chair at time, t. Show at least two complete up-and-down trips.

  b. What is the period of $h(t)$?

4. The heater in Paul's house doesn't have a thermostat; it runs on a timer. It uses 300 watts when it is running. Paul sets the heater to run from 6 a.m. to noon, and again from 4 p.m. to 10 p.m. in the evening.

  a. Sketch a graph of $P(t)$, the power drawn by the heater as a function of time. Show at least two days of heater use.

  b. What is the period of $P(t)$?

5. Match each of the following situations with an appropriate graph from Figure 10.50.

a. When the heart contracts, blood pressure in the arteries rises rapidly to a peak (systolic blood pressure) and then falls off quickly to a minimum (diastolic blood pressure). Blood pressure is a periodic function of time.

b. After an injection is given to a patient, the amount of the drug present in his bloodstream decreases exponentially. The patient receives injections at regular intervals to restore the drug level to the prescribed level. The amount of the drug present is a periodic function of time.

c. The monorail shuttle train between the north and south terminals at Gatwick Airport departs from the south terminal every 12 minutes. The distance from the train to the south terminal is a periodic function of time.

d. Delbert gets a haircut every two weeks. The length of his hair is a periodic function of time.

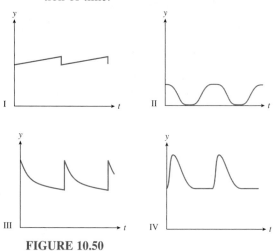

FIGURE 10.50

6. Match each of the following situations with an appropriate graph from Figure 10.51.

a. The number of hours of daylight in Salt Lake City vary from a minimum of 9.6 hours on the winter solstice to a maximum of 14.4 hours on the summer solstice.

b. A weight is 6.5 feet above the floor, suspended from the ceiling by a spring. The weight is pulled down to 5 feet above the floor and released, rising past 6.5 feet in 0.5 seconds before attaining its maximum height of 8 feet. Neglecting the effects of friction, the height of the weight will continue to oscillate between its minimum and maximum height.

c. The voltage used in U.S. electrical current changes from 155 V to −155 V and back 60 times each second.

d. Although the moon is spherical, what we can see from earth looks like a (sometimes only partly visible) disk. The percentage of the moon's disk that is visible varies between 0 (at new moon) to 100 (at full moon).

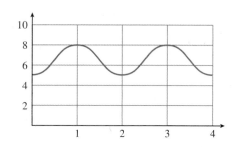

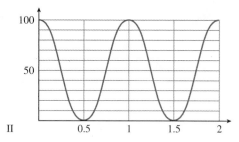

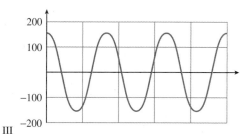

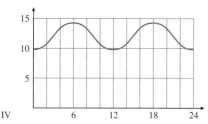

FIGURE 10.51

7. The apparent magnitude of a star is a measure of its brightness as seen from Earth. Smaller values for the apparent magnitude correspond to brighter stars. The graph in Figure 10.52, called a light curve, shows the apparent magnitude of the star Algol as a function of time. Algol is an eclipsing binary star, which means that it is actually a system of two stars, a bright principal star and its dimmer companion, in orbit around each other. As each star passes in front of the other, it eclipses some of the light that reaches Earth from the system.

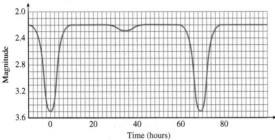

FIGURE 10.52

a. The light curve is periodic. What is its period?

b. What is the range of apparent magnitudes of the Algol system?

c. Explain the large and small dips in the light curve. What is happening to cause the dips?

Source: *Matter, Earth and Sky*, Second Edition by Gamow, © 1965. Reprinted by permission of Prentice-Hall, Inc., Upper Saddle River, NJ. Found in Brandt and Maran, *New Horizon in Astronomy*, 1972.

8. Some stars, called Cepheid variable stars, appear to pulse, getting brighter and dimmer periodically. The graph in Figure 10.53 shows the light curve for the star Delta Cephei.

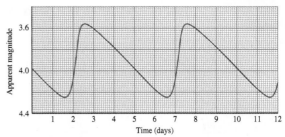

FIGURE 10.53

a. What is the period of the graph?

b. What is the range of apparent magnitudes for Delta Cephei?

c. In 1912 the astronomer Henrietta Leavitt discovered that the absolute magnitude of a Cepheid variable is a function of its period. (Comparing the absolute and apparent magnitudes of a star allows astronomers to calculate its distance from Earth. The Cepheid variables are some of the most distant visible objects.) Use the graph in Figure 10.54 to estimate the absolute magnitude of Delta Cephei.

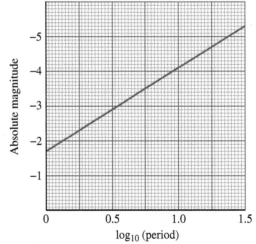

FIGURE 10.54
Source: Ingham, 1997.

Find a linear or quadratic equation that describes each of the following functions.

9.

x	f(x)
1	1
2	4
3	7
4	10
5	13

10.

x	g(x)
1	16
2	12
3	8
4	4
5	0

11.

t	G(t)
1	2
2	5
3	10
4	17
5	26

12.

t	H(t)
1	2
2	8
3	18
4	32
5	50

Find appropriate formulas for each of the following functions.

13. Sketch a rectangle with its base on the *x*-axis and its upper vertices both on the graph of the semicircle $y = \sqrt{16 - x^2}$. (See Figure 10.55.) Note that you can draw many different rectangles, depending on where you choose the vertex (a, b).

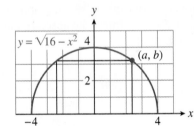

FIGURE 10.55

   a. Express the length of the rectangle's base in terms of *a*.

   b. Express the height of the rectangle in terms of *a*. (*Hint:* Don't forget that the point (a, b) lies on the graph of the semicircle)

   c. Find the area of the rectangle as a function of the *x*-coordinate, *a*, of its upper right corner.

   d. What is the domain of your function? (What values of *a* result in a rectangle?)

   e. Graph your function on its domain. What is the area of the rectangle formed if $a = 2$? What value(s) of *a* result in a rectangle with area 10 square units?

   f. What value of *a* results in the rectangle of largest area?

14. Sketch a rectangle with its base on the *x*-axis and its upper vertices both on the graph of the parabola $y = 25 - x^2$. (See Figure 10.56.) Note that you can draw many different rectangles, depending on where you choose the vertex (a, b).

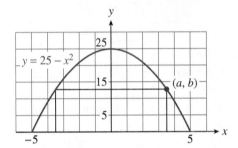

FIGURE 10.56

   a. Express the length of the rectangle's base in terms of *a*.

   b. Express the height of the rectangle in terms of *a*. (*Hint:* Don't forget that the point (a, b) lies on the graph of the parabola.)

   c. Find the area of the rectangle as a function of the *x*-coordinate, *a*, of its upper right corner.

   d. What is the domain of your function? (What values of *a* result in a rectangle?)

   e. Graph your function on its domain. What is the area of the rectangle formed if $a = 4$? What value(s) of *a* result in a rectangle with area 75 square units?

   f. What value of *a* results in the rectangle of largest area?

15. A triangle is formed in the first quadrant as follows: Two of the sides are formed by the *x*- and *y*-axes, the third side is on a line passing through the point (2, 1). Note that there are many different triangles you can draw in this way, depending on the slope of the line through (2, 1).

   a. Let (a, 0) be the *x*-intercept of the third side of the triangle. Express the slope of the third side in terms of a.

   b. Let (0, b) be the *y*-intercept of the third side. Express b in terms of a. (Use your expression for the slope.)

   c. Express the area of the triangle as a function of a.

   d. What is the domain of your function? (What values of a result in a triangle?)

   e. Graph your function on the domain (2, 20.8]. What is the area of the triangle if $a = 3$? What value(s) of a result in a triangle of area 7.2 square units?

   f. What value of a results in the triangle with the smallest area? Describe what happens to the area of the triangle as a increases from 2 to larger values.

16. A triangle is formed in the first quadrant as follows: Two of the sides are formed by the *x*- and *y*-axes, the third side is on a line passing through the point (4, 2).

   a. Let (0, b) be the *y*-intercept of the third side of the triangle. Express the slope of the third side in terms of b.

   b. Let (a, 0) be the *x*-intercept of the third side. Express a in terms of b. (Use your expression for the slope.)

   c. Express the area of the triangle as a function of b.

   d. What is the domain of your function? (What values of b result in a triangle?)

   e. Graph your function on the domain (2, 11.4]. What is the area of the triangle if $b = 3$? What value(s) of b result in a triangle of area 25 square units?

   f. What value of b results in the triangle with the smallest area? Describe what happens to the area of the triangle as the values of b increase.

17. Suppose you would like to measure the radius of a circular pond in the park. It will be easier to measure the circumference of the pond, and then calculate the radius.

   a. Express the radius of a circle in terms of its circumference.

   b. Express the area of a circle as a function of its circumference.

   c. What type of function is your answer to part (b)? What is the domain of your function?

   d. What is the area of a pond whose circumference is 100 yards?

18. The hypotenuse of a right triangle is 12 centimeters long.

   a. If the length of one leg of the triangle is x, what is the length of the other leg?

   b. Express the area of the triangle as a function of x.

   c. What is the domain of your function? Graph your function on its domain.

   d. What is the area of the triangle if the shortest side is $\sqrt{6} \approx 2.45$ inches long?

10.5 Joint Variation

We have often used tables to show a relationship between two variables. In this section we use tables relating three variables. The column headings and row headings show the values of two input variables, and the table entries show the values of the third or output variable.

EXAMPLE 1

Table 10.14 shows the "wind-chill factor" for various combinations of temperature and wind speed.

TABLE 10.14 Wind-chill factors

		\multicolumn{8}{c}{Temperature (° F)}							
		35	30	25	20	15	10	5	0
	5	33	27	21	16	12	7	0	−5
Wind	10	22	16	10	3	−3	−9	−15	−22
speed	15	16	9	2	−5	−11	−18	−25	−31
(mph)	20	12	4	−3	−10	−17	−24	−31	−39
	25	8	1	−7	−15	−22	−29	−36	−44

a. What is the wind-chill factor when the temperature is 15° F and the wind is blowing at 20 mph?

b. At what temperature is the wind-chill factor −15 during a 10 mph wind?

c. What wind speed is needed for a temperature of 30° F to have a wind-chill factor of 1?

Solutions

a. We look in the column for 15° and the row for 20 mph. The associated wind-chill factor is −17.

b. We look in the row for 10 mph until we find the wind-chill factor of −15. The column heading is 5, so the associated temperature is 5° F.

c. In the 30° F column, we find the wind-chill factor of 1 in the 25 mph row, so the wind speed is 25 mph. ∎

EXERCISE 1 Under the company retirement plan, an employee contributes a fixed amount each year until retirement. The amount accumulated (including 8% annual interest) depends on the amount of the annual contribution and on the number of years the employee contributes to the plan.

Amount in retirement fund

Annual contribution	\multicolumn{5}{c}{Number of years of contributions}				
	10	20	30	40	50
500	7243	22,881	56,642	129,528	286,885
1000	14,487	45,762	113,283	259,057	573,770
1500	21,730	68,643	169,925	388,585	860,655
2000	28,973	91,524	226,566	518,113	1,147,540
2500	43,460	137,286	339,850	777,170	1,721,310
3000	50,703	160,167	396,491	906,698	2,008,196

a. If an employee contributes $500 each year, how much will have accumulated after 40 years?

b. How much must you contribute each year in order to accumulate $573,770 after 50 years?

c. If you make annual payments of $2500, how many years will it take to accumulate $137,286?

EXAMPLE 2

A study on the death rates at two hospitals collected information about a particular kind of surgery. Patients were classified as being in good or poor condition before the surgery. The data are shown in Table 10.15.

TABLE 10.15 Patients after a surgery at two hospitals

	Good condition		Poor condition	
	Hospital A	Hospital B	Hospital A	Hospital B
Survived	937	918	1824	563
Died	10	12	77	29

a. How many patients received the surgery at Hospital A? How many at Hospital B?

b. What percentage of patients at Hospital A did not survive the surgery? At Hospital B? Which hospital had the higher death rate?

c. Among patients at Hospital A who went into surgery in good condition, what percent did not survive the surgery? Among the patients at Hospital B?

d. Among patients at Hospital A who went into surgery in poor condition, what percent did not survive the surgery? Among the patients at Hospital B?

Solutions

a. At Hospital A, a total of $937 + 10 + 1824 + 77 = 2848$ patients received the surgery. At Hospital B, there were $918 + 12 + 563 + 29 = 1522$ patients.

b. At Hospital A, $10 + 77 = 87$ patients died in surgery. This represents 87/2848 or 3.1% of the patients who underwent the surgery. At Hospital B, $12 + 29 = 41$ patients died, or $41/1522 \approx 2.7\%$ of the total. Hospital A had the higher death rate.

c. At Hospital A there were $937 + 10 = 947$ patients in good condition before surgery. The fraction of these patients who died was $10/947 \approx 1.1\%$. At Hospital B there were $918 + 12 = 930$ patients in good condition. The fraction who died was $12/930 \approx 1.3\%$. Among patients in good condition, Hospital B had the higher death rate.

d. At Hospital A there were $1824 + 77 = 1901$ patients who went into surgery in poor condition. The fraction who died was $77/1901 \approx 4.1\%$. At Hospital B there were a total of $563 + 29 = 592$ patients in poor condition. The fraction who did not survive was $29/592 \approx 4.9\%$. Among patients in poor condition, Hospital B had the higher death rate. ■

In Example 2, Hospital A has a lower death rate than Hospital B both for patients in good condition and for patients in poor condition, yet Hospital A has the higher death rate overall. This apparently contradictory situation is an example of **Simpson's Paradox.** Notice that for either hospital, the death rate is higher for the patients in poor condition than for the patients in good condition. Because the majority of the patients for Hospital A began in poor condition, the death rate at Hospital A appears worse than the death rate at Hospital B, where the majority of the patients began in good condition.

EXERCISE 2 Delbert is comparing two stock market advisors based on their records from the years 1990 and 2000. Each advisor recommended certain stocks, some of which were profitable and others were not. The results are summarized in the table below.

Recommendations from two stock market advisors

	1990		2000	
	Advisor A	**Advisor B**	**Advisor A**	**Advisor B**
Profitable	32	36	45	14
Not profitable	28	28	23	6

a. How many stocks were recommended by advisor A in the two years? How many by advisor B?

b. How many of the stocks recommended by advisor A were profitable? How many of the stocks recommended by advisor B?

c. What percent of all the stocks recommended by advisor A were profitable? By advisor B? Which advisor recommended the higher percentage of profitable stocks?

d. What percent of the stocks recommmended by advisor A in 1990 were profitable? What was advisor B's record in 1990?

e. What percent of the stocks recommended by advisor A in 2000 were profitable? What was advisor B's record in 2000?

Joint Variation

Sometimes we can find patterns relating the entries in a table.

EXAMPLE 3

Rectangular beams of a given length can support a load that depends on both the width and the depth of the beam. Table 10.16 shows some of the values.

TABLE 10.16 Maximum load (kilograms)

		Width (centimeters)					
		0	**1**	**2**	**3**	**4**	**5**
	0	0	0	0	0	0	0
	1	0	10	20	30	40	50
Depth	**2**	0	40	80	120	160	200
(cm)	**3**	0	90	180	270	360	450
	4	0	160	320	480	640	800
	5	0	250	500	750	1000	1250

a. How great a load can a beam support if it has a width of 2 centimeters and a depth of 5 centimeters?

b. How great a load can a beam support if it has a width of 5 centimeters and a depth of 2 centimeters?

c. Consider the row corresponding to a depth of 3 centimeters. How does the load depend on the width?

d. Consider the column corresponding to a width of 3 cm. How does the load depend on the depth?

Solutions

a. From the column for 2 centimeters and the row for 5 centimeters, we find that the maximum load is 500 kilograms.

b. From the column for 5 centimeters and the row for 2 centimeters, we find that the maximum load is 200 kilograms.

c. The loads for different widths and a depth of 3 centimeters describe a function of width, as shown in the table taken from the fourth row of Table 10.16. For each increase of 1 centimeter in the width of the beam, the maximum load increases by 90 kilograms. When we graph load versus width, the points will lie on a straight line, as shown in Figure 10.57.

Width	Load
0	0
1	90
2	180
3	270
4	360
5	450

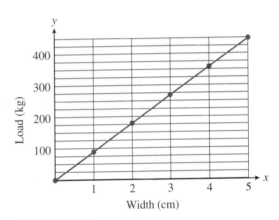

FIGURE 10.57

Because this line passes through the points $(0, 0)$ and $(1, 90)$, its slope is 90 and its y-intercept is $(0, 0)$. Thus, the equation $y = 90x$ relates the maximum load to the width of the beam, where x is the width in centimeters and y is the load in kilograms.

d. We can record the loads from the column for width 3 centimeters in a new table showing the relationship between load and depth. We notice that the increase in load for each increase of 1 centimeter in width is not a constant, so the graph in Figure 10.58 is not a straight line.

The curve does pass through the origin, so perhaps the data describe direct variation with a power of depth. If we

Depth	Load
0	0
1	30
2	120
3	270
4	480
5	750

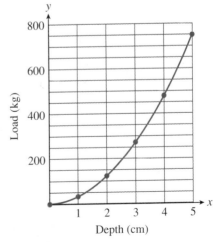

FIGURE 10.58

try the form $y = kx^2$ and use the point $(1, 30)$, we find that $30 = k \cdot 1^2$, so $k = 30$. The equation $y = 30x^2$ does fit the different depths $x = 0, 2, 3, 4$, and 5. ∎

In Example 3, the load varies directly with the width when the depth is 3 centimeters. In fact, the load varies directly with the width for any fixed depth. Also, the load varies with the square of depth, not only when the width is 3 centimeters, but for any fixed width. Consequently, we can find a constant k such that

$$\text{load} = k \cdot \text{width} \cdot \text{depth}^2$$

This relationship between variables is an example of **joint variation.** Whenever

$$z = kxy$$

we say that z **varies jointly** with x and y. If

$$z = k\frac{x}{y}$$

we say that z varies directly with x and inversely with y.

EXAMPLE 4   Find an equation relating the load, width, and depth for the data in Example 3.

Solution

We will use the variable L for the load, w for the width, and d for the depth. The equation we want will have the form

$$L = kwd^2$$

for some value of k, which we must determine. Use the fact, from Table 10.16, that $L = 10$ when, $w = 1 = d$. Then

$$10 = k(1)(1^2)$$
$$10 = k$$

The equation relating load, width, and depth is

$$L = 10wd^2$$

You can check that this formula works for all the values in the table. ∎

EXERCISE 3   The cost of tiling a rectangular floor depends on the length and width of the floor. The table shows the costs in dollars for various floors.

		Length (feet)					
		5	**6**	**7**	**8**	**9**	**10**
	5	400	480	560	640	720	800
	6	480	576	672	768	864	960
Width	**7**	560	672	784	896	1008	1120
(feet)	**8**	640	768	896	1024	1152	1280
	9	720	864	1008	1152	1296	1440
	10	800	960	1120	1280	1440	1600

a. How much does it cost to tile a floor that is 7 feet by 8 feet?

b. How much does it cost to tile a floor that is 5 feet by 9 feet?

c. Suppose you are tiling a floor that is 6 feet wide. How does the cost depend on the length of the floor?

d. Suppose you are tiling a floor that is 10 feet long. How does the cost depend on the width of the floor?

e. The cost varies jointly with the length and width of the floor. Find a formula for the cost of tiling the floor in terms of its length and width.

EXAMPLE 5

The gravitational attraction between a planet and its satellite varies directly with the mass of the satellite and inversely with the square of the distance between them, as shown in Table 10.17.

a. Find an equation relating gravitational force to mass and distance for the data given in the table.

b. Use the equation to predict the gravitational force when the mass is 5 and the distance is 0.1.

TABLE 10.17 Force

		Distance			
		10	**15**	**20**	**25**
Mass	1	0.1	$0.0\overline{4}$	0.025	0.016
	2	0.2	$0.0\overline{8}$	0.050	0.032
	3	0.3	$0.1\overline{3}$	0.075	0.048
	4	0.4	$0.1\overline{7}$	0.100	0.064

Solutions

a. We will use the variable F for the gravitational force, m for the mass, and d for the distance. The equation should have the form

$$F = k\frac{m}{d^2}$$

The variable m is in the numerator because it varies directly with force, and d is in the denominator because force varies inversely with d^2. To determine the value of the constant of proportionality, we can use any entry in the table. For example, $F = 0.1$ when $m = 1$ and $d = 10$. Thus

$$0.1 = k\frac{1}{10^2}$$

$$10 = k$$

and the equation is

$$F = 10\frac{m}{d^2}$$

b. We substitute $m = 5$ and $d = 0.1$ to obtain

$$F = \frac{10(5)}{(0.1)^2} = 5000$$

The force is 5000 when the mass is 5 and the distance 0.1. ∎

> **EXERCISE 4** Use your formula from Exercise 3 to predict the cost of tiling a rectangular floor that is 12 feet wide and 15 feet long.

READING QUESTIONS

1. How can you use a table to show the dependence of an output variable on two input variables?
2. What is joint variation?
3. For the equation in Example 5, describe the graph of force versus distance for a satellite of mass 1.

ANSWERS TO 10.5 EXERCISES

1a. $129,528   **b.** $1000   **c.** 20 years

2a. A: 128; B: 84   **b.** A: 77; B: 50

c. A: 60.2%; B: 59.5%; A has the higher percentage.

d. A: 53.3% profitable; B: 56.3% profitable

e. A: 66.2%; B: 70%

3a. $896   **b.** $720   **c.** cost = 96 · length   **d.** cost = 160 · width

e. cost = 16 · length · width

4. $2,880

HOMEWORK 10.5

Problems 1–4 show examples of Simpson's Paradox.

1. A college is being investigated for discrimination against women. Table 10.18 shows the number of men and women candidates who were accepted into the English and the Engineering graduate programs.

TABLE 10.18 Candidates accepted into graduate school

	English department		Engineering department	
	Men	Women	Men	Women
Accepted	5	17	63	5
Denied	20	62	9	0

a. Calculate the total number of men who applied to graduate school and the percentage of men who were accepted.

b. Calculate the total number of women who applied to graduate school and the percentage of women who were accepted.

c. Of the men who applied to the English program, what percent were accepted?

d. Of the women who applied to the English program, what percent were accepted?

e. Of the men who applied to the Engineering program, what percent were accepted?

f. Of the women who applied to the Engineering program, what percent were accepted?

g. In each department, did men or women have a better success rate? Which group had a better success rate overall?

2. One of the arguments against the death penalty in the United States is that it is imposed in a discriminatory fashion. Table 10.19 shows data compiled in 1981 indicating whether or not convicted murderers were sentenced to death. The factors considered were the race of the defendant and the race of the victim.

TABLE 10.19   Sentences for murder

	Black defendant		White defendant	
	Black victim	White victim	Black victim	White victim
Death	6	11	0	19
Not	97	52	9	132

a. Calculate the total number of black defendants and the percentage of them who were sentenced to death.

b. Calculate the total number of white defendants and the percentage of them who were sentenced to death.

c. Of the black defendants who murdered black victims, what percent were sentenced to death?

d. Of the white defendants who murdered black victims, what percent were sentenced to death?

e. Of the black defendants with white victims, what percent were sentenced to death?

f. Of the white defendants with white victims, what percent were sentenced to death?

g. Was a black defendant or a white defendant more likely to be sentenced to death for killing a black victim? For killing a white victim?

3. The starting salaries of graduates from Mid-State Tech are shown in Table 10.20. The salaries are classified according to the type of job accepted and the level of the graduate's degree.

TABLE 10.20   Salaries of graduates from Mid-State Tech

	Bachelors degree		Masters degree	
	Teaching	Industry	Teaching	Industry
Starting salary	$16,000	$25,000	$23,000	$31,000
Number	5	55	36	4

a. Who has the higher starting salary: teachers with bachelors degrees or teachers with masters degrees?

b. Who has the higher starting salary: industry employees with bachelors degrees or industry employees with masters degrees?

c. What is the average starting salary for graduates with bachelors degrees? (*Hint:* First compute the sum of the starting salaries for all the bachelors degree graduates.)

d. What is the average starting salary for graduates with masters degrees? (*Hint:* First compute the sum of the starting salaries for all the masters degree graduates.)

e. Who has the higher average starting salary: graduates with bachelors degrees or graduates with masters degrees?

4. A professional baseball player named Joe wants to negotiate a new contract based on his batting performance. Joe is a switch-hitter, that is, he bats both left- and right-handed. Table 10.21 shows his batting record for two seasons.

TABLE 10.21 Joe's batting record over two seasons

	Last season		This season	
	Left-handed	**Right-handed**	**Left-handed**	**Right-handed**
Hits	26	55	49	22
At-bats	97	218	190	93

a. Compute Joe's batting average last season. (Batting average is the fraction of "at-bats" that result in "hits.")

b. Compute Joe's batting average this season.

c. What was Joe's left-handed batting average last season?

d. What was Joe's right-handed batting average last season?

e. What is Joe's left-handed batting average this season?

f. What is Joe's right-handed batting average this season?

g. Has Joe's left-handed batting average improved? His right-handed batting average? His overall batting average?

5. The cost of insulating the ceiling in a building depends on the thickness of the insulation and the area of the ceiling. Table 10.22 shows some insulation costs.

TABLE 10.22 Cost (dollars)

		Area (square meters)					
		100	**200**	**300**	**400**	**500**	**600**
	4	72	144	216	288	360	432
	5	90	180	270	360	450	540
Thickness (cm)	6	108	216	324	432	540	648
	7	126	252	378	504	630	756
	8	144	288	432	576	720	864
	9	162	324	486	648	810	972

a. What does it cost to insulate a ceiling with an area of 500 square meters with 5 centimeters of insulation?

b. What thickness can you buy for $864 if the ceiling has an area of 600 square meters?

c. Consider the row corresponding to a thickness of 4 cm. How does the cost of insulating the ceiling depend on the area of the ceiling?

d. Consider the column corresponding to an area of 100 square meters. How does the cost depend on the thickness of the insulation?

e. The cost varies jointly with the thickness of the insulation and the area of the ceiling. Write an equation relating the cost to the area and thickness of insulation.

f. Use your formula from part (e) to determine the cost of insulating a building with 10 centimeters of insulation if the area of the ceiling is 800 square meters.

6. The volume of a certain quantity of helium is related both to the absolute temperature and the pressure of the gas. Table 10.23 shows volumes at some temperatures and pressures.

TABLE 10.23 Volume (cubic meters)

		Absolute temperature (Kelvin)					
		100	**150**	**200**	**250**	**300**	**350**
	1	18	27	36	45	54	63
Pressure (atmospheres)	2	9	13.5	18	22.5	27	31.5
	3	6	9	12	15	18	21
	4	4.5	6.75	9	11.25	13.5	15.75

a. What is the volume of helium when the pressure is 4 atmospheres and the temperature is 350 K?

b. What temperature is necessary to produce a volume of 15 cubic meters at a pressure of 3 atmospheres?

c. Consider the row corresponding to 2 atmospheres. How is the volume related to the absolute temperature?

d. Consider the column corresponding to 300 K. How is the volume related to the pressure?

e. The volume of the gas varies directly with the absolute temperature and inversely with the pressure. Write an equation relating the volume, absolute temperature, and pressure.

f. Use your formula from part (e) to determine the volume of the helium if the absolute temperature is 50 K and the pressure is 0.4 atmosphere.

7. The kinetic energy of a moving object varies jointly with the mass and the square of the velocity of the object. Suppose that a 1.5 kilogram mass moving at 2 meters per second has a kinetic energy of 3 joules.

   a. Write an equation relating the kinetic energy of an object to its mass and velocity.

   b. Use your equation to make a table of kinetic energy values for objects of mass 1.0, 1.5, 2.0, and 2.5 kilograms moving at velocities of 1.0, 2.0, 3.0, and 4.0 meters per second.

8. The cost of paint for a cylindrical grain silo varies jointly with the height and the radius of the silo. Suppose that the paint for a silo that is 40 feet tall with a radius of 20 feet costs $160.

   a. Write an equation relating the cost of the paint with the height and radius of the silo.

   b. Use your equation to make a table of paint costs for silos of height 30, 40, 50, and 60 feet tall with radii of 20, 25, and 30 feet.

9. Ammonia has many uses in industry and agriculture, including the production of fertilizers. It is produced in the laboratory from nitrogen and hydrogen, but the process requires high pressure and temperature for significant yield. The graph in Figure 10.59 illustrates the relationship.

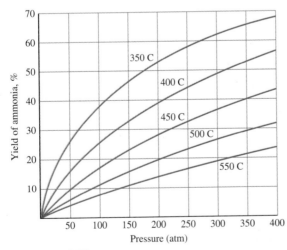

FIGURE 10.59

a. Complete the table showing the yield of ammonia, as a percent of the gas mixture leaving the reactor, at various pressures and temperatures.

	Pressure (atmospheres)							
Temperature (°C)	50	100	150	200	250	300	350	400
350								
400								
450								
500								
550								

b. What happens to the yield of ammonia if the pressure is held constant but the temperature is increased beyond 350°C?

c. Sketch a graph of the yield of ammonia as a function of temperature when the pressure is 300 atmospheres.

Source: © Andrew Hunt and Alan Sykes, Chemistry, 1984. Reprinted by permission of Pearson Education Limited.

10. The graph in Figure 10.60 shows the *heat index*, which combines air temperature and relative humidity to determine an "apparent" temperature, or what the temperature actually feels like.

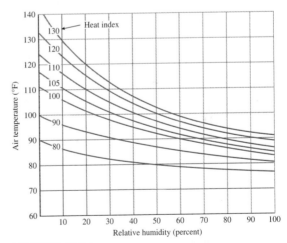

FIGURE 10.60

   a. Complete the table showing the heat index for various combinations of air temperature and relative humidity.

	Relative Humidity (%)					
Air temperature (°F)	0	20	40	60	80	100
80						
90						
100						
110						
120						

Air temperature (°F)	80	90	100	110	120
Relative humidity (%)					

c. Sketch a graph of the heat index as a function of air temperature if the relative humidity is 70%

Source: Ahrens, 1998.

b. Complete the table showing the relative humidity at which the heat index is equal to the actual air temperature.

CHAPTER 10 REVIEW

For Problems 1–10, (a) describe each function as a transformation of a basic function, (b) sketch a graph of the basic function and the given function on the same axes.

1. $g(x) = |x| + 2$

2. $F(t) = \dfrac{1}{t} - 2$

3. $f(s) = \sqrt{s} + 3$

4. $g(u) = \sqrt{u + 2} - 3$

5. $G(t) = |t + 2| - 3$

6. $H(t) = \dfrac{1}{(t - 2)^2} + 3$

7. $h(s) = -2\sqrt{s}$

8. $g(s) = \dfrac{1}{2}|s|$

9. $Q(t) = 2^{t+4}$

10. $N(t) = -\dfrac{1}{4}(2^t)$

Give an equation for the function graphed.

11.

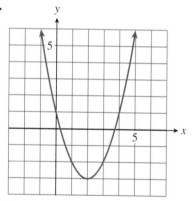

FIGURE 10.61

12.

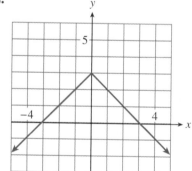

FIGURE 10.62

For Problems 13–18, (a) find a formula for the inverse f^{-1} of each function, (b) graph the function and its inverse on the same set of axes, along with the graph of $y = x$.

13. $f(x) = x + 4$

14. $f(x) = \dfrac{x - 2}{4}$

15. $f(x) = x^3 - 1$

16. $f(x) = \dfrac{1}{x + 2}$

17. $f(x) = \dfrac{1}{x} + 2$

18. $f(x) = \sqrt[3]{x} - 2$

19. If $F(t) = \dfrac{3}{4}t + 2$, find $F^{-1}(2)$.

20. If $G(x) = \dfrac{1}{x} - 4$, find $G^{-1}(3)$.

Evaluate each function.

21. $H(t) = t^2 + 2t$,
$\quad H(2a)$ and $\quad H(a + 1)$.

22. $F(x) = 2 - 3x$,
$\quad F(2) + F(3)$ and $\quad F(2 + 3)$.

23. $f(x) = 2x^2 - 4$,
$\quad f(a) + f(b)$ and $\quad f(a + b)$.

24. $h(x) = 2^x$,
$\quad h(a + 3)$ and $\quad h(3a)$

Graph each piecewise-defined function.

25. $f(x) = \begin{cases} x + 1 & \text{if } x \le 0 \\ x^2 & \text{if } x > 0 \end{cases}$

26. $g(x) = \begin{cases} x - 1 & \text{if } x \le 1 \\ x^3 & \text{if } x > 1 \end{cases}$

27. $H(x) = \begin{cases} x^2 & \text{if } x \le 0 \\ \sqrt{x} & \text{if } x > 0 \end{cases}$

28. $F(x) = \begin{cases} |x| & \text{if } x \le 0 \\ \dfrac{1}{x} & \text{if } x > 0 \end{cases}$

29. $S(x) = \begin{cases} x^3 & \text{if } x \le 1 \\ |x| & \text{if } x > 1 \end{cases}$

30. $T(x) = \begin{cases} \dfrac{1}{x^2} & \text{if } x < 0 \\ \sqrt{x} & \text{if } x \ge 0 \end{cases}$

31. Sketch an equilateral triangle whose sides have length s.

  a. Use the Pythagorean theorem to find the altitude of the triangle in terms of s.

  b. Express the area of the triangle as a function of s.

  c. Sketch the graph of your function.

  d. Find the area of an equilateral triangle whose side is 4 centimeters long.

  e. Find the side of an equilateral triangle whose area is 2.7 square feet.

32. Sketch a rectangle formed in the first quadrant as follows: The base and left side lie on the x- and y-axes, respectively. The upper right vertex (a, b) lies on the line $y = 10 - \frac{1}{2}x$.

  a. Express the length and width of the rectangle in terms of a.

  b. Express the area of the rectangle as a function of a.

  c. What is the domain of your function?

  d. Graph your function on its domain.

  e. Express the perimeter of the rectangle as a function of a.

33. An amortization table shows the monthly payments for a loan or mortgage. The table gives monthly payments for a loan of $100,000. The column headings show the length of the loan in years, and the row headings show the annual interest rate.

  a. You would like to borrow $100,000 for 20 years. What interest rate, to the nearest percent, can you accept if your monthly payments must be no more than $800? What interest rate can you accept if the loan is for 30 years?

  b. Suppose you borrow $100,000 for 10 years at 10% interest. Which would cause a greater reduction in your monthly payment: Reducing the interest rate by 5% or increasing the length of the loan by 5 years?

Monthly payment

	Length of loan (years)					
Interest rate	5	10	15	20	25	30
0.05	1879	1056	788	657	582	535
0.06	1924	1105	840	713	641	597
0.07	1969	1154	894	771	703	661
0.08	2014	1205	949	831	767	729
0.09	2060	1257	1007	893	833	799
0.10	2107	1311	1066	957	901	870

  c. At a fixed interest rate of 8%, is the monthly payment a linear function of the length of the loan?

d. For a fixed loan period of 25 years, is the monthly payment a linear function of interest rate?

e. Does a 1% increase in the interest rate have a greater affect on the monthly payment for a 15-year loan or a 30-year loan?

34. Warmer air can hold more moisture than cooler air. *Relative humidity* is the amount of moisture in the air, as a fraction of the saturation level at the current temperature. A common measure of humidity is the *dewpoint:* the temperature at which the current humidity would saturate the air, so that dew forms. The table gives dewpoints (in °F) for various values of relative humidity and temperature.

Dewpoint

Temperature, °F	Relative humidity, %					
	1	10	20	30	40	50
30	−60	−20	−6	2	9	14
40	−53	−12	2	11	18	23
50	−47	−4	11	20	27	32
60	−41	3	19	29	36	41
70	−35	11	27	37	45	50
80	−29	19	35	46	54	60
90	−23	26	43	54	62	69

Dewpoint

Temperature, °F	Relative humidity, %				
	60	70	80	90	100
30	18	21	25	27	30
40	28	31	34	37	40
50	37	40	44	47	50
60	46	50	54	57	60
70	55	60	64	67	70
80	65	69	73	77	80
90	74	79	83	87	90

a. Estimate the relative humidity if the temperature is 70°F and the dewpoint is 40°F.

b. Does the dewpoint rise or fall with temperature? (*Hint:* Consider any column in the table, and notice how dewpoint changes with increasing temperature.)

c. Does the dewpoint rise or fall with humidity? (*Hint:* Consider any row in the table, and notice how dewpoint changes with increasing humidity.)

d. Suppose that the temperature is 70°F and the relative humidity is 70%. Which would cause a larger change in dewpoint: a rise in temperature to 80°F or an increase in humidity to 80%?

e. Does dewpoint change more rapidly with temperature when the humidity is low or when the humidity is high?

35. In his hiking guidebook, *Afoot and Afield in Los Angeles County,* Jerry Schad notes that the number of people on a wilderness trail is inversely proportional to "the square of the distance and the cube of the elevation gain from the nearest road."

a. Choose variables and write a formula for this relationship.

b. On a sunny Saturday afternoon, you count 42 people enjoying the Rock Pool at Malibu Creek State Park. The Rock Pool is 1.5 miles from the main parking lot, and the trail includes an elevation gain of 250 feet. Calculate the constant of variation in your formula from part (a). (*Hint:* Convert the elevation gain to miles.)

c. Lookout Trail leads 1.9 miles from the parking lot, and involves an elevation gain of 500 feet. How many people would you expect to encounter at the end of the trail?

36. A company's monthly production, P, depends on the capital, C, they have invested and the amount of labor, L, available each month. The *Cobb-Douglas* model for production assumes that P varies jointly with C^a and L^b, where a and b are positive constants less than 1. The Aztech Chip Company invested 625 units of capital and hired 256 workers, and produces 8000 computer chips each month.

a. Suppose that $a = 0.25$, $b = 0.75$. Find the constant of variation and a formula giving P in terms of C and L.

b. If Aztech increases its labor force to 300 workers, what production level can they expect?

c. If Aztech maintains its labor force at 256 workers, what amount of capital outlay would be required for monthly production to reach 16,000 computer chips?

37. a. Is the function shown in Figure 10.63 periodic? If so, what is its period? If not, explain why not.

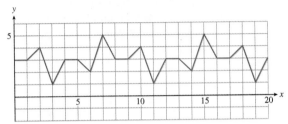

FIGURE 10.63

b. Compute the difference between the maximum and minimum function values. The *midline* of the graph is the horizontal line at the average of the maximum and minimum values. Sketch in the midline of the graph.

c. Find the smallest positive value of k for which $f(x) = f(x + k)$ for all x.

d. Find the smallest positive values of a and b for which $f(b) - f(a)$ is a maximum.

38. a. Find the period, the maximum and minimum values, and the midline of the graph of $y = f(x)$ shown in Figure 10.64. (See Problem 37 for the definition of midline.)

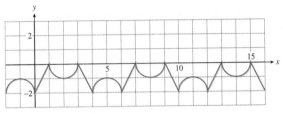

FIGURE 10.64

b. Sketch a graph of $y = 2f(x)$.

c. Sketch a graph of $y = 2 + f(x)$.

d. Modify the graph of $f(x)$ so that the period is twice its current value.

39. Figure 10.65 is a tide chart for Los Angeles for the week of December 17–23, 2000. The horizontal axis shows time in hours, with $t = 12$ corresponding to noon on December 17. The vertical axis shows the height of the tide in feet above mean sea level.

Tide levels in Los Angeles, Dec. 17-23, 2000

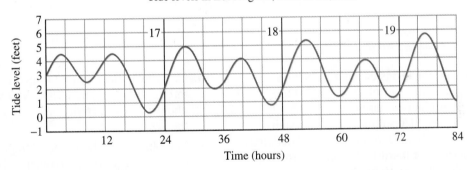

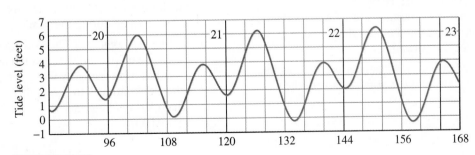

FIGURE 10.65

Source: Tidelines

a. High tides occurred at 3:07 a.m. and 2:08 p.m. on December 17, and low tides at 8:41 a.m. and 9:02 p.m. Estimate the heights of the high and low tides on that day.

b. Is tide height a periodic function of time? Use the information from part (a) to justify your answer.

c. Make a table showing approximate times and heights for the high tides throughout the week. Make a similar table for the low tides.

d. Describe the trend in the heights of the high tides over the week. Describe the trend in the heights of the low tides.

e. What is the largest height difference between consecutive high and low tides during the week shown? When does it occur?

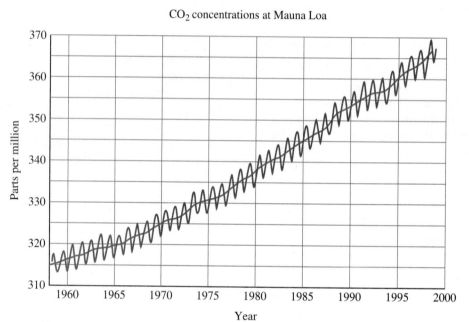

FIGURE 10.66

Source: Scripps Institute of Oceanography

40. Carbon dioxide (CO_2) is called a greenhouse gas because it traps part of the earth's outgoing energy. Animals release CO_2 into the atmosphere, and plants remove CO_2 through photosynthesis. In modern times, deforestation and the burning of fossil fuels both contribute to CO_2 levels. Figure 10.66 shows atmospheric concentrations of CO_2, in parts per million, measured at the Mauna Loa Observatory in Hawaii.

a. The red curve shows annual oscillations in CO_2 levels. Is it the graph of a periodic function?

b. Can you explain why CO_2 levels vary throughout the year? (*Hint:* Why would photosynthesis activity vary throughout the year?)

c. The blue curve shows the average annual CO_2 readings. By approximately how much does the CO_2 level vary from its average value during the year?

d. In 1960 the average CO_2 level was 316.75 parts per million, and the average level has been rising by 0.4% per year. If the level continues to rise at this rate, what CO_2 readings can we expect in the year 2100?

More About Graphing

INVESTIGATION 20

Global Positioning System

The Global Positioning System (GPS) is used by hikers, pilots, surveyors, and even some automobiles to determine their location (latitude, longitude, and elevation) anywhere on the surface of the earth. The system depends on a collection of satellites in orbit around the earth. Each GPS satellite transmits its own position and the current time at regular intervals. A person with a GPS receiver on earth can calculate his or her distance from the satellite by comparing the time of transmission with the time when it receives the signal. Of course, there are many points at the same distance from the satellite—in fact, the set of all points at a certain distance r from the satellite lie on a sphere centered at the satellite. That is why there are several satellites: You calculate your position by finding the intersection point of several such spheres centered on different satellites.

We will consider a simplified, two-dimensional model of a GPS system in which the satellites and the receiver all lie in the xy-plane, instead of in three-dimensional space. In this model we'll need data from two GPS satellites. The satellites are orbiting along a circle of radius 100 meters centered at the origin. You have a receiver inside that circle, and would like to know the coordinates of your position within the circle. To make the computations simpler, we will also assume that the satellite transmissions travel at 5 meters per second.

1. A signal from Satellite A arrives 18 seconds after it was transmitted. How far are you from Satellite A?

2. The signal says that Satellite A was located at (100, 0) at the time of transmission. Use a compass to sketch a graph showing your possible positions relative to Satellite A. (See Figure 11.1.)

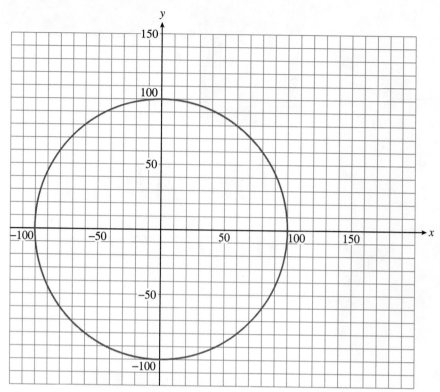

FIGURE 11.1

3. Find an equation for the graph you sketched in part 2.

4. A signal from Satellite B arrives 8.4 seconds after it was transmitted. How far are you from Satellite B?

5. The signal says that Satellite B was located at (28, 96) at the time of transmission. Use a compass to sketch a graph showing your possible positions relative to Satellite B.

6. Find an equation for the graph you sketched in part 5.

7. Your position must lie at the intersection point, P, of your two graphs. Estimate the coordinates of your position from the graph. (Remember that you are within the orbits of the satellites.)

8. Later in this chapter you will learn how to find the coordinates of P algebraically by solving a system of equations. Verify that the ordered pairs (28, 54) and (68.32, 84.24) both satisfy the equations you wrote in part 3 and part 6. What are the coordinates of P?

There are important extra considerations for the real Global Positioning System. The receiver's position is not restricted to the xy-plane, so it is given by three coordinates instead of two. Also, the receiver's clock may not be synchronized with the clocks on the satellites, so there is a fourth variable corresponding to the possible error in the receiver's clock. Thus the GPS receiver must solve a system of four quadratic equations in four unknowns. Each equation comes from a different satellite, so there must be at least four satellites "visible" to the receiver at any given time.

11.1 Conic Sections

In Chapter 1 we saw that the graph of any first-degree equation in two variables,

$$Ax + By = C$$

is a line (as long as A and B are not both 0). We now turn our attention to second-degree equations in two variables. The most general form of such an equation is

$$Ax^2 + Bxy + Cy^2 + Dx + Ey + F = 0$$

where A, B, and C cannot all be zero (because in that case the equation would not be second degree). The graphs of such equations are curves called **conic sections** because they are formed by the intersection of a plane and a cone, as shown in Figure 11.2. Except for a few special cases called *degenerate* conics, which we will describe later, the conic sections fall into four categories called **circles, ellipses, hyperbolas,** and **parabolas.**

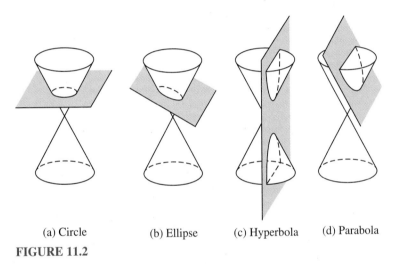

| (a) Circle | (b) Ellipse | (c) Hyperbola | (d) Parabola |

FIGURE 11.2

In this section we consider conic sections whose centers (or vertices, in the case of parabolas) are located at the origin. Such curves are called **central conics.**

Circles and Ellipses

The circle is the most familiar of the conic sections. In Chapter 6 we used the distance formula to derive the standard equation for a circle of radius r centered at the point (h, k).

$$(x - h)^2 + (y - k)^2 = r^2$$

From this equation we can see that a circle whose center is the origin has equation

$$x^2 + y^2 = r^2$$

We can also write this equation in the form

$$\frac{x^2}{r^2} + \frac{y^2}{r^2} = 1$$

Then the denominator of the x-squared term is the square of the x-intercepts, and the denominator of the y-squared term is the square of the y-intercepts. (Do you see why?) Of course, in this case the x- and y-intercepts are equal because the graph is a circle.

If the denominators of the x-squared and y-squared terms are not equal, the graph is called an **ellipse.** For example, the graph of

$$\frac{x^2}{9} + \frac{y^2}{4} = 1$$

is shown in Figure 11.3. When $y = 0$, we have $\frac{x^2}{9} = 1$, so the x-intercepts are $(3, 0)$ and $(-3, 0)$. When $x = 0$, we have $\frac{y^2}{4} = 1$, so the y-intercepts are $(0, 2)$ and $(0, -2)$.

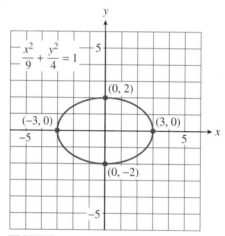

FIGURE 11.3

In general, we define an ellipse as follows.

An **ellipse** is the set of points in the plane, the sum of whose distances from two fixed points (the **foci**) is a constant.

We can visualize the definition in the following way. Drive two nails into a board to represent the two foci. Attach the two ends of a piece of string to the two nails, and stretch the string taut with a pencil. Trace around the two nails, keeping the string taut, as illustrated in Figure 11.4. The figure described will be an ellipse because the sum of the distances from each point to the two foci is the length of the string, which is constant.

Ellipses appear in a variety of applications. The orbits of the planets about the sun and of satellites about the earth are ellipses. The arches in some bridges are elliptical in shape, and whispering domes, such as the ceiling of the Mormon Tabernacle in Salt Lake City, are made from ellipses.

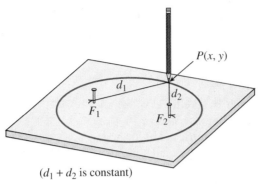

$P(x, y)$

d_1

d_2

F_1

F_2

$(d_1 + d_2 \text{ is constant})$

FIGURE 11.4

Using the distance formula and the definition on page 682, we can show that the equation of an ellipse centered at the origin has the standard form

$$\frac{x^2}{a^2} + \frac{y^2}{b^2} = 1 \tag{1}$$

By setting y equal to zero in Equation (1), we find that the x-intercepts of this ellipse are a and $-a$; by setting x equal to zero, we find that the y-intercepts are b and $-b$.

The line segment that passes through the foci and ends on the ellipse is called the **major axis.** When $a > b$, the foci lie on the x-axis, as in Figure 11.5a. The x-intercepts are the endpoints of the major axis, so its length is $2a$. The vertical segment with length $2b$ is called the **minor axis.** The endpoints of the major axis are the **vertices** of the ellipse and the endpoints of the minor axis are the **covertices.**

When $a < b$, the foci are on the y-axis. In this case the y-intercepts of the ellipse are the endpoints of the major axis, which has length $2b$. (See Figure 11.5b.) The minor axis is horizontal and has length $2a$.

The standard form of the equation for an ellipse gives us enough information to sketch its graph.

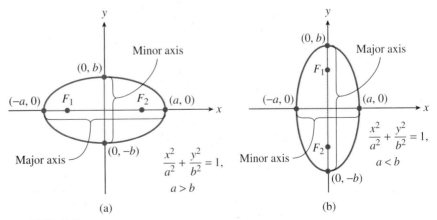

(a)

Minor axis

$(0, b)$

$(-a, 0)$

F_1

F_2

$(a, 0)$

$(0, -b)$

Major axis

$\dfrac{x^2}{a^2} + \dfrac{y^2}{b^2} = 1,$

$a > b$

(b)

$(0, b)$

Major axis

F_1

$(-a, 0)$

$(a, 0)$

$\dfrac{x^2}{a^2} + \dfrac{y^2}{b^2} = 1,$

$a < b$

Minor axis

F_2

$(0, -b)$

FIGURE 11.5

EXAMPLE 1

Graph

$$\frac{x^2}{8} + \frac{y^2}{25} = 1$$

Solution

x	y
0	± 5
$\pm 2\sqrt{2}$	0

The graph is an ellipse with major axis on the y-axis. Since $a^2 = 8$ and $b^2 = 25$, the vertices are located at $(0, 5)$ and $(0, -5)$, and the covertices lie $\sqrt{8} = 2\sqrt{2}$ units to the right and left of the center, or approximately at $(2.8, 0)$ and $(-2.8\ 0)$. (See Figure 11.6.) To sketch the ellipse, first locate the vertices and covertices. Then draw a smooth curve through the points.

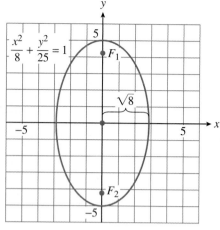

FIGURE 11.6 ∎

The equation of any central ellipse may be written as

$$Ax^2 + By^2 = C$$

where A, B, and C have like signs. The features of the graph are easier to identify if we first convert the equation to standard form.

EXAMPLE 2

Graph $4x^2 + y^2 = 12$

Solution

First convert the equation to standard form: Divide through by the constant term, 12, to obtain

$$\frac{x^2}{3} + \frac{y^2}{12} = 1$$

Because $a^2 = 3$ and $b^2 = 12$, the vertices are $(0, \pm\sqrt{12}) = (0, \pm 2\sqrt{3})$ and the covertices are $(\pm\sqrt{3}, 0)$. Plot points at approximately $(0, \pm 3.4)$ and $(\pm 1.7, 0)$, then draw a central ellipse through the points. See Figure 11.7.

x	y
0	$\pm 2\sqrt{3}$
$\pm\sqrt{3}$	0

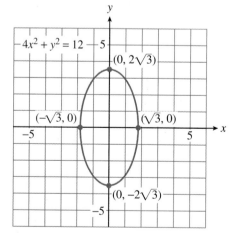

FIGURE 11.7 ∎

EXERCISE 1   Graph $9x^2 + 8y^2 = 16$.

We can find coordinates of other points on an ellipse by substituting a value for one variable and solving for the other variable.

EXAMPLE 3

a. Find the exact coordinates of any points with y-coordinate 2 on the ellipse $4x^2 + y^2 = 12$. Plot and label those points on the ellipse.

b. Solve the equation $4x^2 + y^2 = 12$ when $y = -4$. What do the solutions tell you about the graph of the ellipse?

Solutions

a. We substitute $y = 2$ in the equation and solve for x.

$$4x^2 + (2)^2 = 12$$
$$4x^2 = 8$$
$$x^2 = 2$$
$$x = \pm\sqrt{2}$$

There are two points with $y = 2$, namely $(\sqrt{2}, 2)$ and $(-\sqrt{2}, 2)$. See Figure 11.8.

b. We substitute $y = -4$ and solve for x.

$$4x^2 + (-4)^2 = 12$$
$$4x^2 = -4$$
$$x^2 = -1$$
$$x = \pm i$$

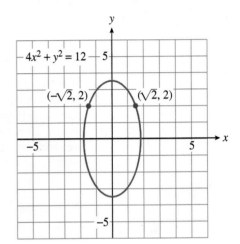

FIGURE 11.8

Because there are no real solutions, there are no points on the ellipse with $y = -4$. ∎

EXERCISE 2 Find the exact coordinates of all points with y-coordinate -1 on the ellipse $9x^2 + 8y^2 = 16$.

Hyperbolas

If a cone is cut by a plane parallel to its axis, the intersection is a **hyperbola**, the only conic section made of two separate pieces, or **branches**. (Look back at Figure 11.2.) Hyperbolas occur in a number of applied settings. The navigational system called LORAN (long-range navigation) uses radio signals to locate a ship or plane at the intersection of two hyperbolas. Satellites moving with sufficient speed will follow an orbit that is a branch of a hyperbola; for example, a rocket sent to the moon must be fitted with retrorockets to reduce its speed in order to achieve an elliptical, rather than hyperbolic, orbit about the moon.

The hyperbola is defined as follows.

> A **hyperbola** is the set of points in the plane, the difference of whose distances from two fixed points (the **foci**) is a constant.

If the origin is the center of the hyperbola and the foci lie on the x-axis, we can show that its equation may be written as

$$\frac{x^2}{a^2} - \frac{y^2}{b^2} = 1 \tag{2}$$

The two branches of the hyperbola open left and right, so the graph has x-intercepts at a and $-a$ but no y-intercepts. (See Figure 11.9a.) The segment joining the x-intercepts is the **transverse axis,** and its length is $2a$. The segment of length $2b$ is called the **conjugate axis.** The endpoints of the transverse axis are the **vertices** of the hyperbola.

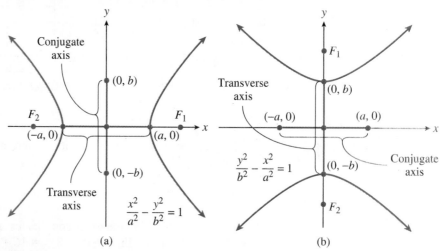

FIGURE 11.9

A hyperbola centered at the origin with foci on the y-axis has the equation

$$\frac{y^2}{b^2} - \frac{x^2}{a^2} = 1 \tag{3}$$

In this case the graph has y-intercepts at b and $-b$ but no x-intercepts—the two branches open up and down. (See Figure 11.9b.) Here the y-intercepts are the vertices, so the transverse axis is vertical and has length $2b$. The conjugate axis has length $2a$.

Asymptotes of Hyperbolas

The branches of the hyperbola approach two straight lines that intersect at its center. These lines are asymptotes of the graph, and they are useful as guidelines for sketching the hyperbola. We can obtain the asymptotes by first forming a rectangle (called the "central rectangle") whose sides are parallel to and the same length as the transverse and conjugate axes. The asymptotes are determined by the diagonals of this rectangle, so their slopes are $\pm\dfrac{b}{a}$.

EXAMPLE 4

Graph $\dfrac{y^2}{9} - \dfrac{x^2}{4} = 1$.

Solution

The graph is a hyperbola with center at the origin. The equation is in form (3), so the branches of the hyperbola open upward and downward. $a^2 = 4$ and $b^2 = 9$, so $a = 2$ and $b = 3$, and the vertices are $(0, 3)$ and $(0, -3)$. There are no x-intercepts.

Construct the central rectangle with dimensions $2a = 4$ and $2b = 6$, as shown in Figure 11.10. Draw the asymptotes through the diagonals of the rectangle. The asymptotes have slopes $\pm \frac{3}{2}$. Finally, sketch the branches of the hyperbola through the vertices and approaching the asymptotes.

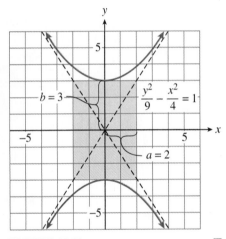

FIGURE 11.10 ∎

EXERCISE 3 Graph $\dfrac{x^2}{9} - \dfrac{y^2}{16} = 1$.

Find a and b, and plot the vertices.
Sketch the central rectangle.
Draw the asymptotes.
Sketch the hyperbola.

The equation of a central hyperbola may be written as

$$Ax^2 + By^2 = C$$

where A and B have *opposite* signs and $C \neq 0$. (For example, $2x^2 - 3y^2 = 6$ and $-4x^2 + y^2 = 3$ are equations of hyperbolas.) As with ellipses, it is best to rewrite the equation in standard form in order to graph it.

EXAMPLE 5

Write the equation $4y^2 - x^2 = 16$ in standard form and describe the important features of its graph.

Solution

Divide each side by 16 to obtain

$$\frac{y^2}{4} - \frac{x^2}{16} = 1 \qquad \text{or} \qquad \frac{y^2}{2^2} - \frac{x^2}{4^2} = 1$$

The graph is a central hyperbola with y-intercepts 2 and -2. The slopes of the asymptotes are given by

$$\frac{\pm b}{a} = \frac{\pm 2}{4} = \pm\frac{1}{2}$$

so the equations of the asymptotes are $y = \frac{1}{2}x$ and $y = \frac{-1}{2}x$. (See Figure 11.11.)

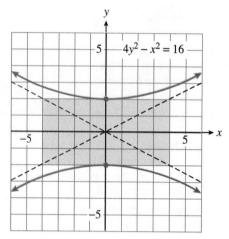

FIGURE 11.11 ■

EXERCISE 4

a. Write the equation $4x^2 = y^2 + 25$ in standard form.

b. Find the vertices of the graph, and the equations of the asymptotes.

We can find exact coordinates of points on a hyperbola by substituting a value for one variable and solving for the other variable.

EXAMPLE 6
🎞

Find the exact coordinates of any points with x-coordinate $x = 2$ on the hyperbola with equation $4y^2 - x^2 = 16$. Plot and label those points on the hyperbola.

Solution

We substitute $x = 2$ in the given equation.

$$4y^2 - (2)^2 = 16$$
$$4y^2 = 20$$
$$y^2 = 5$$
$$y = \pm \sqrt{5}$$

There are two points with $x = 2$, namely $(2, \sqrt{5})$ and $(2, -\sqrt{5})$. See Figure 11.12.

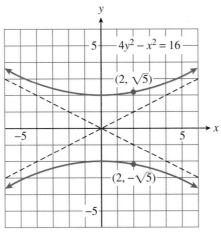

FIGURE 11.12

EXERCISE 5 Solve the equation $4y^2 - x^2 = 16$ when $y = 1$. What is the significance of the solutions to the graph of the hyperbola?

Parabolas

If a cone is cut by a plane parallel to the side of the cone, the intersection is a parabola. (Refer to Figure 11.2.) In Section 4.1 we graphed parabolas whose equations were of the form

$$y = ax^2 + bx + c$$

These parabolas opened either up or down and had vertical axes of symmetry. However, it is also possible to have parabolas that open to the left or right. In general, we define a parabola as follows.

A **parabola** is the set of points in the plane whose distances from a fixed line l and a fixed point F are equal.

The fixed line in the definition is called the **directrix,** and the fixed point is the **focus.** The **axis** of a parabola is the line running through the focus of the parabola and perpendicular to the directrix. (See Figure 11.13.)

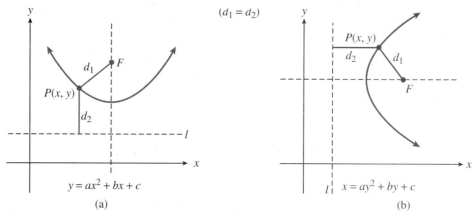

FIGURE 11.13

Parabolas have many applications in optics and communications. Parabolic mirrors are used in telescopes because light waves received reflect off the surface and form an image at the focus of the parabola. For similar reasons radio antennae and television dish receivers are parabolic in shape. The parabolic mirrors in searchlights and automobile headlights reflect light from the focus into beams of parallel rays.

We first consider parabolas with vertex at the origin. Using the distance formula and the definition above, it can be shown that parabolas that open upward and those that open downward have equations of the form

$$y = \frac{x^2}{4p} \qquad \text{and} \qquad y = \frac{-x^2}{4p}$$

respectively, where p is the distance between the vertex of the parabola and its focus or the distance between the vertex and the directrix. (See Figure 11.14.) Parabolas that open to the right and those that open to the left have equations of the forms

$$x = \frac{y^2}{4p} \qquad \text{and} \qquad x = \frac{-y^2}{4p}$$

respectively.

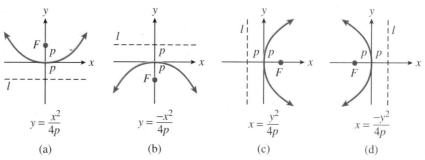

FIGURE 11.14

A line drawn through the focus and parallel to the directrix will intersect the parabola at two points. These points are $2p$ units from the directrix, so they must also be $2p$ units from the focus. (Recall that by definition the points of a parabola are equidistant from the focus and the directrix. See Figure 11.15.) By locating these two "guide points" and the vertex, we can make a reasonable sketch of the parabola, as illustrated in Example 7.

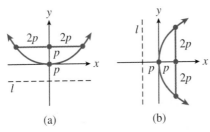

(a)   (b)

FIGURE 11.15

EXAMPLE 7

Graph $x = \dfrac{-y^2}{6}$.

Solution

The parabola has its vertex at the origin and opens to the left, so its axis is the x-axis. Also, $6 = 4p$, so $p = \frac{3}{2}$, and the focus is the point $\left(-\frac{3}{2}, 0\right)$. The directrix is the vertical line $x = \frac{3}{2}$.

To graph the parabola, draw a line segment of length $4p = 6$ perpendicular to the axis and centered at the focus; because $2p = 3$, the endpoints of the segment are $\left(-\frac{3}{2}, 3\right)$ and $\left(-\frac{3}{2}, -3\right)$. Sketch the parabola through the vertex and the two guide points to obtain the graph shown in Figure 11.16.

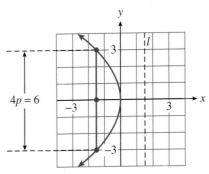

FIGURE 11.16

EXERCISE 6   Graph $y^2 - 8x = 0$.

We summarize the important facts about central conics in Table 11.1 on page 693.

EXAMPLE 8

Write each equation in standard form and describe its graph.

a. $x^2 = 9y^2 - 9$

b. $4x^2 + y = 0$

c. $x^2 = 6y^2 + 8$

d. $x^2 = \dfrac{4 - y^2}{2}$

Solutions

a. The equation $x^2 = 9y^2 - 9$ is equivalent to

$$9y^2 - x^2 = 9 \quad \text{or} \quad \frac{y^2}{1^2} - \frac{x^2}{3^2} = 1$$

The graph is a hyperbola opening up and down.

b. The equation $4x^2 + y = 0$ is equivalent to

$$y = -4x^2 \quad \text{or} \quad y = \frac{-x^2}{1/4}$$

The graph is a parabola that opens downward.

c. The equation $x^2 = 6y^2 + 8$ is equivalent to

$$x^2 - 6y^2 = 8 \quad \text{or} \quad \frac{x^2}{(\sqrt{8})^2} - \frac{y^2}{\left(\dfrac{2}{\sqrt{3}}\right)^2} = 1$$

The graph is a hyperbola that opens left and right.

d. The equation $x^2 = \dfrac{4 - y^2}{2}$ is equivalent to

$$2x^2 + y^2 = 4 \quad \text{or} \quad \frac{x^2}{(\sqrt{2})^2} + \frac{y^2}{2^2} = 1$$

The graph is an ellipse with major axis on the y-axis because $2 > \sqrt{2}$. ∎

Conic Sections Using Graphing Calculators

Graphing calculators are designed to graph functions. Circles, ellipses, hyperbolas, and some parabolas are not the graphs of functions because some values of x correspond to two different values of y. This means we must do a little more work in order to obtain these graphs on a calculator.

EXAMPLE 9   Use a calculator to graph each conic section.

a. $x^2 + y^2 = 9$ **b.** $x^2 - y^2 = 4$ **c.** $x - y^2 = 1$

Solutions

a. We first solve for y in terms of x.

$$x^2 + y^2 = 9$$
$$y^2 = 9 - x^2$$
$$= \pm \sqrt{9 - x^2}$$

We see that there are two values of y associated with a single value of x. That is, the graph of the circle is actually the graph of two functions,

$$Y_1 = \sqrt{9 - x^2} \quad \text{and} \quad Y_2 = -Y_1$$

Enter these expressions and press $\boxed{\text{ZOOM}}\ \boxed{6}$ for the standard graphing window. The graph appears to be part of an ellipse, not a circle, as

TABLE 11.1

Description of conic	Standard form of equation	Graph
circle	$x^2 + y^2 = r^2$	
ellipse (a) major axis on *x*-axis ("wide and short")	$\dfrac{x^2}{a^2} + \dfrac{y^2}{b^2} = 1$ $(a > b)$	
(b) major axis on *y*-axis ("tall and narrow")	$(a < b)$	
hyperbola (a) transverse axis on *x*-axis (opens left and right)	$\dfrac{x^2}{a^2} - \dfrac{y^2}{b^2} = 1$	
(b) transverse axis on *y*-axis (opens up and down)	$\dfrac{y^2}{b^2} - \dfrac{x^2}{a^2} = 1$	
parabola (a) opens up	$y = \dfrac{x^2}{4p}$	
(b) opens down	$y = \dfrac{-x^2}{4p}$	
(c) opens right	$x = \dfrac{y^2}{4p}$	
(d) opens left	$x = \dfrac{-y^2}{4p}$	

shown in Figure 11.17a. This distortion in shape occurs because the scales on the two axes are different. One way to remedy this distortion is to use the "square" window setting: Press ⟨ZOOM⟩ ⟨5⟩ to get Figure 11.17b. We now see that the conic section is a circle. (What is the radius of the circle?)

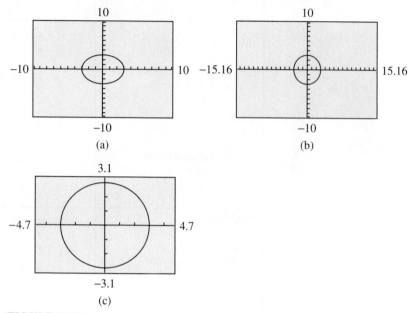

(a)　(b)

(c)

FIGURE 11.17

If we look closely, we see gaps at the left and right of the circle. These gaps occur because the calculator plots only a selection of points (depending on the window chosen) and then tries to "connect the dots." We get a better picture with the **ZDecimal** window (⟨ZOOM⟩ ⟨4⟩). (See Figure 11.17c.)

b.　Begin by solving for y in terms of x.

$$x^2 - y^2 = 4$$
$$x^2 - 4 = y^2$$
$$\pm\sqrt{x^2 - 4} = y$$

Enter

$$Y_1 = \sqrt{x^2 - 4} \qquad \text{and} \qquad Y_2 = -Y_1$$

Press ⟨GRAPH⟩ to see the hyperbola in Figure 11.18a.

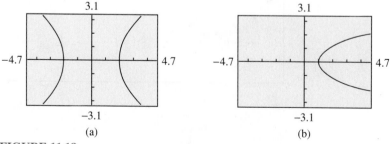

(a)　(b)

FIGURE 11.18

c. Solve for y in terms of x.

$$x - y^2 = 1$$
$$x - 1 = y^2$$
$$\pm\sqrt{x - 1} = y$$

Enter

$$Y_1 = \sqrt{x - 1} \qquad \text{and} \qquad Y_2 = -Y_1$$

The graph is a parabola that opens to the right, as shown in Figure 11.18b. ■

1. Name four types of conic sections. Why are they called conic sections?
2. Define the following terms related to ellipses: major and minor axes, vertices and covertices.
3. What is the standard form for the equation of an ellipse?
4. Define the following terms related to hyperbolas: transverse and conjugate axes, vertices, asymptotes.
5. What is the standard form for the equation of a hyperbola?
6. What are the directrix and focus of a parabola?

ANSWERS TO 11.1 EXERCISES

1.

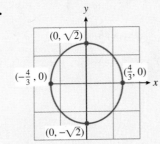

4a. $\dfrac{x^2}{\left(\frac{5}{2}\right)^2} - \dfrac{y^2}{5^2} = 1$

b. $\left(\dfrac{5}{2}, 0\right)\left(\dfrac{-5}{2}, 0\right);\ y = 2x,\ y = -2x$

5. There are no solutions when $y = 1$ because there are no points on the graph with y-coordinate 1.

2. $\left(\dfrac{2\sqrt{2}}{3}, -1\right), \left(\dfrac{-2\sqrt{2}}{3}, -1\right)$

3.

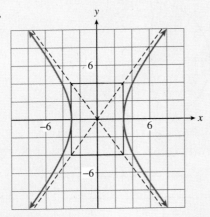

6.

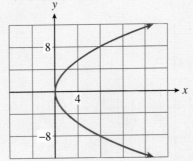

Graph each circle or ellipse in Problems 1–10.

1. $4x^2 = 16 - 4y^2$ **2.** $2x^2 = 18 - 2y^2$ **3.** $\dfrac{x^2}{16} + \dfrac{y^2}{4} = 1$ **4.** $\dfrac{x^2}{9} + \dfrac{y^2}{16} = 1$

5. $\dfrac{x^2}{10} + \dfrac{y^2}{25} = 1$ **6.** $3x^2 + 4y^2 = 36$ **7.** $x^2 + \dfrac{y^2}{14} = 1$ **8.** $x^2 = 36 - 9y^2$

9. $3y^2 = 30 - 2x^2$ **10.** $5y^2 = 30 - 3x^2$

Graph each hyperbola in Problems 11–18.

11. $\dfrac{x^2}{25} - \dfrac{y^2}{9} = 1$ **12.** $\dfrac{y^2}{4} - \dfrac{x^2}{16} = 1$ **13.** $\dfrac{y^2}{12} - \dfrac{x^2}{8} = 1$ **14.** $y^2 - 9x^2 = 36$

15. $9x^2 - 4y^2 = 36$ **16.** $3x^2 = 4y^2 + 24$ **17.** $\dfrac{1}{2}x^2 = y^2 - 12$ **18.** $y^2 = \dfrac{1}{2}x^2 - 16$

Graph each parabola in Problems 19–26.

19. $x^2 = 2y$ **20.** $y^2 = 4x$ **21.** $x = \dfrac{-1}{16}y^2$ **22.** $y^2 = 12x$

23. $4x^2 = 3y$ **24.** $x^2 + 8y = 0$ **25.** $2y^2 - 3x = 0$ **26.** $3x^2 + 5y = 0$

The graphs of the equations in Problems 27–36 are circles, ellipses, hyperbolas, or parabolas. (a) Name the graph and describe its main features. (b) find the coordinates of all points on the graph with the given x- or y-coordinate.

27. $y^2 = 4 - x^2$, $x = -1$ **28.** $y^2 = 6 - 4x^2$, $y = 2$

29. $4y^2 = x^2 - 8$, $y = 2$ **30.** $x^2 + 2y - 4 = 0$, $y = -3$

31. $4x^2 = 12 - 2y^2$, $x = 4$ **32.** $6x^2 = 8 - 6y^2$, $x = \sqrt{2}$

33. $4x^2 = 6 + 4y$, $y = -2$ **34.** $2x^2 = 5 + 4y^2$, $x = -\sqrt{3}$

35. $6 + \dfrac{x^2}{4} = y^2$, $y = -\sqrt{5}$ **36.** $y^2 = 6 - \dfrac{2x^2}{3}$, $y = \sqrt{7}$

For Problems 37–40, (a) give an equation for the conic graphed, (b) use your equation to complete the table.

37.

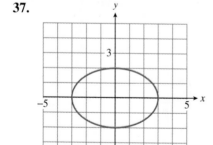

FIGURE 11.19

x	± 3		-2	
y		± 2		1

38.

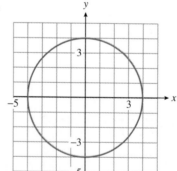

FIGURE 11.20

x	± 4		3	
y		± 4		-2

39.

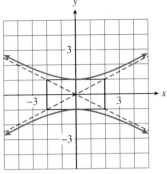

FIGURE 11.21

x	0		4	
y		0		-2

40.

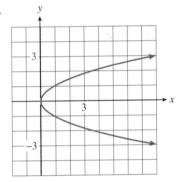

FIGURE 11.22

x	0		8	
y		± 2		3

41. The arch of a bridge forms the top half of an ellipse with a horizontal major axis. (See Figure 11.23.) The arch is 7 feet high and 20 feet wide.

a. Find an equation for the ellipse.

b. How high is the arch at a distance of 8 feet from the peak?

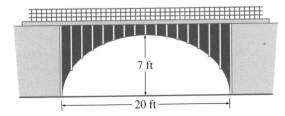

FIGURE 11.23

42. A doorway is topped by a semi-elliptical arch. (See Figure 11.24.) The doorway is 230 centimeters high at its highest point and 200 centimeters high at its lowest point. It is 80 centimeters wide.

a. Find an equation for the ellipse.

b. How high is the doorway 8 centimeters from the left side?

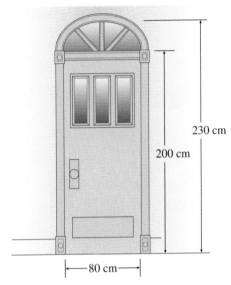

FIGURE 11.24

43. The edge of a sailboat's keel is elliptical in shape, with a major axis of 360 centimeters. The minor axis is 100 centimeters, but part of the ellipse has been cut off parallel to the major axis. This cut edge is 330 centimeters long. (See Figure 11.25.)

a. Find an equation for the ellipse.

b. What is the width of the keel at its widest point? Round your answer to two decimal places.

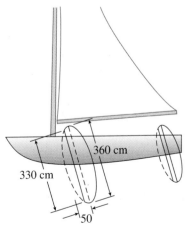

FIGURE 11.25

44. The wing of a World War II British Spitfire is an ellipse whose major axis is 48 feet. The minor axis is 16 feet, but part of the ellipse is cut off parallel to the major axis. This cut edge is 46 feet long. (See Figure 11.26.)

a. Find an equation for the ellipse.

b. How wide is the wing at its center? Round your answer to two decimal places.

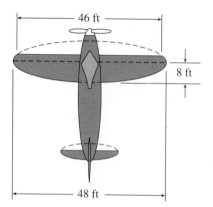

FIGURE 11.26

45. In a reflecting telescope a parabolic mirror is used to collect light, and the image forms at the focus of the parabola. Suppose you want to build a parabolic mirror that is 72 inches in diameter and 3 inches deep, as shown in Figure 11.27.

a. Find an equation for the parabola.

b. How far is the focus from the vertex of the parabola?

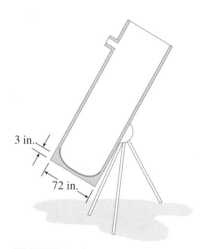

FIGURE 11.27

46. A sonar dish is parabolic in shape, with the receiver located at the focus of the parabola, as shown in Figure 11.28. The dish is 12 inches wide, and the receiver is 12 inches from the vertex of the parabola.

a. Find an equation for the parabola.

b. How deep is the dish?

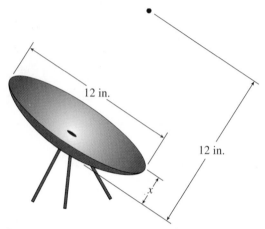

FIGURE 11.28

47. Tien-Ying plans to build a parabolic satellite dish. The dish will have a diameter of 60 centimeters and will be 18 centimeters deep, as shown in Figure 11.29. She must place the receiver at the focus of the parabola.

a. Find an equation for the parabola.

b. How far is the receiver from the vertex of the parabola?

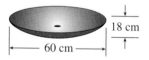

FIGURE 11.29

48. A flashlight has a parabolic mirror to focus a beam of light. The mirror is 8 centimeters in diameter and 4 centimeters deep.

a. Find an equation for the parabola.

b. The light bulb must be positioned at the focus of the parabola. How far is the bulb from the vertex of the parabola?

Problems 49–52 deal with the cooling tower at an electricity-generating facility, as shown in Figure 11.30. The shape of the tower, called a hyperboloid, is obtained by revolving a portion of the hyperbola

$$\frac{x^2}{100^2} - \frac{y^2}{150^2} = 1 \text{ around the } y\text{-axis.}$$

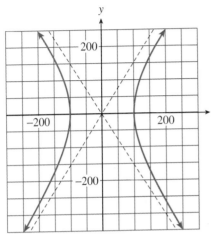

FIGURE 11.30

49. The base of the cooling tower is 360 feet below the center of the hyperbola. What is the diameter of the base?

50. The top of the cooling tower is 200 feet above the center of the hyperbola. What is the diameter of the top?

51. The diameter of the tower first decreases with height, and then increases again. There are two heights at which the tower's diameter is 250 feet. Find the greater of the two heights.

52. Find the height at which the tower's diameter is 200 feet.

53. Graph $x^2 - y^2 = 0$. (*Hint:* Factor the left side of the equation and use the zero-factor principle to write two equations, then graph them.) Describe your graph.

54. Graph $4x^2 - y^2 = 0$. (*Hint:* See the hint for Problem 53.) Describe your graph.

55. Graph $x^2 - y^2 = 4$, $x^2 - y^2 = 1$, and $x^2 - y^2 = 0$ on the same set of axes. What do you observe?

56. Graph $4x^2 - y^2 = 16$, $4x^2 - y^2 = 4$, and $4x^2 - y^2 = 0$ on the same set of axes. What do you observe?

11.2 Translated Conics

In this section we consider ellipses and hyperbolas that are not centered at the origin, and parabolas whose vertices are not at the origin. These curves are **translations** of the central conics we studied in Section 11.1.

Ellipses

In Section 6.4 we used the distance formula to derive the equation of a circle centered at the point (h, k). The standard form for this equation is

$$(x - h)^2 + (y - k)^2 = r^2$$

The standard form for the equation of an ellipse centered at any point (h, k) can also be derived from the distance formula.

$$\frac{(x - h)^2}{a^2} + \frac{(y - k)^2}{b^2} = 1 \qquad (4)$$

The horizontal axis of the ellipse has length $2a$, and the vertical axis $2b$. When $a > b$, the major axis is horizontal and the ellipse is short and wide. When $a < b$, the major axis is vertical and the ellipse is tall and narrow. See Figure 11.31.

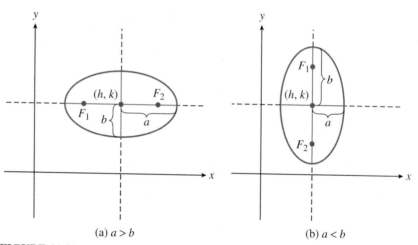

(a) $a > b$ (b) $a < b$

FIGURE 11.31

EXAMPLE 1

a. Graph $\dfrac{(x + 2)^2}{16} + \dfrac{(y - 1)^2}{5} = 1$.

b. Find the exact coordinates of the intercepts of the graph.

Solutions

a. The graph is an ellipse with center at $(-2, 1)$. We have $a^2 = 16$ and $b^2 = 5$, and the major axis is parallel to the x-axis because $a > b$. The vertices lie four units to the left and right of the center, at $(-6, 1)$ and $(2, 1)$. The covertices lie $\sqrt{5}$ units above and below the center, at approximately $(-2, 3.2)$ and $(-2, -1.2)$. The graph is shown in Figure 11.32.

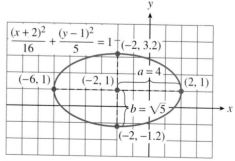

FIGURE 11.32

b. We set $y = 0$ and solve the resulting equation to find the x-intercepts.

$$\frac{(x + 2)^2}{16} + \frac{(0 - 1)^2}{5} = 1 \qquad \text{Subtract } \tfrac{1}{5} \text{ from both sides.}$$

$$\frac{(x + 2)^2}{16} = \frac{4}{5} \qquad \text{Multiply both sides by 16.}$$

$$(x + 2)^2 = \frac{64}{5} \qquad \text{Extract roots.}$$

$$x + 2 = \pm \sqrt{\frac{64}{5}}$$

$$x = -2 \pm \frac{8\sqrt{5}}{5}$$

The x-intercepts are $\left(-2 \pm \frac{8\sqrt{5}}{5}, 0\right)$ or approximately $(1.6, 0)$ and $(-5.6, 0)$. We set $x = 0$ to find the y-intercepts.

$$\frac{(0 + 2)^2}{16} + \frac{(y - 1)^2}{5} = 1 \qquad \text{Subtract } \tfrac{1}{4} \text{ from each side.}$$

$$\frac{(y - 1)^2}{5} = \frac{3}{4} \qquad \text{Multiply both sides by 5.}$$

$$(y - 1)^2 = \frac{15}{4} \qquad \text{Extract roots.}$$

$$y - 1 = \pm \sqrt{\frac{15}{4}}$$

$$y = 1 \pm \frac{\sqrt{15}}{2}$$

The y-intercepts are $\left(0, 1 \pm \frac{\sqrt{15}}{2}\right)$ or approximately $(0, 2.9)$ and $(0, -0.9)$. ∎

EXERCISE 1

a. Graph $\dfrac{(x - 5)^2}{15} + \dfrac{(y + 3)^2}{8} = 1$.

b. Find the coordinates of the vertices.

Second-degree equations in which the coefficients of x^2 and y^2 have the *same* sign can be written in one of the standard forms for an ellipse by completing the square. As with circles, the equation can be graphed easily from the standard form. (You may want to review completing the square for circles in Section 6.4.)

EXAMPLE 2

a. Write the equation in standard form.

$$4x^2 + 9y^2 - 16x - 18y - 11 = 0$$

b. Graph the equation.

c. Find the coordinates of the intercepts.

Solutions

a. First prepare to complete the square in both x and y. Begin by factoring out the coefficients of x^2 and y^2.

$$4(x^2 - 4x \quad\;) + 9(y^2 - 2y \quad\;) = 11$$

Complete the square in x by adding $4 \cdot 4$, or 16, to each side of the equation, and complete the square in y by adding $9 \cdot 1$, or 9, to each side to obtain

$$4(x^2 - 4x + 4) + 9(y^2 - 2y + 1) = 11 + 16 + 9$$

Write each term on the left side as a perfect square to get

$$4(x - 2)^2 + 9(y - 1)^2 = 36 \quad \text{Divide both sides by 36.}$$

$$\frac{(x - 2)^2}{9} + \frac{(y - 1)^2}{4} = 1$$

b. The graph is an ellipse with center at $(2, 1)$, $a^2 = 9$, and $b^2 = 4$. The vertices lie three units to the right and left of center at $(5, 1)$ and $(-1, 1)$, and the covertices lie two units above and below the center at $(2, 3)$ and $(2, -1)$. The graph is shown in Figure 11.33.

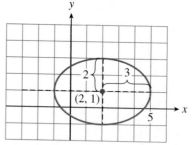

FIGURE 11.33

c. To find the x-intercepts, substitute $y = 0$ into the original equation to obtain

$$4x^2 + 9(0)^2 - 16x - 18(0) - 11 = 0$$
$$4x^2 - 16x - 11 = 0$$

Use the quadratic formula to solve for x.

$$x = \frac{16 \pm \sqrt{432}}{8} = 2 \pm \frac{3\sqrt{3}}{2}$$

The x-intercepts are $\left(2 \pm \frac{3\sqrt{3}}{2}, 0\right)$, or approximately $(4.6, 0)$ and $(-0.6, 0)$. To find the y-intercepts, substitute $x = 0$ into the given equation to obtain

$$4(0)^2 + 9y^2 - 16(0) - 18y - 11 = 0$$
$$9y^2 - 18y - 11 = 0$$

Use the quadratic formula to solve for y.

$$y = \frac{18 \pm \sqrt{720}}{18} = 1 \pm \frac{2\sqrt{5}}{3}$$

The y-intercepts are $\left(0, 1 \pm \frac{2\sqrt{5}}{3}\right)$ or approximately $(0, 2.5)$ and $(0, -0.5)$. ■

a. Write the equation $x^2 + 4y^2 + 4x - 16y + 4 = 0$ in standard form.

b. Graph the equation.

Finding the Equation of an Ellipse

To write the equation of an ellipse from a description of its properties, we must find the center of the ellipse and the lengths of its axes. We can then substitute this information into the standard form.

EXAMPLE 3

Find the equation of the ellipse with vertices at $(3, 3)$ and $(3, -5)$ and covertices at $(1, -1)$ and $(5, -1)$.

Solution

To find the center of the ellipse, find the midpoint of the major (or minor) axis. (The midpoint formula is discussed in Section 6.4.)

$$h = \bar{x} = \frac{1 + 5}{2} = 3$$

$$k = \bar{y} = \frac{3 - 5}{2} = -1$$

Thus, the center is the point $(3, -1)$. The covertices lie on the horizontal axis, so a is the distance between the center and either covertex, say $(5, -1)$, and

$$a = |5 - 3| = 2$$

The value of b is the distance from the center to one of the vertices, say $(3, 3)$, so

$$b = |3 - (-1)| = 4$$

The equation of the ellipse has the form

$$\frac{(x - h)^2}{a^2} + \frac{(y - k)^2}{b^2} = 1$$

with $h = 3$, $k = -1$, $a = 2$, $b = 4$. Thus, the equation is

$$\frac{(x - 3)^2}{4} + \frac{(y + 1)^2}{16} = 1$$

If we clear this equation of fractions and expand the powers, we obtain the general form.

$$4(x - 3)^2 + (y + 1)^2 = 16$$
$$4(x^2 - 6x + 9) + (y^2 + 2y + 1) = 16$$
$$4x^2 - 24x + 36 + y^2 + 2y + 1 = 16$$
$$4x^2 + y^2 - 24x + 2y + 21 = 0$$

∎

Hyperbolas

The following standard forms for equations of hyperbolas centered at the point (h, k) can be derived using the distance formula and the definition of hyperbola.

$$\frac{(x-h)^2}{a^2} - \frac{(y-k)^2}{b^2} = 1 \tag{5}$$

$$\frac{(y-k)^2}{b^2} - \frac{(x-h)^2}{a^2} = 1 \tag{6}$$

Equation (5) describes a hyperbola whose transverse axis is parallel to the x-axis, so that the branches open left and right. Equation (6) describes a hyperbola whose transverse axis is parallel to the y-axis, so that the branches open up and down. See Figure 11.34.

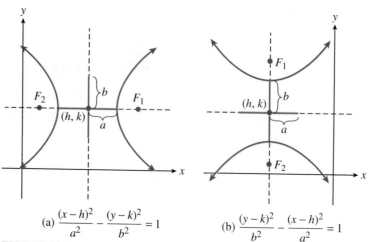

(a) $\dfrac{(x-h)^2}{a^2} - \dfrac{(y-k)^2}{b^2} = 1$ (b) $\dfrac{(y-k)^2}{b^2} - \dfrac{(x-h)^2}{a^2} = 1$

FIGURE 11.34

EXAMPLE 4

a.   Graph $\dfrac{(x-3)^2}{8} - \dfrac{(y+2)^2}{10} = 1.$

b.   Find the equations of the asymptotes.

Solutions

a.   The graph is a hyperbola with $(h, k) = (3, -2)$, $a = \sqrt{8} = 2\sqrt{2}$, and $b = \sqrt{10}$. Because the equation is in standard form (5), the branches open left and right. The coordinates of the vertices are thus $(3 + 2\sqrt{2}, -2)$ and $(3 - 2\sqrt{2}, -2)$, or approximately $(5.8, -2)$ and $(0.2, -2)$. The ends of the conjugate axis are $(3, -2 + \sqrt{10})$ and $(3, -2 - \sqrt{10})$, or approximately $(3, 1.2)$ and $(3, -5.2)$.

The central rectangle is centered at the point $(3, -2)$ and extends to the vertices in the horizontal direction and to the ends of the conjugate axis in the vertical direction as shown in Figure 11.35. Draw the asymptotes through the opposite corners of the central rectangle, and sketch the hyperbola through the vertices and approaching the asymptotes to obtain the graph in Figure 11.35.

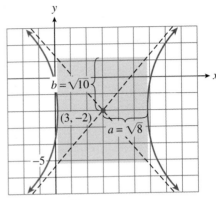

FIGURE 11.35

b. Both asymptotes pass through the center of the hyperbola, $(3, -2)$. Their slopes are $\dfrac{\sqrt{10}}{\sqrt{8}} = \dfrac{\sqrt{5}}{2}$ and $\dfrac{-\sqrt{5}}{2}$. Substitute these values into the point-slope formula to find the equations

$$y + 2 = \frac{\sqrt{5}}{2}(x - 3) \qquad \text{and} \qquad y + 2 = \frac{-\sqrt{5}}{2}(x - 3) \qquad \blacksquare$$

EXERCISE 4

a. Graph $\dfrac{y^2}{9} - \dfrac{(x + 4)^2}{12} = 1$.

b. Find the equations of the asymptotes.

Quadratic equations in which the x^2 term and the y^2 term have *opposite* signs describe hyperbolas. We can write such equations in one of the standard forms (5) or (6) by completing the squares in x and y.

EXAMPLE 5

a. Write the equation in standard form.
$$y^2 - 4x^2 + 4y - 8x - 9 = 0$$

b. Graph the equation.

Solutions

a. First prepare to complete the square by writing
$$(y^2 + 4y \quad) - 4(x^2 + 2x \quad) = 9$$

Complete the square in y by adding 4 to each side of the equation, and complete the square in x by adding $-4 \cdot 1$, or -4 to each side to get

$$(y^2 + 4y + 4) - 4(x^2 + 2x + 1) = 9 + 4 - 4$$

or

$$(y + 2)^2 - 4(x + 1)^2 = 9$$

Divide each side by 9 to obtain the standard form

$$\frac{(y + 2)^2}{9} - \frac{(x + 1)^2}{\frac{9}{4}} = 1$$

b. The graph is a hyperbola with center at $(-1, -2)$. Because the equation is in standard form (6), the transverse axis is parallel to the y-axis, and

$$b^2 = 9, \qquad a^2 = \frac{9}{4}$$

Thus $b = 3$ and $a = \frac{3}{2}$, and the vertices are $(-1, 1)$ and $(-1, -5)$. The ends of the conjugate axis are $\left(-\frac{5}{2}, -2\right)$ and $\left(\frac{1}{2}, -2\right)$.

The central rectangle is centered at $(-1, -2)$ and passes through the vertices and the ends of the conjugate axis, as shown in Figure 11.36. Draw the asymptotes through the corners of the rectangle, then sketch the hyperbola by starting at the vertices and approaching the asymptotes to obtain the graph in Figure 11.36.

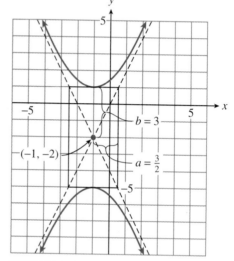

FIGURE 11.36

EXERCISE 5

a. Write the equation $4x^2 - 6y^2 + 16x - 12y - 14 = 0$ in standard form.

b. Graph the equation.

Parabolas

The equation of a parabola whose vertex is located at the point (h, k) has one of the following standard forms.

$$y - k = a(x - h)^2 \quad \begin{array}{l} \text{opens upward if} \quad\quad a > 0 \\ \text{opens downward if} \quad a < 0 \end{array} \quad\quad (7)$$

$$x - h = a(y - k)^2 \quad \begin{array}{l} \text{opens to the right if} \quad a > 0 \\ \text{opens to the left if} \quad\;\; a < 0 \end{array} \quad\quad (8)$$

The constant a is equal to $\dfrac{\pm 1}{4p}$ where p is the distance between the vertex and the focus of the parabola. It controls the width of the parabola; the smaller the absolute value of a, the wider the parabola. You may recognize Equation (7) as the vertex form for a parabola, which we studied in Chapter 4.

EXAMPLE 6

Describe the graph of $y - 3 = \frac{-1}{8}(x + 2)^2$.

Solution

The graph is a parabola that opens downward, because $a < 0$. The vertex of the parabola is the point $(-2, 3)$. Because $|a| = \frac{1}{8} < 1$, the parabola is relatively wide. To improve our description, we can locate one or two points besides the vertex. For example, if $x = 2$, then

$$y = \frac{-1}{8}(2 + 2)^2 + 3 = 1$$

so $(2, 1)$ lies on the graph. By symmetry, there is another point with the same y-coordinate at an equal distance from the axis of the parabola. This point is $(-6, 1)$, as shown on the graph in Figure 11.37.

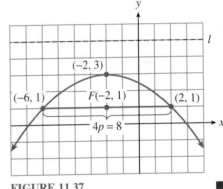

FIGURE 11.37

A quadratic equation that includes either an x^2 term or a y^2 term but not both can be put into one of the standard forms for a parabola by completing the square.

EXAMPLE 7

a. Write the equation in standard form.

$$y^2 + 2y + 8x - 31 = 0$$

b. Graph the equation.

Solutions

a. First prepare to complete the square in y by writing

$$(y^2 + 2y \quad\quad) = -8x + 31$$

Complete the square by adding 1 to each side of the equation to get

$$(y^2 + 2y + 1) = -8x + 32$$

or

$$(y + 1)^2 = -8(x - 4)$$

b. The graph is a parabola that opens to the left. Its vertex is the point $(4, -1)$. To find another point on the graph, choose a value for y, say $y = 3$. Solving for x yields

$$(3 + 1)^2 = -8(x - 4) \quad \text{Divide both sides by } -8.$$

$$\frac{16}{-8} = x - 4 \qquad \text{Add 4 to both sides.}$$

$$2 = x$$

Thus, $(2, 3)$ lies on the graph. Because of symmetry, there is another point on the graph with the same x-coordinate and at an equal distance from the parabola's axis, which is the line $y = -1$. This point is $(2, -5)$, as shown on the graph in Figure 11.38.

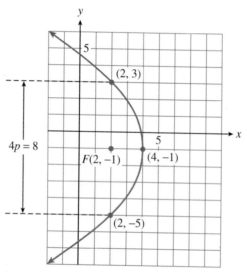

FIGURE 11.38

■

E X E R C I S E 6

a. Write $x = y^2 - 8y + 13$ in standard form.

b. Graph the equation.

General Quadratic Equation in Two Variables

We have considered graphs of second-degree equations in two variables,

$$Ax^2 + Bxy + Cy^2 + Dx + Ey + F = 0$$

for which B, the coefficient of the xy term, is zero. Such graphs are conic sections with axes parallel to one or both of the coordinate axes. If B does not equal zero, the axes of the conic section are rotated with respect to the coordinate axes. The graphing of such equations is taken up in more advanced courses in analytic geometry.

The graph of a second-degree equation can also be a point, a line, a pair of lines, or no graph at all, depending on the values of the coefficients A through F. Such graphs are called **degenerate conics.**

Given an equation of the form

$$Ax^2 + Bxy + Cy^2 + Dx + Ey + F = 0$$

we can determine the nature of the graph from the coefficients of the quadratic terms. If the graph is not a degenerate conic, the following criteria apply.

The graph of $Ax^2 + Cy^2 + Dx + Ey + F = 0$ is

1. a circle if $A = C$.
2. a parabola if $A = 0$ or $C = 0$ (but not both).
3. an ellipse if A and C have the same sign.
4. a hyperbola if A and B have opposite signs.

EXAMPLE 8

Name the graph of each equation assuming that the graph is not degenerate.

a. $3x^2 + 3y^2 - 2x + 4y - 6 = 0$ c. $4x^2 - 6y^2 + x - 2y = 0$

b. $4y^2 + 8x^2 - 3y = 0$ d. $y + x^2 - 4x + 1 = 0$

Solutions

a. The graph is a circle because the coefficients of x^2 and y^2 are equal.
b. The graph is an ellipse because the coefficients of x^2 and y^2 are both positive.
c. The graph is a hyperbola because the coefficients of x^2 and y^2 have opposite signs.
d. The graph is a parabola because y is of first degree and x is of second degree. ∎

The coefficients D, E, and F do not figure in determining the *type* of conic section the equation represents. They do, however, determine the *position* of the graph relative to the origin. Once we recognize the form of the graph, it is helpful to write the equation in the appropriate standard form in order to discover more specific information about the graph. The standard forms for the conic sections are summarized in Table 11.2. For the parabola, (h, k) is the vertex of the graph, whereas for the other conics, (h, k) is the center.

TABLE 11.2

Name of curve	Standard form of equation	Graph
Circle	$(x - h)^2 + (y - k)^2 = r^2$	
Ellipse (a) Major axis parallel to x-axis (b) Major axis parallel to y-axis	$\dfrac{(x - h)^2}{a^2} + \dfrac{(y - k)^2}{b^2} = 1$ $a > b$ $a < b$	
Hyperbola (a) Transverse axis parallel to x-axis (b) Transverse axis parallel to y-axis	$\dfrac{(x - h)^2}{a^2} - \dfrac{(y - k)^2}{b^2} = 1$ $\dfrac{(y - k)^2}{b^2} - \dfrac{(x - h)^2}{a^2} = 1$	
Parabola (a) Opens upward (b) Opens downward (c) Opens to the right (d) Opens to the left	$(y - k) = a(x - h)^2, a > 0$ $(y - k) = a(x - h)^2, a < 0$ $(x - h) = a(y - k)^2, a > 0$ $(x - h) = a(y - k)^2, a < 0$	

Using a Graphing Calculator

As we saw in Section 11.1, the graph of a conic section (except for parabolas that open up or down) is the graph of two functions. We find these two functions by solving the quadratic equation that defines the conic for y in terms of x.

EXAMPLE 9

Use a calculator to graph $y^2 - 3y + x^2 = 0$

Solution

The equation $y^2 - 3y + x^2 = 0$ describes a circle. To solve for y in terms of x, we will use the quadratic formula. Because we are regarding y (not x) as the variable, we have $a = 1$, $b = -3$, and $c = x^2$. Thus

$$y = \frac{-(-3) \pm \sqrt{(-3)^2 - 4(1)(x^2)}}{2(1)} = \frac{3 \pm \sqrt{9 - 4x^2}}{2}$$

The two functions defined by this equation are

$$Y_1 = \frac{3 + \sqrt{9 - 4x^2}}{2} \quad \text{and} \quad Y_2 = \frac{3 - \sqrt{9 - 4x^2}}{2}$$

Use the **ZDecimal** window to obtain the graph shown in Figure 11.39. By tracing values on the two curves, we can see that this circle has a diameter of 3.

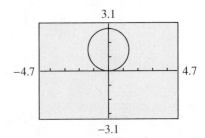

FIGURE 11.39

1a.

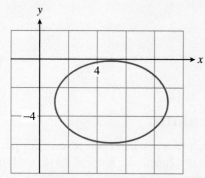

b. $(5 - \sqrt{15}, -3), (5 + \sqrt{15}, -3),$
$(5, -3 + \sqrt{8}), (5, -3 - \sqrt{8})$

3. $\dfrac{(x + 4)^2}{9} + \dfrac{(y - 1)^2}{4} = 1$

4a.

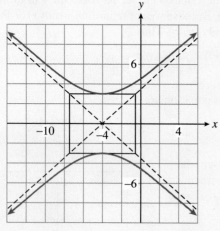

b. $y = \dfrac{\sqrt{3}}{2}x + 4, \ y = \dfrac{-\sqrt{3}}{2}x + 4$

6a. $x + 3 = (y - 4)^2$

b.

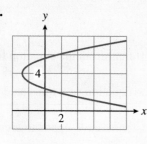

2a. $\dfrac{(x + 2)^2}{16} + \dfrac{(y - 2)^2}{4} = 1$

b.

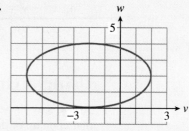

5a. $\dfrac{(x + 2)^2}{6} - \dfrac{(y + 1)^2}{4} = 1$

b.

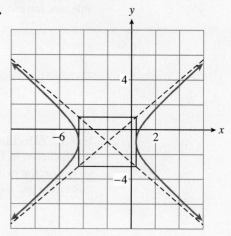

7a. A hyperbola centered at the origin with vertices $(0, 1)$ and $(0, -1)$ and asymptotes $y = \frac{1}{3}x$ and $y = \frac{-1}{3}x$

b. A parabola opening upward from the vertex $(1, -5)$

c. An ellipse centered at $(-2, 1)$ with horizontal major axis, with $a = 2$ and $b = \frac{2}{3}$

For each equation in Problems 1–10, (a) graph the ellipse, (b) give the exact coordinates of any four points on the ellipse.

1. $\dfrac{(x-3)^2}{16} + \dfrac{(y-4)^2}{9} = 1$

2. $\dfrac{(x-2)^2}{4} + \dfrac{(y-5)^2}{25} = 1$

3. $\dfrac{(x+2)^2}{6} + \dfrac{(y-5)^2}{12} = 1$

4. $\dfrac{x^2}{16} + \dfrac{(y+4)^2}{6} = 1$

5. $9x^2 + 4y^2 - 16y = 20$

6. $x^2 + 16y^2 + 6x = 7$

7. $9x^2 + 16y^2 - 18x + 96y + 9 = 0$

8. $6x^2 + 5y^2 - 12x + 20y - 4 = 0$

9. $8x^2 + y^2 - 48x + 4y + 68 = 0$

10. $x^2 + 10y^2 + 4x + 20y + 4 = 0$

In Problems 11–16, write an equation for the ellipse with the properties given.

11. Center at $(1, 6)$, $a = 3$, $b = 2$

12. Center at $(2, 3)$, $a = 4$, $b = 3$

13. Vertices at $(3, 2)$ and $(-7, 2)$, minor axis of length 6

14. Covertices at $(3, 7)$ and $(3, -1)$, major axis of length 10

15. Vertices at $(-4, 9)$ and $(-4, -3)$, covertices at $(-7, 3)$ and $(-1, 3)$

16. Vertices at $(-3, -5)$ and $(9, -5)$, covertices at $(3, 0)$ and $(3, -10)$

For each equation in Problems 17–26, (a) graph the hyperbola, (b) give the exact coordinates of any four points on the hyperbola.

17. $\dfrac{(x-4)^2}{9} - \dfrac{(y+2)^2}{16} = 1$

18. $\dfrac{(y+4)^2}{25} - \dfrac{(x-3)^2}{4} = 1$

19. $\dfrac{x^2}{4} - \dfrac{(y-3)^2}{8} = 1$

20. $\dfrac{(y+2)^2}{6} - \dfrac{(x+2)^2}{10} = 1$

21. $16y^2 - 4x^2 + 32x - 128 = 0$

22. $9x^2 - 4y^2 - 36x - 24y - 36 = 0$

23. $4x^2 - 6y^2 - 32x - 24y + 16 = 0$

24. $9y^2 - 8x^2 + 72y + 16x + 64 = 0$

25. $12x^2 - 3y^2 + 24y - 84 = 0$

26. $10y^2 - 5x^2 + 30x - 95 = 0$

In Problems 27–30, write an equation for the hyperbola with the properties given.

27. Center at $(-1, 5)$, $a = 8$, $b = 6$, opening up and down

28. Center at $(6, -2)$, $a = 1$, $b = 4$, opening left and right

29. One vertex at $(-1, 3)$, one end of the horizontal conjugate axis at $(-5, 1)$

30. One vertex at $(1, -2)$, one end of the vertical conjugate axis at $(-5, 1)$

For each equation in Problems 31–38, (a) graph the parabola, (b) give the exact coordinates of any four points on the parabola.

31. $x = \dfrac{(y+3)^2}{2}$

32. $y = \dfrac{(x-2)^2}{-3}$

33. $-6(y+4) = (x-3)^2$

34. $4(x-5) = (y+1)^2$

35. $y^2 - 4y + 8x + 6 = 0$

36. $4x^2 - 4x = 8y - 5$

37. $9y^2 = 6y + 12x - 1$

38. $9y^2 + 12y - 12x = 0$

Name the graph of each equation and describe its main features.

39. $y^2 = 6 - 4x^2$

40. $4y^2 = x^2 - 8$

41. $x^2 + 2y - 4 = 0$

42. $2x^2 = 5 + 4y^2$

43. $6 + \dfrac{x^2}{4} = y^2$

44. $y^2 = 6 - \dfrac{2x^2}{3}$

45. $\dfrac{1}{2}y^2 - x = 4$

46. $\dfrac{(y - 2)^2}{4} - \dfrac{(x + 3)^2}{8} = 1$

47. $\dfrac{(x + 3)^2}{5} + \dfrac{y^2}{12} = 1$

48. $y - 2 = \dfrac{(x + 4)^2}{4}$

49. $2x^2 + y^2 + 4x = 2$

50. $y^2 - 4x^2 + 2y - x = 0$

11.3   Systems of Quadratic Equations

In Section 4.3 we considered systems of quadratic equations involving two parabolas. Those systems can have one, two, or no solutions, depending on the graphs of the two equations. Systems involving other conics may have up to four solutions.

EXAMPLE 1

Find the intersection points of the graphs of

$$x^2 + y^2 = 5$$
$$xy = 2$$

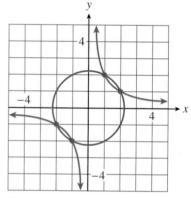

FIGURE 11.40

Solution

We will use substitution to solve the system. Solve the "easier" of the two equations (the second equation) for y to obtain

$$y = \frac{2}{x}$$

Substitute $\dfrac{2}{x}$ for y in the first equation to find

$$x^2 + \left(\frac{2}{x}\right)^2 = 5 \qquad \text{or} \qquad x^2 + \frac{4}{x^2} = 5$$

This equation has only one variable, x, and we solve it by first clearing fractions. Multiply both sides by x^2, and then subtract $5x^2$ to obtain

$$x^4 - 5x^2 + 4 = 0$$

Factor the left side to get

$$(x^2 - 1)(x^2 - 4) = 0$$

and apply the zero-factor principle to find

$$x^2 - 1 = 0 \qquad \text{or} \qquad x^2 - 4 = 0$$

Solve each of these equations to find

$$x = 1, \qquad x = -1, \qquad x = 2, \qquad \text{or} \qquad x = -2$$

Finally, substitute each of these values into $y = \frac{2}{x}$ to find the y-components of each solution. The intersection points of the two graphs are $(1, 2)$, $(-1, -2)$, $(2, 1)$, and $(-2, -1)$. The graph of the system is shown in Figure 11.40. ■

EXERCISE 1 Find the intersection points of the graphs of

$$x^2 - y^2 = 35$$
$$xy = 6$$

Using a Graphing Calculator

If we want to graph the two equations in Example 1 on the calculator, we solve each equation for y in terms of x. However, the circle $x^2 + y^2 = 5$ is not the graph of a function. When we solve for y, we find

$$y = \pm \sqrt{5 - x^2}$$

We can enter this equation into the calculator as *two* functions.

$$Y_1 = \sqrt{5 - x^2}$$
$$Y_2 = -Y_1 = -\sqrt{5 - x^2}$$

The first equation gives the top half of the circle, and the second equation gives the bottom half. Enter Y_1, Y_2, and $Y_3 = \frac{2}{x}$ in the window

$$\text{Xmin} = -4.7 \quad \text{Xmax} = 4.7$$
$$\text{Ymin} = -3.1 \quad \text{Ymax} = 3.1$$

to obtain the graph shown in Figure 11.41. (The two halves of the circle may appear not to touch the x-axis because the calculator does not plot the points where x is exactly $\pm \sqrt{5}$.)

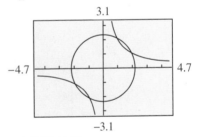

FIGURE 11.41

Solving Systems by Elimination

We used substitution in Example 1 to solve the system. If both equations are of the form

$$ax^2 + by^2 = c$$

elimination of variables is more efficient.

EXAMPLE 2

Find the solutions to the following system of the equations.

$$x^2 - 2y^2 = 1$$
$$\frac{x^2}{15} + \frac{y^2}{10} = 1$$

Verify the solutions on a graph.

Solution

Multiply the first equation by 3 and the second by 60 to obtain the new system

$$3x^2 - 6y^2 = 3$$
$$4x^2 + 6y^2 = 60$$

Adding these two equations, we have

$$7x^2 = 63$$
$$x^2 = 9$$
$$x = \pm 3$$

Substitute these values for x into any of the equations involving y and solve to find the solutions $(3, 2)$, $(3, -2)$, $(-3, 2)$ and $(-3, -2)$.

The two original equations describe a hyperbola and an ellipse. We can obtain graphs on our calculator by solving each equation for y to get

$$y = \pm \sqrt{\frac{x^2 - 1}{2}}$$

$$y = \pm \sqrt{10\left(1 - \frac{x^2}{15}\right)}$$

Enter these equations as shown in Figure 11.42a. Using the window

$$\text{Xmin} = -7.05 \quad \text{Xmax} = 7.05$$
$$\text{Ymin} = -4.65 \quad \text{Ymax} = 4.65$$

we obtain the graph shown in Figure 11.42b. The solutions of the system are the intersection points of the graphs.

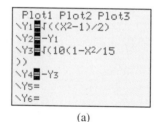

(a)   (b)

FIGURE 11.42

EXERCISE 2 Find the intersection points of the graphs of

$$y^2 - x^2 = 5$$
$$x^2 + y^2 = 13$$

For some quadratic systems, a combination of elimination of variables and substitution can be used.

EXAMPLE 3   Find the intersection of the circles given by the equations.

$$x^2 - 4x + y^2 + 2y = 20 \qquad (1)$$
$$x^2 - 12x + y^2 + 10y = -12 \qquad (2)$$

Solution

Subtract Equation (2) from Equation (1) to obtain

$$8x - 8y = 32$$

Solving for x we have

$$x = y + 4 \qquad (3)$$

Substitute $y + 4$ for x into either of the original equations. Using Equation (1), we find

$$(y + 4)^2 - 4(y + 4) + y^2 + 2y = 20$$
$$(y^2 + 8y + 16) - 4y - 16 + y^2 + 2y = 20$$
$$2y^2 + 6y - 20 = 0$$
$$y^2 + 3y - 10 = 0$$
$$(y + 5)(y - 2) = 0$$

Thus $y = -5$ or $y = 2$. From Equation (3) we find that when $y = -5$, $x = -1$, and when $y = 2$, $x = 6$. Thus the two circles intersect at $(-1, -5)$ and $(6, 2)$ as shown in Figure 11.43.

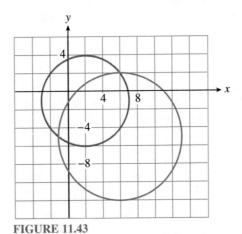

FIGURE 11.43

EXERCISE 3   Find the intersection points of the graphs of

$$x^2 - 8x + y^2 + 2y = 23$$
$$x^2 - 12x + y^2 - 6y = -25$$

Applications

The Global Positioning System (GPS) we discussed in Investigation 20 is very accurate, but it depends on perfectly synchronized atomic clocks. An older navigational system, called LORAN (Long Range Navigation), is still used by sailors to locate their position at sea. The system works by finding the intersection of two hyperbolas. Radio transmitters at known positions broadcast signals simultaneously, and the navigator notes the *difference* in the arrival times of the two signals. (This difference is calculated using only the navigator's clock, and does not depend on the time when the signals were transmitted.) From the time difference the navigator computes the difference in the distances to the two transmitters, using the speed of radio waves.

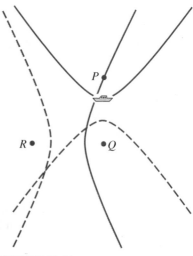

In Section 11.1 we saw that a hyperbola is the set of all points in the plane for which the difference of the distances from two fixed points is a constant. Thus, the navigator's position lies on a hyperbola with the two transmitters as the foci. In Figure 11.44, transmitters located at P and Q determine the hyperbola shown in blue. Transmitters at Q and R determine a second hyperbola, shown in red. The navigator's position must be at an intersection of the two hyperbolas.

FIGURE 11.44

EXAMPLE 4

Your ship receives LORAN signals from transmitters located at $A(0, 28)$, $B(0, -28)$, and $C(-120, -28)$. You calculate that you are 14 kilometers closer to A than to B, and 72 kilometers closer to B than to C. It can be shown that your position must lie on the hyperbolas

$$\frac{y^2}{49} - \frac{x^2}{735} = 1$$
$$\frac{(x + 60)^2}{36^2} - \frac{(y + 28)^2}{48^2} = 1$$

a. Plot and label the positions of the three transmitters. Describe the orientation of the two hyperbolas.

b. Use a graphing calculator to graph the two hyperbolas on the window

$$\text{Xmin} = -130 \quad \text{Xmax} = 130$$
$$\text{Ymin} = -130 \quad \text{Ymax} = 40$$

Locate the point P corresponding to the position of your ship.

c. Find the coordinates of your ship. Verify that your position is 14 kilometers closer to A than B and 72 kilometers closer to B than to C.

Solutions

a. Plot points at A, B, and C, as shown in Figure 11.45. The foci of the first parabola, A and B, lie on a vertical line segment, so the hyperbola opens up and down. Because your position is closer to A than to B, your ship is located on the upper branch of that hyperbola. The foci of the second hyperbola, B and C, lie on a horizontal line, so the second hyperbola opens left and right. Because your position is closer to B than to C, your ship is located on the right branch.

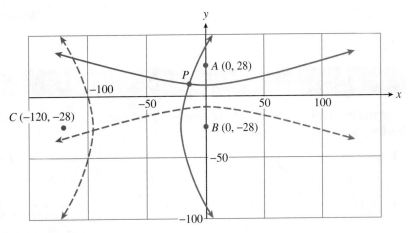

FIGURE 11.45

b. Solve each equation for y in terms of x to find

$$y = \pm 7 \sqrt{1 + \frac{x^2}{735}}$$

$$y = -28 \pm 48 \sqrt{\frac{(x + 60)^2}{36^2} - 1}$$

Graph these four functions to see the two hyperbolas shown in Figure 11.45. Although there are four intersection points, only point P is closer to A than to B and closer to B than to C.

c. Use the **Intersect** or **Value** feature of the calculator to find that the coordinates of P are $(-15, 8)$. Evaluate the distance formula to verify that point P satisfies the conditions of the problem.

$$AP = \sqrt{(-15 - 0)^2 + (8 - 28)^2} = \sqrt{625} = 25$$
$$BP = \sqrt{(-15 - 0)^2 + (8 + 28)^2} = \sqrt{1521} = 39$$
$$CP = \sqrt{(-15 + 120)^2 + (8 + 28)^2} = \sqrt{12321} = 111$$

We see that AP is 14 less than BP, and BP is 72 less than CP. ■

READING QUESTIONS

1. Name two algebraic techniques that are useful in solving systems of quadratic equations.

2. In Example 1, why did we substitute the values of x into the second equation, rather than the first equation?

3. How many solutions can a system of two ellipses have? Illustrate your answer with sketches.

4. Can two parabolas intersect in four points? Illustrate with a sketch.

HOMEWORK 11.3

Find the intersection points of the graphs by solving a system of equations. Verify your solutions by graphing.

1. $xy = 4$
$x^2 + y^2 = 8$

2. $x^2 - 2y^2 = -4$
$xy = -4$

3. $x^2 - y^2 = 16$
$xy = 6$

4. $xy = -2$
$x^2 + y^2 = 5$

5. $x^2 + y^2 = 10$
$x^2 - y^2 = 8$

6. $5y^2 - x^2 = 1$
$x^2 + y^2 = 5$

7. $\dfrac{x^2}{35} - \dfrac{y^2}{10} = 1$
$x^2 + 7y^2 = 77$

8. $\dfrac{x^2}{14} + \dfrac{y^2}{35} = 1$
$x^2 + 2y^2 = 54$

9. $\dfrac{x^2}{2} - \dfrac{y^2}{16} = 1$
$y^2 - x^2 = 12$

10. $15x^2 - y^2 = 6$
$y^2 - 3x^2 = 6$

11. $x^2 - 6y^2 = 10$
$28y^2 - x^2 = 12$

12. $x^2 - 8y^2 = 4$
$10y^2 - x^2 = 4$

13. $x^2 + y^2 + 2y = 19$
$x^2 - 2x + y^2 + 8y = 33$

14. $x^2 - 4x + y^2 + 2y = 5$
$x^2 - 6x + y^2 - 2y = 15$

15. $(x + 1)^2 + (y - 2)^2 = 5$
$(x - 5)^2 + (y + 1)^2 = 50$

16. $(x - 1)^2 + (y - 6)^2 = 26$
$(x - 4)^2 + (y - 8)^2 = 65$

17. Use algebra to solve the system from Investigation 20.

$$(x - 100)^2 + y^2 = 90^2$$
$$(x - 28)^2 + (y - 96)^2 = 42^2$$

18. Verify that the solutions you found in Problem 17 are both 90 meters from the point $(100, 0)$ and 42 meters from $(28, 96)$.

19. Refer to Investigation 20 for a simplified version of GPS. Suppose that you receive two signals that are both 7 seconds old. One is from $(60, 80)$ and the other is from $(93.6, 35.2)$. Calculate your position.

20. Verify that the solutions you found in Problem 19 are 35 meters from both the point $(60, 80)$ and $(93.6, 35.2)$.

Use a system of equations to solve each problem.

21. The area of a rectangle is 216 square feet. If the perimeter is 60 feet, find the dimensions of the rectangle.

22. Leon flies his plane 840 miles in the same time that Marlene drives her automobile 210 miles. Suppose that Leon flies 180 miles per hour faster than Marlene drives. Find the rate of each.

23. At a constant temperature, the pressure, P, and the volume, V, of a gas are related by the equation $PV = K$. The product of the pressure (in pounds per square inch) and the volume (in cubic inches) of a certain gas is 30 inch-pounds. If the temperature remains constant as the pressure is increased by 4 pounds per square inch, the volume is decreased by

2 cubic inches. Find the original pressure and volume of the gas.

24. Kristen drove 50 miles to her sister's house, traveling 10 miles in heavy traffic to get out of the city and then 40 miles in less congested traffic. Her average speed in the city was 20 miles per hour less than her speed in light traffic. What was each rate if her trip took 1 hour and 30 minutes?

25. Your ship receives LORAN signals from transmitters located at $A(0, 15)$, $B(0, -15)$, and $C(0, 29)$. You calculate that your distance from A is 24 kilometers less than your distance from B, and 2 kilometers less than your distance from C. It follows that your position must lie on the two hyperbolas

$$\frac{y^2}{144} - \frac{x^2}{81} = 1$$

$$(y - 22)^2 - \frac{x^2}{48} = 1$$

a. Plot and label the positions of the three transmitters. Sketch portions of the two hyperbolas with their correct orientations.

b. Use a graphing calculator to graph the two hyperbolas on the window

$$\text{Xmin} = -20 \quad \text{Xmax} = 20$$
$$\text{Ymin} = -20 \quad \text{Ymax} = 30$$

If you are east of A, locate the point P corresponding to the position of your ship.

c. Find the coordinates of your ship. Verify that your position is 24 kilometers closer to A than B and 2 kilometers closer to A than to C.

26. Your ship receives LORAN signals from transmitters at $A(150, 0)$, $B(116, 0)$, and $C(-150, 0)$. You calculate that your distance from A is 2 kilometers less than your distance from B, and 240 kilometers less than your distance from C. It follows that your position must lie on the two hyperbolas

$$(x - 133)^2 - \frac{y^2}{288} = 1$$

$$\frac{x^2}{120^2} - \frac{y^2}{90^2} = 1$$

a. Plot and label the positions of the three transmitters. Sketch portions of the two hyperbolas with their correct orientations.

b. Use a graphing calculator to graph the two hyperbolas on the window

$$\text{Xmin} = 120 \quad \text{Xmax} = 140$$
$$\text{Ymin} = -60 \quad \text{Ymax} = 60$$

If you are north of A, locate the point P corresponding to the position of your ship.

c. Find the coordinates of your ship. Verify that your position is 2 kilometers closer to A than to B and 240 kilometers closer to A than to C.

27. Your ship receives LORAN signals from transmitters at $A(28, 0)$, $B(-28, 0)$, and $C(-28, -90)$. You calculate that your distance from A is 14 kilometers less than your distance from B, and your distance from B is 72 kilometers less than your distance from C. It follows that your position must lie on the two hyperbolas

$$\frac{x^2}{49} - \frac{y^2}{735} = 1$$

$$\frac{(y + 45)^2}{36^2} - \frac{(x + 28)^2}{27^2} = 1$$

a. Plot and label the positions of the three transmitters. Sketch portions of the two hyperbolas with their correct orientations.

b. Use a graphing calculator to graph the two hyperbolas on the window

$$\text{Xmin} = -40 \quad \text{Xmax} = 40$$
$$\text{Ymin} = -100 \quad \text{Ymax} = 50$$

Locate the point P corresponding to the position of your ship.

c. Find the coordinates of your ship. Verify that your position is 14 kilometers closer to A than to B and 72 kilometers closer to B than to C.

28. Your ship receives LORAN signals from transmitters at $A(50, 0)$, $B(-50, 0)$, and $C(-50, -450)$. You calculate that your distance from A is 40 kilometers less than your distance from B, and your distance from B is 360 kilometers less than your distance from C. It follows that your position must lie on the two hyperbolas

$$\frac{x^2}{20^2} - \frac{y^2}{2100} = 1$$

$$\frac{(y + 225)^2}{180^2} - \frac{(x + 50)^2}{135^2} = 1$$

a. Plot and label the positions of the three transmitters. Sketch portions of the two hyperbolas with their correct orientations.

b. Use a graphing calculator to graph the two hyperbolas on the window

$$\text{Xmin} = -50 \quad \text{Xmax} = 50$$
$$\text{Ymin} = -100 \quad \text{Ymax} = 100$$

Locate the point P corresponding to the position of your ship.

c. Find the coordinates of your ship. Verify that your position is 40 kilometers closer to A than to B and 360 kilometers closer to B than to C.

MIDCHAPTER REVIEW

Graph each ellipse. Label the vertices with their coordinates.

1. $4x^2 + y^2 = 25$

2. $2x^2 + 9y^2 = 36$

3. Find the exact coordinates of all points with y-coordinate -4 on the ellipse $4x^2 + y^2 = 25$.

4. Find the exact coordinates of all points with x-coordinate 3 on the ellipse $2x^2 + 9y^2 = 36$.

Graph each hyperbola. Find the equations of the asymptotes.

5. $9y^2 = 4 + x^2$

6. $x^2 = 6y^2 + 8$

Graph each parabola.

7. $x = 4y^2$

8. $12x + 4y^2 = 0$

a. Name the conic and describe its main features.

b. Find the coordinates of all points on the graph with the given coordinate.

9. $\dfrac{1}{2}x^2 - y = 4, \quad y = \dfrac{-5}{2}$

10. $\dfrac{x^2}{4} = 4 + 6y^2, \quad x = \sqrt{17}$

Graph each ellipse or hyperbola.

11. $\dfrac{(x-5)^2}{15} + \dfrac{y^2}{25} = 1$

12. $\dfrac{(x-4)^2}{5} - \dfrac{(y-4)^2}{8} = 1$

13. $4y^2 - 3x^2 - 24y - 24x - 24 = 0$

14. $5x^2 + 8y^2 - 20x + 16y - 12 = 0$

Find the equation of the conic with the given properties.

15. An ellipse with center at $(-5, -2)$ and tangent to both axes.

16. A hyperbola with one vertex at $(1, -4)$ and whose asymptotes are the lines $y + 4 = \dfrac{\pm 3}{2}(x - 3)$.

Graph the parabola.

17. $(x + 3)^2 - 2 = y$

18. $x = y^2 + 4y$

Solve each system algebraically, and verify with a graph.

19. $x^2 + y^2 = 17$
$\quad 2xy = 23$

20. $3x^2 - y^2 = 2$
$\quad 2x^2 + y^2 = 43$

21. $y^2 - x^2 = 24$
$\quad 2x^2 - y^2 = 1$

22. $x^2 - 7y^2 = 1$
$\quad 27y^2 - 10x^2 = 1$

23. $x^2 + 4x + y^2 - 4y = 26$
$x^2 - 12x + y^2 + 2 = 0$

24. $x^2 - 2x + y^2 + 4y = 20$
$x^2 + 10x + y^2 - 8y = -28$

11.4 Linear Programming

The term **linear programming** was coined in the late 1940s. It describes a relatively young branch of mathematics, if you compare it with other subjects such as Euclidean geometry, where the major ideas were already well understood 23 centuries ago. (The Greek mathematician Euclid wrote what can be considered the first geometry textbook about 300 BC.) Business managers must routinely solve linear programming problems for purchasing and marketing strategy, so it is possible that linear programming affects your daily life as much as any other branch of mathematics.

The Objective Function and Constraints

Suppose TrailGear produces two kinds of hiking boots, a Weekender model, on which they make $8 profit per pair, and a Sierra model, on which they make $10 profit per pair. TrailGear would like to know how many of each model to produce each week in order to maximize their profit. If we let x represent the number of Weekender boots and y the number of Sierra boots they produce, then the total weekly profit is given by

$$P = 8x + 10y$$

This expression for P is called the **objective function.** The goal of a linear programming problem is to maximize or minimize such an objective function, subject to one or more constraints.

If TrailGear had infinite resources and an infinite market, there would be no limit to the profit they could earn by producing more and more hiking boots. However, every business has to consider many factors, including its supplies of labor and materials, overhead and shipping costs, and the size of the market for its product. To keep things simple, we will concentrate on just two of these factors.

Each pair of Weekender boots requires 3 man-hours of labor to produce, and each pair of Sierra boots requires 6 man-hours. TrailGear has available 2400 man-hours of labor per week. Thus, x and y must satisfy the inequality

$$3x + 6y \leq 2400$$

In addition, suppose that their suppliers can provide at most 1000 ounces of foam rubber each week, with each pair of Weekenders using 2 ounces and each Sierra model using 1 ounce. This means that

$$2x + y \leq 1000$$

Of course, we will also require that $x \geq 0$ and $y \geq 0$. These four inequalities are called the **constraints** of the problem.

Solution by Graphing

We have formulated the original problem into an objective function

$$P = 8x + 10y$$

and a system of inequalities called the constraints.

$$3x + 6y \leq 2400$$
$$2x + y \leq 1000$$
$$x \geq 0, y \geq 0$$

Our goal is to find values for x and y that satisfy the constraints and produce the maximum value for P.

We begin by graphing the solutions to the constraint inequalities. These solutions are shown in the shaded region in Figure 11.46. The points in this region are called **feasible solutions** because they are the only values we can consider while looking for the maximum value of the objective function P.

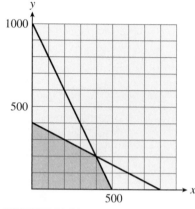

FIGURE 11.46

EXAMPLE 1

a. Verify that the points (300, 100) and (200, 300) represent feasible solutions for the problem above. Show that (300, 400) is not a feasible solution.

b. Find the values of the objective function $P = 8x + 10y$ at the two feasible solutions in part (a).

Solutions

a. The two points (300, 100) and (200, 300) lie within the shaded region in Figure 11.46, but (300, 400) does not. We can also verify that the coordinates of (300, 100) and (200, 300) satisfy each of the constraint inequalities.

b. For (300, 100), we have

$$P = 8(300) + 10(100) = 3400$$

For (200, 300), we have

$$P = 8(200) + 10(300) = 4600 \qquad \blacksquare$$

We cannot check all of the feasible solutions to see which one results in the largest profit. Fortunately there is a simple way to find the optimal solution.

Let's look at the objective function,

$$P = 8x + 10y$$

If TrailGear would like to make $2000 on hiking boots, it could produce 200 pairs of Sierra boots, or 250 pairs of Weekenders. Or it could produce some of each; for example, 50 pairs of Weekenders and 160 pairs of Sierra boots. In fact, every point on the line

$$8x + 10y = 2000$$

represents a combination of Weekenders and Sierra boots that will yield a profit of $2000. This line is labeled $P = 2000$ in Figure 11.47.

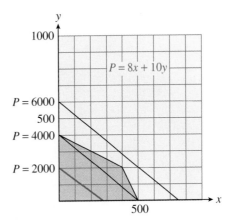

FIGURE 11.47

If TrailGear would like to make $4000 on boots, they should choose a point on the line labeled $P = 4000$. Similarly, all the points on the line labeled $P = 6000$ will yield a profit of $6000, and so on. Different values of P correspond to parallel lines on the graph. Smaller values of P correspond to lines near the origin, and larger values of P have lines farther from the origin. Here is another example.

<table>
<tr><td>

EXAMPLE 2

</td><td>

Figure 11.48 shows the feasible solutions for another linear programming problem. The objective function is $C = 3x + 5y$.

</td><td>

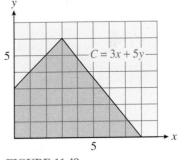

FIGURE 11.48

</td></tr>
</table>

    a. Find the value of C at the point $(0, 3)$. Are there other feasible solutions that give the same value of C?

    b. Find all feasible solutions that result in an objective value of 30.

    c. How many feasible solutions result in an objective value of 39?

    d. Is it possible for a feasible solution to result in an objective value of 45?

Solutions

    a. The objective value at the point $(0, 3)$ is

$$C = 3(0) + 5(3) = 15$$

Another point with the same objective value is $(5, 0)$. In fact, all points on the line $3x + 5y = 15$ have an objective value of 15. This line intersects the set of feasible solutions in a line segment, as shown in Figure 11.49. Thus, there are infinitely many feasible solutions with objective value 15.

b. Points that give an objective value of $C = 30$ lie on the line $3x + 5y = 30$, as shown in the figure. There are infinitely many feasible solutions that lie on this line; for example, one such point is $(5, 3)$.

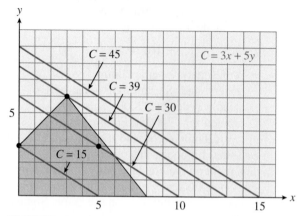

FIGURE 11.49

c. The line $3x + 5y = 39$ intersects the set of feasible solutions in only one point, the point $(3, 6)$. This is the only feasible solution that yields an objective value of 39.

d. The line $3x + 5y = 45$ includes all points for which $C = 45$. This line does not intersect the set of feasible solutions, as we see in the figure. Thus, there are no feasible solutions that result in an objective value of 45. ∎

We are only allowed to choose points from the set of feasible solutions. Imagine the parallel lines representing different values of the objective function sweeping across the graph of the feasible solutions. The objective values increase as the lines sweep up across the graph. What is the last feasible solution the lines intersect before leaving the shaded region? If you study the examples above, perhaps you can see that the largest (and smallest) values of the objective function will occur at corner points of the set of feasible solutions. We haven't proved this fact, but it is true.

> The maximum and the minimum values of the objective function always occur at vertices of the graph of feasible solutions.

Depending on the exact formula for the objective function, the maximum and minimum values may occur at *any* of the vertices of the shaded region.

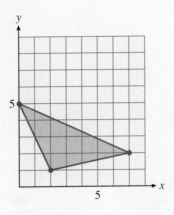
We can now formulate a strategy for solving problems by linear programming.

To Solve a Linear Programming Problem

1. Represent the unknown quantities by variables. Write the objective function and the constraints in terms of the variables.

2. Graph the solutions to the constraint inequalities.

3. Find the coordinates of each vertex of the solution set.

4. Evaluate the objective function at each vertex.

5. The maximum and minimum values of the objective function occur at vertices of the set of feasible solutions.

In Example 3 the set of feasible solutions is an unbounded region.

EXAMPLE 3

Each week the Healthy Food Store buys both granola and muesli in bulk from 2 cereal companies. The store requires at least 12 kilograms of granola and 9 kilograms of muesli. Company A charges $15 for a package that contains 2 kilograms of granola and 1 kilogram of muesli. Company B charges $25 for a package of 3 kilograms of granola and 3 kilograms of muesli. How much should the Healthy Food Store purchase from each company in order to minimize its costs and still meet its needs for granola and muesli? What is the minimum cost?

Solution

Step 1 Number of packages purchased from Company A: x
Number of packages purchased from Company B: y

First write the objective function. The store would like to minimize its cost, so

$$C = 15x + 25y$$

Next, write the constraints. These will be a system of inequalities. It may help to organize the information into a table.

TABLE 11.3

	Company A	Company B	Required
Granola (kg)	$2x$	$3y$	12
Muesli (kg)	x	$3y$	9

The Healthy Food Store will have $2x$ kilograms of granola and x kilograms of muesli from Company A, and $3y$ kilograms of granola and $3y$ kilograms of muesli from Company B. The store's requirements are that

$$2x + 3y \geq 12$$
$$x + 3y \geq 9$$

Because the store cannot purchase negative quantities, we also have

$$x \geq 0, \qquad y \geq 0$$

Step 2 Graph the solutions to the constraint system. The feasible solutions form the shaded region shown in Figure 11.50. Any ordered point on this graph corresponds to a way to purchase granola and muesli that meets the store's needs, but some of these choices cost more than others.

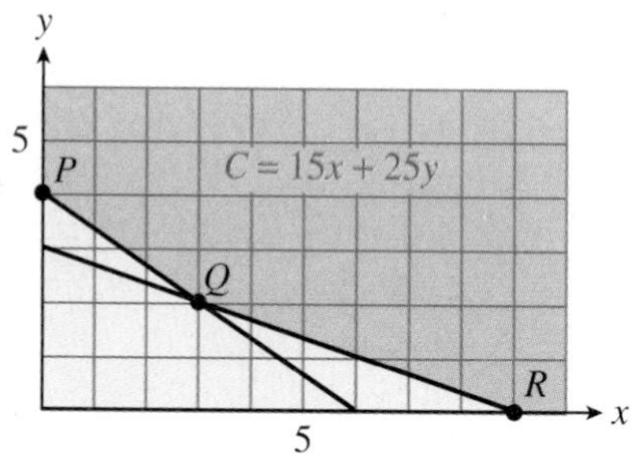

FIGURE 11.50

Step 3 We know that the minimum cost will occur at one of the vertex points, which are labeled in the figure. The coordinates of P and R are easy to see. To find the coordinates of Q, we notice that it is the intersection of the lines $2x + 3y = 12$ and $x + 3y = 9$. Thus, we must solve the system

$$2x + 3y = 12$$
$$x + 3y = 9$$

Subtracting the second equation from the first, we find that $x = 3$. Substituting this value into either of the original two equations we find that $y = 2$. Thus the point Q has coordinates $(3, 2)$.

Step 4 Now we evaluate the objective function at each of the three vertices.

At $P(0, 4)$: $C = 15(0) + 25(4) = 100$

At $Q(3, 2)$: $C = 15(3) + 25(2) = 95$ Minimum cost

At $R(9, 0)$: $C = 15(9) + 25(0) = 135$

The minimum cost occurs at point Q.

Step 5 The Healthy Food Store should buy 3 packages from Company A and 2 packages from Company B. It will pay \$95 for its stock of granola and muesli. ∎

Using a Graphing Calculator

You can use your graphing calculator to solve the problem in Example 3. Set your WINDOW values for the "friendly" window

$$Xmin = 0 \qquad Xmax = 9.4$$
$$Ymin = 0 \qquad Ymax = 6.2$$

Next, graph the set of feasible solutions. We have already taken care of the constraints $x \geq 0$ and $y \geq 0$ by setting Xmin and Ymin to zero. Solve each of the other constraints for y to get

$$y \geq (12 - 2x)/3$$

and

$$y \geq (9 - x)/3$$

To graph these two constraints, press Y = and enter the inequalities as shown in Figure 11.51.

The set of feasible solutions lies above each of the boundary lines because in each constraint y is *greater* than the expression in x. To shade the regions above the graphs of Y_1 and Y_2, move the cursor onto the backslash in front of the equations and press ENTER twice, then press GRAPH. Your display should look like Figure 11.52.

FIGURE 11.51

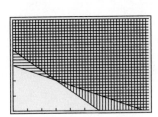 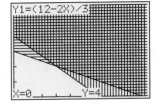

FIGURE 11.52

The feasible solutions lie in the "crosshatched" region that is shaded with both the vertical and horizontal lines. Use the TRACE (or **value** or **intersect** feature) to find the coordinates of one of the vertices, say $(0, 4)$. We'll use the calculator to evaluate the objective function at that vertex.

First press [2nd] [QUIT] to get back to the Home screen. We will enter the formula for the objective function by keying in

$$15X + 25Y$$

(We enter Y by pressing [ALPHA] [1].) Your calculator has stored the values $x = 0$ and $y = 4$ from the [TRACE] key (or **value** feature), so all you have to do now is press [ENTER], and the calculator returns 100 for the value of C. Thus, when $x = 0$ and $y = 4$, $C = 100$.

Now we'll evaluate the objective function at the other vertices. Press [TRACE] to get the graph back, and move the bug to another vertex point, say $(9, 0)$. Then [QUIT] to get back to the **Home** screen. Press [ENTER], and the calculator evaluates the objective function at $(9, 0)$ to get 135. Thus, when $x = 9$ and $y = 0$, $C = 135$. Repeat the process to evaluate the objective function at the last vertex: Press [TRACE] and position the bug at the intersection of the two boundary lines, $(3, 2)$. Then press [2nd] [QUIT] [ENTER] to see that when $x = 3$ and $y = 2$, $C = 95$.

As before, we find that the minimum cost of $95 occurs when $x = 3$ and $y = 2$.

READING QUESTIONS

1. Explain the terms objective function, constraints, and feasible solution.
2. Explain how to solve a linear programming problem by graphing.
3. How can you find the vertices of the set of feasible solutions?
4. Where do the maximum and minimum values of the objective function appear?

ANSWERS TO 11.4 EXERCISES

1a.

$R = x + 5y$

b. At $(2, 1)$, $R = 7$; at $(0, 5)$, $R = 25$; at $(7, 2)$ $R = 17$.

c. The maximum value of R occurs at $(0, 5)$. The maximum value is 25.

d. The minimum value of R occurs at $(2, 1)$. The minimum value is 7.

For Problems 1–4, the objective function for cost

$$C = 3x + 4y$$

is subject to the following constraints.

$$x + y \geq 10, \quad x \leq 8, \quad y \leq 7, \quad x \geq 0, \quad y \geq 0$$

The graph of the feasible solutions is shown in Figure 11.53.

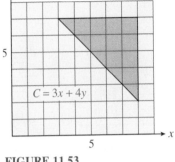

FIGURE 11.53

1. Use a graph to explain why it is impossible in this situation to have a cost as low as $12. (*Hint:* Draw the graph of $12 = 3x + 4y$ together with the graph of the feasible solutions.)

2. Use a graph to explain why the cost will not be as great as $60. (*Hint:* Draw the graph of $60 = 3x + 4y$ together with the graph of the feasible solutions.)

3. Use a graph to determine which vertex of the shaded region will correspond to the minimum cost. What is the minimum cost?

4. Use a graph to determine which vertex of the shaded region will correspond to the maximum cost. What is the maximum cost?

For Problems 5–8, the objective function for profit

$$P = 4x - 2y$$

is subject to the following constraints.

$$5x - y \geq -2, \quad x + y \leq 8, \quad x \geq 0, \quad y \geq 0$$

The graph of the feasible solutions is shown in Figure 11.54.

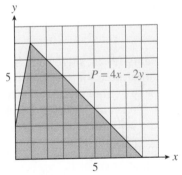

FIGURE 11.54

5. Graph the line that corresponds to a profit of $8. Find the coordinates of at least one feasible solution that gives a profit of $8.

6. Graph the line that corresponds to a profit of $22. Find the coordinates of at least one feasible solution that gives a profit of $22.

7. **a.** Which line is farther from the origin, the line for a profit of $8 or the line for the profit of $22?

   b. Use a graph to determine which vertex corresponds to a maximum profit.

   c. Find the maximum profit.

8. **a.** Use a graph to determine which vertex corresponds to a minimum profit.

   b. Find the minimum profit.

For problems 9–12, objective functions and the graphs of the feasible solutions are given. (a) Use the graph to find the vertex that yields the minimum value of the objective function, and find the minimum value; (b) use the graph to find the vertex that yields the maximum value of the objective function, and find the maximum value.

9. $C = 3x + y$

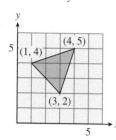

FIGURE 11.55

10. $C = x + 4y$

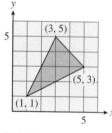

FIGURE 11.56

11. $C = 5x - 2y$

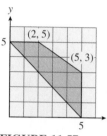

FIGURE 11.57

12. $C = 2x - y$

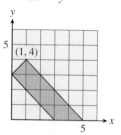

FIGURE 11.58

For problems 13–20, (a) graph the set of feasible solutions; (b) find the vertex that gives the minimum of the objective function, and find the minimum value; and (c) find the vertex that gives the maximum of the objective function, and find the maximum value.

13. Objective function $C = 3x + 2y$ with constraints $x \geq 0$, $y \geq 0$, $2x + y \leq 8$, $4x + 6y \leq 24$

14. Objective function $C = -2x + y$ with constraints $x \geq 0$, $y \geq 0$, $x - 2y \geq -10$, $2x + y \leq 10$

15. Objective function $C = 3x - y$ with constraints $x \geq 0$, $y \geq 0$, $x + y \leq 14$, $5x + y \leq 50$

16. Objective function $C = 5x + 4y$ with constraints $x \geq 0$, $y \geq 0$, $2x + y \leq 10$, $x - 3y \geq -3$

17. Objective function $C = 200x - 20y$ with constraints $x \geq 0$, $y \geq 0$, $3x + 2y \leq 24$, $x + y \leq 9$, $x + 2y \leq 16$

18. Objective function $C = 54x + 24y$ with constraints $x \geq 0$, $y \geq 0$, $3x + 2y \leq 24$, $3x - y \leq 15$, $3x - 4y \geq -12$

For Problems 19–26, solve each linear programming problem by graphing.
a. Write a formula for the objective function.
b. Write a system of inequalities for the constraints.
c. Graph the set of feasible solutions.
d. Find the optimum solution.

19. The math club is selling tickets for a show by a "mathemagician." Student tickets will cost \$1 and faculty tickets will cost \$2. The ticket receipts must be at least \$250 to cover the fee for the performer. An alumna promises to donate one calculator for each student ticket sold and three calculators for each faculty ticket sold. What is the minimum number of calculators that the alumna will donate?

20. The math department is having a book sale of unwanted textbooks to raise funds for $300 in scholarships. Paperback textbooks will be sold for $2 and the hardcover textbooks will be sold for $5. If paperback texts weigh 2 pounds each and hardcover books weigh 3 pounds each, find the minimum weight of textbooks the department must sell in order to raise its required funds.

21. Jeannette has 180 acres of farmland for growing wheat or soy. Each acre of wheat requires two hours of labor at harvest time and each acre of soy needs one hour of labor. She will have 240 hours of labor available at harvest time. Find the maximum profit Jeannette can make from her two crops if she can get a profit of $36 per acre for wheat and $24 per acre for soy.

22. Vassilis has at most $10,000 to invest in 2 banks. Alpha Bank will pay 6% annual interest and Bank Beta pays 5% annual interest. Alpha Bank will only insure up to $6000, so Vassilis will invest no more than that with Alpha. What is the maximum amount of interest Vassilis can earn in 1 year?

23. Gary's pancake recipe includes corn meal and whole wheat flour. Corn meal has 2.4 grams of linoleic acid and 2.5 milligrams of niacin per cup. Whole wheat flour has 0.8 grams of linoleic acid and 5.2 milligrams of niacin per cup. These two dry ingredients do not exceed 3 cups total. They combine for at least 3.2 grams of linoleic acid and at least 10 milligrams of niacin. Minimize the number of calories possible in the recipe if corn meal has 433 calories per cup and whole wheat flour has 400 calories per cup.

24. Cho requires 1 hour of cutting and 2 hours of sewing to make a Batman costume. He requires 2 hours of cutting and 1 hour of sewing to make a Wonder Woman costume. At most 10 hours per day are available for cutting and at most 8 hours per day are available for sewing. At least one costume must be made each day to stay in business. Find Cho's maximum income from selling one day's costumes if a Batman costume costs $68 and a Wonder Woman costume costs $76.

25. Problem 43 of Homework 2.5 described the diet of the Colombian ground squirrel. Refer to the system of constraints and the set of feasible solutions you found in that Problem.

  a. Write a formula for the squirrel's energy consumption. What is the optimum diet for the squirrel if it wants to maximize its energy consumption? Round your answers to one decimal place.

  b. Write a formula for the squirrel's foraging time. What is the optimum diet for the squirrel if it wants to minimize its time foraging? Round your answer to one decimal place.

  c. The actual observed diet of the squirrel consists of approximately 100 grams of forb and 25 grams of grass. Does the squirrel appear to be trying to maximize its caloric intake or to minimize its time eating? (Because 70% of the food available to the squirrel is grass, its diet is not determined by convenience.)

Source: Belovsky, 1986.

26. Problem 44 of Homework 2.5 described the diet of the Isle Royale moose. Refer to the system of constraints and the set of feasible solutions you found in that Problem.

  a. Write a formula for the moose's energy consumption. What is the optimum diet for the moose if it wants to maximize its energy consumption? Round your answers to one decimal place.

  b. Write a formula for the moose's foraging time. What is the optimum diet for the moose if it wants to minimize its time foraging? Round your answer to one decimal place.

  c. The actual observed diet of the moose consists of approximately 868 grams of aquatic plants and 3437 grams of leaves. Does the moose appear to be trying to maximize its caloric intake or to minimize its time eating?

Source: Belovsky, 1978.

For Problems 27–32, use a graphing calculator to find approximate values for the maximum and minimum of the objective function.

27. Objective function $C = 8.7x - 4.2y$ with constraints $x \geq 0$, $y \geq 0$, $1.7x - 4.5y \geq -9$, $14.3x + 10.9y \leq 28.6$

28. Objective function $C = -142x + 83y$ with constraints $x \geq 0$, $y \geq 0$, $21x - 49y \geq -147$, $19x + 21y \leq 171$

29. Objective function $C = 312x + 92y$ with constraints $x \geq 0$, $y \geq 0$, $18x + 17y \leq 284$, $51x + 11y \leq 656$

30. Objective function $C = 5.3x + 4.2y$ with constraints $x \geq 0$, $y \geq 0$, $2.5x + 1.7y \leq 20.1$, $0.09x - 0.31y \geq -0.39$

31. Objective function $C = 202x + 220y$ with constraints $x \geq 0$, $y \geq 0$, $38x + 24y \leq 294$, $35x + 34y \leq 310$, $13x + 29y \leq 197$

32. Objective function $C = 54x + 24y$ with constraints $x \geq 0$, $y \geq 0$, $43x + 32y \leq 333$, $23x - 9y \leq 152$, $73x - 94y \geq -296$

11.5 Displaying Data

We can summarize data effectively using a picture. We have already used graphs on the coordinate plane to represent data. In this section we consider other ways to display data.

Pie Charts

One of the simplest ways to display data is to use a **pie chart** or **circle graph**. The pie chart in Figure 11.59 shows which math courses students at a local college take after algebra class.

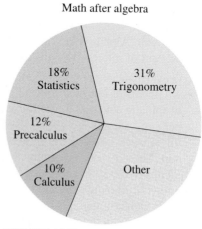

Math after algebra

FIGURE 11.59

EXAMPLE 1

Use Figure 11.59 to answer the following questions.

a. What is the ratio of students who take business calculus to those who take statistics immediately after algebra?

b. If the pie chart represents the results of a study of 1200 students, how many took trigonometry immediately after algebra?

c. How large is the angle for the sector representing "Other"?

Solutions

a. Because 10% of the students take business calculus and 18% take statistics, the ratio is $\frac{10}{18} = \frac{5}{9}$.

b. 31% of 1200 is 0.31(1200) or 372, so 372 students took trigonometry right after algebra.

c. The "Other" section represents

$$100\% - (31 + 18 + 12 + 10)\% = 29\%$$

of the circle. A complete circle is 360°, so the "Other" section has
0.29(360) = 104.4 degrees. ∎

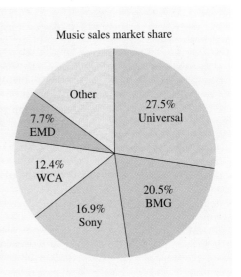

Music sales market share

Histograms

A **histogram** is a type of bar chart. The height of each bar shows the number
of data points at that score or value. A histogram shows us how the data are
distributed.

EXAMPLE 2

The histogram in Figure 11.60a shows the results of a quiz in a physics class.
Find the mode, mean, and median. (See Appendix A.7 to review these terms.)

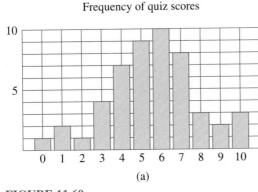

FIGURE 11.60

Solution

From the histogram we can immediately see that the mode is 6, because that is
the score with the tallest bar. The mode has a frequency of 10. In fact, we can
read off all the scores (and their frequencies):

Score	0	1	2	3	4	5	6	7	8	9	10
Frequency	1	2	1	4	7	9	10	8	3	2	3

Entering these data into the calculator, we find that the mean is $\bar{x} = 5.54$, and the median is 6. ∎

The histogram in Figure 11.60b has precisely the same shape as the histogram in Figure 11.60a, but the vertical scale is different. The heights of the bars in Figure 11.60a correspond to the *frequencies* of the different scores. The heights of the bars in Figure 11.60b correspond to the *fraction* of all the quiz scores. For example, the height corresponding to the score 6 is 0.2, which means that 0.2 or 20% of all the scores were 6. In other words, the score 6 has a **relative frequency** of 20%.

The **relative frequency** of a score is the number of times the score occurs divided by the total number of all scores.

The histogram in Figure 11.60b is called a **relative frequency histogram.**

EXERCISE 2 The histogram shows predicted EPS (earnings-per-share), in dollars, for companies with the greatest annual growth.

a. How many companies are depicted in the histogram?
b. Find the mode, mean, and median of EPS for these companies.
c. What is the relative frequency of an EPS of $0.65?

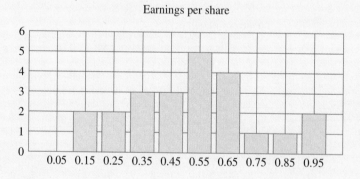

Earnings per share

Source: *LA Times*, April 20, 2000

Quartiles and Boxplots

The mean, median, and mode are all measures of central tendency; they indicate where a typical data point lies. We may also want to know how the data are spread out, or distributed. One simple graphical display of spread is called a **boxplot** or a **box-and-whiskers display.** See Figure 11.61.

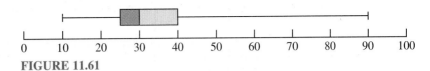

FIGURE 11.61

The boxplot in Figure 11.61 summarizes the annual income (in thousands of dollars) of professors at Quail Community College. The outside ends of the "whiskers" extend to the lowest and highest values of the data, so the incomes range from a low of $10,000 to a high of $90,000. The left edge of the box lies at the **first quartile** (or **lower quartile**). The first quartile is the median of the lower half of the scores, so one-fourth of all the salaries are below $25,000. The right edge of the box is the **third quartile** (or **upper quartile**). The third quartile is the median of the upper half of the scores, so three-fourths of the salaries are below $40,000. The line dividing the box in two shows the median of all the salaries, which is $30,000.

EXAMPLE 3

Use the boxplot in Figure 11.61 to answer the following questions.

a. What fraction of the incomes fall between $25,000 and $40,000?

b. What percent of the incomes are greater than $30,000?

c. At a new school most professors are at the lower end of the salary scale, while older schools have more established faculty with incomes near the upper end. Does this college appear to be new or older?

Solutions

a. Because $40,000 is the upper quartile, three-fourths of the scores fall below $40,000. One-quarter of the incomes fall below $25,000 (which is the lower quartile), so one-half of the incomes fall between $25,000 and $40,000.

b. Because $30,000 is the median, about one-half or 50% lie above $30,000.

c. Although there is one salary at $90,000, only one-quarter of the faculty have salaries exceeding $40,000. Three-fourths of the salaries are concentrated from $10,000 to $40,000, so this is probably a newer school. ∎

EXERCISE 3 The boxplot depicts the spread of earnings-per-share (EPS) among a sample of 50 companies.

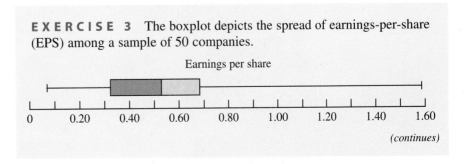

Earnings per share

(continues)

a. Approximately what percentage of the companies have an EPS exceeding $0.32?

b. If these companies are representative of companies on a stock exchange, what fraction of the companies on the exchange have an EPS between $0.32 and $0.68?

Source: *Los Angeles Times*, April 20, 2000

Using a Graphing Calculator to Draw Boxplots and Histograms

A graphing calculator has built-in features to produce statistical plots.

EXAMPLE 4

Use the calculator to make a boxplot and a histogram for the following heights of women in a health club.

Height (inches)	57	58	59	60	61	62	63	64	65	66	67	68	69	70	71	72
Frequency	1	6	14	26	47	69	90	96	89	70	46	27	13	5	2	1

Solution

Press **STAT** **ENTER** to access the statistics lists and enter the data: The heights go in the L_1 column and the frequencies go in the L_2 column. Access the **Stat Plot** menu by pressing **2nd** [Stat Plot], and activate the first plot by pressing **ENTER** **ENTER**. Select the boxplot icon for the **Type,** then select L_1 for the **Xlist** and L_2 for the **Freq,** as shown in Figure 11.62a.

Next adjust the window settings. For this plot we'll use

$$Xmin = 56.5 \quad Ymin = 0$$
$$Xmax = 72.5 \quad Ymax = 100$$
$$Xscl = 1 \quad\quad Yscl = 1$$

Clear out any functions defined in the **Y =** menu. Then press **2nd** [Format] and choose **AxesOff.** Finally, press **GRAPH** to obtain the boxplot shown in Figure 11.62b. Notice that you can use the **TRACE** feature and the arrow keys to display the maximum and minimum values, the median, and the quartiles. From the width of the box, we see that half of the heights are concentrated within the 4-inch range from 62 to 66 inches.

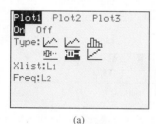

(a)

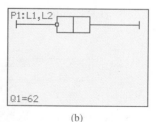

(b)

FIGURE 11.62

To see a histogram, we access the **Stat Plot** menu and choose the bar graph icon for **Type** under **Plot 1,** as shown in Figure 11.63a. The keystrokes are

Figure 11.63b shows a fairly symmetrical histogram for the data: The region to the left of the 64-inch height is nearly a mirror image of the region to the right. The peak at 64 corresponds to the mode. Because of the symmetry of the histogram, this peak also corresponds to the mean and the median.

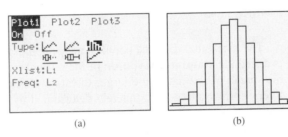

(a)	(b)

FIGURE 11.63

The window settings are important for the histogram. The width of the bars is determined by **Xscl.** We chose **Xscl = 1** because the data were rounded to the nearest inch.

When you finish using the statistical features of the calculator, be sure to turn off the statistics plots (2nd [Stat Plot] 4 ENTER) and set **AxesOn** in the window format menu (2nd [Format]).

Normal Curves and the Standard Deviation

The histogram in Figure 11.63b is roughly bell-shaped. This bell shape becomes more evident if we replace the histogram with a **line graph.** To draw a line graph, we plot a single point at the top of each bar and connect the dots. Choose the line graph icon from the **Stat Plot** menu. Use the keying sequence

The menu screen and the line graph for the data in Example 4 are shown in Figure 11.64.

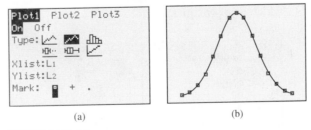

(a)	(b)

FIGURE 11.64

The bell shape identifies a family of line graphs called **normal curves.** A normal curve is symmetrical about the vertical line through its peak, and the x-value of this peak is the mean of the data. Moving away from the peak in either direction, the curve first bends downward, then begins to flatten out and approaches the *x*-axis asymptotically, as shown in Figure 11.65.

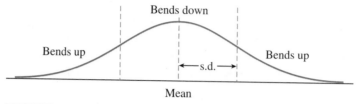

FIGURE 11.65

The horizontal distance between the peak and the steepest point of the curve is called the **standard deviation** of the normal curve. This distance determines the spread of the data—the larger the standard deviation, the wider the curve. Every normal curve is completely determined by its mean and standard deviation; in other words, if two normal curves have the same mean and standard deviation, they are identical curves.

If the relative frequency histogram follows a normal curve, we say that the data are **normally distributed.** Many data sets are approximately normal. For example, the heights of adult American women are (approximately) normally distributed. The mean height is 64 inches and the standard deviation is 2.5 inches.

We already know how to compute the mean of a set of data. It is also possible to compute a standard deviation. We will not discuss the details of the computation but will let the calculator's built-in features compute standard deviations for us.

EXAMPLE 5

Compute the standard deviation for the heights of women from the data in Example 4.

Solution

After entering the data we select **1-Var Stats** by pressing [STAT] [▷] 1, and choose the appropriate data lists: [2nd] [**L₁**] [,] [2nd] [**L₂**] [ENTER]. The results are shown in Figure 11.66. We see that the mean $\bar{x}$ is very close to 64. (The next two lines on the display show intermediate steps used to calculate the standard deviation.) **Sx** and **σx** are different versions of the standard deviation, which is approximately 2.48.

```
1-Var Stats
 x̄=64.00996678
 Σx=38534
 Σx²=2470262
 Sx=2.481861051
 σx=2.479798848
↓n=602
```

```
1-Var Stats
↑n=602
 minX=57
 Q1=62
 Med=64
 Q3=66
 maxX=72
```

FIGURE 11.66

The 68-95-99.7 Rule

Normally distributed data exhibit a remarkable property: You can predict what fraction of the data falls between any two values. We know that most of the data fall near the mean. For normally distributed data, 68% of the data values fall within one standard deviation of the mean. For example, because the mean height of adult American women is 64 inches and the standard deviation is 2.5 inches, we can conclude that 68% of the women are between 61.5 and 66.5 inches tall.

Furthermore, 95% of normally distributed data are no farther from the mean than twice the standard deviation, and 99.7% are no farther away than 3 times the standard deviation. (We say that 95% of the data are within 2 standard deviations of the mean and that 99.7% are within 3 standard deviations of the mean.) We conclude that 95% of adult American women are between 59 and 69 inches tall, and 99.7% are between 56.5 and 71.5 inches tall. (See Figures 11.67 and 11.68.)

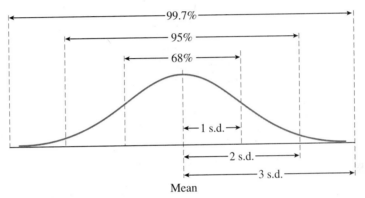

FIGURE 11.67

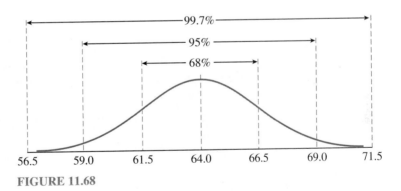

FIGURE 11.68

EXAMPLE 6

Assume that the heights of women in your school are normally distributed with a mean of 64 inches and a standard deviation of 2.5 inches.

    a. What percent of the women in your school are shorter than 4′ 11″?

    b. What percent of the women are between 5′ 6½″ and 5′ 9″ tall?

Solutions

a. 4' 11" is the same as 59 inches, which is 2 standard deviations from the mean. Therefore, 95% of the women are between 59 and 69 inches tall, so 5% have heights outside that range. This 5% is split evenly between the women shorter than 59 inches and the women taller than 69 inches, so 2.5% of the women are shorter than 4' 11". See Figure 11.68.

b. First note that 5' 6½" = 66.5" and 5' 9" = 69". Of the 68% of the women who are between 61.5 and 66.5 inches in height, half, or 34%, are between 64 and 66.5 inches tall. Similarly, of the 95% of the women who are between 59 and 69 inches tall, half, or 47.5%, are between 64 and 69 inches tall. We subtract 34% from 47.5% to conclude that 13.5% of the women are between 5' 6½" and 5' 9" tall. See Figure 11.69.

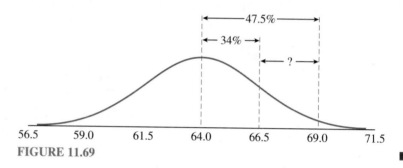

FIGURE 11.69

By following the method in Example 5b, we can use the "68-95-99.7" rule to predict what fraction of the data falls within 1, 2, or 3 standard deviations of the mean. It is also possible to find the fraction between *any* 2 values using statistical tables or a calculator, but we will not do that in this textbook.

The standard deviation can be computed for any set of data. It still gives a measure of the spread of the data. However, the "68-95-99.7" rule is valid only when the data are normally distributed.

EXERCISE 4 The contents of pretzel packages have normally distributed weights. The mean weight is 12 ounces and the standard deviation is 0.05 ounce.

a. What fraction of the pretzel packages contain between 11.9 ounces and 12.1 ounces?

b. What fraction of the packages contain more than 12.15 ounces?

READING QUESTIONS

1. Name four different types of graphs for displaying data discussed in this section.

2. What five numbers can be read from a boxplot?

3. How do you locate the mode of a set of data from a histogram (when each bar represents a different score)?

4. What is the shape of a line graph for normally distributed data?

5. How do you estimate the standard deviation from a line graph for normally distributed data?

6. What is the 68-95-99.7 rule?

ANSWERS TO 11.5 EXERCISES

1a. $\frac{25}{7}$ **b.** 15%, 54°

2a. 23   **b.** mode = 0.55; mean ≈ 0.52; median = 0.55

       c. $\frac{4}{23}$ ≈ 17.39%

3a. 75%   **b.** $\frac{1}{2}$

4a. 95%   **b.** 0.15%

HOMEWORK 11.5

Solve the following problems involving pie charts.

1. An algebra section consists of 23 freshmen, 16 sophomores, 7 juniors, and 4 seniors. Draw a pie chart showing each class level as a fraction of the whole section.

2. During a vacation in AuGres, Valerie saw 27 black squirrels, 9 gray squirrels, and 4 brown squirrels. Draw a pie chart showing each kind of squirrel as a fraction of all the squirrels Valerie saw.

3. The forensics club consists of 15 women and 60 men. Draw a pie chart showing the relationship between the number of women and men in the club. If the ratio of women to men in the club stays constant, how many men will there be when there are 23 women in the club?

4. For Mark Becker Night at Dodger Stadium last year, 54 of Mark's friends won door prizes and 246 did not. Draw a pie chart showing the relationship between the number of winners and nonwinners. If the ratio of winners to nonwinners stays constant, how many nonwinners will there be when there are 63 winners?

5. The pie chart in Figure 11.70 shows the results of a poll of Zanner's Ice Creame Shoppe customers. After the pie chart was created, Helen discovered that the employee who made the pie chart mistakenly used 44 instead of 144 for the number of customers who prefer vanilla. Make a pie chart for the correct percentages. (*Hint:* First determine what the employee mistakenly used for the total number of customers polled, then determine the numbers for chocolate, ices, and the correct total.)

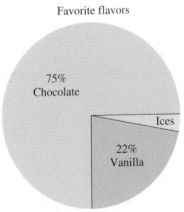

Favorite flavors

FIGURE 11.70

6. The pie chart in Figure 11.71 shows the percentages of registered Democrats, Republicans, and Independents in Whitney Township as reported in the *Whitney Daily Breeze*. However, the Breeze incorrectly used the number 312 instead of 112 for the number of Democrats. Make a pie chart for the correct percentages. (See the hint for Problem 5.)

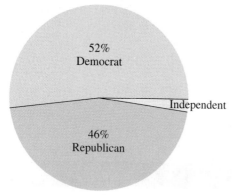

Whitney township voters

FIGURE 11.71

7. The pie chart in Figure 11.72 shows the proposed split for funding a new computer facility by school funds, private donations, and local businesses. The school administration revised its initial proposal for funds, and provided only one-tenth of its original proposal. The private donors and local businesses did meet their original pledges, and a less expensive facility was built. Make a pie chart showing how the funding was actually shared.

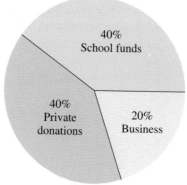

Funding for computers

FIGURE 11.72

8. The pie chart in Figure 11.73 from a VitaSnak package shows what percent of the calories in one candy bar come from carbohydrates, fat, and protein. Although the producers correctly measured the calories from carbohydrates and from protein, their measurement of the number of calories from fat is only half the true value. Make a corrected pie chart for the source of the calories in one candy bar.

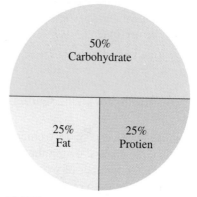

Calories per serving

FIGURE 11.73

a. What are the minimum, maximum, median, and quartiles for the two neighborhoods?

b. Which neighborhood has the youngest child?

c. Which neighborhood has the oldest?

d. Which neighborhood has the wider range of children's ages?

e. Which neighborhood has the larger median age?

Solve the following problems involving boxplots.

9. The boxplots in Figure 11.74 show the spread of ages among children in neighborhoods A and B.

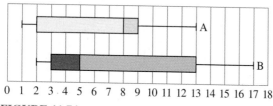

FIGURE 11.74

f. Assume that both neighborhoods have the same total number of children. Stefanie wants to have lots of play friends within a year of her age. Stefanie is 4 years old. Which neighborhood seems to be a better choice? Explain.

10. The boxplots in Figure 11.75 show the price range of bicycles at three different bicycle shops.

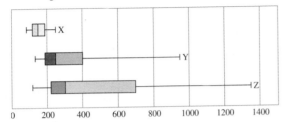

FIGURE 11.75

a. What are the minimum, maximum, median, and quartiles for the three shops?

b. Which shop has the least expensive bicycle?

c. Which shop has the widest range of prices?

d. Which shop offers the greatest fraction of its bicycles at prices below $200?

e. Which shop offers the greatest fraction of its bicycles at prices above $700?

f. Assume that each of the three shops has the same number of bicycles. Which shop seems to have the most bicycles available in the range from $200 to $400? Explain.

For Problems 11–16, match the given verbal description with a pie chart and a histogram in Figure 11.76.

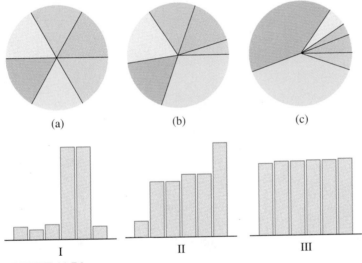

FIGURE 11.76

11. A die is rolled many times, and the frequency of each of the six faces is recorded. The die is not balanced properly, so that in particular 6 occurs too often and 1 occurs too seldom.

12. A fair die is rolled many times, and the six different outcomes occur with close to the same frequency.

13. The results of an election were tallied, and each of the six candidates received roughly the same number of votes.

14. The election requires a runoff because the top two candidates received about the same number of votes.

15. There were many math and English majors, but relatively few physics, biology, linguistics, or political science majors.

16. There are noticeably more people who prefer "Typhoon" shampoo than any of the other brands.

For Problems 17–26, match the given verbal description with a boxplot and a histogram in Figure 11.77.

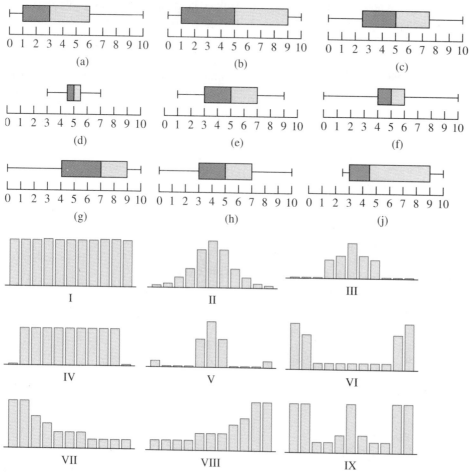

FIGURE 11.77

17. The quiz scores indicated that most students either understood almost all the material or almost none of the material. (The boxplot shows the range of quiz scores, and the histogram shows the frequency of different quiz scores.)

18. The quiz scores indicated that most students understood little of the material, with many students scoring zero, and three-fourths of the students scoring 6 or less. (The boxplot shows the range of quiz scores, and the histogram shows the frequency of different quiz scores.)

19. Because of a flaw in the manufacturing process, many lightbulbs failed immediately, with fewer bulbs lasting for a long time. (The boxplot shows the range of years in the light-bulbs' lifetimes, and the histogram shows the number of bulbs that failed in different time intervals.)

20. The number of chocolate chips per cookie in a package of cookies varied little. There were always at least a few chips, and most cookies had 4–6 chips. (The boxplot shows the numbers of chips in different cookies, and the histogram shows the frequency of different counts of chips.)

21. A movie was rated on a scale from 0 to 10 by many viewers. There was absolutely no agreement in rating, and the movie received each possible rating nearly equally often. (The boxplot shows the range of ratings, and the histogram shows the frequency of each rating.)

22. Because of extremely lax quality control, the weights of packages indicating a content of 5 ounces varied as much as 4 ounces from the proper weight. Within the range of measured weights, each weight occurred as often as any other. (The boxplot shows the range of weights of the packages, and the histogram shows the number of packages with different weights.)

23. The quiz scores range from 0 to 10, with most scores in the middle. The frequency of scores drops off the farther from the middle we move. (The boxplot shows the range of quiz scores, and the histogram shows the frequency of different quiz scores.)

24. A movie was rated on a scale from 0 to 10 by many viewers. Although it received all possible ratings, most of the scores were on the high end, with relatively few viewers giving low scores. (The boxplot shows the range of ratings, and the histogram shows the frequency of each rating.)

25. The quiz scores range from 0 to 10, but not all possible scores were recorded. Instead the scores were bunched in the middle, with just a few at the extremes. (The boxplot shows the range of quiz scores, and the histogram shows the frequency of different quiz scores.)

26. Subjects were tested to see if they could stay balanced on a narrow beam for 10 minutes. Many could pass the test, and more subjects stayed balanced for longer periods of time than for shorter. (The boxplot shows the range of times that the subjects stayed on the beam, and the histogram shows the frequency of different periods of time.)

Solve the following problems involving histograms.

27. Each student in a statistics class tossed the same coin ten times and recorded the number of times it landed heads. The histogram in Figure 11.78 shows the frequency for the number of heads obtained by each student.

   a. Find the mean, median, and mode for the number of heads per student.

   b. Make a boxplot for the data.

   c. How many students are in the class? Make a relative frequency histogram for the data.

   d. Does there seem to be anything unusual about this coin?

favor of raising student fees in order to build a new football stadium. The histogram in Figure 11.79 shows how many "yes" responses were recorded by each pollster.

   a. Find the mean, median, and mode for the number of "yes" responses per pollster.

   b. Make a boxplot for the data.

   c. How many students are in the psychology class? Make a relative frequency histogram for the data.

   d. Based on the results from all the students, what do you estimate for the percentage of students on campus who favor the fee increase?

Results of coin tosses

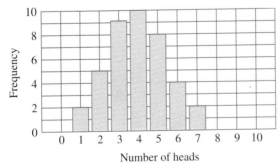

FIGURE 11.78

28. Each student in a psychology class asked ten other students on campus if they were in

Results of student polls

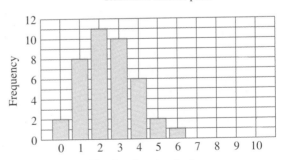

FIGURE 11.79

Solve the following problems using normal curves.

29. Assume that the heights of adult American men are normally distributed with a mean of 70" and a standard deviation of 2.5".

   a. What fraction of adult American men are taller than 5' 10"?

   b. What percentage of adult American men are between 5' $7\frac{1}{2}$" and $6'\frac{1}{2}$" tall?

   c. What percentage of adult American men are between 5' 5" and 6' 3" tall?

   d. What percentage of the men are between 5' $2\frac{1}{2}$" and 6' $5\frac{1}{2}$" tall?

   e. What percentage of the men are between 5' 10" and 6' $\frac{1}{2}$" tall?

   f. What percentage of the men are taller than 5' $7\frac{1}{2}$"?

   g. What percentage of the men are between 5' 5" and 6' $5\frac{1}{2}$" tall?

30. Assume that IQ scores are normally distributed with a mean of 100 and a standard deviation of 10.

   a. What fraction of the population has IQ's below 100?

   b. What percent of the population has IQ's between 90 and 110?

   c. What percent of the population has IQ's between 80 and 120?

   d. What percent of the population has IQ's between 70 and 130?

   e. What percent of the population has IQ's between 90 and 100?

   f. What percent has IQ's less than 90?

   g. What percent has IQ's exceeding 130?

31. A rice packager knows that she cannot attain perfect consistency in the weight of rice per package. In fact, the weights are normally distributed with a standard deviation of 0.5 ounces. She can choose the mean for the weights by adjusting her equipment. How should she set the mean weight if she wants $97\frac{1}{2}$% of her packages to contain at least 16 ounces of rice?

32. A lumber company has been asked to provide a large number of planks each 10 feet long. It is not critical if a plank is too long, but the company faces a stiff penalty for each plank that is too short. The saw cuts wood with lengths that are normally distributed. The mean can be set, but the standard deviation is 1/8 inch. What should the company choose as the mean length if it wants no more than 0.15% of the planks coming out too short?

CHAPTER 11 REVIEW

Graph each conic section and describe its main features.

1. $x^2 + y^2 = 9$

2. $\dfrac{x^2}{9} + y^2 = 1$

3. $\dfrac{x^2}{4} + \dfrac{y^2}{16} = 1$

4. $\dfrac{y^2}{6} - \dfrac{x^2}{8} = 1$

5. $4(y - 2) = (x + 3)^2$

6. $(x - 2)^2 + (y + 3)^2 = 16$

7. $\dfrac{(x - 2)^2}{4} + \dfrac{(y + 3)^2}{9} = 1$

8. $\dfrac{(x + 4)^2}{12} + \dfrac{(y - 2)^2}{6} = 1$

9. $\dfrac{(x - 2)^2}{4} - \dfrac{(y + 3)^2}{9} = 1$

10. $(x - 2)^2 + 4y = 4$

For Problems 11–22, (a) write the equation of each conic section in standard form, (b) identify the conic and describe its main features.

11. $x^2 + y^2 - 4x + 2y - 4 = 0$

12. $x^2 + y^2 - 6y - 4 = 0$

13. $4x^2 + y^2 - 16x + 4y + 4 = 0$

14. $8x^2 + 5y^2 + 16x - 20y - 12 = 0$

15. $x^2 - 8x - y + 6 = 0$

16. $y^2 + 6y + 4x + 1 = 0$

17. $x^2 + y = 4x - 6$

18. $y^2 = 2y + 2x + 2$

19. $2y^2 - 3x^2 - 16y - 12x + 8 = 0$

20. $9x^2 - 4y^2 - 72x - 24y + 72 = 0$

21. $2x^2 - y^2 + 6y - 19 = 0$

22. $4y^2 - x^2 + 8x - 28 = 0$

23. An ellipse centered at the origin has a vertical major axis of length 16 and a horizontal minor axis of length 10.

  a. Find the equation of the ellipse.

  b. What are the values of y when $x = 4$?

24. An ellipse centered at the origin has a horizontal major axis of length 26 and a vertical minor axis of length 18.

  a. Find the equation of the ellipse.

  b. What are the values of y when $x = 12$?

Write an equation for the conic section with the given properties.

25. Circle: center at $(-4, 3)$, radius $2\sqrt{5}$

26. Circle: endpoints of a diameter at $(-5, 2)$, and $(1, 6)$

27. Ellipse: center at $(-1, 4)$, $a = 4$, $b = 2$, major axis horizontal

28. Ellipse: vertices at $(3, 6)$ and $(3, -4)$, covertices at $(1, 1)$ and $(5, 1)$

29. Hyperbola: center $(2, -3)$, $a = 4$, $b = 3$, transverse axis horizontal

30. Hyperbola: one vertex at $(-4, 1)$, one end of the vertical conjugate axis at $(-3, 4)$

31. Parabola: vertex at $(0, 0)$ and focus at $(0, 4)$

32. Parabola: focus at $(0, 0)$ and directrix line $x = 2$

Graph the system of equations and state the solutions of the system.

33. $4x^2 + y^2 = 25$
    $x^2 - y^2 = -5$

34. $4x^2 - 9y^2 + 132 = 0$
    $x^2 + 4y^2 - 67 = 0$

35. $x^2 + 3y^2 = 13$
    $xy = -2$

36. $x^2 + y^2 = 17$
    $2xy = 17$

Use an equation or a system of equations to solve each problem.

37. Moia drives 180 miles in the same time that Fran drives 200 miles. Find the speed of each if Fran drives 5 miles per hour faster than Moia.

38. The perimeter of a rectangle is 26 inches and the area is 12 square inches. Find the dimensions of the rectangle.

39. The perimeter of a rectangle is 34 centimeters and the area is 70 square centimeters. Find the dimensions of the rectangle.

40. A rectangle has a perimeter of 18 feet. If the length is decreased by 5 feet and the width is increased by 12 feet, the area is doubled. Find the dimensions of the original rectangle.

41. Norm takes a commuter train 10 miles to his job in the city. The evening train returns him home at a rate 10 miles per hour faster than the morning train takes him to work. If Norm spends a total of 50 minutes per day commuting, what is the rate of each train?

42. Hattie's annual income from an investment is $32. If she had invested $200 more and the rate had been 1/2% less, her annual income would have been $35. What are the amount and rate of Hattie's investment?

43. Your ship receives LORAN signals from transmitters at $A(0, 100)$, $B(0, 412)$, and $C(0, -100)$. You calculate that your distance from A is 120 kilometers less than your distance from B, and 120 kilometers less than your distance from C. It follows that your position must lie on the hyperbolas

$$\frac{(y - 256)^2}{60^2} - \frac{x^2}{20736} = 1$$

$$\frac{y^2}{60^2} - \frac{x^2}{80^2} = 1$$

a. Plot and label the positions of the three transmitters. Sketch portions of the two hyperbolas with their correct orientations.

b. Use a graphing calculator to graph the two hyperbolas on the window

$$\text{Xmin} = -240 \quad \text{Xmax} = 240$$
$$\text{Ymin} = 0 \quad \text{Ymax} = 450$$

You are east of the transmitters. Locate the point P corresponding to the position of your ship.

c. Find the coordinates of your ship. Verify that your position is 120 kilometers closer to A than to B and 120 kilometers closer to A than to C.

44. Your ship receives LORAN signals from transmitters at $A(1368, 0)$, $B(507, 0)$, and $C(-507, 0)$. You calculate that your distance from A is 189 kilometers less than your distance from B, and your distance from B is 936 kilometers less than your distance from C. It follows that your position must lie on the two hyperbolas

$$\frac{(2x - 1875)^2}{189^2} - \frac{y^2}{420^2} = 1$$

$$\frac{x^2}{468^2} - \frac{y^2}{195^2} = 1$$

a. Plot and label the positions of the three transmitters. Sketch portions of the two hyperbolas with their correct orientations.

b. Use a graphing calculator to graph the two hyperbolas on the window

$$\text{Xmin} = 400 \quad \text{Xmax} = 1500$$
$$\text{Ymin} = -450 \quad \text{Ymax} = 450$$

You are north of the transmitters. Locate the point P corresponding to the position of your ship.

c. Find the coordinates of your ship. Verify that your position is 189 kilometers closer to A than to B and 936 kilometers closer to B than to C.

For Problems 45 and 46, (a) graph the set of feasible solutions; (b) find the vertex that gives the minimum of the objective function, and find the minimum value; (c) find the vertex that gives the maximum of the objective function, and find the maximum value.

45. Objective function $C = 18x + 48y$ with constraints $x \geq 0$, $y \geq 0$, $3x + y \geq 3$, $2x + y \leq 12$, $x + 5y \leq 15$

46. Objective function $C = 10x - 8y$ with constraints $x \geq 0$, $y \geq 0$, $5x - y \geq 2$, $x + 2y \leq 18$, $x - y \leq 3$

47. Ruth wants to provide cookies for the customers at her video rental store. It takes 20 minutes to mix the ingredients for each batch of peanut butter cookies and 10 minutes to bake them. Each batch of granola cookies takes 8 minutes to mix and 10 minutes to bake. Ruth does not want to use the oven more than 2 hours a day,

or to spend more than 2 hours a day mixing ingredients.

a. Write a system of inequalities for the number of batches of peanut butter cookies and granola cookies Ruth can make in one day, and graph the solutions.

b. Ruth decides to sell the cookies. If she charges 25¢ per peanut butter cookie and 20¢ per granola cookie, she will sell all the cookies she bakes. Each batch contains 50 cookies. How many batches of each type of cookie should she bake to maximize her income, and what is the maximum income?

48. A vegetarian recipe calls for at least 32 ounces of a combination of tofu and tempeh. Tofu provides 2 grams of protein per ounce and tempeh provides 1.6 grams of protein per ounce. Graham would like the dish to provide at least 56 grams of protein.

    a. Write a system of inequalities for the amount of tofu and the amount of tempeh for the recipe, and graph the solutions.

    b. Suppose that tofu costs 12¢ per ounce and tempeh costs 16¢ per ounce. What is the least expensive combination of tofu and tempeh Graham can use for the recipe, and how much will it cost?

49. A supermarket selected 40 customers at random and recorded their weekly grocery expenses. The histogram in Figure 11.80 shows the distribution of grocery bills.

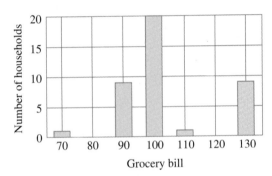

FIGURE 11.80

    a. In which interval do most of the grocery bills fall?

    b. Estimate the mean grocery bill. (*Hint:* Assume that all customers within a given range have as their grocery bill the midpoint of that interval.)

    c. Estimate the first quartile.

    d. Draw a boxplot for the data.

    e. Is it possible for a boxplot to be symmetrical about the median when the histogram is not? Justify your answer.

50. The table summarizes the results of a survey about commuting distances of students at a small college.

Distance (km)	Number of students
$0 \leq d < 5$	80
$5 \leq d < 10$	70
$10 \leq d < 15$	50
$15 \leq d < 20$	20
$20 \leq d < 25$	5

    a. Make a histogram for the data, with the edges of each bar corresponding to the endpoints of the intervals given in the table.

    b. How many students were in the survey?

    c. Which interval contains the median commuting distance?

    d. Estimate the mean of the commuting distances. (*Hint:* Assume that all students within a given interval have a commute equal to the midpoint of that interval.)

51. The weight of the hamburger patties at Mick's Burgers is approximately normally distributed. The line graph in Figure 11.81 was obtained by connecting the tops of the bars of a histogram.

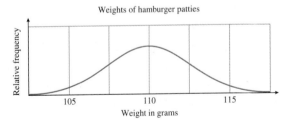

FIGURE 11.81

    a. Estimate the mean weight of a hamburger patty. Explain why the median weight should be the same as the mean.

    b. Estimate the standard deviation of the distribution.

    c. Explain why the first quartile is more than 107.5 grams. Give a rough estimate of the first quartile.

d. Draw a boxplot for the data. Put the ends of the whiskers 3 standard deviations from the mean.

52. The boxplot in Figure 11.82 summarizes test scores for a hundred students. Assume that the distribution is approximately normal.

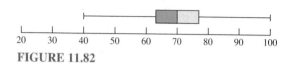

FIGURE 11.82

a. What is the median test score? Explain why the mean should be the same as the median.

b. What are the highest and lowest values?

c. Explain why the standard deviation of the test scores must be more than 7.5.

d. Sketch a bell-shaped curve to correspond to the data. You need not define the vertical scale.

Review Topics

A.1 Algebraic Expressions

From your previous algebra courses you are familiar with the use of letters or **variables** to stand for unknown numbers in equations or formulas. Variables are also used to represent numerical quantities that change over time or in different situations. For example, p might stand for the atmospheric pressure at different heights above the earth's surface. Or N might represent the number of people infected with cholera t days after the start of an epidemic.

An **algebraic expression** is any meaningful combination of numbers, variables, and symbols of operation. Algebraic expressions are used to express relationships between variable quantities.

EXAMPLE 1   Loren makes $6 an hour working at the campus bookstore.

a. Choose a variable for the number of hours Loren works per week.

b. Write an algebraic expression for the amount of Loren's weekly earnings.

Solutions

a. Let h stand for the number of hours Loren works per week.

b. The amount Loren earns is given by

$$6 \times \text{(number of hours Loren worked)}$$

or $6 \cdot h$. Loren's weekly earnings can be expressed as $6h$.   ■

We say that the algebraic expression $6h$ represents the amount of money Loren earns *in terms of* the number of hours she works. If we substitute specific values for the variable or variables in an algebraic expression, we can find a numerical value for the expression. This is called **evaluating** the expression.

EXAMPLE 2 If Loren from Example 1 works for 16 hours in the bookstore this week, how much will she earn?

Solution

Evaluate the expression $6h$ for $h = 16$.

$$6h = 6(16) = 96$$

Loren will make $96. ∎

Equations

If we also assign a variable to Loren's weekly earnings, say w, we can write an equation relating the two variables.

$$w = 6h$$

An **equation** is just a mathematical statement that two expressions are equal. Equations relating two variables are particularly useful. If we know the value of one of the variables, we can find the corresponding value of the other variable by solving the equation.

EXAMPLE 3 How many hours does Loren need to work next week if she wants to earn $225?

Solution

In mathematical language, we know that $w = 225$, and we would like to know the value of h. We substitute the value for w into our equation, and then solve for h.

$$w = 6h \qquad \text{Substitute 225 for } w$$
$$225 = 6h \qquad \text{Divide both sides by 6.}$$
$$\frac{225}{6} = \frac{6h}{6} \qquad \text{Simplify.}$$
$$37.5 = h$$

Loren must work 37.5 hours in order to earn $225. In reality, Loren will probably have to work for 38 hours, because most employers do not pay for portions of an hour's work. Thus, Loren needs to work for 38 hours. ∎

Order of Operations

Algebraic expressions often involve more than one operation. To resolve any ambiguities over how such expressions should be evaluated, we follow the **order of operations.**

Order of Operations

1. Simplify any expressions within grouping symbols (parentheses, brackets, square root bars, or fraction bars). Start with the innermost grouping symbols and work outward.

2. Evaluate all powers.

3. Perform multiplications and divisions in order from left to right.

4. Perform additions and subtractions in order from left to right.

EXAMPLE 4

April sells environmentally friendly cleaning products. Her income consists of $200 per week plus a commission of 9% of her sales.

    a. Choose variables to represent the unknown quantities, and write an algebraic expression for April's weekly income in terms of her sales.

    b. Find April's income for a week in which she sells $350 worth of cleaning products.

Solutions

    a. Let I represent April's total income for the week, and let S represent the total amount of her sales. We can translate the information from the problem into mathematical language as follows:

$$\text{``Her income consists of \$200} \ldots \text{plus} \ldots \text{9\% of her sales.''}$$
$$I \qquad = \qquad 200 \qquad + \qquad 0.09 \; S$$

Thus, $I = 200 + 0.09S$.

    b. We want to evaluate our expression from part (a) with $S = 350$. We substitute **350** for S to find

$$I = 200 + 0.09(\mathbf{350})$$

According to the order of operations, we should perform the multiplication before the addition. Thus, we begin by computing $0.09(350)$.

$$I = 200 + 0.09(350) \qquad \text{Multiply } 0.09(350) \text{ first.}$$
$$= 200 + 31.5 \qquad\qquad \text{Add.}$$
$$= 231.50$$

April's income for the week is $231.50.   ■

On a scientific or a graphing calculator, we can enter the expression from Example 4b just as it is written:

$$200 \boxed{+} 0.09 \boxed{\times} 350 \boxed{\textbf{ENTER}}$$

The calculator will perform the operations in the correct order—multiplication first.

Parentheses and Fraction Bars

We can use parentheses to "override" the multiplication-first rule. Compare the two expressions below.

The sum of four times x and ten	$4x + 10$
Four times the sum of x and ten	$4(x + 10)$

In the first expression we perform the multiplication $4 \times x$ first, but in the second expression we perform the addition $x + 10$ first, because it is enclosed in parentheses.

EXAMPLE 5

Economy Parcel Service charges $2.80 per pound to deliver a package from Pasadena to Cedar Rapids. Andrew wants to mail a painting that weighs 8.3 pounds, plus whatever packing material he uses.

a. Choose variables to represent the unknown quantities and write an expression for the cost of shipping Andrew's painting.

b. Find the shipping cost if Andrew uses 2.9 pounds of packing material.

Solutions

a. Let C stand for the shipping cost and let w stand for the weight of the packing material. Andrew must find the total weight of his package *first*, and then multiply by the shipping charge. The total weight of the package is $8.3 + w$ pounds. We use parentheses around this expression to show that it should be computed first, and the sum should be multiplied by the shipping charge of $2.80 per pound. Thus,

$$C = 2.80(8.3 + w)$$

b. Evaluate the formula from part (a) with $w = 2.9$.

$$C = 2.80(8.3 + 2.9) \qquad \text{Add inside parentheses.}$$
$$= 2.80(11.2) \qquad \text{Multiply.}$$
$$= 31.36$$

The cost of shipping the painting is $31.36. ∎

On a calculator, we enter the expression for C in the order it appears, including the parentheses. (Experiment to see whether your calculator requires you to enter the $\boxed{\times}$ symbol after 2.80.) The keying sequence

$$2.80 \; \boxed{\times} \; \boxed{(} \; 8.3 \; \boxed{+} \; 2.9 \; \boxed{)} \; \boxed{\text{ENTER}}$$

gives the correct result, 31.36.

COMMON ERROR

If we omit the parentheses, the calculator will perform the multiplication before the addition. Thus, the keying sequence

$$2.80 \; \boxed{\times} \; 8.3 \; \boxed{+} \; 2.9$$

gives an incorrect result for Example 5b. (The sequence

$$8.3 \; \boxed{+} \; 2.9 \; \boxed{\times} \; 2.80$$

does not work either!) ∎

The location (or absence) of parentheses can drastically alter the meaning of an algebraic expression. In the following example, note how the location of the parentheses changes the value of the expression.

EXAMPLE 6

a. $5 - 3 \cdot 4^2$

$\quad = 5 - 3 \cdot 16$

$\quad = 5 - 48 = -43$

c. $(5 - 3 \cdot 4)^2$

$\quad = (5 - 12)^2$

$\quad = (-7)^2 = 49$

b. $5 - (3 \cdot 4)^2$

$\quad = 5 - 12^2$

$\quad = 5 - 144 = -139$

d. $(5 - 3) \cdot 4^2$

$\quad = 2 \cdot 16$

$\quad = 32$ ∎

In the expression $5 - 12^2$, which appears in Example 6b, the exponent 2 applies only to 12, not to -12. Thus, $5 - 12^2 \neq 5 + 144$. ∎

The order of operations mentions other "grouping devices" besides parentheses: fraction bars and square root bars. Notice how the placement of the fraction bar affects the expressions in the next example.

EXAMPLE 7

a. $\dfrac{1 + 2}{3 \cdot 4} = \dfrac{3}{12}$

$\qquad = \dfrac{1}{4}$

c. $\dfrac{1 + 2}{3} \cdot 4 = \dfrac{3}{3} \cdot 4$

$\qquad = 1 \cdot 4 = 4$

b. $1 + \dfrac{2}{3 \cdot 4} = 1 + \dfrac{2}{12}$

$\qquad = 1 + \dfrac{1}{6} = \dfrac{7}{6}$

d. $1 + \dfrac{2}{3} \cdot 4 = 1 + \dfrac{8}{3}$

$\qquad = \dfrac{3}{3} + \dfrac{8}{3} = \dfrac{11}{3}$ ∎

Using a Calculator

Because your calculator cannot use a fraction bar as a grouping symbol, *you must insert parentheses around any expression that appears above or below a fraction bar.* Here are keying sequences for each expression in Example 7.

a. $\dfrac{1 + 2}{3 \cdot 4}$ 　 (1 + 2) ÷ (3 × 4)

b. $1 + \dfrac{2}{3 \cdot 4}$ 　 1 + 2 ÷ (3 × 4)

c. $\dfrac{1 + 2}{3} \cdot 4$ 　 (1 + 2) ÷ 3 × 4

d. $1 + \dfrac{2}{3} \cdot 4$ 　 1 + 2 ÷ 3 × 4

Similarly, your calculator does not use a square root bar as a grouping symbol. When using a calculator, *you must insert parentheses around any expression that appears under a square root symbol.*

EXAMPLE 8

If an object falls from a height of h meters, then the time t it will take to reach the ground is given in seconds by the formula $t = \sqrt{\dfrac{h}{4.9}}$. How long will it take a marble to fall to the ground from the top of the Sears Tower in Chicago, which is 443.2 meters tall?

Solution

We evaluate the expression for $h = 443.2$.

$$t = \sqrt{\frac{443.2}{4.9}}$$

To simplify the expression, we perform the division *first,* then take the square root of the quotient. On a graphing calculator, we use the keying sequence

and the calculator returns the result, 9.510466844. The marble takes approximately 9.51 seconds to reach the ground. ■

COMMON ERROR

The calculator displays an open parenthesis with the radical symbol, and we must insert a close parenthesis to mark the end of the radicand before entering any other calculations. For example,

$$\sqrt{(\ 100\ \div\ 4\ )} \quad \text{means} \quad \sqrt{\frac{100}{4}} = \sqrt{25} = 5$$

$$\sqrt{(\ 100\ )\ \div\ 4} \quad \text{means} \quad \frac{\sqrt{100}}{4} = \frac{10}{4} = 2.5$$

■

EXAMPLE 9

Use a calculator to simplify each expression. Then explain how your calculator performed the operations.

a. $-8(6.2)^2$
b. $\dfrac{\sqrt{72}}{4 + 2.5}$

Solutions

a. Use the keying sequence

$$(-)\ 8\ \times\ 6.2\ x^2\ \text{ENTER}$$

(Use the gray negative key, $(-)$, not the blue subtraction key, $-$, to enter -8.) Your calculator performs the power first, to get $(6.2)^2 = 38.44$, then multiplies the result by -8 to get $-8(38.44) = -307.52$.

b. Use the keying sequence

$$\sqrt{(\ 72\ )\ \div\ (\ 4\ +\ 2.5\ )}\ \text{ENTER}$$

We must insert parentheses around the sum $4 + 2.5$. Your calculator first performs the operations below the fraction bar.

$$\frac{\sqrt{72}}{4 + 2.5} = \frac{\sqrt{72}}{6.5} \qquad \text{Approximate the square root.}$$

$$\approx \frac{8.485281374}{6.5} \qquad \text{Divide.}$$

$$\approx 1.305427904 \approx 1.31$$

■

Write algebraic expressions to describe each situation, and evaluate for the given values.

1. Jim was 27 years old when Ana was born.

   a. Choose variables to represent the unknown quantities, and write an expression for Jim's age in terms of Ana's age.

   b. Use your expression to find Jim's age when Ana is 22 years old.

2. Rani wants to replace the wheels on her in-line skates. New wheels cost $6.59 each.

   a. Choose variables to represent the unknown quantities, and write an expression for the total cost of new wheels in terms of the number of wheels Rani must replace.

   b. Use your expression to find the total cost if Rani must replace 8 wheels.

3. Helen decides to drive to visit her father. The trip is a distance of 1260 miles.

   a. Choose variables to represent the unknown quantities, and write an expression for the total number of hours Helen must drive in terms of her average driving speed.

   b. Use your expression to find how long Helen must drive if she averages 45 miles per hour.

4. Ben will inherit one million dollars on his twenty-first birthday.

   a. Choose variables to represent the unknown quantities, and write an expression for the number of years before Ben gets his inheritance in terms of his present age.

   b. Use your expression to calculate how many more years Ben must wait after he turns 13 years old.

5. Dress fabric is sold (from bolts with a standard width) for $5.79 per yard.

   a. Write an expression for the number of yards of fabric in terms of the number of feet of the fabric. (There are 3 feet in a yard.)

   b. Write an expression for the cost of the fabric in terms of the number of feet.

   c. Find the cost of 10 feet of fabric.

6. Yoon can maintain a running speed of 12 miles per hour.

   a. Write an expression for Yoon's running time (in hours) in terms of the distance run.

   b. Write an expression for Yoon's time in *minutes* in terms of the distance run.

   c. How many minutes will it take Yoon to run 2.7 miles?

7. The area of a circle is equal to π times the square of its radius.

   a. Choose variables to represent the unknown quantities, and write an expression for the area of a circle in terms of its radius.

   b. Find the area of a circle whose radius is 5 centimeters.

8. To estimate the speed (in miles per hour) at which an automobile was moving on a dry concrete road when the brakes were applied, multiply the length (in feet) of its skid mark by 24 and take the square root of the resulting product.

   a. Choose variables to represent the unknown quantities, and write an expression for the speed of the car in terms of the length of the skid mark.

   b. How fast was a car traveling if its skid mark was 160 feet long?

9. Farshid wants to compute the average of his homework scores. He has a total of 198 points from the 20 assignments he turned in. However, he missed some of the assignments entirely (and therefore has no points for those assignments).

   a. Write an expression for the total number of assignments in terms of how many Farshid missed.

   b. Write an expression for Farshid's average homework score in terms of the number of assignments he missed.

   c. Find Farshid's average score if he missed five assignments.

10. Roma and four friends together bought a single lottery ticket and agreed to split any winnings evenly. The ticket was a $10,000 winner and in addition qualified for the Big Spin, a cash prize depending on the outcome of a spin of a prize wheel.

    a. Write an expression for their total winnings in terms of the prize from the Big Spin.

    b. Write an expression for Roma's share of the winnings in terms of the prize from the Big Spin.

    c. Find Roma's share if the prize from the Big Spin is $50,000.

11. The sales tax in the city of Preston is 7.9%.

    a. Choose variables to represent the unknown quantities, and write an expression for the amount of sales tax in terms of the price of the item.

    b. Find the total bill for an item (price plus tax) in terms of the price of the item.

    c. Find the total bill for an item whose price is $490.

12. A savings account pays (at the end of one year) 6.4% interest on the amount deposited.

    a. Choose variables to represent the unknown quantities, and write an expression for the amount of interest earned in one year in terms of the amount deposited.

    b. Find the total amount (initial deposit plus interest) in the account after one year in terms of the amount deposited.

    c. Find the total amount in the account after one year if $350 was deposited.

13. Your best friend moves to another state and the long distance phone call costs $1.97 plus $0.39 for each minute.

    a. Choose variables to represent the unknown quantities, and write an expression for the cost of a long distance phone call in terms of the number of minutes of the call.

    b. Find the cost of a 27-minute phone call.

14. Arenac Airlines charges 47 cents per pound on its flight from Omer to Pinconning, both for passengers and for luggage. Mr. Owsley wants to take the flight with 15 pounds of luggage.

    a. Choose variables to represent the unknown quantities, and write an expression for the cost of the flight in terms of Mr. Owsley's weight.

    b. Find the cost if Mr. Owsley weighs 162 pounds.

15. Juan buys a 50-pound bag of rice and consumes about 0.4 pounds per week.

    a. Write an expression for the amount of rice Juan has consumed in terms of the number of weeks since he bought the bag.

    b. Write an expression for the amount of rice Juan has left in terms of the number of weeks since he bought the bag.

    c. Find the amount of rice Juan has left after 6 weeks.

16. Trinh is bicycling down a mountain road that loses 500 feet in elevation for each 1 mile of road. She started at an elevation of 6300 feet.

    a. Write an expression for the elevation Trinh has lost in terms of the distance she has cycled.

    b. Write an expression for Trinh's elevation in terms of the number of miles she has cycled.

    c. Find Trinh's elevation after she has cycled 9 miles.

17. Leon's truck gets 20 miles per gallon of gasoline. The gas tank holds 14.6 gallons of gasoline.

    a. Write an expression for the amount of gasoline used in terms of the number of miles driven.

    b. Write an expression for the amount of gasoline remaining in the tank in terms of the number of miles driven since the tank was last filled.

    c. Find the amount of gasoline left in the tank when Leon has driven 110 miles.

18. As part of a charity fundraising event, Sanaz found sponsors to pay her $16 for each hour that she skates. In addition, Sanaz will win a scholarship if she can raise $500.

    a. Write an expression for the number of hours Sanaz skates in terms of the amount of money she raises.

b. Write an expression for the number of hours Sanaz has left to skate to earn the scholarship in terms of the amount of money she has already raised.

c. Find the number of hours she has left to skate if she has already raised $380.

Simplify each expression according to the order of operations.

19. $\dfrac{3(6-8)}{-2} - \dfrac{6}{-2}$ **20.** $\dfrac{5(3-5)}{2} - \dfrac{18}{-3}$

21. $6[3 - 2(4 + 1)]$ **22.** $5[3 + 4(6 - 4)]$

23. $(4 - 3)[2 + 3(2 - 1)]$

24. $(8 - 6)[5 + 7(2 - 3)]$

25. $64 \div (8[4 - 2(3 + 1)])$

26. $27 \div (3[9 - 3(4 - 2)]$

27. $5[3 + (8 - 1)] \div (-25)$

28. $-3[-2 + (6 - 1)] \div 9$

29. $[-3(8 - 2) + 3] \cdot [24 \div 6]$

30. $[-2 + 3(5 - 8)] \cdot [-15 \div 3]$

31. -5^2 **32.** $(-15)^2$

33. $(-3)^4$ **34.** -3^4

35. -4^3

36. $(-4)^3$

37. $(-2)^5$

38. -2^5

39. $\dfrac{4 \cdot 2^3}{16} + 3 \cdot 4^2$

40. $\dfrac{4 \cdot 3^2}{6} + (3 \cdot 4)^2$

41. $\dfrac{3^2 - 5}{6 - 2^2} - \dfrac{6^2}{3^2}$

42. $\dfrac{3 \cdot 2^2}{4 - 1} + \dfrac{(-3)(2)^3}{6}$

43. $\dfrac{(-5)^2 - 3^2}{4 - 6} + \dfrac{(-3)^2}{2 + 1}$

44. $\dfrac{7^2 - 6^2}{10 + 3} - \dfrac{8^2 \cdot (-2)}{(-4)^2}$

Use a calculator to simplify each expression.

45. $\dfrac{-8398}{26 \cdot 17}$

46. $\dfrac{-415.112}{8.58 + 18.73}$

47. $\dfrac{112.78 + 2599.124}{27.56}$

48. $\dfrac{202462 - 9510}{356}$

49. $\sqrt{24 \cdot 54}$

50. $\sqrt{\dfrac{1216}{19}}$

51. $\dfrac{116 - 35}{215 - 242}$

52. $\dfrac{842 - 987}{443 - 385}$

53. $\sqrt{27^2 + 36^2}$

54. $\sqrt{13^2 - 4 \cdot 21 \cdot 2}$

55. $\dfrac{-27 - \sqrt{27^2 - 4(4)(35)}}{2 \cdot 4}$

56. $\dfrac{13 + \sqrt{13^2 - 4(5)(-6)}}{2 \cdot 5}$

Write an algebraic expression that corresponds to each graphing calculator keying sequence.

57. a. 2 [+] 3 [÷] 4 [ENTER]

b. [( ] 2 [+] 3 [ )] [÷] 4 [ENTER]

58. a. 72 [÷] [( ] 6 [×] 2 [ )] [ENTER]

b. 72 [÷] 6 [×] 2 [ENTER]

59. a. [(−)] 23 [x^2] [ENTER]

b. [( ] [(−)] 23 [ )] [x^2] [ENTER]

60. a. 96 [÷] 2 [∧] 3 [ENTER]

b. [( ] 96 [÷] 2 [ )] [∧] 3 [ENTER]

61. a. [√( ] 9 [+] 16 [ )] [ENTER]

b. [√( ] 9 [ )] [+] 16 [ENTER]

62. a. [(−)] 5 [×] 2 [x^2] [ENTER]

b. [( ] [(−)] 5 [×] 2 [ )] [x^2] [ENTER]

Evaluate each expression for the given values of the variable. Use your calculator where appropriate.

63. $\dfrac{5(F - 32)}{9}$; $F = 212$

64. $\dfrac{a - 4s}{1 - r}$; $r = 2, s = 12$, and $a = 4$

65. $P + Prt$; $P = 1000, r = 0.04$, and $t = 2$

66. $R(1 + at)$; $R = 2.5, a = 0.05$, and $t = 20$

67. $\frac{1}{2}gt^2 - 12t$; $g = 32$ and $t = \frac{3}{4}$

68. $\frac{Mv^2}{g}$; $M = \frac{16}{3}, v = \frac{3}{2}$, and $g = 32$

69. $\frac{32(V - v)^2}{g}$; $V = 12.78, v = 4.26$, and $g = 32$

70. $\frac{32(V - v)^2}{g}$; $V = 38.3, v = -6.7$, and $g = 9.8$

A.2 Linear Equations

Solving an equation often involves generating simpler equations that have the same solutions. Equations that have identical solutions are called **equivalent equations.** For example,

$$3x - 5 = x + 3 \tag{1}$$

and

$$2x = 8$$

are equivalent equations because the solution to each equation is 4. Often we can find simpler equivalent equations by "undoing" in reverse order the operations performed on the variable.

Solving Linear Equations

A great variety of practical problems can be solved with **linear,** or first-degree, equations. These are equations that can be written so that every term is either a constant or a constant times the variable. Equation (1) above is an example of a linear equation. Recall the following rules for solving linear equations.

To generate equivalent equations:

1. We can add or subtract the *same* number on *both* sides of an equation.
2. We can multiply or divide *both* sides of an equation by the *same* number (except zero).

Applying either of these rules produces a new equation equivalent to the old one, and thus preserves the solution. We apply the rules with the goal of "isolating" the variable on one side of the equation.

EXAMPLE 1

Solve the equation $3x - 5 = x + 3$.

Solution

We first collect all the variable terms on one side of the equation, and the constant terms on the other side.

$$3x - 5 - x = x + 3 - x \quad \text{Subtract } x \text{ from both sides.}$$
$$2x - 5 = 3 \quad \text{Simplify.}$$
$$2x - 5 + 5 = 3 + 5 \quad \text{Add 5 to both sides.}$$
$$2x = 8 \quad \text{Simplify.}$$
$$\frac{2x}{2} = \frac{8}{2} \quad \text{Divide both sides by 2.}$$
$$x = 4 \quad \text{Simplify.}$$

The solution is 4. (You can check the solution by substituting 4 into the original equation and showing that a true statement results.) ■

The following steps should enable you to solve any linear equation. Of course, you may not need all the steps for a particular equation.

To solve a linear equation:

1. First, simplify each side of the equation separately.
   a. Apply the distributive law to remove parentheses.
   b. Collect like terms.
2. By adding or subtracting appropriate terms on both sides of the equation, get all the variable terms on one side, and all the constant terms on the other.
3. Divide both sides of the equation by the coefficient of the variable.

EXAMPLE 2

Solve $\quad 3(2x - 5) - 4x = 2x - (6 - 3x)$

Solution

Begin by simplifying each side of the equation.

$$3(2x - 5) - 4x = 2x - (6 - 3x) \quad \text{Apply the distributive law.}$$
$$6x - 15 - 4x = 2x - 6 + 3x \quad \text{Combine like terms on each side.}$$
$$2x - 15 = 5x - 6$$

Next, get all the variable terms on the left side of the equation, and all the constant terms on the right side.

$$2x - 15 - 5x + 15 = 5x - 6 - 5x + 15 \quad \text{Add } -5x + 15 \text{ to both sides.}$$
$$-3x = 9$$

Finally, divide both sides of the equation by the coefficient of the variable.

$$-3x = 9 \qquad \text{Divide both sides by } -3.$$
$$x = -3$$

The solution is -3.

Problem Solving

Problem solving often involves translating a "real-life" problem into a computer programming language, or, in our case, into algebraic expressions. We can then use algebra to solve the mathematical problem, and interpret the solution in the context of the original problem. Learning to extract the relevant information from a situation and to create a "model" for the problem are among the most important mathematical skills you can acquire.

Here are some guidelines for problem solving with algebraic equations.

Guidelines for Problem Solving

Step 1 Identify the unknown quantity and assign a variable to represent it.

Step 2 Find some quantity that can be expressed in two different ways, and write an equation.

Step 3 Solve the equation.

Step 4 Interpret your solution to answer the question in the problem.

In Step 1, it is a good idea to begin by writing an English phrase to describe the quantity you are looking for. Be as specific as possible—if you are going to write an equation about this quantity, you must understand its properties! Remember that your variable must represent a *numerical* quantity. For example, x can represent the *speed* of a train, but not just "the train."

Writing an equation is the hardest part of the problem. Note that the "quantity" mentioned in Step 2 will probably *not* be the same unknown quantity you are looking for, but the algebraic expressions you write *will* involve your variable. For example, if your variable represents the *speed* of a train, your equation might be about the *distance* the train traveled.

Supply and Demand

In economics, we often hear about the law of supply and demand. The higher the price of a product, the more its manufacturers will be willing to supply, but the lower the price, the more consumers will buy. The price at which the demand for a product equals the supply is called the *equilibrium price*.

EXAMPLE 3

The Coffee Connection finds that when it charges p dollars for a pound of coffee, it can sell $800 - 60p$ pounds per month. On the other hand, at a price of p dollars a pound, International Food and Beverage will supply the Connection with $175 + 40p$ pounds of coffee per month. What price should the Coffee

Connection charge for a pound of coffee so that its monthly inventory will sell out?

Solution

Step 1 We are looking for the equilibrium price, p.

Step 2 The Coffee Connection would like the demand for its coffee to equal its supply. We equate the expressions for supply and for demand to obtain the equation

$$800 - 60p = 175 + 40p$$

Step 3 Solve the equation. We want all terms containing the variable, p, on one side of the equation, and all constant terms on the other side. To accomplish this, we add $60p$ to both sides of the equation, and subtract 175 from both sides to obtain

$$800 - 60p + 60p - 175 = 175 + 40p + 60p - 175$$
$$625 = 100p \quad \text{Divide both sides by 100.}$$
$$6.25 = p$$

Step 4 The Coffee Connection should charge $6.25 per pound for their coffee. ■

Percent Problems

Recall the basic formula for computing percents.

$$P = rW$$

the **P**art (or percent) = the percentage **r**ate $\times$ the **W**hole Amount

A *percent increase* or *percent decrease* is calculated as a fraction of the *original* amount. For example, suppose you make $6.00 an hour now, but next month you are expecting a 5% raise. Your new salary should be

$$\$6.00 + 0.05(\$6.00) = \$6.30$$
$$\underset{\text{original salary}}{\$6.00} \quad \underset{\text{increase}}{} \quad \underset{\text{new salary}}{}$$

EXAMPLE 4

The price of housing in urban areas increased 4% over the past year. If a certain house costs $100,000 today, what was its price last year?

Solution

Step 1 Let c represent the cost of the house last year.

Step 2 Express the current price of the house in two different ways. During the past year, the price of the house increased by 4%, or $0.04c$. Its current price is thus

$$\underset{\text{Original cost}}{(1)c} \;+\; \underset{\text{Price increase}}{0.04c} \;=\; c(1 + 0.04) = 1.04c$$

This expression is equal to the value given for current price of the house.

$$1.04c = 100{,}000$$

Step 3 To solve this equation, we divide both sides by 1.04 to find

$$c = \frac{100{,}000}{1.04} = 96{,}153.846$$

Step 4 To the nearest cent, the cost of the house last year was \$96,153.85. ∎

COMMON ERROR   In Example 3 it would be incorrect to calculate last year's price by subtracting 4% of \$100,000 from \$100,000 to get \$96,000. (Do you see why?) ∎

Weighted Averages

We find the *average* or *mean* of a set of values by adding up the values and dividing the sum by the number of values. Thus, the average, $\bar{x}$, of the numbers x_1, $x_2, \ldots, x_n$ is given by

$$\bar{x} = \frac{x_1 + x_2 + \cdots + x_n}{n}$$

In a **weighted average** the numbers being averaged occur with different frequencies, or are "weighted" differently in their contribution to the average value. For instance, suppose a biology class of 12 students takes a 10-point quiz. Two students receive 10's, three receive 9's, five receive 8's, and two receive scores of 6. The average score earned on the quiz is then

$$\bar{x} = \frac{2(10) + 3(9) + 5(8) + 2(6)}{12} = 8.25$$

The numbers in color are called the *weights*—in this example they represent the number of times each score was counted. Note that n, the total number of scores, is equal to the sum of the weights.

$$12 = 2 + 3 + 5 + 2$$

EXAMPLE 5   Kwan's grade in his accounting class will be computed as follows: Tests count for 50% of the grade, homework counts for 20%, and the final exam counts for 30%. If Kwan has an average of 84 on tests and 92 on homework, what score does he need on the final exam to earn a grade of 90?

Solution

Step 1 Let x represent the final exam score Kwan needs.

Step 2 Kwan's grade is the weighted average of his test, homework, and final exam scores.

$$\frac{0.50(84) + 0.20(92) + 0.30x}{1.00} = 90 \qquad (1)$$

(The sum of the weights is 1.00, or 100% of Kwan's grade.) Multiply both sides of the equation by 1.00 to get

$$0.50(84) + 0.20(92) + 0.30x = 1.00(90) \qquad (2)$$

Step 3 Solve the equation. Simplify the left side first.

$$60.4 + 0.30x = 90 \qquad \text{Subtract 60.4 from both sides.}$$
$$0.30x = 29.6 \qquad \text{Divide both sides by 0.30.}$$
$$x = 98.7$$

Step 4 Kwan needs a score of 98.7 on the final exam to earn a grade of 90. ∎

In Example 5, we rewrote the formula for a weighted average, Equation (1), in a simpler form, Equation (2). In this form, the formula says

$$w_1 x_1 + w_2 x_2 + \cdots + w_n x_n = W \overline{x}$$

where W is the sum of the weights, or

The sum of the weighted values equals the sum of the weights times the average value.

This form is particularly useful for solving problems involving mixtures.

EXAMPLE 6

The vet advised Delbert to feed his dog Rollo with kibble that is no more than 8% fat. Rollo likes JuicyBits, which are 15% fat. LeanMeal is much more expensive, but is only 5% fat. How much LeanMeal should Delbert mix with 50 pounds of JuicyBits to make a mixture that is 8% fat?

Solution

Step 1 Let p represent the number of pounds of LeanMeal needed.

Step 2 In this problem we want the weighted average of the fat contents in the two kibbles to be 8%. The "weights" are the number of pounds of each kibble we use. It is often useful to summarize the given information in a table.

	% Fat	Total pounds	Pounds of fat
Juicy Bits	15%	50	0.15(50)
LeanMeal	5%	p	0.05p
Mixture	8%	$50 + p$	0.08(50 + p)

The amount of fat in the mixture must come from adding the amounts of fat in the two ingredients. This gives us an equation,

$$0.15(50) + 0.05p = 0.08(50 + p)$$

This equation is an example of the formula for weighted averages.

Step 3 Simplify each side of the equation, then solve.

$$7.5 + 0.05p = 4 + 0.08p$$
$$3.5 = 0.03p$$
$$p = 116.\overline{6}$$

Step 4 Delbert must mix $116\frac{2}{3}$ pounds of LeanMeal with 50 pounds of JuicyBits to make a mixture that is 8% fat. ∎

Formulas

A formula is any equation that relates several variable quantities. For example, the equation

$$P = 2l + 2w \tag{1}$$

is a formula that gives the perimeter of a rectangle in terms of its length and width.

Suppose we have a fixed amount of wire fencing to enclose an exercise area for rabbits, and we would like to see what dimensions are possible for different rectangles with that perimeter. In this case it would be more useful to have a formula for, say, the length of the rectangle in terms of its perimeter and its width. We can find such a formula by solving Equation (1) for l in terms of P and w.

$$2l + 2w = P \qquad \text{Subtract } 2w \text{ from both sides.}$$
$$2l = P - 2w \qquad \text{Divide both sides by 2.}$$
$$l = \frac{P - 2w}{2}$$

The result is a new formula that gives the length of a rectangle in terms of its perimeter and its width.

EXAMPLE 7

The formula $5F = 9C + 160$ relates the temperature in degrees Fahrenheit, F, to the temperature in degrees Celsius, C. Solve the formula for C in terms of F.

Solution

Begin by isolating the term that contains C.

$$5F = 9C + 160 \qquad \text{Subtract 160 from both sides.}$$
$$5F - 160 = 9C \qquad \text{Divide both sides by 9.}$$
$$\frac{5F - 160}{9} = C$$

The formula for C in terms of F is $C = \dfrac{5F - 160}{9}$, or $C = \dfrac{5}{9}F - \dfrac{160}{9}$. ∎

Solve each linear equation.

1. $3x + 5 = 26$ 2. $2 + 5x = 37$

3. $3(z + 2) = 37$ 4. $2(z - 3) = 15$

5. $3y - 2(y - 4) = 12 - 5y$

6. $5y - 3(y + 1) = 14 + 2y$

7. $0.8w - 2.6 = 1.4w + 0.3$

8. $4.8 - 1.3w = 0.7w + 2.1$

9. $0.25t + 0.10(t - 4) = 11.60$

10. $0.12t + 0.08(t + 10,000) = 12,000$

Solve each exercise by writing and solving an equation.

11. Celine's boutique carries a line of jewelry made by a local artists' co-op. If she charges p dollars for a pair of earrings, she finds that she can sell $200 - 5p$ pairs per month. On the other hand, the co-op will provide her with $56 + 3p$ pairs of earrings when she charges p dollars per pair. What price should Celine charge so that the demand for earrings will equal her supply?

12. Curio Electronics sells garage door openers. If they charge p dollars per unit, they sell $120 - p$ openers per month. The manufacturer will supply $20 + 2p$ openers at a price of p dollars each. What price should Curio Electronics charge so that their monthly supply will meet their demand?

13. Roger sets out on a bicycle trip at an average speed of 16 miles per hour. Six hours later his wife finds his patch kit on the dining room table. If she heads after him in the car at 45 miles per hour, how long will it be before she catches him?

   a. What are we asked to find in this problem? Assign a variable to represent it.

   b. Write an expression in terms of your variable for the distance Roger's wife drives.

   c. Write an expression in terms of your variable for the distance Roger has cycled.

   d. Write an equation and solve it.

14. Kate and Julie set out in their sailboat on a straight course at 9 miles per hour. Two hours later their mother becomes worried and sends their father after them in the speedboat. If their father travels at 24 miles per hour, how long will it be before he catches them?

   a. What are we asked to find in this problem? Assign a variable to represent it.

   b. Write an expression in terms of your variable for the distance Kate and Julie sailed.

   c. Write an expression in terms of your variable for the distance their father traveled.

   d. Write an equation and solve it.

15. The reprographics department has a choice of 2 new copying machines. One sells for $20,000 and costs $0.02 per copy to operate. The other sells for $17,500, but its operating costs are $0.025 per copy. The repro department decides to buy the more expensive machine. How many copies must they make before the higher price is justified?

   a. What are we asked to find in this problem? Assign a variable to represent it.

   b. Write expressions in terms of your variable for the total cost incurred by each machine.

   c. Write an equation and solve it.

16. Annie needs a new refrigerator and can choose between 2 models of the same size. One model sells for $525 and costs $0.08 per hour to run. A more energy-efficient model sells for $700 but runs for $0.05 per hour. If Annie buys the more expensive model, how long will it be before she starts saving money?

   a. What are we asked to find in this problem? Assign a variable to represent it.

   b. Write expressions in terms of your variable for the total cost incurred by each refrigerator.

   c. Write an equation and solve it.

17. The population of Midland has been growing at an annual rate of 8% over the past 5 years. Its present population is 135,000.

a. Assuming the same rate of growth, what do you predict for the population of Midland next year?

b. What was the population of Midland last year?

18. For the past 3 years the annual inflation rate has been 6%. This year a steak dinner at Benny's costs $12.

   a. Assuming the same rate of inflation, what do you predict for the price of a steak dinner next year?

   b. What did a steak dinner cost last year?

19. Virginia took a 7% pay cut when she changed jobs last year. What percent pay increase must she receive this year in order to match her old salary of $24,000? (*Hint:* What was Virginia's salary after the pay cut?)

20. Clarence W. Networth took a 16% loss in the stock market last year. What percent gain must he realize this year in order to restore his original holdings of $85,000? (*Hint:* What was the value of Clarence's stock holdings after the loss?)

21. Delbert's test average in algebra is 77. If the final exam counts for 30% of the grade and the test average counts for 70%, what must Delbert score on the final exam to have a term average of 80?

22. Harold's batting average for the first 8 weeks of the baseball season is 0.385. What batting average must he maintain over the last 18 weeks so that his season average will be 0.350 (assuming he continues the same number of "at-bats" per week)?

23. A horticulturist needs a fertilizer that is 8% potash, but she can find only fertilizers that contain 6% and 15% potash. How much of each should she mix to obtain 10 pounds of 8% potash fertilizer?

   a. What are we asked to find in this problem? Assign a variable to represent it.

   b. Write algebraic expressions in terms of your variable for the amounts of each fertilizer the horticulturist uses. Use the table.

   c. Write expressions for the amount of potash in each batch of fertilizer.

d. Write two different expressions for the amount of potash in the mixture. Now write an equation and solve it.

Pounds of fertilizer	% potash	Pounds of potash

24. A sculptor wants to cast a bronze statue from an alloy that is 60% copper. He has 30 pounds of a 45% alloy. How much 80% copper alloy should he mix with it to obtain the 60% copper alloy?

   a. What are we asked to find in this problem? Assign a variable to represent it.

   b. Write algebraic expressions in terms of your variable for the amounts of each alloy the sculptor uses. Use the table.

   c. Write expressions in terms of your variable for the amount of copper in each batch of alloy.

   d. Write two different expressions for the amount of copper in the mixture. Now write an equation and solve it.

Pounds of alloy	% copper	Pounds of copper

25. Lacy's Department Store wants to keep the average salary of its employees under $19,000 per year. If the downtown store pays its 4 managers $28,000 per year and its 12 department heads $22,000 per year, how much can it pay its 30 clerks?

   a. What are we asked to find in this problem? Assign a variable to represent it.

   b. Write algebraic expressions for the total amounts Lacy's pays its managers, its department heads, and its clerks.

   c. Write two different expressions for the total amount Lacy's pays in salaries each year.

d. Write an equation and solve it.

26. Federal regulations require that 60% of all vehicles manufactured next year comply with new emission standards. Major Motors can bring 85% of their small trucks in line with the standards, but only 40% of their automobiles. If Major Motors plans to manufacture 20,000 automobiles next year, how many trucks will they have to produce in order to comply with the federal regulations?

a. What are we asked to find in this problem? Assign a variable to represent it.

b. Write algebraic expressions for the number of trucks and the number of cars that will meet emission standards.

c. Write two different expressions for the total number of vehicles that will meet the standards.

d. Write an equation and solve it.

Solve each formula for the specified variable.

27. $v = k + gt$, for t

28. $S = 3\pi d + \pi a$, for d

29. $S = 2w(w + 2h)$, for h

30. $A = P(1 + rt)$, for r

31. $P = a + (n - 1)d$, for n

32. $R = 2d + h(a + b)$, for b

33. $A = \pi rh + \pi r^2$, for h

34. $A = 2w^2 + 4lw$, for l

A.3   Facts from Geometry

In this section we review some information you will need from geometry. You are already familiar with the formulas for area and perimeter of common geometric figures; you can find these formulas in the reference section at the front of the book.

Right Triangles and the Pythagorean Theorem

A **right triangle** is a triangle in which one of the angles is a right angle, or 90°. Because the sum of the three angles in *any* triangle is 180°, this means that the other two angles in a right triangle must have a sum of 180° − 90°, or 90°. For instance, if we know that one of the angles in a right triangle is 37°, then the remaining angle must be 90° − 37°, or 53°, as shown in Figure A.1.

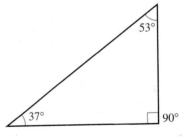

FIGURE A.1

EXAMPLE 1

In a right triangle, the medium-sized angle is 15° less than twice the smallest angle. Find the sizes of the three angles in Figure A.2.

Solution

Step 1 Let x stand for the size of the smallest angle. Then the medium-sized angle must be $2x - 15$.

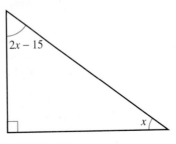

FIGURE A.2

Step 2 Because the right angle is the largest angle, the sum of the smallest and medium-sized angles must be the remaining 90°. Thus

$$x + (2x - 15) = 90$$

Step 3 Solve the equation. Begin by simplifying the left side.

$$3x - 15 = 90 \quad \text{Add 15 to both sides.}$$
$$3x = 105 \quad \text{Divide both sides by 3.}$$
$$x = 35$$

Step 4 The smallest angle is 35°, and the medium-sized angle is $2(35°) - 15°$, or 55°. ∎

In a right triangle (see Figure A.3), the longest side is opposite the right angle and is called the **hypotenuse.** Ordinarily, even if we know the lengths of two sides of a triangle it is not easy to find the length of the third side (to solve this problem, we need trigonometry), but for the special case of a right triangle, there is an equation that relates the lengths of the three sides. This property of right triangles was known to many ancient cultures, and we know it today by the name of a Greek mathematician, Pythagoras, who provided a proof of the result.

Pythagorean Theorem

In a right triangle, if c stands for the length of the hypotenuse and a and b stand for the lengths of the other two sides, then

$$a^2 + b^2 = c^2$$

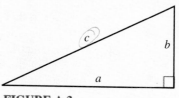

FIGURE A.3

EXAMPLE 2 The hypotenuse of a right triangle is 15 feet long. The third side is twice the length of the shortest side. Find the lengths of the other 2 sides.

Solution

Step 1 Let x represent the length of the shortest side, so that the third side has length $2x$. (See Figure A.4.)

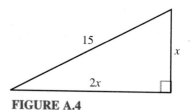

FIGURE A.4

Step 2 Substituting these expressions into the Pythagorean Theorem, we find

$$x^2 + (2x)^2 = 15^2$$

Step 3 This is a quadratic equation with no linear term, so we simplify and then isolate x^2.

$$x^2 + 4x^2 = 225 \quad \text{Combine like terms.}$$
$$5x^2 = 225 \quad \text{Divide both sides by 5.}$$
$$x^2 = 45$$

Taking square roots of both sides yields

$$x = \pm\sqrt{45} \approx \pm 6.708203932$$

Step 4 Because a length must be a positive number, the shortest side has length approximately 6.71 feet, and the third side has length 2 (6.71), or approximately 13.42 feet. ■

Isosceles and Equilateral Triangles

Recall also that an **isosceles** triangle is one that has at least two sides of equal length. In an isosceles triangle, the angles opposite the equal sides, called the **base** angles, are equal in measure, as shown in Figure A.5a. In an **equilateral** triangle (Figure A.5b), all three sides have equal length, and all three angles have equal measure.

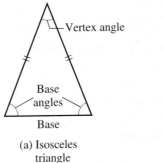 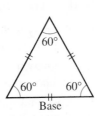

(a) Isosceles triangle

(b) Equilateral triangle

FIGURE A.5

Inequalities and the Triangle Inequality

The longest side in a triangle is always opposite the largest angle, and the shortest side is opposite the smallest angle. It is also true that the sum of the lengths of any two sides of a triangle must be greater than the third side, or else the two sides will not meet to form a triangle! This fact is called the **triangle inequality.**

In Figure A.6, we must have that

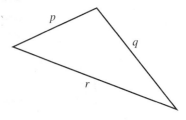

FIGURE A.6

$$p + q > r$$

where p, q, and r are the lengths of the sides of the triangle. Recall that the symbol $>$ is called an **inequality symbol,** and the statement $p + q > r$ is called an **inequality.** There are four basic inequality symbols:

$$> \quad \text{"is greater than"}$$
$$< \quad \text{"is less than"}$$
$$\geq \quad \text{"is greater than or equal to"}$$
$$\leq \quad \text{"is less than or equal to"}$$

Inequalities that include the symbols $>$ or $<$ are called **strict inequalities**; those that include $\geq$ or $\leq$ are called **nonstrict.**

If we multiply or divide both sides of an inequality by a negative number, the direction of the inequality must be reversed. For example, if we multiply both sides of the inequality

$$2 < 5$$

by -3, we get

$$-3(2) > -3(5) \quad \text{Inequality symbol changes from } < \text{ to } >.$$

or

$$-6 > -15$$

Because of this property the rules for solving linear equations must be revised slightly for solving linear inequalities.

Rules for Solving Linear Inequalities

1. We may add or subtract the same number on both sides of an inequality without changing its solutions.

2. We may multiply or divide both sides of an inequality by a *positive* number without changing its solutions.

3. If we multiply or divide both sides of an inequality by a *negative* number, we must *reverse the direction of the inequality symbol.*

EXAMPLE 3

Solve the inequality $4 - 3x \geq -17$.

Solution

Use the rules above to "isolate" x on one side of the inequality.

$$4 - 3x \geq -17 \quad \text{Subtract 4 from both sides.}$$
$$-3x \geq -21 \quad \text{Divide both sides by } -3.$$
$$x \leq 7$$

Notice that we reversed the direction of the inequality when we divided by -3. Any number less than or equal to 7 is a solution of the inequality. ■

Now we can use the Triangle Inequality to discover information about the sides of a triangle.

EXAMPLE 4

Two sides of a triangle have lengths 7 inches and 10 inches. What can you say about the length of the third side (see Figure A.7)?

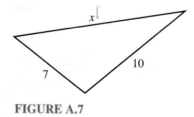

FIGURE A.7

Solution

Let x represent the length of the third side of the triangle. By the Triangle Inequality, we must have that

$$x < 7 + 10, \quad \text{or} \quad x < 17$$

Looking at another pair of sides, we must also have that

$$10 < x + 7, \quad \text{or} \quad x > 3$$

Thus the third side must be greater than 3 inches but less than 17 inches long. ∎

Similar Triangles

Two triangles are said to be **similar** if their corresponding angles are equal. This means that the two triangles will have the same shape, but not necessarily the same size. One of the triangles will be an enlargement or a reduction of the other; so their corresponding sides are proportional. In other words, for similar triangles, the ratios of the corresponding sides are equal. (See Figure A.8.)

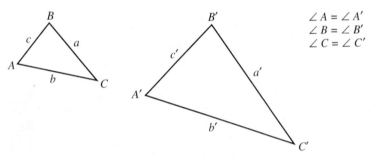

FIGURE A.8

If any two pairs of corresponding angles of two triangles are equal, then the third pair must also be equal, because in both triangles the sum of the angles is 180°. Thus, to show that two triangles are similar, we need only show that two pairs of angles are equal.

EXAMPLE 5

The roof of an A-frame ski chalet forms an isosceles triangle with the floor (see Figure A.9). The floor of the chalet is 24 feet wide, and the ceiling is 20 feet tall at the center. If a loft is built at a height of 8 feet from the floor, how wide will the loft be?

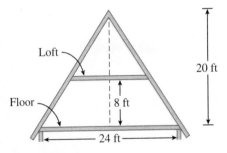

FIGURE A.9

Solution

From Figure A.10 on page 776 we can show that $\triangle ABC$ is similar to $\triangle ADE$. Both triangles include $\angle A$, and because $\overline{DE}$ is parallel to $\overline{BC}$, $\angle ADE$ is equal to $\angle ABC$. Thus, the triangles have two pairs of equal angles, and are therefore similar triangles.

Step 1 Let w stand for the width of the loft.

Step 2 First note that if $FG = 8$, then $AF = 12$. Because $\triangle ABC$ is similar to ADE, the ratios of their corresponding sides (or corresponding altitudes) are equal. In particular,

$$\frac{w}{24} = \frac{12}{20}$$

Step 3 Solve the proportion for w. Multiply both sides by 24.

$$w = \frac{(12)(24)}{20} \qquad \text{Simplify.}$$

$$w = \frac{288}{20} = 14.4 \qquad \text{Divide by 20.}$$

Step 4 The floor of the loft will be 14.4 feet wide.   ■

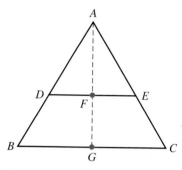

FIGURE A.10

Volume and Surface Area

The **volume** of a three-dimensional object measures its capacity, or how much space it encloses. Volume is measured in cubic units, such as cubic inches or cubic meters. The volume of a rectangular prism, or box, is given by the product of its length, width, and height. For example, the volume of the box of length 4 inches, width 3 inches, and height 2 inches shown in Figure A.11 is

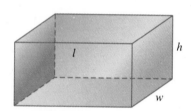

FIGURE A.11

$$V = lwh = (4)(3)(2) = 24 \text{ cubic inches}$$

Formulas for the volumes of other common objects can be found in the reference section.

EXAMPLE 6

A cylindrical can must have a height of 6 inches, but can have any reasonable radius.

a. Write an algebraic expression for the volume of the can in terms of its radius.

b. If the volume of the can should be approximately 170 cubic inches, what should its radius be?

Solutions

a. The formula for the volume of a right circular cylinder is $V = \pi r^2 h$. If the height of the cylinder is 6 inches, then $V = \pi r^2(6)$, or $V = 6\pi r^2$. (See Figure A.12.)

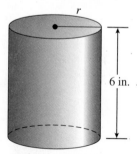

FIGURE A.12

b. Substitute 170 for V and solve for r.

$$170 = 6\pi r^2 \quad \text{Divide both sides by } 6\pi.$$

$$r^2 = \frac{170}{6\pi} \quad \text{Take square roots.}$$

$$r = \sqrt{\frac{170}{6\pi}} \approx 3.00312$$

Thus, the radius of the can should be approximately 3 inches. A calculator keying sequence for the expression above is

$$\boxed{\sqrt{\ (} \ 170 \ \boxed{\div} \ 6 \ \boxed{\pi} \ \boxed{)} \ \boxed{\text{ENTER}}$$

The **surface area** of a solid object is the sum of the areas of all the exterior faces of the object. It measures the amount of paper that would be needed to cover the object entirely. Because it is an area, it is measured in square units.

EXAMPLE 7

Write a formula for the surface area of a closed box in terms of its length, width, and height. (See Figure A.13.)

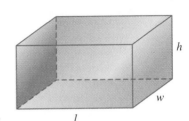

FIGURE A.13

Solution

The box has six sides; we must find the area of each side and add them together. The top and bottom of the box each have area lw, so together they contribute $2lw$ to the surface area. The back and front of the box each have area lh, so they contribute $2lh$ to the surface area. Finally, the left and right sides of the box each have area wh, so they add $2wh$ to the surface area. Thus, the total surface area is

$$S = 2lw + 2lh + 2wh$$

Formulas for the surface areas of other common solids can be found inside the front cover.

EXERCISE A.3

Solve each exercise involving properties of triangles.

1. One angle of a triangle is 10° larger than another, and the third angle is 29° larger than the smallest. How large is each angle?

2. One angle of a triangle is twice as large as the second angle, and the third angle is 10° less than the larger of the other two. How large is each angle?

3. One acute angle of a right triangle is twice the other acute angle. How large is each acute angle?

4. One acute angle of a right triangle is 10° less than three times the other acute angle. How large is each acute angle?

5. The vertex angle of an isosceles triangle is 20° less than the sum of the equal angles. How large is each angle?

6. The vertex angle of an isosceles triangle is 30° less than one of the equal angles. How large is each angle?

7. The perimeter of an isosceles triangle is 42 centimeters and its base is 12 centimeters long. How long are the equal sides?

8. The altitude of an equilateral triangle is $\frac{\sqrt{3}}{2}$ times its base. If the perimeter of an isosceles triangle is 18 inches, what is its area?

Solve each inequality.

9. $3x - 2 > 1 + 2x$

10. $2x + 3 \le x - 1$

11. $\dfrac{-2x - 6}{-3} > 2$

12. $\dfrac{-2x - 3}{2} \le -5$

13. $\dfrac{2x - 3}{3} \le \dfrac{3x}{-2}$

14. $\dfrac{3x - 4}{-2} > \dfrac{-2x}{5}$

15. If two sides of a triangle are 6 feet and 10 feet long, what can you say about the length of the third side?

16. If one of the equal sides of an isosceles triangle is 8 millimeters long, what can you say about the length of the base?

Solve each exercise involving similar triangles.

17. A 6-foot man stands 12 feet from a lamppost. His shadow is 9 feet long. How tall is the lamppost? (See Figure A.14.)

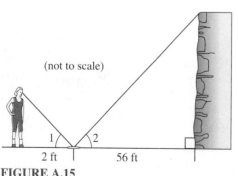

6 ft

12 ft   9 ft

FIGURE A.14

18. A rock climber estimates the height of a cliff she plans to scale as follows. She places a mirror on the ground so that she can just see the top of the cliff in the mirror while she stands straight. (The angles 1 and 2 formed by the light rays are equal, as shown in Figure A.15.) She then measures the distance to the mirror (2 feet) and the distance from the mirror to the base of the cliff. If she is 5 feet 6 inches tall, how high is the cliff?

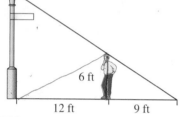

(not to scale)

1   2

2 ft   56 ft

FIGURE A.15

19. A conical tank is 12 feet deep and the diameter of the top is 8 feet. If the tank is filled with water to a depth of 7 feet as shown in Figure A.16, what is the area of the exposed surface of the water?

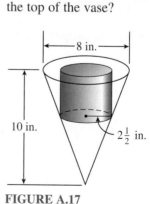

8 ft

12 ft

7 ft

FIGURE A.16

20. A florist fits a cylindrical piece of foam into a conical vase that is 10 inches high and measures 8 inches across the top, as shown in Figure A.17. If the radius of the foam cylinder is $2\frac{1}{2}$ inches, how tall should it be just to reach the top of the vase?

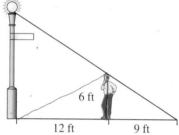

8 in.

10 in.

$2\frac{1}{2}$ in.

FIGURE A.17

21. To measure the distance across the river shown in Figure A.18, stand at A and sight across the river to a convenient landmark at B. Then measure the distances AC, CD, and DE. If AC = 20 feet, CD = 13 feet, and DE = 58 feet, how wide is the river?

22. To measure the distance EC across the lake shown in Figure A.19, stand at A and sight point C across the lake, then mark point B. Then sight to point E and mark point D so that DB is parallel to CE. If AD = 25 yards, AE = 60 yards, and BD = 30 yards, how wide is the lake?

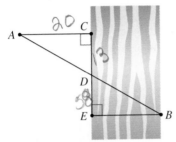

FIGURE A.18

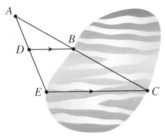

FIGURE A.19

Use the formulas in the reference section to find the volumes and surface areas of the given objects.

23. a. How much helium (in cubic meters) is needed to inflate a spherical balloon to a radius of 1.2 meters?

b. How much gelatin (in square centimeters) is needed to coat a spherical pill whose radius is 0.7 centimeters?

24. a. How much storage space is there in a rectangular box whose length is 12.3 inches, whose width is 4 inches, and whose height is 7.3 inches?

b. How much marine sealer will be needed to paint a rectangular wooden storage locker with length 6.2 feet, width 5.8 feet, and height 2.6 feet?

25. a. How much grain can be stored in a cylindrical silo whose radius is 6 meters and whose height is 23.2 meters?

b. How much paint is needed to cover a cylindrical storage drum whose radius is 15.3 inches and whose height is 4.5 inches?

26. a. A conical pile of sand is 8.1 feet high and has a radius of 4.6 feet. How much sand is in the pile?

b. How much plastic is needed to line a conical funnel with a radius of 16 centimeters and a slant height of 42 centimeters?

Use the formulas for surface area and volume to answer the questions.

27. A cardboard packing box must have length 20 inches and width 16 inches.

a. Write a formula for the surface area of the box in terms of its height.

b. What is the maximum possible height for the box if its surface area should be no more than 1216 square inches?

28. We can find the area of a trapezoid by computing the product of its height and the average of its two bases.

a. Write a formula for the area of a trapezoid whose height is 30 centimeters.

b. If the area of the trapezoid in part (a) is 1020 square centimeters and one base is 36 centimeters long, find the length of the other base.

A.4 Graphs

Graphs are one of the most useful tools for studying mathematical relationships. Constructing a graph provides an overview of a quantity of data, and helps us identify trends or unexpected occurrences. Interpreting the graph can help us answer questions about the data.

The data in Table A.1 show the atmospheric pressure, on a certain day, at different altitudes. Meteorologists regularly collect such data by sending up a weather balloon with a device called a radiosonde, which is equipped with a barometer and a radio transmitter. Altitudes are given in feet, and atmospheric pressures are given in inches of mercury.

TABLE A.1

Altitude	0	5000	10,000	20,000	30,000	40,000	50,000
Pressure	29.7	24.8	20.5	14.6	10.6	8.5	7.3

We observe a generally decreasing trend in pressure as the altitude increases, but it is difficult to say anything more precise about the relationship between pressure and altitude. A clearer picture emerges if we plot the data on a graph. To do this, we use two perpendicular number lines called **axes.** We use the horizontal axis for the values of the first variable, altitude, and the vertical axis for the values of the second variable, pressure.

The entries in Table A.1 are called **ordered pairs,** in which the **first component** is the altitude and the **second component** is the atmospheric pressure measured at that altitude. For example, the first two entries can be represented by (0, 29.7) and (5000, 24.8). We plot the points whose **coordinates** are given by the ordered pairs, as shown in Figure A.20a.

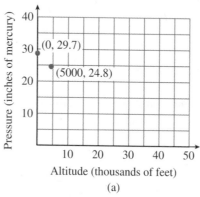
(a)

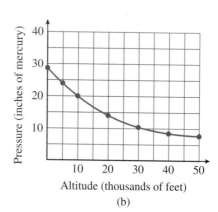
(b)

FIGURE A.20

If we plot all the ordered pairs from Table A.1, we can connect them with a smooth curve as shown in Figure A.20b. In doing this we are actually "estimating" the pressures that correspond to altitudes between those given, such as 15,000 feet or 37,000 feet. For many physical situations, variables are related so that one changes "smoothly" with respect to the other, and a smooth curve will thus serve as a good model. We will assume that this is the case in most of the modeling we do.

EXAMPLE 1

From the graph in Figure A.20b, estimate

   a. the atmospheric pressure measured at an altitude of 15,000 feet,

   b. the altitude at which the pressure is 12 inches of mercury.

Solutions

   a. The point with first coordinate 15,000 on the graph in Figure A.21 has a second coordinate of approximately 17.4. Hence, we estimate the pressure at 15,000 feet to be 17.4 inches of mercury.

   b. The point on the graph with second coordinate 12 has a first coordinate of approximately 25,000, so an atmospheric pressure of 12 inches of mercury occurs at about 25,000 feet.

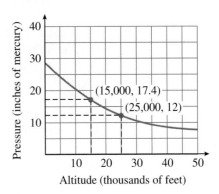

FIGURE A.21

By using the graph in Figure A.20b, we can obtain information about the relationship between altitude and pressure that would be difficult to obtain from the data alone.

EXAMPLE 2

   a. For what altitudes is the pressure less than 18 inches of mercury?

   b. How much does the pressure decrease as the altitude increases from 15,000 feet to 25,000 feet?

   c. For which 10,000-foot increase in altitude does the pressure change most rapidly?

Solutions

   a. From the graph in Figure A.20b we see that the pressure has dropped to 18 inches of mercury at about 14,000 feet, and that it continues to decrease as the altitude increases. Therefore, the pressure is less than 18 inches of mercury for altitudes greater than 14,000 feet.

   b. At 15,000 feet the pressure is approximately 17.4 inches of mercury, and at 25,000 feet it is 12 inches. This represents a decrease in pressure of 17.4 − 12, or 5.4, inches of mercury.

   c. By studying the graph we see that the pressure decreases most rapidly at low altitudes, so we conclude that the greatest drop in pressure occurs between 0 and 10,000 feet.

Graphs of Equations

In Example 1 we used a graph to illustrate a collection of data given by a table. Graphs can also help us analyze models given by equations. Let's first review some facts about solutions of equations in two variables.

An equation in two variables, such as $y = 2x + 3$, is said to be *satisfied* if the variables are replaced by a pair of numbers that make the statement true. The pair of numbers is called a **solution** of the equation and is usually written as an ordered pair (x, y). (The first number in the pair is the value of x and the second number is the value of y.)

To find a solution of a given equation, we assign a number to one of the variables and then solve the resulting equation for the second variable. By choosing different values for x, we can find many different solutions to the equation. For example, to obtain solutions to the equation

$$y = 2x + 3 \qquad\qquad (1)$$

we might choose the values -2, 0, and 1 for x. When we substitute these x-values into the equation, we find a corresponding y-value for each.

$$\text{When } x = -2, \qquad y = 2(-2) + 3 = -1$$
$$\text{when } x = 0, \qquad y = 2(0) + 3 = 3$$
$$\text{when } x = 1, \qquad y = 2(1) + 3 = 5$$

Thus, the ordered pairs $(-2, -1)$, $(0, 3)$, and $(1, 5)$ are three solutions of $y = 2x + 3$.

We can also substitute values for y and see if there are any x-values that make the equation true. For example, if we let $y = 10$, we have

$$10 = 2x + 3 \quad \text{\small Subtract 3 from both sides.}$$

Solving this equation for x, we find

$$7 = 2x$$

or $x = 3.5$. This means that the ordered pair $(3.5, 10)$ is another solution of Equation (1). Because we could have used any value for x (or for y), there are actually infinitely many solutions for Equation (1).

An equation in two variables may have infinitely many solutions, so we cannot list them all. However, we can display the solutions on a graph. For this we use a **Cartesian (or rectangular) coordinate system,** as shown in Figure A.22. Every point in the plane can be located by its coordinates, and every ordered pair of coordinates corresponds to a unique point in the plane, called its **graph.**

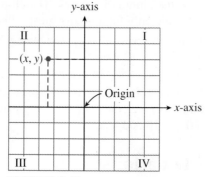

FIGURE A.22

The graph of an equation is the graph of all the solutions of the equation. Thus, a particular point is included in the graph if its coordinates satisfy the equation. If the coordinates of a point do not satisfy the equation, then the point is not part of the graph. We can think of the graph of an equation as a picture of the solutions of the equation. For example, the graph of Equation (1),

$$y = 2x + 3$$

is shown in Figure A.23.

The graph does not display *all* the solutions of Equation (1), but shows

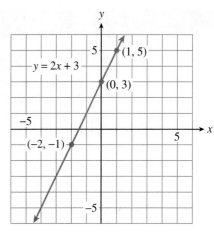

FIGURE A.23

important features such as the intercepts on the x- and y-axes, and enough of a picture for us to infer how the rest of the curve looks. (In this case, a straight line.) Because there is a solution corresponding to every real number x, the graph extends infinitely in either direction, as indicated by the arrows.

EXAMPLE 3

Use the graph of $y = 0.5x^2 - 2$ in Figure A.24 to decide whether the given ordered pairs are solutions of the equation. Verify your answers algebraically.

a. $(-4, 6)$ **b.** $(3, 0)$

Solutions

a. Because the point $(-4, 6)$ does lie on the graph, the ordered pair $x = -4$, $y = 6$ is a solution of $y = 0.5x^2 - 2$. We can verify this by substituting -4 for x and 6 for y.

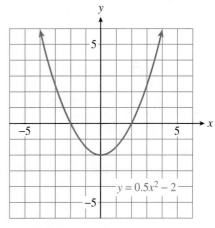

FIGURE A.24

$$0.5(-4)^2 - 2 = 0.5(16) - 2$$
$$= 8 - 2 = 6$$

b. Because the point $(3, 0)$ does not lie on the graph, the ordered pair $x = 3$, $y = 0$ is not a solution of $y = 0.52x^2 - 2$. We substitute 3 for x and 0 for y to verify this.

$$0.5(3)^2 - 2 = 0.5(9) - 2$$
$$= 4.5 - 2 = 2.5 \neq 0$$

■

Answer the questions about each graph.

1. Figure A.25 shows the graph of the temperatures recorded during a winter day in Billings, Montana.

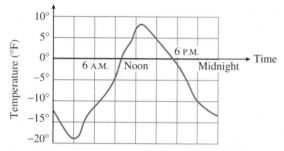

FIGURE A.25

a. What were the high and low temperatures recorded during the day?

b. During what time intervals is the temperature above 5°F? Below −5°F?

c. Estimate the temperatures at 7 a.m. and 2 p.m. At what time(s) is the temperature approximately 0°F? Approximately −12°F?

d. How much did the temperature increase between 3 a.m. and 6 a.m.? Between 9 a.m. and noon? How much did the temperature decrease between 6 p.m. and 9 p.m.?

e. During which 3-hour interval did the temperature increase most rapidly? Decrease most rapidly?

2. Figure A.26 shows a graph of the altitude of a commercial jetliner during its flight from Denver to Los Angeles.

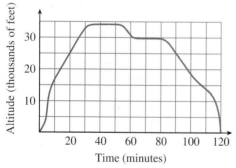

FIGURE A.26

a. What was the highest altitude the jet achieved? At what time(s) was this altitude recorded?

b. During what time intervals was the altitude greater than 10,000 feet? Below 20,000 feet?

c. Estimate the altitudes 15 minutes into the flight and 35 minutes into the flight. At what time(s) was the altitude approximately 16,000 feet? 32,000 feet?

d. How many feet did the jet climb during the first 10 minutes of flight? Between 20 minutes and 30 minutes? How many feet did the jet descend between 100 minutes and 120 minutes?

e. During which 10-minute interval did the jet ascend most rapidly? Descend most rapidly?

3. Figure A.27 shows the gas mileage achieved by an experimental model automobile at different speeds.

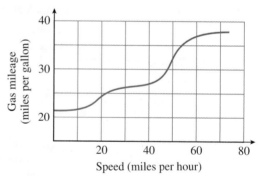

FIGURE A.27

a. Estimate the gas mileage achieved at 43 miles per hour.

b. Estimate the speed at which a gas mileage of 34 miles per gallon is achieved.

c. At what speed is the best gas mileage achieved? Do you think that the gas mileage will continue to improve as the speed increases? Why or why not?

d. The data illustrated by the graph were collected under ideal test conditions. What factors might affect the gas mileage if the car were driven under more realistic conditions?

4. Figure A.28 shows the average height of young women aged 0 to 18 years.

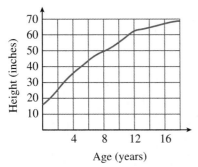

FIGURE A.28

a. Estimate the average height of 5-year-old girls.

b. Estimate the age at which the average young woman is 50 inches tall.

c. At what age does the average woman achieve her maximum height? Do you think that height will continue to increase as age increases? Why or why not?

d. The data recorded in the graph reflect the average heights for young women at given ages. What factors might affect the heights of specific individuals?

5. Figure A.29 shows the speed of a car during an hour-long journey.

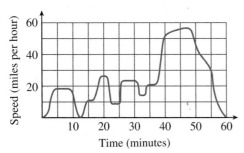

FIGURE A.29

a. When did the car stop at a traffic signal?

b. During what time interval did the car drive in "stop-and-go" city traffic?

c. During what time interval did the car travel on the freeway?

6. Figure A.30 shows the fish population of a popular fishing pond.

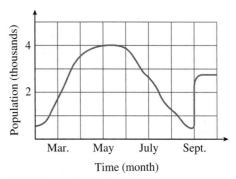

FIGURE A.30

a. During what months do the young fish hatch?

b. During what months is fishing allowed?

c. When does the park service restock the pond?

For Exercises 7–14, (a) use the graph to find the missing component in each solution of the equation, (b) verify your answers algebraically.

7. $s = 2t + 4$

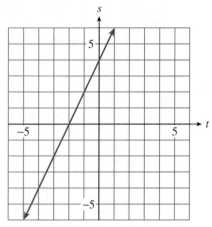

FIGURE A.31

$(-3, ?), (1, ?), (?, 0), (?, 4)$

9. $w = v^2 + 2$

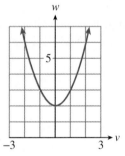

FIGURE A.33

$(-2, ?), (2, ?), (?, 3), (?, 2)$

8. $s = -2t + 4$

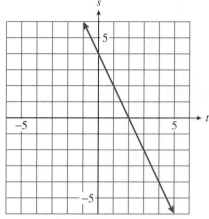

FIGURE A.32

$(-2, ?), (3, ?), (?, 0), (?, 4)$

10. $w = v^2 - 4$

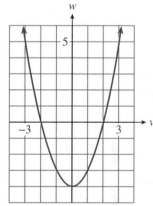

FIGURE A.34

$(-1, ?) (3, ?), (?, 0), (?, -4)$

11. $p = \dfrac{1}{m-1}$

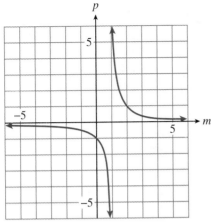

FIGURE A.35

$(-1, ?), \left(\dfrac{1}{2}, ?\right), \left(?, \dfrac{1}{3}\right), (?, -1)$

12. $p = \dfrac{1}{m+1}$

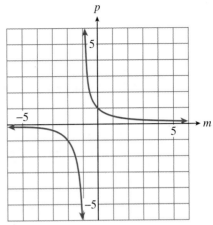

FIGURE A.36

$\left(\dfrac{-3}{2}, ?\right), (3, ?), (?, -1), (?, 2)$

13. $y = x^3$

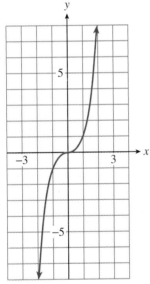

FIGURE A.37

$(-2, ?), \left(\dfrac{1}{2}, ?\right), (?, 0), (?, -1)$

14. $y = \sqrt{x+4}$

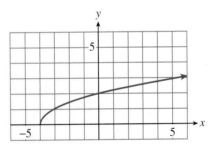

FIGURE A.38

$(0, ?), (5, ?), (?, 0), (?, 1)$

A.5 Laws of Exponents

In this section we review the rules for performing operations on powers.

Product of Powers

Consider a product of two powers with the same base.

$$(a^3)(a^2) = aaa \cdot aa = a^5$$

because a occurs as a factor five times. The number of a's in the product is the *sum* of the number of a's in each factor.

> ### First Law of Exponents
>
> To multiply two powers with the same base, add the exponents and leave the base unchanged.
>
> $$a^m \cdot a^n = a^{m+n}$$

EXAMPLE 1

Add exponents.

a. $5^3 \cdot 5^4 = 5^{3+4} = 5^7$

Same base

Add exponents.

b. $x^4 \cdot x^2 = x^{4+2} = x^6$

Same base

Here are some mistakes to avoid.

COMMON ERRORS

1. Note that we do not *multiply* the exponents when simplifying a product. For example,

$$b^4 \cdot b^2 \neq b^8$$

You can check this with your calculator by choosing a value for b, for instance, $b = 3$.

$$3^4 \cdot 3^2 \neq 3^8$$

2. In order to apply the first law of exponents, the bases must be the same. For example,

$$2^3 \cdot 3^5 \neq 6^8$$

(Check this on your calculator.)

3. We do not multiply the bases when simplifying a product. In Example 1a, note that

$$5^3 \cdot 5^4 \neq 25^7$$

4. Although we can simplify the product x^2x^3 as x^5, we cannot simplify the *sum $x^2 + x^3$*, because x^2 and x^3 are not like terms. ∎

EXAMPLE 2

Multiply $(-3x^4z^2)(5x^3z)$.

Solution

Rearrange the factors to group together the numerical coefficients and the powers of each base. Apply the first law of exponents.

$$(-3x^4z^2)(5x^3z) = (-3)(5)x^4x^3z^2z$$
$$= -15x^7z^3 \qquad \blacksquare$$

Quotients of Powers

We can reduce fractions by "canceling," that is, by dividing both numerator and denominator by any common factors.

$$\frac{x^7}{x^4} = \frac{x x x \cancel{x}\cancel{x}\cancel{x}\cancel{x}}{\cancel{x}\cancel{x}\cancel{x}\cancel{x}} = \frac{x^3}{1} = x^3$$

We can obtain the quotient by *subtracting* the exponent of the denominator from the exponent of the numerator.

$$\frac{x^7}{x^4} = x^{7-4} = x^3$$

What if the larger power occurs in the denominator of the fraction?

$$\frac{x^4}{x^7} = \frac{\cancel{x}\cancel{x}\cancel{x}\cancel{x}}{\cancel{x}\cancel{x}\cancel{x}\cancel{x}xxx} = \frac{1}{x^3}$$

We subtract the exponent of the numerator from the exponent of the denominator, that is

$$\frac{x^4}{x^7} = \frac{1}{x^{7-4}} = \frac{1}{x^3}$$

These examples suggest the following law.

Second Law of Exponents

To divide two powers with the same base, subtract the smaller exponent from the larger one, and keep the same base.

a. If the larger exponent occurs in the numerator, put the power in the numerator.

$$\text{If} \quad m > n, \quad \text{then} \quad \frac{a^m}{a^n} = a^{m-n} \qquad (a \neq 0)$$

b. If the larger exponent occurs in the denominator, put the power in the denominator.

$$\text{If} \quad m < n, \quad \text{then} \quad \frac{a^m}{a^n} = \frac{1}{a^{n-m}} \qquad (a \neq 0)$$

EXAMPLE 3

a. $\dfrac{3^8}{3^2} = 3^{8-2} = 3^6$ Subtract exponents: $8 > 2$.

b. $\dfrac{w^3}{w^6} = \dfrac{1}{w^{6-3}} = \dfrac{1}{w^3}$ Subtract exponents: $3 < 6$.   ■

EXAMPLE 4   Divide $\dfrac{3x^2y^4}{6x^3y}$

Solution

Consider the numerical coefficients and the powers of each variable separately. Use the second law of exponents to simplify each quotient of powers.

$$\dfrac{3x^2y^4}{6x^3y} = \dfrac{3}{6} \cdot \dfrac{x^2}{x^3} \cdot \dfrac{y^4}{y}$$

$$= \dfrac{1}{2} \cdot \dfrac{1}{x^{3-2}} \cdot y^{4-1}$$ Subtract exponents.

$$= \dfrac{1}{2} \cdot \dfrac{1}{x} \cdot \dfrac{y^3}{1} = \dfrac{y^3}{2x}$$ ■

Power of a Power

Consider the expression $(a^4)^3$, the third power of a^4.

— Add exponents.

$$(a^4)^3 = (a^4)(a^4)(a^4) = a^{4+4+4} = a^{12}$$

We can obtain the same result by multiplying the exponents together.

$$(a^4)^3 = a^{4 \cdot 3} = a^{12}$$

Third Law of Exponents

To raise a power to a power, keep the same base and multiply the exponents.

$$(a^m)^n = a^{mn}$$

— Multiply exponents.

EXAMPLE 5   **a.** $(4^3)^5 = 4^{3 \cdot 5} = 4^{15}$

— Multiply exponents.

b. $(y^5)^2 = y^{5 \cdot 2} = y^{10}$ ■

COMMON ERROR   Notice the difference between the expressions

$$(x^3)(x^4) = x^{3+4} = x^7$$

and

$$(x^3)^4 = x^{3 \cdot 4} = x^{12}$$

The first expression is a product, so we add the exponents. The second expression raises a power to a power, so we multiply the exponents. ■

Power of a Product

To simplify the expression $(5a)^3$, we use the associative and commutative laws to regroup the factors as follows.

$$\begin{aligned}
(5a)^3 &= (5a)(5a)(5a) \\
&= 5 \cdot 5 \cdot 5 \cdot a \cdot a \cdot a \\
&= 5^3 a^3
\end{aligned}$$

Thus, to raise a product to a power, we can simply raise each factor to the power.

Fourth Law of Exponents

A power of a product is equal to the product of the powers of each of its factors.

$$(ab)^n = a^n b^n$$

EXAMPLE 6

a. $(5a)^3 = 5^3 a^3 = 125a^3$ Cube each factor.

b. $(-xy^2)^4 = (-x)^4(y^2)^4$ Raise each factor to the fourth power.

$\quad\quad = x^4 y^8$ Apply the third law of exponents. ■

COMMON ERRORS

1. Compare the two expressions $3a^2$ and $(3a)^2$; they are not the same. In the expression $3a^2$ only the factor a is squared, but in $(3a)^2$ both 3 and a are squared. Thus

$$3a^2 \text{ cannot be simplified,}$$

but

$$(3a)^2 = 3^2 a^2 = 9a^2$$

2. Compare the two expressions $(3a)^2$ and $(3 + a)^2$. The fourth law of exponents applies to the *product* $3a$, but not to the *sum* $3 + a$. Thus

$$(3 + a)^2 \neq 3^2 + a^2$$

In order to simplify $(3 + a)^2$, we must expand the binomial product:

$$(3 + a)^2 = (3 + a)(3 + a) = 9 + 6a + a^2$$ ■

Power of a Quotient

To simplify the expression $\left(\dfrac{x}{3}\right)^4$, we multiply together 4 copies of the fraction $\dfrac{x}{3}$. That is,

$$\left(\frac{x}{3}\right)^4 = \frac{x}{3} \cdot \frac{x}{3} \cdot \frac{x}{3} \cdot \frac{x}{3} = \frac{x \cdot x \cdot x \cdot x}{3 \cdot 3 \cdot 3 \cdot 3}$$

$$= \frac{x^4}{3^4} = \frac{x^4}{81}$$

In general, we have the following rule.

Fifth Law of Exponents

To raise a quotient to a power, raise both the numerator and denominator to the power.

$$\left(\frac{a}{b}\right)^n = \frac{a^n}{b^n} \qquad (b \neq 0)$$

For reference, we state all of the laws of exponents together. All the laws are valid when a and b are not equal to zero, and when the exponents m and n are whole numbers.

I. $a^m \cdot a^n = a^{m+n}$ **III.** $(a^m)^n = a^{mn}$

IIa. $\dfrac{a^m}{a^n} = a^{m-n} \qquad (m > n)$ **IV.** $(ab)^n = a^n b^n$

b. $\dfrac{a^m}{a^n} = \dfrac{1}{a^{n-m}} \qquad (m < n)$ **V.** $\left(\dfrac{a}{b}\right)^n = \dfrac{a^n}{b^n}$

EXAMPLE 7

Simplify $5x^2 y^3 (2xy^2)^4$.

Solution

According to the order of operations, we should perform any powers before multiplications. Thus, we begin by simplifying $(2xy^2)^4$. We apply the fourth law.

$$5x^2 y^3 (2xy^2)^4 = 5x^2 y^3 \cdot 2^4 x^4 (y^2)^4 \qquad \text{Apply the third law.}$$
$$= 5x^2 y^3 \cdot 2^4 x^4 y^8$$

Finally, multiply powers with the same base. Apply the first law.

$$5x^2 y^3 \cdot 2^4 x^4 y^8 = 5 \cdot 2^4 x^2 x^4 y^3 y^8 = 80 x^6 y^{11} \qquad \blacksquare$$

EXAMPLE 8

Simplify $\left(\dfrac{2x}{z^2}\right)^3$

Solution

Begin by applying the fifth law.

$$\left(\frac{2x}{z^2}\right)^3 = \frac{(2x)^3}{(z^2)^3}$$

Apply the fourth law to the numerator and the third law to the denominator.

$$= \frac{2^3 x^3}{z^6} = \frac{8x^3}{z^6}$$

∎

Simplify by applying the appropriate law of exponents.

1. a. $b^4 \cdot b^5$ **b.** $b^2 \cdot b^8$ **c.** $(q^3)(q)(q^5)$ **d.** $(p^2)(p^4)(p^4)$

2. a. $\dfrac{w^6}{w^3}$ **b.** $\dfrac{c^{12}}{c^4}$ **c.** $\dfrac{z^6}{z^9}$ **d.** $\dfrac{b^4}{b^8}$

3. a. $2^7 \cdot 2^2$ **b.** $6^5 \cdot 6^3$ **c.** $\dfrac{2^9}{2^4}$ **d.** $\dfrac{8^6}{8^2}$

4. a. $(d^3)^5$ **b.** $(d^4)^2$ **c.** $(5^4)^3$ **d.** $(4^3)^3$

5. a. $(6x)^3$ **b.** $(3y)^4$ **c.** $(2t^3)^5$ **d.** $(6s^2)^2$

6. a. $\left(\dfrac{w}{2}\right)^6$ **b.** $\left(\dfrac{5}{u}\right)^4$ **c.** $\left(\dfrac{-4}{p^5}\right)^3$ **d.** $\left(\dfrac{-3}{q^4}\right)^5$

7. a. $\left(\dfrac{h^2}{m^3}\right)^4$ **b.** $\left(\dfrac{n^3}{k^4}\right)^8$ **c.** $(-4a^2b^4)^4$ **d.** $(-5ab^8)^3$

8. a. $\dfrac{ab^2}{(ab)^2}$ **b.** $\dfrac{(x^2y)^2}{x^2y^2}$ **c.** $\dfrac{(2mp)^3}{2m^3p}$ **d.** $\dfrac{4^2rt^4}{2^4r^4t}$

Simplify if possible.

9. a. $w + w$ **b.** $w(w)$ **10. a.** $m^2 - m^2$ **b.** $m^2(-m^2)$

11. a. $4z^2 - 6z^2$ **b.** $4z^2(-6z^2)$ **12. a.** $t^3 + 3t^3$ **b.** $t^3(3t^3)$

13. a. $4p^2 + 3p^3$ **b.** $4p^2(3p^3)$ **14. a.** $2w^2 - 5w^4$ **b.** $(2w^2)(-5w^4)$

15. a. $3^9 \cdot 3^8$ **b.** $3^9 + 3^8$ **16. a.** $(-2)^7(-2)^5$ **b.** $-2^7 - 2^5$

Multiply.

17. a. $(4y)(-6y)$ **b.** $(-4z)(-8z)$ **18. a.** $(2wz^3)(-8z)$ **b.** $(4wz)(-9w^2z^2)$

19. a. $-4x(3xy)(xy^3)$ **b.** $-5x^2(2xy)(5x^2)$ **20. a.** $-7ab^2(-3ab^3)$ **b.** $-4a^2b(-3a^3b^2)$

Divide.

21. a. $\dfrac{2a^3b}{8a^4b^5}$ **b.** $\dfrac{8a^2b}{12a^5b^3}$ **22. a.** $\dfrac{-12qw^4}{8qw^2}$ **b.** $\dfrac{-12rz^6}{20rz}$

Multiply or divide.

23. a. $\dfrac{-15bc(b^2c)}{-3b^3c^4}$ **b.** $\dfrac{-25c(c^2d^2)}{-5c^8d^2}$ **24. a.** $-2x^3(x^2y)(-4y^2)$

     b. $3xy^3(-x^4)(-2y^2)$

Simplify by applying the laws of exponents.

25. a. $b^3(b^2)^5$ **b.** $b(b^4)^6$ **26. a.** $(p^2q)^3(pq^3)$ **b.** $(p^3)^4(p^3q^4)$

27. a. $(2x^3y)^2(xy^3)^4$ **b.** $(3xy^2)^3(2x^2y^2)^2$ **28. a.** $-a^2(-a)^2$ **b.** $-a^3(-a)^3$

Simplify by applying the laws of exponents.

29. a. $\left(\dfrac{-2x}{3y^2}\right)^3$ **b.** $\left(\dfrac{-x^2}{2y}\right)^4$ **30. a.** $\dfrac{(4x)^3}{(-2x^2)^2}$ **b.** $\dfrac{(5x)^2}{(-3x^2)^3}$

31. a. $\dfrac{(xy)^2(-x^2y)^3}{(x^2y^2)^2}$ **b.** $\dfrac{(-x)^2(-x^2)^4}{(x^2)^3}$ **32. a.** $\left(\dfrac{-2x}{y^2}\right)\left(\dfrac{y^2}{3x}\right)^2$ **b.** $\left(\dfrac{x^2z}{2}\right)\left(\dfrac{-2}{x^2z}\right)^3$

Evaluate each function for the given expressions, and simplify.

33. $f(x) = x^3$
     a. $f(a^2)$ **b.** $a^3 \cdot f(a^3)$ **c.** $f(ab)$ **d.** $f(a + b)$

34. $g(x) = x^4$
     a. $g(a^3)$ **b.** $a^4 \cdot g(a^4)$ **c.** $g(ab)$ **d.** $g(a + b)$

35. $F(x) = 3x^5$
     a. $F(2a)$ **b.** $2F(a)$ **c.** $F(a^2)$ **d.** $[F(a)]^2$

36. $G(x) = 4x^3$
     a. $G(3a)$ **b.** $3G(a)$ **c.** $G(a^4)$ **d.** $[G(a)]^4$

A.6   Products and Factoring

In Section A.5 we used the first law of exponents to multiply two or more monomials. In this section we review techniques for multiplying and factoring polynomials of several terms.

Polynomials

A **polynomial** is a sum of terms in which all the exponents on the variables are whole numbers and no variables appear in the denominator or under a radical. The expressions

$$0.1R^4, \qquad d^2 + 32d - 21, \qquad \text{and} \qquad 128x^3 - 960x^2 + 8000$$

are all examples of polynomials in one variable.

An algebraic expression consisting of one term of the form cx^n, where c is a constant and n is a whole number, is called a **monomial.** For example,

$$y^3, \qquad -3x^8, \qquad \text{and} \qquad 0.1R^4$$

are monomials. A polynomial is just a sum of one or more monomials.

A polynomial with exactly two terms, such as $\frac{1}{2}n^2 + \frac{1}{2}n$, is called a **binomial.** A polynomial with exactly three terms, such as

$$d^2 + 32d - 21 \text{ or } 128x^3 - 960x^2 + 8000$$

is called a **trinomial.** We have no special names for polynomials with more than three terms.

EXAMPLE 1

Which of the following expressions are polynomials?

a. πr^2

b. $23.4s^6 - 47.9s^4$

c. $\dfrac{2}{3}w^3 - \dfrac{7}{3}w^2 + \dfrac{1}{3}w$

d. $7 + m^{-2}$

e. $\dfrac{x-2}{x+2}$

f. $\sqrt[3]{4y}$

Solutions

The first three are all polynomials. In fact, (a) is a monomial, (b) is a binomial, and (c) is a trinomial. The last three are not polynomials. The variable in (d) has a negative exponent, the variable in (e) occurs in the denominator, and the variable in (f) occurs under a radical. ■

In a polynomial containing only one variable, the greatest exponent that appears on the variable is called the **degree** of the polynomial. If there is no variable at all, then the polynomial is called a **constant,** and the degree of a constant is zero.

EXAMPLE 2

Give the degree of each polynomial.

a. $b^3 - 3b^2 + 3b - 1$

b. 10^{10}

c. $-4w^3$

d. $s^2 - s^6$

Solutions

a. This is a polynomial in the variable b, and because the greatest exponent on b is 3, the degree of this polynomial is 3.

b. This is a constant polynomial, so its degree is 0. (The exponent on a *constant* does not affect the degree!)

c. This monomial has degree 3.

d. This is a binomial of degree 6. ■

We can evaluate a polynomial just as we evaluate any other algebraic expression: We replace the variable with a number and simplify the result.

EXAMPLE 3

Let $P = -2x^2 + 3x - 1$. Evaluate P in each of the following.

a. $x = 2$

b. $x = -1$

c. $x = t$

d. $x = t + 3$

Solutions

In each case we replace the x by the given value.

a. $P = -2(2)^2 + 3(2) - 1 = -8 + 6 - 1 = -3$

b. $P = -2(-1)^2 + 3(-1) - 1 = -2 + (-3) - 1 = -6$

c. $P = -2(t)^2 + 3(t) - 1 = -2t^2 + 3t - 1$

d. $P = -2(t + 3)^2 + 3(t + 3) - 1$
$$= -2(t^2 + 6t + 9) + 3(t + 3) - 1$$
$$= -2t^2 - 9t - 10 \qquad \blacksquare$$

Products of Polynomials

To multiply polynomials we use a generalized form of the distributive property.

$$a(b + c + d + \cdots) = ab + ac + ad + \cdots$$

To multiply a polynomial, by a monomial, we multiply each term of the polynomial by the monomial.

EXAMPLE 4

a. $3x(x + y + z) = 3x(x) + 3x(y) + 3x(z)$
$$= 3x^2 + 3xy + 3xz$$

b. $-2ab^2(3a^2 - ab + 2b^2) = -2ab^2(3a^2) - 2ab^2(-ab) - 2ab^2(2b^2)$
$$= -6a^3b^2 + 2a^2b^3 - 4ab^4 \qquad \blacksquare$$

Products of Binomials

Products of binomials occur so frequently that it is worthwhile to learn a shortcut for this type of multiplication. We can use the following scheme to perform the multiplication mentally. (See Figure A.39.)

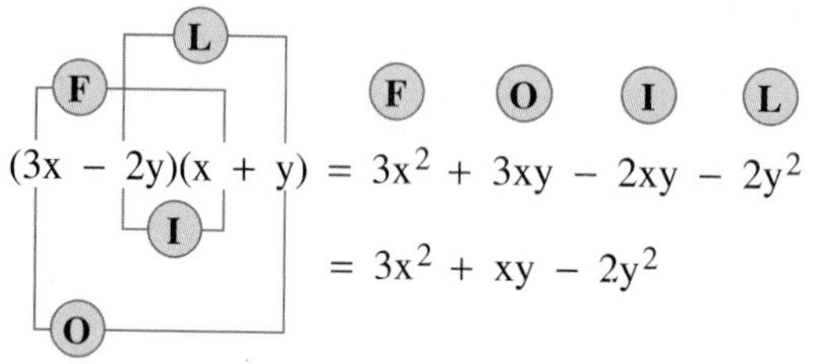

$$(3x - 2y)(x + y) = 3x^2 + 3xy - 2xy - 2y^2$$
$$= 3x^2 + xy - 2y^2$$

FIGURE A.39

This process is sometimes called the **FOIL** method, where FOIL represents

the product of the **F**irst terms
the product of the **O**uter terms
the product of the **I**nner terms
the product of the **L**ast terms

EXAMPLE 5

$$(2x - 1)(x + 3) = 2x^2 + 6x - x - 3$$
$$= 2x^2 + 5x - 3 \qquad \blacksquare$$

Factoring

We sometimes find it useful to write a polynomial as a single *term* composed of two or more *factors*. This process is the reverse of multiplication and is called **factoring**. For example, observe that

$$3x^2 + 6x = 3x(x + 2)$$

We will consider only factorization in which the factors have integer coefficients.

Common Factors

We can factor a common factor from a polynomial by using the distributive property in the form

$$ab + ac = a(b + c)$$

We first identify the common factor. For example, each term of the polynomial

$$6x^3 + 9x^2 - 3x$$

contains the monomial $3x$ as a factor, therefore

$$6x^3 + 9x^2 - 3x = 3x \,(\underline{\qquad})$$

Next we insert the proper polynomial factor within the parentheses. This factor can be determined by inspection. We ask ourselves for monomials that, when multiplied by $3x$, yield $6x^3$, $9x^2$, and $-3x$, respectively, and obtain

$$6x^3 + 9x^2 - 3x = 3x(2x^2 + 3x - 1)$$

We can check the result of factoring an expression by multiplying the factors. In the example above,

$$3x(2x^2 + 3x - 1) = 6x^3 + 9x^2 - 3x$$

EXAMPLE 6

a. $18x^2y - 24xy^2 = 6xy(? - ?)$
$$= 6xy(3x - 4y)$$

because

$$6xy(3x - 4y) = 18x^2y - 24xy^2$$

b. $y(x - 2) + z(x - 2) = (x - 2)(? - ?)$
$$= (x - 2)(y + z)$$

because

$$(x - 2)(y + z) = y(x - 2) + z(x - 2) \qquad \blacksquare$$

Opposite of a Binomial

It is often useful to factor -1 from the terms of a binomial.

$$a - b = (-1)(-a + b)$$
$$= (-1)(b - a)$$
$$= -(b - a)$$

Hence we have the important relationship

$$a - b = -(b - a)$$

that is, $a - b$ and $b - a$ are opposites or negatives of each other.

EXAMPLE 7

a. $3x - y = -(y - 3x)$

b. $a - 2b = -(2b - a)$ ∎

Factoring Quadratic Trinomials

Consider the trinomial

$$x^2 + 6x + 16 \tag{1}$$

Can we find two binomial factors,

$$(x + a)(x + b) \tag{2}$$

whose product is the given trinomial? The product of the binomials is

$$(x + a)(x + b) = x^2 + (a + b)x + ab$$

We really need to find only the two numbers, a and b. By comparing Equations (1) and (2), we see that $a + b = 6$ and $ab = 16$. That is, the sum of the two numbers is the coefficient of the linear term, 6, and their product is the constant term, 16. To find the numbers we list all the possible integer factorizations of 16, namely $1 \cdot 16$, $2 \cdot 8$, and $4 \cdot 4$, and we see that only one combination gives the correct linear term: 8 and 2. These are the numbers a and b, so

$$x^2 + 6x - 16 = (x + 8)(x + 2)$$

In Example 8 we factor quadratic trinomials in which one or more of the coefficients is negative.

EXAMPLE 8

Factor

a. $x^2 - 7x + 12$ **b.** $x^2 - x - 12$

Solutions

a. Find two numbers whose product is 12 and whose sum is -7. Because the product is positive and the sum is negative, the two numbers must both be negative. The possible factors of 12 are -1 and -12, -2 and -6, or -3 and -4. Only -4 and -3 have the correct sum, -7. Hence

$$x^2 - 7x + 12 = (x - 4)(x - 3)$$

b. Find two numbers whose product is -12 and whose sum is -1. Because the product is negative, the two numbers must be of opposite sign and their sum must be -1. By listing the possible factors of -12, we find that the two numbers are -4 and 3. Hence

$$x^2 - x - 12 = (x - 4)(x + 3)$$ ■

If the coefficient of the quadratic term is not 1, we must also consider its factors.

EXAMPLE 9

Factor $8x^2 - 9 - 21x$.

Solution

Step 1 Write the trinomial in decreasing powers of x

$$8x^2 - 21x - 9$$

$(8x \quad 9)(x \quad 1)$

$(8x \quad 1)(x \quad 9)$

$(8x \quad 3)(x \quad 3)$

$(4x \quad 9)(2x \quad 1)$

$(4x \quad 1)(2x \quad 9)$

$(4x \quad 3)(2x \quad 3)$

FIGURE A.40

Step 2 List the possible factors for the quadratic term.

$$(8x \qquad)(x \qquad)$$
$$(4x \qquad)(2x \qquad)$$

Step 3 Consider possible factors for the constant term: 9 may be factored as $9 \cdot 1$ or as $3 \cdot 3$. Form all possible pairs of binomial factor using these factorizations. (See Figure A.40.)

Step 4 Select the combinations of the products ① and ② whose sum or difference could be the linear term, $-21x$.

$$(8x \quad 3)(x \quad 3)$$

Step 5 Insert the proper signs:

$$(8x + 3)(x - 3)$$ ■

With practice, you can usually factor trinomials of the form $Ax^2 + Bx + C$ mentally. The following observations may help.

1. If A, B, and C are all positive, both signs in the factored form are positive. For example, as a first step in factoring $6x^2 + 11x + 4$ we could write

$$(\underline{\quad} + \underline{\quad})(\underline{\quad} + \underline{\quad})$$

2. If A and C are positive and B is negative, both signs in the factored form are negative. Thus as a first step in factoring $6x^2 - 11x + 4$ we could write

$$(\underline{\quad} - \underline{\quad})(\underline{\quad} - \underline{\quad})$$

3. If C is negative, the signs in the factored form are opposite. Thus as a first step in factoring $6x^2 - 5x - 4$ we could write

$$(\underline{\quad} + \underline{\quad})(\underline{\quad} - \underline{\quad}) \quad \text{or} \quad (\underline{\quad} - \underline{\quad})(\underline{\quad} + \underline{\quad})$$

EXAMPLE 10

a. $6x^2 + 5x + 1 = (\underline{\quad} + \underline{\quad})(\underline{\quad} + \underline{\quad})$
$\qquad\qquad = (3x + 1)(2x + 1)$

b. $6x^2 - 5x + 1 = (\underline{\quad} - \underline{\quad})(\underline{\quad} - \underline{\quad})$
$\qquad\qquad = (3x - 1)(2x - 1)$

c. $6x^2 - x - 1 = (\underline{\quad} + \underline{\quad})(\underline{\quad} - \underline{\quad})$
$\qquad\qquad = (3x + 1)(2x - 1)$

d. $6x^2 - xy - y^2 = (\underline{\quad} + \underline{\quad})(\underline{\quad} - \underline{\quad})$
$\qquad\qquad = (3x + y)(2x - y)$ ∎

Special Products and Factors

The products below are special cases of the multiplication of binomials. They occur so often that you should learn to recognize them on sight.

I. $(a + b)^2 = (a + b)(a + b) = a^2 + 2ab + b^2$
II. $(a - b)^2 = (a - b)(a - b) = a^2 - 2ab + b^2$
III. $(a + b)(a - b) = a^2 - b^2$

Notice that in (I) $(a + b)^2 \neq a^2 + b^2$, and that in (II) $(a - b)^2 \neq a^2 - b^2$. ∎

EXAMPLE 11

a. $3(x + 4)^2 = 3(x^2 + 2 \cdot 4x + 4^2)$
$\qquad\qquad = 3x^2 + 24x + 48$

b. $(y + 5)(y - 5) = y^2 - 5^2$
$\qquad\qquad = y^2 - 25$

c. $(3x - 2y)^2 = (3x)^2 - 2(3x)(2y) + (2y)^2$
$\qquad\qquad = 9x^2 - 12xy + 4y^2$ ∎

Each of the special products above, when viewed from right to left, also represents a special case of factoring quadratic polynomials.

I. $a^2 + 2ab + b^2 = (a + b)^2$
II. $a^2 - 2ab + b^2 = (a - b)^2$
III. $a^2 - b^2 = (a + b)(a - b)$

The trinomials in (I) and (II) are sometimes called **perfect-square trinomials** because they are squares of binomials. The expression $a^2 + b^2$ cannot be factored.

EXAMPLE 12

Factor

a. $x^2 + 8x + 16$ **c.** $4a^2 - 12ab + 9b^2$

b. $y^2 - 10y + 25$ **d.** $25m^2n^2 + 20mn + 4$

Solutions

a. Because 16 is equal to 4^2 and 8 is equal to $2 \cdot 4$,

$$x^2 + 8x + 16 = x^2 - 2 \cdot 4x + 4^2$$
$$= (x + 4)^2$$

b. Because $25 = 5^2$ and $10 = 2 \cdot 5$,

$$y^2 - 10y + 25 = y^2 - 2 \cdot 5y + 5^2$$
$$= (y - 5)^2$$

c. Because $4a^2 = (2a)^2$, $9b^2 = (3b)^2$, and $12ab = 2(2a)(3b)$,

$$4a^2 - 12ab + 9b^2 = (2a)^2 - 2(2a)(3b) + (3b)^2$$
$$= (2a - 3b)^2$$

d. Because $25m^2n^2 = (5mn)^2$, $4 = 2^2$, and $20mn = 2(5mn)(2)$,

$$25m^2n^2 + 20mn + 4 = (5mn)^2 + 2(5mn)(2) + 2^2$$
$$= (5mn + 2)^2 \qquad ■$$

Binomials of the form $a^2 - b^2$ are often called the **difference of two squares.**

EXAMPLE 13

Factor if possible

a. $x^2 - 81$ **b.** $4x^2 - 9y^2$ **c.** $x^2 + 81$

Solutions

a. The expression $x^2 - 81$ is the difference of two squares, $x^2 - 9^2$, and thus can be factored according to (III) above.

$$x^2 - 81 = (x + 9)(x - 9) = x^2 - 9^2$$

b. Because $4x^2 - 9y^2$ can be written as $(2x)^2 - (3y)^2$,

$$4x^2 - 9y^2 = (2x)^2 - (3y)^2$$
$$= (2x + 3y)(2x - 3y)$$

c. The expression $x^2 + 81$, or $x^2 + 0x + 81$, is *not* factorable, because no two real numbers have a product of 81 and a sum of 0.   ■

The factors $x + 9$ and $x - 9$ in Example 13(a) are called **conjugates** of each other. In general, any binomials of the form $a + b$ and $a - b$ are called a **conjugate pair.**

COMMON ERROR

$x^2 + 81 \neq (x + 9)(x + 9)$, which you can verify by multiplying.

$$(x + 9)(x + 9) = x^2 + 18x + 81 \qquad ■$$

Identify each polynomial as a monomial, a binomial, or a trinomial. Give the degree of each polynomial.

1. $2x^3 - x^2$ **2.** $x^2 - 2x + 1$ **3.** $5n^4$ **4.** $3n + 1$

5. $3r^2 - 4r + 2$ **6.** r^3 **7.** $y^3 - 2y^2 - y$ **8.** $3y^2 + 1$

Which of the following are *not* polynomials?

9. a. $1 - 0.04t^2$ **b.** $3x^2 - 4x + \dfrac{2}{x}$ **c.** $2\sqrt{z} - 7z^3 + 2$ **d.** $\sqrt{2}w^3 + \dfrac{3}{4}w^2 - w$

10. a. $\sqrt{3}p^2 - 7p + 2$ **b.** $2h^{4/3} + 6h^{1/3} - 2$ **c.** $\dfrac{2}{x^2 - 6x + 5}$ **d.** $\dfrac{1}{4}y^{-2} + 3y^{-1} + 4$

11. a. $\dfrac{1}{m^2 + 3}$ **b.** $v^2 - 16 + 2^v$ **c.** $\sqrt{x^3 - 4x}$ **d.** $\dfrac{m^4}{12}$

12. a. $3^t - 5t^3 + 2$ **b.** $\dfrac{q + 3}{q - 1}$ **c.** $c^{1/2} - c$ **d.** $\sqrt[3]{d + 1}$

Evaluate each polynomial function for the given values of the variable.

13. $P = x^3 - 3x^2 + x + 1$

     a. $x = 2$ **b.** $x = -2$ **c.** $x = 2b$

14. $P = 2x^3 + x^2 - 3x + 4$

     a. $x = 3$ **b.** $x = -3$ **c.** $x = -a$

15. $Q = t^2 + 3t + 1$

     a. $t = \dfrac{1}{2}$ **b.** $t = -\dfrac{1}{3}$ **c.** $t = -w$

16. $Q = 2t^2 - t + 1$

     a. $t = \dfrac{1}{4}$ **b.** $t = -\dfrac{1}{2}$ **c.** $t = 3v$

17. $R = 3z^4 - 2z^2 + 3$

     a. $z = 1.8$ **b.** $z = -2.6$ **c.** $z = k - 1$

18. $R = z^4 + 4z - 2$

     a. $z = 2.1$ **b.** $z = -3.1$ **c.** $z = h + 2$

19. $N = a^6 - a^5$

     a. $a = -1$ **b.** $a = -2$ **c.** $a = \dfrac{m}{3}$

20. $N = a^5 - a^4$

     a. $a = -1$ **b.** $a = -2$ **c.** $a = \dfrac{q}{2}$

Write each product as a polynomial and simplify.

21. $4y(x - 2y)$ **22.** $3x(2x + y)$ **23.** $-6x(2x^2 - x + 1)$ **24.** $-2y(y^2 - 3y + 2)$

25. $a^2b(3a^2 - 2ab - b)$ **26.** $ab^3(-a^2b^2 + 4ab - 3)$

27. $2x^2y^3(4xy^4 - 2xy - 3x^3y^2)$ **28.** $5x^2y^2(3x^4y^2 + 3x^2y - xy^6)$

29. $(n + 2)(n + 8)$ **30.** $(r - 1)(r - 6)$ **31.** $(r + 5)(r - 2)$ **32.** $(z - 3)(z + 5)$

33. $(2z + 1)(z - 3)$ **34.** $(3t - 1)(2t + 1)$

35. $(4r + 3s)(2r - s)$ **36.** $(2z - w)(3z + 5w)$

37. $(2x - 3y)(3x - 2y)$ **38.** $(3a + 5b)(3a + 4b)$

39. $(3t - 4s)(3t + 4s)$ **40.** $(2x - 3z)(2x + 3z)$

41. $(2a^2 + b^2)(a^2 - 3b^2)$ **42.** $(s^2 - 5t^2)(3s^2 + 2t^2)$

Factor completely. Check your answers by multiplying factors.

43. $4x^2z + 8xz$ **44.** $3x^2y + 6xy$ **45.** $3n^4 - 6n^2 + 12n$

46. $2x^4 - 4x^2 + 8x$ **47.** $15r^2s + 18rs^2 - 3r$ **48.** $2x^2y^2 - 3xy + 5x^2$

49. $3m^2n^4 - 6m^3n^3 + 14m^3n^2$

51. $15a^4b^3c^4 - 12a^2b^2c^5 + 6a^2b^3c^4$

53. $a(a + 3) + b(a + 3)$

55. $y(y - 2) - 3x(y - 2)$

57. $4(x - 2)^2 - 8x(x - 2)^3$

59. $x(x - 5)^2 - x^2(x - 5)^3$

50. $6x^3y - 6xy^3 + 12x^2y^2$

52. $14xy^4z^3 + 21x^2y^3z^2 - 28x^3y^2z^5$

54. $b(a - 2) + a(a - 2)$

56. $2x(x + 3) - y(x + 3)$

58. $6(x + 1) - 3x(x + 1)^2$

60. $x^2(x + 3)^3 - x(x + 3)^2$

Supply the missing factors or terms.

61. $3m - 2n = -(\quad ?\quad)$

63. $-2x + 2 = -2(\quad ?\quad)$

65. $-ab - ac = ?(b + c)$

67. $2x - y + 3z = -(\quad ?\quad)$

62. $2a - b = -(\quad ?\quad)$

64. $-6x - 9 = -3(\quad ?\quad)$

66. $-a^2 + ab = ?(a - b)$

68. $3x + 3y - 2z = -(\quad ?\quad)$

Factor completely.

69. $x^2 + 5x + 6$

72. $y^2 - 7y + 10$

75. $2x^2 + 3x - 2$

78. $1 - 5x + 6x^2$

81. $10u^2 - 3 - u$

84. $24x^2 - 29x + 5$

87. $12 - 53x + 30x^2$

90. $48t^2 - 122t + 39$

93. $15x^2 - 4xy - 4y^2$

96. $24u^2 - 20v^2 + 17uv$

99. $10a^2b^2 - 19ab + 6$

102. $54x^2y^2 + 3xy - 2$

70. $x^2 + 5x + 4$

73. $x^2 - 6 - x$

76. $3x^2 - 7x + 2$

79. $9y^2 - 21y - 8$

82. $8u^2 - 3 + 5u$

85. $5a + 72a^2 - 12$

88. $39x + 80x^2 - 20$

91. $3x^2 - 7ax + 2a^2$

94. $12x^2 + 7xy - 12y^2$

97. $12a^2 - 14b^2 - 13ab$

100. $12a^2b^2 - ab - 20$

103. $22a^2z^2 - 21 - 19az$

71. $y^2 - 7y + 12$

74. $x^2 - 15 - 2x$

77. $7x + 4x^2 - 2$

80. $10y^2 - 3y - 18$

83. $21x^2 - 43x - 14$

86. $-30a + 72a^2 - 25$

89. $-30t - 44 + 54t^2$

92. $9x^2 + 9ax - 10a^2$

95. $18u^2 + 20v^2 - 39uv$

98. $24a^2 - 15b^2 - 2ab$

101. $56x^2y^2 - 2xy - 4$

104. $26a^2z^2 - 24 + 23az$

Write each expression as a polynomial.

105. $(x + 3)^2$

108. $(3x + 2)^2$

111. $(3t - 4s)(3t + 4s)$

114. $(4u + 5v)(4u + 5v)$

106. $(y - 4)^2$

109. $(x + 3)(x - 3)$

112. $(2x + a)(2x - a)$

115. $(8xz + 3)(8xz + 3)$

107. $(2y - 5)^2$

110. $(x - 7)(x + 7)$

113. $(5a - 2b)(5a - 2b)$

116. $(7yz - 2)(7yz - 2)$

Factor completely.

117. $x^2 - 25$

121. $x^2 - 4y^2$

125. $9u^2 - 30uv + 25v^2$

127. $4a^2 - 25b^2$

131. $9x^2y^2 + 6xy + 1$

135. $(x + 2)^2 - y^2$

118. $x^2 - 36$

122. $9x^2 - y^2$

128. $16a^2 - 9b^2$

132. $4x^2y^2 + 12xy + 9$

119. $x^2 - 24x + 144$

123. $4x^2 + 12x + 9$

126. $16s^2 - 56st + 49t^2$

129. $x^2y^2 - 81$

133. $16x^2y^2 - 1$

136. $x^2 - (y - 3)^2$

120. $x^2 + 26x + 169$

124. $4y^2 + 4y + 1$

130. $x^2y^2 - 64$

134. $64x^2y^2 - 1$

A.7 Summarizing Data

The purpose of statistics is to make sense of information stored in numbers. Numerical information is called *data,* and the practice of statistics includes gathering data, presenting data, and drawing conclusions based on data. Statistics affects you in more ways than you probably realize. It determines the cost of your insurance, the television programs that are broadcast, the quality of your athletic shoes, the type of medical treatment you receive, and many other aspects of your daily life.

Suppose that your instructor has handed back your first exam. You may be curious about how your score compares with those of your classmates. It would be too tedious to compare your score with each of the other scores. It would be easier if you could look at just one or two meaningful numbers. In this section we consider ways to give a lot of information about a set of data with just a few numbers.

Measures of Center: Mean, Median, Mode

You might be satisfied to know if your test score is better or worse than the average score in your class. But exactly what do we mean by "average"? In this situation, the average is a single number that summarizes the class scores. The mean, median, and mode are three different types of "average" used in statistics.

Consider the following test scores from an intermediate algebra class of 15 students.

$$72, 77, 80, 82, 84, 84, 86, 88, 90, 90, 90, 97, 98, 99, 100$$

The **mean** of the test scores is the value obtained by adding all the scores together and dividing by the number of students. For this set of data, the mean is

$$\frac{72 + 77 + 80 + 82 + 84 + 84 + 86 + 88 + 90 + 90 + 90 + 97 + 98 + 99 + 100}{15} = 87.8$$

The mean is what most people think of as the "average."

The **median** of the scores is the middle score when they are listed in increasing order. For our data, the median is 88. Half the scores are at or below the median, and half the scores are at or above the median. There is always a middle score if the number of scores is odd. If the number of scores is even, the median is the value halfway between the two middle scores.

The **mode** of the scores is the score that occurs most frequently. For our data set, the mode is 90. When there is no score that occurs more often than the rest, we say that there is no mode.

Each of these three values gives a measure of the middle, or of what is typical, in a set of numbers.

EXAMPLE 1   There are seven houses for sale in a particular neighborhood. Their listed prices are shown at right. What are the mean, median, and mode of the prices?

House	Price in dollars
A	190,000
B	100,000
C	110,000
D	2,500,000
E	100,000
F	170,000
G	120,000

Solution

We find the mean by adding the prices and dividing by the number of houses. Thus the mean is

$$\frac{190{,}000 + 100{,}000 + 110{,}000 + 2{,}500{,}000 + 100{,}000 + 170{,}000 + 120{,}000}{7} = 470{,}000$$

The mean price is $470,000.

The median is the middle price when they are listed in order. From lowest to highest, the prices (in thousands) are 100, 100, 110, 120, 170, 190, and 2500. The middle score is 120, so the median price is $120,000.

The mode is the most frequent score. Here the mode is $100,000. ■

Notice that the three types of "average" are all different values in this example. If you were interested in buying a house in this neighborhood, the mean price might not be particularly helpful to you, because the mean is much more than any but the single most expensive house. This example shows a weakness of using the mean as a measurement of the "typical" value: It can be greatly affected by one single large (or small) value. In Example 1, you should check that if the price of house D were to change to $400,000, neither the median nor the mode would change, but the mean price would drop to $170,000.

The mode in Example 1 is not a particularly good indication of the typical house price either, because it corresponds to the lowest price. But if you were a building contractor and needed to know what price for a house would be most popular, you might find the mode more useful than the mean or median.

EXAMPLE 2

Find the mean, the median, and the mode for each set of data.

a. 6, 3, 5, 5, 7, 4

b. 9.5, 0, 5, 0.5, 10, 5

c. 50, −20, 30, 50, 60, 10

d. 35, 45, 50, 0, 0, 50

e. 1, 2, 3, 4

f. 2, 2, 3, 4, 4, 5

Solutions

a. $\frac{6 + 3 + 5 + 5 + 7 + 4}{6} = 5$, so 5 is the mean. The middle scores of 3, 4, 5, 5, 6, 7 are both 5, so 5 is the median.
The mode is 5, which is the only score that shows up more than once.

b. $\frac{9.5 + 0 + 5 + 0.5 + 10 + 5}{6} = 5$, so 5 is the mean.
The middle scores of 0, 0.5, 5, 5, 9.5, 10 are both 5, so 5 is the median.
The mode is 5, the only score that shows up more than once.

c. $\frac{50 - 20 + 30 + 50 + 60 + 10}{6} = 30$; the mean is 30.
The middle scores of −20, 10, 30, 50, 50, 60 are 30 and 50, so the median is $\frac{30 + 50}{2} = 40$.
The mode is 50.

d. $\frac{35 + 45 + 50 + 0 + 0 + 50}{6} = 30$; the mean is 30.
The middle scores of 0, 0, 35, 45, 50, 50 and 35 and 45, so the median is $\frac{35 + 45}{2} = 40$.
The mode is 50.

e. $\frac{1+2+3+4}{4} = 2.5$; the mean is 2.5.

The middle values are 2 and 3, so the median is $\frac{2+3}{2} = 2.5$.
Because no value occurs more often than the rest, there is no mode.

f. $\frac{2+2+3+4+4+5}{6} = \frac{10}{3}$; the mean is $\frac{10}{3}$ or approximately 3.33.

The middle values are 3 and 4, so the median is $\frac{3+4}{2} = 3.5$.
No value occurs more often than the rest, so there is no mode. (Some statisticians, however, call this a "bimodal" data set because two values, 2 and 4, occur more often than the others.) ∎

The mean, median, and mode of a data set can all be equal, as in Example 2a and 2b. Notice that the data sets in Examples 2c and 2d are different, but they have the same mean (30), median (40), and mode (50). These examples show us that the summary numbers alone do not describe the original data completely.

Measures of Spread: Extremes, Range, Quartiles, and the Five-Number Summary

We often want information about how the data are spread out, or distributed. The most obvious measure of spread is the **range** of values, the difference between the smallest and greatest values. We call the smallest and greatest values the **extremes,** so that the range is the difference between the extremes.

If you know both the median and the extremes, then you know the middle, low, and high scores. Between the median and the low score is the **first quartile** (or **lower quartile**), and between the median and high score is the **third quartile** (or **upper quartile**). One-fourth of all the scores are below the lower quartile, and three-fourths are below the upper quartile.

The combination of the extremes, the median, and the first and third quartile give a lot of information about the spread of the scores. They are called the **five-number summary.**

EXAMPLE 3 Find the five-number summary for the following exam scores in a history class.

0, 10, 20, 30, 38, 42, 48, 54, 60, 68, 72, 76, 80, 84, 88, 92, 94, 96, 98, 100

Solution

The twenty scores are already in increasing order. The lowest and highest scores are 0 and 100, respectively. The median is the average of the two middle scores, 68 and 72, so the median is 70. The first quartile can be found by determining the median of the first 10 scores. The 2 middle scores are 38 and 42, so the first quartile is 40. The third quartile is the median of the last half of all the scores, so it is the average of 88 and 92, namely 90. Thus the 5-number summary is 0, 40, 70, 90, 100. ∎

Notice that there is a spread of 40 points between the lowest score and the first quartile but a spread of only 10 points between the third quartile and the highest score. This reflects the fact that there are relatively few students with low scores and many students with high scores.

EXAMPLE 4

A sporting goods store conducts a study about the number of hours its customers watch television each week. The five-number summary for the average number of hours spent watching television per week is 0, 1.5, 7, 9.5, 78.

    a. What are the extremes of weekly viewing hours?

    b. What fraction of the customers in the study watch more than 9.5 hours?

    c. What fraction of the customers watch from 1.5 to 9.5 hours per week?

Solutions

    a. The viewing hours ranged from a low of 0 to a high of 78 hours per week.

    b. Because 9.5 is the upper quartile, three-fourths of the customers watch less than 9.5 hours, hence one-fourth of the customers watch more than 9.5 hours each week.

    c. Because one-quarter watch less than 1.5 hours per week and one-quarter watch more than 9.5 hours per week, half of the customers watch from 1.5 to 9.5 hours per week. ■

Using the Calculator

The graphing calculator can compute most of the summarizing numbers we have discussed.

EXAMPLE 5

Use a graphing calculator to find the mean, the median, and the five-number summary for the following bowling scores.

$$99, 158, 221, 182, 169, 155, 157, 162, 201,$$
$$125, 141, 170, 148, 192, 197, 170, 106, 127$$

Solution

Clear the home screen (press **CLEAR**) before entering the data. Press **STAT** **ENTER** to begin entering data. If there are some unwanted data already in the table, use the arrow keys to highlight the column heading and press **CLEAR** **ENTER** to clear out the values. Put the bowling scores into the first column, using either **ENTER** or ▽ to move to the next row. If you enter a value incorrectly, you can use the arrow keys △ or ▽ to highlight the incorrect value and then key in the correct value. (See Figure A.41.)

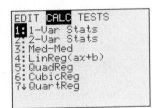

FIGURE A.41   **FIGURE A.42**

Access the statistics calculation submenu by pressing [STAT] [▷]. Use key: [1] (see Figure A.42). Then compute the summary values by pressing [STAT] [▷] [1] [ENTER].

The TI-83 denotes the mean by $\bar{x}$, shown on the second line on the display (see Figure A.43). For our data the mean is 160. The third line shows the sum of all the bowling scores. (The Greek letter Σ is used to indicate a summation, as discussed in Chapter 9.) The last line shows that there were 18 bowling scores. We will ignore the other lines for now.

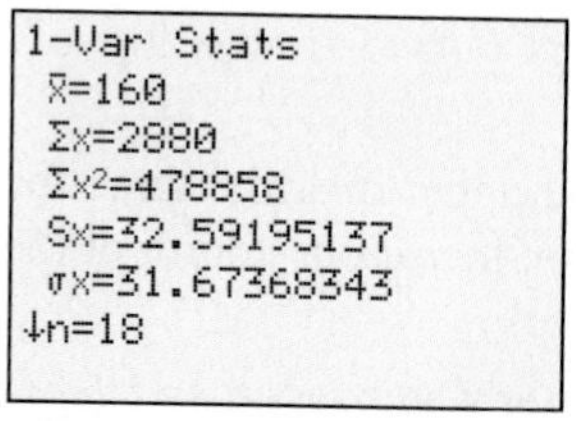

```
1-Var Stats
 x̄=160
 Σx=2880
 Σx²=478858
 Sx=32.59195137
 σx=31.67368343
↓n=18
```

FIGURE A.43

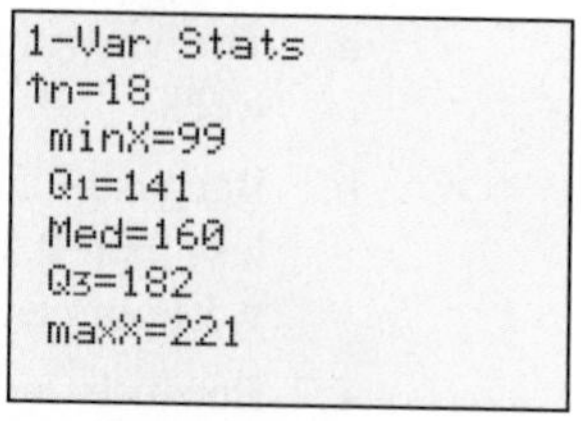

```
1-Var Stats
↑n=18
 minX=99
 Q₁=141
 Med=160
 Q₃=182
 maxX=221
```

FIGURE A.44

Press the down arrow for more summary values. (See Figure A.44.) We see that median is 160, and in fact the five-number summary is 99, 141, 160, 182, 221.

∎

You may have noticed that the command we used to compute our summary values was **"1-Var Stats."** Any number used to summarize information about a set of measurements is called a **statistic.** The mean, median, mode, range, extreme values, and quartiles are all examples of statistics. These are *one-variable* statistics because there is only one measurement or score per person. (The coefficients of the least-squares regression line discussed in Section 1.5 are an example of two-variable statistics.)

Statisticians often receive data that are already partially organized. For example, instead of showing 12 separate listings of a score of 50, the data might simply indicate that the score 50 occurs 12 times. In other words, the score 50 has a **frequency** of 12.

EXAMPLE 6

An advertising agency conducted a study to discover how many magazine subscriptions a typical college student maintains. From a survey of 337 students, 17 had no magazine subscriptions, 89 had 1 subscription, 98 had 2 subscriptions, 71 had 3, 47 had 4, 12 had 5, 1 had 6, 1 had 8, and 1 had 12. Find the mean, mode, and the five-number summary for the number of subscriptions maintained by students in the survey.

Solution

We will not enter the 337 scores into the calculator individually. Instead, we enter the values 0, 1, 2, 3, 4, 5, 6, 8, and 12 just once each, and then enter the frequencies, which tell how often to count each of the scores.

First enter the nine different scores in the L_1 column. In the L_2 column enter the corresponding frequencies as in Figure A.45. Then press [STAT] [▷] [1] to select **1-Var Stats**, and press [2nd] [1] [,] [2nd] [2], as shown in Figure A.46, for values in L_1 and frequencies in L_2. Finally press [ENTER] to get the summary numbers (see Figures A.47 and A.48). The mean is approximately 2.3, and the five-number summary is 0, 1, 2, 3, 12. The mode is 2 because it is the number of subscriptions with the highest frequency (98).

```
L1      L2      L3
0       17      ------
1       89
2       98
3       71
4       47
5       12
6        1
L2(7)= 1
```

FIGURE A.45

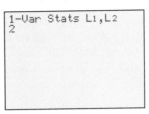

FIGURE A.46

```
1-Var Stats
 x̄=2.290801187
 Σx=772
 Σx²=2416
 Sx=1.388195861
 σx=1.386134693
 ↓n=337
■
```

FIGURE A.47

```
1-Var Stats
 ↑n=337
 minX=0
 Q1=1
 Med=2
 Q3=3
 maxX=12
```

FIGURE A.48

EXERCISE A.7

For Exercises 1–4, find the mean, median, and mode.

1. 7, 2, 3, 4, 6, 4, 5, 1, 4

2. 8, 23, 29, 15, 42, 10, 23

3. 110, 93, 27, 44, 87, 77, 66

4. 7.5, 2.1, 9.6, 5.8, 6.4, 7.2, 8.3

For Exercises 5–10, answer the questions about each data set.

5. There are only seven acting roles in a new movie. The salaries of the seven players are shown in the table at right.

Player	Salary in 1000 dollars
A	1200
B	100
C	20
D	10
E	7
F	7
G	7

   a. Find the mean, median, and mode of the salaries.

   b. The producer thinks the mean salary is too high. Change the salary of Player A (and no one else's) so that the mean salary becomes $100,000. (*Hint:* Let x be Player A's new salary. Set up and solve an appropriate equation.)

6. Sims Township decides to hold a raffle. First prize will be $100,000. There will also be one prize each of $1000, $500, and $100, and 21 consolation prizes of $20.

   a. Find the mean, median, and mode of the prizes.

   b. The county decides that it has sufficient funds to raise the mean prize to $6000. They decide to put all the extra money into the first prize. How large will the new first prize be? (*Hint:* Let x be the number of dollars in the new first prize. Set up and solve an appropriate equation.)

7. Find 7 numbers so that the mean is 0, the median is 10, and the mode is 20. Assume that two of the numbers are both 5. (*Hint:* Begin by considering how many scores must be greater than the median, then determine the frequency of the score 20. Set up an equation and solve it to find the remaining unknown number.)

8. Find 7 scores so that the mean is 15, the median is 20, and the mode is 0. Assume that

2 of the scores are both 25. (*Hint:* Begin by considering how many scores must be greater than the median, then determine the frequency of the score 0. Set up an equation and solve it to find the remaining unknown number.)

9. Find a set of data so that the median is less than the mean, which in turn in less than the mode.

10. Find a set of data so that the median is less than the mode, which in turn in less than the mean.

For Exercises 11–14, determine the five-number summary for each set of data from Exercises 1–4.

For Exercises 15 and 16, determine the mean, mode, and the five-number summary for each set of data.

15. On a 10-point quiz, 1 student received a 0, 1 received a 2, 3 received a 4, 3 received a 5, 8 received a 6, 14 received a 7, 12 received an 8, 6 received a 9, and 2 had scores of 10.

16. A restaurant asks its patrons to rate its service on a score of 1 to 5. Last year they received 97 scores of 1, 95 scores of 2, 53 scores of 3, 38 scores of 4, and 47 scores of 5.

For Exercises 17–20, answer the questions about each data set.

17. The five-number summary for your class's first exam is 28, 64, 73, 81, 98.

   a. What was the range of scores?

   b. What was the upper quartile in the class?

   c. What percent of the class scored at or below 73?

   d. What percent of the class scored at or above 81?

18. The five-number summary for used car prices advertised in a local paper is 1, 11, 50, 60, 300. The summary is in hundreds of dollars.

   a. What was the range of prices?

   b. What was the lower quartile of used car prices in the paper?

   c. What percent of the cars were priced at or above $6000?

   d. What percent of the cars were priced at or above $5000?

19. The five-number summary for another section's first exam is 3, 74, 82, 91, 94. Compare this with the five-number summary for your class in Exercise 17.

   a. Which class has the wider range of scores?

   b. Which class has the larger fraction of students with scores below 82? Explain.

   c. Which class has the greater fraction of students with scores above 90? Explain.

   d. Which class do you believe did better overall? Explain.

20. The five-number summary for used car prices on a used car lot is 20, 30, 60, 90, 150. The summary is in hundreds of dollars. Compare this with the five-number summary of used car prices in Exercise 18.

   a. Which has the wider range of prices?

   b. Which has the larger fraction of cars selling below $6000? Explain.

   c. Which has the larger fraction of cars below $15,000? Explain.

   d. Which has the larger fraction of cars between $5000 and $6000? Explain.

21. Which of the three statistics—mean, median, and mode—must always be one of the values in the data set? Give examples in which the other two statistics are not in the data set.

22. Which of the values in the five-number summary must always be in the data set? Give examples in which the other statistics are not in the data set.

B

The Number System

B.1 Real Numbers

Subsets of the Real Numbers

The numbers associated with points on a number line are called the **real numbers.** The set of real numbers is denoted by $\mathbb{R}$. You are already familiar with several types, or subsets, of real numbers.

- The set $\mathbb{N}$ of **natural,** or **counting numbers,** as its name suggests, consists of the numbers 1, 2, 3, 4, . . . , where " . . . " indicates that the list continues without end.

- The set $\mathbb{W}$ of **whole numbers** consists of the natural numbers and zero: 0, 1, 2, 3, ...

- The set $\mathbb{Z}$ of **integers** consists of the natural numbers, their negatives, and zero: . . . , -3, -2, -1, 0, 1, 2, 3, ...

All of these numbers are subsets of the rational numbers.

Rational Numbers

A number that can be expressed as the quotient of two integers $\dfrac{a}{b}$, where $b \neq 0$, is called a **rational number.** The integers are rational numbers, and so are common fractions. Some examples of rational numbers are 5, -2, 0, $\frac{2}{9}$, $\sqrt{16}$, and $\frac{-4}{17}$. The set of rational numbers is denoted by $\mathbb{Q}$.

Every rational number has a decimal form that either terminates or repeats a pattern of digits. For example,

$$\frac{3}{4} = 3 \div 4 = 0.75, \text{ a terminating decimal,}$$

and

$$\frac{9}{37} = 9 \div 37 = 0.243243243\ldots$$

where the pattern of digits 243 is repeated endlessly. We use the "repeater bar" notation to write a repeating decimal fraction.

$$\frac{9}{37} = 0.\overline{243}$$

Irrational Numbers

Some real numbers *cannot* be written in the form $\frac{a}{b}$ where a and b are integers. For example, the number $\sqrt{2}$ is not equal to any common fraction. Such numbers are called **irrational numbers.** Examples of irrational numbers are $\sqrt{15}$, π, and $-\sqrt[3]{7}$.

The decimal form of an irrational number never terminates, and its digits do not follow a repeating pattern, so it is impossible to write down an exact decimal equivalent for an irrational number. However, we can obtain decimal *approximations* correct to any desired degree of accuracy by rounding off. A graphing calculator gives the decimal representation of π as 3.141592654. This is not the *exact* value of π, but for most calculations it is quite adequate.

Some n^{th} roots are rational numbers and some are irrational numbers. For example,

$$\sqrt{49}, \qquad \sqrt[3]{\frac{27}{8}}, \qquad \text{and} \qquad 81^{1/4}$$

are rational numbers because they are equal to 7, $\frac{3}{2}$, and 3, respectively. On the other hand,

$$\sqrt{5}, \qquad \sqrt[3]{56}, \qquad \text{and} \qquad 7^{1/5}$$

are irrational numbers. We can use a calculator to obtain decimal approximations for each of these numbers.

$$\sqrt{5} \approx 2.236, \qquad \sqrt[3]{56} \approx 3.826, \qquad \text{and} \qquad 7^{1/5} \approx 1.476$$

The subsets of the real numbers are related as shown in Figure B.1. Every natural number is also a whole number, every whole number is an integer, every integer is a rational number, and every rational number is real. Also, every real number is either rational or irrational.

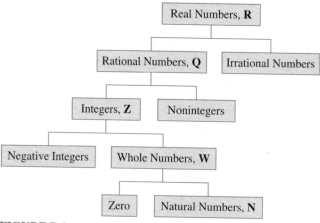

FIGURE B.1

EXAMPLE 1

a. 2 is a natural number, a whole number, an integer, a rational number, and a real number.

b. $\sqrt{15}$ is an irrational number and a real number.

c. The number π, whose decimal representation begins 3.14159 . . . is irrational and real.

d. 3.14159 is a rational and real number (which is close but not exactly equal to π). ■

Properties of the Real Numbers

The real numbers have several useful properties. If a, b, and c represent real numbers, then each of the following equations is true.

$$a + b = b + a \qquad \text{Commutative properties}$$
$$ab = ba$$
$$(a + b) + c = a + (b + c) \qquad \text{Associative properties}$$
$$(ab)c = a(bc)$$
$$a(b + c) = ab + ac \qquad \text{Distributive property}$$
$$a + 0 = a \qquad \text{Identity properties}$$
$$a \cdot 1 = a$$

These properties do not mention subtraction or division. But we can define subtraction and division in terms of addition and multiplication. For example we can define the difference $a - b$ as follows.

$$a - b = a + (-b)$$

where $-b$, the **additive inverse** (or **opposite**) of b, is the number that satisfies

$$b + (-b) = 0$$

Similarly, we can define the quotient $\dfrac{a}{b}$.

$$\frac{a}{b} = a\left(\frac{1}{b}\right) \qquad (b \neq 0)$$

where $\dfrac{1}{b}$, the **multiplicative inverse** (or **reciprocal**) of b, is the number that satisfies

$$b \cdot \frac{1}{b} = 1 \qquad (b \neq 0)$$

Division by zero is not defined.

EXERCISE B.1

Name the subsets of the real numbers to which each of the numbers in Exercises 1–12 belongs.

1. $-\dfrac{5}{8}$

2. 137

3. $\sqrt{8}$

4. 2.71828 . . .

5. -36

6. $\sqrt{49}$

7. 0

8. $0.\overline{357}$

9. $13.\overline{289}$

10. $\sqrt{\dfrac{4}{9}}$

11. 2π

12. $\dfrac{13}{7}$

Write each rational number in decimal form. Does the decimal terminate or does it repeat a pattern?

13. $\dfrac{3}{8}$ **14.** $\dfrac{5}{6}$ **15.** $\dfrac{2}{7}$ **16.** $\dfrac{43}{11}$

17. $\dfrac{7}{16}$ **18.** $\dfrac{5}{12}$ **19.** $\dfrac{11}{13}$ **20.** $\dfrac{25}{6}$

B.2   Complex Numbers

In Chapter 3 we saw that the square root of a negative number, such as $\sqrt{-4}$, is not a real number. We said that such a radical was "undefined." However, for some applications square roots of negative numbers are so important that a new kind of number was invented in order to work with these radicals.

Imaginary Numbers

Recall that we defined a new number, i, whose square is -1. Thus,

$$i^2 = -1 \qquad \text{or} \qquad i = \sqrt{-1}$$

The letter i used in this way is not a variable, it is the name of a specific number, and hence is a constant.

    The square root of any negative number can be written as the product of a real number and i. For example,

$$\sqrt{-4} = \sqrt{-1 \cdot 4}$$
$$= \sqrt{-1} \sqrt{4} = i \cdot 2$$

or $\sqrt{-4} = 2i$. Any number that is the product of i and a real number is called an **imaginary number.** Examples of imaginary numbers are

$$3i, \quad \frac{7}{8}i, \quad -38i, \qquad \text{and} \qquad \sqrt{5}i$$

Complex Numbers

Consider the quadratic equation

$$x^2 - 2x + 5 = 0 \tag{1}$$

We can use the quadratic formula to solve Equation (1) as follows.

$$x = \frac{-(-2) \pm \sqrt{(-2)^2 - 4(1)(5)}}{2}$$
$$= \frac{2 \pm \sqrt{-16}}{2}$$

If we now replace $\sqrt{-16}$ by $4i$, we have

$$x = \frac{2 \pm 4i}{2} = 1 \pm 2i$$

The two solutions of Equation (1) are $1 + 2i$ and $1 - 2i$. These numbers are examples of **complex numbers.**

> A **complex number** is a number that can be written in the form $a + bi$, where a and b are real numbers.

Thus,

$$3 - 5i, \quad 2 + \sqrt{7}i, \quad \frac{4 - i}{3}, \quad 6i, \quad \text{and} \quad -9$$

are all complex numbers. In the complex number $a + bi$, a is called the **real part,** and b is called the **imaginary part.** All real numbers are also complex numbers (with imaginary part equal to zero), and all imaginary numbers are complex numbers with real part equal to zero.

All the properties of real numbers listed in section B.1 are also true of complex numbers. We can also carry out arithmetic operations with complex numbers.

Sums and Differences of Complex Numbers

We add and subtract complex numbers by combining their real and imaginary parts separately. For example,

$$(4 + 5i) + (2 - 3i) = (4 + 2) + (5 - 3)i$$
$$= 6 + 2i$$

In general, we have

> $$(a + bi) + (c + di) = (a + c) + (b + d)i$$
> $$(a + bi) - (c + di) = (a - c) + (b - d)i$$

EXAMPLE 1

Subtract $(8 - 6i) - (5 + 2i)$.

Solution

Combine the real and imaginary parts.

$$(8 - 6i) - (5 + 2i) = (8 - 5) + (-6 - 2)i$$
$$= 3 + (-8)i$$
$$= 3 - 8i$$ ∎

Products of Complex Numbers

To find the product of two imaginary numbers, we use the fact that $i^2 = -1$. For example,

$$(3i) \cdot (4i) = 3 \cdot 4i^2$$
$$= 12(-1) = -12$$

To find the product of two complex numbers, we use the FOIL method, as if the numbers were binomials. For example,

$$(2 + 3i)(3 - 5i) = 6 - 10i + 9i - 15i^2$$

Because $i^2 = -1$, the last term, $-15i^2$, can be replaced by $-15(-1)$, or 15, to obtain

$$6 - 10i + 9i + 15$$

Finally, we combine the real parts and imaginary parts to obtain

$$(6 + 15) + (-10i + 9i) = 21 - i$$

EXAMPLE 2

Multiply $(7 - 4i)(-2 - i)$.

Solution

$$(7 - 4i)(-2 - i) = -14 - 7i + 8i + 4i^2 \quad \text{Replace } i^2 \text{ by } -1.$$
$$= -14 - 7i + 8i - 4 \quad \text{Combine real parts and imaginary parts.}$$
$$= -18 + i \qquad\qquad\qquad ■$$

Quotients of Complex Numbers

To find the quotient of two complex numbers, we use the technique of rationalizing the denominator that we learned in Section 6.5. For example, consider the quotient

$$\frac{3 + 4i}{2i}$$

Because i is really a radical (remember that $i = \sqrt{-1}$), we multiply the numerator and denominator of the quotient by i to obtain

$$\frac{(3 + 4i) \cdot i}{2i \cdot i} = \frac{3i + 4i^2}{2i^2} \quad \text{Apply the distributive law to the numerator. Recall that } i^2 = -1.$$
$$= \frac{3i - 4}{-2}$$

Finally, we divide -2 into each term of the numerator to get

$$\frac{3i}{-2} - \frac{4}{-2} = \frac{-3}{2}i + 2$$

EXAMPLE 3

Divide $\dfrac{10 - 15i}{5i}$.

Solution

Multiply numerator and denominator by i.

$$\frac{10 - 15i}{5i} = \frac{(10 - 15i) \cdot i}{5i \cdot i}$$

$$= \frac{10i - 15i^2}{5i^2} = \frac{10i + 15}{-5} \qquad \text{Replace } i^2 \text{ by } -1.$$

$$= \frac{10i}{-5} + \frac{15}{-5} = -2i - 3 \qquad \text{Divide } -5 \text{ into each term of numerator.} \quad \blacksquare$$

If the divisor has both a real and an imaginary part, we multiply numerator and denominator by the conjugate of the denominator. The **conjugate** of a complex number $a + bi$ is $a - bi$.

EXAMPLE 4

Divide $\dfrac{2 + 3i}{4 - 2i}$.

Solution

Multiply numerator and denominator by the conjugate of the denominator, $4 + 2i$.

$$\frac{2 + 3i}{4 - 2i} = \frac{(2 + 3i)(4 + 2i)}{(4 - 2i)(4 + 2i)} \qquad \text{Expand numerator and denominator.}$$

$$= \frac{8 + 4i + 12i + 6i^2}{16 + 8i - 8i - 4i^2} \qquad \text{Replace } i^2 \text{ by } -1.$$

$$= \frac{8 + 16i - 6}{16 - (-4)} \qquad \text{Combine like terms.}$$

$$= \frac{2 + 16i}{20} \qquad \text{Divide 20 into each term of numerator.}$$

$$= \frac{2}{20} + \frac{16i}{20} = \frac{1}{10} + \frac{4i}{5} \qquad \blacksquare$$

EXERCISE B.2

Add or subtract.

1. $(11 - 4i) - (-2 - 8i)$

2. $(7i - 2) + (6 - 4i)$

3. $(2.1 + 5.6i) + (-1.8i - 2.9)$

4. $\left(\dfrac{1}{5}i - \dfrac{2}{5}\right) - \left(\dfrac{4}{5} - \dfrac{3}{5}i\right)$

Multiply.

5. $5i(2 - 4i)$

6. $-7i(-1 + 4i)$

7. $(4 - i)(-6 + 7i)$

8. $(2 - 3i)(2 - 3i)$

9. $(7 + i\sqrt{3})^2$

10. $(5 - i\sqrt{2})^2$

11. $(7 + i\sqrt{3})(7 - i\sqrt{3})$

12. $(5 - i\sqrt{2})(5 + i\sqrt{2})$

Evaluate each polynomial for the given value of the variable.

13. $z^2 + 9$

    a. $z = 3i$ **b.** $z = -3i$

15. $x^2 - 2x + 2$

    a. $x = 1 - i$ **b.** $x = 1 + i$

17. $2y^2 - y + 2$

    a. $y = 2 - i$ **b.** $y = -2 - i$

14. $3w^2 + 5$

    a. $w = 2i$ **b.** $w = -2i$

16. $q^2 + 4q + 13$

    a. $q = -2 + 3i$ **b.** $q = -2 - 3i$

18. $v^2 + 2v + 3$

    a. $v = 1 + i$ **b.** $v = -1 + i$

Divide.

19. $\dfrac{12 + 3i}{-3i}$

20. $\dfrac{12 + 4i}{8i}$

21. $\dfrac{10 + 15i}{2 + i}$

22. $\dfrac{4 - 6i}{1 - i}$

23. $\dfrac{5i}{2 - 5i}$

24. $\dfrac{-2i}{7 + 2i}$

25. $\dfrac{\sqrt{3}}{\sqrt{3} + i}$

26. $\dfrac{2\sqrt{2}}{1 - i\sqrt{2}}$

27. $\dfrac{1 + i\sqrt{5}}{1 - i\sqrt{5}}$

28. $\dfrac{\sqrt{2} - i}{\sqrt{2} + i}$

29. $\dfrac{3 + 2i}{2 - 3i}$

30. $\dfrac{4 - 6i}{-3 - 2i}$

Expand each product of polynomials.

31. $(2z + 7i)(2z - 7i)$

33. $[x + (3 + i)][x + (3 - i)]$

35. $[v - (4 + i)][v - (4 - i)]$

37. For what values of x will $\sqrt{x - 5}$ be real? Imaginary?

39. Simplify. (*Hint:* $i^2 = -1$ and $i^4 = 1$.)

    a. i^6 **b.** i^{12}

    c. i^{15} **d.** i^{102}

32. $(5w + 3i)(5w - 3i)$

34. $[s - (1 + 2i)][s - (1 - 2i)]$

36. $[Z + (2 + i)][Z + (2 - i)]$

38. For what values of x will $\sqrt{x + 3}$ be real? Imaginary?

40. Express with a positive exponent and simplify.

    a. i^{-1} **b.** i^{-2}

    c. i^{-3} **d.** i^{-6}

Answers to Odd-Numbered Problems

Homework 1.1

1.

w	0	4	8	12	16
A	250	190	130	70	10

a. $A = 250 - 15w$

b.

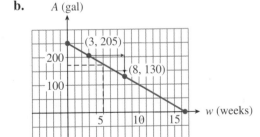

A (gal)

(3, 205)

(8, 130)

w (weeks)

c. 75 gal. **d.** Up to the fifth week

3.

t	0	5	10	15	20
P	−800	−600	−400	−200	0

a. $P = -800 + 40t$

b.

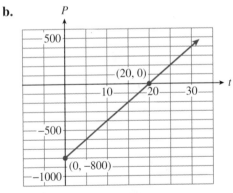

P

(20, 0)

(0, −800)

t

c. The *P*-intercept, −800, is the initial ($t = 0$) value of the profit. Phil and Ernie start out 800 in debt. The *t*-intercept, 20, is the number of hours required for Phil and Ernie to break even.

5a. (8, 0), (0, 4)

b.

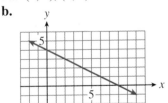

y

x

7a. $(4, 0)$, $(0, -3)$

b.

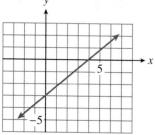

9a. $(9, 0)$, $(0, -4)$

b.

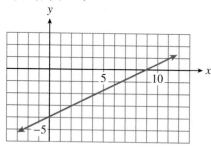

11a. $\left(\dfrac{3}{2}, 0\right)$, $\left(0, \dfrac{11}{3}\right)$

b.

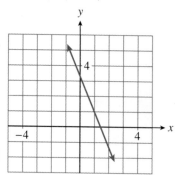

13a. $(-2250, 0)$, $(0, 1500)$

b.

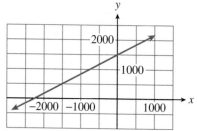

15a. $(12, 0)$, $(0, 4)$

b.

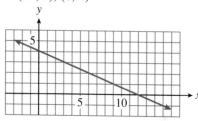

17a. $\$0.6x$, $\$0.8y$ **b.** $0.60x + 0.80y = 4800$

c.

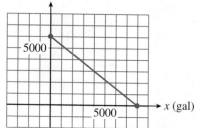

d. The y-intercept, 6000 gallons, is the amount of premium that the gas station owner can buy if he buys no regular. The x-intercept, 8000 gallons, is the amount of regular he can buy if he buys no premium.

19.

s	200	500	800	1200	1500
I	16	25	34	46	55

a. $I = 10{,}000 + 0.03s$

b.

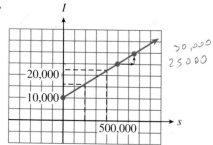

c. Sales between \$200,000 and \$400,000

d. Her salary will increase by \$6000.

Homework 1.2

1a. $y = 6 - 2x$

b.

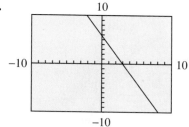

c.

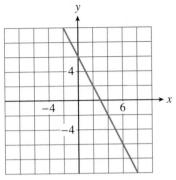

3a. $y = \dfrac{3}{4}x - 300$

b.

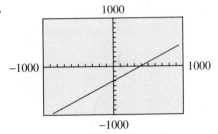

c.

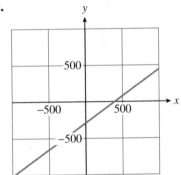

5a. $y = 0.02 - 0.04x$

b.

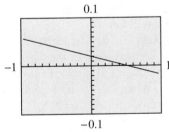

c.

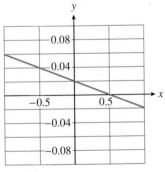

7a. $y = 210 - 35x$

b.

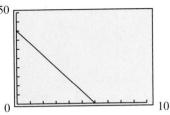

c.

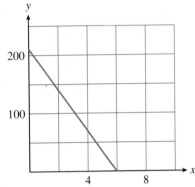

9a. $(100, 0), (0, 100)$ **b.** $y = -x + 100$
c. Xmin $= -20$, Xmax $= 120$, Ymin $= -20$,
Ymax $= 120$ **d.**

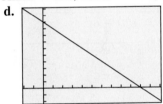

11a. $(0.04, 0), (0, -0.028)$
b. $y = (25x - 1)/36$ **c.** Xmin $= -0.1$,
Xmax $= 0.1$, Ymin $= -0.1$, Ymax $= 0.1$

d.

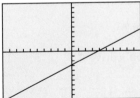

13a. $(-47, 0) (0, 12)$ **b.** $y = 12(x/47 + 1)$
c. Xmin $= -50$, Xmax $= 10$, Ymin $= -5$,
Ymax $= 15$ **d.**

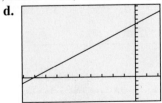

15a. $(-42, 0), (0, -28)$ **b.** $y = -\dfrac{2}{3}x - 28$

c. Xmin $= -50$, Xmax $= 10$, Ymin $= -30$,
Ymax $= 5$

d.

17a. i) $x = -3$ ii) $x < -3$ iii) $x > -3$
b. i) $x = -3$ ii) $x < -3$ iii) $x > -3$
c. The answers to parts (a) and (b) are the same because, for points on the graph of $y = -2x + 6$, the values of the y-coordinates are the same as $-2x + 6$. **19a.** $x = 0.6$ **b.** $x = -0.4$
c. $x > 0.6$ **d.** $x < -0.4$ **21a.** $x = 4$
b. $x = -5$ **c.** $x > 1$ **d.** $x < 14$
23a. $x = 11$ **b.** $x = -10$ **c.** $x \geq -5$
d. $x \leq 8$ **25a.** $x = 4$ **b.** $x < 22$
27a. $x = 20$ **b.** $x \leq 7$
29a.

Time on treadmill (minutes)	0	1	2	5	10
Blood pressure (mm Hg)	120	126	132	150	180

b. $p = 120 + 6t$
c. 200

d. 141 **e.** At 7.5 minutes

Homework 1.3

1. Anthony **3.** Bob's driveway **5.** -1
7. $-\dfrac{2}{3}$ **9a.** $\dfrac{3}{2}$ **b.** $6, \dfrac{3}{2}$ **c.** Yes
d. $-9, \dfrac{3}{2}$ **e.** 27 **11a.** 5 **b.** $\dfrac{-5}{2}$
c. 15 **d.** $-6\dfrac{1}{4}$ **13.** 14.29 ft
15. 3375 m
17a.

b. $m = \dfrac{3}{4}$

19a.

b. $m = -3$

21a.

b. $m = \dfrac{8}{5}$

23a. Yes. The slope between any two data points is the same. **b.** 0.5 grams per degree Celsius
25. (a) **27a.**

t	2	5	6	8
S	16	40	48	64

b. 70

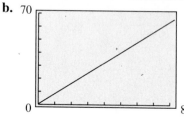

c. $m = 8$ dollars per hour **d.** The slope gives the typist's rate of pay, in dollars per hour.
29a. $m = 1250$ barrels per day **b.** Rate of pumping **31a.** $m = -6$ liters per day
b. Rate of water consumption **33a.** $m = 12$ inches per foot **b.** Conversion rate from feet to inches **35a.** $m = 4$ dollars per kilogram
b. Unit cost of coffee beans, per kilogram
37a. Yes **b.** 2π **39a.** About 1750 km
b. 3.5 km and 5.5 km, 2 km
c. 0.0017, 0.07, 0.0085 **d.** The slopes appear steeper because the vertical scale of the figure has been stretched by a factor of 100 relative to the horizontal scale. **41a.** 7.9 **b.** 8.35 km
c. 2 hours and 5 minutes **43.** $(-2, 5)$, $(1, 7)$
(Other answers are possible.)

Midchapter 1 Review

1.

h	0	3	6	9	10
T	65	80	95	110	115

a. $T = 65 + 5h$ **b.**

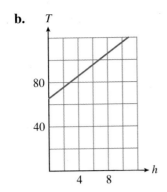

c. 95°   **d.** 3 p.m.

3a.

t	A
0	300
10	250
20	200
30	150
40	100

b. $A = 300 - 5t$

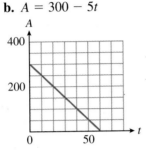

c. The t-intercept shows that Delbert reached the ground after 60 minutes; the A-intercept shows that Delbert began at an altitude of 300 meters.

5a. $(200, 0)$, $(0, -300)$ **b.**

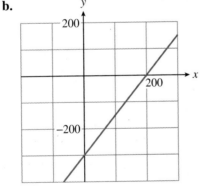

c. $y = \dfrac{3}{2}x - 300$ **7a.** $(-1.5, 0)$, $(0, 280)$

b.

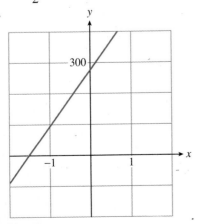

c. $y = \dfrac{560}{3}x + 280$ **9a.** $x = 6$ **c.** $x > 6$

11a. $x = 4$ **b.** $x < -4$ **c.** $x \geq 2$

d. $x \leq -1$ **13.** -200 **15.** $\dfrac{1}{20}$

17. $(-6, 0)$, $(0, 10)$, $m = \dfrac{5}{3}$

19. $(3, 0)$, $(0, -120)$, $m = 40$

21a. $m = 1.5$ feet/second   **b.** Francine is rising at a speed of 1.5 feet per second.

Homework 1.4

1a. $y = -\dfrac{3}{2}x + \dfrac{1}{2}$ **b.** $m = -\dfrac{3}{2}, b = \dfrac{1}{2}$

3a. $y = -\dfrac{1}{6}x + \dfrac{1}{9}$ **b.** $m = -\dfrac{1}{6}, b = \dfrac{1}{9}$

5a. $y = 14x - 22$ **b.** $m = 14, b = -22$

7a. $y = -29$ **b.** $m = 0, b = -29$

9a. $y = -\dfrac{5}{3}x + 16\dfrac{1}{3}$ **b.** $m = -\dfrac{5}{3}, b = 16\dfrac{1}{3}$

11a.

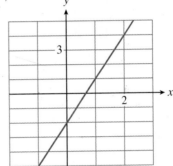

b. $y = 3x - 2$ **c.** $x = \dfrac{2}{3}$

13a.

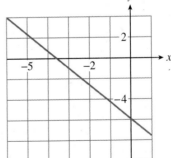

b. $y = -\dfrac{5}{3}x - 6$ **c.** $x = \dfrac{-18}{5}$

15a. $a = 100 + 150t$ **b.** $(0, 100)$, $m = 150$. The skier's starting altitude is 100 feet and she rises at 150 feet per minute.

17a. $G = 25 + 12.5t$ **b.** $(0, 25)$, $m = 12.5$. There were 25 tons in the dump before the regulations, and the dump is filling at 12.5 tons per year.

19a. $M = 7000 - 400w$ **b.** $(0, 7000)$, $m = -400$. Tammy had \$7000 saved, and she spends \$400 per week.   **21a.** $50°F$

b. $-20°C$ **c.**

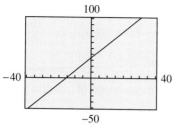

d. The Fahrenheit temperature increases $\frac{9}{5}$ of a degree for every degree increase in the Celsius temperature.   **e.** The C-intercept gives the Celsius temperature at $0°F$, and the F-intercept gives the Fahrenheit temperature at $0°C$.

23a.

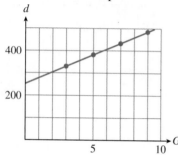

b. $m = 25, b = 250$ **c.** $d = 25G + 250$

25a.

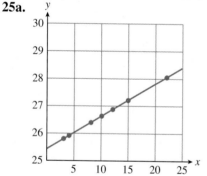

                     $b = 25.4$

b. $y = 0.12x + 25.4$ **c.** 18 kg

27. $-\dfrac{9}{19}$ **29.** 0.387 **31.** Undefined

33a.

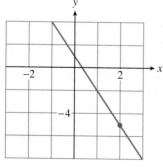

b. $y + 5 = -3(x - 2)$ **c.** $y = -3x + 1$

35a.

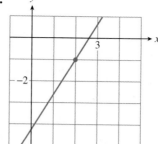

b. $y + 1 = \dfrac{5}{3}(x - 2)$ **c.** $y = \dfrac{5}{3}x - \dfrac{13}{3}$

37a. II   **b.** III   **c.** I   **d.** IV

39. $m = 2, (6, -1)$ **41.** $m = -\dfrac{4}{3}, (-5, 3)$

43a. $y + 3.5 = -0.25(x + 6.4)$

b. $y = -0.25x - 5.1$

c.

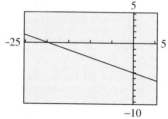

45a. $y = 250 = 2.4(x - 80)$

b. $y = 2.4x - 442$

c.

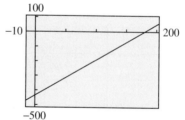

47a. $m = 4, b = 40$ **b.** $y = 4x + 40$

49a. $m = -80, b = -2000$

b. $P = -80t - 2000$ **51a.** $m = \dfrac{1}{4}, b = 0$

b. $V = \dfrac{1}{4}d$ **53a.** $m = 2.5$ **b.** $b = 6.25$

55a. $m = -8.4$ **b.** $b = 63$

Homework 1.5

1a. 12 sec   **b.** 39

c.

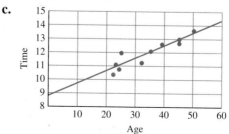

d. 11.6 sec   **e.** $y = 0.09x + 9.1$ **f.** 12.7 sec, 10.18 sec. The predicted time for the 40-year-old is reasonable, but the predicted time for the 12-year-old is not.   **3b.** 129 lb, 145 lb
c. $y = 2.\overline{6}x - 44.\overline{3}$ **d.** 137 lb
5a. $y = 2.84x - 55.74$ **b.** 137.33 lb
7a. 10 meters every 29 years, or about 0.34 meter per year   **b.** $D = 0.34a$ **c.** Over 1305 years old   **9a.**

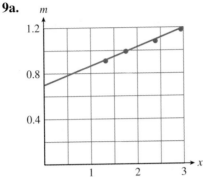

b. $m = 0.68 + 0.17x$ **c.** $k = 0.17; m_0 = 0.68$
11a. 22 km per million years
b.

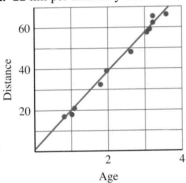

19 km per million years   **c.** 19 mm per year
13a. E
b.

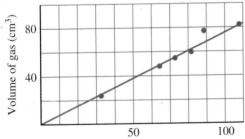

c. $y = \dfrac{4}{3}x$ **d.** 1330 mg   **e.** Oxygen

15a.

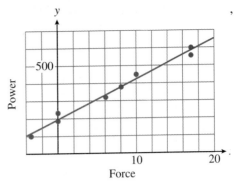

$y \approx 23.048x + 195.3$ **b.** 541 watts
c. -8.5 newtons   **d.** 3.4 watts   **e.** approximately 1.7%   **17a.**

x	50	125
C	9000	15,000

b. $C = 80x + 5000$
c.

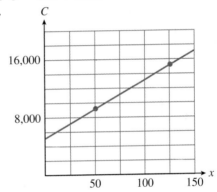

d. $m = 80$ dollars is the cost of making each bike.
19a.

g	12	5
d	312	130

b. $d = 26g$ **c.**

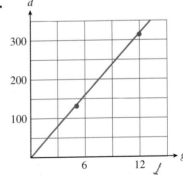

d. $m = 26$ miles per gallon gives the car's fuel efficiency.   **21a.**

C	15	-5
F	59	23

b. $F = \dfrac{9}{5}C + 32$

c.

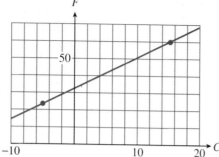

d. $m = \dfrac{9}{5}$ degrees Fahrenheit per degree Celsius compares the size of a degree in the two temperature scales.   **23a.** $75°$ **b.** The slope gives the rate of change of the temperature in degrees per hour.   **25a.** 20 mph   **b.** The slope gives the acceleration of the car in miles per hour per second.   **27.** 2 min: $21°C$; 2 hr: $729°C$. The first answer is reasonable; the second is not.
29. 128 lb
31a.

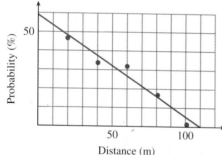

b. 31.7%   **c.** 90 meters   **d.** The regression line gives a negative probability, which is not reasonable.

Homework 1.6

1a.

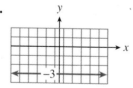

b. $m = 0$

3a.

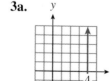

b. m is undefined.

5a.

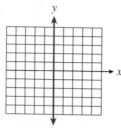

b. m is undefined.   **7.** $x = -5$ **9.** $y = 0$
11. $y = 9$ **13a.** l_1 negative, l_2 negative, l_3 positive, l_4 zero   **b.** l_1, l_2, l_4, l_3
15a.

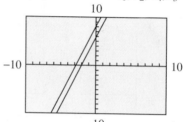

b. $m = 3$, $m = 3.1$. No.   **c.** $y = 68$ for both lines. The lines intersect at $(20, 68)$.   **17.** parallel: a, g, h; perpendicular: c, f   **19a.** parallel
b. neither   **c.** neither   **d.** parallel

21b. Slope $\overline{AB}$: -1, slope $\overline{BC}$: 1, slope $\overline{AC}$: $\dfrac{1}{4}$. Hence $\overline{AB} \perp \overline{AC}$, so the triangle is a right triangle.

23. Slope $\overline{PQ}$: 4, slope $\overline{QR}$: $-\dfrac{7}{2}$, slope $\overline{RS}$: 4, slope $\overline{SP}$: $-\dfrac{7}{2}$. Hence $\overline{PQ} \parallel \overline{RS}$ and $\overline{QR} \parallel \overline{SP}$, so the points are the vertices of a parallelogram.

25. Slope $\overline{AB} = \dfrac{7}{6} = $ slope $\overline{BC}$, so A, B, and C lie on the same line.   **27a.** $y = \dfrac{1}{2}x - \dfrac{5}{2}$

b. $\dfrac{1}{2}$ **d.** $y = \dfrac{1}{2}x - 2$ **29a.** $y = \dfrac{3}{2}x + \dfrac{5}{2}$

b. $-\dfrac{2}{3}$ **d.** $y = -\dfrac{2}{3}x + \dfrac{14}{3}$

31a. $y = -2x - 8$ **b.** $y = \dfrac{1}{2}x - 3$

33b. $m = \dfrac{4}{5}$ **c.** $m = -\dfrac{5}{4}$

d. $y = -\dfrac{5}{4}x - \dfrac{9}{2}$ **35.** $y = \dfrac{3}{2}x + \dfrac{15}{2}$

37a. Right angles are equal.   **b.** Alternate interior angles are equal.   **c.** Two angles of one triangle equal two angles of the other.
d. Definition of slope   **e.** Corresponding sides of similar triangles are proportional.

Chapter 1 Review

1a.

n	100	500	800	1200	1500
C	4000	12,000	18,000	26,000	32,000

b. $C = 20n + 2000$

c.

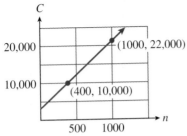

d. $22,000   **e.** 400

3a.

t	5	10	15	20	25
R	1560	1460	1360	1260	1160

b. $R = 1660 - 20t$ **c.** $(0, 1660), (83, 0)$

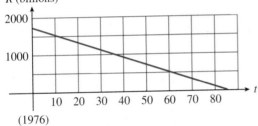

d. In 1976, R was 1660. The world's oil reserves will be gone in 2059.   **5a.** $5A + 2C = 1000$

b.

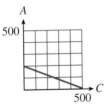

c. 200 children's tickets   **d.** C-intercept (500): only children's tickets are sold. A-intercept (200): only adults' tickets are sold.

7.

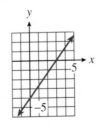

9.

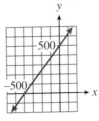

11.

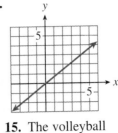

13.

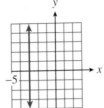

15. The volleyball

17. Highway 33   **19a.** $B = 800 - 5t$

b.

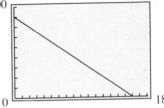

c. $m = -5$ barrels per minute is the rate at which oil is leaking.   **21a.** $F = 500 + 0.10C$

b.

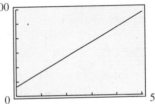

c. $m = 0.10$. The decorator charges 10% of the cost of the job (plus a flat $500 fee).   **23.** $-\dfrac{3}{2}$

25. -0.4 **27.** Neither is linear.

29. $d = 1, V = 4.2$ **31.** 80 ft

33. $m = \dfrac{1}{2}, b = -\dfrac{5}{4}$ **35.** $m = -4, b = 3$

37a.

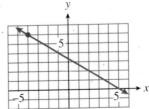

b. $y = -\dfrac{2}{3}x + \dfrac{10}{3}$ **39a.** $T = 62 - 0.0036h$

b. $-46°F$; $108°F$ **c.** $-71°F$

41. $y = -\dfrac{9}{5}x + \dfrac{2}{5}$

43a.

t	0	15
P	4800	6780

b. $P = 4800 + 132t$

c. $m = 132$ people per year gives the rate of population growth.   **45a.** $m = -2$, $b = 3$

b. $y = -2x + 3$ **47.** $\dfrac{3}{5}$ **49a.** $m = \dfrac{3}{2}$

b. $(4, 2)$, No   **c.** $(6, 5)$ **51.** Parallel

53. $y = -\dfrac{2}{3}x + \dfrac{14}{3}$ **55.** $y = \dfrac{2}{3}x - \dfrac{26}{3}$

57. 6   **59a.** 45 cm   **b.** 87 cm

c. $y = 1.2x - 3$ **d.** 69 cm

e. $y = 1.197x - 3.660$, 68.16 cm

Homework 2.1

1. $(3, 0)$ **3.** $(50, 70)$ **5a.** The median age

b.

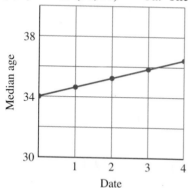

Date

c. The median age for women is increasing by 0.3 year per year.

d.

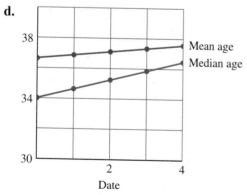

Date

e. Approximately 2004   **f.** There are more young women than old women, with some women much older than the mean.   **7.** $(-2, 3)$

9. $(2, 3)$ **11.** Inconsistent   **13.** Consistent

15. Inconsistent   **17a.** $D = 10 + 0.09x$

b. $F = 15 + 0.05x$

c.

x	0	30	60	90	120	150
Dash	10	12.70	15.40	18.10	20.80	23.50
Friendly	15	16.50	18.00	19.50	21.00	22.50

e. 125 min   **19a.** $y = 50x$ **b.** $y = 2100 - 20x$

d. 30 cents per bushel, 1500 bushels

21a. $C = 200 + 4x$ **b.** $R = 12x$

c.

x	5	10	15	20	25	30
C	220	240	260	280	300	320
R	60	120	180	240	300	360

d. 25 pendants

23a.

	Number of tickets	Cost per ticket	Revenue
Adults	x	7.50	$7.50x$
Students	y	4.25	$4.25y$
Total	$x + y$	—	$7.50x + 4.25y$

b. $x + y = 82$ **c.** $7.50x + 4.25y = 465.50$

d. 36 adults, 46 students   **25.** $(-4.6, 52)$

27. $(7.15, 4.3)$

Homework 2.2

1. $(2, 1)$ **3.** $(1, 2)$ **5.** $(1, -2)$

7. $(6, 0)$ **9.** $(1, 2)$ **11.** $(1, 2)$

13. Inconsistent   **15.** Consistent

17. Dependent   **19a.** $y = 34 + 0.3x$

b. $y = 36.58 + 0.115x$ **c.** 13.95

21a. $M = 4 - \dfrac{16}{75}s$ **b.** $F = -\dfrac{4}{75}s$

c.

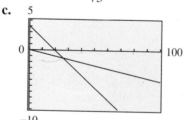

d. The salinity is 25% and the freezing point is $-1\frac{1}{3}°C$.

23a.

	Principal	Interest rate	Interest
Bonds	x	0.10	$0.10x$
Certificate	y	0.08	$0.08y$
Total	$x + y$	—	$0.10x + 0.08y$

b. $x + y = 2000$ **c.** $0.10x + 0.08y = 184$
d. $800 at 8\%$, $1200 at 10\%$
25a.

	Pounds	% silver	Amount of silver
First alloy	x	0.45	$0.45x$
Second alloy	y	0.60	$0.60y$
Mixture	$x + y$	—	$0.45x + 0.60y$

b. $x + y = 40$ **c.** $0.45x + 0.60y = 0.48(40)$
d. 32 lb   **27.** True-false: 2 points; fill-ins:
5 points.
29a.

	Rate	Time	Distance
Detroit to Denver	$x - y$	4	1120
Denver to Detroit	$x + y$	3.5	1120

b. $4(x - y) = 1120$ **c.** $3.5(x + y) = 1120$
d. Airplane: 300 mph; wind: 20 mph
31a.

	Cups	Calories per cup	Calories
Oat flakes	x	310	$310x$
Wheat flakes	y	290	$290y$
Mixture	$x + y$	—	$310x + 290y$

b. $x + y = 1$ **c.** $310x + 290y = 302$
d. 0.6 cup oats, 0.4 cup wheat   **33a.** $S = 35x$,
$D = 1700 - 15x$ **b.** $34 per pair, 1190 pairs
35. $(2.3, 1.6)$ **37.** $(182, 134)$

Homework 2.3

1. $(1, 2, -1)$ **3.** $(2, -1, -1)$
5. $(4, 4, -3)$ **7.** $(2, -2, 0)$ **9.** $(0, -2, 3)$
11. $(-1, 1, -2)$ **13.** $\left(\dfrac{1}{2}, \dfrac{2}{3}, -3\right)$
15. $(4, -2, 2)$ **17.** $(1, 1, 0)$
19. $\left(\dfrac{1}{2}, -\dfrac{1}{2}, \dfrac{1}{3}\right)$ **21.** Inconsistent
23. $\left(\dfrac{1}{2}, 0, 3\right)$ **25.** $(-1, 3, 0)$
27. Inconsistent   **29.** $\left(\dfrac{1}{2}, \dfrac{1}{2}, 3\right)$
31. 60 nickels, 20 dimes, 5 quarters
33. $x = 40$ in., $y = 60$ in., $z = 55$ in.
35. 0.3 cup carrots, 0.4 cup green beans, 0.3 cup
cauliflower   **37.** 40 score only, 20 evaluation,

80 narrative report   **39.** 20 tennis, 15 Ping-
Pong, 10 squash

Midchapter 2 Review

1. $(3, 5)$ **3.** $(-80, 70)$ **5.** dependent
7a. $460x + 120y = 7520$ **b.** $4x = y$
c. $(8, 32)$ Etienne can buy 8 tables and 32 chairs.
9. $(2, -4)$ **11.** $\left(\dfrac{-2}{3}, \dfrac{1}{3}\right)$ **13.** Inconsistent

15a.

	R	T	D
P	5.4	x	y
S	3	$x + 90$	y

b. $y = 3(x + 90)$ **c.** $y = 5.4x$
d. $(112.5, 607.5)$ The seismograph is 607.5 miles
from the earthquake.   **17.** $(2, 6, -3)$
19. $(-14, -4, 4)$ **21.** dependent   **23.** He
should use 0.4 quarts of cranberry juice, 0.2 quarts
of apricot nectar, and 0.4 quarts of club soda.

Homework 2.4

1. $\begin{bmatrix} -2 & 1 & | & 0 \\ -9 & 3 & | & -6 \end{bmatrix}$ **3.** $\begin{bmatrix} 1 & -3 & | & 6 \\ 0 & -2 & | & 11 \end{bmatrix}$

5. $\begin{bmatrix} 1 & 0 & -2 & | & 5 \\ 2 & 6 & -1 & | & 4 \\ 0 & -3 & 2 & | & -3 \end{bmatrix}$

7. $\begin{bmatrix} 1 & 2 & 1 & | & -5 \\ 0 & 4 & -2 & | & 3 \\ 0 & -9 & 2 & | & 12 \end{bmatrix}$ **9.** $\begin{bmatrix} 1 & -3 & | & 2 \\ 0 & 7 & | & 0 \end{bmatrix}$

11. $\begin{bmatrix} 2 & 6 & | & -4 \\ 4 & 0 & | & 3 \end{bmatrix}$ **13.** $\begin{bmatrix} 1 & -2 & 2 & | & 1 \\ 0 & 7 & -5 & | & 4 \\ 0 & 9 & -11 & | & -1 \end{bmatrix}$

15. $\begin{bmatrix} -1 & 4 & 3 & | & 2 \\ 3 & 0 & -5 & | & 14 \\ -3 & 0 & -3 & | & 8 \end{bmatrix}$

17. $\begin{bmatrix} -2 & 1 & -3 & | & -2 \\ 0 & 4 & -6 & | & -2 \\ 0 & 0 & -4 & | & -5 \end{bmatrix}$ **19.** $(2, 3)$

21. $(-2, 1)$ **23.** $(3, -1)$ **25.** $\left(-\dfrac{7}{3}, \dfrac{17}{3}\right)$
27. $(1, 2, 2)$ **29.** $\left(2, -\dfrac{1}{2}, \dfrac{1}{2}\right)$
31. $(-3, 1, -3)$ **33.** $\left(\dfrac{5}{4}, \dfrac{5}{2}, -\dfrac{1}{2}\right)$

Homework 2.5

1.

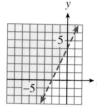

3.

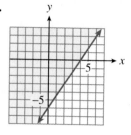

5.

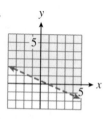

7.

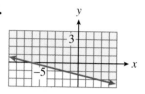

9.

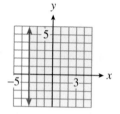

11.

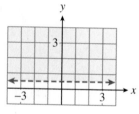

13.

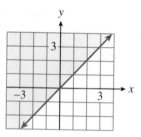

15.

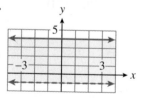

17.

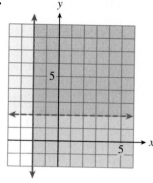

19.

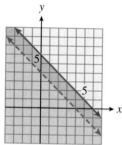

21.

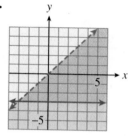

23.

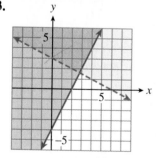

25.

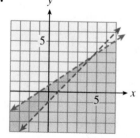

27.

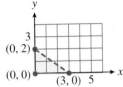

29.

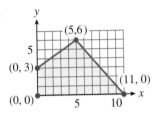

31.

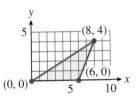

33.

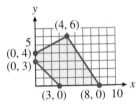

35.

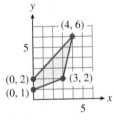

37.

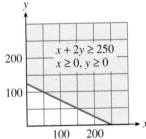

$$x + 2y \geq 250$$
$$x \geq 0, y \geq 0$$

39.

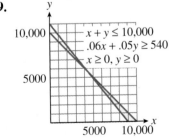

$$x + y \leq 10,000$$
$$.06x + .05y \geq 540$$
$$x \geq 0, y \geq 0$$

41.

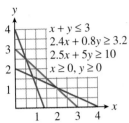

$$x + y \leq 3$$
$$2.4x + 0.8y \geq 3.2$$
$$2.5x + 5y \geq 10$$
$$x \geq 0, y \geq 0$$

43a. $2.05x + 5.21y \leq 342$
$2.67x + 1.64y \leq 314$
$2.44x + 2.26y \geq 100$
$x \geq 0, y \geq 0$

b.

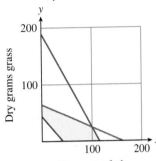

c. Many answers are possible.   **d.** At least 44.25 grams of grass; at least 40.98 grams of forb.

Chapter 2 Review

1. $x = -1, y = 2,$

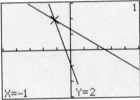

3. $x = \dfrac{1}{2}, y = \dfrac{7}{2}$ **5.** $x = 12, y = 0$

7. Consistent   **9.** Dependent   **11.** $x = 2,$
$y = 0, z = -1$ **13.** $x = 2, y = -5, z = 3$
15. $x = -2, y = 1, z = 3$ **17.** $x = 3, y = -1$
19. $x = 4, y = 1$ **21.** $x = -1, y = 0, z = 2$
23. 26   **25.** \$3181.82 at 8%, \$1818.18 at
13.5%   **27.** 5 cm, 12 cm, 13 cm
29.

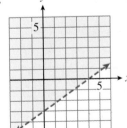

31.

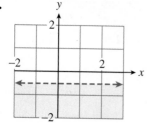

33.

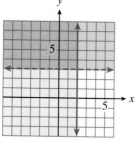

35.

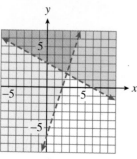

37.

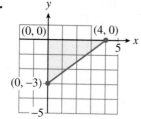

Vertices: $(0, 0)$, $(0, -3)$, $(4, 0)$

39.

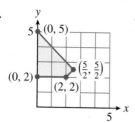

Vertices: $(0, 2)$, $(0, 5)$, $(2, 2)$, $\left(\dfrac{5}{2}, \dfrac{5}{2}\right)$

41.

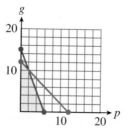

$$20p + 8g \le 120$$
$$10p + 10g \le 120$$

Homework 3.1

1. $\dfrac{\pm 5}{3}$ **3.** $\pm\sqrt{6}$ **5.** $\pm\sqrt{6}$

7. ± 2.65 **9.** ± 5.72 **11.** ± 5.73

13. $\pm\sqrt{\dfrac{Fr}{m}}$ **15.** $\pm\sqrt{\dfrac{2s}{g}}$ **17a.** $V = 8.8r^2$

b.

r	1	2	3	4	5	6	7	8
V	8.8	35.2	79.2	140.8	220	316.8	431.2	563.2

The volume increases by a factor of four.

c. 5.86 cm

d.

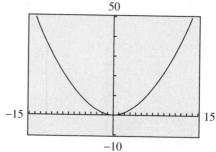

19. 21 in.   **21.** 42.4 m   **23.** 11.3 in.

25a.

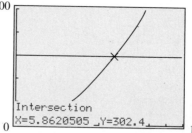

b. ± 12

27a.

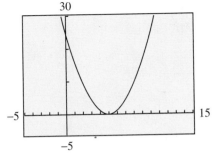

b. 1, 9

29a.

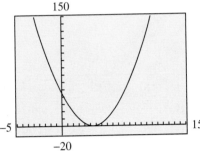

b. 10, −2   **31.** 5, −1   **33.** $\dfrac{5}{2}, \dfrac{-3}{2}$

35. $-2 \pm \sqrt{3}$ **37.** $\dfrac{1}{2} \pm \dfrac{\sqrt{3}}{2}$

39. $\dfrac{-2}{9}, \dfrac{-4}{9}$ **41.** $\dfrac{7}{8} \pm \dfrac{\sqrt{8}}{8}$

43a. $B = 5000(1 + r)^2$

b.

r	0.02	0.04	0.06	0.08
B	5202	5408	5616	5832

c. 11.8%

d. 10,000

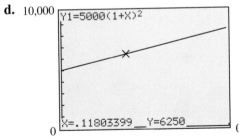

0.3

45. 8%   **47.** 7.98 mm   **49a.** 1.73 sq cm, 6.93 sq cm, 43.3 sq cm

b. 60

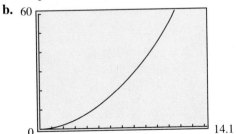

14.1

c. An equilateral triangle with side 5.1 cm has area 11.26 sq cm.   **d.** About 6.8 cm

e. 6.8 cm   **f.** 20 cm   **51.** $\pm\sqrt{\dfrac{bc}{a}}$

53. $a \pm 4$ **55.** $\dfrac{-b \pm 3}{a}$

57a,c.

v	0	1	2	3	4	5	6	7	8	9	10	11
J	0	0.05	0.20	0.46	0.82	1.28	1.84	2.5	3.27	4.13	5.1	6.17
H	0.9	0.95	1.10	1.36	1.72	2.18	2.74	3.4	4.17	5.03	6.0	7.07

b,c.

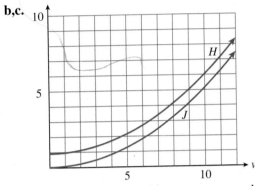

d. 5.5 meters   **e.** 10.14 meters per second

Homework 3.2

1a.

Height	Base	Area	Height	Base	Area
1	34	34	10	16	160
2	32	64	11	14	154
3	30	90	12	12	144
4	28	112	13	10	130
5	26	130	14	8	112
6	24	144	15	6	90
7	22	154	16	4	64
8	20	160	17	2	34
9	18	162	18	0	0

b.

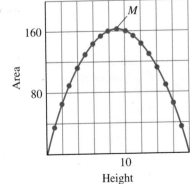

c. 162 sq in, 9 in by 18 in   **d.** Base: $36 - 2x$, area: $36x - 2x^2$

f. 170

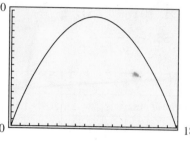

g. Two possible rectangles, one with height 11.5 in. and the other with height 6.5 in.

3a.

t	0	1	2	3	4	5
h	300	304	276	216	124	0

b. 350

Y1=-16X²+20X+300
X=.625   Y=306.25

c. 306.25 ft at 0.625 sec   **d.** 1.25 sec
e. 5 sec   **5.** $3x^2 - 15x$ **7.** $2b^2 + 9b - 18$
9. $16w^2 - 24w + 9$ **11.** $6p^3 - 33p^2 + 45p$
13. $-50 - 100r - 50r^2$
15. $12q^4 - 36q^3 + 27q^2$ **17.** $(x - 5)(x - 2)$
19. $(x - 15)(x + 15)$ **21.** $(w - 8)(w + 4)$
23. $(2z - 5)(z + 8)$ **25.** $(3n + 4)^2$
27. $3a^2(a + 1)^2$ **29.** $4h^2(h - 3)(h + 3)$
31. $-10(u + 13)(u - 3)$ **33.** $6t^2(4t^2 + 1)$

Homework 3.3

1. $x = 2, -2.5$ **3.** $x = 0, -\dfrac{10}{3}$

5. $x = -\dfrac{3}{4}, -8$ **7.** $x = 4, 4$

9. $a = \dfrac{1}{2}, -3$ **11.** $x = 0, 3$ **13.** $y = 1, 1$

15. $x = \dfrac{1}{2}, 1$ **17.** $t = 2, 3$ **19.** $z = -1, 2$

21. $v = -3, 6$ **23 and 25.** The graphs have the same x-intercepts. In general, the graph of $y = ax^2 + bx + c$ has the same x-intercepts as the graph of $y = k(ax^2 + bx + c)$.
27. $x^2 + x - 2 = 0$ **29.** $x^2 + 5x = 0$
31. $2x^2 + 5x - 3 = 0$ **33.** $8x^2 - 10x - 3 = 0$
35. $0.1(x - 18)(x + 15)$
37. $-0.08(x - 18)(x + 32)$

39a. $10^2 + h^2 = (h + 2)^2$ **b.** 24 ft
41a. $h = -16t^2 + 16t + 8$ **b.** 12 ft; 8 ft

c. $11 = -16t^2 + 16t + 8$; at $\dfrac{1}{4}$ sec and $\dfrac{3}{4}$ sec

d. $\Delta\text{Tbl} = 0.25$

e. 15

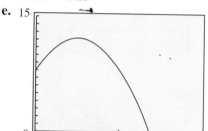

f. 1.37 sec

43a.

Width	10	20	30	40	50
Length	170	160	150	140	130
Area	1700	3200	4500	5600	6500

Width	60	70	80
Length	120	110	100
Area	7200	7700	8000

b. $l = 180 - x$, $A = 180x - x^2$; 80 yd by 100 yd
c. $180x - x^2 = 8000$, 80 yd by 100 yd, or 100 yd by 80 yd. There are two solutions because the pasture can be oriented in two directions.
45a. $l = x - 4$, $w = x - 4$, $h = 2$, $V = 2(x - 4)^2$

b.

x	4	5	6	7	8	9	10
V	0	2	8	18	32	50	72

c. As x increases, V increases.   **d.** 9 in. by 9 in.

e. $2(x - 4)^2 = 50$, $x = 9$ **47a.** $\dfrac{\sqrt{2v}}{g}$ seconds

b. $\dfrac{v^2}{g}$ meters   **c.** 207 meters   **d.** Air resistance decreases the range.

Midchapter 3 Review

1. $x = \pm\sqrt{2}$ **3.** $x = -4 \pm\sqrt{20}$

5. $s = \sqrt{\dfrac{2A}{3\sqrt{3}}}$ **7.** The volume quadruples.

9a. $A = lw$, $P = 2l + 2w$ **b.** (Many answers are possible.) $1'' \times 29''$, $A = 29$ square inches; $14'' \times 16''$, $A = 224$ square inches
11. $-18b^2 + 6b$ **13.** $4t^2 - 34t + 16$
15. $2(x + 7)(x - 3)$ **17.** $(7a + 2b)^2$
19. $ab = 0$ only if $a = 0$ or $b = 0$.

21. $x = \dfrac{7}{2}, x = -1$ **23.** $x = 0, x = \dfrac{-5}{2}$

25. 1 second

Homework 3.4

1.

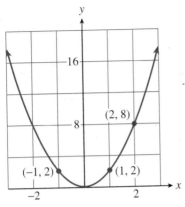

3.

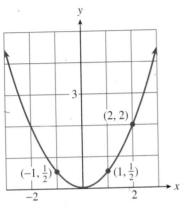

5.

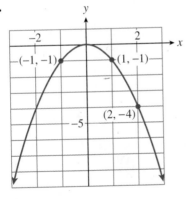

7.

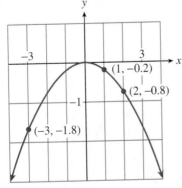

9.

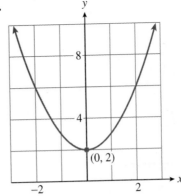

11.

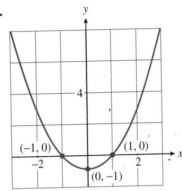

13.

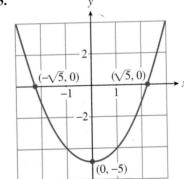

15.

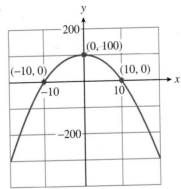

17. $(0, 0), (4, 0); (2, -4)$

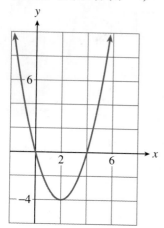

19. $(0, 0), (-2, 0); (-1, -1)$

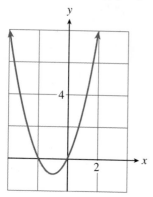

21. $(0, 0), (-2, 0); (-1, -3)$

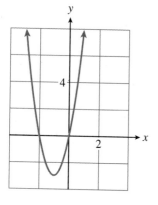

23. $(0, 0), \left(\dfrac{5}{2}, 0\right);$

$\left(\dfrac{5}{4}, 3\dfrac{1}{8}\right)$

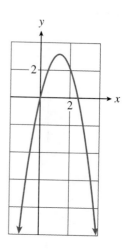

25.

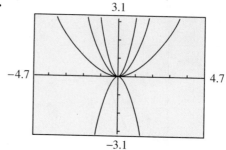

The constants affect the width of the parabola.

27.

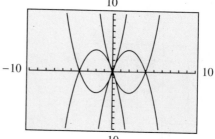

The constants affect the location of the x-intercepts. **29a.** II   **b.** IV   **c.** I   **d.** III **e.** VI   **f.** V   **31a.** At $x = 0$ and $x = 4000$ there is no increase in biomass.
b. (2000, 400). The largest annual increase in biomass, 400 tons, occurs when the biomass is 2000 tons.
c.

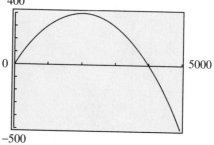

d. The biomass decreases if $4000 < x \le 5000$. If the population becomes too large, its supply of food may be inadequate.

Homework 3.5

1. $(x + 4)^2$ **3.** $\left(x - \dfrac{7}{2}\right)^2$ **5.** $\left(x + \dfrac{3}{4}\right)^2$

7. $\left(x - \dfrac{2}{5}\right)^2$ **9.** $1, 1$ **11.** $-4, -5$

13. $-\dfrac{3}{2} \pm \sqrt{\dfrac{21}{4}}$ **15.** $-1 \pm \sqrt{\dfrac{5}{2}}$

17. $-\dfrac{4}{3}, 1$ **19.** $\dfrac{1}{4} \pm \sqrt{\dfrac{13}{16}}$ **21.** $\dfrac{4}{3}, -1$

23. $\dfrac{2}{5}, -2$ **25.** $-1 \pm \sqrt{1 - c}$

27. $-\dfrac{b}{2} \pm \sqrt{\dfrac{b^2 - 4}{4}}$ **29.** $-\dfrac{1}{a} \pm \sqrt{\dfrac{4a + 1}{a^2}}$

31. $3 \pm \sqrt{\dfrac{V}{\pi h}}$ **33.** $\pm \sqrt{\dfrac{2(E - mgh)}{m}}$

35. $\pm \sqrt{\dfrac{V}{2w} - s^2}$ **37.** $0, x^2$ **39.** $\dfrac{3x \pm 3}{2}$

41. $\pm \sqrt{\dfrac{4x^2 - 36}{9}}$ **43.** $\dfrac{\pm 2x}{5}$

45. $-\dfrac{b}{2} \pm \sqrt{\dfrac{b^2 - 4c}{4}}$

47a. $A = (x + y)^2$ **b.** $A = x^2 + 2xy + y^2$

c. x^2, xy, xy, y^2

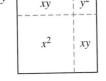

49a. $A = \dfrac{1}{2}(x^2 - y^2)$ **b.** $A = \dfrac{1}{2}(x - y)(x + y)$

c. 18 sq ft

Homework 3.6

1. $1.618, -0.618$ **3.** $1.449, -3.449$

5. $1.695, -0.295$ **7.** $1.434, 0.232$

9. $-5.894, 39.740$

11a.

s	10	20	30	40	50
d	9	27	53	87	129

s	60	70	80	90	100
d	180	239	307	383	467

b.

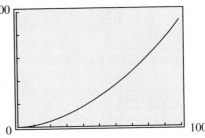

c. $\dfrac{s^2}{24} + \dfrac{s}{2} = 50$; 29.16 mph

13a.

t	0	5	10	15	20	25
h	11,000	10,520	9240	7160	4280	600

b.

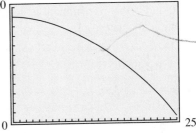

c. $-16t^2 - 16t + 11,000 = 1000$; 24.5 sec

d. 1.2 sec   **15a.** $2l + 4w = 100$

b. $l = 50 - 2w$ **c.** $w(50 - 2w) = 250$;

$w = 6.91, 18.09$ **d.** 12.06 m by 6.91 m, or

4.61 m by 18.09 m   **17a.** 45 mi

b. 1.26 mi   **19a.** $2 - \sqrt{5}$ **b.** $x^2 - 4x - 1 = 0$

21a. $4 + 3i$ **b.** $x^2 - 8x + 25 = 0$

23a.

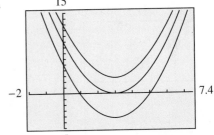

$(5, 0)$ and $(1, 0)$; $(3, 0)$; no x-intercepts

b. $16, 0, -12$. If the discriminant is positive, there are two distinct real roots, and hence two x-intercepts. If the discriminant is zero, there is one repeated real root, and one x-intercept. If the discriminant is negative, there are two complex roots and no x-intercepts.

25. Two complex solutions   **27.** One repeated real solution   **29.** Two distinct real solutions
31. No   **33.** Yes

35. $w = \dfrac{-4l \pm \sqrt{16l^2 + 8A}}{4}$

37. $t = \dfrac{4 \pm \sqrt{16 - 64h}}{32}$

39. $t = \dfrac{v \pm \sqrt{v^2 - 2as}}{a}$

41. $y = \dfrac{-x \pm \sqrt{8 - 11x^2}}{2}$ **43.** $\dfrac{-b}{a}$

45. The discriminant is the quantity under the radical in the quadratic equation. If the discriminant is positive, it has two distinct real square roots; if the discriminant is zero, it has only one square root; and if the discriminant is negative, it has two complex square roots.

Chapter 3 Review

1. $x = 1, x = 4$ **3.** $x = -1, x = 2$
5. $x = -2, x = 3$ **7.** $4x^2 - 29x - 24 = 0$
9.

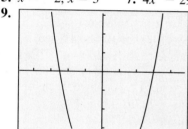

$y = (x - 3)(x + 2.4)$ **11.** $(0, 0)$;

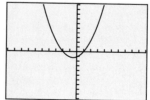

13. $\left(\dfrac{9}{2}, \dfrac{-81}{4}\right)$; $(0, 0)$, $(9, 0)$, $(0, 0)$

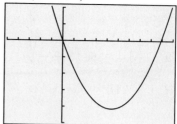

15. $x = 2 \pm \sqrt{10}$ **17.** $x = \dfrac{3}{2} \pm \sqrt{\dfrac{3}{4}}$

19. $x = 1, x = 2$ **21.** $x = 3.41, x = 0.59$

23. $\pm\sqrt{\dfrac{2K}{m}}$ **25.** $\dfrac{6 \pm \sqrt{36 - 12h}}{6}$

27. one real solution   **29.** two complex solutions   **31.** 9   **33.** 11%   **35.** 10 ft by 18 ft or 12 ft by 15 ft   **37a.** $h = 100t - 2.8t^2$
c. 893 ft   **d.** 15.7 sec

39. $A_1 = $ area of square $-$ area of two triangles $= x \cdot x - \left(\dfrac{1}{2}y \cdot y - \dfrac{1}{2}y \cdot y\right) = x^2 - y^2$.
$A_2 = $ (length)(width) $= (x + y)(x - y) = x^2 - y^2$.

Homework 4.1

1. $(1.5, 4.25)$, maximum   **3.** $\left(\dfrac{2}{3}, \dfrac{1}{9}\right)$, minimum

5. $(-4.5, 18.5)$, maximum   **7.** $\left(\dfrac{-1}{2}, 0\right)$, $(4, 0)$, $(0, 4)$; $\left(\dfrac{7}{4}, 10\dfrac{1}{8}\right)$

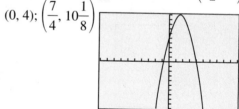

9. $(-2, 0)$, $(1, 0)$, $(0, -1.2)$; $(-0.5, -1.35)$

11. $(0, 7)$; $(-2, 3)$

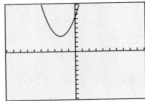

13. $(0.41, 0)$, $(-2.41, 0)$, $(0, -1)$; $(-1, -2)$

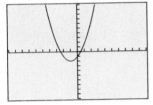

15. $(2.37, 0)$, $(0.63, 0)$, $(0, -3)$; $(1.5, 1.5)$

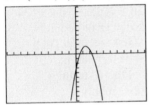

17a. IV   **b.** V   **c.** I   **d.** VII

19a. $y = x^2 + x - 6$ **b.** $y = 2x^2 + 2x - 12$

21a. $(3, 4)$ **b.** $y = 2x^2 - 12x + 22$

23a. $(-4, -3)$ **b.** $y = \dfrac{-1}{2}x^2 - 4x - 11$

25a. $y = a(x + 2)^2 + 6$ **b.** 3

27a. $y = \dfrac{3}{4}x^2 - 3$ **b.** $y = -x^2 - 3$

29. $y = x^2 - 9$ **31.** $y = -2x^2$

33. $y = x^2 - 2x - 15$ **35.** $y = x^2 - 4x + 5$

37a.

t	0	0.5	1.0	1.5	2.0	2.5	3.0	3.5
x	0	8.225	15.9	23.025	29.6	35.625	41.1	46.025
y	0	7.44	12.48	15.12	15.36	13.2	8.64	1.68

b.

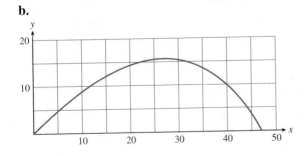

c. $y \approx 15.5$ m   **d.** $x \approx 47$ m   **e.** 3.6 sec

f. $x = 46.94$ m   **g.** $y = 15.55$ m

39a.

No. of price increases	Price of room	No. of rooms rented	Total revenue
0	20	60	1200
1	22	57	1254
2	24	54	1296
3	26	51	1326
4	28	48	1344
5	30	45	1350
6	32	42	1344
7	34	39	1326
8	36	36	1296
10	40	30	1200
12	44	24	1056
16	52	12	624
20	60	0	0

b. Price of a room: $20 + 2x$; rooms rented: $60 - 3x$; revenue: $1200 + 60x - 6x^2$ **d.** 20

e.

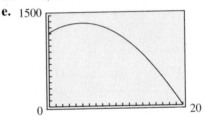

f. \$24; \$36   **g.** \$1350; \$30; 45 rooms

Homework 4.2

1. $a = 3$, $b = 1$, $c = -2$

3a. $P = -0.16x^2 + 7.4x - 71$ **b.** 14%

c.

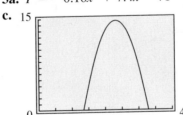

5a. $C = 0.032t^2 + 0.08t + 27.7$ **b.** 44.95 lb

c.

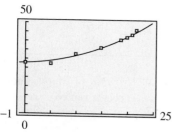

7. $D = \dfrac{1}{2}n^2 - \dfrac{3}{2}n$ **9.** $y = -2(x - 30)^2 + 280$

11a. $y = \dfrac{-1}{40}(x - 80)^2 + 164$ **b.** 160.99 ft

13. $y = 0.00012(x - 2000)^2 + 20$

15a. $y = 8.24x + 38.89$ **b.** 162.5 m

c. 160

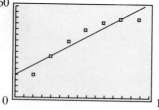

d. $y = -0.81x^2 + 21.2x$ **e.** 135.7 m

f. 160

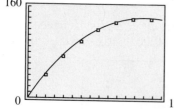

g. Quadratic   **17a.** $y = 0.0051x + 11.325$

b. 13.2 hrs   **c.** 15

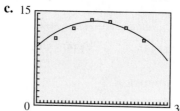

d. $y = -0.00016x^2 + 0.053x + 9.32$ **e.** 7.4 hrs

f. 15

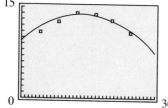

g. 9.8 hrs. Neither model is very good, but the quadratic is better.

19a. $y = -0.023x^2 + 0.44x - 0.157$

b. 3

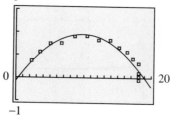

c. (9.6, 1.99). The estimate of 9.6 grams for the weight at which maximum growth occurs is the same as the previous estimate.

Homework 4.3

1. 3 sec, 144 ft   **3a.** Length: $50 - w$; Area: $50w - w^2$ **b.** 625 sq in   **5a.** $300w - 2w^2$
b. 11,250 sq yd   **7a.** Number of people: $16 + x$, Price per person: $2400 - 100x$, Total revenue: $38,400 + 800x - 100x^2$ **b.** 20
9. 100 baskets, $2000   **11a.** 6 starlings, 26 pecks per minute   **b.** Roughly

c. 30

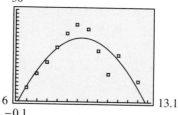

d. 6.4 or 6 starlings, 23 pecks per minute
13. $a = 0.9$, $I = \$865.80$ **15.** $(-1, 12)$, $(4, 7)$
17. $(-2, 7)$ **19.** None   **21.** $(-2, -5)$,
$(5, 16)$ **23.** $(1, 4)$ **25.** $(3, 1)$
27a.

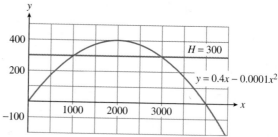

b. Larger, by 75 tons. Smaller, by 125 tons.
c. 1000 tons and 3000 tons   **d.** The fish population will decrease each year until it is completely depleted.

29a.

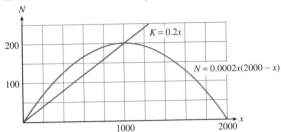

b. $K > N$. The population will decrease by 48 bears.   **c.** The population will increase by 18 bears.   **d.** 1000   **e.** Populations between 0 and 1000 will increase, populations over 1000 will decrease.   **f.** 1000 (unless the population is 0)   **g.** 500 (unless the population is 0)

31a. (200, 2600), (1400, 18,200)

b.

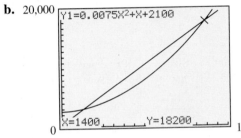

c. $x = 800$ **33a.** (60, 39,000), (340, 221,000)

b.

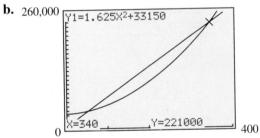

c. $x = 200$

Midchapter 4 Review

1. $\left(\dfrac{5}{4}, \dfrac{17}{8}\right)$, maximum   **3.** Two

5a. $(0, -6), (-3, 0), (2, 0)$; vertex $\left(\dfrac{-1}{2}, \dfrac{-25}{4}\right)$

b.

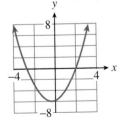

7a. $(-3, -2)$

b. $y = \dfrac{1}{3}x^2 + 2x + 1$

c.

x	-6	0	1
y	1	1	$\frac{10}{3}$

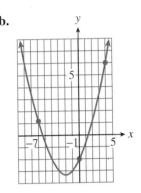

9a. $y = \dfrac{1}{4}x^2 + x - 2$ **b.**

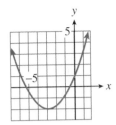

11a. Leaves hand at $(-4, 45)$; vertex $(0, 49)$

b. $h = \dfrac{-1}{4}x^2 + 49$ **c.** 14 feet

13a.

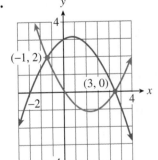

b. $(-0.03, 234.2)$ The velocity is zero at maximum height.   **15a.** 45   **b.** $810

17a. $(-1, 2), (3, 0)$

b.

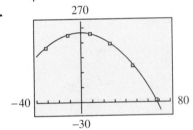

Homework 4.4

1a, b.

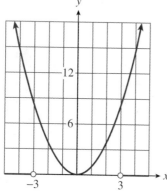

c. $x < -3$ are also solutions.

3a,b.

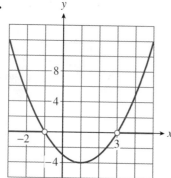

c. $x > 3$ or $x < -1$

b. $x < -12$ or $x > 15$

b. $0.3 \leq x \leq 0.5$

5a. $x = -12, 15$

7a. $x = 0.3, 0.5$

9. $(-5, 3]$

(number line: open circle at −5, closed at 3, with −5, 0, 3 labeled)

11. $[-4, 0]$

(number line: closed dots at −4 and 0)

13. $(-6, \infty)$

(number line: open circle at −6, shading right)

15. $(-\infty, -3) \cup [-1, \infty)$

(number line: −3, −1 0)

17. $[-6, -4) \cup (-2, 0]$

(number line: −6, −4, −2, 0)

19a. $(-\infty, -3) \cup (6, \infty)$ **b.** $(-3, 6)$
c. $[-2, 5]$ **d.** $(-\infty, -2) \cup (5, \infty)$
21a. $(-4, 4)$ **b.** $(-\infty, -4) \cup (4, \infty)$
c. $(-\infty, -3] \cup [3, \infty)$ **d.** $[-3, 3]$

23. $x < -2$ or $x > 3$ **25.** $0 \leq k \leq 4$
27. $p < -1$ or $p > 6$ **29.** $-4.8 < x < 6.2$
31. $x \leq -7.2$ or $x \geq 0.6$ **33.** $x < 5.2$ or
$x > 8.8$ **35.** $x < -3.5$ or $x > 3.5$
37. All x **39.** $-10.6 < x < 145.6$

Homework 4.5

1. $(-3, 4)$ **3.** $[-7, 4]$
5. $\left(-\infty, \dfrac{-1}{2}\right) \cup (4, \infty)$ **7.** $(-8, 8)$
9. $(-2.24, 2.24)$ **11.** $(-\infty, 0.4] \cup [6, \infty)$
13. $\left(-\infty, \dfrac{-2}{3}\right) \cup \left[\dfrac{1}{2}, \infty\right)$
15. $(-\infty, 0.27] \cup [3.73, \infty)$

17. All x **19.** No solution   **21.** $4 < t < 16$
23. $0 \leq x < 100$ or $600 < x \leq 700$
25. $10 < p < 30$ **27.** $5 < r < 12$
29a.

Additional people	Size of group	Price per person	Total income
0	20	600	12,000
5	25	550	13,750
10	30	500	15,000
15	35	450	15,750
20	40	400	16,000
25	45	350	15,750
30	50	300	15,000
35	55	250	13,750
40	60	200	12,000
45	65	150	9750
50	70	100	7000
55	75	50	3750
60	80	0	0

b. Size of group: $20 + x$; Price per person:
$600 - 10x$ **c.** $y = (20 + x)(600 - 10x)$
e. $16,000$; 40   **f.** Between 35 and 45
g.

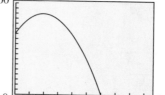

Chapter 4 Review

1. $\left(\dfrac{1}{2}, \dfrac{-49}{4}\right)$; $(-3, 0)$, $(4, 0)$, $(0, -12)$

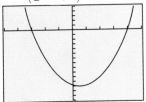

3. $(1, 5)$; $(-1.24, 0)$, $(3.24, 0)$, $(0, 4)$

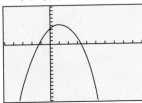

5. $(-\infty, -2) \cup (3, \infty)$ **7.** $\left[-1, \dfrac{3}{2}\right]$

9. $[-2, 2]$ **11a.** $R = p\left(220 - \dfrac{1}{4}p\right)$

b. $400 < p < 480$

13a. $y = 60(4 + x)(32 - 4x)$ **b.** 2

15. $(1, 3)$, $(-1, 3)$

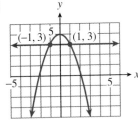

17. $(-1, -4)$, $(5, 20)$

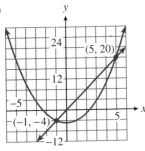

19. $(-9, 155)$, $(5, 15)$

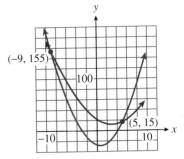

21. $a = 1$, $b = -1$, $c = -6$

23. $y = 0.2(x - 15)^2 - 6$

Homework 5.1

1. Function; the tax is determined by the price of the item   **3.** Not a function; incomes may differ for same number of years of education
5. Function; weight is determined by volume
7. Independent: items purchased; dependent: price of item. Yes, a function.   **9.** Independent: topics; dependent: page or pages on which topic occurs. No, not a function.   **11.** Independent: students' names; dependent: students' scores on quizzes, tests, etc. No, not a function.
13. Independent: person stepping on scales; dependent: person's weight. Yes, a function.
15. No   **17.** Yes   **19.** Yes   **21.** Yes
23. No   **25.** Yes   **27a.** 60   **b.** 37.5
c. 30   **29a.** 15%   **b.** 14%
c. $7010– $9169   **31a.** 1991   **b.** 1 yr
c. 1 yr   **d.** 7000   **33a.** Approximately
$365   **b.** $2 or $8   **c.** $3.25 < d < 6.75$

35a.

Initial clutch size laid	Experimental clutch size				
	4	5	6	7	8
5	0.3	0.7	0.5	0.3	0
6	1.7	1.9	2.8	0.8	1.2
7	3.5	2.3	3.1	3.6	2.4
8	2.5	3.5	3.5	4.3	4.5

b.

Initial clutch size	5	6	7	8
Optimum clutch size	5	6	7	8

c. In each category, the natural clutch size produced the largest number of fledglings. Thus, optimal clutch size seems to depend upon the individual bird.

37a.

Altitude (km)	Percent of Earth's surface	Altitude (km)	Percent of Earth's surface
−7 to −6	1.0	−1 to 0	8.5
−6 to −5	16.4	0 to 1	20.9
−5 to −4	23.2	1 to 2	4.5
−4 to −3	13.9	2 to 3	2.2
−3 to −2	4.8	3 to 4	1.1
−2 to −1	3.0	4 to 5	0.5

b. No   **c.** 4 to 5 km deep and sea level to 1 km elevation   **d.** Most of Earth's surface lies at two levels, the abyssal plains of the ocean floor and low-lying land. Areas of greater elevation are successively less common.
e. Regions of depth 1 to 2 km below sea level.
f. Mount Everest should appear between 8 and 9 km on the horizontal axis. It is not represented because the percentage of Earth's surface at that elevation is so small.   **39a.** 0   **b.** 10
c. -19.4 **d.** $\dfrac{14}{3}$ **41a.** 1   **b.** 6
c. $\dfrac{3}{8}$ **d.** 96.48   **43a.** $\dfrac{5}{6}$ **b.** 9
c. $\dfrac{-1}{10}$ **d.** ≈ 0.923 **45a.** $\sqrt{12}$ **b.** 0
c. $\sqrt{3}$ **d.** ≈ 0.447
47a.

t	2	5	8	10
V	23,520	16,800	10,080	5600

b. $V(12) = 1120$ gives the value of the sports utility vehicle after 12 years.
49a.

p	5000	8,000	10,000	12,000
N	2400	1500	1200	1000

b. $N(6000) = 2000$ is the number of cars that will be sold at a price of $6000.
51a.

d	20	50	80	100
v	15.5	24.5	31.0	34.6

b. $v(250) = 54.8$ is the speed of a car that left 250-foot skid marks.

Homework 5.2

1a. $-2, 0, 5$ **b.** 2   **c.** $h(-2) = 0, h(0) = -2$, $h(1) = 0$ **d.** 5   **e.** 3   **f.** Increasing: $(-4, -2)$ and $(0, 3)$, decreasing: $(-2, 0)$
3a. $-1, 2$ **b.** $3, -1.25$ **c.** $R(-2) = 0$, $R(0) = 4, R(2) = 0, R(4) = 0$ **d.** Maximum: 4, minimum: -5 **e.** Maximum at $p = 0$, minimum at $p = 5$ **f.** Increasing: $(-3, 0)$ and $(1, 3)$, decreasing: $(0, 1)$ and $(3, 5)$
5a. $0, \dfrac{1}{2}, 0$ **b.** $\dfrac{5}{6}$ **c.** $\dfrac{-5}{6}, \dfrac{-1}{6}, \dfrac{7}{6}, \dfrac{11}{6}$
d. Maximum: 1, minimum: -1 **e.** Maximum at $x = -1.5, 0.5$, minimum at $x = -0.5, 1.5$
7a. $2, 2, 1$ **b.** $-6 \le s < -4$ or $0 \le s < 2$
c. Maximum: 2, minimum: -1 **d.** Maximum for $-3 \le s < -1$ or $3 \le s < 5$, minimum for $-6 \le s < -4$ or $0 \le s < 2$

9. (a) and (d)
11a.

x	-2	-1	0	1	2
$g(x)$	-4	3	4	5	12

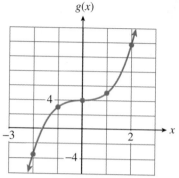

13a.

x	-5	-4	-3	-2	-1
$G(x)$	3	2.8	2.6	2.4	2.2

x	0	1	2	3	4
$G(x)$	2	1.7	1.4	1	0

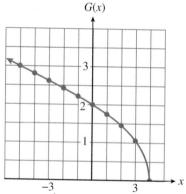

15a.

x	-3	-2	-1	0	1	2	3
$v(x)$	10	-3	-4	1	6	5	-8

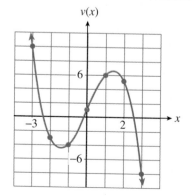

17a. $-1, 1$ **b.** $-3 < x < 2$ or $2 < x < 3$
19a. $-2, 2$ **b.** $-2.8, 0, 2.8$
c. $-2.5 < q < -1.25, 1.25 < q < 2.5$
d. $-2 < q < 0, 2 < q$
21. $(4.8, 3.6), (-4.8, 3.6)$
23. $(-1.6, 4.352), (1.6, -4.352)$ **25a.** No
b. Maximum sediment yield of 800 tons per
square mile occurs when precipitation is 12 inches
per year.   **c.** $12 < p < 60$ **d.** In desert
shrub, increased precipitation causes increased
runoff. In grassland, runoff decreases to 350 tons
per square mile as precipitation increases to 30
inches annually, because vegetation increases. In
forests, runoff decreases slightly with increased
precipitation, but levels off beyond 50 inches at
about 300 tons per square mile.   **27a.** $18°$
b. 0.2 to 0.5 km   **c.** About $-26°$ per kilometer
29a.

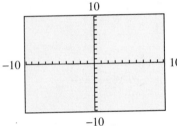

b.

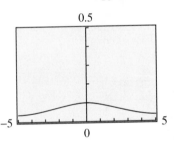

31a.

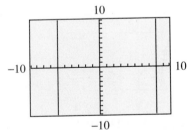

b.

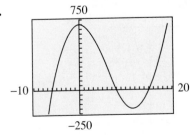

33.

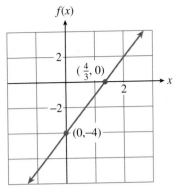

35.

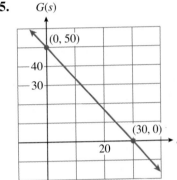

37.

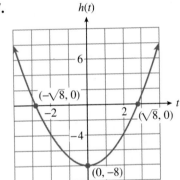

39.

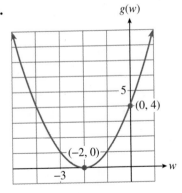

41.

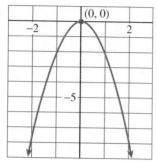

43a. $0, 5$ **b.** $0, \dfrac{-3}{2}$ **c.** $\dfrac{5}{6}$ **d.** $-5, \dfrac{1}{2}$

e.

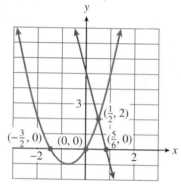

45a. $0, 3$ **b.** $0, 1$ **c.** None   **d.** $-1, 3$

e.

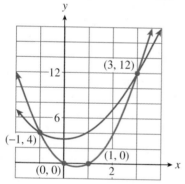

Homework 5.3

1a. 8   **b.** -5 **c.** -0.4 **d.** 1.2

3a. -4 **b.** $\dfrac{-1}{3}$ **5a.** -9 **b.** 9

7a. -4 **b.** 20   **9.** -50 **11.** 144

13. 1

15.

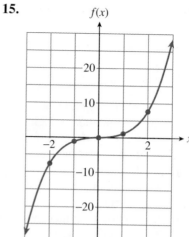

17.

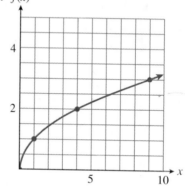

19.

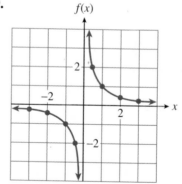

21a. 2.7   **b.** -2.7 **c.** 2.9   **d.** 1.8
e. $-2.3 \le x \le 2.5$ **23a.** 0.3   **b.** -0.4
c. 0.2   **d.** $0.2 \le x \le 3.3$
25.

g is shifted 2 units down from f, and h is shifted 1 unit up.   **27.**

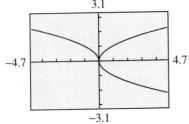

g is shifted 1.5 units to the left from f, and h is shifted 1 unit to the right.
29.

g is reflected about the x-axis from f, and h is reflected about the y-axis.   **31a.** $f(x) = \sqrt{x}$

b. $f(x) = \sqrt[3]{x}$ **c.** $f(x) = |x|$ **d.** $f(x) = \dfrac{1}{x}$

e. $f(x) = x^3$ **f.** $f(x) = \dfrac{1}{x^2}$ **33a.** $x = 12$

b. $x = 18$ **c.** $x < 9$ **d.** $x > 3$
35a. $t = -3$ **b.** $t = 1.5$ **c.** $t < 0.8$
d. $-2.4 < t < 0.4$ **37a.** (vi)   **b.** (ii)
c. (iv)   **d.** (i)   **e.** (v)   **f.** (iii)
39a. 41   **b.** $29 < x \le 61$ **41a.** $-5, 17$
b. $-1 < x < 13$

Midchapter 5 Review

1. At any given time, a person's age is a function of social security number. (There is only one person associated with any given social security number.)   **3.** The table shows $x = 1$ associated with both $y = 9$ and $y = 6$, so y is not a function of x.   **5a.** $s = h(t)$ **b.** The duck was at a height of 7 meters 3 seconds after it was flushed out of the bushes.   **7a.** 315   **b.** When no money is spent on advertising, revenue is 20 thousand dollars.
9.

x	0	1	2	3	4	5	6	7	8	9
$f(x)$	2	1	.586	.268	0	$-.236$	$-.449$	$-.646$	$-.828$	-1

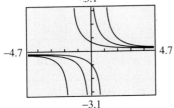

f(x)

11.   p(x)

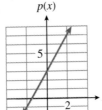

13. 7   **15.** -15

Homework 5.4

1. Domain: $[-5, 3]$, range: $[-3, 7]$
3. Domain: $[-4, 5]$, range: $[-1, 1) \cup [3, 6]$
5. Domain: $[-2, 2]$, range: $[-1, 1]$
7. Domain: $(-5, 5]$, range: $\{-1, 0, 2, 3\}$
9. Domain: $[-2, 5]$, range: $[-4, 12]$
11. Domain: $[-5, 3]$, range: $[-15, 1]$
13. Domain: $[-2, 2]$, range: $[-9, 7]$
15. Domain: $[-1, 8]$, range: $[0, 3]$
17. Domain: $[-1.25, 2.75]$, range: $\left[\dfrac{4}{17}, 4\right]$

19. Domain: $(3, 6]$, range: $\left(-\infty, \dfrac{-1}{3}\right]$
21a. Domain: all real numbers, range: all real numbers   **b.** Domain: all real numbers, range: $[0, \infty)$ **23a.** Domain: all real numbers except zero, range: $(0, \infty)$ **b.** Domain: all real numbers except zero, range: all real numbers except zero   **25a.** Domain: all real numbers except 4, range: $(0, \infty)$ **b.** Domain: all real numbers except zero, range: $(-4, \infty)$ **27a.** Domain: all real numbers, range: all real numbers
b. Domain: all real numbers, range: all real numbers   **29a.** Domain: $[2, \infty)$, range: $[0, \infty)$
b. Domain: $[0, \infty)$, range: $[-2, \infty)$ **31a.** $f(x)$
b. $g(x)$ **33a.** $g(x)$ **b.** $f(x)$
35. Increasing: $x^3, \sqrt{x}, \sqrt[3]{x}, x$; decreasing: $\dfrac{1}{x}$
37. $x^3, |x|, \sqrt{x}, \sqrt[3]{x}, x, x^2$

Homework 5.5

1a. Yes, 6.5%   **b.** $T = 0.065p$

c. T

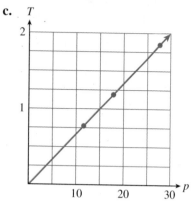

3a. 24   **b.** $L = \dfrac{24}{w}$

c. L

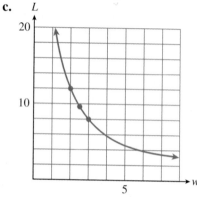

5a. No   **b.** $P = 16 + 2l$ **c.** Yes
d. $A = 8l$ **7.** (b)   **9.** (c)
11a. $m = 0.165w$

w	100	150	200	400
m	16.5	24.75	33	66

b. 19.8 lb   **c.** 303.03 lb
13a. $L = 0.8125T^2$

T	1	5	10	20
L	0.8125	20.3	81.25	325

b. 234.8125 ft   **c.** 0.96 sec
15a. $B = \dfrac{88}{d}$

d	2	4	12	24
B	44	22	7.3	3.7

b. 8.8 milligauss   **c.** More than 20.47 in.
17a. $P = 0.2228w^3$

w	10	20	40	80
P	223	1782	14,259	114,074

b. 752 kilowatts   **c.** 35.54 mph

19a. $y = 0.3x$
b.

x	2	5	8	12	15
y	0.6	1.5	2.4	3.6	4.5

c. y doubles also   **21a.** $y = \dfrac{2}{3}x^2$
b.

x	3	6	9	12	15
y	6	24	54	96	150

c. y is quadrupled   **23a.** $y = \dfrac{120}{x}$
b.

x	4	8	20	30	40
y	30	15	6	4	3

c. y is halved   **25.** (b): $k = 0.5$ **27.** (c)
29. (b): $k = 72$ **31.** (c)   **33a.** $d = 0.005v^2$

b. 50 m   **35a.** $m = \dfrac{8}{p}$ **b.** 0.8 ton

37a. $T = \dfrac{6}{d}$ **b.** 1°C   **39a.** $W = 600d^2$

b. 864 newtons   **41.** One-fourth of original
illumination   **43.** 81% of original resistance
45. If $y_1 = kx_1$, then $y_2 = k(cx_1) = cy_1$

47. If $y = kx^2$, then $\dfrac{y}{x^2} = k$.   **49.** Yes

Homework 5.6

1. (b)   **3.** (a)
5. h

7. d

9. d

11. (b)   **13a.** II   **b.** IV   **c.** I   **d.** III
15a. Table (4), Graph (c)   **b.** Table (3),
Graph (b)   **c.** Table (1), Graph (d)
d. Table (2), Graph (a)   **17a.** III   **b.** III

19a. $k = 50$, $s = 50\sqrt{d}$ **b.** 30 cm/sec

c. 40

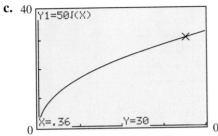
0
0.4

21a. $k = 1.225$, $d = 1.225\sqrt{h}$ **b.** 173.24 mi

c. 200

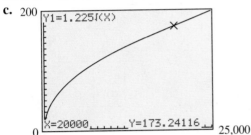
0
25,000

23. $|x| = 6$ **25.** $|p + 3| = 5$ **27.** $|t - 6| \le 3$
29. $|b + 1| \ge 0.5$ **31a.** $|x + 12|$, $|x + 4|$, $|x - 24|$
b. $f(x) = |x + 12|$, $|x + 4|$, $|x - 24|$
c. She should stand at x-coordinate -4.
33. They should find an apartment as close to the health club as possible.
35a. $x = -5$, $x = -1$ **b.** $-7 \le x \le 1$

c. $x < -8$ or $x > 2$ **37.** $x = \dfrac{-3}{2}$, $x = \dfrac{5}{2}$

39. $\dfrac{-9}{2} < x < \dfrac{-3}{2}$ **41.** $x \le -2$ or $x \ge 5$

Chapter 5 Review

1. Function; there is a unique value of the dependent variable for each value of the independent variable.   **3.** Not a function; there is not a unique value of the dependent variable for each value of the independent variable.
5. $N(10) = 7000$; this is the number of barrels of oil 10 days after a new well is opened.
7. $F(0) = 1$, $F(-3) = \sqrt{37} \approx 6.08$
9. $h(8) = -6$, $h(-8) = -14$ **11a.** $P(0) = 5$
b. $x = 1$, $x = 5$ **13a.** $f(-2) = 3$, $f(2) = 5$
b. $t = 1, 3$ **c.** t-intercepts: -3, 4; $f(t)$-intercept: 2   **d.** Maximum value: 5, occurs at $t = 2$
15. Domain: $[-2, 4]$; range: $[-10, 2.25]$

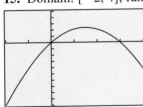

17. Domain: $(-2, 4]$; range: $\left[\dfrac{1}{6}, \infty\right)$

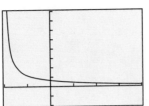

19. Function

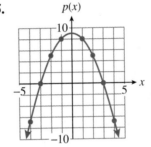

21. Not a function   **23.** $f(t)$

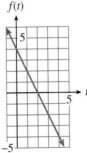

25. $p(x)$

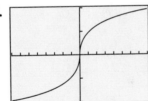

27.

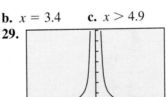

a. $x = 0.5$

b. $x = 3.4$ **c.** $x > 4.9$ **d.** $x \le 2.0$
29.

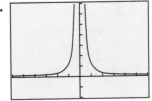

a. $x = \pm 5.8$

b. $x = \pm 0.4$ **c.** $-2.5 < x < 0$ or $0 < x < 2.5$
d. $x \le -0.5$ or $x \ge 0.5$ **31.** $y = 1.2x^2$

33. $y = \dfrac{20}{x}$ **35a.** $s = 1.75t^2$ **b.** 63 cm

37. 480   **39a.** $w = \dfrac{k}{r^2}$

b. *w*

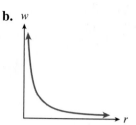

c. $3960\sqrt{3} \approx 6858.92$ mi

41. *w*

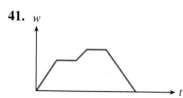

43. $|x| = 4$ **45.** $|p - 7| \leq 4$

47. $\dfrac{-2}{3} < x < 2$ **49.** $y \leq -0.9$ or $y \geq 0.1$

51.

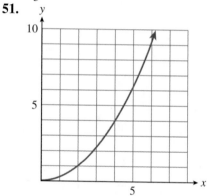

53.

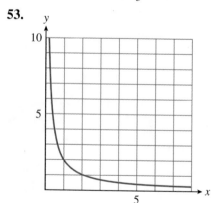

55a. *g(x)*

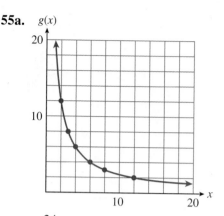

b. $y = \dfrac{24}{x}$

57a.

x	0	4	8	14	16	22
y	24	20	16	10	8	2

b. $y = 24 - x$

59a.

x	0	1	4	9	16	25
y	0	1	2	3	4	5

b. $y = \sqrt{x}$

61a.

x	−3	−2	1	0	1	2
y	5	0	−3	−4	−3	0

b. $y = x^2 - 4$

Homework 6.1

1a. $\dfrac{1}{2}$ **b.** $\dfrac{1}{25}$ **c.** 27   **d.** 16

3a. 320   **b.** $\dfrac{1}{32q^5}$ **c.** $\dfrac{-4}{x^2}$ **d.** $8b^3$

5a. $\dfrac{1}{(m-n)^2}$ **b.** $\dfrac{1}{y^2} + \dfrac{1}{y^3}$ **c.** $\dfrac{2p}{q^4}$

d. $\dfrac{-5x^5}{y^2}$ **7a.** 8   **b.** −8   **c.** $\dfrac{1}{8}$

d. $\dfrac{-1}{8}$ **9a.** $\dfrac{1}{8}$ **b.** $\dfrac{-1}{8}$ **c.** 8   **d.** −8

11a.

x	1	2	4	8	16
f(x)	1	0.25	0.06	0.02	0.00

b. The values of $f(x)$ decrease, because x^{-2} is the reciprocal of x^2.

c.

x	1	0.5	0.25	0.125	0.0625
x^{-2}	1	4	16	64	256

d. The values of $f(x)$ increase towards infinity.
13. (b), (c), and (d) have the same graph, because they represent the same function.

15a. $F(r) = 3r^{-4}$ **b.** $G(w) = \dfrac{2}{5}w^{-3}$

c. $H(z) = \dfrac{1}{9}z^{-2}$ **17a.** $T = kr^2d^4$

b. $k = 0.00625$ **19a.** $E = \dfrac{k}{x^3}$

b. $k = \dfrac{2qL}{4\pi\epsilon}$

21a.

v^2	29.16	43.56	54.76	60.84	75.69	90.25	98.01
F	12.7	15.6	19.0	22.4	23.4	24.4	30.7

v^2	100	110.25	114.49	129.96	139.24	141.61
F	27.3	32.2	30.2	36.1	37.6	35.6

b. $a \approx 0.21$, $b \approx 7.18$

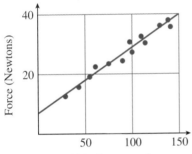

c. 30.7 newtons   **d.** 9.13 meters per second
e.

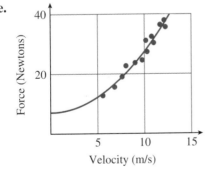

23a.

x	1	2.25	4	6.25	9	12.25
d	0.95	2.16	3.89	6.12	8.80	11.85

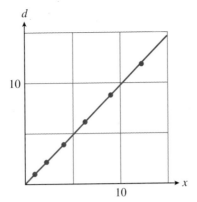

b. $k \approx 0.972$ **c.** $g \approx 9.8$

25a. $\lambda_{max} = \dfrac{2898}{T}$ **b.** $0.50\ \mu m$

c. Green light is a narrow band of wavelengths in the middle of the spectrum. The sun does not appear green because it also radiates light with wavelengths on either side of λ_{max}, that is, blue and red light. When blue, red, and green light are mixed, they appear white. The higher the temperature of a star, the shorter the wavelength at λ_{max}. Thus, hotter stars appear blue and cooler stars appear red.   **d.** $T = \dfrac{2898}{\lambda_{max}}$

e.

Star	R Cygni	Betelgeuse	Arcturus
λ_{max}	1.115	0.966	0.725
Temperature	2600	3000	4000

Star	Polaris	Sirius	Rigel
λ_{max}	0.414	0.322	0.223
Temperature	7000	9000	13,000

f.

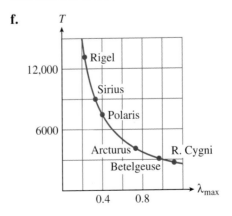

27a. a^5 **b.** $\dfrac{1}{5^7}$ **c.** $\dfrac{1}{p^3}$ **d.** $\dfrac{1}{7^{10}}$

29a. $\dfrac{20}{x^3}$ **b.** $\dfrac{1}{3u^{12}}$ **c.** $5^8 t$ **31a.** $\dfrac{x^4}{9y^6}$

b. $\dfrac{a^6 b^4}{36}$ **c.** $\dfrac{5}{6h^6}$ **33a.** No. Because

$\dfrac{1}{(x+y)^2} \neq \dfrac{1}{x^2} + \dfrac{1}{y^2}$ **b.** $\dfrac{1}{(2+2)^2} = \dfrac{1}{16}$, but

$\dfrac{1}{2^2} + \dfrac{1}{2^2} = \dfrac{1}{2}$ **35a.** $\dfrac{1}{3}x + 3x^{-1}$

b. $\dfrac{1}{4}x^{-2} - \dfrac{3}{2}x^{-1}$ **37a.** $\dfrac{1}{2}x^{-2} + x^{-3} - \dfrac{1}{2}x^{-4}$

b. $\dfrac{2}{3}x^{-2} - \dfrac{1}{9} + \dfrac{1}{6}x^2$ **39.** $x - 3 + 2x^{-1}$

41. $-3 + 6t^{-2} + 12t^{-4}$
43. $-4 - 2u^{-1} + 6u^{-2}$ **45.** $4x^{-2}(x^4 + 4)$
47. $a^{-3}(3 - 3a^4 + a^6)$ **49a.** 2.85×10^2
b. 8.372×10^6 **c.** 2.4×10^{-2}
d. 5.23×10^{-4} **51a.** 240
b. 6,870,000,000,000,000 **c.** 0.005
d. 0.000202 **53a.** 3,000,000 **b.** 112,500
55a. 5.605304×10^{12} **b.** \$20,553.93
57a. 7.9×10^9 km/yr **b.** 250,000 m/sec
59a. 250,000 times **b.** 8×10^{19} meters
61. 1,000,000 years **63a.** 500 picowatts
b. $P = 8000d^{-4}$

c.

d (nautical miles)	4	5	7	10
P (picowatts)	31.3	12.8	3.3	0.8

d. 16.8 nautical miles
e.

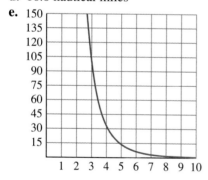

Homework 6.2

1a. 11 **b.** 3 **c.** 5 **3a.** 2 **b.** 2
c. 9 **5a.** 3 **b.** 3 **c.** 2 **7a.** 2
b. $\dfrac{1}{2}$ **c.** $\dfrac{1}{8}$ **9a.** $\sqrt{3}$ **b.** $4\sqrt[3]{x}$

c. $\sqrt[5]{4x}$ **11a.** $\dfrac{1}{\sqrt[3]{6}}$ **b.** $\dfrac{3}{\sqrt[8]{xy}}$ **c.** $\sqrt[4]{x-2}$

13a. $7^{1/2}$ **b.** $(2x)^{1/3}$ **c.** $2z^{1/5}$

15a. $-3(6^{-1/4})$ **b.** $(x - 3y)^{1/4}$
c. $-(1 + 3b)^{-1/3}$ **17a.** 1.414 **b.** 4.217
c. 1.125 **d.** 0.140 **e.** 2.782 **19.** 132.6 km
21a.

r (meters)	0.2	0.4	0.6	0.8	1.0
v (meters per second)	1.4	2	2.4	2.8	3.1

b. 3 meters per second **c.** 2.2 meters per second
d.

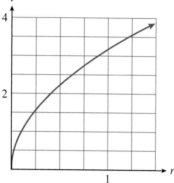

e. 1.9 meters **f.** 1.2 meters per second
23a. 6.5×10^{-13} centimeters; 1.15×10^{-36}
cubic centimeters **b.** 1.8×10^{14} grams per
cubic centimeter
c.

Element	Carbon	Potassium	Cobalt
Mass number, A	14	40	60
Radius, r (10^{-13} cm)	3.1	4.4	5.1

Element	Technetium	Radium
Mass number, A	99	226
Radius, r (10^{-13} cm)	6	7.9

d.

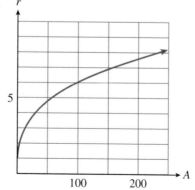

25a. 287; 343   **b.** 1985; 2028

c.

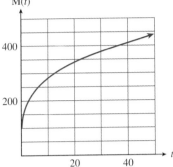

The membership grows rapidly at first, then slows down.   **27a.** I   **b.** III   **c.** II

29a. All the graphs are increasing and concave down, and each flattens out more quickly than the previous one.   **b.** 10, 4.64, 3.16, 2.51

c. 1.58, 1.05, 1.005; $100^{1/n}$ gets closer to 1.

31 and 33. y_1 and y_2 are symmetrical about $y_3 = x$.   **35a.** $x^{1/2}$ **b.** $(x^{1/2})^{1/2}$

c. $\sqrt{\sqrt{x}} = (x^{1/2})^{1/2} = x^{1/4} = \sqrt[4]{x}$ **37a.** 125

b. 2   **c.** 63   **d.** $-2x^7$ **39.** -216

41. 241   **43.** 9.5   **45.** ± 5.477

47. $-5, 12.5$

49a.

a	1	2	4	8
V	3.14	25.13	201.06	1608.5

b. 6.38 m

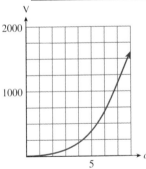

51. $4840°$K   **53.** $\dfrac{4\pi r^3}{3}$ **55.** $\dfrac{8Lvf}{\pi R^4}$

57a.

Type of ball	Bounce height	e
Baseball	1.5	0.50
Basketball	3.375	0.75
Golfball	2.16	0.60
Handball	3.84	0.80
Softball	1.815	0.55
Superball	4.86	0.90
Tennisball	3.286	0.74
Volleyball	3.375	0.75

b. $e = \sqrt{\dfrac{H}{6}}$

c.

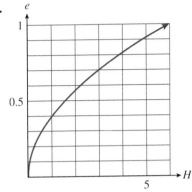

d. Four times as high

59. $\dfrac{1}{4}x^{1/2} - 2x^{-1/2} + \dfrac{1}{\sqrt{2}}x$ **61.** $3x^{-1/3} - \dfrac{1}{2}$

63. $x^{0.5} + x^{-0.25} - 1$

Homework 6.3

1a. 27   **b.** 25   **c.** 125   **3a.** $\dfrac{1}{64}$

b. $\dfrac{1}{16}$ **c.** $\dfrac{1}{256}$ **5a.** $\sqrt[5]{x^4}$ **b.** $\dfrac{1}{\sqrt[6]{b^5}}$

c. $\dfrac{1}{\sqrt[3]{p^2 q^2}}$ **7a.** $3\sqrt[5]{x^2}$ **b.** $\dfrac{4}{\sqrt[3]{z^4}}$

c. $-2\sqrt[4]{xy^3}$ **9a.** $x^{2/3}$ **b.** $2a^{1/5}b^{3/5}$

c. $-4mp^{-7/6}$ **11a.** $(ab)^{2/3}$ **b.** $8x^{3/4}$

c. $\dfrac{1}{3}RT^{-1/2}K^{-5/2}$ **13a.** 8   **b.** -81

c. $2y^3$ **15a.** $-a^4 b^8$ **b.** $2x^3 y^9$

c. $-3a^2 b^3$ **17a.** 7.931   **b.** 10.903

c. 0.090   **d.** 35.142

19a.

t	5	10	15	20
$I(t)$	131	199	254	302

b. 20 days

c. 500

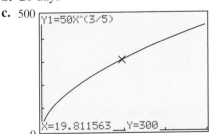

21. All the graphs are increasing and concave up. For $x > 1$, each graph increases more quickly than the previous one.   **23a.** $V = L^3, A = 6L^2$

b. $L = V^{1/3}; L = \left(\dfrac{A}{6}\right)^{1/2}$ **c.** $A = 6V^{2/3}$

d. $\frac{A}{V} = \frac{6}{L}$. As L increases, the surface-to-volume ratio decreases.

25a and b. 5000

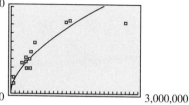

c. 4165 miles   **d.** 341,000 square miles

27a. Home range size: II, Lung volume: III, brain mass: I, Respiration rate: IV   **b.** If $p > 1$, the graph is increasing and concave up. If $0 < p < 1$, the graph is increasing and concave down. If $p < 0$, the graph is decreasing and concave up.

29a. *Tricosanthes* is the snake gourd and *Lagenaria* is the bottle gourd.   **b.** $a \approx 9.5$
c. $a \approx 2$ **d.** Yes.   **31a.** Both graphs bend upwards very slightly, so time is not precisely proportional to distance.   **b.** The slopes of the line segments joining consecutive data points increase slightly as distance increases, so the graphs are not linear.

c.

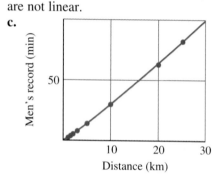

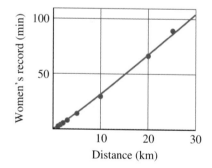

The graph of $W(x)$ bends upwards more than the graph of $M(x)$. Because both exponents are greater than 1, the times increase more sharply as distance increases. This indicates that the runners tire and slow down over longer distances.

33a. 450

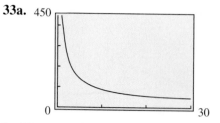

b. 79   **c.** 18.4°C   **d.** An increase in temperature range from 9° to 10° results in a greater drop in the number of species (6) than an increase from 19° to 20° (2 species). The curve is steeper near 10° than near 20°, so a 1° horizontal change on the graph near 10° results in a greater vertical change than a 1° horizontal change near 20°.

35a. The animal's metabolic rate is the rate at which it must consume energy, and the available energy content of the animal's food is proportional to the amount consumed.   **b.** Since the amount of food a particular region can provide is limited, the number of animals that region can support is also limited. The total amount of food consumed by a population of animals cannot exceed the available supply per unit area, C, so for each unit of land area

$$\text{(number of animals)} \times \text{(consumption per animal)} = C$$

or

$$\text{number of animals} = \frac{C}{\text{consumption per animal}}$$

c. By part (b), the population density is inversely proportional to the rate of food consumption per animal, and by part (a) an individual's food consumption is proportional to its metabolic rate, or BMR. Thus

$$\text{population density} = \frac{C'}{\text{BMR}}$$

But Kleiber's rule says that the BMR $= Km^{0.75}$. Therefore,

$$\text{population density} = \frac{C'}{Km^{0.75}} = km^{-0.75}.$$

d. *D*

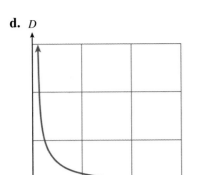

37a. $4a^2$ **b.** $9b^{5/3}$ **39a.** $4w^{3/2}$ **b.** $3z^2$

41a. $\dfrac{1}{2k^{1/4}}$ **b.** $\dfrac{4}{3h^{1/3}}$ **43a.** Wren: 15 days;
greylag goose: 28 days

b. $\dfrac{I(m) \times W(m)}{m} = 0.18m^{-0.04}$ **c.** If $m^{-0.04}$ is
considered to be close to $m^0 = 1$, then the fraction
mass lost over initial mass is 0.18. **45.** 64

47. $\dfrac{1}{243}$ **49.** 2.466 **51a.** $p = K^{1/2}a^{3/2}$

b. 1.88 yr **53.** $\dfrac{13}{3}$ **55.** 0.665

57. $2x^{3/2} - 2x$ **59.** $\dfrac{1}{2}y^{1/3} + \dfrac{3}{2}y^{-7/6}$

61. $2x^{1/2} - x^{1/4} - 1$ **63.** $a^{3/2} - 4a^{3/4} + 4$

65. $x(x^{1/2} + 1)$ **67.** $\dfrac{y - 1}{y^{1/4}}$

69. $\dfrac{a^{2/3} + 3a^{1/3} - 1}{a^{1/3}}$

Midchapter 6 Review

1. $\dfrac{4}{10^3}$ **3.** $t = 10$ **5.** $f(x) = \dfrac{2}{3}x^{-4}$

7a. *v* Yes.

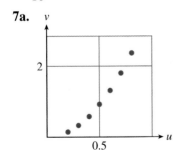

b.

x	.04	.09	.16	.25	.36	.49	.64
v	0.148	0.333	0.592	0.925	1.332	1.813	2.368

c. *v* Yes, $v \approx 3.7x$

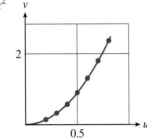

d. $v \approx 3.7u^2$ *v*

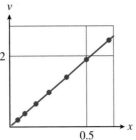

9. $\dfrac{3t^7}{w^3}$ **11.** 1.234×10^6 **13.** 0.3

15. $\sqrt[6]{4096} = 4$ **17.** $\sqrt[5]{0.00243^2} = 0.09$

19. $-\sqrt[4]{256} = -4$ **21.** $\sqrt[3]{-1728} = -12$

23. $-2(5m)^{1/3}$ **25.** $9(2p)^{-3.4}$ **27.** $11\sqrt[5]{h}$

29. $\dfrac{2}{3}\sqrt[5]{c^4}$ **31.** $(7t)^2$ **33.** $\dfrac{k^7}{64}$

35. $8a^2$ **37.** $x = 75$

39. $y = 29{,}524$

41.

x	0	1	5	10	20	50	70	100
f(x)	0	1	1.62	2.00	2.46	3.23	3.58	3.98

f(x)

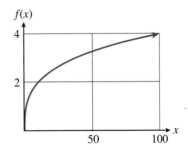

Homework 6.4

1. $5; \left(\dfrac{5}{2}, 3\right)$ **3.** $\sqrt{20}; (0, -2)$ **5.** $5; \left(\dfrac{1}{2}, 5\right)$

7a. 13 miles **b.** $\dfrac{\sqrt{125}}{2} \approx 5.6$ miles

9. $7 + \sqrt{89} + \sqrt{68} \approx 24.7$ **11.** The rectangle is a square because all the sides have length $\sqrt{61}$.   **13.** Both diagonals have length $\sqrt{50} \approx 7.1$.   **15.** $AC = BC = \sqrt{18}$

17. $y = \dfrac{1}{2}x + \dfrac{5}{4}$

19.

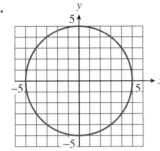

21.

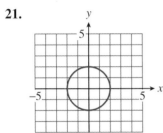

23.

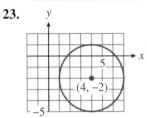

$(4, -2)$

25.

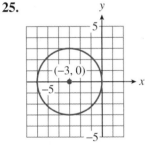

$(-3, 0)$

27. $(x + 1)^2 + (y - 2)^2 = 11; (-1, 2), \sqrt{11}$
29. $(x + 4)^2 + (y - 0)^2 = 20; (-4, 0), \sqrt{20}$
31. $x^2 + y^2 + 4x - 10y + 17 = 0$
33. $x^2 + y^2 - 3x + 8y + 11 = 0$
35. $x^2 + y^2 - 4x - 4y - 2 = 0$
37. $x^2 + y^2 + 6x + 2y + 9 = 0$
39. $x^2 + y^2 + 2x + 2y - 23 = 0$
41. $MD \parallel AC$, therefore $\angle BAC = \angle BMD$ and $\triangle BAC \sim \triangle MAE \sim \triangle BMD$. Hence $\frac{AB}{AM} = \frac{AC}{AE}$. Since M is the midpoint of AB, $AB = 2AM$. Therefore, $AC = 2AE$ as well. Similarly, $\frac{AB}{MB} = \frac{BC}{BD}$. Hence $BC = 2BD$.

Homework 6.5

1a. $3\sqrt{2}$ **b.** $2\sqrt[3]{3}$ **c.** $-2\sqrt[4]{4} = -2\sqrt{2}$
3a. $100\sqrt{6}$ **b.** $10\sqrt[3]{900}$ **c.** $\dfrac{-2}{3}\sqrt[5]{5}$
5a. $x^3\sqrt[3]{x}$ **b.** $3z\sqrt{3z}$ **c.** $2a^2\sqrt[4]{3a}$
7a. $-6s^2$ **b.** $-7h$ **c.** $2\sqrt{4 - x^2}$
9a. $A\sqrt[3]{8 + A^3}$ **b.** $3xy\sqrt{x}$ **c.** $\dfrac{2b\sqrt[3]{b^2}}{a^2}$
11. $5\sqrt{7}$ **13.** $\sqrt{3}$ **15.** $9\sqrt{2x}$
17. $-\sqrt[3]{2}$ **19.** $6 - 2\sqrt{5}$ **21.** $2\sqrt{3} + 2\sqrt{5}$
23. $2\sqrt[3]{5} - 4\sqrt[3]{3}$ **25.** $x - 9$ **27.** $-4 + \sqrt{6}$
29. $7 - 2\sqrt{10}$ **31.** $a - 4\sqrt{ab} + 4b$
33. $(1 + \sqrt{3})^2 - 2(1 + \sqrt{3}) - 3 = 0$
35. $(-3 + 3\sqrt{2})^2 + 6(-3 + 3\sqrt{2}) - 9 = 0$
37. $2\sqrt{3}$ **39.** $\dfrac{\sqrt{14x}}{6}$ **41.** $\dfrac{\sqrt{2ab}}{b}$
43. $\dfrac{\sqrt{6k}}{k}$ **45.** $-2(1 - \sqrt{3})$
47. $\dfrac{x(x + \sqrt{3})}{x^2 - 3}$ **49.** $\dfrac{\sqrt{6}}{2}$
51a. $y = \sqrt{x^2} = |x|$ **b.** $y = \sqrt[3]{x^3} = x$
53a. $2|x|$ **b.** $|x - 5|$ **c.** $|x - 3|$
55. $\dfrac{\sqrt[3]{4x^2}}{2x}$ **57.** $\dfrac{\sqrt[3]{x}}{x}$ **59.** $\dfrac{\sqrt[3]{18y^2}}{3y}$
61. $\dfrac{\sqrt[4]{2xy}}{2y}$

Homework 6.6

1. 64   **3.** -2 **5.** $\dfrac{9}{4}$ **7.** -5
9. 90 ft   **11.** $\dfrac{gT^2}{4\pi^2}$ **13.** $\pm\sqrt{t^2 - r^2}$
15. $\dfrac{-1}{3}$ **17.** $\dfrac{-1}{2}, 2$ **19.** 12   **21.** 5
23. 4   **25.** 5   **27.** 0   **29.** 4
31. $(1, 2)$ **33a.** t miles   **b.** $d = \sqrt{64 + t^2}$
d. 12.7 min   **35a.** $700 - 10t$ **b.** $15t$
d. $D = \sqrt{(15t)^2 + (700 - 10t)^2}$ **e.** 618.5 ft
f. 719.8 ft   **g.** 22 sec   **h.** $x_{\max} = 70$
37a. 1 hr 30 min   **b.** 1 hr 42 min
c. $t = \dfrac{\sqrt{1 + x^2}}{3} + \dfrac{5 - x}{4}$
f. $x = 1.15$ mi; $t = 1$ hr 28.2 min

Chapter 6 Review

1a. $\dfrac{1}{81}$ **b.** $\dfrac{1}{64}$ **3a.** $\dfrac{1}{243m^5}$ **b.** $\dfrac{-7}{y^8}$

5a. $\dfrac{2}{c^3}$ **b.** $\dfrac{99}{z^2}$ **7a.** 1.018×10^{-9} sec
or 0.000000001018 sec **b.** 8 min 20 sec
9a. 5000 seconds, or 83 minutes **b.** 42 years
c. $\dfrac{1}{10}$

11a.

v^2	31.36	56.25	77.44	108.16	139.24
E	0.9	1.8	2.6	3.9	5

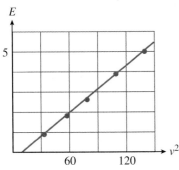

b. $E \approx 0.038v^2 - 0.336$ **c.** The data from
this experiment do not show that E is exactly proportional to v^2. If it were, the vertical intercept of
the graph would be zero. However, there are many
sources of experimental error and statistical variation that could explain the discrepancy.
d. $E \approx 3.1$ calories per kilogram per second

13a. $25\sqrt{m}$ **b.** $\dfrac{8}{\sqrt[3]{n}}$ **15a.** $\dfrac{1}{\sqrt[4]{(3q)^3}}$

b. $7\sqrt{u^3v^3}$ **17a.** $2x^{2/3}$ **b.** $\dfrac{1}{4}x^{1/4}$

19a. $6b^{-3/4}$ **b.** $\dfrac{-1}{3}b^{-1/3}$ **21.** 112 kg

23a. 480 **b.** 498

25a. 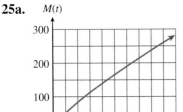 **b.** 283 **c.** 2051

27a.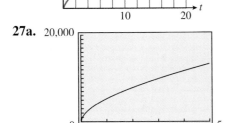

b. \$7114.32 **29a.** It is the cost of producing
the first ship. **b.** $C = \dfrac{12}{\sqrt[8]{x}}$ **c.** About \$11

million; about 8.3%; about 8.3% **d.** About
8.3% **31a.** 4096 **b.** $\dfrac{1}{8}$ **c.** 62.35
d. 400,000 **33.** 169 **35.** 16 **37.** 1, 4
39. 9 **41.** 7 **43.** 5 **45.** $g = \dfrac{2v}{t^2}$
47. $p = \pm 2\sqrt{R^2 - R}$ **49.** 3.07 hr
51. 21.59; yes
53.

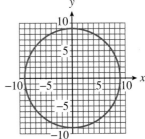

55.

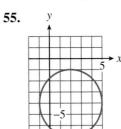

57. $(x - 5)^2 + (y + 2)^2 = 32$
59. $(x - 1)^2 + (y - 4)^2 = 10$
61a. $\dfrac{5p^4}{a^2}\sqrt{5p}$ **b.** $\dfrac{2}{w^2}\sqrt{3v^2}$
63a. $2\sqrt[3]{a^3 - 2b^6}$ **b.** $-4ab^2\sqrt[3]{2}$
65a. $x^2 - 4x\sqrt{x} + 4x$ **b.** $x^2 - 4x$
67a. $\dfrac{7\sqrt{5y}}{5y}$ **b.** $3\sqrt{2d}$

69a. $\dfrac{-3\sqrt{a} + 6}{a - 4}$ **b.** $\dfrac{-3\sqrt{z} - 12}{z - 16}$

Homework 7.1

1a.

Weeks	0	1	2	3	4
Bacteria	300	600	1200	2400	4800

b. $P(t) = 300(2)^t$
c. 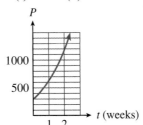 **d.** 76,800; 492

3a.

Weeks	0	6	12	18	24
Bees	20,000	50,000	125,000	312,500	781,250

b. $P(t) = 20,000(2.5)^{t/6}$

c. P (thousands)

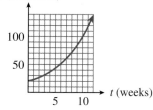

d. 36,840; 424,128

5a.

Years	0	1	2	3	4
Account balance	4000	4320	4665.60	5038.85	5441.96

b. $A(t) = 4000(1.08)^{t}$

c.

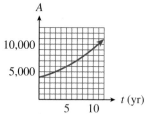

d. $4665.60; $8635.70

7a.

Years since 1963	Value of house
0	20,000
1	21,000
2	22,050
3	23,125.50
4	24,310.13

b. $P(t) = 20,000(1.05)^{t}$

c. P (thousands)

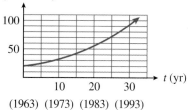

d. $35,917.13; $74,669.13

9a.

Weeks	0	2	4	6	8
Mosquitoes	250,000	187,500	140,625	105,469	79,102

b. $P(t) = 250,000(0.75)^{t/2}$

c. P (thousands)

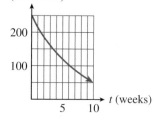

d. 162,380; 79,102

11a.

Feet	0	4	8	12	16
% of light	100	85	72.25	61.41	52.20

b. $L(d) = (0.85)^{d/4}$

c. L (%)   **d.** 44%, 16%

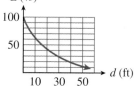

13a.

Years	0	1	2	3	4
Pounds of plutonium-238	50	49.6	49.2	48.8	48.4

b. $P(t) = 50(0.992)^{t}$

c. P (lb)

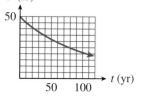

d. 46.1 lb; 22.4 lb   **15a.** To evaluate $P(t)$, we raise 3 to a power and multiply the result by 2. To evaluate $Q(t)$, we raise 6 to the power.

b.

t	0	1	2	3
$P(t)$	2	6	18	54
$Q(t)$	1	6	36	216

17a. 365   **b.** $N(t) = 365(0.356)^{t}$

c.

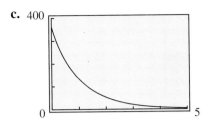

d. 0.03

19a.

n	$-\infty$	1	2	3	4	5	6
$r(n)$	0.4	0.7	1	1.6	2.8	5.2	10

b.

n	$-\infty$	1	2
Planet	Mercury	Venus	Earth

n	3	4	5	6
Planet	Mars	—	Jupiter	Saturn

With the rather contrived strategem of using $-\infty$ for Mercury, Bode's law fits fairly well for the first six planets, except that we must skip $n = 4$ between Mars and Jupiter.   **c.** 19.6 AU
d. Yes, the orbital radius corresponding to $n = 4$ is 2.8 AU, which matches the asteroids' orbital radii.   **e.** 38.8 AU for Neptune and 77.2 for Pluto

21.

x	0	1	2	3	4
Q	20	24	28.8	34.56	41.472

23.

t	0	1	2	3	4
P	8	12	18	27	40.5

25.

w	0	1	2	3	4
N	120	96	76.8	61.44	49.152

27. 4   **29.** 1.2   **31.** 0.14   **33.** 0.04
35a. 39; 1.045   **b.** 36; 1.047   **c.** Species B
37a. $P(t) = 9,579,700(1 + r)^t$ **b.** 1.57%
39a. 3.53%   **b.** 3.53%   **c.** No
d. 3.53%

41a.

t	0	2	4	6	8
$L(t)$	3	6	9	12	15

$L(t) = 1.5t + 6$

b.

t	0	2	4	6	8
$E(t)$	3	6	12	24	48

$E(t) = 3(2)^{t/2}$

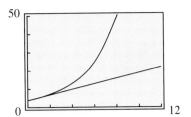

43a. 64 geese per year   **b.** 1.125; 12.5%

Homework 7.2

1a. (0, 26); increasing   **b.** (0, 1.2); decreasing
c. (0, 75); decreasing   **d.** $\left(0, \dfrac{2}{3}\right)$; increasing
3a. power   **b.** exponential   **c.** power
d. neither   **5.**

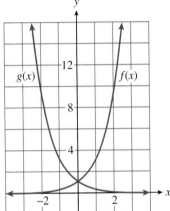

7.

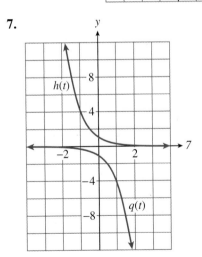

9. (b), (c), and (d) have identical graphs because their formulas are equivalent expressions.

11a.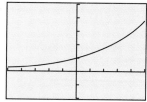

b. [0.27, 3.71]

13a.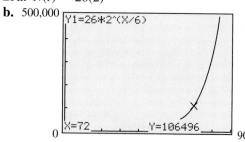

b. [0.33, 3.05] 15. $\dfrac{2}{3}$ 17. $\dfrac{-1}{4}$

19. $\dfrac{1}{7}$ 21. $\dfrac{-5}{4}$ 23. ± 2

25a. $N(t) = 26(2)^{t/6}$

b. 500,000

Y1=26*2^(X/6)

X=72 Y=106496 0 ... 90

c. 72 days later 27a. $V(t) = 700(0.7)^{t/2}$

b. 800

Y1=700*.7^(X/2)

X=4 Y=343 0 ... 20

c. 4 yr 29. 2.26 31. -1.40

33. (a), growth factor; 2; (d), decay factor: $\dfrac{1}{3}$

35a. $P_0 = 300$

b.

x	0	1	2
f(x)	300	600	1200

c. $a = 2$

d. $f(x) = 300(2)^x$ 37a. $S_0 = 150$

b. $a = 0.55$ c. $S(d) = 150(0.55)^d$

39a. $N(x) = 1{,}600{,}000(0.212)^{x-1}$

b. $L(x) = 2.3^{x-1}$ c. About 3,120,000 miles

41a. Reaction time appears to be inversely proportional to a power of the concentration.

b.

Concentration (moles/liter)	0.15	0.12	0.09	0.06	0.03
Reaction time (seconds)	0.023	0.018	0.015	0.0095	0.0041

c. $r = 0.1543c$ d. Yes, the equation for reaction rate as a function of the concentration is of the form $y = kx$. e. $t = \dfrac{6.48}{c}$

43a.

Stage n	S(n)	N(n)	P(n)
0	1	3	3
1	$\frac{1}{3}$	12	4
2	$\frac{1}{9}$	48	$\frac{16}{3}$
3	$\frac{1}{27}$	192	$\frac{64}{9}$

b. $S(n) = \left(\dfrac{1}{3}\right)^n$ c. $N(n) = 3(4)^n$

d. $P(n) = 3\left(\dfrac{4}{3}\right)^n$ e. It becomes longer and longer, eventually becoming infinite in length.

f. The area is finite.

45.

x	$f(x) = x^2$	$g(x) = 2^x$
-2	4	$\frac{1}{4}$
-1	1	$\frac{1}{2}$
0	0	1
1	1	2
2	4	4
3	9	8
4	16	16
5	25	32
6	36	64

a. Range of f: $[0, \infty)$; range of g: $(0, \infty)$

b. Three c. $x = 2, x = 4, x \approx -0.77$

d. $(-0.77, 2) \cup (4, \infty)$ e. $g(x)$ 47a. To evaluate f, we raise 2 to the power $x - 1$; to evaluate g we subtract 1 from 2^x.

b.

x	$y = 2^x$	$f(x)$	$g(x)$
-2	$\frac{1}{4}$	$\frac{1}{8}$	$\frac{-3}{4}$
-1	$\frac{1}{2}$	$\frac{1}{4}$	$\frac{-1}{2}$
0	1	$\frac{1}{2}$	0
1	2	1	1
2	4	2	3

49a. To evaluate f, we multiply 3^x by -1; to evaluate g, we raise 3 to the power $-x$.

b.

x	$y = 3^x$	$f(x)$	$g(x)$
-2	$\frac{1}{9}$	$\frac{-1}{9}$	9
-1	$\frac{1}{3}$	$\frac{-1}{3}$	3
0	1	-1	1
1	3	-3	$\frac{1}{3}$
2	9	-9	$\frac{1}{9}$

51.

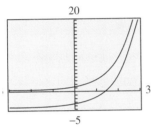

The second graph is shifted 5 units down from the first graph.

53.

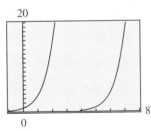

The second graph is shifted 3 units to the right from the first graph.

55.

X	Y_1	Y_2
-3	8	-8
-2	4	-4
-1	2	-2
0	1	-1
1	$.5$	$-.5$
2	$.25$	$-.25$
3	$.125$	$-.125$

$X = -3$

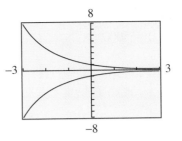

The second graph is reflected about the x-axis.

57.

X	Y_1	Y_2
-3	$.001$	20.1
-2	$.01$	21
-1	$.1$	30
0	1	120
1	10	1020
2	100	10020
3	1000	100020

$Y_2 = 20 + 10^{(X+2)}$

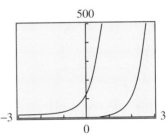

The second graph is shifted 20 units up and 2 units to the left.

Homework 7.3

1a. 2 **b.** 5 **3a.** $\dfrac{1}{2}$ **b.** -1

5a. 1 **b.** 0 **7a.** 5 **b.** 6

9a. -1 **b.** -3 **11.** $\log_t 16 = \dfrac{3}{2}$

13. $\log_{0.8} M = 1.2$ **15.** $\log_x(W - 3) = 5t$

17. $\log_3 2N_0 = -0.2t$ **19a.** $\log_4 2.5$

b. 0.7 **21a.** $\log_{10} 0.003$ **b.** -2.5

23. 0.85 **25.** 3.84 **27a.** 1.7348

b. 3.3670 **c.** -1.1367 **d.** -2.2118

29. -0.23 **31.** 2.53 **33.** 0.77

35. -0.68 **37.** 3.63 **39.** 9.60 in

41. 1.91 mi   **43.** 3.34 mi   **45a.** $19{,}969{,}613$

b. $25{,}372{,}873;\ 32{,}238{,}116;\ 40{,}960{,}915$

c. 1970 **d.** 1987 **47.** $16^w = 256$

49. $b^{-2} = 9$ **51.** $10^{-2.3} = A$

53. $4^{2q-1} = 36$ **55.** $u^w = v$ **57.** 2

59. $\dfrac{1}{2}$ **61.** 100 **63.** 11 **65.** $7^{2/3}$

67. 1 **69.** 0 **71.** 1 **73.** 0

Midchapter 7 Review

1a. $2^{1/4}$ **b.** $P(t) = 5000 \cdot 2^{t/4}$

c. $P(t)$ (thousands)

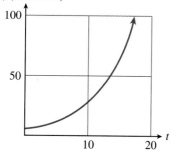

3a.

Years since 1992	0	1	2	3	4
Value of house	135,000	148,500	163,350	179,685	197,654

b. $f(t) = 135,000 \cdot 1.10^t$

c. 200,000

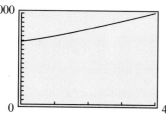

d. \$263,077

5a.

Years since 1995	0	1	2	3	4
Agriculture majors	200	160	128	102	82

b. $f(t) = 200\left(\dfrac{4}{5}\right)^t$

c. 200

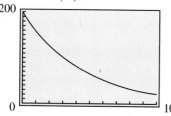

d. 66

7. $f(x)$

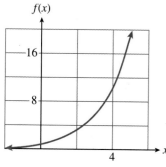

Domain: all real numbers, range: all positive numbers **9.** $x = 7$ **11.** $x \approx 1.58$

13. -2 **15.** $\log_{64} \dfrac{1}{2} = \dfrac{-1}{6}$

17. $\log_{10} 6 \approx 0.7782$ **19a.** $N(t) = 20 \cdot 9^{t/5}$

b. 100,000

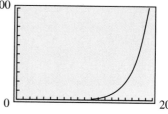

c. 15 weeks **21.** $x = \dfrac{1}{32}$

Homework 7.4

1a.

x	-2	-1	0	1	2
$f(x)$	$\frac{1}{4}$	$\frac{1}{2}$	1	2	4

x	$\frac{1}{4}$	$\frac{1}{2}$	1	2	4
$f^{-1}(x)$	-2	-1	0	1	2

b.

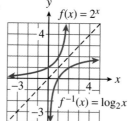

3a.

x	-2	-1	0	1	2
$f(x)$	9	3	1	$\frac{1}{3}$	$\frac{1}{9}$

x	9	3	1	$\frac{1}{3}$	$\frac{1}{9}$
$f^{-1}(x)$	-2	-1	0	1	2

b.

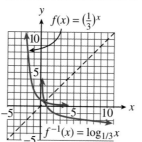

5a. 2.688

b. 0.334 **7a.** Undefined **b.** 8.683

9a. 15.614 **b.** 0.419 **11.** -1.58×10^{-5}

13. 1.76 **15.** 25.70 **17.** 3.31 **19.** 0.05

21. 3.2 **23.** 1.3×10^{-2} **25.** 100 dB

27. 6.31×10^6 W per sq m   **29.** 1000 times
31. 25,119 times   **33.** 4.7
35a.

Distance from bed (feet)	0.2	0.4	0.6	0.8	1.0	1.2
Velocity, Hoback River (ft/sec)	1.91	2.22	2.40	2.53	2.63	2.71
Velocity, Pole Creek (ft/sec)	1.51	1.70	1.82	1.90	1.96	2.01

b. Hoback River, 0.31 ft/sec; Pole Creek, 0.196 ft/sec

c.

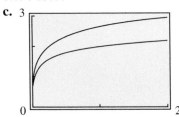

d. 2.3 ft/sec, 1.75 ft/sec   **37a.** 81   **b.** 4
c. $\log_3 x$ and 3^x are inverse functions.   **d.** 1.8
e. a **39a.** IV   **b.** V   **c.** I   **d.** II
e. III   **f.** VI   **41a.** "Take the log base 6 of x."   **b.** "Raise 5 to the power x."
43a. $10^4 = 10,000$ **b.** $10^8 = 100,000,000$
45. $x > 9$
47a.

x	x^2	$\log_{10} x$	$\log_{10} x^2$
1	1	0	0
2	4	0.3010	0.6021
3	9	0.4771	0.9542
4	16	0.6021	1.2041
5	25	0.6990	1.3979
6	36	0.7782	1.5563

b. $2\log_{10} x = \log_{10} x^2$

49.

x	$y = \log_e x$
1	0
2	0.693
4	1.386
16	2.773
$\frac{1}{2}$	-0.693
$\frac{1}{4}$	-1.386
$\frac{1}{16}$	-2.773

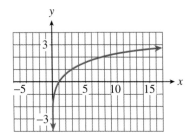

Homework 7.5

1a. $\log_b 2 + \log_b x$ **b.** $\log_b x - \log_b y$
c. $\log_b x + \log_b y - \log_b z$ **d.** $3\log_b x$
3a. $\frac{1}{2}\log_b x$ **b.** $\frac{2}{3}\log_b x$
c. $2\log_b x + 3\log_b y$
d. $\frac{1}{2}\log_b x + \log_b y - 2\log_b z$
5a. $\frac{1}{3}(\log_{10} x + 2\log_{10} y - \log_{10} z)$

b. $\frac{1}{2}\log_{10} 2 + \frac{1}{2}\log_{10} L - \log_{10} R$

c. $\log_{10} 2 + \log_{10} \pi + \frac{1}{2}\log_{10} l - \frac{1}{2}\log_{10} g$

d. $\log_{10} 2 + \frac{2}{3}\log_{10} y + \frac{1}{3}\log_{10} x$

7a. 1.7917   **b.** -0.9163 **9a.** 2.1972
b. 1.956   **11.** 2.8074   **13.** 0.8928
15. ± 1.3977 **17.** -1.6092 **19.** 0.2736
21. -12.4864 **23a.** $W(t) = 0.01(3)^{t/16}$
b. 56.97 hr   **25a.** $C(t) = 0.7(0.8)^t$
b. 2.5 hr later   **27a.** No   **b.** No
29a. No   **b.** No
31.

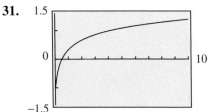

The graphs are the same because the two expressions are equivalent.
33.

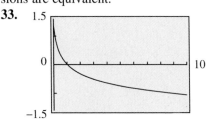

The graphs are the same because the two expressions are equivalent.

35a. $P(t) = P_0(1.037)^t$ **b.** 19.08 yr

c. 10,000

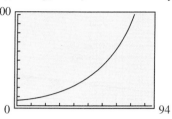

   0   94

37a. 12.49 hr   **b.** 24.97 hr; 37.46 hr

39a. The data points appear to lie on a power function.

b. $Y = \log t = \log (r^k) = k \log r = kX.$

c.

Planet	Mercury	Venus	Earth	Mars	Jupiter
$X = \log r$	−0.412	−0.141	0	0.183	0.716
$Y = \log t$	−0.618	−0.211	0	0.274	1.074

Planet	Saturn	Uranus	Neptune	Pluto
$X = \log r$	0.980	1.283	1.478	1.595
$Y = \log t$	1.469	1.924	2.217	2.340

d. $Y = 1.5X$ **e.** $t = r^{1.5}$. Yes

f. 4.685 years   **41a.** $\log_b 4$ **b.** $\log_b x^2 y^3$

c. $\log_b \dfrac{1}{x^2}$ **43a.** $\log_{10} \sqrt{\dfrac{xy}{z^3}}$ **b.** $\log_b \dfrac{1}{4}$

45. 4   **47.** 11   **49.** 3   **51.** No solution

53a. 20,000

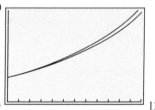

   0   12

b. $15,529.24; $16,310.19   **c.** 6 years later

55. 12.9%   **57.** 10 yr 361 days

59. $t = \dfrac{1}{k} \log_{10} \left(\dfrac{A}{A_0} + 1 \right)$ **61.** $q = \log_v \left(\dfrac{w}{p} \right)$

63. $A = k(10^{t/T} - 1)$ **65a.** $x = b^m, y = b^n$

b. $\log_b (b^m \cdot b^n)$ **c.** $\log_b (b^{m+n})$

d. $m + n$ **e.** $\log_b (xy) = \log_b x + \log_b y$

67a. $x = b^m$ **b.** $\log_b (b^m)^k$ **c.** $\log_b b^{mk}$

d. mk **e.** $\log_b (x^k) = k \log_b x$

Homework 7.6

1.

x	−10	−5	0	5	10	15	20
$f(x)$	0.14	0.37	1	2.72	7.39	20.09	54.60

60

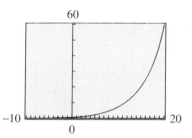

 −10   20
            0

3.

x	−10	−5	0	5	10	15	20
$f(x)$	20.09	4.48	1	0.22	0.05	0.01	0.00

20

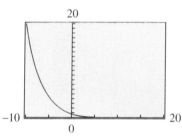

 −10   20
            0

5a. 0.6419   **b.** 3.8067   **c.** −1.2040

7a. 4.137   **b.** 1.878   **c.** 0.0743

9a. $N(t) = 6000e^{0.04t}$

b.

t	0	5	10	15	20	25	30
$N(t)$	6000	7328	8951	10,933	13,353	16,310	19,921

c.

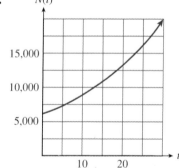

d. 15,670   **e.** 70.3 hrs

11a.

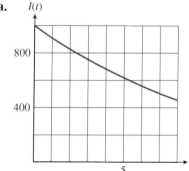

b. 941.8 lumens   **c.** 2.2 cm
13. $P(t) = 20(1.49)^t$; increasing
15. $P(t) = 6500(0.082)^t$; decreasing
17a.

x	0	0.5	1	1.5	2	2.5
e^x	1	1.6487	2.7183	4.4817	7.3891	12.1825

b. Each ratio equals $e^{0.5} \approx 1.6487$.
19a.

x	0	0.6931	1.3863	2.0794	2.7726	3.4657	4.1589
e^x	1	2	4	8	16	32	64

b. Each difference in x-values is approximately 0.6931. Since $e^{0.6931} \approx 2$, and by the first law of exponents, each function value is approximately double the previous.   **21.** 0.8277
23. -2.9720 **25.** 1.6451   **27.** -3.0713
29. $t = \dfrac{1}{k} \ln y$ **31.** $t = \ln\left(\dfrac{k}{k - y}\right)$
33. $k = e^{T/T_0} - 10$
35a.

n	0.39	3.9	39	390
$\ln n$	-0.942	1.361	3.664	5.966

b. Each difference is 2.303. Each x-value is 10 times the previous, so by the first log rule, each function value is $\ln 10 \approx 2.303$ greater than the previous.
37a.

n	2	4	8	16
$\ln n$	0.693	1.386	2.079	2.773

b. Each quotient equals k, where $n = 2^k$. By the third log rule, $\ln 2^k = k \ln 2$, so $\frac{\ln n}{\ln 2} = k$.
39. $N(t) = 100e^{0.6931t}$ **41.** $N(t) = 1200e^{-0.5108t}$
43. $N(t) = 10e^{0.1398t}$ **45a.** 20,000
b. 1.0576   **c.** $N(t) = 20,000e^{0.05596t}$
d. 107,182   **47a.** 0.8775
b. $N(t) = 500e^{-0.1307t}$ **c.** 135.32 mg
49a. $A(t) = 500e^{0.095t}$ **b.** 7.3 yrs   **c.** 7.3 yrs
d.

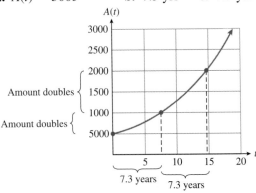

51a. 1921 yr   **b.** 5589.9 yr
53a. $\dfrac{1}{2} N_0$; $\dfrac{1}{4} N_0$; $\dfrac{1}{16} N_0$
b.

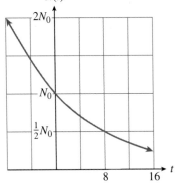

c. $N(t) = N_0 e^{-0.0866t}$ **55a.** $y = 116(0.975)^t$
b. $G(t) = 116e^{-0.025t}$ **c.** 28 minutes
57b. The graph is shifted 2 units to the right.
c. The values of a and b change the height and position of the peak of the curve.   **d.** The area of increased yield is larger, so it appears that hedgerows are good for crop yield.   **e.** The beneficial effects extend to 20 times the height of the hedgerow. The field should be 50 meters by 50 meters.   **59a.** $8^x = 20$
b. $\log_{10} 8^x = \log_{10} 20$
c. $x = \dfrac{\log_{10} 20}{\log_{10} 8} = 1.4406$
61. $\log_8 Q = \dfrac{\log_{10} Q}{\log_{10} 8}$
63. $\ln Q = \dfrac{\log_{10} Q}{\log_{10} e} = 2.3 \log_{10} Q$

Chapter 7 Review

1a.

Years after 1974	0	5	10	15	20
Number of degrees	8	12	18	27	41

b. $N(t) = 8(1.5)^{t/5}$
c.

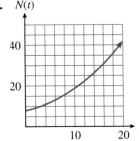

d. 18; 44

3a.

Hours after 8 am	0	1	2	3	4	5
Medication (mg)	100	85	72.25	61.41	52.20	44.37

b. $A(t) = 100(0.85)^t$

c.

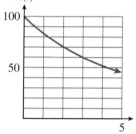

d. 52.20 mg; 19.69 mg

5.

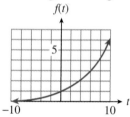

7.

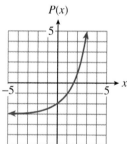

9. $\dfrac{-4}{3}$

11. -11 **13.** $C(T) = C_0(0.5)^{(T-T_0)/10}$, where T is the temperature in degrees Celsius, $C(T)$ is the organic content at temperature T, and C_0 is the organic content at T_0 degrees.   **15.** 4

17. -1 **19.** -3 **21.** $2^{x-2} = 3$

23. $\log_{0.3}(x+1) = -2$ **25.** $y = -1$

27. $b = 4$ **29.** $x = 0.544$ **31.** $x = 0.021$

33. $k = 0.0543$ **35.** $h = 0.195$

37. $\log_b x + \dfrac{1}{3}\log_b y - 2\log_b z$

39. $\dfrac{1}{3}(4\log_{10} x - \log_{10} y)$ **41.** $\log_{10}\sqrt[3]{\dfrac{x}{y^2}}$

43. $\log_{10}\dfrac{1}{8}$ **45.** $\dfrac{9}{4}$ **47.** 31.16

49. 3.77   **51.** -7.278

53. $t = \dfrac{1}{k}\log_{10}\left(\dfrac{N}{N_0}\right)$ **55a.** 238   **b.** 2010

57a. $C(t) = 90(1.06)^t$ **b.** \$94.48
c. About 5 yrs   **59.** 1.548   **61.** 411.58

63. 2.286

65a.

t	0	2	4	6	8	10
$V(t)$	0	63.2	86.5	95.0	98.2	99.3

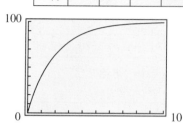

b. The graph is increasing and concave down. In the long run $V(t)$ approaches 100.   **c.** 2.77 sec

67. $t = \dfrac{1}{k}\ln\left(\dfrac{12}{y-6}\right)$ **69.** $N(t) \approx 600e^{-0.9163t}$

71b. $W(t) = 2^t$ cents   **c.** \$327.68; \$10,737,418.24

Homework 8.1

1. $12x^3 - 5x^2 - 8x + 4$

3. $x^3 - 6x^2 + 11x - 6$

5. $6a^4 - 5a^3 - 5a^2 + 5a - 1$

7. $y^4 + 5y^3 - 20y - 16$ **9.** $6 + x + 5x^2$

11. $4 - 7x^2 - 8x^4$ **13.** $0x^2$ **15.** $-8x^3$

17a. 4   **b.** 5   **c.** 7   **19a.** 4   **b.** 5

21. $(x+y)^3 = (x+y)(x+y)^2$
$= (x+y)(x^2 + 2xy + y^2)$
$= x(x^2 + 2xy + y^2) + y(x^2 + 2xy + y^2)$
$= x^3 + 2x^2y + xy^2 + x^2y + 2xy^2 + y^3$
$= x^3 + 3x^2y + 3xy^2 + y^3$

23. $(x+y)(x^2 - xy + y^2)$
$= x(x^2 - xy + y^2) + y(x^2 - xy + y^2)$
$= x^3 - x^2y + xy^2 + x^2y - xy^2 + y^3 = x^3 + y^3$

25a. The formula begins with x^3 and ends with y^3. As you proceed from term to term, the powers on x decrease while the powers on y increase, and on each term the sum of the powers is 3. The coefficients of the two middle terms are both 3.
b. The formula is the same as for $(x+y)^3$, except that the terms alternate in sign.

27. $1 + 6z + 12z^2 + 8z^3$

29. $1 - 15\sqrt{t} + 75t - 125t\sqrt{t}$

31. $x^3 - 1$ **33.** $8x^3 + 1$ **35.** $27a^3 - 8b^3$

37. $(x+3)(x^2 - 3x + 9)$

39. $(a - 2b)(a^2 + 2ab + 4b^2)$

41. $(xy - 1)(x^2y^2 + xy + 1)$

43. $(3a + 4b)(9a^2 - 12ab + 16b^2)$

45. $(5ab - 1)(25a^2b^2 + 5ab + 1)$

47a. $2x^2 + 3_2 x$ **b.** 1224 sq in

49a. $\left(6 - \dfrac{5\pi}{4}\right)x^2$ **b.** 132.67 sq in

51a. $\dfrac{2}{3}\pi r^3 + \pi r^2 h$ **b.** $\dfrac{14}{3}\pi r^3$

53a. $500(1 + r)^2$; $500(1 + r)^3$; $500(1 + r)^4$
b. $500r^2 + 1000r + 500$;
$500r^3 + 1500r^2 + 1500r + 500$;
$500r^4 + 2000r^3 + 3000r^2 + 2000r + 500$
c. $583.20; $629.86; $680.24
55a. length: $16 - 2x$; width: $12 - 2x$; height x
b. $V = x(16 - 2x)(12 - 2x)$ **c.** $0 < x < 6$
d.

x	1	2	3
V	140	192	180

e.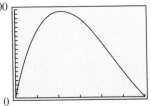

f. $x = 2.26$ in; maximum volume: 194.07 cu in
57a. 0, 9
b. $0 < x < 9$
c.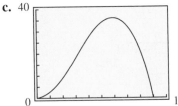

d. $\dfrac{28}{3}$ points   **e.** 36 points
f. 3 ml or 8.2 ml
59a.
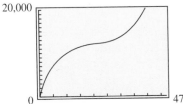

b. 900; 11,145; 15,078
c. 1341; 171; 627   **d.** 1981   **61a.** 20 cm
b. 100 cm   **63a.** $763 < H(t) \le 864$
b.
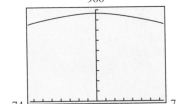

c. 864 min   **d.** 859.8 min
e. $-34 \le t \le 34$ **f.** $t > 66$ or $t < -66$

Homework 8.2

1.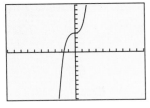

3.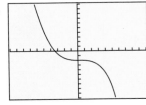

5.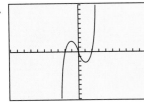

7.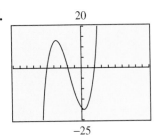

If the coefficient of x^3 is positive, the y-values increase from $-\infty$ to ∞ as x increases. If the coefficient of x^3 is negative, the y-values decrease from ∞ to $-\infty$ as x increases.

9a.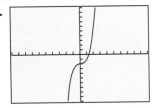

b.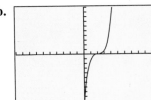

c.

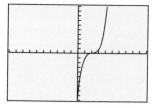

The graphs of (b) and (c) are the same.

11.

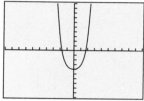

13.

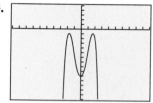

15.

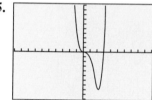

17.

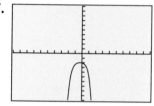

If the coefficient of x^4 is positive, the y-values decrease from ∞ and increase to ∞ at the ends. If the coefficient of x^4 is negative, the y-values increase from $-\infty$ and decrease to $-\infty$ at the ends.

19.

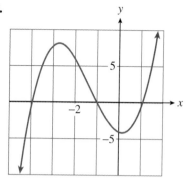

21.

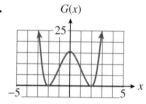

23.

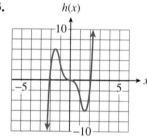

25.

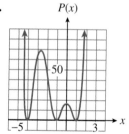

27a. $(-2, 0), (-1, 0), (3, 0)$
b. $P(x) = (x + 2)(x + 1)(x - 3)$
29a. $(-2, 0), (0, 0), (1, 0), (2, 0)$
b. $R(x) = x(x - 2)(x - 1)(x + 2)$
31a. $(-2, 0) + (1, 0), (4, 0)$
b. $p(x) = (x - 4)(x - 1)(x + 2)$
33a. $(-2, 0), (2, 0), (3, 0)$
b. $r(x) = (x + 2)^2(x - 2)(x - 3)$
35a. 0 (multiplicity 2)

b.

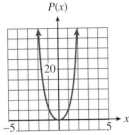

$P(x)$

37a. 0 (multiplicity 2), -2 (multiplicity 2)

b.

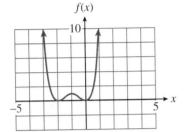

$f(x)$

39a. $0, \pm 2$ **b.**

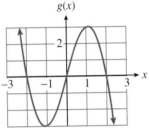

$g(x)$

41a. $\pm\sqrt{2}, \pm\sqrt{8}$ **b.**

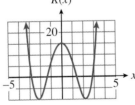

$K(x)$

43a. $\pm 1, -3$ (multiplicity 2)

b.

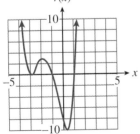

$r(x)$

45. $P(x) = (x + 2)(x - 1)(x - 4)$

47. $P(x) = (x + 3)^2(x - 2)$

49. $P(x) = (x + 2)(x - 2)^3$

51a. $y = x^3 - 4x + 3$, graph is shifted upward three units   **b.** $y = x^3 - 4x - 5$, graph is shifted downward five units

c. $y = (x - 2)^3 - 4(x - 2)$, graph is shifted two units to the right   **d.** $y = (x + 3)^3 - 4(x + 3)$, graph is shifted three units to the left

53a. $y = x^4 - 4x^2 + 6$, graph is shifted six units upward   **b.** $y = x^4 - 4x^2 - 2$, graph is shifted two units downward

c. $y = (x - 4)^4 - 4(x - 4)^2$, graph is shifted four units to the right   **d.** $y = (x + 2)^4 - 4(x + 2)^2$, graph is shifted two units to the left

Homework 8.3

1a. $t = \dfrac{150}{50 - v}$

b.

v	0	5	10	15	20	25	30	35	40	45	50
t	3	3.33	3.75	4.29	5	6	7.5	10	15	30	—

Their travel time increases as the headwind increases.   **c.** 20

Asymptotes:
$x = 50, y = 0$

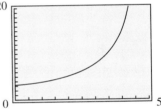

3a. $0 \le p < 100$

b.

p	0	15	25	40	50	75	80	90	100
C	0	12.7	24	48	72	216	288	648	—

c. 2000   60%

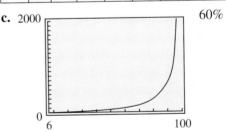

d. $p > 96\%$ **e.** $p = 100$; as the percentage immunized approaches 100, the cost becomes infinitely large.   **5a.** $C = 8 + \dfrac{20{,}000}{n}$

b.

n	100	200	400	500	1000	2000	4000	5000	8000
C	208	108	58	48	28	18	13	12	10.5

c.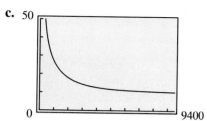

d. 2000   **e.** $n > 5000$ **f.** $C = 8$; as n increases, the average cost per calculator approaches \$8.

7a. Cost of reordering $= 4500 + \dfrac{3000}{x}$; $C = 6x + 4500 + \dfrac{3000}{x}$

b.

x	10	20	30	40	50	60	70	80	90	100
C	4860	4770	4780	4815	4860	4910	4963	5018	5073	5130

c.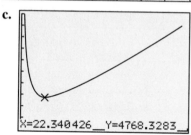

Minimum value for C: \$4768.33   **d.** 22; 14

e.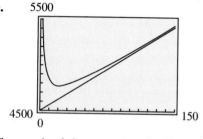

The graph of C approaches the line as an asymptote.   **9a.** Surface area $= 2x^2 + 4xh = 96$, so $h = \dfrac{96 - 2x^2}{4x} = \dfrac{24}{x} - \dfrac{x}{2}$ **b.** $V = 24x - \dfrac{1}{2}x^3$

c.

x	1	2	3	4	5	6	7
h	23.5	11	6.5	4	2.3	1	-0.07
V	23.5	44	58.5	64	57.5	36	-3.5

d.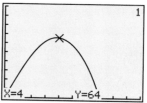

Maximum volume is 64 cm^3 **e.** $x = 4$ cm $h = 4$ cm

f.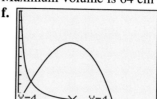

11a.

v	-100	-75	-50	-25	0	25	50	75	100
P	338.15	358.92	382.41	409.19	440	475.83	518.01	568.4	629.66

b.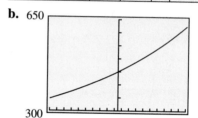

c. -20 m per sec; 68 m per sec   **d.** $v > 12$ (approaching at more than 12 m per sec)

e. $v = 332$; as v approaches 332 m per sec, the pitch becomes infinitely high

13.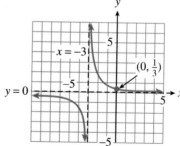

15.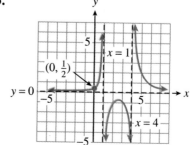

17.

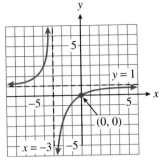

19.

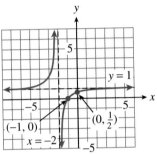

21.

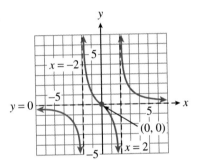

23.

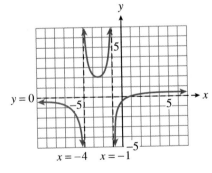

25.

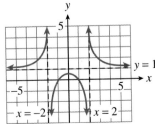

27.

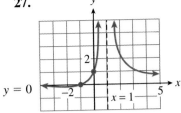

29.

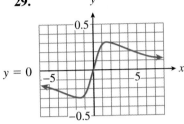

31.

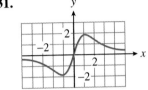

33a. $\dfrac{1}{y} = \dfrac{1}{x} + \dfrac{1}{k} = \dfrac{x + k}{kx}$, $y = \dfrac{kx}{x + k}$

b.

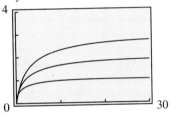

The graphs increase from the origin and approach a horizontal asymptote at $y = k$.

35. $y = \dfrac{12x}{x + 20}$ **37a.** V **b.** $\dfrac{V}{2}$

c. $V = 0.7$, $K = 2.2$ (many answers are possible)

d.

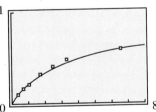

39a. $a = \dfrac{1}{V}$, $b = \dfrac{K}{V}$

b.

s	0.33	0.66	1	1.66	2.50	3.33	6.66
$\frac{s}{v}$	4.1	4.7	5	5.5	6.4	7.2	11.5

c. $\dfrac{s}{v} = 1.13s + 3.74$ **d.** $V = 0.88$, $K = 3.29$

Midchapter 8 Review

1. $t^3 + 3t^2 - 5t - 4$
3. $v^3 - 30v^2 + 300v - 1000$
5. $(y + 3x)(y^2 - 3xy + 9x^2)$
7.

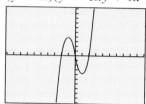

9. $x = 0, x = \pm\sqrt{8}$

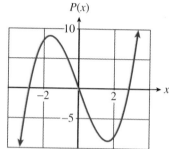

11.

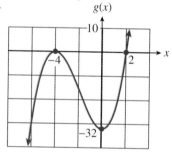

13. All real numbers except -2, 0, and 2
15.

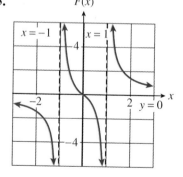

Homework 8.4

1. $\dfrac{-2}{d^2}$ **3.** $\dfrac{2x + 3}{3}$ **5.** $\dfrac{3a^2 - 2a}{2}$

7. $\dfrac{6(1 + t)}{1 - t}$ **9.** $y - 2$ **11.** $\dfrac{-2}{v^2 + 3v + 9}$

13. $\dfrac{2(x - 3)}{3}$ **15.** $\dfrac{y + 3x}{y - 3x}$ **17.** $\dfrac{2x - 3}{x - 1}$

19. $\dfrac{4z^2 + 6z + 9}{2z + 3}$ **21.** (b) **23.** None

25. $\dfrac{-np^2}{2}$ **27.** $\dfrac{5}{ab}$ **29.** 5

31. $\dfrac{a(2a - 1)}{a + 4}$ **33.** $\dfrac{x - 2}{x + 1}$

35. $\dfrac{6x(x - 2)(x - 1)^2}{(x^2 - 8)(x^2 - 2x + 4)}$ **37.** $\dfrac{2}{9}$

39. $\dfrac{a + 1}{a - 2}$ **41.** $3(x^2 - xy + y^2)$

43. $\dfrac{x + 2}{x^2 - 1}$ **45.** $\dfrac{x^2(x - 4)}{(x + 1)}$ **47.** $\dfrac{x + 3}{6y}$

49. $6rs - 5 + \dfrac{2}{rs}$ **51.** $-5s^8 + 7s^3 - \dfrac{2}{s^2}$

53. $2y + 5 + \dfrac{2}{2y + 1}$

55. $x^2 + 4x + 9 + \dfrac{19}{x - 2}$

57. $4z^3 - 2z^2 + 3z + 1 + \dfrac{2}{2z + 1}$

59. $x^3 + 2x^2 + 4x + 8 + \dfrac{15}{x - 2}$

61. $P(x) = (x - 1)(x^2 - x - 1)$ so 1 is a zero of $P(x)$. The other zeros are $\dfrac{1 \pm \sqrt{5}}{2}$.

63. $P(x) = (x + 3)(x^3 - 6x^2 + 8x)$ so -3 is a zero of $P(x)$. The other zeros are 0, 2, and 4.

65a. All real numbers except 2.
b. $y = x + 2, x \neq 2$
c.

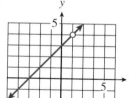

67a. All real numbers except 1 and -1.
b. $y = \dfrac{1}{x - 1}, x \neq -1$
c.

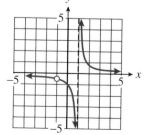

Homework 8.5

1. $\dfrac{x-3}{2}$ **3.** $\dfrac{a+b-5c}{6}$ **5.** $\dfrac{2x-1}{2y}$

7. $\dfrac{-2x+7}{x+2y}$ **9.** $12xy^2(x+y)^2$

11. $(a+4)(a+1)^2$ **13.** $x(x-1)^3$

15. $\dfrac{7x}{6}$ **17.** $\dfrac{y}{12}$ **19.** $\dfrac{7x+1}{6x}$

21. $\dfrac{8x-5}{x(x-1)}$ **23.** $\dfrac{3y-3y^2}{(y+1)(2y-1)}$

25. $\dfrac{y^2-4y+5}{(y+1)(2y-3)}$ **27.** $\dfrac{-4}{15(x-2)}$

29. $\dfrac{3y+1}{y(y-3)(y+2)}$ **31.** $\dfrac{x^2-1}{x}$

33. $\dfrac{x^3-2x^2+2x-2}{(x-1)^2}$ **35a.** $\dfrac{25}{s+8}$

b. $\dfrac{25}{s-8}$ **c.** $\dfrac{50s}{s^2-64}$ **37a.** $\dfrac{900}{400+w}$

b. $\dfrac{900}{400-w}$

c. Orville, by $\dfrac{1800w}{160{,}000-w^2}$ hours

39. $\dfrac{7}{10a+2}$ **41.** $\dfrac{a}{a-2}$ **43.** h

45. $\dfrac{q}{q-1}$ **47.** LC **49.** $\dfrac{-2(x+z)}{xz}$

51a. $\dfrac{1}{f}=\dfrac{1}{q+60}+\dfrac{1}{q}$ **b.** $f=\dfrac{q^2+60q}{2q+60}$

53a. $T_1=\dfrac{d}{r_1},\ T_2=\dfrac{d}{r_2}$ **b.** $D=2d,$

$T=\dfrac{d}{r_1}+\dfrac{d}{r_2}$ **c.** $r=\dfrac{D}{T}=\dfrac{2d}{\frac{d}{r_1}+\frac{d}{r_2}}$

d. $r=\dfrac{2r_1r_2}{r_1+r_2}$ **e.** $r=58\dfrac{1}{3}$ mph

55. $\dfrac{x^2+y^2}{x^2y^2}$ **57.** $\dfrac{8w-1}{4w^2}$ **59.** $\dfrac{b^2-a^2}{ab}$

61. $\dfrac{xy}{x+y}$ **63.** $\dfrac{x^3+1}{x^3}$ **65.** $b+a$

67. $\dfrac{\sqrt7+\sqrt3}{2}$ **69.** $\dfrac{6\sqrt3+7\sqrt2}{5}$

Homework 8.6

1. $\dfrac{-1}{2}$ **3.** $\dfrac{13}{8}$ **5.** $\pm\sqrt{\dfrac{15}{8}}$ **7.** 0.97

9. 37 ft   **11a.** $t=\dfrac{150}{50-v}$

b. $4=\dfrac{150}{50-v},\ v=12.5$ mph

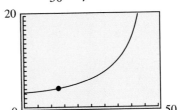

13. $168=\dfrac{72p}{100-p},\ p=70\%$

15.

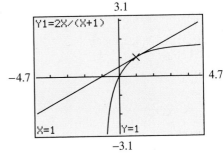

$x=1$

17.

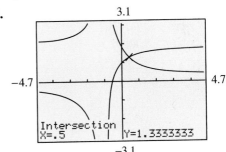

$x=\dfrac{1}{2}$ **19a.** 150

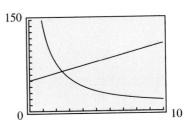

$2.50   **b.** $\dfrac{160}{x}=6x+49,\ x=2.5$

21a. $L=\dfrac{3200}{w}$ **b.** $P=\dfrac{6400}{w}+2w$

c.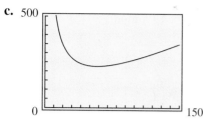

Lowest point: (56.6, 226); the minimum perimeter is 226 ft for a width of 56.6 ft.

d. $240 = \dfrac{6400}{w} + 2w$ **e.** $w = 40$ ft, $l = 80$ ft

23. $b = ka$ and $d = kc$, so $\dfrac{b}{a} = k = \dfrac{d}{c}$ or $\dfrac{b}{a} = \dfrac{d}{c}$.

25. 4 **27.** 40 **29.** \$6187.50 **31.** 45 mi

33. 689 **35a.** 19,882 meters **b.** 0.3%

c. 0.00657 inches **37a.** $AE = 1$,

$ED = x - 1$, $CD = 1$ **b.** $\dfrac{1}{x} = \dfrac{x-1}{1}$

c. $\dfrac{1 + \sqrt{5}}{2}$ **39.** $r = \dfrac{S - a}{S}$

41. $x = \dfrac{Hy}{2y - H}$ **43.** $d = \pm\sqrt{\dfrac{Gm_1m_2}{F}}$

45. $r = \dfrac{2QI}{I + Q}$ **47a.** $v = \dfrac{1}{T}\left(\dfrac{rD}{R} - d\right)$

b. 59.5 km/min **49a.** The effective population equals the actual population, $N = F + M$.

b.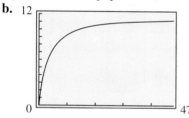

c. Largest population: 11; 33 males

d. 9 males **e.** 3 **51.** 5 **53.** 1

55. $\dfrac{-14}{5}$ **57.** $\dfrac{-1}{6}, \dfrac{-4}{3}$

59a. $t_1 = \dfrac{144}{s - 20}$ **b.** $t_2 = \dfrac{144}{s + 20}$

c.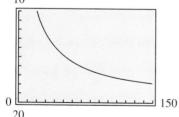

(100, 3): If the airspeed of the plane is 100 mph, the round trip will take 3 hours.

d. $\dfrac{144}{s - 20} + \dfrac{144}{s + 20} = 3$ **e.** 100 mph

Chapter 8 Review

1. $2x^3 - 11x^2 + 19x - 10$

3. $(2x - 3z)(4x^2 + 6xz + 9z^2)$

5a. $\dfrac{1}{6}n^3 - \dfrac{1}{2}n^2 + \dfrac{1}{3}n$

b. 220 **c.** 20 **7.** $x = 0, \pm 2$

9a. $x = 2, -1$ **b.**

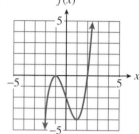

11a. $x = 0, 1, -3$

b.

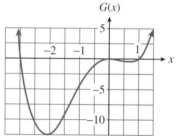

13a. $x = 0, -1, 1$

b.

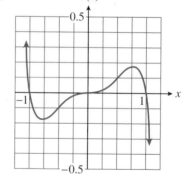

15a. $x = -1, 1$

b.

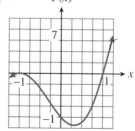

17a. $x = -1, 1, \pm\sqrt{6}$

b.

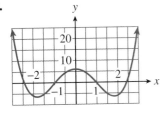

19a. $V = \dfrac{\pi h^3}{4}$

b. $2\pi \approx 6.28 \text{ cm}^3$; $16\pi \approx 50.27 \text{ cm}^3$

c.

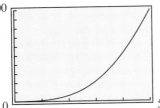

$h \approx 5$

21. The Galaxy is about twice as thick as a compact disk, compared to its diameter.

23a.

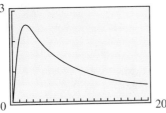

b. 338   **c.** Months 2 and 20   **d.** During month 6. The number of members eventually decreases to zero.   **25a.** Asymptotes: $x = 4$, $y = 0$; y-intercept: $\left(0, \dfrac{-1}{4}\right)$

b.

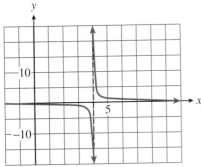

27a. Asymptotes: $x = -3$, $y = 1$; x-intercept: $(2, 0)$, y-intercept: $\left(0, \dfrac{-2}{3}\right)$

b.

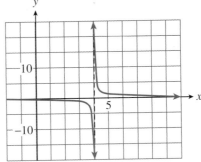

29a. Asymptotes: $x = -2$, $x = 2$, $y = 3$; x- and y-intercept: $(0, 0)$

b.

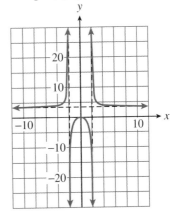

31a. $t_1 = \dfrac{90}{v - 2}$ **b.** $t_2 = \dfrac{90}{v + 2}$

c.

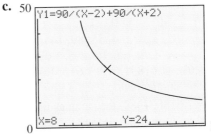

$(8, 24)$ If the club paddles at 8 miles per hour, they will complete the trip after 24 hours of traveling.

d. $\dfrac{90}{v + 2} + \dfrac{90}{v - 2} = 24$; $v = 8$ hours

33. $\dfrac{a}{2(a - 1)}$ **35.** $\dfrac{y^2 - 2x}{2}$

37. $\dfrac{a - 3}{2a + 6}$ **39.** $10ab$ **41.** $\dfrac{6x}{2x + 3}$

43. $\dfrac{a^2 - 2a}{a^2 + 3a + 2}$ **45.** $\dfrac{1}{2x - 1}$

47. $9x^2 - 7 + \dfrac{4}{x^2} - \dfrac{1}{x^4}$

49. $x^2 - 2x - 2 - \dfrac{1}{x - 2}$ **51.** $\dfrac{2}{x}$

53. $\dfrac{3x + 1}{2(x - 3)(x + 3)}$ **55.** $\dfrac{2a^2 - a + 1}{(a - 3)(a - 1)}$

57. $\dfrac{1}{5}$ **59.** $\dfrac{x}{x + 4}$ **61.** -2

63. No solution **65.** $n = \dfrac{Ct}{C - V}$

67. $q = \dfrac{pr}{r - p}$ **69.** $\dfrac{x^3 + y}{x^3 y}$ **71.** $\dfrac{-y}{x}$

73. $\dfrac{-(x - y)^2}{xy}$

Homework 9.1

1. $-4, -3, -2, -1$ **3.** $\dfrac{-1}{2}, 1, \dfrac{7}{2}, 7$

5. $2, 1.5, 1.\overline{3}, 1.25$ **7.** $0, 1, 3, 6$ **9.** $-1, 1,$ $-1, 1$ **11.** $1, 0, \dfrac{-1}{3}, \dfrac{1}{2}$ **13.** $1, 1, 1, 1$

15. $a_1 = \dfrac{4}{3}$ **17.** $a_n = 3a_{n-1}$

19. $a_{n+1} = \dfrac{1}{3}a_{n+2}$ **21.** 58 **23.** 1.415

25. 8.944 **27.** $3, 5, 7, 9, 11$ **29.** $24, -12,$ $6, -3, 1.5$ **31.** $1, 2, 6, 24, 120$ **33.** $100,$ $210, 331, 464.1, 610.51$ **35a.** $14,000; 11,900;$ $10,115; 8597.75$ **b.** $v_1 = 14,000;$ $v_{n+1} = 0.85v_n$ **37a.** $1.55, 2.00, 2.45, 2.90$ **b.** $c_1 = 1.55; c_{n+1} = c_n + 0.45$ **39a.** $50,000;$ $50,000; 50,000; 50,000$ **b.** $v_1 = v_n$ **41a.** $10, 18, 24.4, 29.52$ **b.** $d_1 = 10;$ $d_{n+1} = 0.8d_n + 10$ **43a.** 3 **b.** 6 **c.** $0, 1, 3, 6, 10$ **d.** $L_1 = 0,$ $L_{n+1} = n + L_n$ **45a.** $1, 1, 2, 3, 5, 8, 13, 21, 34, 55, 89, 144, 233,$ $377, 610, 987$ **b.** $1, 2, 1.5, 1.\overline{6}, 1.6, 1.625,$ $1.615, 1.619, 1.618, 1.618, 1.618, 1.618, 1.618,$ $1.618, 1.618.$ The quotients approach a limit near $1.618.$ $\dfrac{1 + \sqrt{5}}{2} \approx 1.618033989,$ the same as the limit above. **47.** a_n approaches $1.4142,$ or $\sqrt{2}$ **49.** c_n approaches $0.64039,$ or $\dfrac{1 + \sqrt{17}}{8}$ **51.** s_n approaches 2
53a.

n	A_n	B_n	C_n	S_n
0	2	0	0	2
1	2	2	0	6
2	4	2	2	14
3	8	4	2	22
4	14	8	4	42
5	26	14	8	78
6	48	26	14	142
7	88	48	26	262
8	162	88	48	482

c. $A_0 = 2, A_n = A_{n-1} + B_{n-1} + C_{n-1};$ $B_0 = 0, B_n = A_{n-1}; C_0 = 0, C_n = B_{n-1}$ **d.** $S_n = A_n + 2B_n + 3C_n$ $= A_n + 2A_{n-1} + 3A_{n-2}.$ (To define S_n completely in terms of A_n, we must also define $A_{-1} = 0$ and $A_{-2} = 0.$)

Homework 9.2

1. Geometric **3.** Arithmetic **5.** Geometric **7.** Neither **9.** Geometric **11.** Geometric

13. $2, 6, 10, 14$ **15.** $\dfrac{1}{2}, \dfrac{3}{4}, 1, \dfrac{5}{4}$ **17.** $2.7,$ $1.9, 1.1, 0.3$ **19.** $5, -10, 20, -40$

21. $9, 6, 4, \dfrac{8}{3}$ **23.** $60, 24, 9.6, 3.84$

25. $15, 19, 23, \ldots, 3 + (n - 1)4$ **27.** $-13, -17, -21, \ldots, -1 - 4(n - 1)$ **29.** $\dfrac{16}{3}, \dfrac{32}{3}, \dfrac{64}{3}, \ldots, \dfrac{2}{3}(2)^{n-1}$ **31.** $\dfrac{-1}{2}, \dfrac{1}{4},$ $\dfrac{-1}{8}, \ldots, 4\left(\dfrac{-1}{2}\right)^{n-1}$ **33.** 7.5 **35.** $\dfrac{3}{128}$ **37.** 3 **39.** 13 **41.** $s_n = 3 + 2(n - 1)$

43. $x_n = -3(n - 1)$ **45.** $d_n = 24\left(\dfrac{-1}{2}\right)^{n-1}$

47. $w_n = 2^{n-1}$ **49b.** 128 **51b.** $\$115$ **53b.** $\$1203.31$ **55b.** 1.15 kg

Homework 9.3

1. 99 **3.** $106\dfrac{2}{3}$ **5.** 141 **7.** 410

9. -95.8125 **11.** 89.88075 **13.** Arithmetic; 2550 **15.** Geometric; 2046 **17.** Neither; 784 **19.** Arithmetic; 1071 **21.** Geometric; 8.996 **23.** 1938 **25.** 78 **27a.** 10.125 ft **b.** 107.25 ft **29.** $\$7,400,000$ **31.** 66.25 sec **33.** $\$6,504,532.78$ **35.** 4549.7 sec **37.** $\$15,269.50$ **39.** $\$10,737,418.23$ **41a.** N **b.** $N + 1$ **c.** $2S = N \cdot (N + 1);$ $S = \dfrac{N(N + 1)}{2}$ **d.** $\dfrac{N(N + 1)}{2}$ **43a.** r **b.** $Ar = r + r^2 + r^3 + \cdots + r^N + r^{N+1}$ **c.** $A - Ar = 1 - r^{N+1}; A(1 - r) = 1 - r^{N+1}$ **d.** $A = \dfrac{1 - r^{N+1}}{1 - r}$ **45a.** $F + (4 - 1)d;$ $F + (9 - 1)d$ **b.** $F + (N - 1)d$ **c.** $\dfrac{N}{2}[2F + (N - 1)d]$

Midchapter 9 Review

1. $B_n = 1.0025B_{n-1} - 200$ **3.** $1, 5, 14, 30$ **5.** $3, 3, 3, 3$ **7a.** $100, 93, 86.49, 80.4357$

b. $c_1 = 100$, $c_{n+1} = 0.93c_n$
c. $c_n = 100(0.93)^{n-1}$ **9.** $-7, -13, -19, -25$
b. $a_1 = -7$, $a_{n+1} = a_n - 6$
c. $a_n = -1 - 6n$ **11.** Arithmetic, common difference -5.3 **13.** Neither **15.** $\dfrac{9}{16}$

17. 6715.61 **19.** 310.5

Homework 9.4

1. $1^2 + 2^2 + 3^2 + 4^2$ **3.** $3 + 4 + 5$
5. $1(2) + 2(3) + 3(4) + 4(5)$
7. $\dfrac{-1}{2} + \dfrac{1}{2^2} - \dfrac{1}{2^3} + \dfrac{1}{2^4}$ **9.** $\displaystyle\sum_{k=1}^{4}(2k-1)$

11. $\displaystyle\sum_{k=1}^{4} 5^{2k-1}$ **13.** $\displaystyle\sum_{k=1}^{5} k^2$ **15.** $\displaystyle\sum_{k=1}^{5} \dfrac{k}{k+1}$

17. $\displaystyle\sum_{k=1}^{6} \dfrac{k}{2k-1}$ **19.** $\displaystyle\sum_{k=1}^{\infty} \dfrac{2^{k-1}}{k}$ **21.** Neither; 97

23. Neither; $\dfrac{25}{12}$ **25.** Neither; 100

27. Geometric; 5,230,176,600
29. Arithmetic; 20,100 **31.** Neither; 441
33. Arithmetic; 1364 **35.** Arithmetic; -520
37. Geometric; 24,414,062 **39.** Geometric:

1074.76 **41.** 1 **43.** 14.12 **45.** $\dfrac{5}{2}$

47. $\dfrac{3}{8}$ **49.** $\dfrac{4}{9}$ **51.** $\dfrac{31}{99}$

53. $2\dfrac{410}{999}$ **55.** $\dfrac{29}{225}$ **57.** 120 in

59. 30 ft

Homework 9.5

1. 51; 101 **3.** 100; 50
5. 11 rows:

```
                    1
                  1   1
                1   2   1
              1   3   3   1
            1   4   6   4   1
          1   5   10  10   5   1
        1   6   15  20  15   6   1
      1   7   21  35  35  21   7   1
    1   8   28  56  70  56  28   8   1
  1   9   36  84 126 126  84  36   9   1
1  10  45 120 210 252 210 120  45  10   1
```

7. $x^5 + 15x^4 + 90x^3 + 270x^2 + 405x + 243$
9. $z^4 - 12z^3 + 54z^2 - 108z + 81$

11. $8x^3 - 6x^2y + \dfrac{3}{2}xy^2 - \dfrac{1}{8}y^3$

13. $x^{14} - 21x^{12} + 189x^{10} - 945x^8 + 2835x^6 - 5103x^4 + 5103x^2 - 2187$

15. $x^5 + 5x^4y + 10x^3y^2 + 10x^2y^3 + 5xy^4 + y^5$
17. $p^4 - 8p^3q + 24p^2q^2 - 32pq^3 - 16q^4$
19. $2 + 150t^2$
21. $z^5 - 5z^3 + 10z - 10z^{-1} + 5z^{-3} - z^{-5}$
23a. $5 \cdot 4 \cdot 3 \cdot 2 \cdot 1 = 120$

b. $\dfrac{9 \cdot 8 \cdot 7 \cdot 6 \cdot 5 \cdot 4 \cdot 3 \cdot 2 \cdot 1}{7 \cdot 6 \cdot 5 \cdot 4 \cdot 3 \cdot 2 \cdot 1} = 72$

c. $\dfrac{(5 \cdot 4 \cdot 3 \cdot 2 \cdot 1)(7 \cdot 6 \cdot 5 \cdot 4 \cdot 3 \cdot 2 \cdot 1)}{12 \cdot 11 \cdot 10 \cdot 9 \cdot 8 \cdot 7 \cdot 6 \cdot 5 \cdot 4 \cdot 3 \cdot 2 \cdot 1}$

$= \dfrac{1}{792}$ **d.** $\dfrac{8 \cdot 7 \cdot 6 \cdot 5 \cdot 4 \cdot 3 \cdot 2 \cdot 1}{(2 \cdot 1)(6 \cdot 5 \cdot 4 \cdot 3 \cdot 2 \cdot 1)} = 28$

25a. 84 **b.** 220 **c.** 190 **d.** 2002
27a. $1 - 14x + 84x^2$ **b.** $64 - 192x + 240x^2$
29. $-2560u^4 - 320u^2 - 2$
31a. $1 - 30c + 375c^2 - 2500c^3$
b. $1 - 34c + 495c^2 - 4000c^3$ **33.** 56
35. 77,520 **37.** $-101,376$ **39.** 1680
41. -84 **43.** 1, 11, 121, 1331, 14, 641. The digits of the terms in the sequence correspond to the numbers in the first five rows of Pascal's triangle. If $11^n = (10 + 1)^n$ is expanded as a binomial, each term is the product of a number from Pascal's triangle times a power of 10 times a power of 1.

Chapter 9 Review

1. $\dfrac{1}{2}, \dfrac{2}{5}, \dfrac{3}{10}, \dfrac{4}{17}$ **3.** $5, 2, -1, -4, -7$
5a. 1584, 1393.92, 1226.65, 1079.45
b. $a_1 = 1584$, $a_{n+1} = 0.88a_n$ **7a.** 30, 3.75, 43.125, 47.34375 **b.** $a_1 = 30$,
$a_{n+1} = 0.75a_n + 15$ **9.** $x_7 = -25$ **11.** 32

13. $\dfrac{-81}{8}$ **15.** 136 **17.** 68

19. Geometric; $\dfrac{-1}{16}, \dfrac{1}{32}, \dfrac{-1}{64}, \dfrac{1}{128}$;

$a_n = (-1)^n \left(\dfrac{1}{2}\right)^{n-1}$ **21.** Arithmetic; -14,
$-19, -24, -29$; $a_n = 11 - 5n$
23. Geometric; $-16, 32, -64, 128$;
$a_n = (-1)^n(-2)^{n-1}$ **25.** Arithmetic; $-1, -5$,
$-9, -13$; $a_n = 7 - 4n$ **27.** Geometric; -48,
$192, -768, 3072$; $a_n = 12(-4)^{n-1}$

29. $\displaystyle\sum_{k=2}^{5} k(k-1) = 2(1) + 3(2) + 4(3) + 5(4)$

31. $\displaystyle\sum_{k=1}^{12}(2^k - 1)$ **33.** Arithmetic; 210

35. Arithmetic; 57　　**37.** Geometric; $\dfrac{121}{243}$

39. Neither; -4　　**41.** Geometric; $\dfrac{-9}{5}$

43. $\dfrac{64}{27}$ ft　　**45a.** 810　　**b.** $\displaystyle\sum_{n=2}^{16} 6n$

47. 84 ft　　**49.** $\dfrac{29}{9}$

51. $x^5 - 10x^4 + 40x^3 - 80x^2 + 80x - 32$
53. 20　　**55.** 21　　**57.** 32　　**59.** -672

Homework 10.1

1a. Shift the graph of $y = |x|$ 2 units down
b.

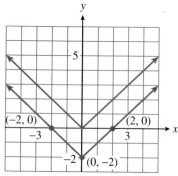

3a. Shift the graph of $y = \sqrt[3]{s}$ 4 units right
b.

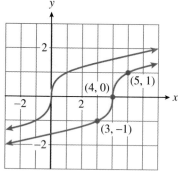

5a. Shift the graph of $y = \dfrac{1}{t^2}$ 1 unit up
b.

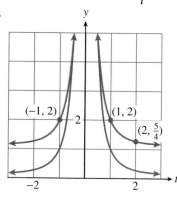

7a. Shift the graph of $y = r^3$ left 2 units
b.

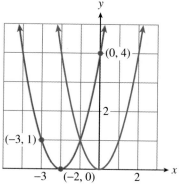

9a. Shift the graph of $y = \sqrt{d}$ down 3 units
b.

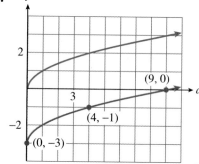

11a. Shift the graph of $y = \dfrac{1}{v}$ left 6 units
b.

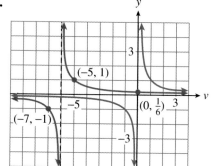

13a. Translate 3 units right and 2 units up.
b.

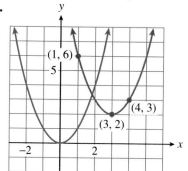

15a. Translate 2 units left and 3 units down

b.

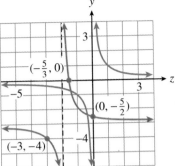

17a. Translate 4 units left and 4 units up

b.

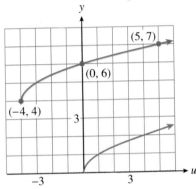

19a. Translate 5 units right and 1 unit down

b.

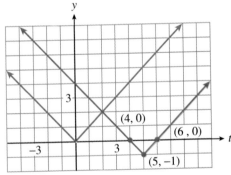

21a. Translate 1 unit right and 6 units up

b.

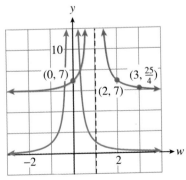

23a. Translate 8 units right and 1 unit down.

b.

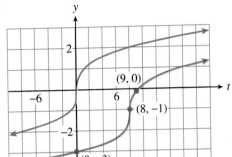

25a. $a = \dfrac{1}{3}$, compresses the basic graph by a factor of $\dfrac{1}{3}$

b.

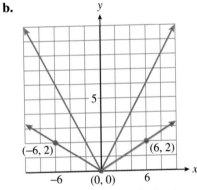

27a. $a = -2$, stretches the basic graph by a factor of 2 and reflects about the x-axis

b.

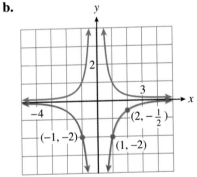

29a. $a = -3$, stretches the basic graph by a factor of 3 and reflects about the x-axis

b.

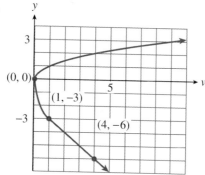

31a. $a = \dfrac{-1}{2}$, compresses the basic graph by a factor of $\dfrac{1}{2}$ and reflects about the x-axis

b.

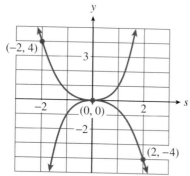

33a. $a = \dfrac{1}{3}$, compresses the basic graph by a factor of $\dfrac{1}{3}$

b.

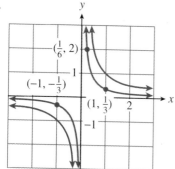

35a. $y = |x|$ shifted 1 unit left and 2 units down
b. $y = |x + 1| - 2$ **37a.** $y = \sqrt{x}$ reflected about the x-axis and shifted 3 units up
b. $y = -\sqrt{x} + 3$ **39a.** $y = x^3$ shifted 3 units right and 1 unit up   **b.** $y = (x - 3)^3 + 1$
41a. $y = 2^x$ reflected about the y-axis and shifted 2 units up   **b.** $y = 2^{-x} + 2$ **43a.** $y = 2^x$ reflected about the x-axis and shifted 10 units up
b. $y = -2^x + 10$ **45a.** $y = (x - 2)^2 + 3$

b.

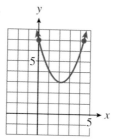

47a. $y = (x + 1)^2 - 4$
b.

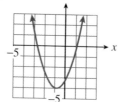

Homework 10.2

1a. $27a^2 - 18a$ **b.** $3a^2 + 6a$
c. $3a^2 - 6a + 2$ **d.** $3a^2 + 6a$
3a. 8   **b.** 8   **c.** 8   **d.** 8   **5a.** $8x^3 - 1$
b. $2x^3 - 2$ **c.** $x^6 - 1$ **d.** $x^6 - 2x^3 + 1$
7a. 11   **b.** 13   **c.** $3a + 3b - 4$
d. $3a + 3b - 2; f(a) + f(b) \neq f(a + b)$
9a. 19   **b.** 28   **c.** $a^2 + b^2 + 6$
d. $a^2 + 2ab + b^2 + 3; f(a) + f(b) \neq f(a + b)$
11a. $\sqrt{3} + 2$ **b.** $\sqrt{6}$ **c.** $\sqrt{a + 1} + \sqrt{b + 1}$
d. $\sqrt{a + b + 1}; f(a) + f(b) \neq f(a + b)$
13a. $\dfrac{-5}{3}$ **b.** $\dfrac{-2}{5}$ **c.** $\dfrac{-2}{a} - \dfrac{2}{b}$
d. $\dfrac{-2}{a + b}; f(a) + f(b) \neq f(a + b)$
15a. $(1, F(1)), (4, F(4)); F(4) - F(1)$
b. $(a, g(a)), (b, g(b)); g(b) - g(a)$
17a. $(a, f(a)), (b, f(b))$ **b.** $m = \dfrac{f(b) - f(a)}{b - a}$
19a. $(x_1, f(x_1)), (x_2, f(x_2))$
b. $m = \dfrac{f(x_2) - f(x_1)}{x_2 - x_1}$
21.

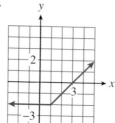

23.

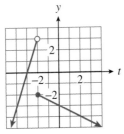

25.

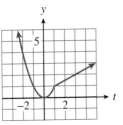

27.

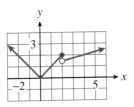

29.

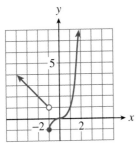

31.

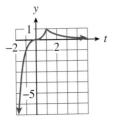

33a.

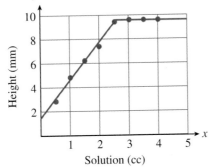

b. Approximately 3.4 mm per cc. For each cc of lead nitrate solution added, the height of the precipitate increases by 3.4 mm.

c. $f(x) = \begin{cases} 3.4x + 1.1, & 0.5 \le x \le 2.5 \\ 9.6, & x > 2.5 \end{cases}$

d. The amount of precipitate increases linearly until 2.5 cc of lead nitrate solution have been added. After that, all the potassium iodide has been used, so no more lead iodide is formed, and the height of the precipitate remains constant.

35a.

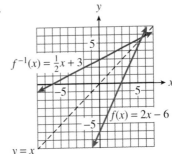

b. The volunteer walked for three minutes, during which time her heart rate was 100 beats per minute. Then she began jogging, and her heart rate rose to 156 beats per minute over the next minute. As her jogging speed increased, her heart rate increased to 185 beats per minute. During the cool-down period, her heart rate decreased rapidly and then leveled off at about 100 beats per minute.

Homework 10.3

1a.

x	0	2	4	6
$f(x)$	-6	-2	2	6

x	-6	-2	2	6
$f^{-1}(x)$	0	2	4	6

b. $f^{-1}(x) = \dfrac{1}{2}x + 3$

c.

3a.

x	-2	-1	0	1
$f(x)$	-7	0	1	2

x	-7	0	1	2
$f^{-1}(x)$	-2	-1	0	1

b. $f^{-1}(x) = \sqrt[3]{x-1}$

c.

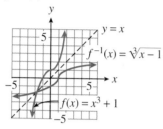

5a.

x	0.1	1	5	10
$f(x)$	-1	0	0.7	1

x	-1	0	0.7	1
$f^{-1}(x)$	0.1	1	5	10

b. $f^{-1}(x) = 10^x$

c.

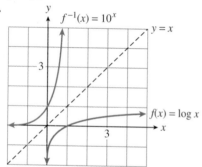

7a.

x	-2	-1	0	2
$f(x)$	$\frac{-1}{3}$	$\frac{-1}{2}$	-1	1

x	$\frac{-1}{3}$	$\frac{-1}{2}$	-1	1
$f^{-1}(x)$	-2	-1	0	2

b. $f^{-1}(x) = \dfrac{x+1}{x}$

c.

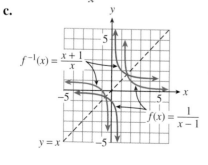

9a.

x	0.1	1	5	10
$f(x)$	-2	-1	-0.3	0

x	-2	-1	-0.3	0
$f^{-1}(x)$	0.1	1	5	10

b. $f^{-1}(x) = 10^{x+1}$

c.

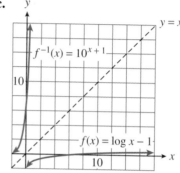

11a.

x	-2	-1	0	1
$f(x)$	0.07	0.7	7	70

x	0.07	0.7	7	70
$f^{-1}(x)$	-2	-1	0	1

b. $f^{-1}(x) = \log\left(\dfrac{x}{7}\right)$

c.

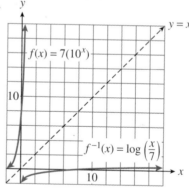

13a. $g(x) = \sqrt[3]{x} + 2$ **b.** $f(4) = (4-2)^3 = 8,$
$g(8) = \sqrt[3]{8} + 2 = 4$
c. $g(-8) = \sqrt[3]{-8} + 2 = 0, f(0) = (0-2)^3$
$= -8$

d.

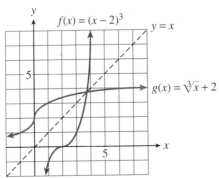

$f(x) = (x - 2)^3$

$y = x$

$g(x) = \sqrt[3]{x} + 2$

15a. $g(x) = \dfrac{1}{2} \log x$ **b.** $f(-1) = 10^{-2} = 0.01,$

$g(0.01) = \dfrac{1}{2} \log 0.01 = -1$

c. $g(100) = \dfrac{1}{2} \log 100 = 1, f(1) = 10^2 = 100$

d.

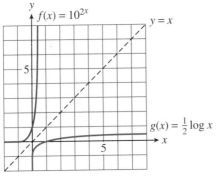

$f(x) = 10^{2x}$

$y = x$

$g(x) = \frac{1}{2} \log x$

17a. Yes   **b.** No   **c.** No   **d.** Yes
19a. Yes   **b.** No   **21a.** Yes   **b.** No
23a. Yes   **b.** No   **25a.** III   **b.** II
c. I   **d.** V   **e.** IV   **f.** None of these

27. 6   **29.** $\dfrac{2}{9}$ **31.** 4

33. 256

35a.

x	−1	0	1	2
$f(x)$	0	1	−2	−1

x	0	1	−2	−1
$f^{-1}(x)$	−1	0	1	2

b. 0   **c.** 2
37a.

x	−1	0	1	2
$f(x)$	−1	1	3	11

x	−1	1	3	11
$f^{-1}(x)$	−1	0	1	2

b. 0   **c.** 1   **39.** No. The graph of f does not pass the horizontal line test.   **41a.** 40°C

b. 98.6°F   **c.** $F = \dfrac{9}{5}C + 32$ **d.** 98.6°F

Midchapter 10 Review

1.

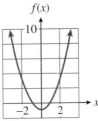

$f(x)$

3.

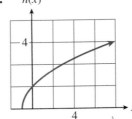

$h(x)$

5.

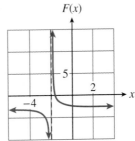

$F(x)$

7.

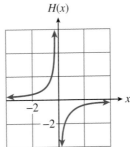

$H(x)$

9. 2

11. $\sqrt[3]{4 + b^2}$ **13a.** 1   **b.** 56
c. $-3(x + 1)^3 + 3$
15.

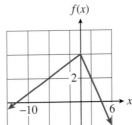

$f(x)$

17. $f^{-1}(4) = 2$, $f^{-1}(6) = 4$

19. $f^{-1}(x) = \dfrac{x^3 - 1}{2}$ **21a.** $f^{-1}(-3) = -14$,
$f(-14) = -3$ **b.** $f(13) = 3$, $f^{-1}(3) = 13$

23.

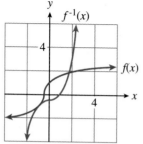

25. (a), (b), (c)

Homework 10.4

1.

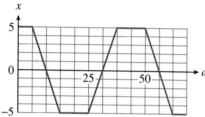

3a.

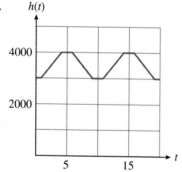

b. 10 min   **5a.** IV   **b.** III   **c.** II   **d.** I
7a. 69 hours   **b.** From 3.5 to 2.2   **c.** At the
large dip, the dimmer star eclipses the brighter
star, and at the small dip, the bright star eclipses
its dimmer companion.   **9.** $f(x) = 3x - 2$
11. $f(x) = x^2 + 1$ **13a.** $2a$ **b.** $\sqrt{16 - a^2}$
c. $f(a) = 2a\sqrt{16 - a^2}$ **d.** $0 < a < 4$

e. $f(a)$

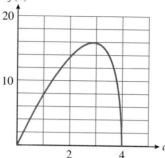

13.86; 1.32, or 3.77   **f.** 2.83   **15a.** $\dfrac{1}{2 - a}$

b. $\dfrac{a}{a - 2}$ **c.** $f(a) = \dfrac{a^2}{2(a - 2)}$ **d.** $a > 2$

e.   15

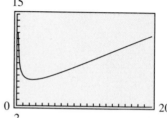

4.5; 2.4 and 12   **f.** The smallest area (4 square
units) occurs when $a = 4$. The area of the triangle
decreases as a increases from 2 to 4. For $a > 4$,
the area is increasing again.   **17a.** $r = \dfrac{C}{2\pi}$

b. $A = \dfrac{C^2}{4\pi}$ **c.** Quadratic. $C > 0$

d. $\dfrac{2500}{\pi} \approx 796$ sq yds

Homework 10.5

1a. 95 applied, 72% were accepted
b. 94 applied, 23% were accepted   **c.** 20%
d. 21.5%   **e.** 90%   **f.** 100%
g. Women did better in each department, but
men had a higher success rate overall.
3a. Master's salary ($23,000) is higher than bach-
elor's ($16,000).   **b.** Master's salary ($31,000)
is higher than bachelor's ($25,000).
c. $24,250   **d.** $23,800   **e.** Graduates with
bachelor's degrees had the higher average starting
salary.   **5a.** $450   **b.** 8 cm
c. $C = 0.72A$ **d.** $C = 18T$

e. $C = 0.18AT$ **7a.** $E = \dfrac{1}{2}mv^2$

b.

	Velocity (m/sec)			
Mass (kg)	**1.0**	**2.0**	**3.0**	**4.0**
1.0	0.5	2.0	4.5	8.0
1.5	0.75	3.0	6.75	12
2.0	1.0	4.0	9.0	16
2.5	1.25	5.0	11.25	20

9a.

	Pressure (atmospheres)							
Temperature (°C)	50	100	150	200	250	300	350	400
350	25	38	46	53	58	62	66	68
400	16	26	33	38	45	48	53	56
450	9	17	23	28	32	37	40	43
500	6	11	16	20	23	27	29	32
550	4	8	11	14	17	19	22	24

b. The yield of ammonia decreases.

c.

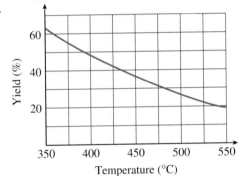

Chapter 10 Review

1a. Shift the graph of $y = |x|$ up 2 units

b.

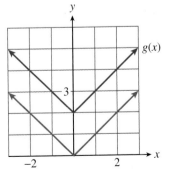

3a. Shift the graph of $y = \sqrt{s}$ up 3 units.

b.

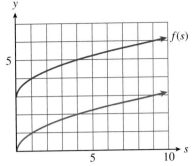

5. Shift the graph of $y = |t|$ left 2 units and down 3 units

b.

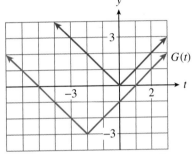

7a. Stretch the graph of $y = \sqrt{s}$ by a factor of 2 and reflect about the x-axis.

b.

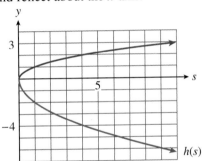

9a. Shift the graph of $y = 2^t$ left 4 units

b.

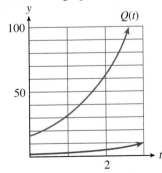

11. $y = (x - 2)^2 - 3$

13a. $f^{-1}(x) = x - 4$
b.

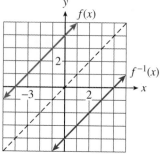

15a. $f^{-1}(x) = \sqrt[3]{x + 1}$
b.

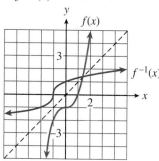

17a. $f^{-1}(x) = \dfrac{1}{x - 2}$
b.

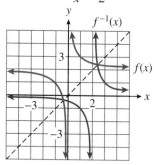

19. 0   **21.** $H(2a) = 4a^2 + 4a$,
$H(a + 1) = a^2 + 4a + 3$
23. $f(a) + f(b) = 2a^2 + 2b^2 - 8$;
$f(a + b) = 2a^2 + 4ab + 2b^2 - 4$
25.

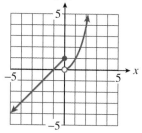

27.

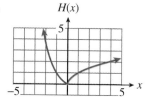

29.

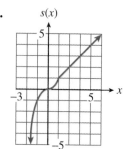

31a. $\dfrac{s\sqrt{3}}{2}$

b. $\dfrac{s^2\sqrt{3}}{4}$ **c.**

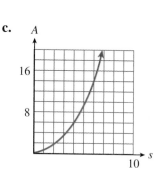

d. $4\sqrt{3}$ sq cm   **e.** 2.5 ft   **33a.** 7%
b. reducing rate by 5%   **c.** no   **d.** no
e. 30-year   **35a.** $N = \dfrac{k}{d^2 E^3}$, where N is
number of people, d is distance in miles from the
road, E is the elevation gain, k is the constant of
variation.   **b.** $k \approx 0.01$ **c.** 3
37a. period is 8.
b.

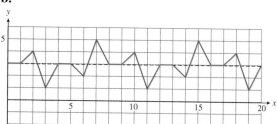

c. $k = 8$ **d.** $a = 3, b = 7$ **39a.** high: 4.5
and 4.4, low: 2.5 and 0.3   **b.** No. Neither high
tide nor low tide values repeat at any constant
time interval.

c.

Time (hours)	3.1	14.1	28.1	39.7	52.9	65.1	77.6
High tide (ft)	4.5	4.4	4.9	4	5.3	3.8	5.6

Time (hours)	90.3	102.2	115.3	126.8	140.1	151.3	164.8
High tide (ft)	3.8	5.9	3.8	6.1	3.8	6.2	3.8

Time (hours)	8.7	21	34.3	46	59.5	70.9	84.4
Low tide (ft)	2.5	0.3	1.9	0.7	1.3	1.1	0.6

Time (hours)	95.7	109.2	120.4	133.8	145	158.4
Low tide (ft)	1.4	0.1	1.6	−0.3	1.9	−0.5

d. The morning high tides increase from 4.5 to 6.2, while the evening high tides decrease from 4 to 3.8. The morning low tides decrease from 2.5 to −0.5, while the evening low tides increase from 0.3 to 1.9.   **e.** 6.7 feet, on Decemaber 23.

Homework 11.1

1.

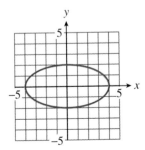

3.

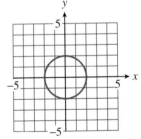

5.

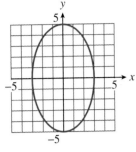

7.

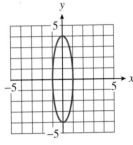

9.

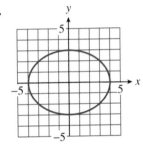

11.

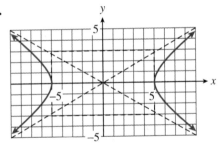

13.

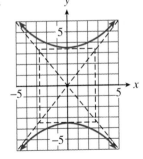

15.

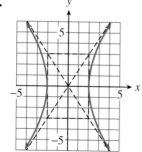

17.

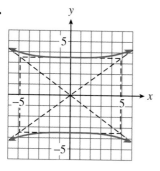

19.

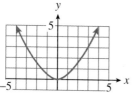

21.

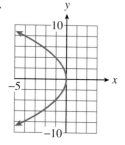

23.

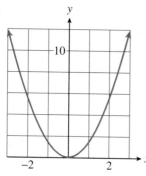

25.

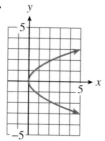

27a. Circle; radius: 2

b. $(-1, \pm\sqrt{3})$ **29a.** Hyperbola; x-intercepts: $\pm\sqrt{8}$ **b.** $(\pm2\sqrt{6}, 2)$ **31a.** Ellipse; x-intercepts: $\pm\sqrt{3}$, y-intercepts: $\pm\sqrt{6}$

b. None   **33a.** Parabola; vertex: $\left(0, \dfrac{-3}{2}\right)$, opens up   **b.** None   **35a.** Hyperbola; y-intercepts: $\pm\sqrt{6}$ **b.** None

37a. $4x^2 + 9y^2 = 36$

b.

x	±3	0	-2	±2.6
y	0	±2	±1.49	1

39a. $4y^2 - x^2 = 4$

b.

x	0	None	4	±3.46
y	±1	0	±2.24	-2

41a. $\dfrac{x^2}{10^2} + \dfrac{y^2}{7^2} = 1$ **b.** 4.2 ft

43a. $\dfrac{x^2}{180^2} + \dfrac{y^2}{50^2} = 1$ **b.** 69.98 cm

45a. $x^2 = 432y$ **b.** 108 in   **47a.** $x^2 = 50y$
b. 12.5 cm   **49.** 520 ft   **51.** 472.5 ft
53.

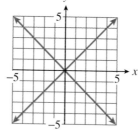

The graph is two straight lines through the origin.
55.

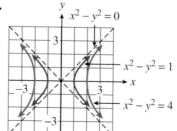

The last graph provides the asymptotes for the first two hyperbolas.

Homework 11.2

1a.

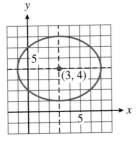

b. $(-1, 4), (7, 4), (3, 1), (3, 7)$

3a.

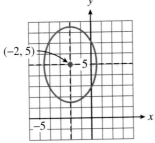

b. $(-2 - \sqrt{6}, 5), (-2 + \sqrt{6}, 5),$
$(-2, 5 - 2\sqrt{3}), (-2, 5 + 2\sqrt{3})$

5a.

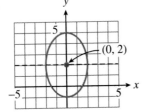

b. $(-2, 2), (2, 2), (0, -1), (0, 5)$

7a.

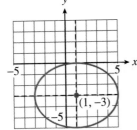

b. $(-3, -3), (5, -3), (1, 0), (1, -6)$

9a.

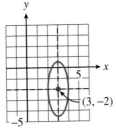

b. $(2, -2), (4, -2), (3, -2 - \sqrt{8}), (3, -2 + \sqrt{8})$

11. $4x^2 + 9y^2 - 8x - 108y + 292 = 0$

13. $9x^2 + 25y^2 + 36x - 100y - 89 = 0$

15. $4x^2 + y^2 + 32x - 6y + 37 = 0$

17a.

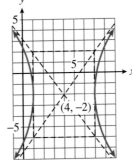

b. $(7, -2), (1, -2), (4 \pm \sqrt{2}, 2)$

19a.

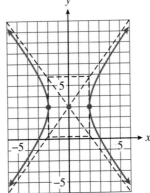

b. $(\sqrt{6}, 1), (\sqrt{6}, 5), (-\sqrt{6}, 1), (-\sqrt{6}, 5)$

21a.

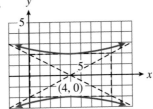

b. $(4, \pm 2), (0, \pm 2\sqrt{2})$

23a.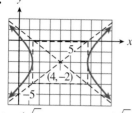

b. $(4 \pm \sqrt{6}, -2), (4 \pm 2\sqrt{3}, 0)$

25a.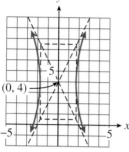

b. $(\pm\sqrt{3}, 4), (\pm 2\sqrt{3}, -2)$

27. $64y^2 - 36x^2 - 640y - 72x - 740 = 0$

29. $4y^2 - x^2 - 8y - 2x - 13 = 0$

31a.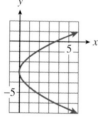

b. $(1, -3 \pm\sqrt{2}), (2, -5), (2, -1)$

33a.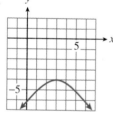

b. $(3 \pm\sqrt{6}, -5), (3 \pm 2\sqrt{3}, -6)$

35a.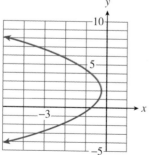

b. $(-1, 2 \pm\sqrt{6}), (-2, 2 \pm\sqrt{14})$

37a.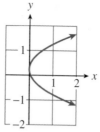

b. $\left(3, \dfrac{7}{3}\right), \left(3, \dfrac{-5}{3}\right), \left(6, \dfrac{1 \pm 6\sqrt{2}}{3}\right)$

39. Ellipse; center $(0, 0)$, major axis vertical, $a^2 = \dfrac{3}{2}, b^2 = 6$ **41.** Parabola; vertex $(0, 2)$, opens downward, $a = \dfrac{-1}{2}$ **43.** Hyperbola; center $(0, 0)$, transverse axis on the y-axis, $a^2 = 24, b^2 = 6$ **45.** Parabola; vertex $(-4, 0)$, opens to the right, $a = \dfrac{1}{2}$ **47.** Ellipse; center $(-3, 0)$, major axis vertical, $a^2 = 5, b^2 = 12$ **49.** Ellipse; center $(-1, 0)$, major axis vertical, $a^2 = 2, b^2 = 4$

Homework 11.3

1. $(2, 2), (-2, -2)$ **3.** $(3\sqrt{2}, \sqrt{2})$, $(-3\sqrt{2}, -\sqrt{2})$ **5.** $(3, 1), (-3, 1), (3, -1)$, $(-3, -1)$ **7.** $(7, 2), (-7, 2), (7, -2)$, $(-7, -2)$ **9.** $(2, 4), (-2, 4), (2, -4)$, $(-2, -4)$ **11.** $(4, 1), (-4, 1), (4, -1)$, $(-4, -1)$ **13.** $(2, 3), (-4, 1)$ **15.** $(0, 4)$, $(-2, 0)$ **17.** $(28, 54), \left(\dfrac{1708}{25}, \dfrac{2106}{25}\right)$

19. $(60, 45)$ **21.** 12 feet by 18 feet

23. Pressure: 6 lb per sq in.; volume: 5 cu in.

25. $(12, 20)$ **27.** $(8, 15)$

Midchapter Review

1.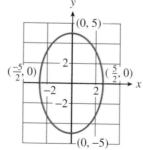

3. $\left(\dfrac{3}{2}, -4\right), \left(\dfrac{-3}{2}, -4\right)$

5.

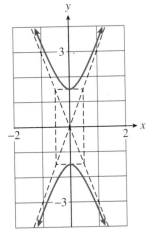

7.

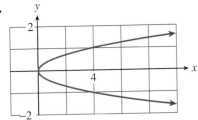

9a. Parabola opening up with vertex at $(0, -4)$

b. $\left(\sqrt{3}, \dfrac{-5}{2}\right), \left(-\sqrt{3}, \dfrac{-5}{2}\right)$

11.

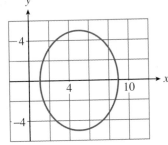

13.

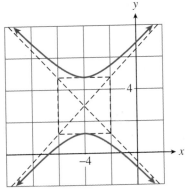

15. $\dfrac{(x + 5)^2}{25} + \dfrac{(y + 2)^2}{4} = 1$

17.

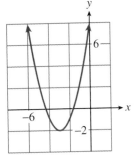

19. No solution

21. $(5, 7), (5, -7), (-5, 7), (-5, -7)$
23. $(1, -3), (3, 5)$

Homework 11.4

1. The graph of $12 = 3x + 4y$ does not intersect the set of feasible solutions. **3.** $(8, 2)$; \$32
5. $(2, 0)$ **7a.** \$22 **b.** $(8, 0)$ **c.** \$32
9a. $(1, 4)$ **b.** 7 **c.** $(4, 5)$ **d.** 17
11a. $(0, 5)$ **b.** -10 **c.** $(5, 0)$ **d.** 25
13b. $(0, 0)$; 0 **c.** $(3, 2)$; 13 **15b.** $(0, 14)$;
-14 **c.** $(10, 0)$; 30 **17b.** $(0, 8)$; -160
c. $(8, 0)$; 1600 **19.** 250 **21.** \$5040
23. 967.7 calories **25a.** $E = 2.44x + 2.26y$;
101.9 grams of forb, 25.5 grams of grass
b. $T = 2.05x + 5.21y$; 41 grams of forb, no grass
c. caloric intake **27.** Maximum: 17.4; minimum: -8.4 **29.** Maximum: 4112; minimum: 0
31. Maximum: 1908; minimum 0

Homework 11.5

1. Algebra section

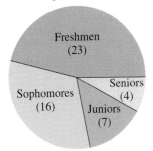

3. Forensics club   92 men

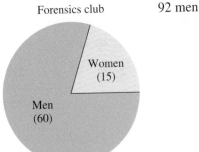

5. Favorite flavors

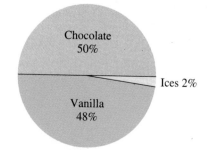

7. Funding for computers

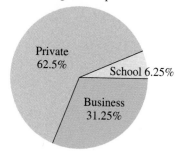

9a. A: 1, 2, 8, 9, 13; B: 2, 3, 5, 13, 17   **b.** A
c. B   **d.** B   **e.** A   **f.** B, since 25% of B
are 3-5 years old   **11.** B; II   **13.** A; III
15. c; I   **17.** b; VI   **19.** a; VII   **21.** c; I
23. f or h; II   **25.** f; V   **27a.** Mean = 3.925;
median = 4; mode = 4

b.

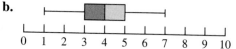

c. 40;

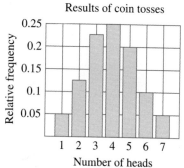

d. The coin lands tails more often than heads.

29a. $\frac{1}{2}$ **b.** 68%   **c.** 95%   **d.** 99.7%
e. 34%   **f.** 84%   **g.** 97.35%
31. Mean = 17 ounces

Chapter 11 Review

1. Circle; center (0, 0), radius 3

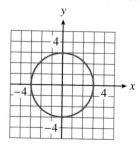

3. Ellipse; center (0, 0), major axis vertical,
$a = 2, b = 4$

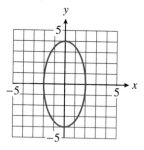

5. Parabola; vertex (−3, 2), opens upward,
$a = \frac{1}{4}$

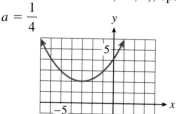

7. Ellipse; center (2, −3), major axis vertical,
$a = 2, b = 3$

9. Hyperbola; center (2, −3), transverse axis horizontal, $a = 2, b = 3$

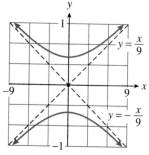

11. Circle; center $(2, -1)$, radius 3, $(x - 2)^2 + (y + 1)^2 = 9$

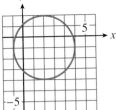

13. Ellipse; center $(2, -2)$, major axis vertical, $a = 2, b = 4, \dfrac{(x - 2)^2}{4} + \dfrac{(y + 2)^2}{16} = 1$

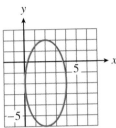

15. Parabola; vertex $(4, 10)$, opens upward, $a = 1$, $y + 10 = (x - 4)^2$

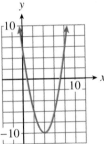

17. Parabola; vertex $(2, -2)$, opens downward, $a = 1, y + 2 = -(x - 2)^2$

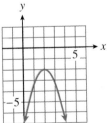

19. Hyperbola; center $(-2, 4)$, transverse axis vertical, $a = 2, b = \sqrt{6}, \dfrac{(y - 4)^2}{6} - \dfrac{(x + 2)^2}{4} = 1$

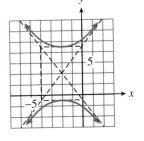

21. Hyperbola; center $(0, 3)$, transverse axis horizontal, $a = \sqrt{5}, b = \sqrt{10}, \dfrac{x^2}{5} - \dfrac{(y - 3)^2}{10} = 1$

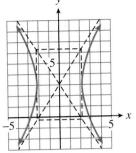

23a. $\dfrac{x^2}{25} + \dfrac{y^2}{64} = 1$ **b.** $\pm\dfrac{24}{5}$

25. $(x + 4)^2 + (y - 3)^2 = 20$

27. $\dfrac{(x + 1)^2}{16} + \dfrac{(y - 4)^2}{4} = 1$

29. $\dfrac{(x - 2)^2}{16} - \dfrac{(y + 3)^2}{9} = 1$ **31.** $y = \dfrac{1}{16}x^2$

33. $(2, 3), (-2, 3), (2, -3), (-2, -3)$

35. $(1, -2), (-1, 2), \left(2\sqrt{3}, \dfrac{-1}{\sqrt{3}}\right), \left(-2\sqrt{3}, \dfrac{1}{\sqrt{3}}\right)$

37. Moia: 45 mph, Fran: 50 mph   **39.** 7 cm by 10 cm   **41.** Morning train: 20 mph, evening train: 30 mph   **43.** $(192, 156)$ **45b.** $(1, 0)$; 18   **c.** $(5, 2)$; 186

47a. $20p + 8g \leq 120, 10p + 10g \leq 120, p \geq 0, g \geq 0$ **b.** 2 batches of peanut butter cookies and 10 batches of granola cookies, for $125 income   **49a.** 95-105   **b.** 104   **c.** 95

d.

e. These data show a boxplot can be symmetrical even with a nonsymmetrical distribution.

51a. The mean is 110 grams. The mean and median are the same for normally distributed data (or for any other symmetrically distributed data set).   **b.** 2.5 grams   **c.** 25% of the data lie between the first quartile and the median, but 34% of the data will lie within one standard deviation below the mean. Q1 ≈ 108 grams

d.

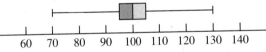

Appendix A.1

1a. $j = a + 27$ **b.** 49   **3a.** $h = \dfrac{1260}{r}$

b. 28 hr   **5a.** $y = \dfrac{f}{3}$ **b.** $c = 5.79\left(\dfrac{f}{3}\right)$

c. $19.30   **7a.** $A = \pi r^2$ **b.** 78.54 sq cm

9a. $n = 20 + m$ **b.** $a = \dfrac{198}{20 + m}$ **c.** 7.92

11a. $t = 0.079p$ **b.** $b = 1.079p$

c. $528.71   **13a.** $C = 1.97 + 0.39m$

b. $12.50   **15a.** $c = 0.4w$

b. $r = 50 - 0.4w$ **c.** 47.6 lb.

17a. $u = \dfrac{m}{20}$ **b.** $g = 14.6 - \dfrac{m}{20}$ **c.** 9.1 gal

19. 6   **21.** -42 **23.** 5   **25.** -2

27. -2 **29.** -60 **31.** -25 **33.** 81

35. -64 **37.** -32 **39.** 50   **41.** -2

43. -5 **45.** -19 **47.** 98.4   **49.** 36

51. -3 **53.** 45   **55.** -5 **57a.** $2 + \dfrac{3}{4}$

b. $\dfrac{2 + 3}{4}$ **59a.** -23^2 **b.** $(-23)^2$

61a. $\sqrt{9 + 16}$ **b.** $\sqrt{9} + 16$ **63.** 100

65. 1080   **67.** 0   **69.** 72.5904

Appendix A.2

1. 7   **3.** $\dfrac{31}{3}$ **5.** $\dfrac{2}{3}$ **7.** $-4.8\overline{3}$

9. 34.29   **11.** $18   **13a.** $t =$ time wife drives   **b.** $45t$ **c.** $16(t + 6)$ **d.** 3.3 hr

15a. $n =$ number of copies

b. $20,000 + 0.02n$; $17,500 + 0.025n$

c. 500,000 copies   **17a.** 145,800

b. 125,000   **19.** 7.53%   **21.** 87

23d. 7.7 lb of 6%; $2.\overline{2}$ lb of 15%

25d. $16,600   **27.** $\dfrac{v - k}{g}$ **29.** $\dfrac{S - 2w^2}{4w}$

31. $\dfrac{P - a + d}{d}$ **33.** $\dfrac{A - \pi r^2}{\pi r}$

Appendix A.3

1. $47°, 57°, 76°$ **3.** $30°, 60°$ **5.** $50°, 50°, 80°$

7. 15 cm   **9.** $x > 3$ **11.** $x > 0$

13. $x \le \dfrac{6}{13}$ **15.** $4 < x < 16$ **17.** 14 ft

19. 17.1 sq ft   **21.** 89.23 ft

23a. 7.24 cu m   **b.** 6.16 sq cm

25a. 2623.86 cu m   **b.** 1903.43 sq in.

27a. $72h + 640$ **b.** 8 in.

Appendix A.4

1a. High: 7°F; low: -19°F   **b.** Above 5°F from noon to 3:00 P.M.; below -5°F from mid-

night to 9:00 A.M. and from 7:00 P.M. to midnight   **c.** 7:00 A.M.: -10°F; 2:00 P.M.: 6°F; 10:00 A.M. and 5:00 P.M.: 0°F; 6:00 A.M. and 10:00 P.M.: -12°F   **d.** Between 3:00 A.M. and 6:00 A.M.: 6°F; between 9:00 A.M. and noon: 10°F; between 6:00 P.M. and 9:00 P.M.: 9°F   **e.** Increased most rapidly: 9:00 A.M. to noon; decreased most rapidly: 6:00 P.M. to 9:00 P.M.   **3a.** 28 mpg   **b.** 50 mph   **c.** Best gas mileage: at 70 mph. The graph seems to be "leveling off" for higher speeds; any improvement in mileage probably would not be appreciable, and the mileage might in fact deteriorate.   **d.** Road condition, weather conditions, traffic, weight in the car   **5a.** 12 min   **b.** First 38 min   **c.** Approximately from 38 min to 55 min   **7.** $(-3, -2)$, $(1, 6)$, $(-2, 0)$, $(0, 4)$ **9.** $(-2, 6)$, $(2, 6)$, $(1, 3)$ or $(-1, 3)$, $(0, 2)$ **11.** $\left(-1, -\dfrac{1}{2}\right)$, $\left(\dfrac{1}{2}, -2\right)$, $\left(4, \dfrac{1}{3}\right)$, $(0, -1)$ **13.** $(-2, -8)$, $\left(\dfrac{1}{2}, \dfrac{1}{8}\right)$, $(0, 0)$, $(-1, -1)$

Appendix A.5

1a. b^9 **b.** b^{10} **c.** q^9 **d.** p^{10} **3a.** 2^9

b. 6^8 **c.** 2^5 **d.** 8^4 **5a.** $216x^3$

b. $81y^4$ **c.** $32t^{15}$ **d.** $36s^4$ **7a.** $\dfrac{h^8}{m^{12}}$

b. $\dfrac{n^{24}}{k^{32}}$ **c.** $256a^8b^{16}$ **d.** $-125a^3b^{24}$

9a. $2w$ **b.** w^2 **11a.** $-2z^2$ **b.** $-24z^4$

13a. Can't be simplified   **b.** $12p^5$

15a. 3^{17} **b.** Can't be simplified

17a. $-24y^2$ **b.** $32z^2$ **19a.** $-12x^3y^4$

b. $-50x^5y$ **21a.** $\dfrac{1}{4ab^4}$ **b.** $\dfrac{2}{3a^3b^2}$

23a. $\dfrac{5}{c^2}$ **b.** $\dfrac{5}{c^5}$ **25a.** b^{13} **b.** b^{25}

27a. $4x^{10}y^{14}$ **b.** $108x^7y^{10}$ **29a.** $\dfrac{-8x^3}{27y^6}$

b. $\dfrac{x^8}{16y^4}$ **31a.** $-x^4y$ **b.** x^4 **33a.** a^6

b. a^{12} **c.** a^3b^3 **d.** $(a + b)^3$ **35a.** $96a^5$

b. $6a^5$ **c.** $3a^{10}$ **d.** $9a^{10}$

Appendix A.6

1. Binomial; 3   **3.** Monomial; 4

5. Trinomial; 2   **7.** Trinomial; 3

9. b and c   **11.** a, b, c   **13a.** -1

b. -21 **c.** $8b^3 - 12b^2 + 2b + 1$

15a. $\dfrac{11}{4}$ **b.** $\dfrac{1}{9}$ **c.** $w^2 - 3w + 1$

17a. 28.0128   **b.** 126.5728
c. $3k^4 - 12k^3 + 16k^2 - 8k + 4$

19a. 2   **b.** 96   **c.** $\dfrac{m^6}{729} - \dfrac{m^5}{243}$

21. $4xy - 8y^2$ **23.** $-12x^3 + 6x^2 - 6x$
25. $3a^4b - 2a^3b^2 - a^2b^2$
27. $8x^3y^7 - 4x^4y^4 - 6x^5y^5$
29. $n^2 + 10n + 16$ **31.** $r^2 + 3r - 10$
33. $2z^2 - 5z - 3$ **35.** $8r^2 + 2rs - 3s^2$
37. $6x^2 - 13xy + 6y^2$ **39.** $9t^2 - 16s^2$
41. $2a^4 - 5a^2b^2 - 3b^4$ **43.** $4xz(x + 2)$
45. $3n^2(n^2 - 2n + 4)$ **47.** $3r(5rs + 6s^2 - 1)$
49. $m^2n^2(3n^2 - 6mn + 14m)$
51. $3a^2b^2c^4(5a^2b - 4c + 2b)$
53. $(a + b)(a + 3)$ **55.** $(y - 3x)(y - 2)$
57. $4(x - 2)^2(-2x^2 + 4x + 1)$
59. $x(x - 5)^2(-x^2 + 5x + 1)$
61. $-(2n - 3m)$ **63.** $-2(x - 1)$
65. $-a(b + c)$ **67.** $-(-2x + y - 3z)$
69. $(x + 2)(x + 3)$ **71.** $(y - 3)(y - 4)$
73. $(x - 3)(x + 2)$ **75.** $(2x - 1)(x + 2)$
77. $(4x - 1)(x + 2)$ **79.** $(3y + 1)(3y - 8)$
81. $(2u + 1)(5u - 3)$ **83.** $(3x - 7)(7x + 2)$
85. $(9a + 4)(8a - 3)$ **87.** $(2x - 3)(15x - 4)$
89. $2(3t + 2)(9t - 11)$ **91.** $(x - 2a)(3x - a)$
93. $(3x - 2y)(5x + 2y)$
95. $(3u - 4v)(6u - 5v)$
97. $(3a + 2b)(4a - 7b)$
99. $(5ab - 2)(2ab - 3)$
101. $2(4xy + 1)(7xy - 2)$
103. $(2az - 3)(11az + 7)$ **105.** $x^2 + 6x + 9$
107. $4y^2 - 20y + 25$ **109.** $x^2 - 9$
111. $9t^2 - 16s^2$ **113.** $25a^2 - 20ab + 4b^2$
115. $64x^2z^2 + 48xz + 9$ **117.** $(x + 5)(x - 5)$
119. $(x - 12)^2$ **121.** $(x + 2y)(x - 2y)$
123. $(2x + 3)^2$ **125.** $(3u - 5v)^2$
127. $(2a + 5b)(2a - 5b)$
129. $(xy + 9)(xy - 9)$ **131.** $(3xy + 1)^2$
133. $(4xy - 1)(4xy + 1)$
135. $(x + 2 - y)(x + 2 + y)$

Appendix A.7

1. mean = 4; median = 4; mode = 4
3. mean = 72; median = 77; no mode
5a. mean = \$193,000; median = \$10,000;
mode = \$7,000   **b.** \$549,000
7. $-80, 5, 5, 10, 20, 20, 20$

9. (many possibilities) $-2, -1, 0, 2, 2$ has
median = 0, mean = $\dfrac{1}{2}$, mode = 2
11. 1, 2.5, 4, 5.5, 7   **13.** 27, 44, 77, 93, 110
15. mean = 6.9; median = 7; mode = 7; 0, 6, 7, 8, 10   **17a.** 28 to 98
b. 81   **c.** 50%   **d.** 25%   **19a.** the second section, (range 3 to 94)   **b.** your section, with more than $\frac{3}{4}$ scoring below 82—only $\frac{1}{2}$ of the other section scored below 82   **c.** the other section, with 25% above 90   **d.** the other section, which had a higher percentage than your section scoring above 74, above 82, and above 91.
21. When the mode exists, it must be one of the values in the data set; 0, 0, 0, 1, 5, 6 has
mean = 2, mode = 0.5.

Appendix B.1

1. Rationals   **3.** Irrationals   **5.** Integers
7. Whole numbers   **9.** Rationals
11. Irrationals   **13.** 0.375, terminates
15. $0.\overline{285714}$, repeats a pattern   **17.** 0.4375,
terminates   **19.** $0.\overline{846153}$, repeats a pattern

Appendix B.2

1. $13 + 4i$ **3.** $-0.8 + 3.8i$ **5.** $20 + 10i$
7. $-17 + 34i$ **9.** $46 + 14\sqrt{3}i$ **11.** 52
13a. 0   **b.** 0   **15a.** 0   **b.** 0
17a. $6 - 7i$ **b.** $10 + 9i$ **19.** $4i - 1$
21. $7 + 4i$ **23.** $\dfrac{-25}{29} + \dfrac{10}{29}i$
25. $\dfrac{3}{4} - \dfrac{\sqrt{3}}{4}i$ **27.** $\dfrac{-2}{3} + \dfrac{\sqrt{5}}{3}i$ **29.** i
31. $4z^2 + 49$ **33.** $x^2 + 6x + 10$
35. $v^2 - 8v + 17$ **37.** Real: $x \geq 5$, imaginary: $x < 5$ **39a.** -1 **b.** 1   **c.** $-i$
d. -1

Index

METRIC CONVERSIONS

Length

1 centimeter	0.394 inch
1 kilometer	0.621 mile
1 inch	2.54 centimeters
1 mile	1.609 kilometers

Volume

1 liter	1.057 quarts
1 milliliter	1 cubic centimeter
1 quart	0.946 liter

Mass (Weight)

1 gram	0.035 ounces
1 kilogram	2.205 pounds
1 ounce	28.350 grams
1 pound	0.454 kilogram